Macintosh® Software

Your own organic chemistry tutor for only $39.95?

Beaker Version 2.1: An Expert System for the Organic Chemistry Student

by Joyce Brockwell, John Werner, Steve Townsend, Nim Tea, and Elizabeth Toon, all of Northwestern University

For use on a Macintosh® 512e, +, SE, or II, with one 800k disk drive

Phenomenal software at a textbook price!

Beaker is sophisticated yet easy-to-use software that allows you to explore organic chemistry principles, study and solve problems, sketch and analyze molecules, do lab simulations, and much more.

Using Beaker, you can quickly draw a molecule, or simply type in an IUPAC name and let Beaker do the drawing. Beaker makes sure the drawing is correct chemistry.

Then you can ask Beaker to: redraw the molecule and show all the atoms ■ give the IUPAC name ■ show all the resonance forms ■ draw all the connectivity isomers ■ create the Newman projections of all the staggered and eclipsed rotamers ■ determine the absolute stereo-chemistry (R,S) at chiral bonds ■ add a reagent and perform a reaction ■ show all the reaction products and the mechanisms that produced them ■ construct an NMR spectrum and much more!

Beaker enhances lectures, individual study, problem solving, and assignments

Beaker contains rules for finding solutions to problems such as naming and carrying out reactions. When asked questions, Beaker computes answers based on the information you input. This means Beaker can answer your questions about diverse areas of chemical structure and reactivity in any order and at varying levels of detail. And, Beaker lets you satisfy your curiosity. You can pose "what if" questions, then immediately see the results displayed on the computer screen.

How to Order

CREDIT CARD ORDERS: We accept VISA, Master-Card, and American Express. To order, use our toll-free number (800) 354-9706 or use the order form. Be sure to include: Type of credit card and account number, expiration date, and your signature. We pay shipping charges on credit card orders. Prices subject to change without notice.

PREPAID ORDERS: Checks or money orders should be made payable to Brooks/Cole Publishing Company. We pay shipping, unless you request special handling. Do not send cash through the mail. We will refund payment for unshipped out-of-stock titles after 120 days and for not-yet-published titles after 180 days unless an earlier date is requested by you.

100% Money-back Guarantee

We back all Beaker™ packages with our 100% money-back guarantee. If, for any reason, you are not satisfied with the product you order from us, simply return it in salable condition within 30 days for a complete refund.

Of course, we replace damaged or defective disks at no cost to you for 30 days after purchase.

Macintosh is a registered trademark of Apple Computer, Inc.

FOLD

Order Form

Please send me (quantity) _____ *BEAKER 2.1 Single User Package(s)* (ISBN: 0-534-15973-7) for $39.95 each **Total:** _____

Please send me (quantity) _____ *BEAKER 2.1 User's Guide(s)* (ISBN: 0-534-11684-1) only for $12.25 each **Total:** _____

Add Tax Due: _____ (Residents of CA, CT, CO, FL, GA, IL, IN, KY, MA, MD, MI, MN, MO, NC, NJ, NY, OH, PA, RI, TN, TX, UT, VA, WA, WI add appropriate sales tax.)

Total Due: _____

Prices subject to change without notice.

Instructors: Beaker is available in Lab Packages and Network Licenses. Call (800) 354-9706 for more information.

Payment Options

_____ Purchase Order enclosed. Please bill me.
_____ Check or money order enclosed.
_____ Charge my _____ VISA _____ MasterCard
_____ American Express

_____   _____
Card Number   Expiration Date

_____
Signature

Please ship to:

_____
Name

_____
Institution

_____
Address

_____
City/State/Zip

_____
Telephone

Billing Address:

_____
Name

_____
Institution

_____
Address

_____
City/State/Zip

_____
Telephone

Secure

Study Guide and Solutions Manual for
Organic Chemistry

Third Edition

Susan McMurry

Brooks/Cole Publishing Company

Pacific Grove, California

Brooks/Cole Publishing Company
A Division of Wadsworth, Inc.

Printed in the United States of America

10   9   8   7   6   5   4   3

ISBN 0-534-16219-3

Sponsoring Editor: *Maureen Allaire*
Editorial Assistant: *Nancy Miaoulis*
Production Coordinator: *Dorothy Bell*
Cover Design: *Michael Rogondino, Vernon T. Boes*
Cover Photo: *MOZO Photo/Design*
Printing and Binding: *Malloy Lithographing, Inc.*

Preface

What enters your mind when you hear the words "organic chemistry"? Some of you may think, "the chemistry of life," or "the chemistry of carbon." Other responses might include "premed," "pressure," "difficult," or "memorization." Although organic chemistry is, formally, the study of the compounds of carbon, the discipline of organic chemistry encompasses many skills that are common to other areas of study. Organic chemistry is as much a liberal art as a science, and mastery of the concepts and techniques of organic chemistry can lead to an enhanced competence in other fields.

As you proceed to solve the problems that accompany the text, you will bring to the task many problem-solving techniques. For example, planning an organic synthesis requires the skills of a chess player; you must plan your moves while looking several steps ahead, and you must keep your plan flexible. Structure-determination problems are like detective problems, in which many clues must be assembled to yield the most likely solution. Naming organic compounds is similar to the systematic naming of biological specimens; in both cases, a set of rules must be learned and then applied to the specimen or compound under study.

The problems in the text fall into two categories: drill problems and complex problems. Drill problems, which appear throughout the text and at the end of each chapter, test your knowledge of one fact or technique at a time. You may need to rely on memorization to solve these problems, which you should work on first. More complicated problems require you to recall facts from several parts of the text and then use one or more of the problem-solving techniques mentioned above. As each major type of problem — synthesis, nomenclature, or structure determination — is introduced in the text, a solution to that type of problem is extensively worked out in this *Solutions Manual*.

Here are several suggestions that may help you with problem solving:

1. The text is organized into chapters that describe individual functional groups. As you study each functional group, *make sure that you understand the structure and reactivity of that group*. In case your memory of a specific reaction fails you, you can rely on your general knowledge of functional groups for help.

2. *Use molecular models*. It is difficult to imagine the three-dimensional structure of an organic molecule when looking at a two-dimensional drawing. Models will help you to appreciate the structural aspects of organic chemistry and are indispensable tools for understanding stereochemistry.

3. Every effort has been made to make this *Solutions Manual* as clear, attractive, and error-free as possible. Nevertheless, you should *use the Solutions Manual in moderation*. The principle use of this book should be to check answers to problems you have already worked out. Also, this *Manual* can give you assistance if you are unable to solve a problem. The *Solutions Manual* should not be used as a substitute for mental effort; at times, struggling with a problem is the only way to teach yourself.

4. *Look through the appendices at the end of the Solutions Manual*. Some of these appendices contain tables that may help you in working problems; others present information related to the history of organic chemistry.

Acknowledgments   First, I would like to thank my husband, John McMurry, for suggesting this project and for supporting my efforts while this book was being written. My appreciation goes to Virginia Severn Goodman, Sonja Erion, and Melba Wallace, all of whom were involved in producing previous editions of this book. Finally, I am most grateful to Sherrie Yourstone for her superb job in the preparation of this manuscript on a Macintosh computer using the programs WordPerfect, ChemConnection, Canvas, and Expressionist.

Contents

Solutions to Problems

Chapter 1   Structure and Bonding   1
Chapter 2   Bonding and Molecular Properties   13
Chapter 3   The Nature of Organic Compounds: Alkanes and Cycloalkanes   27
Chapter 4   Stereochemistry of Alkanes and Cycloalkanes   43
Chapter 5   An Overview of Organic Reactions   61
Chapter 6   Alkenes: Structure and Reactivity   72
Chapter 7   Alkenes: Reactions and Synthesis   88
Chapter 8   Alkynes   106
Chapter 9   Stereochemistry   122
Chapter 10   Alkyl Halides   144
Chapter 11   Reactions of Alkyl Halides: Nucleophilic Substitutions and Eliminations   158
Chapter 12   Structure Determination: Mass Spectroscopy and Infrared Spectroscopy   178
Chapter 13   Structure Determination: Nuclear Magnetic Resonance Spectroscopy   190
Chapter 14   Conjugated Dienes and Ultraviolet Spectroscopy   209
Chapter 15   Benzene and Aromaticity   226
Chapter 16   Chemistry of Benzene: Electrophilic Aromatic Substitution   241
Chapter 17   Alcohols and Thiols   273
Chapter 18   Ethers, Epoxides, and Sulfides   295
Chapter 19   Aldehydes and Ketones: Nucleophilic Addition Reactions   313
Chapter 20   Carboxylic Acids   343
Chapter 21   Carboxylic Acid Derivatives and Nucleophilic Acyl Substitution Reactions   358
Chapter 22   Carbonyl Alpha-Substitution Reactions   383
Chapter 23   Carbonyl Condensation Reactions   407
Chapter 24   Carbohydrates   437
Chapter 25   Aliphatic Amines   457
Chapter 26   Arylamines and Phenols   481
Chapter 27   Amino Acids, Peptides, and Proteins   502
Chapter 28   Lipids   522
Chapter 29   Heterocycles and Nucleic Acids   538
Chapter 30   Orbitals and Organic Chemistry: Pericyclic Reactions   558
Chapter 31   Synthetic Polymers   575

Appendices

Functional-Group Synthesis   590
Functional-Group Reactions   596
Reagents in Organic Chemistry   601
Name Reactions in Organic Chemistry   610
Abbreviations   619
Infrared Absorption Frequencies   622
Proton NMR Chemical Shifts   625
Top 40 Organic Chemicals   626
Nobel Prize Winners in Chemistry   630
Suggested Readings   638

Chapter 1 – Structure and Bonding

1.1 The elements of the periodic table are organized into groups that are based on the number of outer-shell electrons each element has. For example, an element in Group 1A has one outer-shell electron, and an element in Group 5A has five outer-shell electrons. To find the number of outermost-shell electrons for a given element, use the periodic table to locate its group. For part (a), potassium is a member of Group 1A and thus has one electron in its outermost shell.

(b) Aluminum (Group 3A) has three outermost-shell electrons.

(c) Krypton is a noble gas and has eight electrons in its outermost shell.

1.2 a) To find the ground-state electronic configuration of an element, first locate its atomic number. For boron, the atomic number is 5; boron thus has 5 protons and 5 *electrons*. Next we assign the electrons to the proper energy levels, starting with the lowest level:

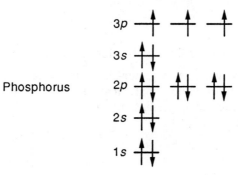

Remember that only two electrons can occupy the same orbital, and that they must be of opposite spin.

A different way to represent the ground-state electron configuration is to simply write down the occupied orbitals and to indicate the number of electrons in each orbital. For example, the electron configuration for boron is $1s^2 2s^2 2p$.

Often, we are interested only in the electrons in the outermost shell. We can then represent all filled levels by the symbol for the inert gas having the same levels filled. In the case of boron, the filled $1s$ energy level is represented by [He], and the *valence shell configuration* is symbolized by $[He]2s^2 2p$.

b) Let's consider an element with many electrons. Phosphorus, with an atomic number of 15, has 15 *electrons*. Assigning these to energy levels:

Notice that the 3p electrons are all in different orbitals. According to *Hund's rule*, we must place one electron into each orbital of the same energy level until all orbitals are half-filled.

The more concise way to represent ground-state electron configuration for phosphorus: $1s^2 2s^2 2p^6 3s^2 3p^3$

Valence shell electron configuration: $[Ne]3s^2 3p^3$

c)   Oxygen (atomic number 8)   d) Chlorine (atomic number 17)

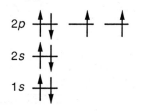

$1s^2 2s^2 2p^4$ $1s^2 2s^2 2p^6 3s^2 3p^5$

$[He]2s^2 2p^4$ $[Ne]3s^2 3p^5$

1.3

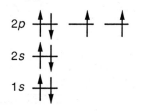

Chloroform

1.4

a)   Carbon (Group 4A) has four electrons in its valence shell and forms four bonds to achieve the noble-gas configuration of neon. Here, a likely formula is CCl_4.

Element	Group	Likely Formula
b) Al	3A	AlH_3
c) C	4A	CH_2Cl_2
d) Si	4A	SiF_4
e) N	5A	CH_3NH_2

1.5   Follow these three steps for drawing the Lewis structure of a molecule.

(1)   Determine the number of valence, or outer-shell electrons for each atom in the molecule. For chloroform, we know that carbon has four valence electrons, hydrogen has one, and each chlorine has seven.

$\cdot\overset{\displaystyle\cdot}{C}\cdot$ 4 x 1 = 4

H·   1 x 1 = 1

$:\overset{\displaystyle\cdot\cdot}{\underset{\displaystyle\cdot\cdot}{Cl}}\cdot$ 7 x 3 = 21

            26   total valence electrons

(2) Next, use two electrons for each single bond.

$$\begin{array}{c} H \\ \cdot\cdot \\ Cl:C:Cl \\ \cdot\cdot \\ Cl \end{array}$$

(3) Finally, use the remaining electrons to achieve an inert gas configuration for all atoms.

Molecule	Lewis structure	Line-bond structure
a) $CHCl_3$	H :Cl:C:Cl: :Cl:	H Cl—C—Cl Cl
b) H_2S 8 valence electrons	H:S: H	H—S H
c) CH_3NH_2 14 valence electrons	H H H:C:N:H H	H H H—C—N—H H
d) BH_3 6 valence electrons	H:B:H H	H—B—H H

Borane can't achieve an inert gas configuration because it has only six valence electrons.

e) NaH 2 valence electrons	Na:H	Na—H
f) CH_3Li 8 valence electrons	H H:C:Li H	H H—C—Li H

1.6 Bonds formed between an electropositive element and an electronegative element are ionic. Bonds formed between an element in the middle of the periodic table and another element are most often covalent, but exceptions can be found.

     Ionic bonds: LiI, KBr, $MgCl_2$
     Covalent bonds: CH_4, CH_2Cl_2, Cl_2

1.7

H H
H :C :C :H   Ethane

H H

H H
H—C—C—H
H H

1.8

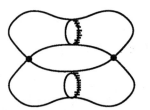

A plane through the middle of a *p–p* pi bond reveals two ovals, one above and one below the bond axis.

1.9

Propane

All carbon atoms are tetrahedral, and all bond angles are approximately 109.5°.

1.10 The two carbons bond to each other by overlap of two sp^3 hybrid orbitals. Six sp^3 hybrid orbitals (three from each carbon) are left over, and they can bond with a maximum of six hydrogens. Thus, a formula such as C_2H_7 is not possible.

1.11

Propene

   The *C3–H bonds* are sigma bonds formed by overlap of an sp^3 orbital of carbon 3 with an *s* orbital of hydrogen.

   The *C2–H and C1–H bonds* are sigma bonds formed by overlap of an sp^2 orbital of carbon with an *s* orbital of hydrogen.

   The *C2–C3 bond* is a sigma bond formed by overlap of an sp^3 orbital of carbon 3 with an sp^2 orbital of carbon 2.

   There are two *C1–C2 bonds*. One is a sigma bond formed by overlap of an sp^2 orbital of carbon 1 with an sp^2 orbital of carbon 2. The other is a pi bond formed by overlap of a *p* orbital of carbon 1 with a *p* orbital of carbon 2. All four atoms connected to the carbon-carbon double bond lie in the same plane, and all bond angles between these atoms are 120°.

1.12

H—C(sp²)=C₄(sp²)—C₃(sp²)... (structure)

All atoms lie in the same plane, and all bond angles are approximately 120°

1.13

H:C:C::O: Acetaldehyde

H—C—C=O

1.14

Propyne

The *C3-H bonds* are sigma bonds formed by overlap of an sp^3 orbital of carbon 3 with an *s* orbital of hydrogen.

The *C1-H bond* is a sigma bond formed by overlap of an *sp* orbital or carbon 1 with an *s* orbital of hydrogen.

The *C2-C3 bond* is a sigma bond formed by overlap of an *sp* orbital of carbon 2 with an sp^3 orbital of carbon 3.

There are three *C1-C2 bonds*. One is a sigma bond formed by overlap of an *sp* orbital of carbon 1 with an *sp* orbital of carbon 2. The other two bonds are pi bonds formed by overlap of two *p* orbitals of carbon 1 with two *p* orbitals of carbon 2.

The three carbon atoms of propyne lie on a straight line; the bond angle is 180°.

1.15

The two nitrogens are triple-bonded. A sigma bond is formed by the overlap of one *sp* orbital from each nitrogen, and two pi bonds are formed by the overlap of two *p* orbitals from one nitrogen atom with two *p* orbitals from the other nitrogen atom. A lone pair of electrons occupies the remaining *sp* orbital of each nitrogen.

1.16

H:C::N:H   Formaldimine

H—C=N—H

The bond between carbon and nitrogen is a double bond. The nitrogen atom is sp^2 hybridized.

1.17

a)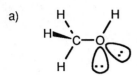

The sp^3–hybridized oxygen atom has tetrahedral geometry.

b)

Pyramidal geometry

c)

Like nitrogen, phosphorus has five outer-shell electrons. PH_3 has pyramidal geometry.

1.18

Element	Atomic Number	Number of valence electrons
a) Magnesium	12	2
b) Sulfur	16	6
c) Bromine	35	7

1.19

Element	Atomic Number	Ground-state Electronic configuration
a) Sodium	11	$1s^2 2s^2 2p^6 3s$
b) Aluminum	13	$1s^2 2s^2 2p^6 3s^2 3p$
c) Silicon	14	$1s^2 2s^2 2p^6 3s^2 3p^2$
d) Calcium	20	$1s^2 2s^2 2p^6 3s^2 3p^6 4s^2$

1.20 (a) $AlCl_3$ (b) CF_2Cl_2 (c) NI_3

1.21

a) H : C ::: C : H   10 valence electrons

b) H : Al : H   6 valence electrons

c) H : C : S : C : H   20 valence electrons

d)
$$H H$$
$$H:C::C:\overset{..}{\underset{..}{Cl}}:$$
18 valence electrons

e)
$$H H H H$$
$$H:C::C:C::C:H$$
22 valence electrons

f)
$$H \overset{..}{\underset{..}{O}}:$$
$$H:\overset{..}{C}:C:\overset{..}{\underset{..}{O}}:H$$
$$H$$
24 valence electrons

1.22

$$H$$
$$H:\overset{..}{C}:C:::N:$$
$$H$$
Acetonitrile

Nitrogen has five electrons in its outer electron shell. Three are used in the carbon-nitrogen triple bond, and two are a non-bonding electron pair.

1.23

a) $CH_3-\overset{..}{\underset{..}{O}}-CH_3$

b) $CH_3-\overset{\overset{:O:}{\|}}{C}-CH_3$

c) $CH_3-\overset{\overset{:O:}{\|}}{C}-\overset{..}{N}H_2$

d) $:\overset{..}{\underset{..}{F}}-\overset{\overset{H}{|}}{\underset{\underset{H}{|}}{C}}-\overset{..}{\underset{..}{Cl}}:$

1.24 In molecular formulas of organic molecules, carbon is listed first, followed by hydrogen. All other elements are listed in alphabetical order.

Compound	Molecular Formula
a) Phenol	C_6H_6O
b) Aspirin	$C_9H_8O_4$
c) Vitamin C	$C_6H_8O_6$
d) Nicotine	$C_{10}H_{14}N_2$
e) Novocain	$C_{13}H_{21}ClN_2O_2$
f) Glucose	$C_6H_{12}O_6$

1.25 In order to work a problem of this sort, you must examine *all* possible structures consistent with the rules of valence. You must systematically consider all possible attachments, including those that have branches, rings and multiple bonds.

a)
$$\overset{\overset{H\ H\ H}{|\ \ |\ \ |}}{H-C-C-C-H}$$
$$\underset{\underset{H\ H\ H}{|\ \ |\ \ |}}{}$$

b)
$$\overset{\overset{H\ H}{|\ \ |}}{H-C-N-H}$$
$$\underset{\underset{H}{|}}{}$$

c)
$$\overset{\overset{H\ \ \ \ H}{|\ \ \ \ |}}{H-C-O-C-H}$$
$$\underset{\underset{H\ \ \ \ H}{|\ \ \ \ |}}{}$$
and
$$\overset{\overset{H\ H}{|\ \ |}}{H-C-C-O-H}$$
$$\underset{\underset{H\ H}{|\ \ |}}{}$$

d)   H–C–C–C–H   and   H–C–C–C–H
(with H H H above and H H Br below for first; H H H above and H Br H below for second)

e)   H–C–C–H   (with H O above, H below, double bond C=O)

(structure) C=C with H, H on left and H, OH on right

(structure) epoxide ring with O at top, C–C, H's attached

f)   H–C–C–C–N–H   (with H H H H above, H H H below)

H–C–C–N–C–H   (with H H H H above, H H H below)

H–C–C–C–H   (with H H H above, H N H below, H H below N)

H–C–N–C–H   (with H H above, H C H below, H H H below)

1.26

a)   H—C—C—C—H   with sp³, sp³, sp³ labels and H's above and below

b)   H—C sp³ (with H, H above), sp² C=C sp², H—C sp³ (with H, H below), H, H

c)   C sp² (with H above), sp² C—C≡C—H (sp, sp labels), H—C sp² (with H below)

d)   H, C sp², C sp³ (with H, H), C sp³, C sp³ (with H, H), C sp², H (with H's attached)

e)   H—C—O—C—H   with sp³, sp³ labels and H's above and below

1.27

(benzene ring structure with H atoms)   Benzene

All carbon atoms of benzene are sp^2–hybridized, and all bond angles of benzene are 120°. Benzene is a planar molecule.

1.28 a) sp^3 b) sp^3 c) sp^2

1.29 All angles are approximate.
    a) 109°   b) 109°   c) 109°   d) 120°

1.30

a)

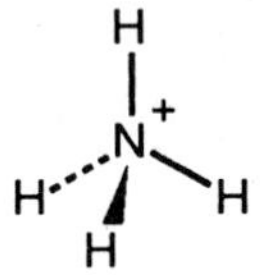

The ammonium ion is tetrahedral because nitrogen is sp^3 hybridized

b)

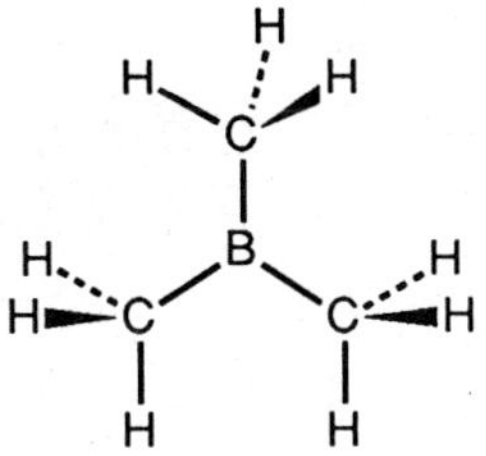

The boron-carbon portion of the molecule is planar because of sp^2 hybridization at boron. The $-CH_3$ portions are tetrahedral.

c)

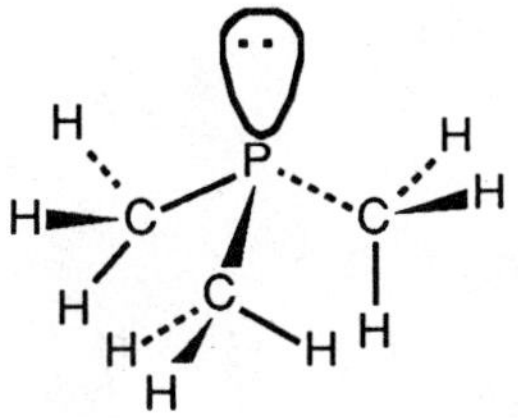

Trimethylphosphine is pyramidal. The $-CH_3$ portions are tetrahedral.

d)

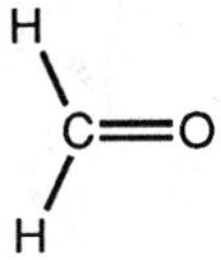

Formaldehyde is planar because carbon is sp^2 hybridized.

1.31

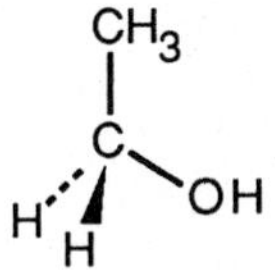

Ethanol

1.32 a) SO_2 has eighteen valence electrons (six from sulfur and six from each oxygen). The oxygen-sulfur-oxygen bond angle is approximately 120°.

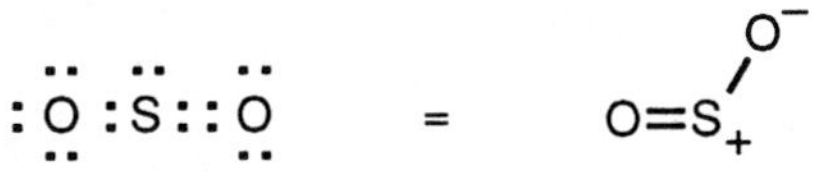

b) SO_3 has 24 valence electrons. SO_3 is a planar molecule.

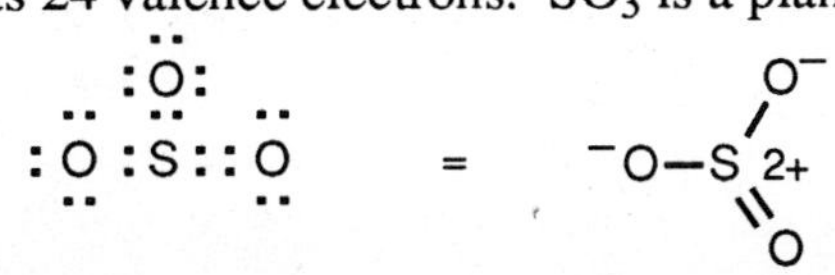

c)   Four oxygens and one sulfur contribute 30 valence electrons. In addition, there are two electrons that give SO_4^{2-} its negative charge. The total number of electrons used to draw the Lewis structure is 32. SO_4^{2-} is a tetrahedral anion.

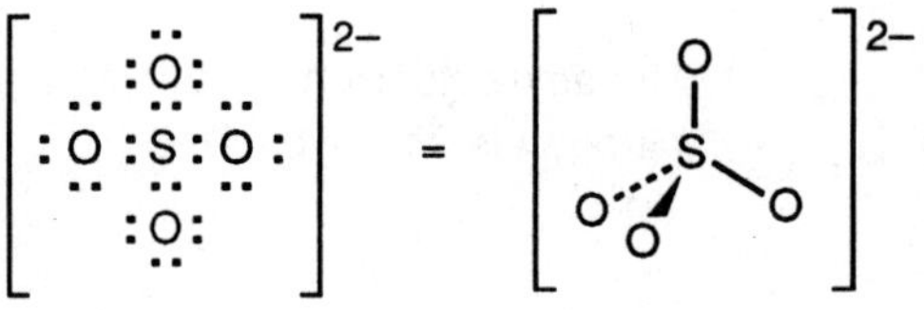

$$\left[\begin{array}{c} :\ddot{O}: \\ :\ddot{O}:S:\ddot{O}: \\ :\ddot{O}: \end{array} \right]^{2-} = \left[\begin{array}{c} O \\ O \cdots S \\ O \end{array} \right]^{2-}$$

1.33

a)

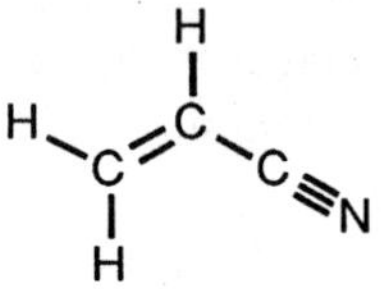

Acrylonitrile

b)

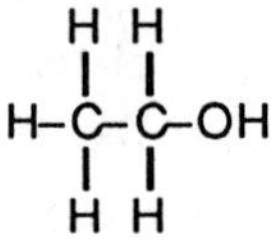

H H
H–C–C–OH
H H

Ethanol

c)

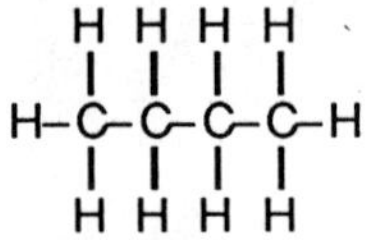

H H H H
H–C–C–C–C–H
H H H H

Butane

1.34

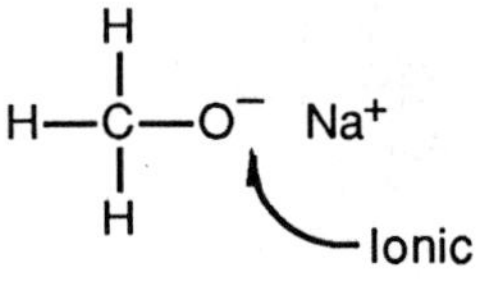

H
|
H—C—O⁻ Na⁺   All other bonds are covalent.
|
H
        ↖— Ionic

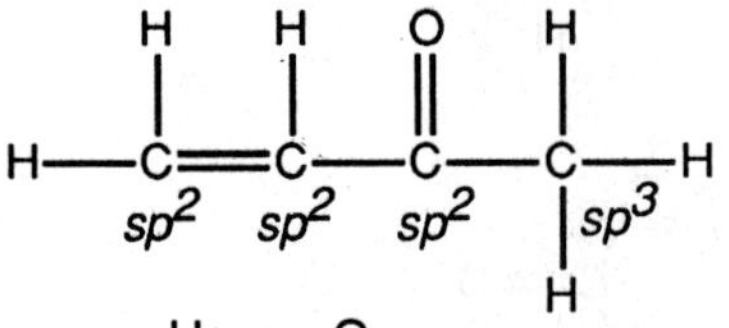

1.35

a) 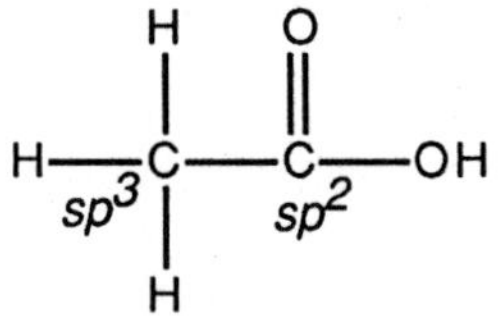   H—C—C—OH
        sp^3 sp^2

b)   H—C=C—C—C—H
        sp^2 sp^2 sp^2 sp^3

c)   H—C=C—C≡N
        sp^2 sp^2 sp

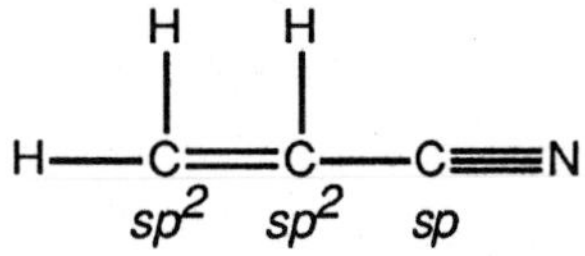

d)

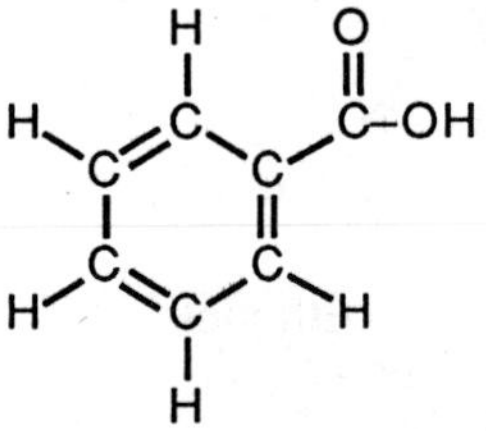

All carbon atoms of benzoic acid
are sp^2 hybridized.

1.36 a) sp^3 b) sp^2 c) sp d) sp^3

1.37 All angles are approximate.
    a) 109°   b) 180°   c) 109°

1.38 Ionic: NaCl
    Covalent: CH_3Cl, Cl_2, HOCl

1.39

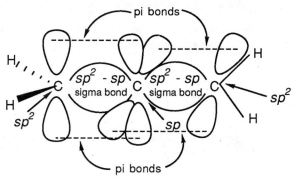

The central carbon of allene forms two sigma bonds and two pi bonds. The central carbon is *sp*–hybridized, and the two terminal carbons are *sp*^{2}–hybridized. The carbon-carbon bond angle is 180°, indicating linear geometry for the carbons.

1.40

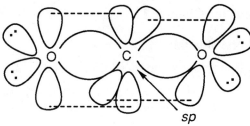

Carbon dioxide is a linear molecule.

1.41 a) The positively charged carbon atom has six valence shell electrons.
   b) A carbocation is *sp*^{2}–hybridized.
   c) A carbocation is planar.
   d) A carbocation is *isoelectronic* with (has the same number of electrons as) a trivalent boron compound.

1.42

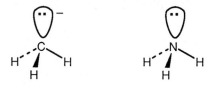

The negatively charged carbanion carbon has eight valence electrons and is *sp*^{3}–hybridized. A carbanion is tetrahedral and is isoelectronic with a trivalent nitrogen compound.

1.43 To answer this problem correctly, we must first remember the Pauli Exclusion Principle, which states that two electrons in the same orbital must have opposite spins. Thus, the two electrons of triplet (spin-unpaired) methylene must occupy different orbitals. In triplet methylene, *sp*–hybridized carbon forms one bond to each of two hydrogens. Each of the two unpaired electrons occupies a *p* orbital. In singlet (spin-paired) methylene the two electrons can occupy the same orbital because they have opposite spins. Including the two C–H bonds, there are a total of three occupied orbitals. We predict sp^2 hybridization and planar geometry for singlet methylene.

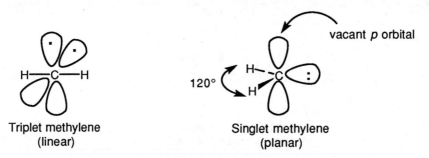

Triplet methylene
(linear)

Singlet methylene
(planar)

Study Guide for Chapter 1

After studying this chapter, you should be able to:

(1) Predict the ground-state electronic configuration of atoms (Problems 1.1, 1.2, 1.18, 1.19).

(2) Draw Lewis electron-dot structures of simple compounds (Problems 1.5, 1.7, 1.13, 1.16, 1.21, 1.22, 1.23, 1.32).

(3) Draw simple organic compounds with the correct three-dimensional geometry (1.3, 1.31).

(4) Identify bonds as ionic or covalent (1.6, 1.34, 1.38).

(5) Predict and describe the hybridization of bonds in simple compounds (1.9, 1.11, 1.12, 1.14, 1.15, 1.26, 1.27, 1.28, 1.35, 1.36, 1.41, 1.42, 1.43).

(6) Predict bond angles and shapes of molecules (1.17, 1.29, 1.30, 1.37, 1.40).

(7) Convert Kekulé structures into molecular formulas and *vice-versa* (1.24, 1.25, 1.33).

Chapter 2 – Bonding and Molecular Properties

2.1 a) This problem will be worked out in detail, using the rules in the text.

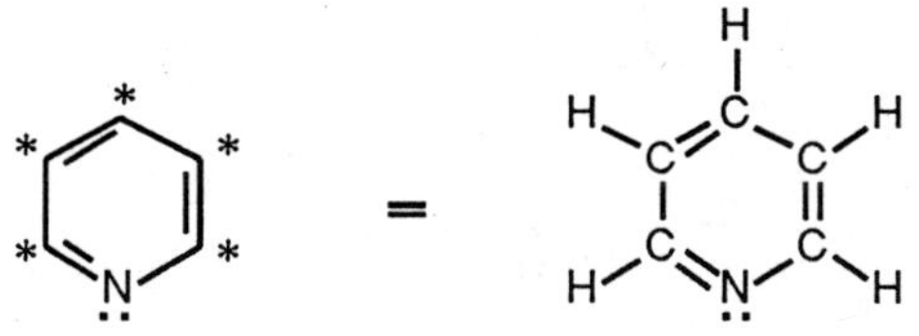

Remember: (1) carbon atoms occur at the intersection of two line-bonds (indicated by * above); (2) to satisfy valency, a hydrogen is implicitly understood to be bonded to each of the above carbons. The molecular formula of pyridine is C_5H_5N.

b) Cyclohexanone: $C_6H_{10}O$

c) Indole: C_8H_7N

2.2 Several possible shorthand structures can satisfy each molecular formula.

a) C_5H_{12}

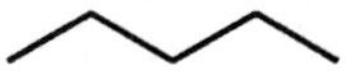

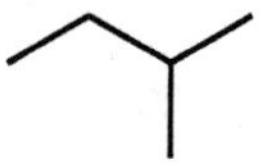

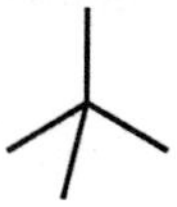

b) C_2H_7N

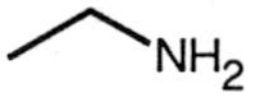

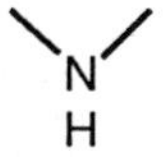

c) C_3H_6O

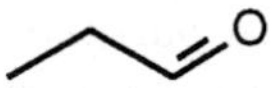

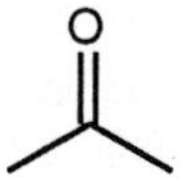

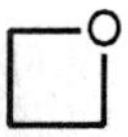

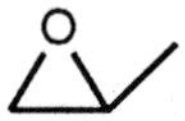

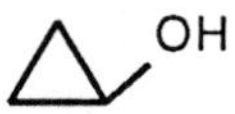

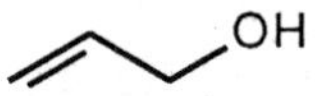

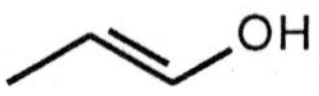

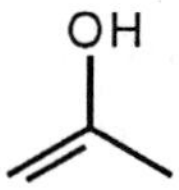

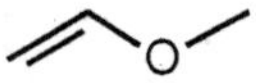

d) C_4H_9Cl

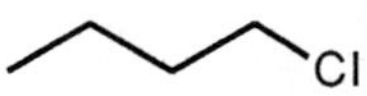

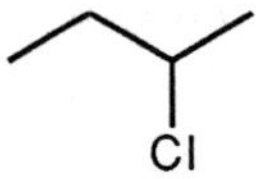

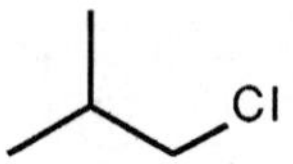

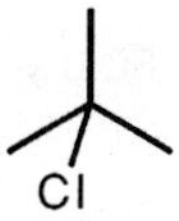

2.4

$$\text{Formal charge (FC)} = \left[\begin{matrix}\text{\# of valence}\\\text{electrons}\end{matrix}\right] - \left[\frac{\text{\# of bonding electrons}}{2}\right] - \left[\begin{matrix}\text{\# non-bonding}\\\text{electrons}\end{matrix}\right]$$

For sulfur, FC $= 6 - \dfrac{6}{2} - 2 = +1$

For oxygen, FC $= 6 - \dfrac{2}{2} - 6 = -1$

2.5

$$\text{Formal charge (FC)} = \left[\begin{matrix}\text{\# of valence}\\\text{electrons}\end{matrix}\right] - \left[\frac{\text{\# of bonding electrons}}{2}\right] - \left[\begin{matrix}\text{\# non-bonding}\\\text{electrons}\end{matrix}\right]$$

a) $H_2C=N=\ddot{N}:$ The Lewis dot structure is $H:\overset{H}{\underset{}{C}}::\overset{1}{N}::\overset{2}{N}:$

Hydrogen   FC $=$ $1 - \dfrac{2}{2} - 0 = 0$

Carbon   FC $=$ $4 - \dfrac{8}{2} - 0 = 0$

Nitrogen 1   FC $=$ $5 - \dfrac{8}{2} - 0 = +1$

Nitrogen 2   FC $=$ $5 - \dfrac{4}{2} - 4 = -1$

Remember: <u>Valence electrons</u> are the electrons particular to a specific element. <u>Bonding electrons</u> are those electrons involved in bonding to other atoms. <u>Non-bonding electrons</u> are those electrons in lone pairs.

b) $CH_3-C\equiv N-O$ $\equiv$ $H:\overset{H}{\underset{H}{\overset{1}{C}}}:\overset{2}{C}:::N:\ddot{O}:$

Hydrogen   FC $=$ $1 - \dfrac{2}{2} - 0 = 0$

Carbon 1   FC $=$ $4 - \dfrac{8}{2} - 0 = 0$

Carbon 2   FC $=$ $4 - \dfrac{8}{2} - 0 = 0$

Nitrogen   FC $=$ $5 - \dfrac{8}{2} - 0 = +1$

Oxygen   FC $=$ $6 - \dfrac{2}{2} - 6 = -1$

c) $\underset{\underset{H}{\overset{\overset{H}{1\,\underset{\cdot\cdot}{\overset{\cdot\cdot}{C}}}}{}}{H:}\!\underset{}{\overset{}{C}}:N:::\overset{2}{C}:$

Hydrogen	FC =	$1 - \dfrac{2}{2} - 0 = 0$
Carbon 1	FC =	$4 - \dfrac{8}{2} - 0 = 0$
Carbon 2	FC =	$4 - \dfrac{6}{2} - 2 = -1$
Nitrogen	FC =	$5 - \dfrac{8}{2} - 0 = +1$

2.6 Use Figure 2.3 to answer this problem. The larger the number, the more electronegative the element.

More electronegative *Less electronegative*

a) H (2.1)   Li (1.0)
b) Br (2.8)   Be (1.5)
c) Cl (3.0)   I (2.5)
d) C (2.5)   H (2.1)

Carbon is slightly more electronegative than hydrogen.

2.7 As in Problem 2.6, use Figure 2.3.

a) $H_3C—Br$ b) $H_3C—NH_2$ c) $H_3C—Li$ d) $H_2N—H$
      δ+ δ−   δ+ δ−   δ− δ+   δ− δ+

e) $H_3C—OH$ f) $H_3C—MgBr$ g) $H_3C—F$
      δ+ δ−   δ− δ+   δ+ δ−

2.8 a) We must look at the polarization of the individual bonds of a molecule to account for an observed dipole moment. In the case of methanol, CH_3OH, the individual bond polarities can be estimated from Figure 2.3. A bond is polarized in the direction of the more electronegative element (the larger numbers in Figure 2.3). In addition, we must take into account the contribution of the two lone pairs of oxygen. Indicating the individual bond polarities by arrows, we can predict the direction of the dipole moment.

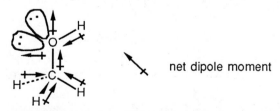

net dipole moment

It would be difficult to calculate the dipole moment from the individual bond moments. It is usually possible, however, to estimate qualitatively the direction and relative magnitude of the dipole moment by estimating the net direction of the bond polarities.

b)

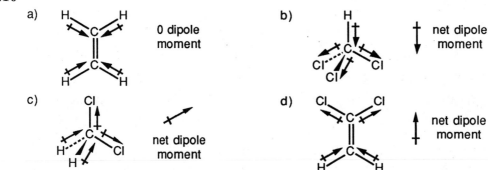

0 dipole moment

Notice that all individual bond moments cancel to give zero net dipole moment.

2.9

$$\delta^-\ \ \delta^+\ \ \delta^-$$
$$:\ddot{O}=C=\ddot{O}:$$

The dipole moment of CO_2 is zero because the bond polarities of the two carbon-oxygen bonds cancel.

2.10

a)   0 dipole moment

b)   net dipole moment

c)   net dipole moment

d)   net dipole moment

2.11 Recall from Section 2.6 that a strong acid has a low pK_a and a weak acid has a high pK_a. Accordingly, picric acid ($pK_a = 0.3$) is a stronger acid than formic acid ($pK_a = 3.7$).

2.12 HO–H is a stronger acid than H_2N–H. Since H_2N^- is a stronger base than HO^-, the conjugate acid of H_2N^- (H_2N–H) is a weaker acid than the conjugate acid of HO^- (HO–H).

2.13

a)   H–CN + $CH_3COO^-Na^+$ ----$\overset{?}{\longrightarrow}$ $Na^+{}^-CN$ + CH_3COOH

$pK_a = 9.2$ $pK_a = 4.7$
Weaker acid   Stronger acid

Remember that the lower the pK_a, the stronger the acid. Thus CH_3COOH, not HCN, is the stronger acid, and the above reaction will not take place in the direction written.

b) CH_3CH_2O-H + $Na^+ {}^-CN$ $\xrightarrow{\text{?}}$ $CH_3CH_2O^- Na^+$ + $H-CN$

$pK_a = 16.0$ $pK_a = 9.2$

Weaker acid   Stronger acid

Using the same reasoning as in part a), we can see that the above reaction will not occur.

2.14

$$\underset{\substack{pK_a = 20 \\ \text{Stronger acid}}}{H_3C-\overset{\overset{\displaystyle O}{\|}}{C}-CH_3} + Na^+ {}^-:NH_2 \longrightarrow \underset{\substack{pK_a = 36 \\ \text{Weaker acid}}}{H_3C-\overset{\overset{\displaystyle O}{\|}}{C}-CH_2:{}^- Na^+} + :NH_3$$

The above reaction will take place as written.

2.15 Enter –9.2 into the calculator and use the INV LOG function to arrive at the answer $K_a = 6 \times 10^{-10}$.

2.16

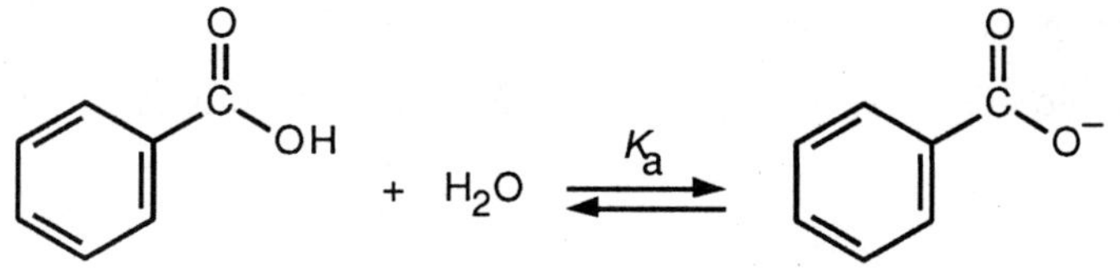

[0.500 M]   [x]   [x]

Benzoic acid   Benzoate anion

$$K_a = \frac{[\text{Benzoate}] [H_3O^+]}{[\text{Benzoic acid}]} = \frac{x^2}{0.500 - x}$$

Because x is small, $0.500 - x$ is approximately equal to 0.500.

Since $pK_a = 4.2$, $K_a = 6 \times 10^{-5}$.

$$6 \times 10^{-5} = \frac{x^2}{0.500}$$

$$30 \times 10^{-6} = x^2$$

$$5.6 \times 10^{-3} = [H_3O^+]; pH = 2.3.$$

2.17 A Lewis base has a nonbonding electron pair to share. A Lewis acid has a vacant orbital to accept an electron pair. Look for a lone electron pair when deciding if a molecule is a Lewis base.

Lewis acids: $MgBr_2$, $B(CH_3)_3$, ${}^+CH_3$

Lewis bases: $CH_3CH_2\ddot{\underset{\cdot\cdot}{O}}H$, $CH_3\ddot{N}HCH_3$, $\ddot{P}(CH_3)_3$

2.18

a)

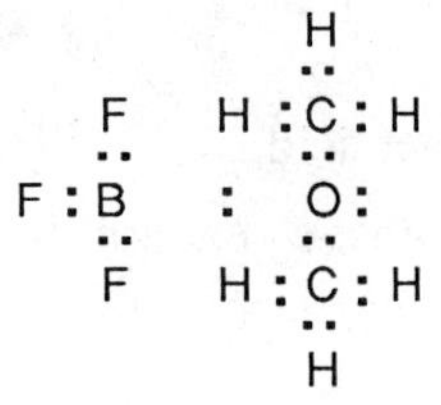

$$FC = 3 - \frac{8}{2} - 0 = -1 \quad \text{boron:}$$

$$FC = 6 - \frac{6}{2} - 2 = +1 \quad \text{oxygen:}$$

The formal charge of -1 for boron indicates that boron has a net negative charge; oxygen has a net positive charge.

b)

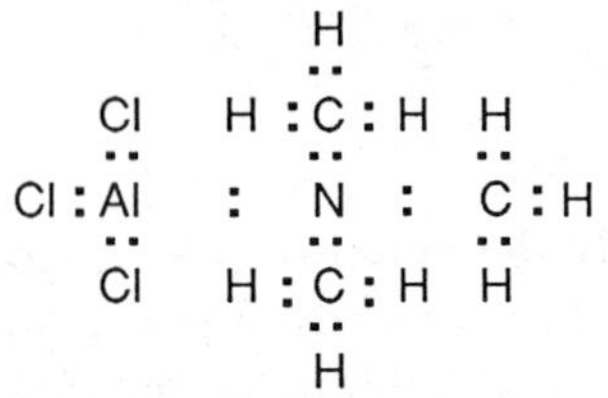

$$\text{aluminum:} \quad FC = 3 - \frac{8}{2} - 0 = -1$$

$$\text{nitrogen:} \quad FC = 5 - \frac{8}{2} - 0 = +1$$

The formal charge of -1 for aluminum indicates that it has a net negative charge; the formal charge $+1$ for nitrogen indicates that it has a net positive charge.

For clarity, electron dots have been left off F and Cl in the above structures.

2.19

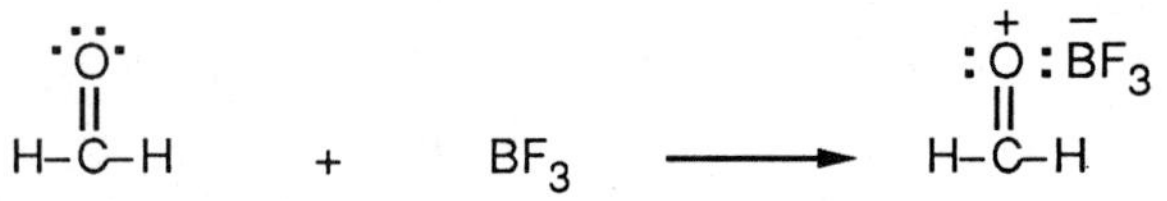

H–C–H   + BF_3 $\longrightarrow$ H–C–H

Lewis base   Lewis acid

2.20

a)   Benzene, C_6H_6

molecular weight $= 6(12.011) + 6(1.008) = 78.11$ g/mol

$$\% \text{ carbon} = \frac{6(12.011)}{78.11} \times 100 = \frac{72.06}{78.11} \times 100 = 92.25\%$$

$$\% \text{ hydrogen} = \frac{6(1.008)}{78.11} \times 100 = \frac{6.048}{78.11} \times 100 = 7.75\%$$

b)   Laetrile, $C_{14}H_{15}NO_7$

molecular weight $= 309.28$ g/mol

$$\% \text{ carbon} = \frac{168.15}{309.28} \times 100 = 54.37\% \qquad \% \text{ nitrogen} = \frac{14.01}{309.28} \times 100 = 4.53\%$$

$$\% \text{ hydrogen} = \frac{15.12}{309.28} \times 100 = 4.89\% \qquad \% \text{ oxygen} = \frac{112.00}{309.28} \times 100 = 36.21\%$$

c) Quinine, $C_{20}H_{24}N_2O_2$

molecular weight = 324.42 g/mol

% carbon $= \dfrac{240.22}{324.42} \times 100 = 74.05\%$ % nitrogen $= \dfrac{28.01}{324.42} \times 100 = 8.63\%$

% hydrogen $= \dfrac{24.19}{324.42} \times 100 = 7.46\%$ % oxygen $= \dfrac{32.00}{324.42} \times 100 = 9.86\%$

d) Diethylstilbesterol, $C_{18}H_{20}O_2$

molecular weight = 268.36 g/mol

% carbon $= \dfrac{216.20}{268.36} \times 100 = 80.56\%$ % oxygen $= \dfrac{32.00}{268.36} \times 100 = 11.92\%$

% hydrogen $= \dfrac{20.16}{268.36} \times 100 = 7.51\%$

2.21 Citral has a percent composition 78.9% carbon, 10.6% hydrogen, 10.5% oxygen. Assume that we have a 100 g sample of citral. It contains 78.9 g carbon, 10.6 g hydrogen, 10.5 g oxygen. If we now divide the weight of each element in the 100 g sample by the atomic weight of the element, we get the *relative* number of moles of each element in the sample.

for C $\dfrac{78.9 \text{ g}}{12.0 \text{ g/mol}} = 6.56$ moles C

for H $\dfrac{10.6 \text{ g}}{1.01 \text{ g/mol}} = 10.51$ moles H

for O $\dfrac{10.5 \text{ g}}{16.0 \text{ g/mol}} = 0.656$ moles O

Next, it is necessary to find the ratio of the numbers of moles of the elements of citral in terms of the smallest whole numbers.

C : H : O = 6.56 : 10.51 : 0.656 = 10 : 16 : 1

So, the empirical formula of citral is $C_{10}H_{16}O$. The formula weight corresponding to $C_{10}H_{16}O$ is 152; since the actual molecular weight given is 152 g/mol, we know that the molecular formula of citral is $C_{10}H_{16}O$.

2.22 The first step in solving this problem is to determine how many mg of carbon and hydrogen are in an 8.00 mg sample of squalene.

mg carbon $= mg\,CO_2 \times \dfrac{\text{at. wt. C}}{\text{m.w. } CO_2} = 25.6 \text{ mg} \times \dfrac{12.0}{44.0} = 6.98$ mg C

mg hydrogen $= mg\,H_2O \times \dfrac{2(\text{at. wt. H})}{\text{m.w. } H_2O} = 8.75 \text{ mg} \times \dfrac{2.02}{18.0} = 0.98$ mg H

6.98 mg + 0.98 mg = 7.96 mg (C + H). Because this number is so close to 8.00 mg, we can assume that squalene contains only C and H.

Next, it is necessary to *calculate the percent* of carbon and hydrogen in the sample.

$$\frac{6.98 \text{ mg}}{8.00 \text{ mg}} \times 100 = 87.2\% \text{ carbon} \qquad \frac{0.98 \text{ mg}}{8.00 \text{ mg}} \times 100 = 12.2\% \text{ hydrogen}$$

Now, as in problem 2.8, we must *find the relative number of moles* of carbon and hydrogen.

$$\frac{87.2 \text{ g}}{12.0 \text{ g/mol}} = 7.26 \text{ moles carbon}$$

$$\frac{12.2 \text{ g}}{1.01 \text{ g/mol}} = 12.1 \text{ moles hydrogen}$$

Finding the nearest whole number ratio, we obtain:

C : H = 7.26 : 12.1 = 3 : 5

Thus, the empirical formula of squalene is C_3H_5. The formula weight of a compound having the formula C_3H_5 is $3(12.0) + 5(1.01) = 41.0$. Since the actual molecular weight of squalene is 410 g/mol, the molecular formula = 10 x (empirical formula), or $C_{30}H_{50}$.

2.23

a)

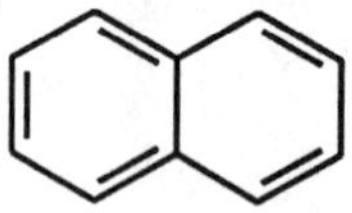

b)

c)

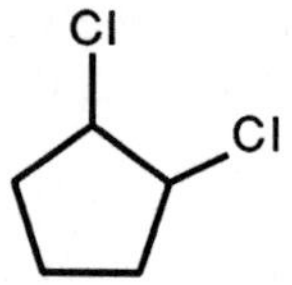

d)

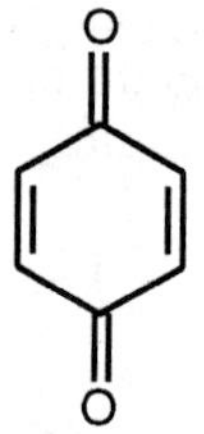

2.24

a)

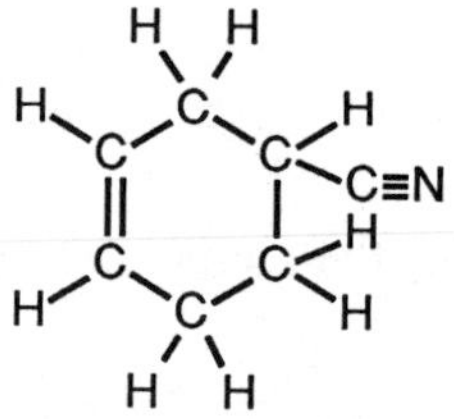

b)

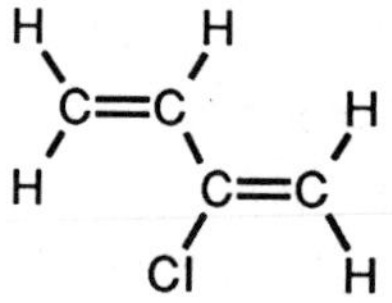

c)

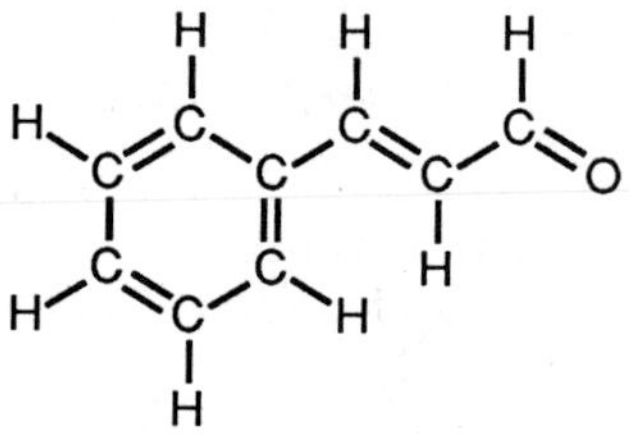

d)

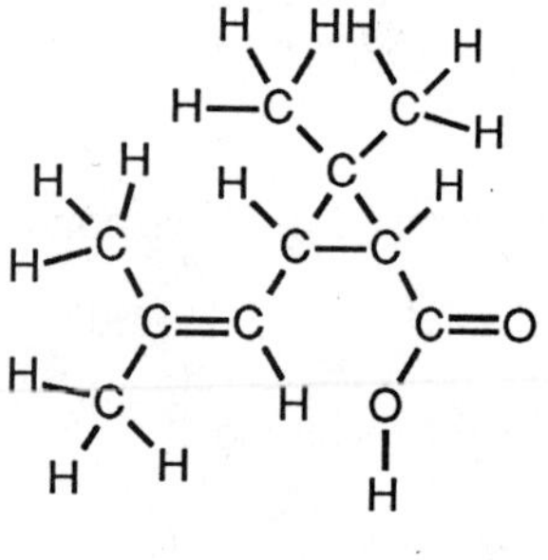

2.25

a) $(CH_3)_2\overset{..}{O}BF_3$

oxygen $FC = 6 - \dfrac{6}{2} - 2 = +1$

boron $FC = 3 - \dfrac{8}{2} - 0 = -1$

b) $H_2\overset{..}{C}-\overset{1}{N}\equiv\overset{2}{N}:$

carbon   FC $= 4 - \dfrac{6}{2} - 2 = -1$

nitrogen 1   FC $= 5 - \dfrac{8}{2} - 0 = +1$

nitrogen 2   FC $= 5 - \dfrac{6}{2} - 2 = 0$

c) $H_2C=\overset{1}{N}=\overset{2}{\underset{..}{N}}:$

carbon   FC $= 4 - \dfrac{8}{2} - 0 = 0$

nitrogen 1   FC $= 5 - \dfrac{8}{2} - 0 = +1$

nitrogen 2   FC $= 5 - \dfrac{4}{2} - 4 = -1$

d) $:\overset{1}{\underset{..}{O}}=\overset{2}{\underset{..}{O}}-\overset{3}{\underset{..}{O}}:$

oxygen 1   FC $= 6 - \dfrac{4}{2} - 4 = 0$

oxygen 2   FC $= 6 - \dfrac{6}{2} - 2 = +1$

oxygen 3   FC $= 6 - \dfrac{2}{2} - 6 = -1$

e) $H_2\overset{..}{C}-\underset{\underset{CH_3}{|}}{\overset{\overset{CH_3}{|}}{P}}-CH_3$

carbon   FC $= 4 - \dfrac{6}{2} - 2 = -1$

phosphorus   FC $= 5 - \dfrac{8}{2} - 0 = +1$

f) (pyridine ring)$N-\overset{..}{\underset{..}{O}}:$

nitrogen   FC $= 5 - \dfrac{8}{2} - 0 = +1$

oxygen   FC $= 6 - \dfrac{2}{2} - 6 = -1$

2.26–2.27

More polar	Less polar

a) $H_3C\xrightarrow{+} Cl$ $\delta^+ \quad \delta^-$ $Cl-Cl$

b) $H\xrightarrow{+} Cl$ $\delta^+ \quad \delta^-$ $H_3C\xleftarrow{+} H$ $\delta^- \quad \delta^+$

c) $HO\xleftarrow{+} CH_3$ $\delta^- \quad \delta^+$ $(CH_3)_3Si\xrightarrow{+} CH_3$ $\delta^+ \quad \delta^-$

d) $Li\xrightarrow{+} OH$ $\delta^+ \quad \delta^-$ $H_3C\xleftarrow{+} Li$ $\delta^- \quad \delta^+$

2.28

a)

b) no dipole moment

c)   Li–H

d) $F_3B \xleftarrow{\quad} \overset{+}{N}(CH_3)_3$

e)

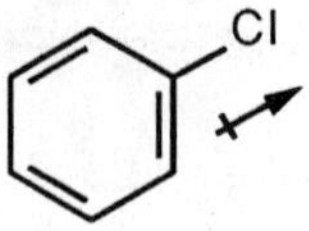

f)

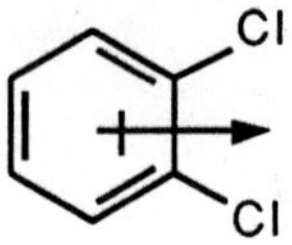

2.29 The magnitude of a dipole moment depends on both charge and distance between atoms. Fluorine is more electronegative than chlorine, but because a C–F bond is shorter than a C–Cl bond, the dipole moment of CH_3F is smaller than that of CH_3Cl.

2.30 Use Figure 2.3 for electronegativities. The most electronegative atom is starred.

a) $\overset{*}{CH_2}FCl$

b) $\overset{*}{F}CH_2CH_2CH_2Br$

c) $H\overset{*}{O}CH_2CH_2NH_2$

d) $CH_3\overset{*}{O}CH_2Li$

2.31

$CH_3\overset{..}{O}H$ + HCl $\rightleftharpoons$ $CH_3\overset{+}{\overset{..}{O}}H_2$ + Cl^-

$CH_3\overset{..}{O}H$ + $Na^+ {}^-\overset{..}{N}H_2$ $\rightleftharpoons$ $CH_3\overset{..}{\underset{..}{O}}{:}Na^+$ + $\overset{..}{N}H_3$

2.32

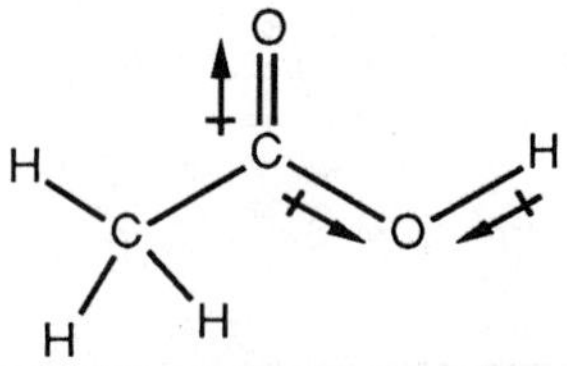

The O–H hydrogen of acetic acid is more acidic than the C–H hydrogens. The –OH oxygen is quite electronegative, and, consequently, the –O–H bond is more highly polarized than the –C–H bonds.

2.33

Lewis acids: $AlBr_3$, BH_3, HF, $TiCl_4$

Lewis bases: $CH_3CH_2\overset{..}{N}H_2$, $CH_3-\overset{..}{\underset{..}{S}}-CH_3$

2.34

a)   :Br:Al:Br:
         :Br:

b)   H H
     H:C:C:N:H
       H H H

c)   H:B:H
         H

d)   H:F:

e)   H   H
     H:C:S:C:H
       H   H

f)   :Cl:
     :Cl:Ti:Cl:
        :Cl:

2.35

a) $H_3C-\overset{\underset{\displaystyle CH_3}{|}}{\underset{\underset{\displaystyle CH_3}{|}}{N^+}}-\overset{..}{\underset{..}{O}}{:}^-$

b) $H_3C-\overset{..}{N}-\overset{+}{N}\equiv N:$

c) $H_3C-\overset{..}{N}=\overset{+}{N}=\overset{..}{\underset{..}{N}}{:}^-$

2.36

Least acidic $\longrightarrow$ Most acidic

$$\underset{\displaystyle pK_a = 20}{CH_3\overset{\overset{\displaystyle O}{\|}}{C}CH_3} \quad < \quad \underset{\displaystyle pK_a = 10}{\text{(phenol)}-OH} \quad < \quad \underset{\displaystyle pK_a = 9}{CH_3\overset{\overset{\displaystyle O}{\|}}{C}CH_2\overset{\overset{\displaystyle O}{\|}}{C}CH_3} \quad < \quad \underset{\displaystyle pK_a = 4.7}{CH_3\overset{\overset{\displaystyle O}{\|}}{C}OH}$$

2.37 To react completely with NaOH, an acid must have a pK_a somewhat lower than the pK_a of H_2O. Thus, all substances in the previous problem except acetone will react completely with NaOH.

2.38 The stronger the acid (lower pK_a), the weaker its conjugate base. Since NH_4^+ is a stronger acid than $CH_3NH_3^+$, NH_3 is a weaker base than CH_3NH_2.

2.39

$$\underset{\substack{pK_a = 15.7 \\ \text{stronger acid}}}{H_3C-\overset{\overset{\displaystyle CH_3}{|}}{\underset{\underset{\displaystyle CH_3}{|}}{C}}-O^-K^+} + H_2O \longrightarrow \underset{\substack{pK_a \approx 18 \\ \text{weaker acid}}}{H_3C-\overset{\overset{\displaystyle CH_3}{|}}{\underset{\underset{\displaystyle CH_3}{|}}{C}}-OH} + KOH$$

The reaction will take place as written because water is a stronger acid than *tert*–butyl alcohol.

2.40 a) Acetone: $K_a = 10^{-20}$ **b)** Formic acid: $K_a = 2.0 \times 10^{-4}$

2.41 a) Nitromethane: $pK_a = 10.30$ **b)** Acrylic acid: $pK_a = 4.20$

2.42

$$\underset{\substack{[0.050 \text{ M}]}}{\text{Formic acid}} + H_2O \underset{}{\overset{K_a}{\rightleftharpoons}} \underset{[x]}{\text{Formate}^-} + \underset{[x]}{H_3O^+}$$

$$K_a = 2.0 \times 10^{-4} = \frac{x^2}{0.050 - x}$$

If you let $0.050 - x = 0.050$, then $x = 3.16 \times 10^{-3}$ and pH = 2.50. If you calculate x exactly, then $x = 3.36 \times 10^{-3}$ and pH = 2.47.

2.43 Only acetic acid will react with sodium bicarbonate.

2.44 Sodium bicarbonate reacts with acetic acid to produce carbonic acid, which breaks down to form CO_2. The resulting CO_2 bubbles indicate the presence of acetic acid. Phenol does not react with sodium bicarbonate.

2.45

a) $CH_3\ddot{O}H$ + H^+ ⟶ $CH_3\overset{+}{\ddot{O}}H_2$

   base   acid

b) $CH_3\ddot{O}H$ + $:\bar{N}H_2$ ⟶ $CH_3\ddot{O}:^-$ + $:NH_3$

   acid   base

c)

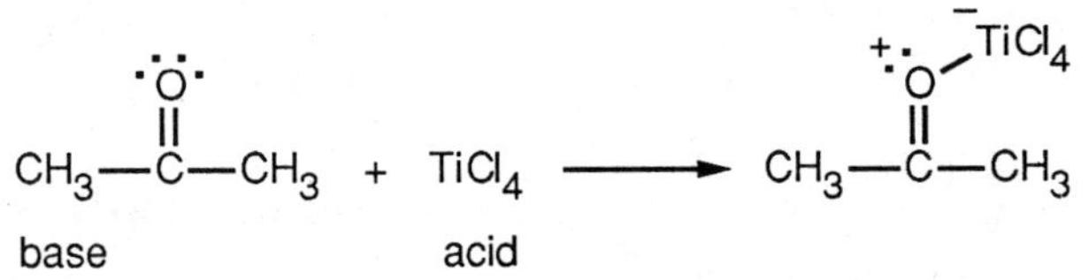

  base   acid

d)

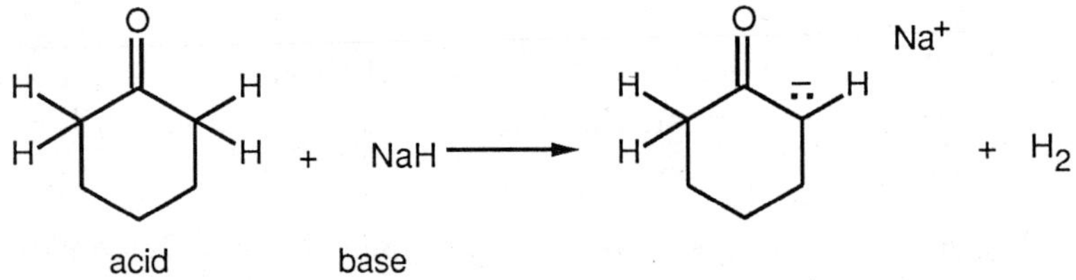

   acid   base

e)

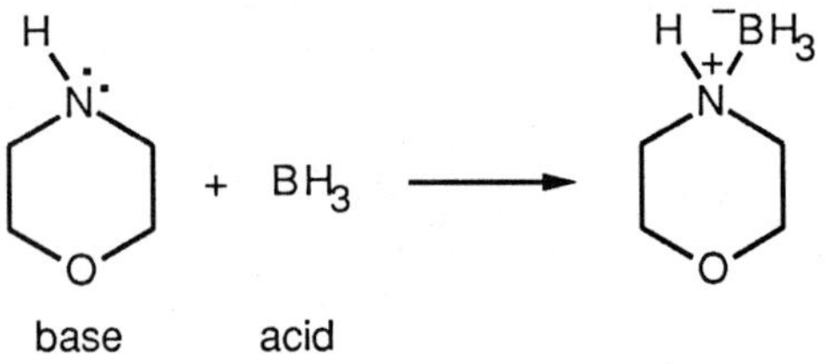

   base   acid

f) $(CH_3)_3\overset{+}{O}\overset{-}{B}F_4$ +

   acid   base

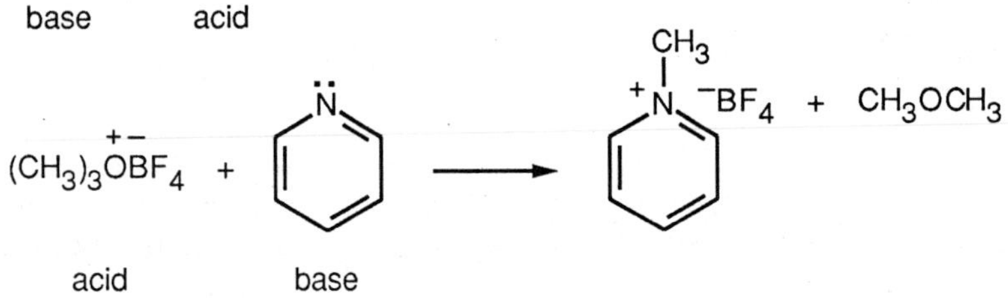

2.46 a) $C_9H_8O_4$ 60.00% C: 4.47% H;   35.52% O
     b) $C_{16}H_{30}O$ 80.61% C;   12.68% H;   6.71% O
     c) $C_{17}H_{19}NO_3$ 71.56% C;   6.71% H;   4.91% N;   16.82% O
     d) $C_{21}H_{22}N_2O_2$ 75.42% C;   6.63% H;   8.38% N;   9.57% O

2.47 The molecular formula of α–pinene is $C_{10}H_{16}$; see problem 2.21 for the method of solution.

2.48 Because the percentages of carbon and hydrogen do not add up to 100%, another element is present in jasmone. The other element is oxygen, present as 9.6% by weight.

As in problem 2.21, assume that we have a 100 g sample. Then we would have

$$80.7 \text{ g carbon} \qquad \frac{80.7 \text{ g}}{12.0 \text{ g/mol}} = 6.72 \text{ mol carbon}$$

$$9.7 \text{ g hydrogen} \qquad \frac{9.7 \text{ g}}{1.01 \text{ g/mol}} = 9.62 \text{ mol hydrogen}$$

$$9.6 \text{ g oxygen} \qquad \frac{9.6 \text{ g}}{16.0 \text{ g/mol}} = 0.60 \text{ mol oxygen}$$

C : H : O = 6.72 : 9.62 : 0.60 = 11.2 : 16.0 : 1 or $C_{11}H_{16}O$.

A compound of this empirical formula has a formula weight of 164. Since this weight agrees with the observed molecular weight, the molecular formula of jasmone is $C_{11}H_{16}O$.

2.49

$$\text{Weight of carbon in sample} = 0.0147 \text{ g} \times \frac{12.0}{44.0} = 0.00401 \text{ g}$$

$$\text{Weight of hydrogen in sample} = 0.0043 \text{ g} \times \frac{2.02}{18.0} = 0.00048 \text{ g}$$

Since these weights do not add up to 0.0050 g, we assume that oxygen is also present.

Weight of oxygen = 0.00500 g − 0.00401 g − 0.00048 g = 0.00051 g oxygen

$$\% \text{ C} = \frac{0.0401 \text{ g}}{0.0500 \text{ g}} = 80.2\% \qquad \text{rel. no. moles:} \quad \text{C} = \frac{80.2}{12.0} = 6.68$$

$$\% \text{ H} = \frac{0.0048 \text{ g}}{0.0500 \text{ g}} = 9.6\% \qquad\qquad\qquad\quad \text{H} = \frac{9.6}{1.01} = 9.50$$

$$\% \text{ O} = \frac{0.0051 \text{ g}}{0.0500 \text{ g}} = 10.2\% \qquad\qquad\qquad\quad \text{O} = \frac{10.2}{16.0} = 0.64$$

C : H : O = 6.68 : 9.50 : 0.64   =   10.45 : 14.84 : 1   =   21 : 30 : 2

A compound of empirical formula $C_{21}H_{30}O_2$ has a formula weight of 314, which is identical to the given molecular weight of progesterone. Thus, the molecular formula of progesterone is $C_{21}H_{30}O_2$.

2.50

$$\text{Weight of carbon} = 0.086 \text{ g } CO_2 \times \frac{12.0}{44.0} = 0.0235 \text{ g C}$$

$$\text{Weight of hydrogen} = 0.051 \text{ g } H_2O \times \frac{2.02}{18.0} = 0.0057 \text{ g H}$$

$$\text{Weight of nitrogen} = \frac{8.6 \text{ mL}}{22.4 \text{ mL/mmol}} \times 0.028 \text{ g/mmol} = 0.0108 \text{ g N}$$

Since these weights add up to 0.040 g, no other elements are present.

$$\% \text{ C} = \frac{0.0235 \text{ g}}{0.040 \text{ g}} \times 100 = 58.8\% \qquad\qquad \frac{58.8 \text{ g}}{12.0 \text{ g/mol}} = 4.90 \text{ mol C}$$

$$\% \text{ H} = \frac{0.0057 \text{ g}}{0.040 \text{ g}} \times 100 = 14.2\% \qquad\qquad \frac{14.2 \text{ g}}{1.01 \text{ g/mol}} = 14.1 \text{ mol H}$$

$$\% \text{ N} = \frac{0.0108 \text{ g}}{0.040 \text{ g}} \times 100 = 27.0\% \qquad\qquad \frac{27.0 \text{ g}}{14.0 \text{ g/mol}} = 1.93 \text{ mol N}$$

C : H : N = 4.90 : 1.93 or approximately 5 : 14 : 2.

The formula weight for $C_5H_{14}N_2 = 102$. Since this weight is equal to the given molecular weight, $C_5H_{14}N_2$ is the molecular formula of cadaverine. Cadaverine has a vile stench.

2.51

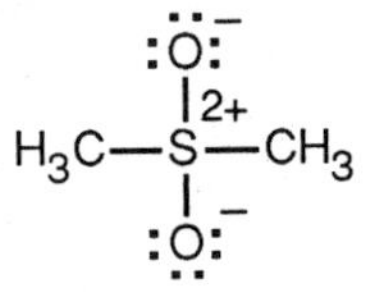

FC of each oxygen $= 6 - \dfrac{2}{2} - 6 = -1$

FC of sulfur $= 6 - \dfrac{8}{2} - 0 = +2$

The presence of formal charges indicates that electrons are not shared equally between S and O; they are strongly attracted to oxygen. The S–O bonds, therefore, are strongly polar. If the geometry of the molecule were planar (as drawn above), the dipole moments of the individual S–O bonds would cancel, resulting in a net dipole moment of zero. Tetrahedral geometry, however, predicts a large dipole moment.

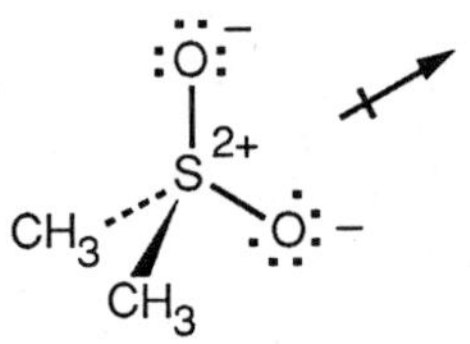

Study Guide for Chapter 2

After studying this chapter, you should be able to:

(1) Draw chemical structures from molecular formulas, and *vice-versa* (2.1, 2.2, 2.23, 2.24).

(2) Calculate the formal charge for atoms in molecules (2.4, 2.5, 2.18, 2.25, 2.35, 2.51).

(3) Predict which of a pair of elements is more electronegative (2.6, 2.7, 2.30).

(4) Predict the direction of polarity of a chemical bond, and predict the dipole moment of a compound (2.8, 2.9, 2.10, 2.26, 2.27, 2.28, 2.29, 2.32).

(5) Predict the relative acid/base strength of Brønsted acids and bases (2.11, 2.12, 2.36, 2.37, 2.38).

(6) Predict the direction of Brønsted acid-base reactions (2.13, 2.14, 2.31, 2.39, 2.43, 2.44).

(7) Identify Lewis acids and bases (2.17, 2.19, 2.33, 2.34, 2.45).

(8) Calculate:

(a) pK_a from K_a, and *vice-versa* (2.15, 2.40, 2.41).

(b) pH of a solution of a weak acid (2.16, 2.42).

(9) Calculate:

(a) Percent composition (2.20, 2.46).

(b) Molecular formula (2.21, 2.22, 2.47, 2.48, 2.49, 2.50).

Chapter 3 – The Nature of Organic Compounds: Alkanes and Cycloalkanes

3.1

a)

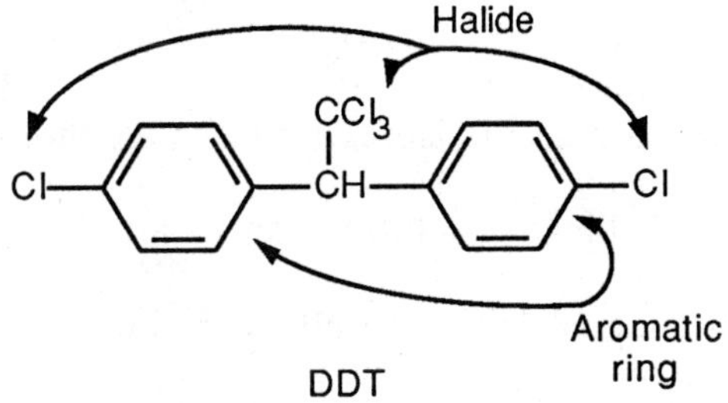

Halide · CCl_3 · Aromatic ring

Cl—⟨ring⟩—CH—⟨ring⟩—Cl

DDT

b)

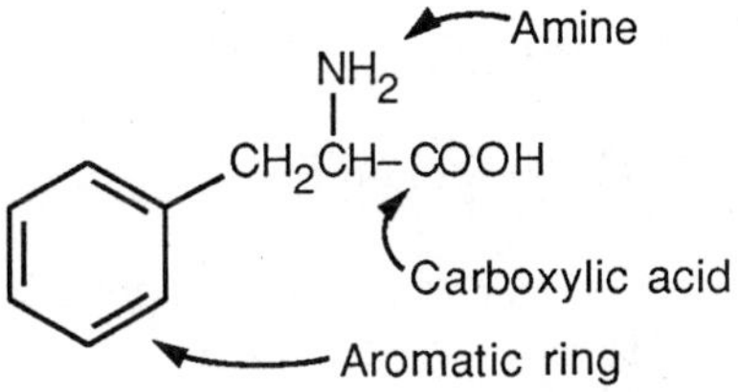

Amine · NH_2

$CH_2CH–COOH$

Carboxylic acid

Aromatic ring

Phenylalanine

c)

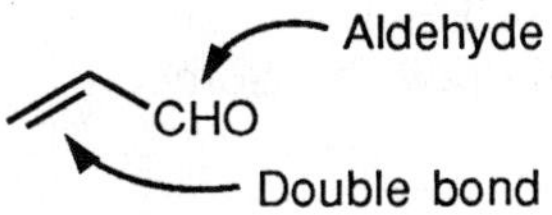

Aldehyde · CHO

Double bond

Acrolein

d)

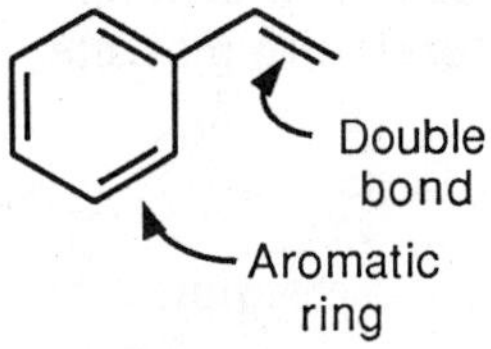

Double bond

Aromatic ring

Styrene

3.2

a) CH_3OH
Methanol

b)

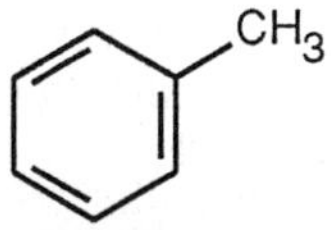

CH_3

Toluene

c) CH_3COOH
Acetic acid

d) CH_3NH_2
Methylamine

e)
$$CH_3\overset{\overset{\displaystyle O}{\|}}{C}CH_2NH_2$$
Aminoacetone

f)

1,3–Butadiene

3.3 We know that carbon forms four bonds and hydrogen forms one. Thus, if you draw all possible six carbon skeletons and add hydrogens so that all carbons have four bonds, you will arrive at the following structures:

$$CH_3CH_2CH_2CH_2CH_2CH_3$$

$$\underset{\displaystyle CH_3CH_2CH_2\overset{\displaystyle \overset{CH_3}{|}}{C}HCH_3}{}$$

$$CH_3CH_2\overset{\displaystyle \overset{CH_3}{|}}{C}HCH_2CH_3$$

$$CH_3CH_2\overset{\displaystyle \overset{CH_3}{|}}{\underset{\displaystyle \underset{CH_3}{|}}{C}}CH_3$$

$$CH_3\overset{\displaystyle \overset{CH_3}{|}}{C}H\overset{\displaystyle \overset{CH_3}{|}}{\underset{\displaystyle \underset{CH_3}{|}}{C}}HCH_3$$

3.4 This problem becomes easier when you realize that the isomers can be alcohols and ethers.

A systematic approach to this type of problem is helpful. Let's start with the alcohol isomers.

1) Draw the simplest long-chain parent alkane. Here, the alkane is butane, $CH_3CH_2CH_2CH_3$.

2) Find the number of different sites to which a functional group may be attached. For butane, two different sites are possible ($-CH_3$ and $-CH_2-$).

3) At each different site, replace an $-H$ by an $-OH$ and draw the isomer.

$$CH_3CH_2CH_2CH_2-OH \quad \text{and} \quad CH_3CH_2\overset{\displaystyle \overset{OH}{|}}{C}HCH_3$$

4) Draw the simplest branched C_4H_{10} alkane.

$$CH_3\overset{\displaystyle \overset{CH_3}{|}}{C}HCH_3$$

5) Find the number of different sites. (There are two for the above alkane.)

6) For each site, replace an $-H$ with an $-OH$ and draw the isomer.

$$CH_3\overset{\displaystyle \overset{CH_3}{|}}{C}HCH_2-OH \quad \text{and} \quad CH_3\overset{\displaystyle \overset{CH_3}{|}}{\underset{\displaystyle \underset{OH}{|}}{C}}CH_3$$

7) Proceed with the next simplest branched C_4H_{10} alkane. In this problem, we have already drawn all alcohol isomers.

For the ethers, start with the $-OCH_3$ isomers. There are two possible sites for attachment of an $-OCH_3$ group to propane ($CH_3CH_2CH_3$) and thus there are two $-OCH_3$ isomers.

$$CH_3CH_2CH_2-OCH_3 \quad \text{and} \quad CH_3\overset{\displaystyle \overset{OCH_3}{|}}{C}HCH_3$$

Finally, there is one $-OCH_2CH_3$ ether, diethyl ether, $CH_3CH_2OCH_2CH_3$.

3.5 a) Nine isomeric esters of formula $C_5H_{10}O_2$ can be drawn.

$$CH_3CH_2CH_2\overset{\displaystyle \overset{O}{\|}}{C}OCH_3$$

$$CH_3\overset{\displaystyle \overset{O}{\|}}{C}H\overset{}{C}OCH_3 \atop \underset{\displaystyle \underset{CH_3}{|}}{}$$

$$CH_3CH_2\overset{\displaystyle \overset{O}{\|}}{C}OCH_2CH_3$$

$$CH_3\overset{\displaystyle \overset{O}{\|}}{C}OCH_2CH_2CH_3$$

$$CH_3\overset{\displaystyle \overset{O}{\|}}{C}O\overset{\displaystyle \overset{CH_3}{|}}{C}HCH_3$$

$$H\overset{\displaystyle \overset{O}{\|}}{C}OCH_2CH_2CH_2CH_3$$

$$\underset{HCO}{\overset{O}{\underset{||}{}}}\underset{CHCH_2CH_3}{\overset{CH_3}{\underset{|}{}}} \qquad \underset{HCO}{\overset{O}{\underset{||}{}}}\underset{CH_2CHCH_3}{\overset{CH_3}{\underset{|}{}}} \qquad \underset{HCO}{\overset{O}{\underset{||}{}}}\underset{\underset{CH_3}{\overset{CH_3}{\underset{|}{C}}}CH_3}{}$$

b) $CH_3CH_2CH_2C{\equiv}N \qquad \underset{CH_3}{\overset{CH_3}{\underset{|}{CH}}C{\equiv}N}$

3.6 a)   Two alcohols have the formula C_3H_8O.

$CH_3CH_2CH_2OH \qquad \underset{CH_3}{\overset{OH}{\underset{|}{CHCH_3}}}$ ... $CH_3\overset{OH}{\underset{|}{CH}}CH_3$

b)   Four bromoalkanes have the formula C_4H_9Br. Refer to the solution of 3.4 if you need help.

$CH_3CH_2CH_2CH_2Br \qquad CH_3CH_2\overset{Br}{\underset{|}{CH}}CH_3 \qquad CH_3\overset{CH_3}{\underset{|}{CH}}CH_2Br \qquad CH_3\overset{CH_3}{\underset{\underset{Br}{|}}{\overset{|}{C}}}CH_3$

3.7

$CH_3CH_2CH_2CH_2CH_2{-}\xi \qquad CH_3CH_2CH_2\overset{}{\underset{CH_3}{\underset{|}{CH}}}{-}\xi \qquad CH_3CH_2\overset{}{\underset{CH_2CH_3}{\underset{|}{CH}}}{-}\xi \qquad CH_3CH_2CHCH_2{-}\xi \; \underset{CH_3}{\underset{|}{}}$

$CH_3\overset{}{\underset{CH_3}{\underset{|}{CH}}}CH_2CH_2{-}\xi \qquad CH_3CH_2\overset{CH_3}{\underset{\underset{CH_3}{|}}{\overset{|}{C}}}{-}\xi \qquad CH_3\overset{CH_3}{\underset{\underset{CH_3}{|}}{\overset{|}{CH}}}CH{-}\xi \qquad CH_3\overset{CH_3}{\underset{\underset{CH_3}{|}}{\overset{|}{C}}}CH_2{-}\xi$

3.8

a) $CH_3\overset{CH_3 \; t}{\underset{\underset{t}{\overset{|}{CH}}}{CHCH_3}}CH_3$ ... $\underset{CH_3}{}$

b) $CH_3CH_2 \boxed{\overset{CH_3}{\underset{|}{CHCH_3}}}$

c) $CH_3CH_2\overset{q \; CH_3}{\underset{\underset{s}{\overset{|}{C}}}{CCH_3}}CH_3$ ... $\underset{CH_3}{}$

3.9

a) $\overset{p}{CH_3}$
$CH_3\underset{p}{\overset{|}{CH}}CH_2CH_2CH_3$
$\underset{t \quad s \quad s \quad p}{}$

b) $\overset{p \; t \; p}{CH_3CHCH_3}$
$CH_3CH_2\underset{s}{CHCH_2CH_3}$
$\underset{p \quad s \quad t \quad s \quad p}{}$

c) $\overset{p}{CH_3} \qquad \overset{p}{CH_3}$
$CH_3CHCH_2{-}\overset{|}{\underset{|}{C}}{-}CH_3$
$\underset{p \quad t \quad s \qquad |q \; p}{} \; CH_3$
$\underset{p}{}$

p = primary; s = secondary; t = tertiary; q = quaternary

3.10

a) $\overset{p}{CH_3}$
$CH_3\overset{|}{CH}CH_2CH_2CH_3$
$\underset{p \quad t \quad s \quad s \quad p}{}$

b) $\overset{p \; t \; p}{CH_3CHCH_3}$
$CH_3CH_2\overset{|}{CH}CH_2CH_3$
$\underset{p \quad s \quad t \quad s \quad p}{}$

c) $\overset{p}{CH_3} \qquad \overset{p}{CH_3}$
$CH_3CHCH_2{-}\overset{|}{\underset{|}{C}}{-}CH_3$
$\underset{p \quad t \quad s^2 \qquad p}{}$
$\underset{p \; CH_3}{}$

3.11

a) $CH_3CH_2CH_2CH_2CH_3$ $CH_3CH_2\overset{\overset{\displaystyle CH_3}{|}}{C}HCH_3$ $CH_3\overset{\overset{\displaystyle CH_3}{|}}{\underset{\underset{\displaystyle CH_3}{|}}{C}}CH_3$

           Pentane   2–Methylbutane   2,2–Dimethylpropane

b)

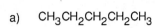

The longest chain is a *hexane*.
The substituents are: 3–methyl, 4–methyl.
The IUPAC name is 3,4–dimethylhexane.

c) $(CH_3)_2CHCH_2\overset{\overset{\displaystyle CH_3}{|}}{C}HCH_3$ d) $(CH_3)_3CCH_2CH_2\overset{\overset{\displaystyle CH_3}{|}}{\underset{\underset{\displaystyle CH_2CH_3}{|}}{C}}H$

    2,4–Dimethylpentane

                                 2,2,5–Trimethylheptane

3.12

a)   3,4–Dimethylnonane $CH_3CH_2CH_2CH_2CH_2\overset{\overset{\displaystyle CH_3}{|}}{C}H\underset{\underset{\displaystyle CH_3}{|}}{C}HCH_2CH_3$

b)   3–Ethyl–4,4–dimethylheptane $CH_3CH_2CH_2\overset{\overset{\displaystyle CH_3}{|}}{\underset{\underset{\displaystyle CH_3}{|}}{C}}\text{———}\underset{\underset{\displaystyle CH_2CH_3}{|}}{C}HCH_2CH_3$

c)   2,2–Dimethyl–4–propyloctane $CH_3CH_2CH_2CH_2\overset{\overset{\displaystyle }{|}}{\underset{\underset{\displaystyle CH_2CH_2CH_3}{|}}{C}}HCH_2C(CH_3)_3$

d)   2,2,4–Trimethylpentane $CH_3\overset{\overset{\displaystyle CH_3}{|}}{C}HCH_2\overset{\overset{\displaystyle CH_3}{|}}{\underset{\underset{\displaystyle CH_3}{|}}{C}}CH_3$

3.13

a) $CH_3CH_2CH_2CH_2\overset{\overset{\displaystyle CH_3}{|}}{\underset{\underset{\displaystyle CH_3}{|}}{C}}H$ The longest chain is a *hexane*.
The correct name is 2–methylhexane.

b) $CH_3CH_2CH_2\overset{\overset{\displaystyle CH_3}{|}}{C}HCH\text{—}CH_3$ The longest chain is an *octane*. The
correct name is 4,5–dimethyloctane.
                $CH_2CH_2CH_3$

c) 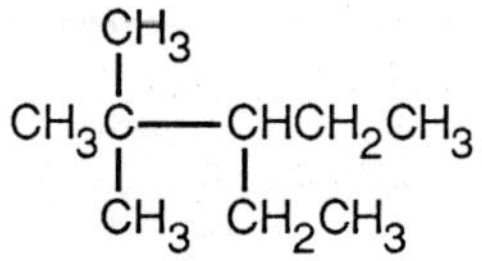

The numbering of substituents should start from the other end of the chain. Also, substituents should be cited in alphabetical order. The correct name is 3–ethyl–2,2–dimethylpentane.

d) 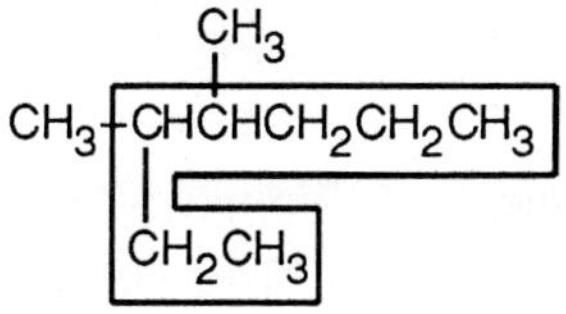

The longest chain is a *heptane;* the numbering of substituents should start from the other end. The correct name is 3,4–dimethylheptane.

e) 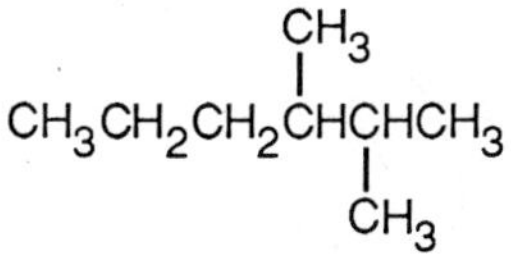

The name must include the prefix "di-" when two substituents are the same. The correct name is 2,3–dimethylhexane.

f) 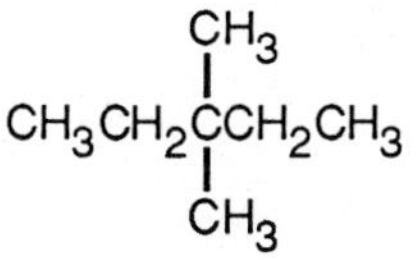

The number "3" must be repeated in the name. The correct name is 3,3–dimethylpentane.

3.14

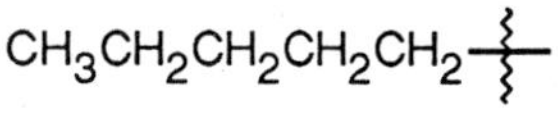

Pentyl

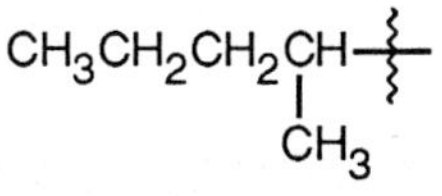

1–Methylbutyl

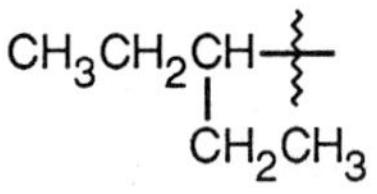

1–Ethylpropyl

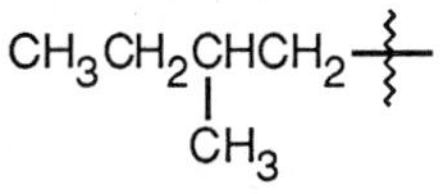

2–Methylbutyl

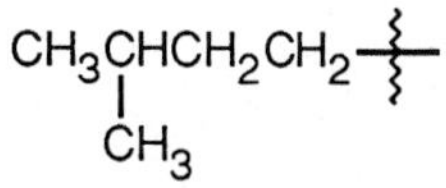

3–Methylbutyl

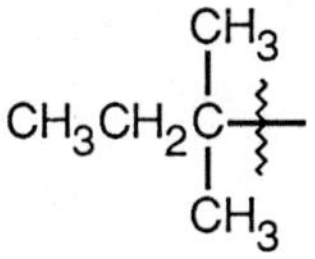

1,1–Dimethylpropyl

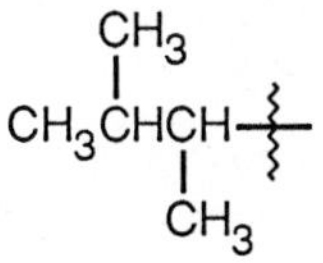

1,2–Dimethylpropyl

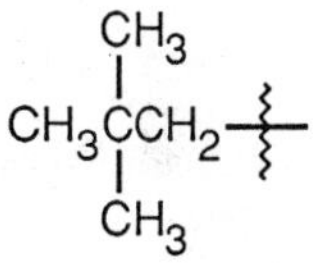

2,2–Dimethylpropyl

3.15

a) 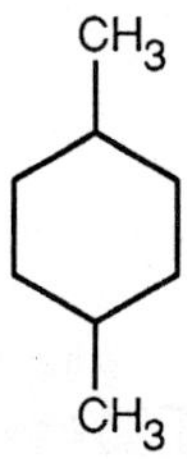

1,4–Dimethylcyclohexane

b) 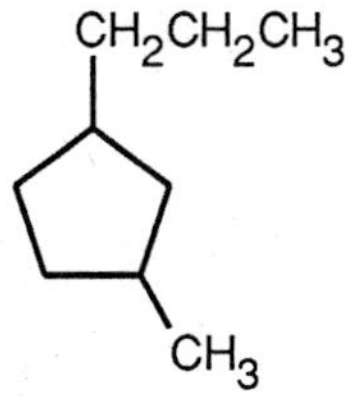

1–Methyl–3–propylcyclopentane

c) 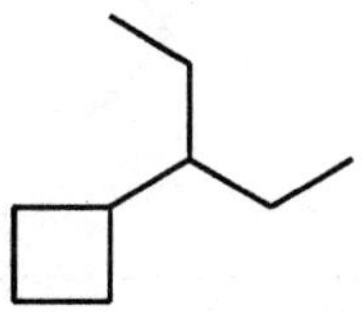

3–Cyclobutylpentane

d)

CH$_2$CH$_3$

Br

1–Bromo–4–ethylcyclodecane

e)

CH$_3$

CH(CH$_3$)$_2$

1–Isopropyl–2–methylcyclohexane

f)

Br

CH$_3$

C(CH$_3$)$_3$

4–Bromo–1–*tert*–butyl–2–
methylcycloheptane

3.16

a)

CH$_3$

CH$_3$

1,1–Dimethylcyclooctane

b)

3–Cyclobutylhexane

c)

Cl

Cl

1,2–Dichlorocyclopentane

d)

CH$_3$

Br   Br

1,3–Dibromo–5–methylcyclohexane

3.17

a)

CH$_3$

H

H

1

Br

trans–1–Bromo–3–methylcyclohexane

b)

CH$_3$

CH$_3$

H   1

H

cis–1,2–Dimethylcyclopentane

c)

CH$_2$CH$_3$

H

H   1

C(CH$_3$)$_3$

trans–1–*t*–Butyl–2–ethylcyclohexane

3.18

a)

OH

hydroxyl

aromatic
ring

Phenol

b)

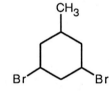

ketone

double
bond

2–Cyclohexenone

c)

NH$_2$ amine

CH$_3$CHCOOH

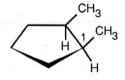

carboxylic
acid

Alanine

d)

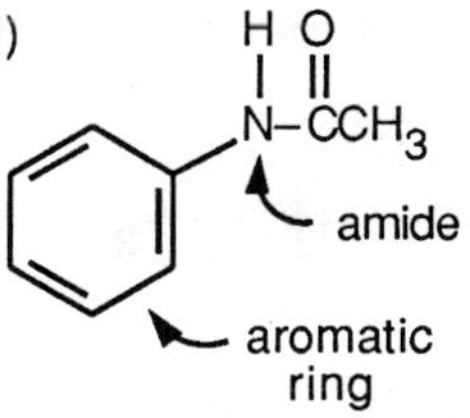

Acetanilide

e)

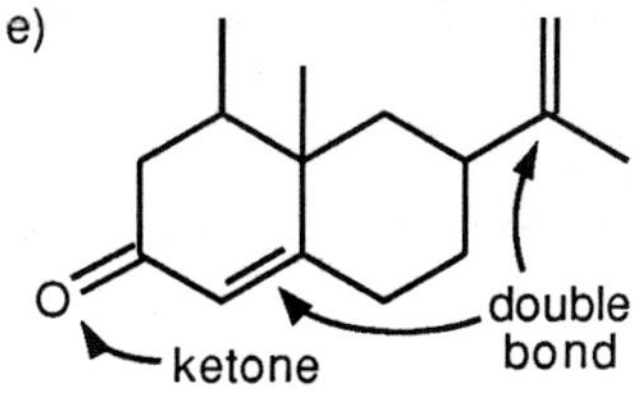

Nootkatone

f)

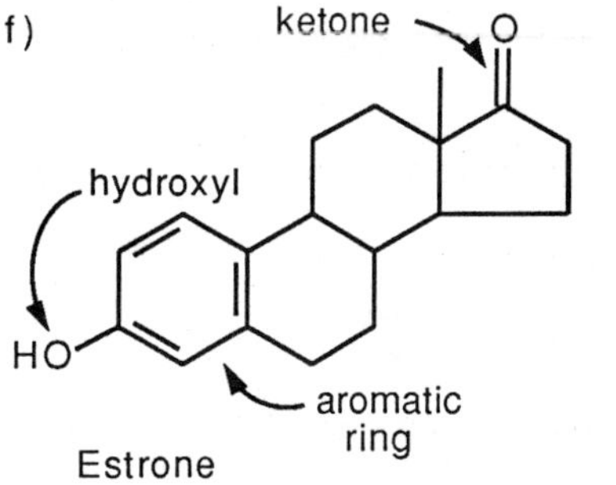

Estrone

g)

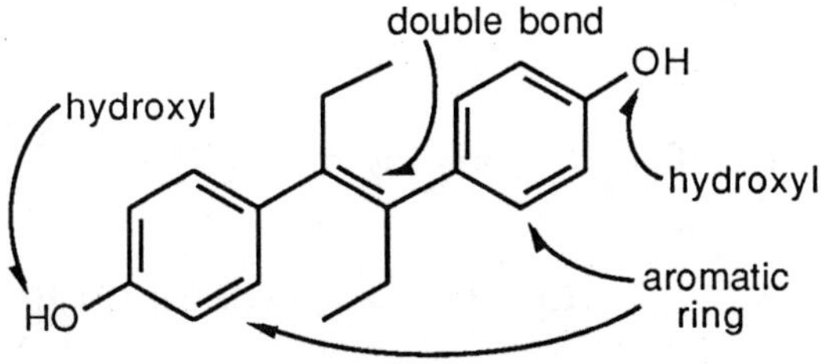

Diethylstilbesterol

h)

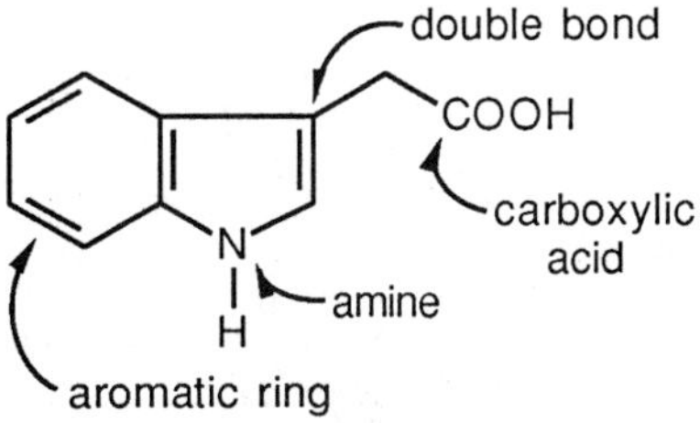

3-Indoleacetic acid

3.19 a) Eighteen isomers have the formula C_8H_{18}. Three of them are pictured.

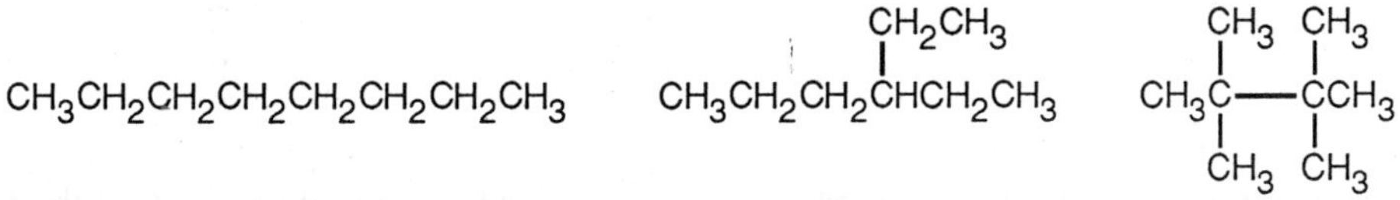

$$CH_3CH_2CH_2CH_2CH_2CH_2CH_2CH_3$$

b) Structures with the formula $C_4H_8O_2$ may represent esters, carboxylic acids or many other complicated molecules. Two possibilities:

3.20

$$CH_3CH_2CH_2CH_2CH_2CH_2CH_3$$

Heptane

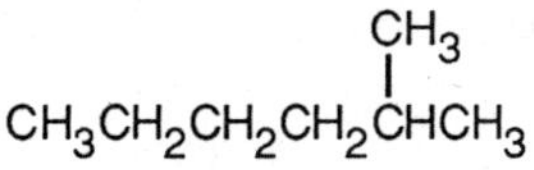

2-Methylhexane

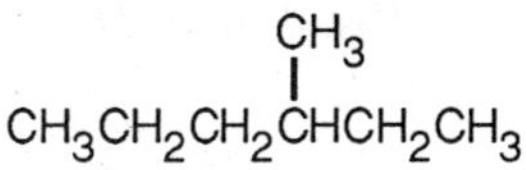

3-Methylhexane

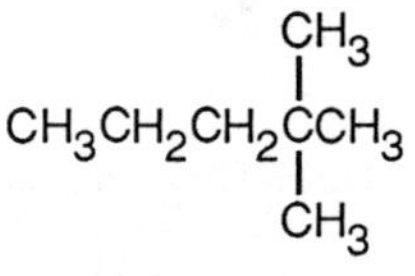

2,2-Dimethylpentane

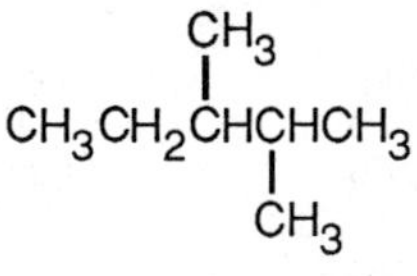

2,3-Dimethylpentane

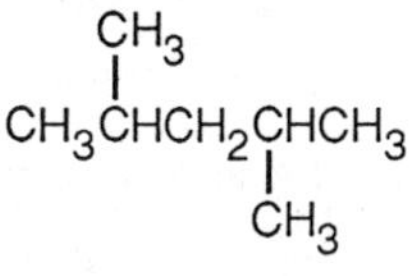

2,4-Dimethylpentane

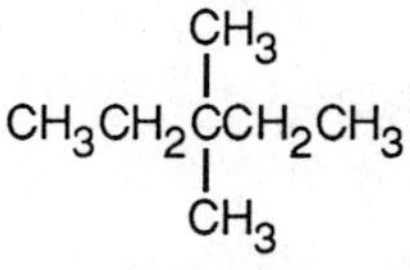

3,3-Dimethylpentane

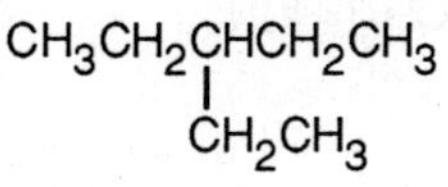

3-Ethylpentane

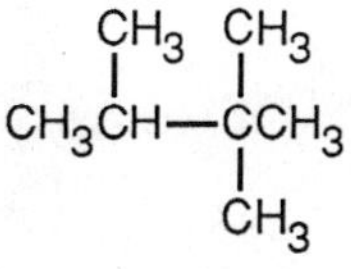

2,2,3-Trimethylbutane

3.21 Other answers to this problem and to problem 3.22 are acceptable.

a) CH₃CH₂C(=O)CH₂CH₃

b) CH₃C(=O)NHCH₂CH₃

c) CH₃C(=O)OCH₂CH₂CH₃

d)   (benzene ring with CH=O substituent)

e) CH₃C(=O)CH₂C(=O)OCH₂CH₃

f) H₂NCH₂CH₂OH

3.22

a) C₄H₈O : CH₃C(=O)CH₂CH₃

b) C₅H₉N : CH₃CH₂CH₂CH₂C≡N

c) C₄H₆O₂ : HC(=O)CH₂CH₂C(=O)H

d) C₆H₁₁Br : CH₃CH₂CH=CHCH₂CH₂Br

e) C₆H₁₄ : CH₃CH₂CH₂CH₂CH₂CH₃

f) C₆H₁₂ : (cyclohexane ring)

g) C₅H₈ : CH₃CH=CHCH=CH₂

h) C₅H₈O : H₂C=CHC(=O)CH₂CH₃

3.23 First, draw all straight-chain isomers. Then proceed to the simplest branched structure.

a) CH₃CH₂CH₂CH₂OH   CH₃CH₂CH(OH)CH₃   CH₃CH(CH₃)CH₂OH   CH₃C(CH₃)(OH)CH₃

There are *4 alcohol isomers* of C₄H₁₀O.

b) CH₃CH₂CH₂CH₂CH₂NH₂   CH₃CH₂CH₂CH(NH₂)CH₃   CH₃CH₂CH(NH₂)CH₂CH₃

CH₃CH₂CH(CH₃)CH₂NH₂   CH₃CH₂C(CH₃)(NH₂)CH₃   CH₃CH(CH₃)CH(NH₂)CH₃   H₂NCH₂CH₂CH(CH₃)CH₃

CH₃C(CH₃)(CH₃)CH₂NH₂   CH₃CH₂CH₂CH₂NHCH₃   CH₃CH₂CH₂NHCH₂CH₃

$$\underset{\text{CH}_3}{\overset{\overset{\text{CH}_3}{|}}{\text{CH}_3\text{CH}_2\text{CHNHCH}_3}}
\quad
\overset{\overset{\text{CH}_3}{|}}{\text{CH}_3\text{CHCH}_2\text{NHCH}_3}
\quad
\underset{\underset{\text{CH}_3}{|}}{\overset{\overset{\text{CH}_3}{|}}{\text{CH}_3\text{CNHCH}_3}}
\quad
\overset{\overset{\text{CH}_3}{|}}{\text{CH}_3\text{CH}_2\text{NHCHCH}_3}$$

$$\overset{\overset{\text{CH}_3}{|}}{\text{CH}_3\text{CH}_2\text{CH}_2\text{NCH}_3}
\quad
\overset{\overset{\text{CH}_3}{|}}{\text{CH}_3\text{CH}_2\text{NCH}_2\text{CH}_3}
\quad
\underset{\underset{\text{CH}_3}{|}}{\overset{\overset{\text{CH}_3}{|}}{\text{CH}_3\text{CHNCH}_3}}$$

There are *17 isomers* of $C_5H_{13}N$. Nitrogen can be bonded to one, two or three alkyl groups.

c) $\quad \text{CH}_3\text{CH}_2\text{CH}_2\overset{\overset{\text{O}}{||}}{\text{C}}\text{CH}_3 \qquad \text{CH}_3\text{CH}_2\overset{\overset{\text{O}}{||}}{\text{C}}\text{CH}_2\text{CH}_3 \qquad \underset{\underset{\text{CH}_3}{|}}{\text{CH}_3\text{CH}\overset{\overset{\text{O}}{||}}{\text{C}}\text{CH}_3}$

There are *3 ketone isomers* with the formula $C_5H_{10}O$.

d) $\quad \text{CH}_3\text{CH}_2\text{CH}_2\text{CH}_2\overset{\overset{\text{O}}{||}}{\text{CH}} \qquad \underset{\underset{\text{CH}_3}{|}}{\text{CH}_3\text{CHCH}_2\overset{\overset{\text{O}}{||}}{\text{CH}}} \qquad \underset{\underset{\text{CH}_3}{|}}{\text{CH}_3\text{CH}_2\text{CH}\overset{\overset{\text{O}}{||}}{\text{CH}}} \qquad \underset{\underset{\text{CH}_3}{|}}{\text{CH}_3\overset{\overset{\text{CH}_3}{|}}{\text{C}}\text{—}\overset{\overset{\text{O}}{||}}{\text{CH}}}$

There are *4 isomeric aldehydes* with the formula $C_5H_{10}O$. Remember that the aldehyde functional group can occur only at the end of a chain.

e) $\quad \text{CH}_3\text{CH}_2\overset{\overset{\text{O}}{||}}{\text{C}}\text{OCH}_3 \qquad \text{CH}_3\overset{\overset{\text{O}}{||}}{\text{C}}\text{OCH}_2\text{CH}_3 \qquad \text{H}\overset{\overset{\text{O}}{||}}{\text{C}}\text{OCH}_2\text{CH}_2\text{CH}_3 \qquad \text{H}\overset{\overset{\text{O}}{||}}{\text{C}}\text{O}\underset{\underset{\text{CH}_3}{|}}{\text{CH}}\text{CH}_3$

There are *4 esters* with the formula $C_4H_8O_2$.

f) $\quad \text{CH}_3\text{CH}_2\text{OCH}_2\text{CH}_3 \qquad \text{CH}_3\text{OCH}_2\text{CH}_2\text{CH}_3 \qquad \underset{\underset{\text{CH}_3}{|}}{\text{CH}_3\text{OCHCH}_3}$

There are *3 ethers* with the formula $C_4H_{10}O_2$.

3.24

a) $\quad \text{CH}_3\text{CH}_2\text{OH}$

b) $\quad \underset{\underset{\text{CH}_3}{|}}{\overset{\overset{\text{CH}_3}{|}}{\text{CH}_3\text{CC}\equiv\text{N}}}$

c) $\quad \overset{\overset{\text{Br}}{|}}{\text{CH}_3\text{CHCH}_3}$

d) $\quad \overset{\overset{\text{OH}}{|}}{\text{CH}_3\text{CHCH}_2\text{OH}}$

e) $\quad \overset{\overset{\text{CH}_3}{|}}{\text{CH}_3\text{CHOCH}_3}$

f) $\quad \underset{\underset{\text{CH}_3}{|}}{\overset{\overset{\text{CH}_3}{|}}{\text{CH}_3\text{CCH}_3}}$

3.25

$\text{CH}_3\text{CH}_2\text{CH}_2\text{CH}_2\text{CH}_2\text{Br}$
1–Bromopentane

$\overset{\overset{\text{Br}}{|}}{\text{CH}_3\text{CH}_2\text{CH}_2\text{CHCH}_3}$
2–Bromopentane

$\overset{\overset{\text{Br}}{|}}{\text{CH}_3\text{CH}_2\text{CHCH}_2\text{CH}_3}$
3–Bromopentane

3.26

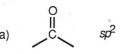

CH₃CHCH₂CH₂CHCH₂Cl

CH₃CHCH₂CH₂CCH₃

CH₃CHCH₂CHCHCH₃

1–Chloro–2,5–dimethylhexane   2–Chloro–2,5–dimethylhexane   3–Chloro–2,5–dimethylhexane

3.27

a) sp^2

b) —C≡N sp

c) sp^2

d) sp^3

3.28

a)   CH₃CH₂CH₂CH₂CH₂CHCH₃

   2–Methylheptane

b)   CH₃CH₂CH—CH₂CCH₃

   4–Ethyl–2,2–dimethylhexane

c)   CH₃CH₂CH₂CH₂C—CHCH₂CH₃

   4–Ethyl–3,4–dimethyloctane

d)   CH₃CH₂CH₂CCH₂CHCH₃

   2,4,4–Trimethylheptane

e)   CH₃CH₂CH₃CH₂CHCH₂C——CHCH₃

   3,3–Diethyl–2,5–dimethylnonane

f)   CH₃CH₂CH₂CHCHCH₂CH₃

   4–Isopropyl–3–methylheptane

3.29

a)   CH₃CHCH₃

b)

c)   CH₃CH₂CH₂CH₂CH₂CH₃

3.30

a)

b)   CH₃CH—CHCH₃

3.31

a) $CH_3CH_2CH_2CH_2CH_2Br$

b)   [cyclobutane with OCH_2CH_3 substituent]

c) $CH_3\overset{\underset{\displaystyle CH_3}{|}}{CH}C\equiv N$

d)   [cyclopentane with $-CH_2OH$ substituent]

e)   There are no aldehyde isomers. However,

$$CH_3\overset{\displaystyle O}{\overset{\displaystyle ||}{C}}CH_3$$

is a ketone isomer.

f)   [benzene ring with COOH and CH_3 substituents]

3.32 The purpose of this problem is to teach you to recognize identical structures when they are drawn slightly differently.

a)   1 and 2 are the same.   b) All structures are the same.
c)   1 and 2 are the same.   d) 1 and 3 are the same.
e)   1 and 3 are the same.

3.33

a)   [cyclopentane ring with Br at top, H and Br markings]

trans–1,3–Dibromocyclopentane

b)   [cyclohexane ring with CH_3CH_2, CH_2CH_3, and H substituents]

cis–1,4–Diethylcyclohexane

c)   [cycloheptane ring with CH_3, H, H, and $CH(CH_3)_2$ substituents]

trans–1–Isopropyl–3–methylcycloheptane

d)   [two cyclohexane rings connected by $-CH_2-$]

Dicyclohexylmethane

3.34

a) $\overset{\underset{\displaystyle 1°}{\overset{\displaystyle CH_3}{|}}}{CH_3}\underset{1°}{CH_3}\underset{3°}{CH}\underset{2°}{CH_2}\underset{1°}{CH_3}$

b) $\underset{1°}{(CH_3)_2}\underset{3°}{CH}\underset{3°}{CH}(\underset{2°}{CH_2}\underset{1°}{CH_3})_2$

c) $\underset{1°}{(CH_3)_3}\underset{4°}{C}\underset{2°}{CH_2}\underset{2°}{CH_2}\underset{}{CH}\overset{\overset{\displaystyle 1°}{\displaystyle CH_3}}{|}$
$\underset{1°}{\underset{\displaystyle CH_3}{|}}$

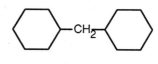

d)   [cyclohexane ring labeled: 2°, 2°, 2°, 2°, 3°, with 1° CH₃]

e)   [fused bicyclic ring system labeled with 2°, 3°, 2°, 2°, 2°, 2°, 3°, 2°]

f)   [bicyclic (norbornane-type) ring labeled 3°, 2°, 2°, 2°, 2°, 2°, 4°, with 1° CH₃]

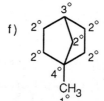

3.35 a)   2–Methylpentane   b) 2,2–Dimethylbutane
c)   2,3,3–Trimethylhexane   d) 5–Ethyl–2–methylheptane

e) 3,3,5–Trimethyloctane   f) 2,2,3,3–Tetramethylhexane

g) 5–Ethyl–3,5–dimethyloctane

3.36

$$CH_3CH_2CH_2CH_2CH_2CH_3$$

Hexane

$$CH_3CH_2CH_2\overset{\overset{\displaystyle CH_3}{|}}{C}HCH_3$$

2–Methylpentane

$$CH_3CH_2\overset{\overset{\displaystyle CH_3}{|}}{C}HCH_2CH_3$$

3–Methylpentane

$$CH_3CH_2\overset{\overset{\displaystyle CH_3}{|}}{\underset{\underset{\displaystyle CH_3}{|}}{C}}CH_3$$

2,2–Dimethylbutane

$$CH_3\overset{\overset{\displaystyle CH_3}{|}}{C}H\overset{\overset{\displaystyle CH_3}{|}}{C}HCH_3$$

2,3–Dimethylbutane

3.37

a) $CH_3\overset{\overset{\displaystyle CH_2CH_3}{|}}{C}HCH_2CH_2CH_2\overset{\overset{\displaystyle CH_3}{|}}{\underset{\underset{\displaystyle CH_3}{|}}{C}}CH_3$ The longest chain is an octane.

correct name: 2,2,6–Trimethyloctane

b) $CH_3\overset{\overset{\displaystyle CH_3}{|}}{C}H\overset{}{\underset{\underset{\displaystyle CH_2CH_3}{|}}{C}}HCH_2CH_2CH_3$ The longest chain is a hexane; numbering should start from the other end.

correct name: 3–Ethyl–2–methylhexane

c) $CH_3CH_2\overset{\overset{\displaystyle CH_3}{|}}{\underset{\underset{\displaystyle CH_3}{|}}{C}}\!\!-\!\!\overset{}{\underset{\underset{\displaystyle CH_2CH_3}{|}}{C}}HCH_2CH_3$ Numbering should start from the other end.

correct name: 4–Ethyl–3,3–dimethylhexane

d) $CH_3CH_2\overset{\overset{\displaystyle CH_3}{|}}{C}H\!\!-\!\!\overset{\overset{\displaystyle CH_3}{|}}{\underset{\underset{\displaystyle CH_3}{|}}{C}}CH_2CH_2CH_2CH_3$ Numbering should start from the other end.

correct name: 3,4,4–Trimethyloctane

e) $CH_3CH_2CH_2\overset{\overset{\displaystyle CH_3}{|}}{C}H\overset{}{\underset{\underset{\displaystyle CH_3CHCH_3}{|}}{C}}H_2CHCH_3$ The longest chain is an octane.

correct name: 2,3,5–Trimethyloctane

f)

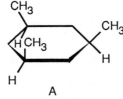

The substituents should have the lowest possible numbers.

correct name: *cis*–1,3–Dimethylcyclohexane

3.38

a)

CH₃

CH₃

1,1–Dimethylcyclooctane

b)

$$CH_3 \quad CH_2CH_3$$
$$CH_3CCH_2CCH_2CH_3$$
$$CH_3 \quad CH_2CH_3$$

4,4–Diethyl–2,2–dimethylhexane

c)

CH₃

CH₃

CH₃

1,1,2–Trimethylcyclohexane

d)

$$CH_3$$
$$CH_2CH_2CHCH_3$$
$$CH_3CH_2CH_2CH_2CH_2CHCH_2CH_2CH_2CH_3$$

6–(3–Methylbutyl)–undecane

Remember that you must choose an alkane whose principal chain is long enough so that the substituent does not become part of the principal chain.

3.39

CH₃

CH₃

H CH₃

H

H

A

CH₃

H

H CH₃

CH₃

H

B

Two cis–trans isomers of 1,3,5–trimethylcyclohexane are possible. In one isomer (A), all methyl groups are cis; in B, one methyl group is trans to the other two.

3.40

a)

Br

Br

H

H

cis–1,3–Dibromocyclohexane

H

Br

Br

H

trans–1,4–Dibromocyclohexane

constitutional isomers

b)

$$CH_3$$
$$CH_3CH_2CH_2CHCHCH_3$$
$$CH_3$$

2,3–Dimethylhexane

$$CH_3 \quad CH_3$$
$$CH_3CHCH_2CH_2CHCH_3$$

2,5,5–Dimethylpentane
(correct name: 2,5–Dimethylhexane)

constitutional isomers

c)

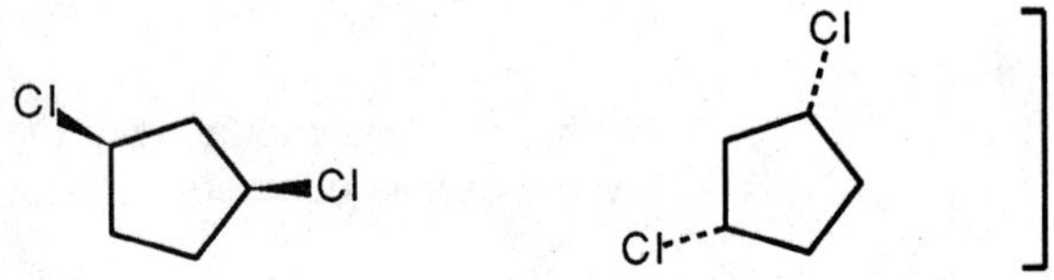

identical

3.41

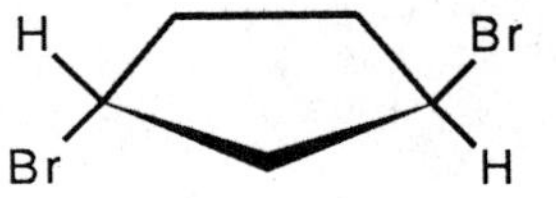

trans–1,3–Dibromocyclopentane    *cis*–1,3–Dibromocyclopentane

3.42

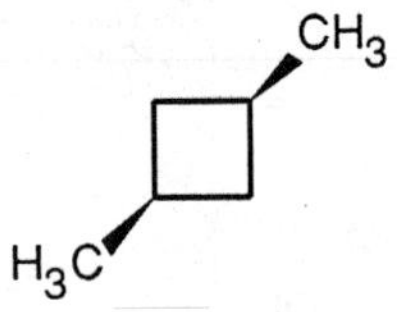

cis–1,3–Dimethylcyclobutane

3.43 Because malic acid has 2–COOH groups, the formula for the rest of the molecule is C_2H_4O. Possible structures for malic acid are:

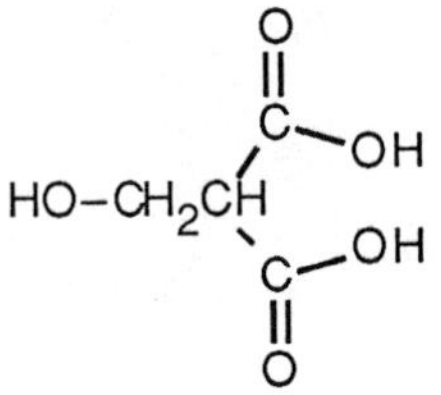

primary alcohol 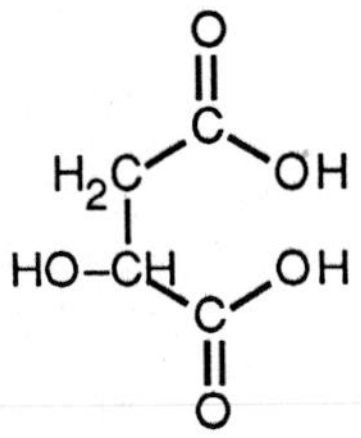   secondary alcohol 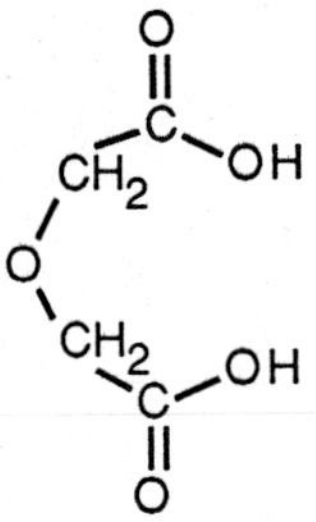   ether

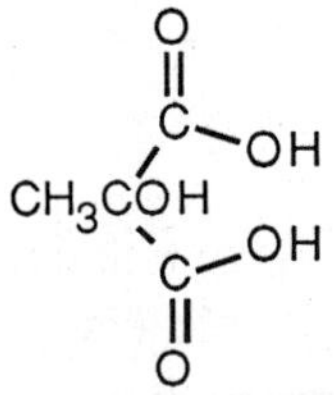

tertiary alcohol 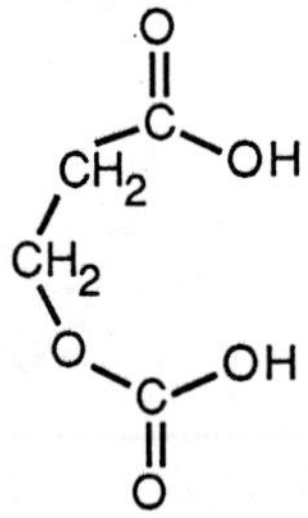   ester 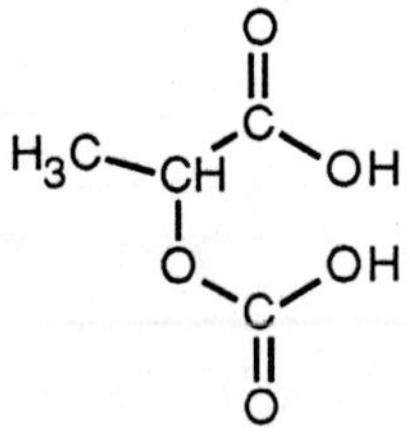   ester

Because only one of these compounds (the second one) is also a secondary alcohol, it must be malic acid.

3.44

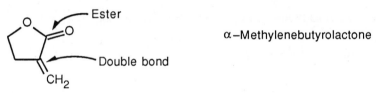

If   [CH$_2$Br / CH$_2$ / CH$_2$Br]   —2 Na→   [CH$_2$ / CH$_2$ / CH$_2$]   +   2 NaBr

Then   [BrCH$_2$ CH$_2$Br / C / BrCH$_2$ CH$_2$Br]   —4 Na→   [CH$_2$ CH$_2$ / C / CH$_2$ CH$_2$]   +   4 NaBr

The two rings are perpendicular in order to keep the geometry of the central carbon as close to tetrahedral as possible.

3.45

Ester

Double bond

CH$_2$

α–Methylenebutyrolactone

3.46 Many students do not know where to begin when they are assigned this type of problem. To start, read the problem carefully, word for word. Then try to interpret parts of the problem. For example:

1)   Formaldehyde is an aldehyde, H–C–H.
           $$\overset{\displaystyle O}{\underset{\displaystyle \parallel}{}}$$
2)   It trimerizes -- that is, 3 formaldehydes come together to form a compound $C_3H_6O_3$. No atoms are eliminated, so all of the original atoms are still present.
3)   There are no carbonyls. This means that trioxane cannot contain any –C=O functional groups. If you look back to Table 3.1, you can see that the only oxygen functional groups that can be present are either ethers or alcohols.
4)   A monobromo derivative is a compound in which one of the H's has been replaced by a Br. Because only one monobromo derivative is possible, we know that there can only be one type of hydrogen in trioxane. The only possibility for trioxane is:

CH$_2$

O   O

H$_2$C   CH$_2$

O

Trioxane

3.47

The two *trans*–1,2–dimethylcyclopentanes are mirror images.

Study Guide for Chapter 3

After studying this chapter, you should be able to:

(1) Identify functional groups in molecules, and draw molecules containing a given functional group (3.1, 3.2, 3.18, 3.21, 3.22, 3.23, 3.24, 3.31, 3.45).

(2) Systematically draw all possible isomers of a given molecular formula (3.3, 3.4, 3.5, 3.6, 3.19, 3.20, 3.23, 3.25, 3.26, 3.43).

(3) Name and draw alkanes and alkyl groups (3.7, 3.11, 3.12, 3.13, 3.14, 3.28, 3.35, 3.36, 3.37, 3.38).

(4) Identify carbon and hydrogen as being primary, secondary or tertiary (3.8, 3.9, 3.10, 3.29, 3.30, 3.34).

(5) Name and draw cycloalkanes, indicating *cis-trans* geometry (if required) (3.15, 3.16, 3.17, 3.33, 3.39, 3.40, 3.41, 3.42, 3.47).

4.2

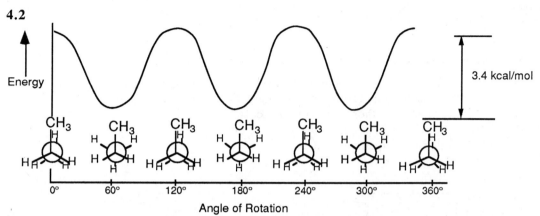

4.3

a)

This conformer of 2,3–dimethylbutane is the most stable since it is staggered and has the fewest $CH_3 - CH_3$ *gauche* interactions.

4.4

a)

The *most stable* conformer occurs at 60° 180°, and 300°.

b)

1.4 kcal/mol

1.4 kcal/mol

1.0 kcal/mol

The *least stable* conformer occurs at 0°, 120°, 240°, and 360°.

c,d)

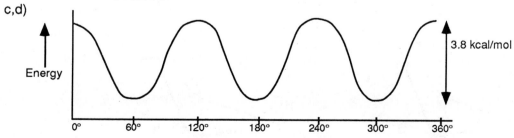

4.5

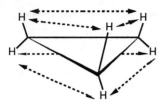

All hydrogen atoms on the same side of the cyclopropane ring are eclipsed by each other. If we draw each hydrogen-hydrogen interaction, we count six eclipsing interactions. Since each of these interactions "costs" 1.0 kcal/mol, all six cost 6.0 kcal/mol. (6.0 kcal/mol ÷ 27.6 kcal/mol) = 0.22; thus, 22% of the total strain energy of cyclopropane is due to eclipsing strain.

4.6

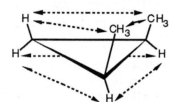 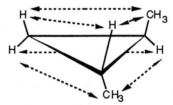

eclipsing interaction	energy cost (kcal/mol)	cis isomer		trans isomer	
		# of interactions	total energy cost (kcal/mol)	# of interactions	total energy cost (kcal/mol)
H–H	1.0	3	3.0	2	2.0
H–CH$_3$	1.4	2	2.8	4	5.6
CH$_3$–CH$_3$	2.5	1	2.5	0	0
			8.3		7.6

The added energy cost of eclipsing interactions causes *cis*–1,2–dimethylcyclopropane to be of higher energy, and to be less stable than the *trans* isomer. Since the *cis* isomer is of higher energy, its heat of combustion is also greater.

4.7 Cyclopropane would be a more efficient fuel because its heat of combustion per –CH$_2$– group is greater than that of propane.

4.8

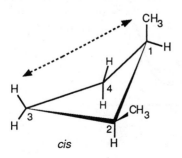

Two types of interaction are present in *cis*–1,2–dimethylcyclobutane. One interaction occurs between the two methyl groups, which are almost eclipsed. The other is an across-the-ring interaction between methyl group at position 1 of the ring and a hydrogen at position 3. Because neither of these interactions are present in *trans* isomer, it is more stable than the *cis* isomer.

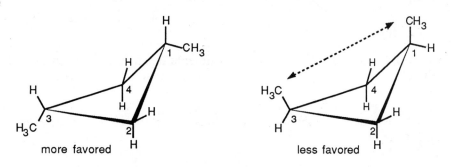

cis   *trans*

In *trans*–1,3–dimethylcyclobutane an across-the-ring interaction occurs between the methyl group at position 1 of the ring and a hydrogen at position 3. Because no interactions are present in the *cis* isomer, it is more stable than the *trans* isomer.

4.9

more favored   less favored

4.10

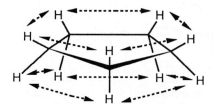

If cyclopentane were planar, it would have ten hydrogen–hydrogen interactions with a total energy cost of 10.0 kcal/mol. The measured total strain energy of 6.5 kcal/mol indicates that 3.5 kcal/mol of eclipsing strain in cyclopentane has been relieved by puckering.

4.11

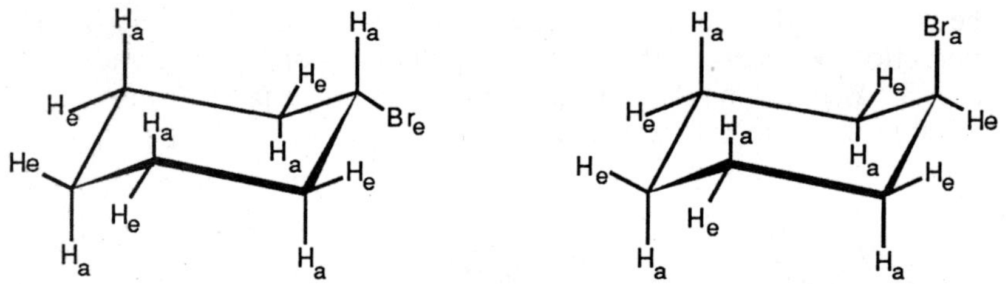

The conformation with bromine in the equatorial position is more stable.

4.12 Make a model of *cis*–1,2–dichlorocyclohexane. Take note of the fact that all *cis* substituents are on the same side of the ring and that two adjacent *cis* substituents have an axial-equatorial relationship. Now, perform a ring-flip on the cyclohexane.

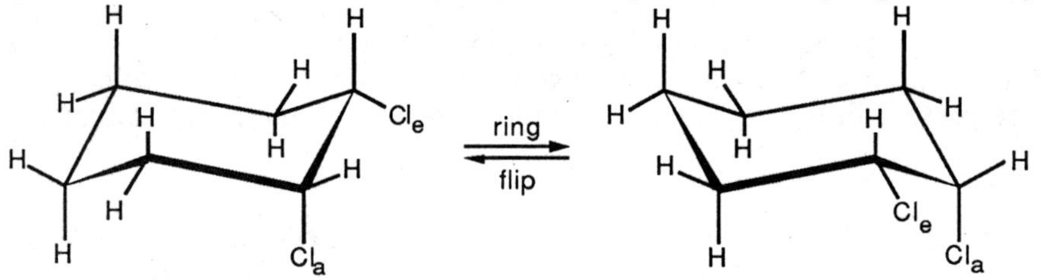

After the ring-flip, the relationship of the two substituents is still axial-equatorial. No two adjacent *cis* substituents can be converted to being both axial or both equatorial without breaking bonds.

4.13 For a *trans*–1,2–disubstituted cyclohexane, two adjacent substituents must be either both axial or both equatorial.

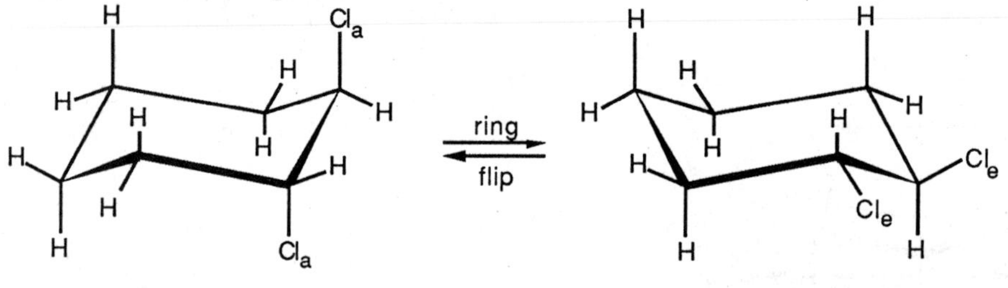

A ring flip converts two adjacent axial substituents to equatorial substituents, and *vice versa*. As in Problem 4.12, no two adjacent *trans* substituents can be converted to an axial-equatorial relationship without bond breaking.

4.14

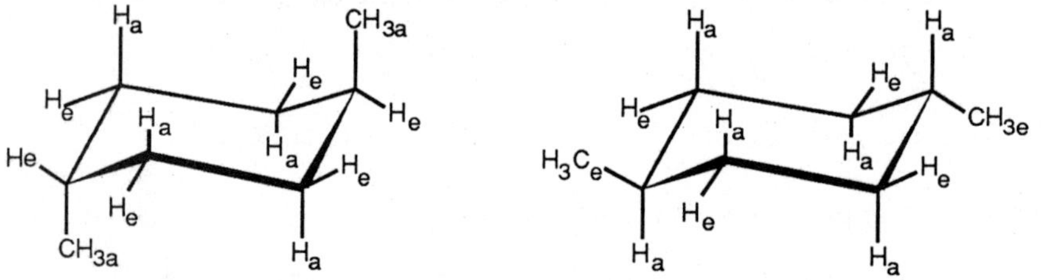

trans–1,4–Dimethylcyclohexane

4.15

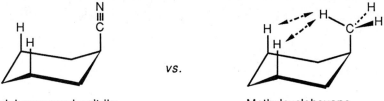

The most stable conformation of axial *tert*–butylcyclohexane is pictured. One methyl group is positioned above the ring and competes for space with two axial ring protons. In the other axial alkylcyclohexanes, this methyl group is replaced by hydrogen, which has a much smaller space requirement. The steric strain caused by an axial *tert*–butyl group is therefore higher than the strain caused by axial methyl, ethyl or isopropyl groups.

4.16 The energy difference between an axial and an equatorial cyano group is very small because there are no 1–3 diaxial interactions for a cyano group.

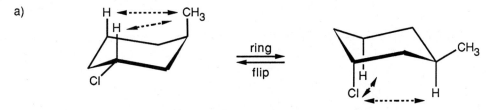

Cyclohexanecarbonitrile   *vs.*   Methylcyclohexane

4.17 Table 4.4 shows that an axial *tert*–butyl group causes 2 x 2.7 kcal/mol of steric strain. Thus, the energy difference between axial and equatorial *tert*–butyl–cyclohexane is 5.4 kcal/mol. According to Table 4.3, this energy difference corresponds to a > 99.9 : 1 ratio of more stable : less stable conformer. Thus, more than 99.9% of *tert*–butylcyclohexane molecules are in the equatorial conformation, and fewer than 0.1% are in the axial conformation at any given moment.

4.18

a)

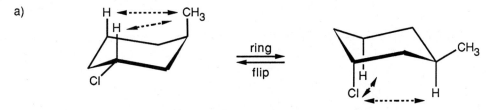

trans–1–Chloro–3–methylcyclohexane

2 (H–CH₃) = 1.8 kcal/mol   2 (H–Cl) = 0.5 kcal/mol

The second conformation is more stable than the first.

b)

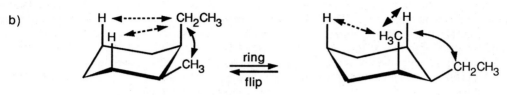

cis–1–Ethyl–2–methylcyclohexane

one CH_3–CH_2CH_3 gauche
interaction = 0.9 kcal/mol
2 (H–CH_2CH_3) = 1.9 kcal/mol
———————————————
Total = 2.8 kcal/mol

one CH_3–CH_2CH_3 gauche
interaction = 0.9 kcal/mol
2 (H–CH_3) = 1.8 kcal/mol
———————————————
Total = 2.7 kcal/mol

The second conformation is more stable than the first.

c)

cis–1–Bromo–4–ethylcyclohexane

2 (H–CH_2CH_3) = 1.9 kcal/mol   2 (H–CH_3) = 0.5 kcal/mol

The second conformation is more stable than the first.

d)

cis–1–*tert*–Butyl–4–ethylcyclohexane

2 [H–C(CH$_3$)$_3$] = 5.4 kcal/mol   2 (H–CH_2CH_3) = 1.9 kcal/mol

The second conformation is more stable than the first.

4.19

trans–1,3–Di–*tert*–butylcyclohexane

In the chair conformation of *trans*–1,3–di–*tert*–butylcyclohexane, one *tert*–butyl group is axial and one is equatorial. The 1,3–diaxial interactions of the axial *tert*–butyl group make the chair form of *trans*–1,3–di–*tert*–butylcyclohexane 5.4 kcal/mol less stable than a cyclohexane with no axial substituents. Since a twist boat conformation is 5.5 kcal/mol less stable than a chair, it is almost as likely that the compound will assume a twist-boat conformation as a chair conformation. The twist-boat removes the 1,3–diaxial interaction present in the chair form of *trans*–1,3–di–*tert*–butylcyclohexane.

4.20

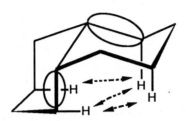

Trans–decalin is more stable than *cis*–decalin. Three 1,3–diaxial interactions cause *cis*–decalin to be of higher energy than *trans*–decalin. You may be able to visualize these interactions by thinking of the circled parts of *cis*–decalin as similar to axial methyl groups. The *gauche* interactions that occur with axial methyl groups also occur in *cis*–decalin.

4.21 a) *Angle strain* is the strain caused by the deformation of bond angles from the normal tetrahedral value.

b) *Steric strain* is the repulsive interaction caused by atoms attempting to occupy the same space.

c) *Torsional strain* is the repulsive interaction between two bonds as they rotate past each other. Torsional strain is responsible for the barrier to rotation in ethane and makes the eclipsed form higher in energy than the staggered form.

d) The *heat of combustion* of an organic compound is the heat liberated when the compound burns completely in oxygen to form CO_2 and H_2O. The heat of combustion for two compounds having the same formula is greater for a strained compound, indicating that the strained compound is of higher energy.

e) A *conformation* is one of the many possible arrangements of atoms caused by rotation about a single bond.

f) A *staggered* conformation is the conformation in which all groups on two adjacent carbons are as far from each other as possible.

g) An *eclipsed* conformation is the conformation in which all groups on two adjacent carbons are as close to each other as possible.

h) *Gauche butane* is the conformation of butane in which the C1 and C4 methyl groups are 60° apart. Gauche butane is of higher energy than anti butane because of steric strain.

4.22

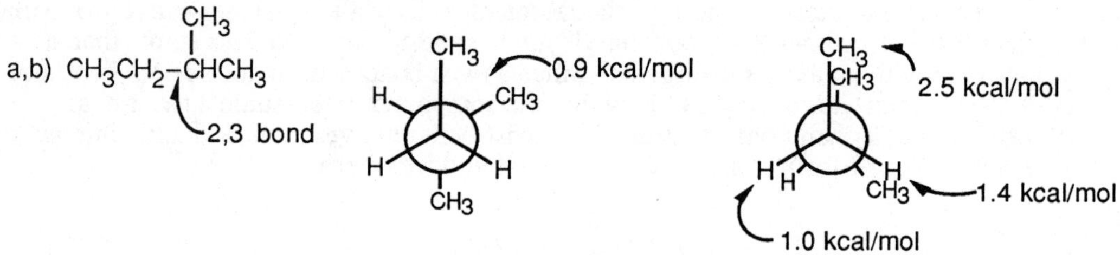

a,b) CH$_3$CH$_2$–CHCH$_3$ (with CH$_3$ above the CH, and "2,3 bond" indicated below)

0.9 kcal/mol

2.5 kcal/mol
1.4 kcal/mol
1.0 kcal/mol

2–Methylbutane   Most stable conformation   Least stable conformation

The energy difference between the two conformations is $(2.5 + 1.4 + 1.0) - 0.9 = 4.0$ kcal/mol.

c) Consider the least stable conformation to be at zero degrees. Keeping the "front" of the projection unchanged, rotate the "back" by 60° to obtain each conformation.

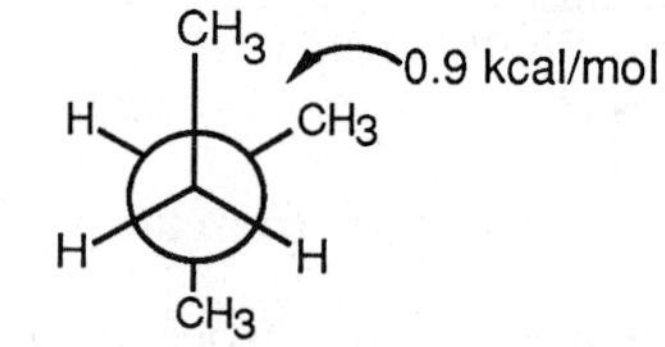

0.9 kcal/mol

at *60°* energy = 0.9 kcal/mol

1.4 kcal/mol   1.4 kcal/mol
1.4 kcal/mol

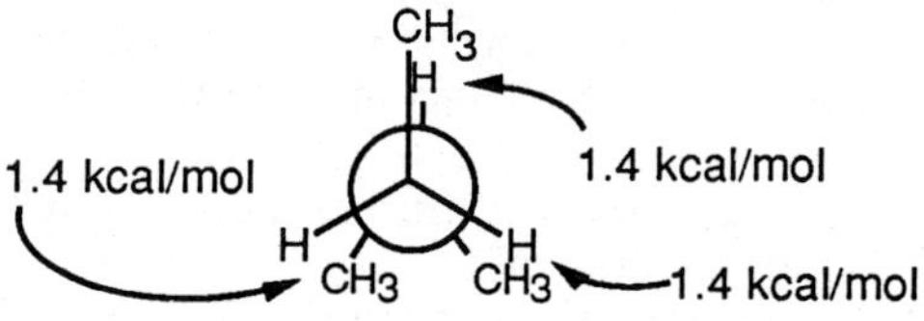

at *120°* energy = 4.2 kcal/mol

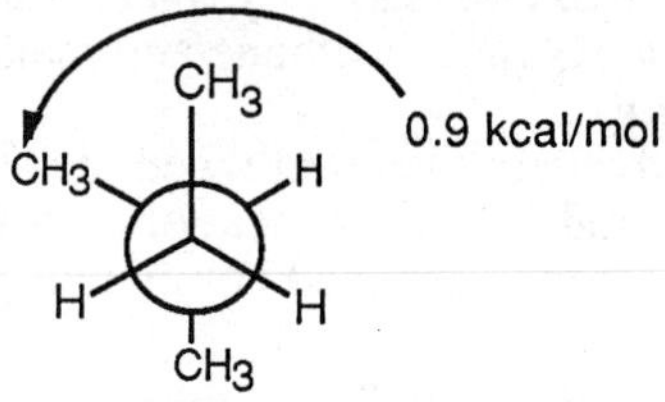

0.9 kcal/mol

at *180°* energy = 0.9 kcal/mol

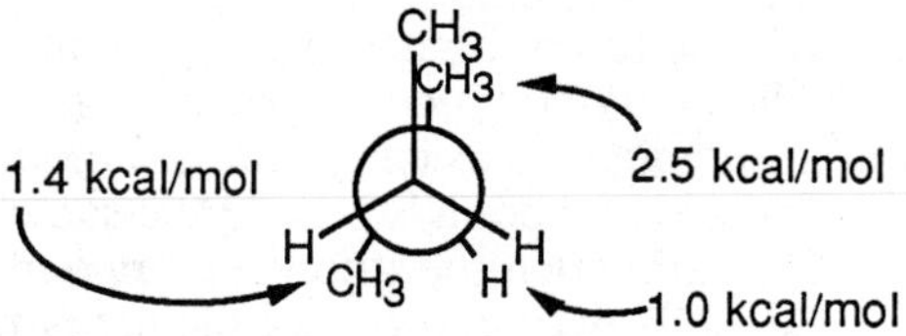

1.4 kcal/mol   2.5 kcal/mol
1.0 kcal/mol

at *240°* energy = 4.9 kcal/mol

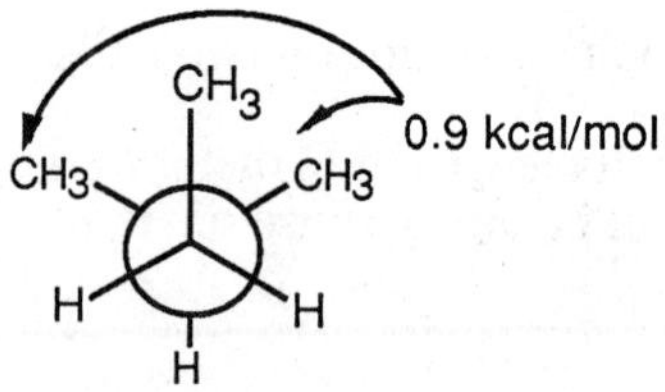

0.9 kcal/mol

at *300°* energy = 1.8 kcal/mol

Use the lowest energy conformation as the energy minimum. The highest energy conformation is 4.0 kcal/mol higher in energy than the lowest energy conformation.

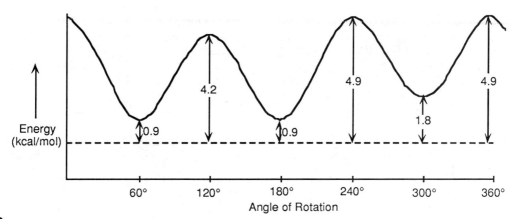

4.23

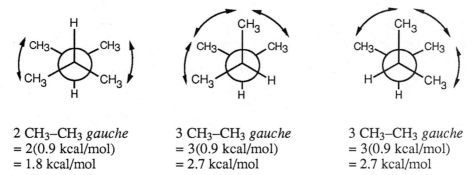

2 CH$_3$–CH$_3$ *gauche*
= 2(0.9 kcal/mol)
= 1.8 kcal/mol

3 CH$_3$–CH$_3$ *gauche*
= 3(0.9 kcal/mol)
= 2.7 kcal/mol

3 CH$_3$–CH$_3$ *gauche*
= 3(0.9 kcal/mol)
= 2.7 kcal/mol

4.24 Since we are not told the values of the interactions for 1.2–dibromoethane, the diagram can only be qualitative.

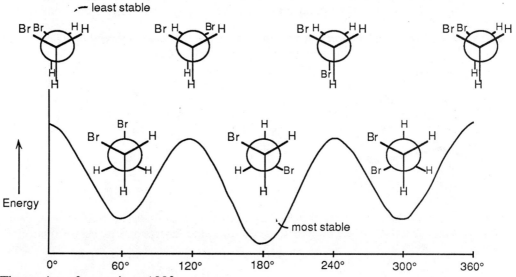

The *anti* conformer is at 180°.
The *gauche* conformers are at 60°, 300°.

4.25 The *anti* conformer has no dipole moment because the net dipole moments of the individual bonds cancel. The *gauche* conformer, however, has a dipole moment. Because the observed dipole moment is 1.0 D at room temperature, a mixture of conformers must be present.

4.26 The highest energy conformation of bromoethane is 3.6 kcal/mol. Since this includes two H–H eclipsing interactions of 1.0 kcal/mol each, the value of an H–Br eclipsing interaction is 3.6 – 2(1.0) = 1.6 kcal/mol.

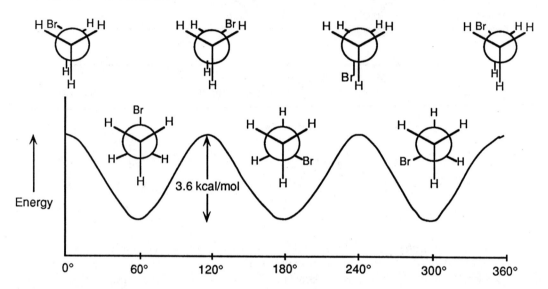

4.27 a)   An *axial bond* of a cyclohexane ring is a bond that is perpendicular to the plane of the ring.

b)   An *equatorial bond* of a cyclohexane ring lies more or less in the plane of the ring.

c)   The *chair conformation* of a cyclohexane ring is the "puckered" conformation that allows all carbon–carbon bond angles to have a value very close to the ideal tetrahedral angle.

d)   A *1,3–diaxial interaction* is a type of steric strain between an axial functional group on C1 of a cyclohexane ring and an axial group on C3 (or C5).

4.28

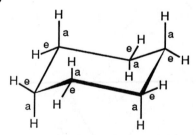

4.29

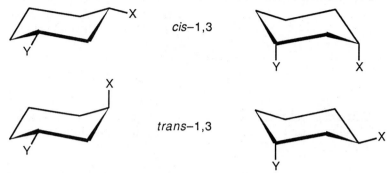

A *cis*–1,3–disubstituted isomer exists almost exclusively in the diequatorial conformation, which has no 1,3–diaxial interactions. The *trans* isomer must have one group axial, leading to 1,3–diaxial interactions. Thus, the *trans* isomer is less stable than the *cis* isomer.

4.30

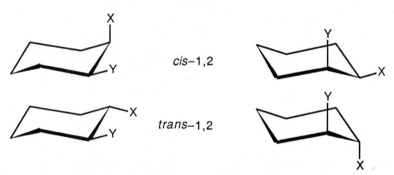

Reasoning similar to that in the previous problem can be used to show that the *trans*–1,2 disubstituted isomer, which exists principally in the diequatorial conformation, is more stable than the *cis* isomer.

4.31

The *trans*–1,4–isomer is more stable.

4.32 Since the methyl group of *N*–methylpiperidine prefers an equatorial conformation, the steric requirements of a methyl group must be greater than those of an electron lone pair.

4.33

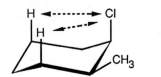 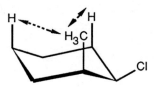

Use Table 4.4 to find the values of 1,3–diaxial interactions. For the first conformation, the steric strain is 2 x 0.25 kcal/mol = 0.5 kcal/mol. The steric strain of the second conformation is 2 x 0.9 kcal/mol, or 1.8 kcal/mol. The first conformation is more stable than the second conformation by 1.3 kcal/mol.

4.34

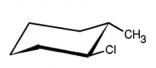 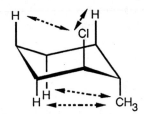

no 1,3–diaxial
interactions

2 x 0.90 kcal/mol = 1.8 kcal/mol
2 x 0.25 kcal/mol = 0.5 kcal/mol
_____
2.3 kcal/mol

The first conformation is more stable than the second conformation by 2.3 kcal/mol.

4.35

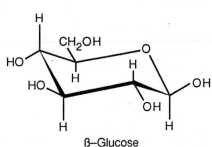

β–Glucose

4.36 To solve this problem; (1) Find the energy cost of a 1–3 diaxial interaction by using Table 4.4.

(2) Convert this energy difference into a percent by using Table 4.3.

a)

2 (H–CH(CH$_3$)$_2$) = 2.2 kcal/mol

$$\frac{\%\ \text{equatorial}}{\%\ \text{axial}} = \frac{97.5}{2.5}$$

2.5 % axial

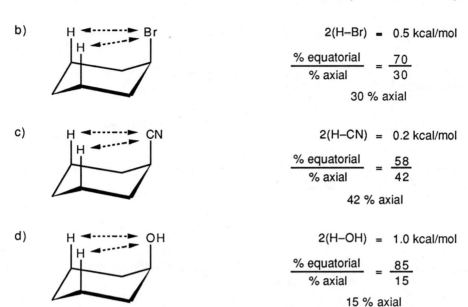

b)

H ◀┈┈┈▶ Br

2(H–Br) = 0.5 kcal/mol

$$\frac{\% \text{ equatorial}}{\% \text{ axial}} = \frac{70}{30}$$

30 % axial

c)

H ◀┈┈┈▶ CN

2(H–CN) = 0.2 kcal/mol

$$\frac{\% \text{ equatorial}}{\% \text{ axial}} = \frac{58}{42}$$

42 % axial

d)

H ◀┈┈┈▶ OH

2(H–OH) = 1.0 kcal/mol

$$\frac{\% \text{ equatorial}}{\% \text{ axial}} = \frac{85}{15}$$

15 % axial

4.37 Be aware of the distinction between axial–equatorial and *cis–trans*. Axial substituents are parallel to the axis of the ring; equatorial substituents lie around the "equator" of the ring. *Cis* substituents are on the same side of the ring; *trans* substituents are on opposite side of the ring.

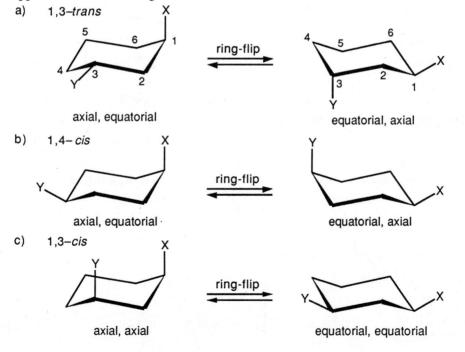

a) 1,3–*trans*

ring-flip

axial, equatorial   equatorial, axial

b) 1,4–*cis*

ring-flip

axial, equatorial   equatorial, axial

c) 1,3–*cis*

ring-flip

axial, axial   equatorial, equatorial

d)   1,5–*trans* is the same as 1,3–*trans*
e)   1,5–*cis* is the same as 1,3–*cis*

f)   1,6–*trans*

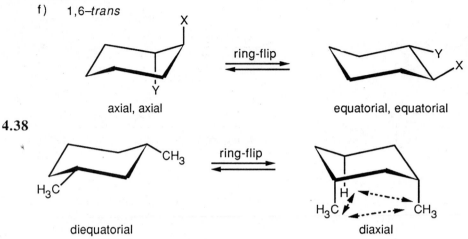

axial, axial   equatorial, equatorial

4.38

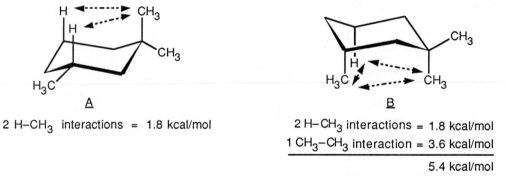

diequatorial   diaxial

The large energy difference between conformations is due to the severe 1,3–diaxial interaction between the two methyl groups.

4.39 Diaxial *cis*–1,3–dimethylcyclohexane contains three 1,3–diaxial interactions -- two H–CH_3 interactions (0.9 kcal/mol each) and one CH_3–CH_3 interaction. If the diaxial conformation is 5.4 kcal/mol less stable than the diequatorial, 5.4 – 2(0.9) = 3.6 kcal/mol of this strain energy must be due to the CH_3–CH_3 interaction.

4.40

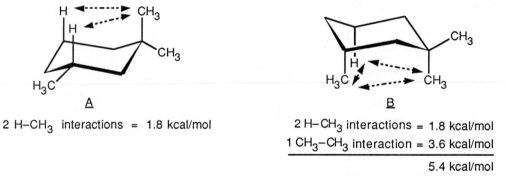

A   B

2 H–CH_3 interactions = 1.8 kcal/mol   2 H–CH_3 interactions = 1.8 kcal/mol
                                              1 CH_3–CH_3 interaction = 3.6 kcal/mol
                                              _____
                                              5.4 kcal/mol

Conformation A is favored because it is 3.6 kcal/mol lower in energy than conformation B.

4.41

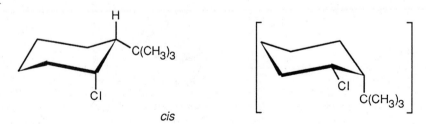

cis

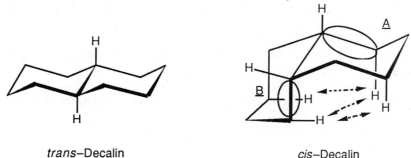

trans

All four conformations of the two isomers are illustrated. The second conformation of each pair has a high degree of steric strain, and thus each isomer adopts the first conformation. Since only the *cis* isomer has chlorine in the necessary axial position, it is more reactive than the *trans* isomer.

4.42 Note: In working with decalins, it is essential to use models. Many structural features of decalins that are obvious with models are not easily visualized with drawings.

trans–Decalin   *cis*–Decalin

No 1,3–diaxial interactions are present in *trans*–decalin.

At the ring junction of *cis*–decalin, one ring acts as an axial substituent of the other (see circled bonds). The circled part of ring <u>B</u> has two 1,3–diaxial interactions with ring <u>A</u> (indicated by arrows). Similarly, the circled part of ring <u>A</u> has two 1,3–diaxial interactions with ring <u>B</u>; one of these interactions is the same as an interaction of part of the <u>B</u> ring with ring <u>A</u>. These three 1,3–diaxial interactions have a total energy cost of 2.7 kcal/mol. *Cis*–decalin is therefore less stable than *trans*–decalin by 2.7 kcal/mol.

4.43

ring-flip   ≡

A ring-flip converts an axial substituent into an equatorial substituent and *vice versa*. At the ring junction of *trans*–decalin, each ring is a *trans–trans* diequatorial substituent of the other. If a ring-flip were to occur, the two rings would become axial substituents of each other. You can see with models that a diaxial ring junction is impossibly strained. Consequently, *trans*–decalin does not ring-flip.

The rings of *cis*–decalin are joined by an axial bond and an equatorial bond. After a ring-flip, the rings are still linked by an equatorial and an axial bond. No additional strain or interaction is introduced by a ring-flip of *cis*–decalin.

4.44

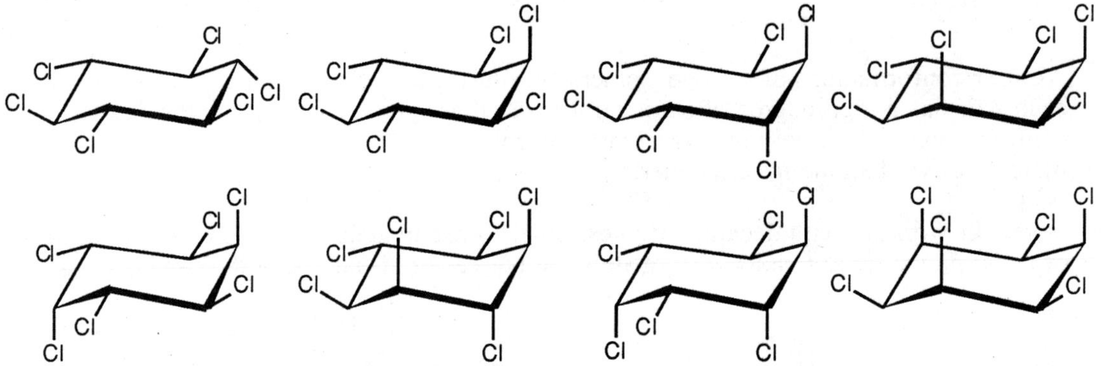

The first isomer is the most stable because all chlorine atoms can assume an equatorial configuration.

4.45 Draw the four possible isomers of 4–*tert*–butylcyclohexane–1,3–diol. Make models of these isomers also.

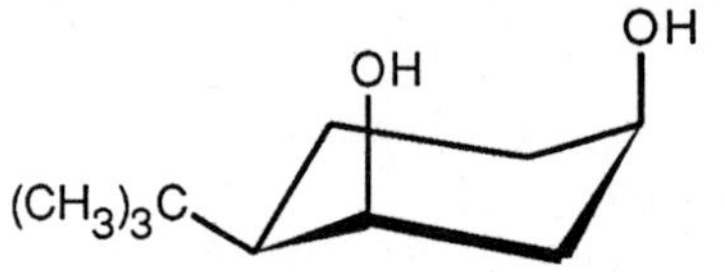

1

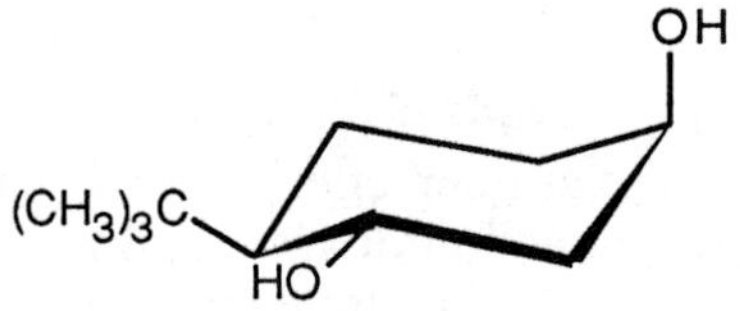

2

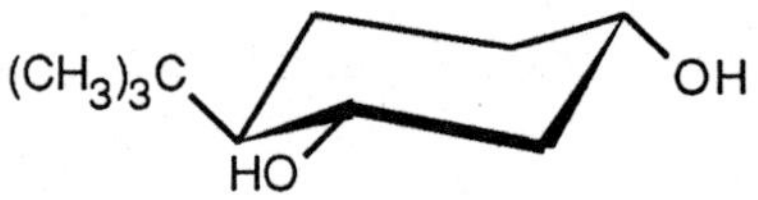

3

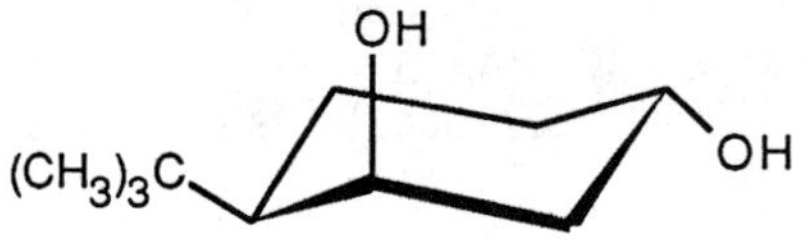

4

Only when the two hydroxyl groups are *cis*–diaxial (structure 1) can the acetal ring form. In any other conformation, the oxygen atoms are too far apart to be incorporated into a six-membered ring.

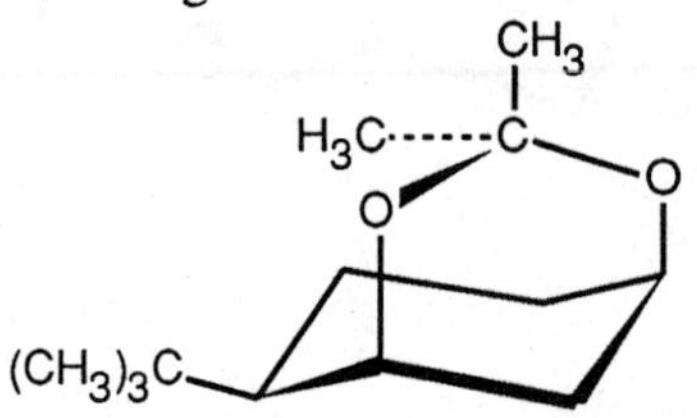

4.46

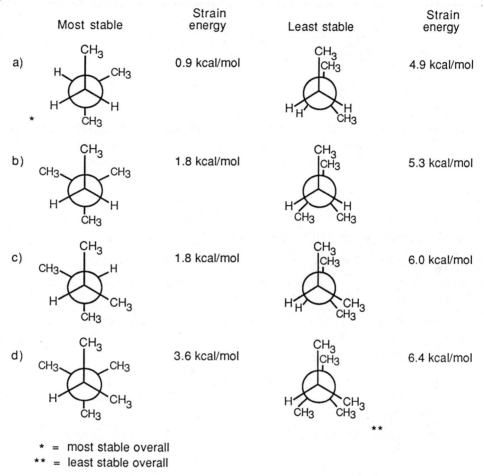

	Most stable	Strain energy	Least stable	Strain energy
a)		0.9 kcal/mol		4.9 kcal/mol
b)		1.8 kcal/mol		5.3 kcal/mol
c)		1.8 kcal/mol		6.0 kcal/mol
d)		3.6 kcal/mol		6.4 kcal/mol

* = most stable overall
** = least stable overall

4.47

A   B

Conformation A of *cis*–1–chloro–3–methylcyclohexane has no 1,3—diaxial interactions and is the more stable conformation. Steric strain in B is due to one CH_3–H interaction (0.9 kcal/mol), one Cl–H interaction (0.25 kcal/mol) and one CH_3–Cl interaction. Since the total-strain energy of B is 3.7 kcal/mol, 3.7 – 0.9 - 0.25 = 2.55 kcal/mol of strain is caused by a CH_3–Cl interaction.

4.48

If you build a model of 1–norbornene, you will find that it is almost impossible to form the bridgehead double bond. sp^2–Hybridization at the double bond requires all carbons bonded to the starred carbon to lie in a plane, yet the angle strain caused by sp^2 hybridization is too severe to allow a bridgehead double bond to exist.

Study Guide for Chapter 4

After studying this chapter, you should be able to:

(1) Draw energy vs. angle of rotation graphs for single bond conformations (4.2, 4.24, 4.26).

(2) Draw Newman projections of bond conformation and predict their relative stability (4.3, 4.4, 4.22, 4.23, 4.46).

(3) Understand the geometry of, and predict the stability of cycloalkanes having fewer than six carbons (4.5, 4.6, 4.7, 4.8, 4.9, 4.10).

(4) Draw and name substituted cyclohexanes (4.11, 4.12, 4.13, 4.14, 4.28, 4.35, 4.37, 4.45).

(5) Predict the stability of substituted cyclohexanes by estimating steric interactions (4.15, 4.16, 4.17, 4.18, 4.19, 4.20, 4.29, 4.30, 4.31, 4.32, 4.33, 4.34, 4.36, 4.38, 4.39, 4.40, 4.41, 4.42, 4.43, 4.44, 4.47, 4.48).

(6) Define important terms relating to alkane and cycloalkane stereochemistry (4.21, 4.27).

5.1

a) $CH_3Br + KOH \longrightarrow CH_3OH + KBr$ substitution

b) $CH_3CH_2OH \longrightarrow H_2C=CH_2 + H_2O$ elimination

c) $H_2C=CH_2 + H_2 \longrightarrow CH_3CH_3$ addition

5.2 Irradiation initiates the chlorination reaction by producing chlorine radicals. Although these radicals are consumed in the propagation steps, new Cl· radicals are formed to carry on the reaction. After irradiation stops, chlorine radicals are still present to carry on the propagation steps, but, as time goes on, radicals combine with each other in termination reactions that remove radicals from the reaction mixture. Because the number of radicals decreases, fewer propagation cycles occur, and the reaction gradually slows down and stops.

5.3

$$CH_3CH_2CH_2\overset{\overset{\displaystyle CH_3}{|}}{C}HCH_2Cl \quad + \quad CH_3CH_2CH_2\overset{\overset{\displaystyle CH_3}{|}}{\underset{\underset{\displaystyle Cl}{|}}{C}}CH_3 \quad +$$

1–Chloro–2–methylpentane   2–Chloro–2–methylpentane

$$CH_3CH_2CH_2\overset{\overset{\displaystyle CH_3}{|}}{C}HCH_3 \quad \underset{h\nu}{\overset{Cl_2}{\longrightarrow}} \quad CH_3CH_2\overset{\overset{\displaystyle CH_3}{|}}{C}H\underset{\underset{\displaystyle Cl}{|}}{C}HCH_3 \quad + \quad CH_3\overset{\overset{\displaystyle CH_3}{|}}{\underset{\underset{\displaystyle Cl}{|}}{C}}HCH_2\overset{\overset{\displaystyle CH_3}{|}}{C}HCH_3 \quad +$$

2–Methylpentane   3–Chloro–2–methylpentane   2–Chloro–4–methylpentane

$$ClCH_2CH_2CH_2\overset{\overset{\displaystyle CH_3}{|}}{C}HCH_3$$

1–Chloro–4–methylpentane

5.4 Pentane has three types of hydrogen atoms, $\overset{a}{C}H_3\overset{b}{C}H_2\overset{c}{C}H_2\overset{b}{C}H_2\overset{a}{C}H_3$. Although monochlorination produces $CH_3CH_2CH_2CH_2CH_2Cl$, it is not possible to avoid producing $CH_3CH_2CH_2CHClCH_3$ and $CH_3CH_2CHClCH_2CH_3$ as well. Since neopentane has only one type of hydrogen, monochlorination yields a single product.

5.5

a)

$$CH_3\overset{\overset{\displaystyle O^{\delta-}}{\|}}{\underset{\delta+}{C}}CH_3$$

ketone

b)

$$H_2C=CH\overset{\overset{\displaystyle {}^{\delta-}O}{\|}}{\underset{\delta+\ \ \delta-}{C}}OCH_2CH_3 \quad\nearrow \text{ester}$$

double bond

c)

$$\overset{\delta-}{Cl} \leftarrow \text{halide}$$
$$H_2C=\underset{\delta+}{CH}$$

double bond

d)

$$\overset{\delta-}{C}H_2CH_3$$
$$CH_3CH_2\underset{\delta-}{\overset{\delta+\ \ \delta-}{-Pb-}}CH_2CH_3$$
$$\underset{\delta-}{C}H_2CH_3$$

organometallic

5.6   To identify electrophiles, look for Lewis acids, positively charged species, or positively polarized atoms. To identify nucleophiles, look for Lewis bases or negatively charged species.

Electrophiles: H^+, $:\!\overset{..}{\underset{..}{Br}}{}^+$, Mg^{2+}, $\overset{\delta-\ \ \delta+\ \ \delta-}{O=C=O}$ Nucleophiles: $H\overset{..}{\underset{..}{O}}:^-$, $:NH_3$

5.7

$$+\ \ H-Br \longrightarrow$$

5.8   According to Table 5.3, a negative $\Delta G°$ indicates that a reaction is favorable. Thus, a reaction with $\Delta G° = -11$ kcal/mol is more favorable than a reaction with $\Delta G° = +11$ kcal/mol.

5.9   From the expression $\Delta G° = -RT\ln K_{eq}$, we can see that a large positive K_{eq} is related to a large negative $\Delta G°$ and thus a large negative $\Delta H°$ (if $\Delta S°$ is small). Consequently, a reaction with $K_{eq} = 1000$ is more exothermic than a reaction with $K_{eq} = 0.001$.

5.10 $\Delta G° = -RT\ln K_{eq}$; $R = 1.986$ cal/K mol; $T = 298$ K

$\Delta G° = -1.986$ cal/K mol x 298 K x $\ln K_{eq} = -592$ cal/mol x $\ln K_{eq}$

If $K_{eq} = 1000$, then $\ln K_{eq} = 6.91$, and $\Delta G° = -592$ cal/mol x 6.91 = 4090 cal/mol = 4.1 kcal/mol

If $K_{eq} = 1$, then $\ln K_{eq} = 0$ and $\underline{\Delta G° = 0}$;

If $K_{eq} = 0.001$, then $\Delta G° = -4$ kcal/mol

$\Delta G° = -RT\ln K_{eq}$; $\ln K_{eq} = -\Delta G° \div RT$; $T = 298$ K

$\ln K_{eq} = -\Delta G° \div 592$ cal/mol $= -\Delta G° \div 0.592$ kcal/mol

If $\Delta G° = -10$ kcal/mol, then $\ln K_{eq} = -[-10$ kcal/mol $\div 0.592$ kcal/mol], and $\ln K_{eq} = 16.90$; $\underline{K_{eq} = 2.2\ \ x\ 10^7}$

If $\Delta G^\circ = 0$, then $\underline{K_{eq} = 1}$

If $\Delta G^\circ = +10$ kcal/mol, then $\ln K_{eq} = -10$ kcal/mol $\div 0.592$ kcal/mol $= -16.90$; $\underline{K_{eq} = 4.6 \times 10^{-8}}$

5.11

a) $Cl_2 \longrightarrow 2Cl\cdot$

Bond broken	ΔH°
Cl—Cl	58 kcal/mol

$\Delta H_a^\circ = +58$ kcal/mol

b) $CH_4 + Cl\cdot \longrightarrow \cdot CH_3 + HCl$

Bond broken	ΔH°	Bond formed	ΔH°
CH_3—H	+104 kcal/mol	H—Cl	+103 kcal/mol

$\Delta H_b^\circ = \Delta H^\circ_{\substack{bond\\broken}} - \Delta H^\circ_{\substack{bond\\formed}} = 104$ kcal/mol $- 103$ kcal/mol $= +1$ kcal/mol

c) $\cdot CH_3 + Cl_2 \longrightarrow CH_3Cl + \cdot Cl$

Bond broken	ΔH°	Bond formed	ΔH°
Cl—Cl	+58 kcal/mol	CH_3—Cl	+84 kcal/mol

$\Delta H_c^\circ = \Delta H^\circ_{\substack{bond\\broken}} - \Delta H^\circ_{\substack{bond\\formed}} = 58$ kcal/mol $- 84$ kcal/mol $= -26$ kcal/mol

d) $\Delta H^\circ_{overall} = \Delta H_b^\circ + \Delta H_c^\circ = -26$ kcal/mol $+ 1$ kcal/mol $= -25$ kcal/mol

The overall reaction between chlorine and methane is exothermic.

5.12

a) $CH_3CH_2OCH_3 + HI \longrightarrow CH_3CH_2OH + CH_3I$

Bonds broken	ΔH°	Bonds formed	ΔH°
CH_3CH_2O—CH_3	81 kcal/mol	CH_3CH_2O—H	103 kcal/mol
H—I	71 kcal/mol	CH_3–I	56 kcal/mol
$\Delta H^\circ_{\substack{bonds\\broken}} = 152$ kcal/mol		$\Delta H^\circ_{\substack{bonds\\formed}} = 159$ kcal/mol	

$\Delta H^\circ_{overall} = \Delta H^\circ_{\substack{bonds\\broken}} - \Delta H^\circ_{\substack{bonds\\formed}} = 152$ kcal/mol $- 159$ kcal/mol $= \underline{-7 \text{ kcal/mol}}$

b) CH_3Cl + NH_3 $\longrightarrow$ CH_3NH_2 + HCl

Bonds broken	$\Delta H°$	Bonds formed	$\Delta H°$
CH_3-Cl	84 kcal/mol	CH_3-NH_2	80 kcal/mol
NH_2-H	103 kcal/mol	$H-Cl$	103 kcal/mol

$\Delta H°_{bonds\ broken}$ = 187 kcal/mol $\Delta H°_{bonds\ formed}$ = 183 kcal/mol

$\Delta H°_{overall}$ = $\Delta H°_{bonds\ broken}$ − $\Delta H°_{bonds\ formed}$ = 187 kcal/mol − 183 kcal/mol = +4 kcal/mol

5.13 A reaction with $\Delta G^{\ddagger}$ = 45 kJ/mol is faster than a reaction with $\Delta G^{\ddagger}$ = 70 kJ//mol. It is not possible to measure the size of K_{eq} from $\Delta G^{\ddagger}$ because $\Delta G^{\ddagger}$ measures the energy difference between reactant and transition state. $\Delta G^{\ddagger}$ does not deal with the energy difference between reactant and *product*, which is described by $\Delta G°$, and thus also by K_{eq}.

5.14

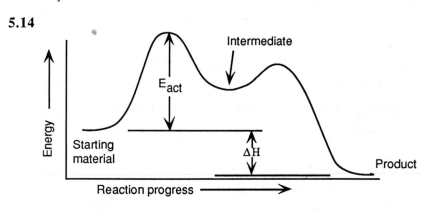

5.15

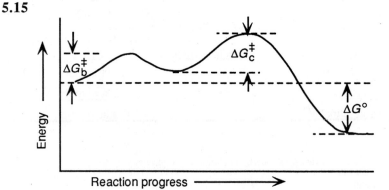

Refer to problem 5.11 for a description of steps b and c of the chlorination reaction. Since $\Delta H°_{overall}$ is negative, $\Delta G°$ is also negative.

5.16 – 5.17

a) $\overset{\delta+}{C}H_3CH_2\overset{\delta-}{C}{\equiv}N$

nitrile

b)

ether

c)

ketone ester

d) ketone

double bonds

e) double bond   amide

f) aldehyde

aromatic ring

5.18 a) substitution   b) elimination   c) addition   d) substitution

5.19 An *addition reaction* takes place when two reactants form a single product.
An *elimination reaction* takes place when one reactant splits apart to give two products.
A *substitution reaction* occurs when two reactants exchange parts to yield two different products.
A *rearrangement reaction* occurs when a reactant undergoes a reorganization of bonds to give a different product.

5.20 a) *Polar reactions* are processes that involve unsymmetrical bond breaking and formation. In a polar reaction, electron-rich sites in the functional groups of one molecule react with electron-poor sites in the functional groups of another molecule.
b) *Heterolytic bond breakage* occurs when both bonding electrons leave with one fragment
c) *Homolytic bond breakage* occurs when one bonding electron leaves with each fragment.
d) A *radical reaction* is a reaction in which odd-electron species are produced or consumed
e) A *functional group* is a group of atoms that has a characteristic reactivity.
f) *Polarization* is the temporary change in the electron distribution of atoms or functional groups due to interactions with reagents or solvent.

5.21

a) $HO:^-$

b) H^+

c), d) $CH_3Br + HO:^- \longrightarrow CH_3OH + :Br:^-$

e)   HO—H $\longrightarrow$ HO:$^-$ + H$^+$

f)   Cl—Cl $\longrightarrow$ 2 :Cl·

5.22

Nucleophiles: :Cl:$^-$, HO:$^-$, CH$_3$NH$_2$

Electrophile: BF$_3$

5.23

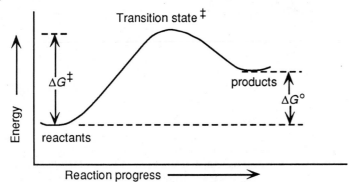

$\Delta G°$ is positive.

5.24

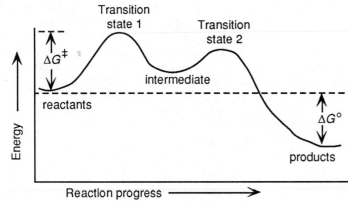

$\Delta G°$ is negative.

5.25 A transition state represents a structure occurring at an energy maximum. An intermediate occurs at an energy minimum between two transition states. Even though an intermediate may be of such high energy that it cannot be isolated, it is still of lower energy than a transition state. An intermediate may be stable and isolate, however.

5.26 Problem 5.24 shows a reaction energy diagram of a two-step exothermic reaction. Step 2 is faster than step 1 because $\Delta G_2^{\ddagger} < \Delta G_1^{\ddagger}$.

5.27

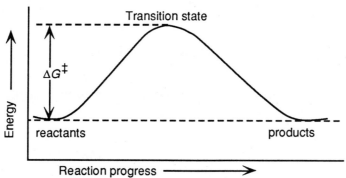

A reaction with $K_{eq} = 1$ has $\Delta G^{\circ} = 0$.

5.28

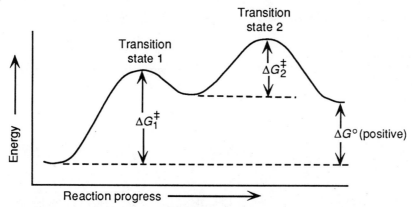

There are two steps in the reaction; step 2 is faster since $\Delta G_2^{\ddagger} < \Delta G_1^{\ddagger}$. There are also two transition states; they are indicated on the diagram.

5.29

a) $CH_3{-}OH$ + $H{-}Br$ $\longrightarrow$ $CH_3{-}Br$ + $H{-}OH$

   91 kcal/mol   88 kcal/mol   70 kcal/mol   119 kcal/mol

   ΔH° = −10 kcal/mol

b) $CH_3CH_2O{-}H$ + $CH_3{-}Cl$ $\longrightarrow$ $CH_3CH_2O{-}CH_3$ + $H{-}Cl$

   103 kcal/mol   84 kcal/mol   81 kcal/mol   103 kcal/mol

   ΔH° = +3 kcal/mol

5.30

a) $CH_3CH_2{-}H$ + Cl_2 $\longrightarrow$ $CH_3CH_2{-}Cl$ + $H{-}Cl$

   98 kcal/mol   58 kcal/mol   81 kcal/mol   103 kcal/mol

   ΔH° = −28 kcal/mol

b) CH_3CH_2-H + Br_2 $\longrightarrow$ CH_3CH_2-Br +   H–Br

   98 kcal/mol   46 kcal/mol   68 kcal/mol   88 kcal/mol

   ΔH° = –12 kcal/mol

c) CH_3CH_2-H + I_2 $\longrightarrow$ CH_3CH_2-I +   H–I

   98 kcal/mol   36 kcal/mol   53 kcal/mol   71 kcal/mol

   ΔH° = +10 kcal/mol

Of the three halogenation reactions, chlorination proceeds the most readily; iodination is the most difficult reaction.

5.31

Bonds broken	ΔH°	Bonds formed	ΔH°
CH_3-CH_3	88 kcal/mol	CH_3-Br	70 kcal/mol
Br_2	46 kcal/mol	CH_3-Br	70 kcal/mol
	ΔH° = 134 kcal/mol		ΔH° = 140 kcal/mol

$\Delta H^\circ_{overall}$ = $\Delta H^\circ_{bonds\ broken}$ – $\Delta H^\circ_{bonds\ formed}$ = 134 kcal/mol – 140 kcal/mol = –6 kcal/mol

ΔH° for bromoethane formation is –12 kcal/mol; ΔH° for bromomethane formation is –6 kcal/mol. Although both of these reactions have negative ΔH°, the reaction that forms bromoethane is more favorable.

5.32, 5.33

	Product		ΔH°
	$CH_3CH_2\overset{\underset{\mid}{CH_3}}{C}HCH_2Cl$ + HCl		–28 kcal/mol
	1–Chloro–2–methylbutane		
	$ClCH_2CH_2\overset{\underset{\mid}{CH_3}}{C}HCH_3$ + HCl		–28 kcal/mol
$CH_3CH_2\overset{\underset{\mid}{CH_3}}{C}HCH_3 + Cl_2 \xrightarrow{light}$	1–Chloro–3–methylbutane		
	$CH_3CHCl\overset{\underset{\mid}{CH_3}}{C}HCH_3$ + HCl		–30 kcal/mol
	2–Chloro–3–methylbutane		
	$CH_3CH_2\overset{\underset{\mid}{CH_3}}{C}ClCH_3$ + HCl		–33 kcal/mol
	2–Chloro–2–methylbutane		

Formation of the tertiary chloride is favored, but a mixture of products is expected since all three ΔH° values are quite close to one another.

5.34 The following compounds yield single monohalogenation products since each has only one kind of hydrogen atom.

C_2H_6,   ⬡   , $CH_3C \equiv CCH_3$

5.35 For the first series of steps:
a) $\Delta H° = +58$ kcal/mol
b) $\Delta H° = +\ 1$ kcal/mol ⎤
c) $\Delta H° = -26$ kcal/mol ⎦ $\Delta H°_{overall} = -25$ kcal/mol

For the alternate series:
a) $\Delta H° = +58$ kcal/mol
b) $\Delta H° = +20$ kcal/mol ⎤
c) $\Delta H° = -45$ kcal/mol ⎦ $\Delta H°_{overall} = -25$ kcal/mol

For both series of reactions, the radical-producing initiation step has $\Delta H° = +58$ kcal/mol.
    In both series, $\Delta H°$ for the propagation steps b + c is –25 kcal/mol. In series 2, however, one step has $\Delta H° = +20$ kcal/mol; this step is energetically much less favorable than any step in series 1 and disfavors the second series as a whole. The first route for chlorination of methane is thus more likely to occur.

5.36 a)   The reaction is a polar rearrangement.

b)

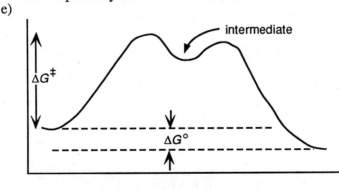

c) $K_{eq} = \dfrac{[Products]}{[Reactants]} = \dfrac{.70}{.30} = 2.3$

d)   Section 5.9 states that reactions that occur spontaneously have $\Delta G^{\ddagger}$ of less than 20 kcal/mol at room temperature. Since this reaction proceeds slowly at room temperature, $\Delta G^{\ddagger}$ is probably close to 20 kcal/mol.

e)

intermediate

Note that it is not possible to determine the energy of the intermediate.

5.37

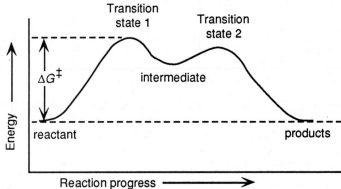

reactant   intermediate   products

$\Delta G^{\ddagger}$ is approximately 20 kcal/mol. ΔG° is approximately zero.

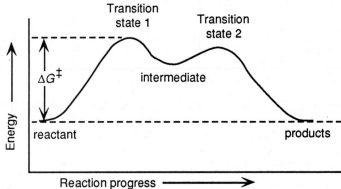

5.38 $\Delta G^{\circ} = \Delta H^{\circ} - T\Delta S^{\circ}$

$= -18 \text{ kcal/mol} - 298 \text{ K} \times 13 \text{ cal/mol K}$

$= -18 \text{ kcal/mol} - 3.9 \text{ kcal/mol}$

$= -21.9 \text{ kcal/mol}$

The reaction is exothermic because ΔG° is less than zero.

5.39 $\Delta G^{\circ} = -RT\ln K_{eq}$

$-21.9 \text{ kcal/mol} = -1.986 \text{ cal/mol K} \times 298 \text{ K} \times \ln K_{eq}$

$-21.9 \text{ kcal/mol} \div -0.592 \text{ kcal/mol} = \ln K_{eq}$

$37.0 = \ln K_{eq}$

$1.2 \times 10^{16} = K_{eq}$

Study Guide for Chapter 5

After studying this chapter, you should be able to:

(1) Identify reactions as polar, radical, substitution, elimination, addition or rearrangement reactions, and define these terms (5.1, 5.18, 5.19, 5.20).

(2) Understand the mechanism of radical reactions (5.2, 5.3, 5.4, 5.34).

(3) Predict the direction of bond polarity for a given functional group (5.5, 5.17).

(4) Identify reagents as electrophiles or nucleophiles (5.6, 5.21, 5.22).

(5) Understand the concepts of equilibrium and rate (5.8, 5.9, 5.13, 5.25).

(6) Calculate $\Delta G°$ and K_{eq} of reactions (5.10, 5.36, 5.38, 5.39).

(7) Use bond dissociation energies to calculate $\Delta H°$ of simple reactions (5.11, 5.12, 5.29, 5.30, 5.31, 5.32, 5.35).

(8) Draw reaction energy diagrams and label them properly (5.14, 5.15, 5.23, 5.24, 5.26, 5.27, 5.28, 5.36, 5.37).

6.1 "Degree of unsaturation" refers to the number of rings, double bonds, or triple bonds that a compound may contain.

As an example, consider the hydrocarbon C_8H_{14} in part (a). The formula of a C_8 *alkane* is C_8H_{18}. C_8H_{14}, which contains four fewer (or two pairs fewer) hydrogens than C_8H_{18}, may have two double bonds, or two rings, or one of each, or one triple bond. C_8H_{14} thus has a degree of unsaturation of 2.

Compound	Degree of Unsaturation
a) C_8H_{14}	2
b) C_5H_6	3
c) $C_{12}H_{20}$	3
d) $C_{20}H_{32}$	5
e) $C_{40}H_{56}$	13

6.2

Compound	Degree of Unsaturation	Structures
a) C_4H_8	1	$CH_3CH_2CH=CH_2$, $CH_3CH=CHCH_3$, $(CH_3)_2C=CH_2$

$H_2C—CH_2$ / $H_2C—CH_2$, H_2C $\diagup$ $CHCH_3$ / H_2C

b) C_4H_6	2	$CH_2=CHCH=CH_2$, $CH_3CH=C=CH_2$, $CH_3C\equiv CCH_3$

$CH_3CH_2C\equiv CH$, $HC—CH_2$ / $H_2C—CH$, $HC—CH_2$ / $HC—CH_2$,

HC / HC $\diagup$ $CHCH_3$, HC / H_2C $\diagup$ CCH_3, H_2C / H_2C $\diagup$ $C=CH_2$

c) C_3H_4	2	$CH_2=C=CH_2$, $CH_3C\equiv CH$, HC / HC $\diagup$ CH_2

6.3 a)   Subtract one hydrogen for each nitrogen present to find the base formula C_6H_4. Compared to the alkane C_6H_{14}, the compound of formula C_6H_4 has ten fewer hydrogens, or five fewer hydrogen pairs, and contains five degrees of unsaturation.

b) $C_6H_5NO_2$ also contains five degrees of unsaturation because oxygen does not affect the base formula of a compound.

c)   A halogen atom is equivalent to a hydrogen atom in the calculation of the base formula. Here, the "base formula" is C_8H_{12}. $C_8H_9Cl_3$ has three degrees of unsaturation.

d) $C_9H_{16}Br_2$ – one degree of unsaturation.
e) $C_{10}H_{12}N_2O_3$ – six degrees of unsaturation.
f) $C_{20}H_{32}O_2$ – five degrees of unsaturation.

6.4

a) $\underset{\text{H}_2\text{C}=\text{CHCH}-\text{CCH}_3}{}$

with H$_3$C and CH$_3$ groups, numbered 1 2 3 4 5, and CH$_3$ below

1) Find the longest carbon chain containing the double bond, and name the parent compound. Here the longest chain contains five carbons, and the compound is a *pentene*.
2) Number the carbon atoms, giving to the double bond the lowest possible number.
3) Name the compound: 3,4,4–trimethyl–1–pentene.

b) $CH_3CH_2CH=CCH_2CH_3$ (with CH$_3$ above)   3–Methyl–3–hexene

c) $CH_3CH=CHCHCH=CHCHCH_3$ (with CH$_3$ and CH$_3$ above)   4,7–Dimethyl–2,5–octadiene

6.5

a) $CH_2=CHCH_2CH_2C=CH_2$ (with CH$_3$ above)

2–Methyl–1,5–hexadiene

b) $CH_3CH_2CH_2CH=CC(CH_3)_3$ (with CH$_2$CH$_3$ above)

3–Ethyl–2,2–dimethyl–3–heptene

c) $CH_3CH=CHCH=CHC-\!-C=CH_2$ (with CH$_3$ above; CH$_3$ CH$_3$ below)

2,3,3–Trimethyl–1,4,6–octatriene

d)

3,4–Diisopropyl–2,5–dimethyl–3–hexene

e) $CH_3CH_2CH_2CHCH_2CHCH_3$ (with C(CH$_3$)$_3$ above; CH$_3$ below)

4–*tert*–Butyl–2–methylheptane

6.6

a)

1,2–Dimethylcyclohexene

b)

4,4–Dimethylcycloheptene

c)

3–Isopropylcyclopentene

6.7 Compounds (c), (e), and (f) can exist as pairs of *cis–trans* isomers.

c) $CH_3CH_2CH=CHCH_3$

e) $ClCH=CHCl$

f) $BrCH=CHCl$

6.8 Models are essential here! A model of cyclohexene shows that a six-membered ring is too small to contain a *trans* double bond without causing severe strain to the ring. A ten-membered ring is flexible enough to accommodate either a *cis* or a *trans* double bond, although the *cis* isomer has less ring strain than the *trans* isomer.

6.9 Review the sequence rules of Section 6.6. In summary:

Rule 1: A high-atomic-weight atom has priority over a low-atomic weight atom.

Rule 2: If a decision can't be reached by Rule 1, look at the second, third or fourth atom out until a decision can be made.

Rule 3: Multiple-bonded atoms are considered to be equivalent to the same number of single bonded atoms.

	High	Low	Rule
a)	$-Br$	$-H$	1
b)	$-Br$	$-Cl$	1
c)	$-CH_2CH_3$	$-CH_3$	2
d)	$-OH$	$-NH_2$	1
e)	$-CH_2OH$	$-CH_3$	2
f)	$-CH=O$	$-CH_2OH$	3

6.10

Highest priority ⟶ Lowest priority

a)   −Cl, −OH, −CH₃, −H

b)   −CH₂OH, −CH=CH₂, −CH₂CH₃, −CH₃

c)   −COOH, −CH₂OH, −CN, −CH₂NH₂

d)   −CH₂OCH₃, −CN, −C≡CH, −CH₂CH₃

6.11

a) $\underline{Z}$

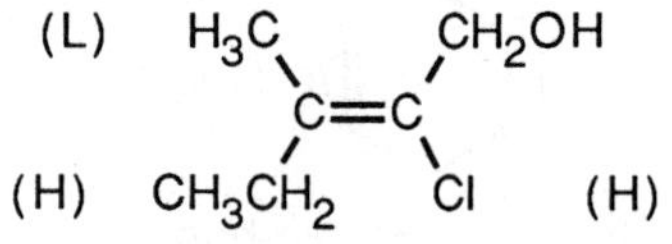

First, consider substituents on the right side of the double bond. −Cl ranks higher than −CH₂OH by Cahn-Ingold-Prelog rules. On the left side, −CH₂CH₃ ranks higher than −CH₃. The isomer has $\underline{Z}$ configuration because the higher priority groups are on the same side of the double bond.

b) $\underline{E}$

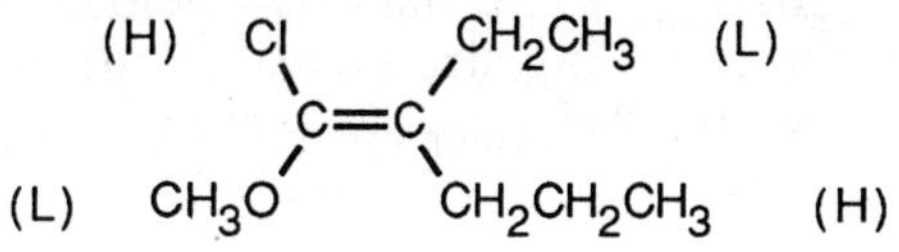

c) $\underline{Z}$

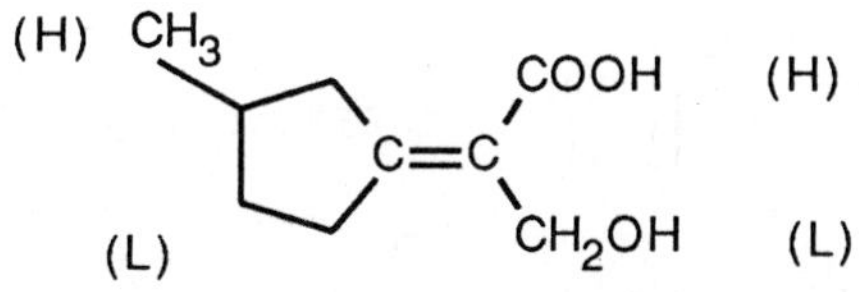

Notice that in ranking substituents on the left side of the bond, the upper substituent is of higher priority because of the methyl group attached to the ring.

d) $\underline{E}$

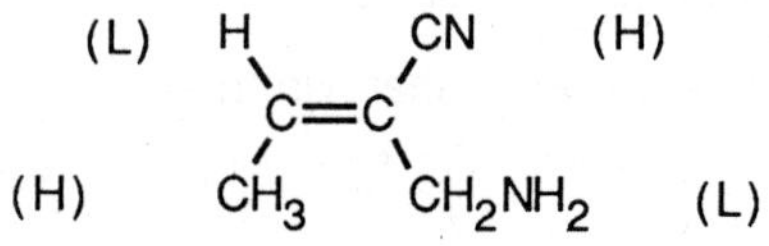

6.12

	More stable	Less stable
a)		
	disubstituted double bond	monosubstituted double bond

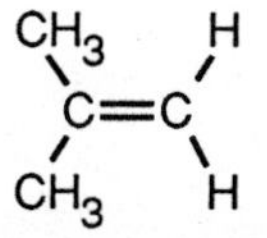

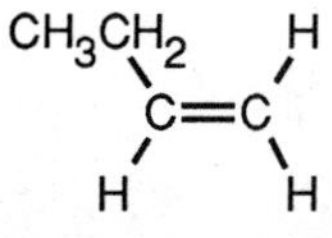

b)   E

no steric strain

CH$_3$CH$_2$CH$_2$ CH$_3$
          C=C   Z
          H   H

steric strain of groups on
same side of double bond

c)

trisubstituted double bond

disubstituted double bond

6.13 In all of these examples, ΔH_{hydrog} for the *trans* cycloalkenes is higher than ΔH_{hydrog} for the *cis* compounds, indicating that the *trans*–cycloalkenes are less stable than *cis*–cycloalkenes. Build models of the two cyclooctenes and notice the large amount of strain in *trans*–cyclooctene relative to *cis*–cyclooctene. This strain causes the *trans* isomer to be of higher energy and to have a ΔH_{hydrog} larger than *cis*–cyclooctene. Use models to construct the other four cycloalkenes. As ring size increases, the problem of strain for the *trans* rings becomes less severe, and ΔH_{hydrog} decreases.

6.14

a)   +   HCl   ⟶   Chlorocyclohexane

Since the starting material is symmetrical, only one product is possible.

b)   (CH$_3$)$_2$C=CHCH$_2$CH$_3$ $\xrightarrow{\text{HBr}}$ (CH$_3$)$_2$CCH$_2$CH$_2$CH$_3$ (with Br)

2–Bromo–2–methylpentane

c)   CH$_3$CH$_2$CH$_2$CH=CH$_2$ $\xrightarrow{\text{H}_3\text{PO}_4,\ \text{KI}}$ CH$_3$CH$_2$CH$_2$CHCH$_3$ (with I)

2–Iodopentane

d)   +   HBr   ⟶

1–Bromo–1–methylcyclohexane

6.15

a)

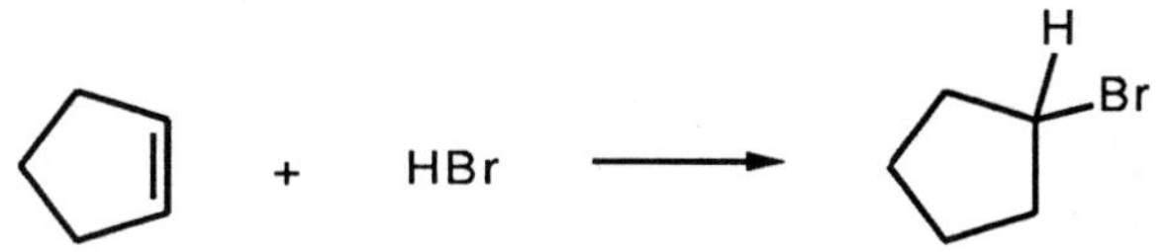

Cyclopentene

b) $CH_3CH_2CH=CHCH_2CH_3$ + HBr $\longrightarrow$ $CH_3CH_2CHBrCH_2CH_2CH_3$

   3–Hexene

c)

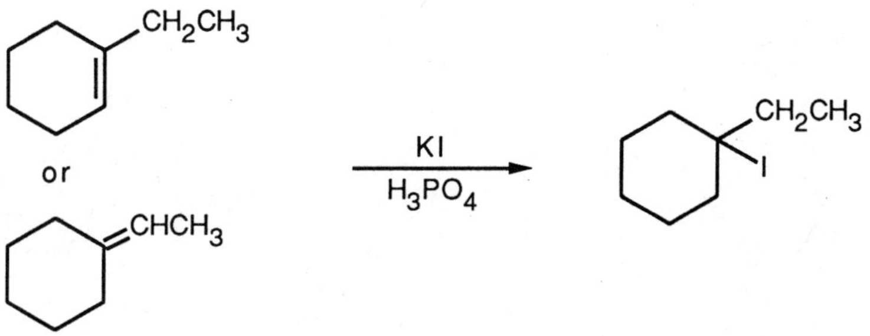

or

$\xrightarrow[H_3PO_4]{KI}$

d)

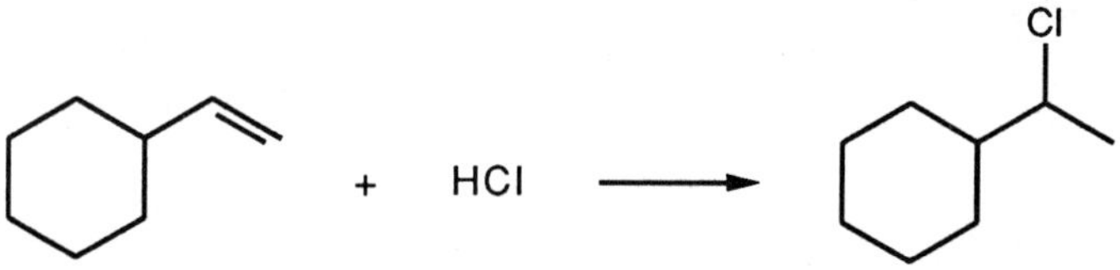

+   HCl $\longrightarrow$

6.16

a) $\underset{\overset{|}{CH_3}}{CH_3CH_2C}=\underset{\overset{|}{CH_3}}{CHCHCH_3}$ + HBr $\longrightarrow$ $\left[\underset{+}{CH_3CH_2\underset{\overset{|}{CH_3}}{C}-CH_2\underset{\overset{|}{CH_3}}{CHCH_3}}\right]$ $\longrightarrow$ $CH_3CH_2\underset{\overset{|}{Br}}{\overset{\overset{CH_3}{|}}{C}}-CH_2\underset{\overset{|}{CH_3}}{CHCH_3}$

carbocation intermediate

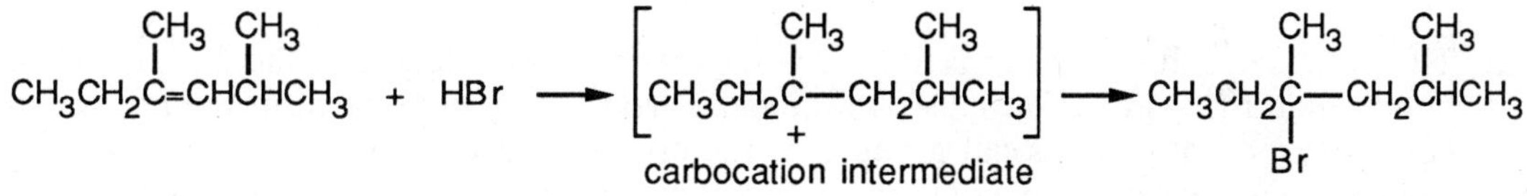

b)

=CHCH₃ + HI $\longrightarrow$ [+—CH₂CH₃] $\longrightarrow$

carbocation intermediate

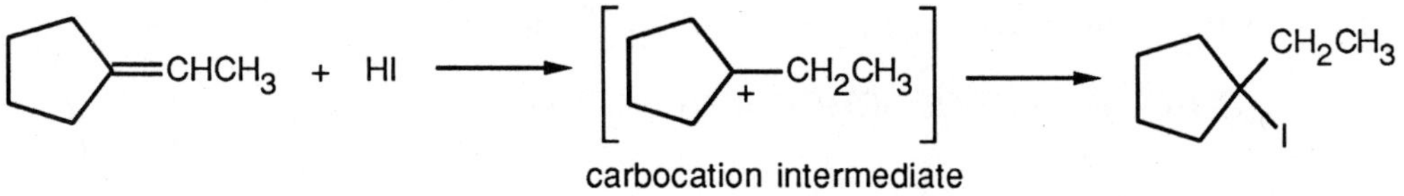

6.17 The second step in the electrophilic addition of HCl to alkenes is exothermic. According to the Hammond postulate, the transition state should resemble the carbocation intermediate.

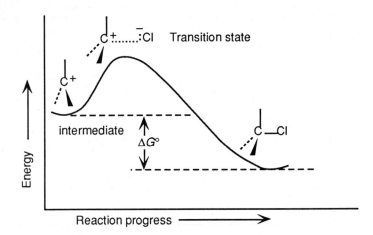

6.18

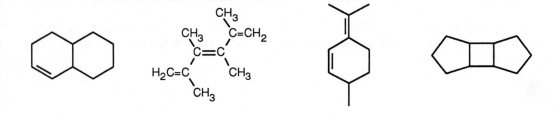

a 2°
carbocation

a 3°
carbocation

6.19 *Formula   Degree of Unsaturation*

a)	C_6H_6	4
b)	C_6H_{10}	2
c)	$C_{10}H_{16}$	3
d)	$C_6H_6Cl_6$	1
e)	C_5H_5N	4
f)	$C_{10}H_{10}O_2$	6

6.20 The purpose of this problem is to give you experience in calculating the number of double bonds and/or rings in a formula. Additionally, you will gain practice in writing structures containing various functional groups. Remember that any formulas that satisfy the rules of valency are acceptable. Try to identify functional groups in the formulas you draw.

a) $C_{10}H_{16}$ -- three degrees of unsaturation. Examples:

$$CH_3CH_2CH=CHCH=CHCH=CHCH_2CH_3$$

b) C_8H_8O. The parent hydrocarbon is C_8H_8; each structure contains five degrees of unsaturation. Examples:

c)   This compound has C_7H_{12} as its base formula. $C_7H_{10}Cl_2$ has two degrees of unsaturation. Examples:

d) $C_{10}H_{16}O_2$ -- three degrees of unsaturation.

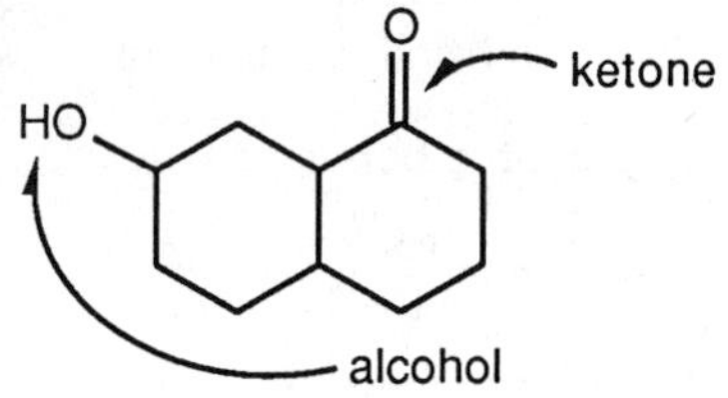

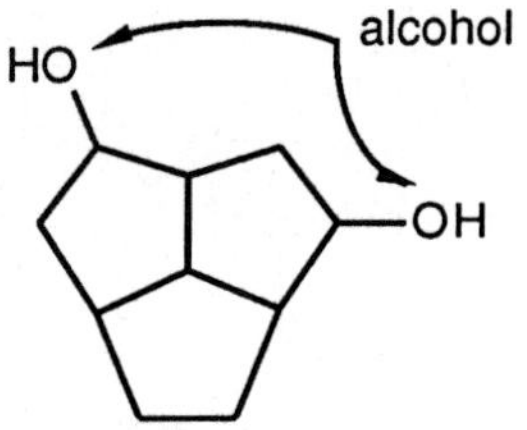

ketone

alcohol

ketone

HO— alcohol

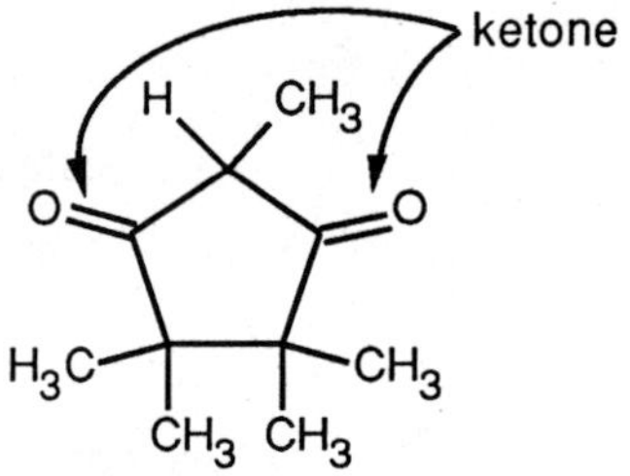

HO— alcohol

—OH

H   CH₃

O   O

H₃C—   —CH₃

CH₃ CH₃

e)   C₅H₉NO₂ -- two degrees of unsaturation.

ketone

O   O amide

$CH_3CH_2CCH_2CNH_2$

double bond   O   nitro group

$CH_2=CHCH_2CH_2CH_2N^+-O^-$

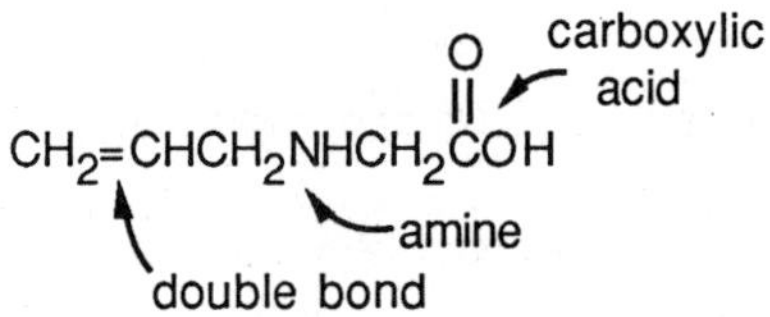

carboxylic acid

O

$CH_2=CHCH_2NHCH_2COH$

amine

double bond

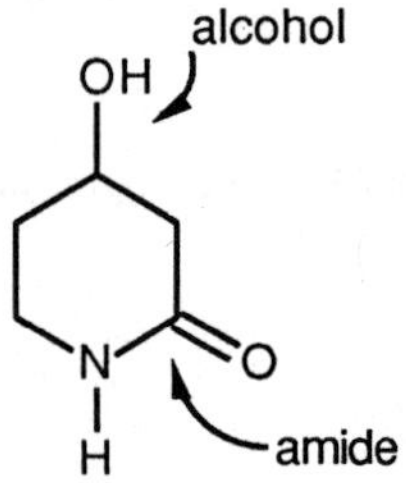

alcohol

OH

amide

N
H

ketone

O

NH₂

amine

ether

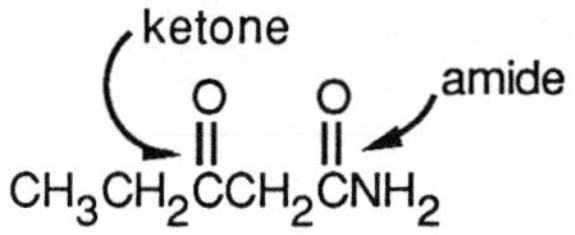

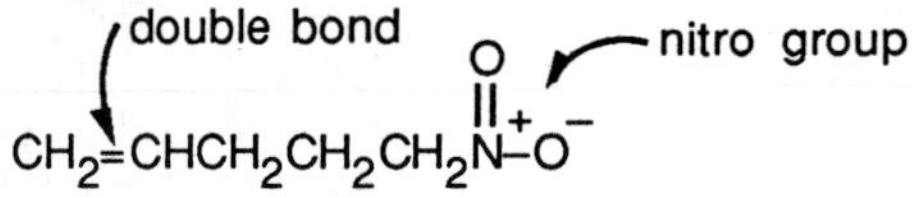

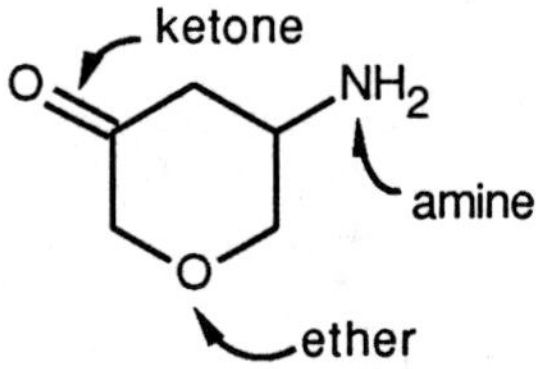

f)   C₈H₁₀ClNO -- four degrees of unsaturation.

halide   O amide

$ClCH=CHCH=CHCH=CHCH_2CNH_2$

double bonds

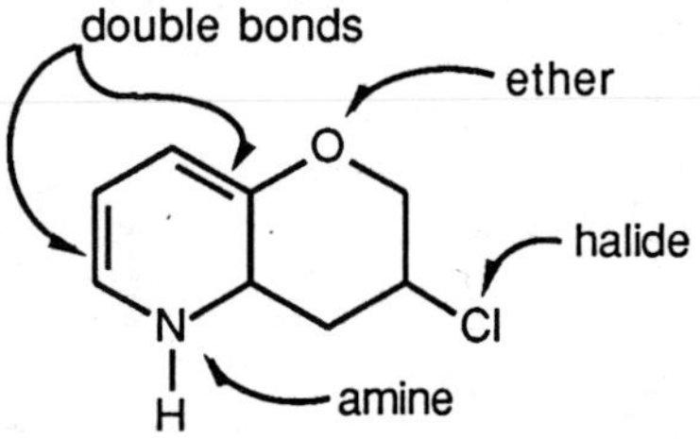

double bonds ether

O

halide

Cl

N
H   amine

alcohol   CH₃   aromatic ring

amine

HO— CH_2NH_2

Cl

halide

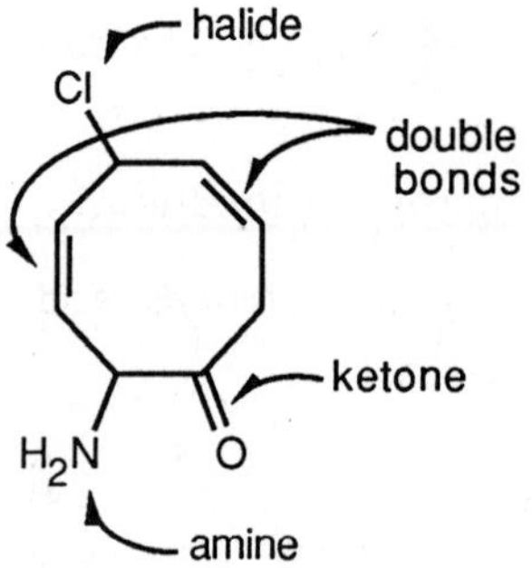

halide

Cl   double bonds

ketone

H_2N O   ketone

amine

H   amine
N

O

ketone

halide

Cl

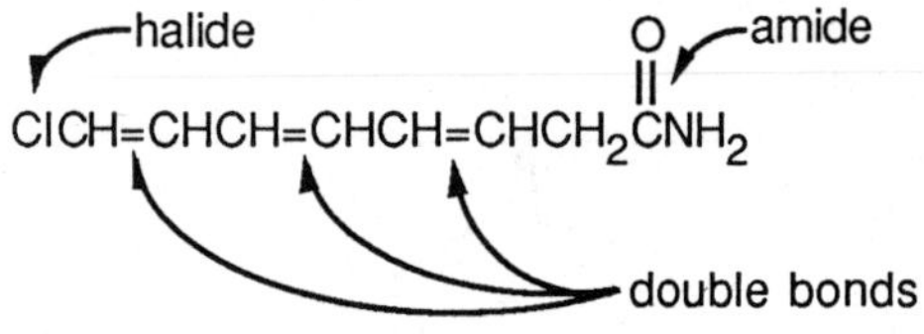

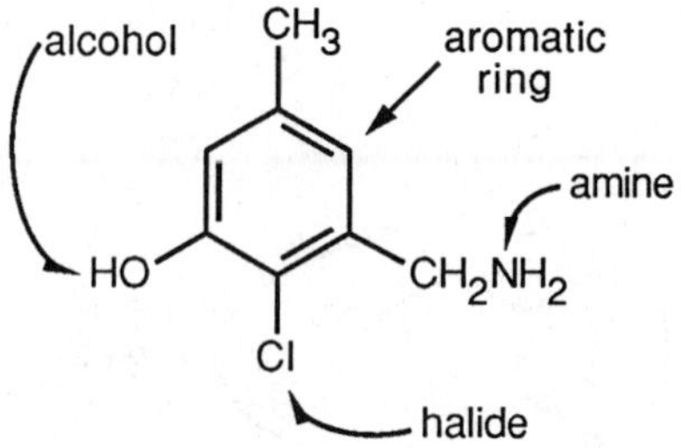

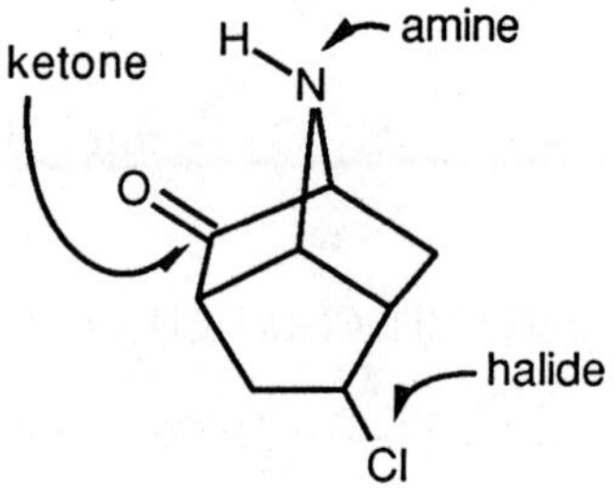

6.21 Interpreting problems of this sort is often difficult. To start, you should train yourself to *read every word* of the problem. Then you should try to solve the problem phrase by phrase. For example, "A compound of formula $C_{10}H_{14}$" describes a compound having four degrees of unsaturation ($C_{10}H_{14}$ has four fewer hydrogen pairs than a C_{10} acyclic alkene). The phrase "undergoes catalytic hydrogenation" means that H_2 is added to the double bonds. "Absorbs only two equivalents of H_2" means that only *two* of the degrees of unsaturation are double bonds (or a triple bond). The other two must be rings.

6.22 a) 4–Methyl–2–hexene   b) 4–Butyl–7–methyl–2–octene
c) 2–Ethyl–1–butene   d) 3,4–Dimethyl–1,5–heptadiene
e) 4–Methyl–1,3–hexadiene   f) 1,2–Butadiene
g) 3,3–Dimethyl–1–butene   h) 2,2,5,5–Tetramethyl–3–hexene

6.23 Because the longest carbon chain contains 8 carbons, and because there are three double bonds present, ocimene is an <u>octatriene</u>. Start numbering at the end that will give the lowest number to the first double bond (1,3,6 is lower than 2,5,7). Number the methyl substituents and, finally, name the compound.

(3*E*)–3,7–Dimethyl–1,3,6–octatriene

6.24, 6.25

(3*E*, 6*E*)–3,7,11–Trimethyl–1,3,6,10–dodecatetraene

6.26

a) $CH_3CH=CCH_2C=CH_2$ (with CH_3 and CH_3 substituents)

b) $CH_3CH_2CH=CHCHC(CH_3)_2CH=CH_2$ (with $CH_2CH_2CH_3$ substituent)

c) $CH_3CHCH=C=CH_2$ (with CH_3 substituent)

d) $H_2C=CHC=CHCH=CHC=CH_2$ (with CH_3 and CH_3 substituents)

e) $CH_3CH_2CH_2CH_2C=CHCH_3$ (with $CH_2CH_2CH_2CH_3$ substituent)

f) $(CH_3)_3CCH=CHC(CH_3)_3$

6.27

Menthene

6.28

a) $CH_2=CHCH=C(CH_3)_2$

Correct name: 4–Methyl–1,3–pentadiene
Numbering must start at the other end.

b) $CH_3CH_2\overset{\overset{\displaystyle CH_2}{\|}}{C}CH=CH_2$

Correct name: 2–Ethyl–1,3–butadiene
The parent chain must contain *both*
double bonds.

c) $CH_3CH=CHCH_2CH=CHCH_2CH_3$

Correct name: 2,5–Octadiene
Numbering must start at the other end.

d) $CH_3CH_2CH_2\overset{\overset{\displaystyle CH_2CH_3}{|}}{C}=CHCH_2CH_2CH_3$

Correct name: 4–Ethyl–4–octene
Numbering must start at the other end.

e) $CH_3CH_2CH_2CH=\overset{\overset{\displaystyle CH_2CH_2CH_3}{|}}{C}CH_2CH_3$

Correct name: 4–Ethyl–4–octene
The longest chain containing the double
bond is an *octene*.

f) $H_2C=CHCH_2CH=CH_2$

Correct name: 1,4–Pentadiene
The longest chain must contain both
double bonds.

6.29

$CH_3CH_2CH_2CH=CH_2$
1–Pentene

$CH_3CH_2CH=CHCH_3$
2–Pentene

$CH_3CH_2\overset{\overset{\displaystyle CH_3}{|}}{C}=CH_2$
2–Methyl–1–butene

$CH_3\overset{\overset{\displaystyle CH_3}{|}}{C}HCH=CH_2$
3–Methyl–1–butene

$CH_3CH=\overset{\overset{\displaystyle CH_3}{|}}{C}CH_3$
2–Methyl–2–butene

6.30

$CH_3CH_2CH_2CH_2CH=CH_2$
1–Hexene

$CH_3CH_2CH_2CH=CHCH_3$
2–Hexene

$CH_3CH_2CH=CHCH_2CH_3$
3–Hexene

$CH_3CH_2CH_2\overset{\overset{\displaystyle CH_3}{|}}{C}=CH_2$
2–Methyl–1–pentene

$CH_3CH_2\overset{\overset{\displaystyle CH_3}{|}}{C}HCH=CH_2$
3–Methyl–1–pentene

$CH_3\overset{\overset{\displaystyle CH_3}{|}}{C}HCH_2CH=CH_2$
4–Methyl–1–pentene

$CH_3CH_2CH=\overset{\overset{\displaystyle CH_3}{|}}{C}CH_3$
2–Methyl–2–pentene

$CH_3CH_2\overset{\overset{\displaystyle CH_3}{|}}{C}=CHCH_3$
3–Methyl–2–pentene

$CH_3\overset{\overset{\displaystyle CH_3}{|}}{C}HCH=CHCH_3$
4–Methyl–2–pentene

CH₃
|
CH₃CHC=CH₂
|
CH₃

2,3–Dimethyl–1–butene

CH₃
|
CH₃CCH=CH₂
|
CH₃

3,3–Dimethyl–1–butene

CH₃   CH₃
\   /
C=C
/   \
CH₃   CH₃

2,3–Dimethyl–2–butene

CH₂CH₃
|
CH₃CH₂C=CH₂

2–Ethyl–1–butene

6.31 In problem 6.29, only 2–pentene shows *cis–trans* isomerism. In problem 6.30, 2–hexene, 3–hexene, 3–methyl–2–pentene, and 4–methyl–2–pentene show *cis–trans* isomerism.

6.32 As expected, the two *trans* compounds are more stable than their *cis* counterparts. The *cis–trans* difference is much more extreme for the tetramethyl compound, however. Build a model of *cis*–2,2,5,5–tetramethyl–3–hexene and notice the extreme crowding of the methyl groups. Steric interference makes the *cis* isomer much less stable than the *trans* isomer and causes *cis* ΔH_{hydrog} to be much larger than *trans* ΔH_{hydrog}.

6.33 The central carbon of allene forms two sigma bonds and two pi bonds. The central carbon is *sp* hybridized, and the carbon-carbon bond angle is 180°, indicating linear geometry for the carbons of allene.

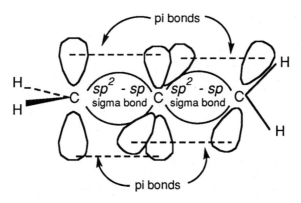

6.34 Because its heat of hydrogenation is so much larger, 1,2–propadiene must be much less stable than 1,4–pentadiene. This instability may be due to strain encountered when one carbon must form a double bond to each of two different carbons.

6.35

a)   CH₃CH₂CH=CCH₂CH₃   +   HCl   ⟶   CH₃CH₂CH₂CCH₂CH₃
                | |
                CH₃   CH₃
                                                             |
                                                             Cl

b)

$$\text{(cyclopentene with CH}_2\text{CH}_3\text{)} + \text{HBr} \longrightarrow \text{(cyclopentane with Br and CH}_2\text{CH}_3\text{)}$$

c) $CH_3CH_2\overset{\underset{\displaystyle |}{CH_3}}{C}=CHC(CH_3)_3$ + HI $\longrightarrow$ $CH_3CH_2\overset{\underset{\displaystyle |}{CH_3}}{\underset{\displaystyle |}{C}}CH_2C(CH_3)_3$

d) $CH_2=CHCH_2CH_2CH_2CH=CH_2$ + 2HCl $\longrightarrow$ $CH_3\overset{\underset{\displaystyle |}{Cl}}{C}HCH_2CH_2CH_2\overset{\underset{\displaystyle |}{Cl}}{C}HCH_3$

e)

Both products are likely to be formed.

6.36

Highest priority $\longrightarrow$ Lowest priority

a) $-I, -Br, -CH_3, -H$

b) $-OCH_3, -OH, -COOH, -H$

c) $-COOCH_3, -COOH, -CH_2OH, -CH_3$

d) $-COCH_3, -CH_2CH_2OH, -CH_2CH_3, -CH_3$

e) $-CH_2Br, -C\equiv N, -CH_2NH_2, -CH=CH_2$

f) $-CH_2OCH_3, -CH_2OH, -CH=CH_2, -CH_2CH_3$

6.37

a)   (H) $HOCH_2$ CH_3 (H)

(L) CH_3 H   (L)   **Z**

b)   (L) HOOC   H   (L)

(H) Cl OCH_3 (H)   **Z**

c)   (H) NC CH_3 (L)

(L) CH_3CH_2 CH_2OH (H)   **E**

d)   (H) CH_3O_2C $CH=CH_2$ (H)

(L) HO_2C CH_2CH_3 (L)   **Z**

6.38

a)   3–Methylcyclohexene   b) 1,5–Dimethylcyclopentene
c)   Ethylcyclobutadiene   d) 1,2–Dimethyl–1,4–cyclohexadiene
e)   5–Methyl–1,3–cyclohexadiene   f) 1,5–Cyclooctadiene

6.39

a) 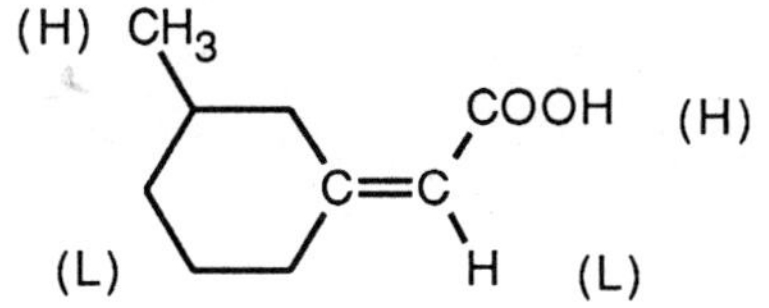
(H) CH3
COOH (H)
C=C
(L)
H (L)

Z (correct)

b) 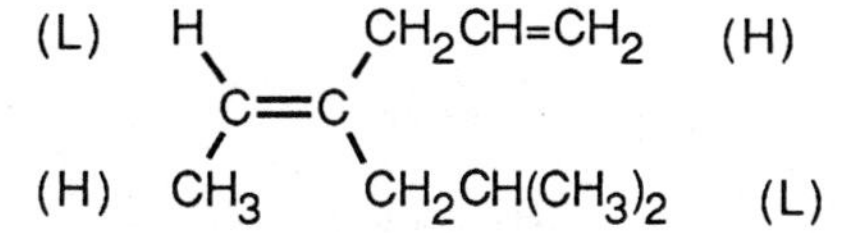
(L) H
CH2CH=CH2 (H)
C=C
(H) CH3
CH2CH(CH3)2 (L)

E (correct)

c) 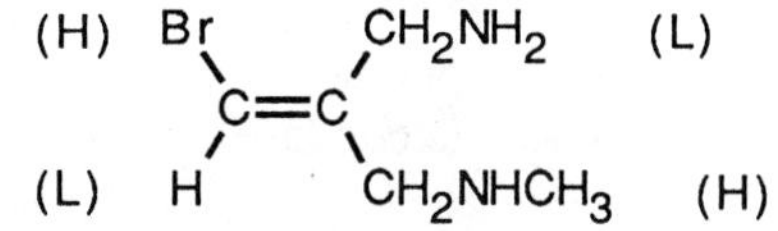
(H) Br
CH2NH2 (L)
C=C
(L) H
CH2NHCH3 (H)

E (incorrect)

d) 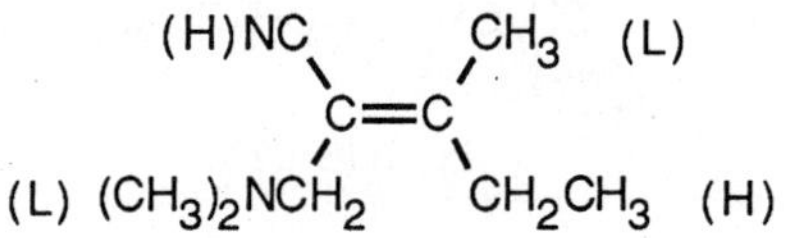
(H)NC
CH3 (L)
C=C
(L) (CH3)2NCH2
CH2CH3 (H)

E (correct)

e) 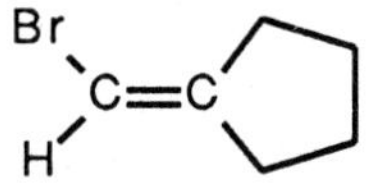
Br
C=C
H

This compound does not exhibit *E–Z* isomerism.

f) 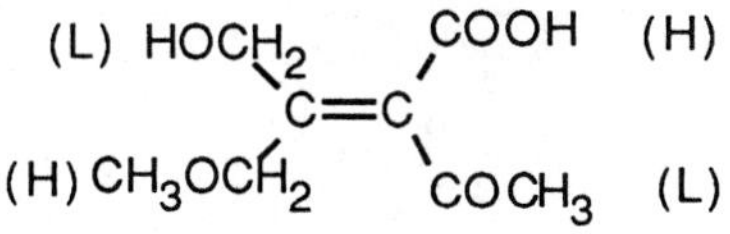
(L) HOCH2
COOH (H)
C=C
(H) CH3OCH2
COCH3 (L)

E (correct)

6.40

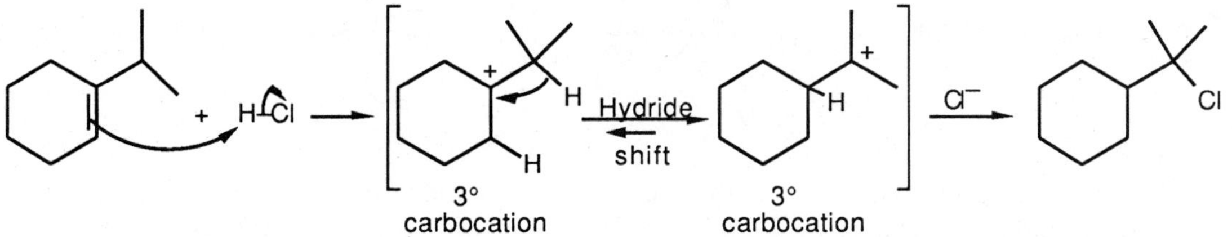
+ H–Cl
H
H
Hydride
shift
H
Cl⁻
Cl
3°
carbocation
3°
carbocation

6.41

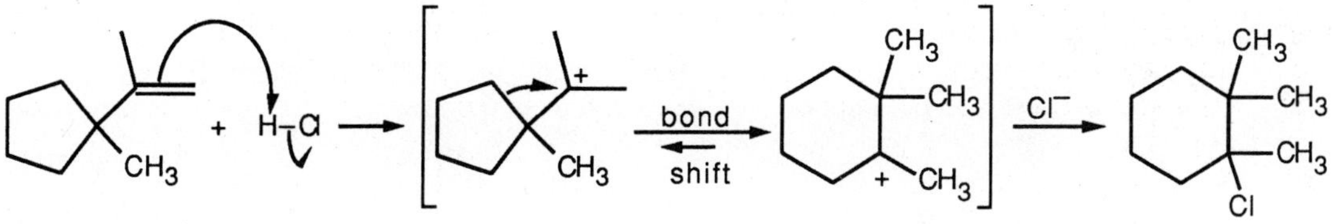
CH3
+ H–Cl
+
CH3
bond
shift
CH3
CH3
+ CH3
Cl⁻
CH3
CH3
CH3
Cl

6.42 a) $C_{27}H_{46}O$ five degrees of unsaturation
   b) $C_{14}H_9Cl_5$ eight degrees of unsaturation
   c) $C_{20}H_{34}O_5$ four degrees of unsaturation
   d) $C_8H_{10}N_4O_2$ six degrees of unsaturation
   e) $C_{21}H_{28}O_5$ eight degrees of unsaturation
   f) $C_{17}H_{23}NO_3$ seven degrees of unsaturation

6.43

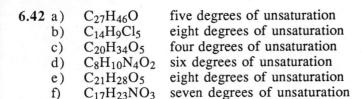

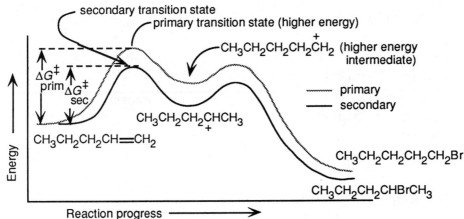

6.44

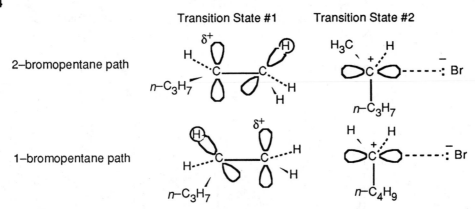

The first step (protonation) for both reaction paths is endothermic, and both transition states resemble the carbocation intermediate. Transition states for the exothermic second step also resemble the carbocation intermediate. Transition state #1 for 1–bromopentane is more like the carbocation intermediate than is transition state #1 for 2–bromopentane.

Study Guide for Chapter 6

After studying this chapter, you should be able to:

(1) Calculate the degree of unsaturation of any compound, including those containing N, O and halogen (6.1, 6.2, 6.3, 6.19, 6.20, 6.21, 6.42).

(2) Name cyclic and acyclic alkenes, and draw structures corresponding to given names (6.4, 6.5, 6.6, 6.22, 6.23, 6.24, 6.26, 6.27, 6.28, 6.29, 6.30, 6.38).

(3) Draw *cis–trans* isomers of alkenes (6.7, 6.8, 6.32).

(4) Assign priorities to double-bond substituents according to sequence rules (6.9, 6.10, 6.36).

(5) Assign _E_, _Z_–configurations to double bonds (6.11, 6.25, 6.37, 6.39).

(6) Predict the relative stability of alkene double bonds (6.12, 6.13, 6.34).

(7) Understand the mechanism of electrophilic addition reactions (6.14, 6.15, 6.16, 6.18, 6.35, 6.40, 6.41).

(8) Understand the Hammond postulate (6.17, 6.43, 6.44).

7.1

$$CH_3CH_2\overset{\overset{\displaystyle Br}{|}}{C}(CH_3)_2 \xrightarrow{\text{KOH}} CH_3CH=C(CH_3)_2 \quad + \quad CH_3CH_2\overset{\overset{\displaystyle CH_3}{|}}{C}=CH_2$$

Dehydrobromination can occur in two directions to yield a mixture of products.

7.2

$$CH_3CH_2\overset{\overset{\displaystyle OH}{|}}{\underset{\underset{\displaystyle CH_3}{|}}{C}}CH_2CH_2CH_3 \xrightarrow[\text{H}_2\text{O}]{\text{H}_2\text{SO}_4,}$$

$$\begin{matrix} \underset{CH_3}{\overset{CH_3CH_2}{\diagdown}}C=C\underset{H}{\overset{CH_2CH_3}{\diagup}} \end{matrix}$$

(Z)–3–Methyl–3–hexene

+

$$\begin{matrix} \underset{CH_3}{\overset{CH_3CH_2}{\diagdown}}C=C\underset{CH_2CH_3}{\overset{H}{\diagup}} \end{matrix}$$

(E)–3–Methyl–3–hexene

+

$$\begin{matrix} \underset{H}{\overset{CH_3}{\diagdown}}C=C\underset{CH_3}{\overset{CH_2CH_2CH_3}{\diagup}} \end{matrix}$$

(Z)–3–Methyl–2–hexene

+

$$\begin{matrix} \underset{H}{\overset{CH_3}{\diagdown}}C=C\underset{CH_2CH_2CH_3}{\overset{CH_3}{\diagup}} \end{matrix}$$

(E)–3–Methyl–2–hexene

+

$$CH_3CH_2\overset{\overset{\displaystyle CH_2}{\|}}{C}CH_2CH_2CH_3$$

2–Ethyl–1–pentene

Five alkene products, including E, Z isomers, might be obtained by dehydration of 3–methyl–3– hexanol.

7.3

1,2–Dimethylcyclohexene

$\xrightarrow{Cl_2}$

trans–1,2–Dichloro-
1,2–dimethylcyclohexane

The chlorines are *trans* to one another in the product, as are the methyl groups.

7.4

Addition of hydrogen halides involves formation of an open carbocation, not a cyclic halonium ion intermediate. The carbocation, which is sp^2 hybridized and planar, can be attacked by chloride from either top or bottom, yielding products in which the two methyl groups can be either *cis* or *trans* to each other.

7.5

Br and –OH are *trans* in the product.

7.6 NBS is the source of the electrophilic Br^+ ion in bromohydrin formation. Attack of the alkene pi electrons on Br^+ forms a cyclic bromonium ion. When this bromonium ion is opened by water, a partial positive charge develops at the carbon whose bond to bromine is being cleaved.

vs.

less favorable

Since a secondary carbon can stabilize this charge better than a primary carbon, opening of the bromonium ion occurs at the secondary carbon to yield Markovnikov product.

7.7 Addition of IN_3 to the alkene yields a product in which I is bonded to the primary carbon and N_3 is bonded to the secondary carbon. If addition occurs with Markovnikov orientation, I^+ must be the electrophile, and the reaction must proceed through an iodonium ion intermediate. Opening of the iodonium ion gives Markovnikov product for the reasons discussed in Problem 7.6. The bond polarity of iodine azide is:

$$CH_3CH_2CH=CH_2 \xrightarrow{\qquad} \left[CH_3CH_2CH\text{———}CH_2 \right] \xrightarrow{\qquad} CH_3CH_2\overset{\displaystyle N_3}{\underset{\displaystyle |}{CH}}CH_2I$$

7.8 Keep in mind that oxymercuration corresponds to Markovnikov addition of H_2O to an alkene.

a) $CH_3CH_2CH_2CH=CH_2 \xrightarrow[\text{2. NaBH}_4]{\text{1. Hg(OAc)}_2, \text{H}_2\text{O}} CH_3CH_2CH_2\overset{\displaystyle OH}{\underset{\displaystyle |}{CH}}CH_3$

b) $CH_3CH_2CH=\overset{\displaystyle CH_3}{\underset{\displaystyle |}{C}}CH_3 \xrightarrow[\text{2. NaBH}_4]{\text{1. Hg(OAc)}_2, \text{H}_2\text{O}} CH_3CH_2CH_2\overset{\displaystyle CH_3}{\underset{\displaystyle |}{\underset{\displaystyle |}{C}}}CH_3$
(with OH below)

7.9

a) $CH_3\overset{}{\underset{\displaystyle |}{\underset{\displaystyle CH_3}{C}}}=CHCH_2CH_2CH_3$

or

$CH_3\overset{\displaystyle ||}{\underset{\displaystyle CH_2}{C}}CH_2CH_2CH_2CH_3$

$\xrightarrow[\text{2. NaBH}_4]{\text{1. Hg(OAc)}_2, \text{H}_2\text{O}}$ $CH_3\overset{\displaystyle OH}{\underset{\displaystyle CH_3}{\underset{\displaystyle |}{C}}}CH_2CH_2CH_2CH_3$

b)

$\xrightarrow[\text{2. NaBH}_4]{\text{1. Hg(OAc)}_2, \text{H}_2\text{O}}$ (cyclohexyl-CH(OH)CH$_3$)

Oxymercuration occurs with Markovnikov orientation.

7.10 Recall the mechanism of hydroboration and note that the hydrogen added to the double bond comes from borane. The product of hydroboration with BD_3 has deuterium bonded to the more substituted carbon; –D and –OH are *cis* to one another.

$\xrightarrow[\text{2. H}_2\text{O}_2, \ ^-\text{OH}]{\text{1. BD}_3, \text{THF}}$

7.11

a) $(CH_3)_2CHCH=CH_2 \xrightarrow[\text{2. H}_2\text{O}_2, \ ^-\text{OH}]{\text{1. BH}_3, \text{THF}} (CH_3)_2CHCH_2CH_2OH$

b) $(CH_3)_2C=CHCH_3 \xrightarrow[\text{2. H}_2\text{O}_2, \ ^-\text{OH}]{\text{1. BH}_3, \text{THF}} (CH_3)_2CHCH\overset{}{\underset{\displaystyle OH}{\underset{\displaystyle |}{C}}}H_3$

c)

$$\xrightarrow[\text{2. H}_2\text{O}_2, \ ^-\text{OH}]{\text{1. BH}_3, \text{THF}}$$

7.12

a) $\underset{\overset{|}{\underset{\text{CH}_3}{}}}{\text{CH}_3\text{CH}_2\text{C}}=\text{CH}_2$ $\xrightarrow{\text{HBr}}$ $\underset{\text{CH}_3\text{CH}_2\text{C(CH}_3)_2}{\overset{\text{Br}}{|}}$

2–Methyl–1–butene

b) $\text{CH}_3\text{CH}_2\text{CH}=\text{CH}_2$ $\xrightarrow[\text{peroxides}]{\text{HBr}}$ $\text{CH}_3\text{CH}_2\text{CH}_2\text{CH}_2\text{Br}$

1–Butene

c) $\underset{\overset{|}{\underset{\text{CH}_3}{}}}{\text{CH}_3\text{CH}_2\text{C}}=\text{CHCH}_2\text{CH}_3$ $\xrightarrow[\text{peroxides}]{\text{HBr}}$ $\underset{\overset{|}{\underset{\text{Br}}{}}}{\text{CH}_3\text{CH}_2}\underset{\overset{|}{\underset{}{}}}{\text{CHCHCH}_2\text{CH}_3}$ $\overset{\text{CH}_3}{}$

3–Methyl–3–hexene

7.13

1. $= \Delta G^{\ddagger}_{sec}$

2. $= \Delta G^{\ddagger}_{tert}$

3. $= \Delta G^{\circ}_{sec}$

4. $= \Delta G^{\circ}_{tert}$

The reaction producing the more stable radical $(\text{CH}_3\text{CHBr}\dot{\text{C}}(\text{CH}_3)_2)$ has a lower ΔG° for the first step. (Remember that a larger negative ΔG° indicates that a reaction is more favorable.) According to the Hammond Postulate, the more stable radical also forms faster, and $\Delta G^{\ddagger}$ for its formation is also lower.

7.14

a) $(\text{CH}_3)_2\text{C}=\text{CHCH}_2\text{CH}_3$ $\xrightarrow[\text{Ethanol}]{\text{H}_2, \text{Pd/C}}$ $(\text{CH}_3)_2\text{CHCH}_2\text{CH}_2\text{CH}_3$

2–Methyl–2–pentene   2–Methylpentane

b)

$\xrightarrow[\text{Ethanol}]{\text{H}_2, \text{Pd/C}}$

3,3–Dimethylcyclopentene   1,1-Dimethylcyclopentane

7.15

a)

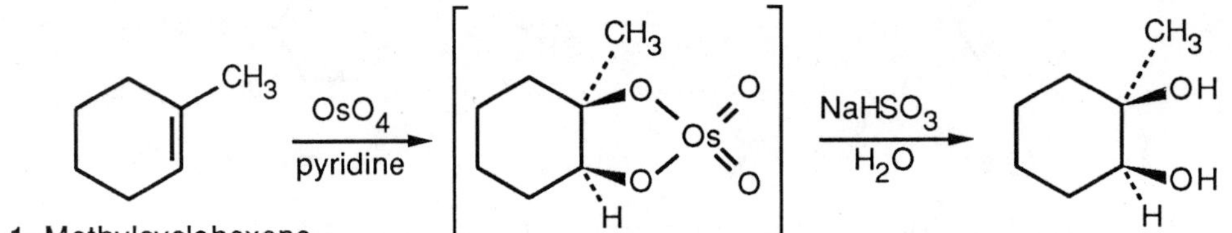

1–Methylcyclohexene

b) $CH_3CH_2CH=C(CH_3)_2$ $\xrightarrow[\text{2. NaHSO}_3, \text{H}_2\text{O}]{\text{1. OsO}_4, \text{pyridine}}$ $CH_3CH_2\overset{\underset{|}{HO}}{C}H-\overset{\underset{|}{OH}}{\underset{|}{C}}CH_3$
$\quad\quad\quad\quad\quad\quad\quad\quad\quad\quad\quad\quad\quad\quad\quad\quad\quad\quad\quad\quad\quad\quad\quad CH_3$

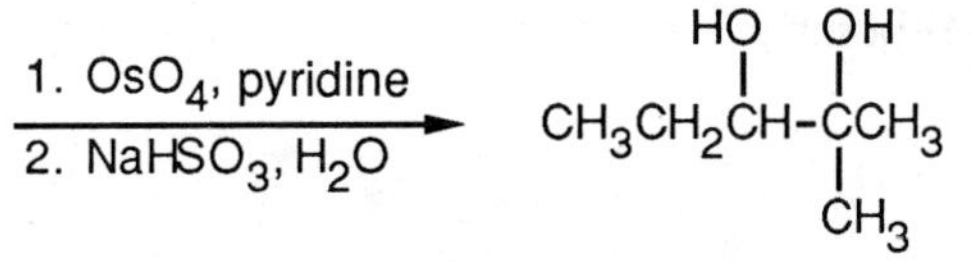

2–Methyl–2–pentene

c) $CH_2=CHCH=CH_2$ $\xrightarrow[\text{2. NaHSO}_3, \text{H}_2\text{O}]{\text{1. 2 OsO}_4, \text{pyridine}}$ $HOCH_2\overset{\underset{|}{HO}}{C}H\overset{\underset{|}{OH}}{C}HCH_2OH$

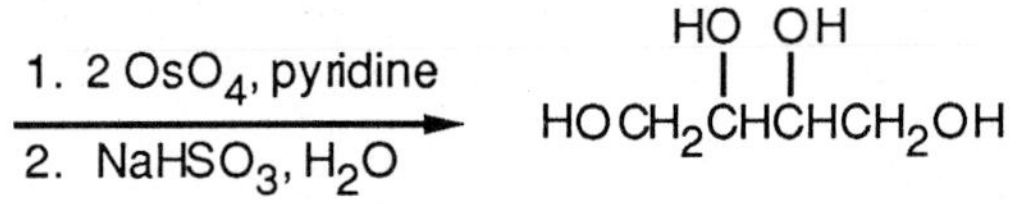

1,3–Butadiene

Alternatively, aqueous alkaline $KMnO_4$ can be used to carry out hydroxylation of the above alkenes.

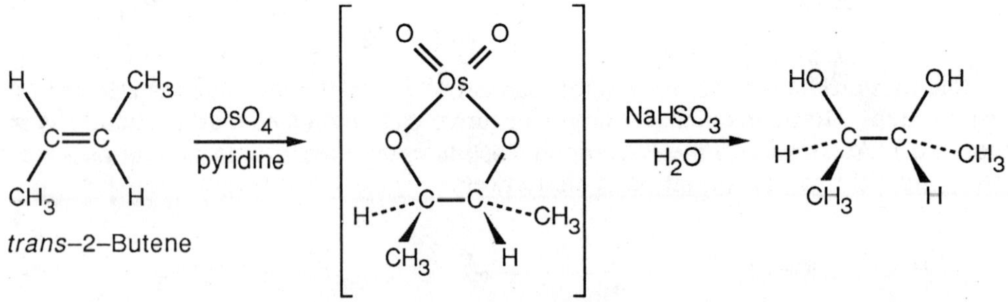

7.16

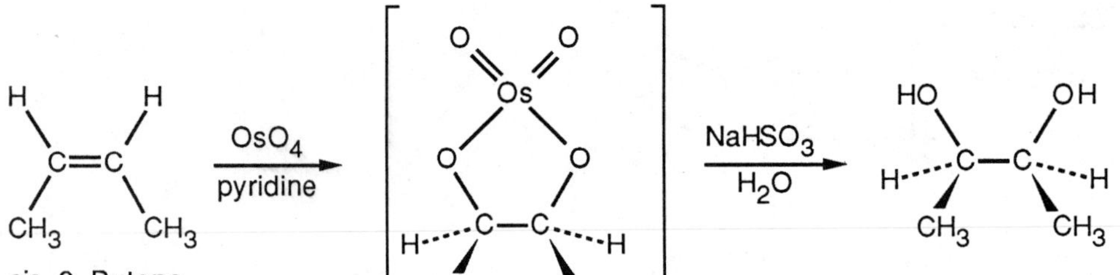

Formation of the cyclic osmate, which occurs with *syn* stereochemistry, retains the *cis-trans* stereochemistry of the double bond, since osmate formation is a single-step reaction. Treatment of the osmate ester with $NaHSO_3$ does not affect the stereochemistry of the carbon-oxygen bond. The diol produced from *cis*–2–butene is isomeric with the diol produced from *trans*–2–butene.

7.17

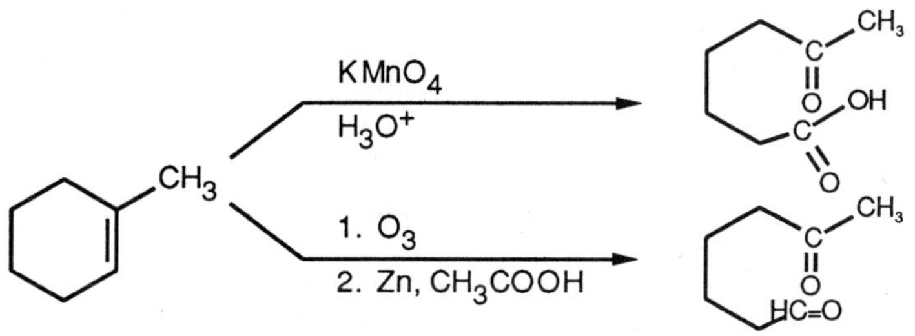

7.18

a) $(CH_3)_2C=CH_2$ $\xrightarrow[\text{2. Zn, }H_3O^+]{\text{1. }O_3}$ $(CH_3)_2C=O$ + $O=CH_2$

b) $CH_3CH_2CH=CHCH_2CH_3$ $\xrightarrow[\text{2. Zn, }H_3O^+]{\text{1. }O_3}$ $CH_3CH_2CH=O$ + $O=CHCH_2CH_3$

7.19

a)

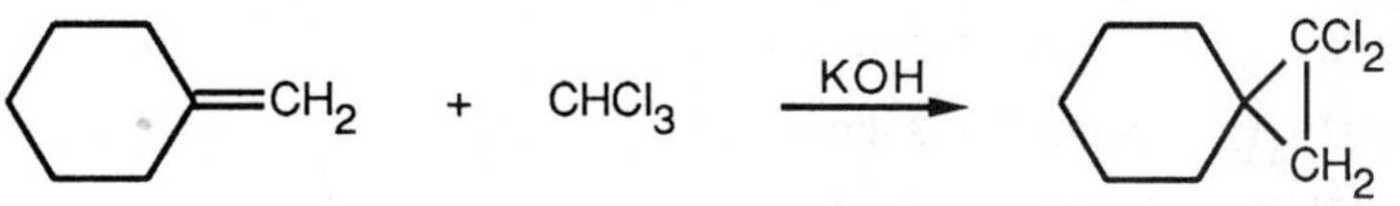

b) 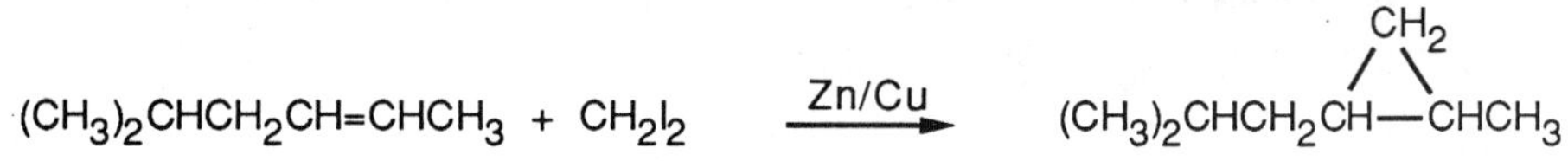 $(CH_3)_2CHCH_2CH=CHCH_3 + CH_2I_2$ $\xrightarrow{Zn/Cu}$

7.20

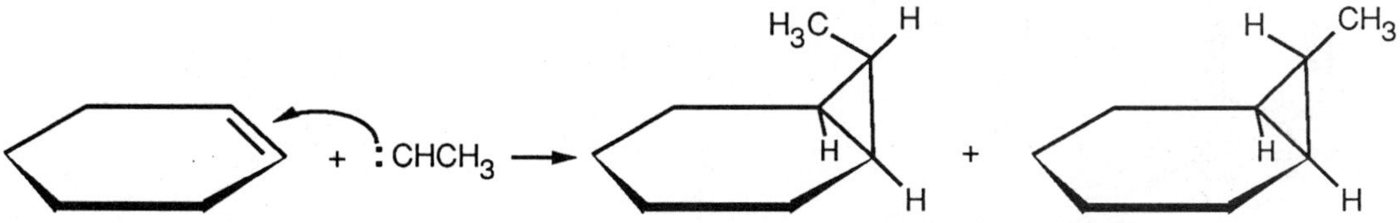

Focus on the stereochemistry of the three-membered ring. Simmons-Smith reaction of 1,1–diiodoethane with the double bond occurs with *syn* stereochemistry and can produce two isomers. In one of these isomers (<u>A</u>), the methyl group is on the same side of the three-membered ring as the cyclohexane ring carbons. In <u>B</u>, the methyl group is on the side of the three-membered ring opposite to the cyclohexane ring carbons.

7.21

a)

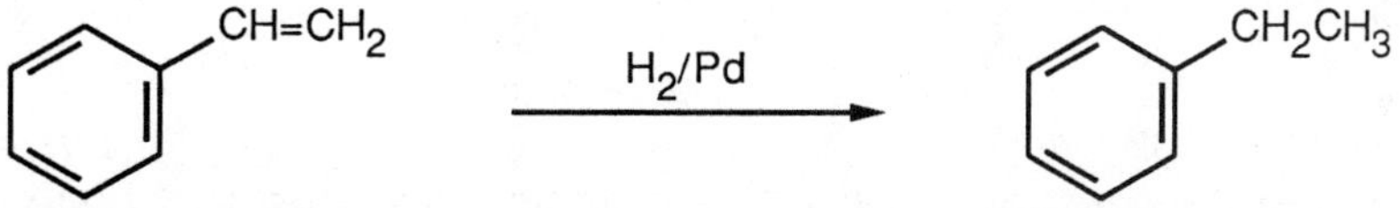

b)

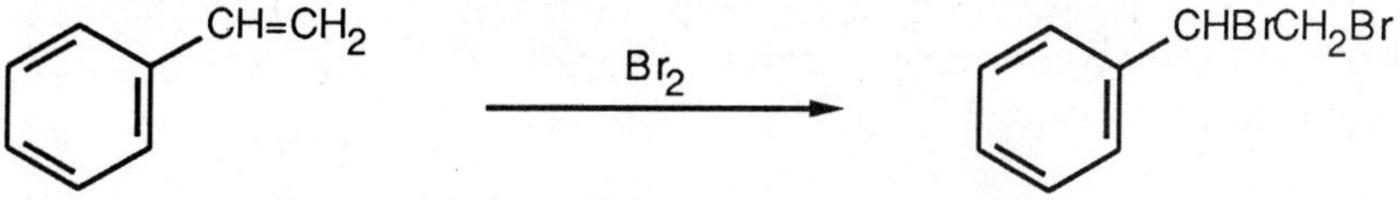

c)

$$\xrightarrow{\text{HBr}}$$

d)

$$\xrightarrow[\text{2. NaHSO}_3]{\text{1. OsO}_4}$$

e)

$$\xrightarrow{\text{D}_2/\text{Pd}}$$

7.22

a)

$$CH_3CH_2CH_2CH_2\overset{\overset{\displaystyle CH_3}{|}}{C}=CH_2 \quad \text{2-Methyl-1-hexene}$$

$$CH_3CH_2CH_2CH=C(CH_3)_2 \quad \text{2-Methyl-2-hexene}$$

$$CH_3CH_2CH=CHCH(CH_3)_2 \quad \text{2-Methyl-3-hexene}$$

$$CH_3CH=CHCH_2CH(CH_3)_2 \quad \text{5-Methyl-2-hexene}$$

$$CH_2=CHCH_2CH_2CH(CH_3)_2 \quad \text{5-Methyl-1-hexene}$$

$$\xrightarrow{\text{H}_2/\text{Pd}} \quad CH_3CH_2CH_2CH_2CH(CH_3)_2$$

2-Methylhexane

b)

3,3-Dimethylcyclohexene

4,4-Dimethylcyclohexene

$$\xrightarrow{\text{H}_2/\text{Pd}}$$

1,1-Dimethylcyclohexane

c) $CH_3CH=CHCH_2CH(CH_3)_2$

5-Methyl-2-hexene

$$\xrightarrow{\text{Br}_2/\text{CCl}_4}$$

$CH_3CHBrCHBrCH_2CH(CH_3)_2$

2,3-Dibromo-5-methylhexane

d) $CH_3CH_2CH_2CH=CH_2$

1-Pentene

$$\xrightarrow[\text{2. NaBH}_4]{\text{1. Hg(OAc)}_2, \text{H}_2\text{O}}$$

$CH_3CH_2CH_2CH(OH)CH_3$

2-Pentanol

e) $CH_3CH_2CH_2CH_2\overset{\overset{\displaystyle }{|}}{C}=CHCH_3$
CH_3

3-Methyl-2-heptene

$$\xrightarrow[\text{peroxides}]{\text{HBr}}$$

$CH_3CH_2CH_2CH_2\overset{\overset{\displaystyle Br}{|}}{C}HCHCH_3$
CH_3

2-Bromo-3-methylheptane

f) 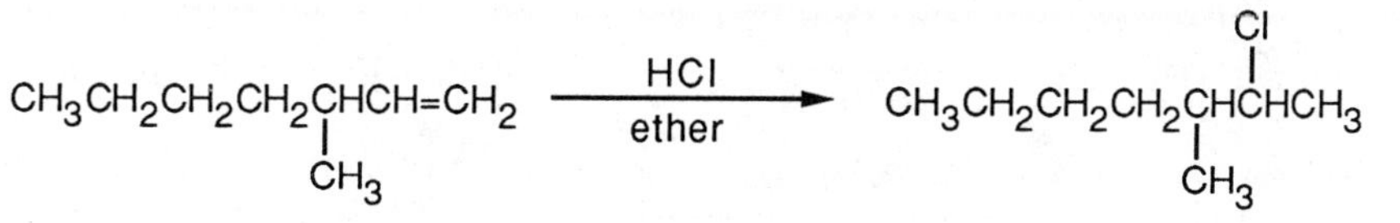
$CH_3CH_2CH_2CH_2CHCH=CH_2$
CH_3
HCl
ether
Cl
$CH_3CH_2CH_2CH_2CHCHCH_3$
CH_3

3–Methyl–1–heptene 2–Chloro–3–methylheptane

7.23

a) 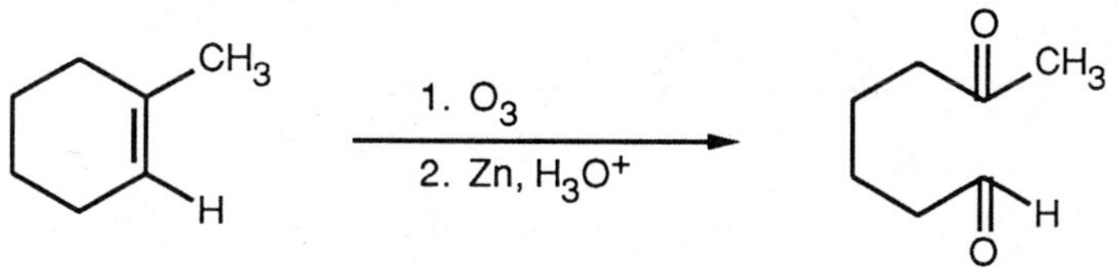
CH_3
H
1. O_3
2. Zn, H_3O^+
O
CH_3
H
O

b) 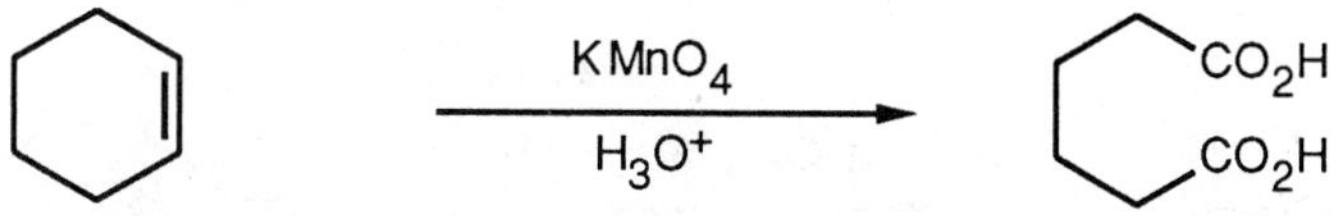
$KMnO_4$
H_3O^+
CO_2H
CO_2H

c) 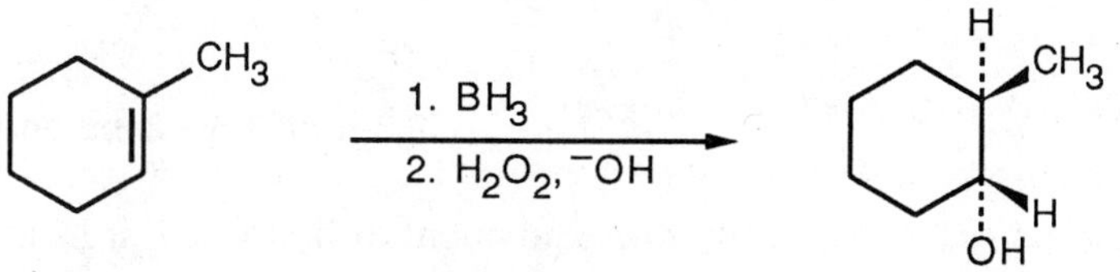
CH_3
1. BH_3
2. H_2O_2, ^-OH
H
CH_3
H
OH

Remember that –H and –OH add *syn* across the double bond.

d) 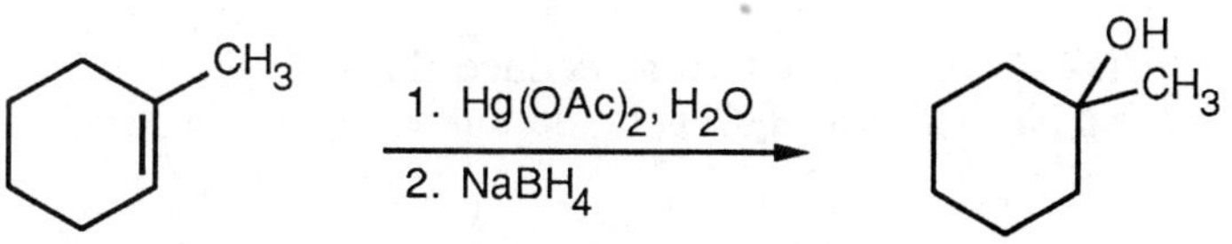
CH_3
1. $Hg(OAc)_2$, H_2O
2. $NaBH_4$
OH
CH_3

7.24

a) 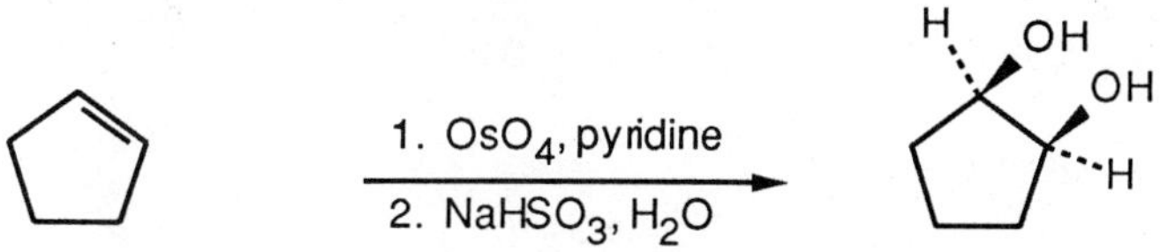
1. OsO_4, pyridine
2. $NaHSO_3$, H_2O
H
OH
OH
H

b) 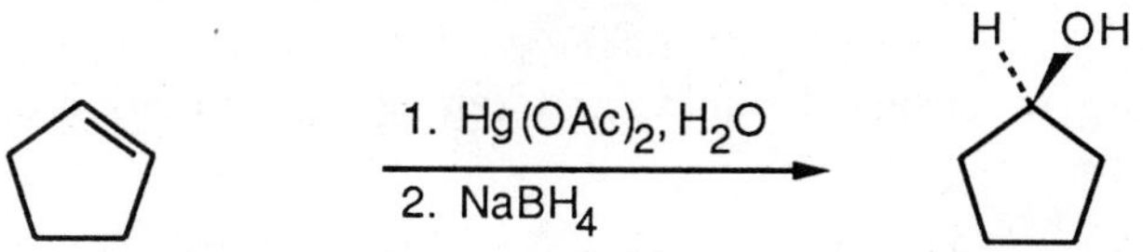
1. $Hg(OAc)_2$, H_2O
2. $NaBH_4$
H
OH

Hydroboration/oxidation or hydration are other routes to this product.

c) 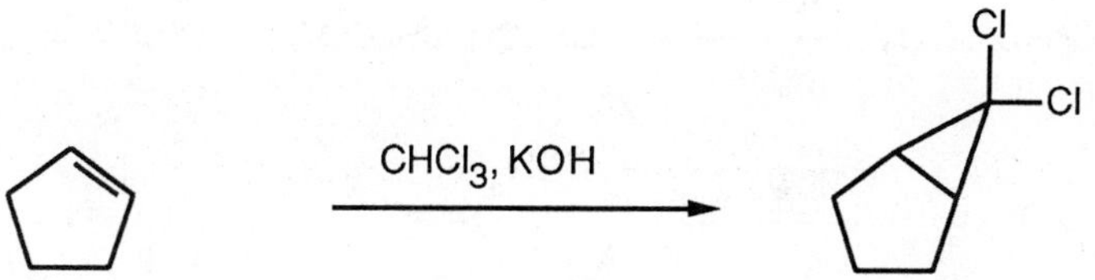
$CHCl_3$, KOH
Cl
Cl

d) $CH_3CH=CHCH(CH_3)_2$ $\xrightarrow[\text{2. Zn, }H_3O^+]{\text{1. }O_3}$ CH_3CHO + $(CH_3)_2CHCHO$

e) $(CH_3)_2C=CH_2$ $\xrightarrow[\text{2. }H_2O_2,\ ^-OH]{\text{1. }BH_3}$ $(CH_3)_2CHCH_2OH$

f)

$\xrightarrow[\text{heat}]{H_2SO_4, H_2O}$

7.25 Because ozonolysis gives only <u>one</u> product, we can assume that the alkene is symmetrical.

$\xleftarrow[\text{2. Zn, }H_3O^+]{\text{1. }O_3}$

2,3–Dimethyl–2–butene

7.26 a) If the hydrocarbon reacts with only one equivalent of hydrogen, it has only one double bond.

  b) If only one type of aldehyde is produced on ozonolysis, the alkene must be symmetrical.

Putting these two facts together allows us to deduce that
$CH_3CH_2CH_2CH_2CH=CHCH_2CH_2CH_2CH_3$ is the unknown hydrocarbon.

$CH_3CH_2CH_2CH_2CH=CHCH_2CH_2CH_2CH_3$ $\xrightarrow[\text{2. Zn, }H_3O^+]{\text{1. }O_3}$ $2\ CH_3CH_2CH_2CH_2\overset{\displaystyle O}{\overset{\displaystyle \|}{C}}H$

5-Decene   Pentanal

$\downarrow H_2/Pd$

$CH_3CH_2CH_2CH_2CH_2CH_2CH_2CH_2CH_2CH_3$

Decane

7.27 Remember that alkenes give carboxylic acids and ketones on oxidative cleavage with $KMnO_4$ in acidic solution.

a) $CH_3CH_2CH=CH_2$ $\xrightarrow[H_2O]{KMnO_4}$ CH_3CH_2COOH + CO_2

b) $CH_3CH_2CH_2CH=C(CH_3)_2$ $\xrightarrow[H_2O]{KMnO_4}$ $CH_3CH_2CH_2COOH$ + $(CH_3)_2C=O$

c) $\xrightarrow[H_2O]{KMnO_4}$ $=O$ + $(CH_3)_2C=O$

7.28 Compound $\underline{A}$ has three degrees of unsaturation. Because compound $\underline{A}$ contains only one double bond, the other two degrees of unsaturation must be rings.

Other structures having two fused rings are possible.

7.29 Don't get discouraged by the amount of information in this problem. Read slowly and interpret piece by piece. We know the following.
1) Hydrocarbon $\underline{A}$ (C_6H_{12}) has one double bond or ring.
2) Because $\underline{A}$ reacts with one equivalent of H_2, it has one double bond and no ring.
3) Compound $\underline{A}$ forms a diol when reacted with OsO_4.
4) When alkenes are oxidized with $KMnO_4$ they give either carboxylic acids or ketones, depending on the substitution pattern of the double bond.
    a)   A ketone is produced from what was originally a disubstituted carbon in the double bond.
    b)   A carboxylic acid is produced from what was originally a monosubstituted carbon in the double bond.
5) One fragment from $KMnO_4$ oxidation is a carboxylic acid, CH_3CH_2COOH.
    a)   This fragment was $CH_3CH_2CH=$ (a monosubstituted double bond) in compound $\underline{A}$.
    b)   It contains three of the six carbons of compound $\underline{A}$.
6) a)   The other fragment contains three carbons.
    b)   It is a ketone.
    c)   The only three carbon ketone is acetone, $O=C(CH_3)_2$.
    d)   This fragment was $=C(CH_3)_2$ in compound $\underline{A}$.
7) If we join the fragment in 5a with the one in 6d, we get:

$$CH_3CH_2CH=C(CH_3)_2 \qquad\qquad C_6H_{12}$$
$$\underline{A}$$

The complete scheme:

7.30 The oxidative cleavage reaction of alkenes with O_3, followed by Zn in acid, produces aldehyde and ketone functional groups at sites where double bonds used to be. On ozonolysis, these two dienes yield only aldehydes because all double bonds are monosubstituted.

Because the other diene is symmetrical, only one dialdehyde, $OCHCH_2CHO$, is produced.

7.31 Try to solve this problem phrase by phrase.
1) $C_{10}H_{18}O$ has *two* double bonds and/or rings.
2) $C_{10}H_{18}O$ must be an alcohol because it undergoes reaction with H_2SO_4.
3) When $C_{10}H_{18}O$ is treated with dilute H_2SO_4, a mixture of alkenes of the formula $C_{10}H_{16}$ is produced.
4) Since the major alkene product $\underline{B}$ yields only one product, C_5H_8O, on ozonolysis, $\underline{B}$ and $\underline{A}$ contain two rings. $\underline{A}$ therefore has no double bonds.

7.32

Addition to 1–methylcyclohexene occurs at a faster rate. Positive charge generated during the reaction is better stabilized by a tertiary carbocation intermediate than by a secondary carbocation intermediate.

7.33

a) $CH_3CH=CHCH_3$ $\xrightarrow{\text{HBr}}$ $CH_3CH_2\overset{\overset{\displaystyle Br}{|}}{C}HCH_3$

    2–Butene

b) $3 \ CH_3CH=CHCH_3 \xrightarrow[THF]{BH_3} (CH_3CH_2\overset{\overset{\displaystyle CH_3}{|}}{C}H)_3B \xrightarrow[^-OH]{H_2O_2} 3 \ CH_3CH_2\overset{\overset{\displaystyle OH}{|}}{C}HCH_3$

2–Butene

c) $(CH_3)_2C=CH_2 \xrightarrow[peroxides]{HBr} (CH_3)_2CHCH_2Br$

2–Methylpropene

d) $CH_3CH=C(CH_3)_2 \xrightarrow[peroxides]{HI} CH_3CH_2\overset{\overset{\displaystyle |}{|}}{C}(CH_3)_2$

2–Methyl–2–butene

Addition of HI always gives Markovnikov products.

7.34

$\xleftarrow{2H_2/Pd}$ $\xrightarrow[2. \ Zn, \ H_3O^+]{1. \ O_3} 2 \ H\overset{\overset{\displaystyle O}{||}}{C}CH_2CH_2\overset{\overset{\displaystyle O}{||}}{C}H$

Cyclooctane   1,5–Cyclooctadiene

7.35 a) Non-Markovnikov addition using peroxides succeeds only with HBr. Other hydrogen halides yield products of Markovnikov addition.
b) Hydroxylation of double bonds produces *cis*, not *trans* diols.
c) Ozone reacts with <u>both</u> double bonds of 1,4–cyclohexadiene.
d) Because hydroboration is a *syn* addition, the –H and the –OH added to the double bond must be *cis* to each other.

7.36 a) This alcohol can't be synthesized selectively by hydroboration//oxidation. Consider the two possible starting materials.

1. $CH_3CH_2CH_2CH=CH_2 \xrightarrow[2. \ H_2O_2, \ ^-OH]{1. \ BH_3} CH_3CH_2CH_2CH_2CH_2OH$

1–Pentene yields only the primary alcohol.

2. $CH_3CH_2CH=CHCH_3 \xrightarrow[2. \ H_2O_2, \ ^-OH]{1. \ BH_3} CH_3CH_2\overset{\overset{\displaystyle OH}{|}}{C}HCH_2CH_3 \ + \ CH_3CH_2CH_2\overset{\overset{\displaystyle OH}{|}}{C}HCH_3$

2–Pentene yields a mixture of alcohols.

b) $(CH_3)_2C=C(CH_3)_2 \xrightarrow[2. \ H_2O_2, \ ^-OH]{1. \ BH_3} (CH_3)_2CH\overset{\overset{\displaystyle OH}{|}}{C}H(CH_3)_2$

2,3–Dimethyl–2–butene yields the desired alcohol exclusively.

c) This alcohol can't be formed by a hydroboration reaction. The –H and –OH added to a double bond must be *cis* to each other; in the product shown, they are *trans*.

d) The product shown is not a hydroboration product; hydroboration yields an alcohol in which ¯OH is bonded to the *less* substituted carbon.

7.37

a) 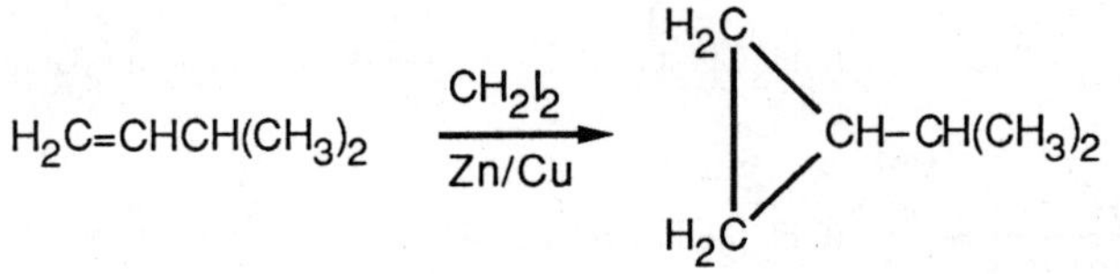 $H_2C=CHCH(CH_3)_2$ $\xrightarrow[\text{Zn/Cu}]{CH_2I_2}$

b)

+ $CHCl_3$ $\xrightarrow{KOH}$

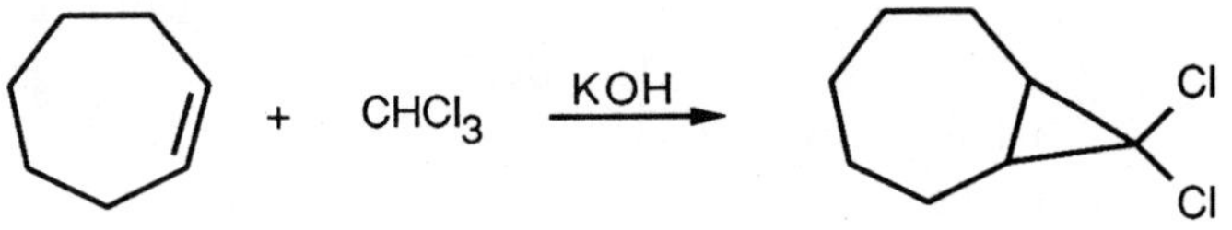

Cycloheptene

7.38

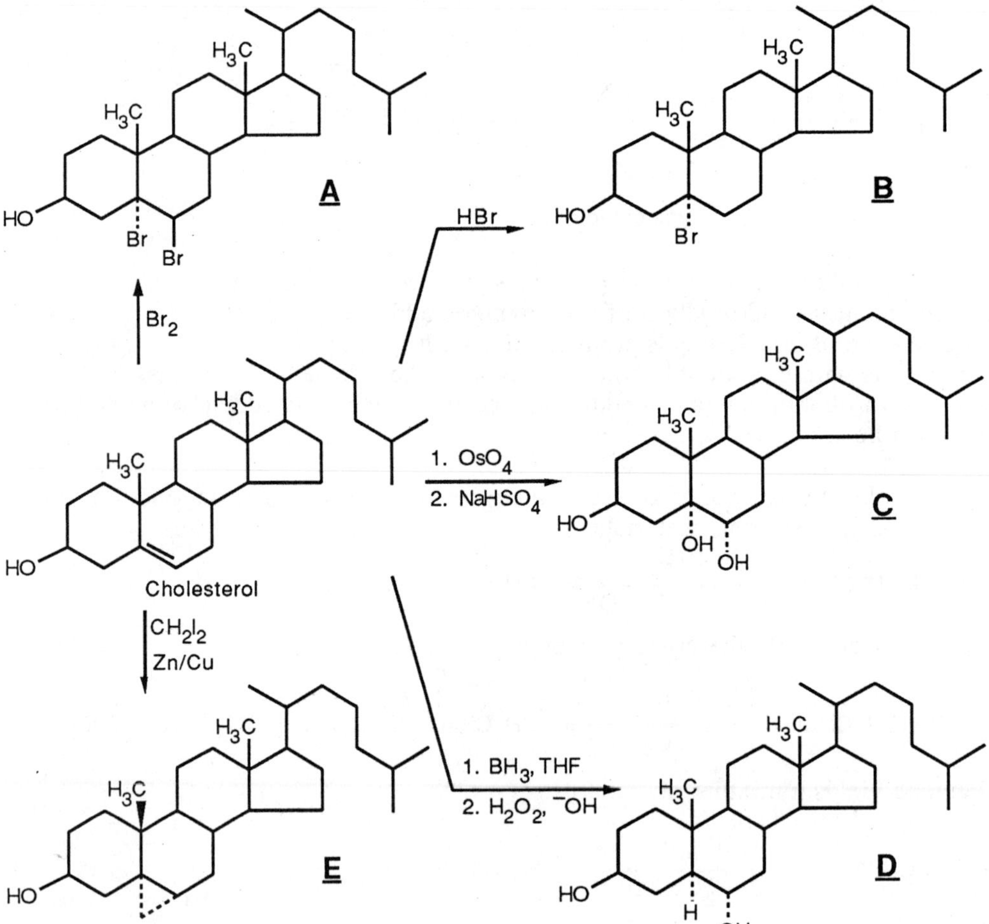

7.39

$CH_3(CH_2)_{12}CH=CH(CH_2)_7CH_3$ $\xrightarrow[\text{H}_3\text{O}^+]{KMnO_4}$ $CH_3(CH_2)_{12}COOH$ + $CH_3(CH_2)_7COOH$

7.40 C_8H_8 has five double bonds and/or rings. One of these double bonds reacts with H_2/Pd. Stronger conditions cause the uptake of four equivalents of H_2. C_8H_8 thus contains four double bonds, three of which are in an aromatic ring, and one C=C double bond. A good guess for C_8H_8 at this point is:

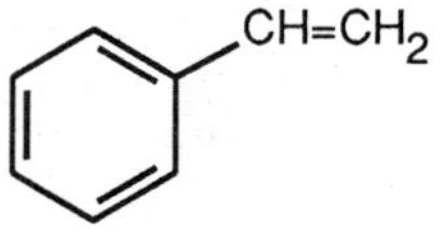

Reaction of a double bond with $KMnO_4$ yields cleavage products of the highest possible degree of oxidation. In this case, the products are $CO_2 + C_6H_5CO_2H$.

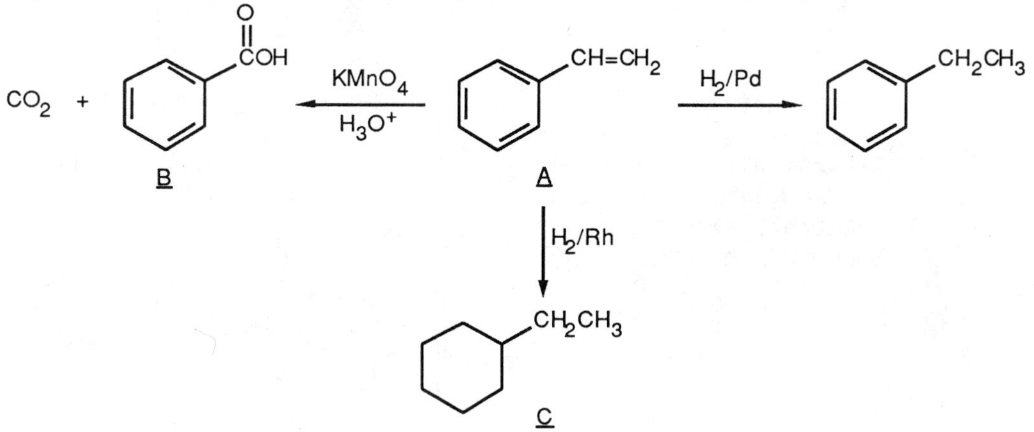

7.41

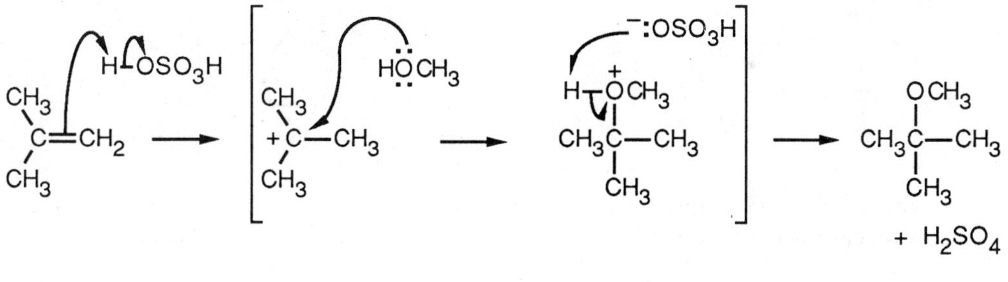

7.42

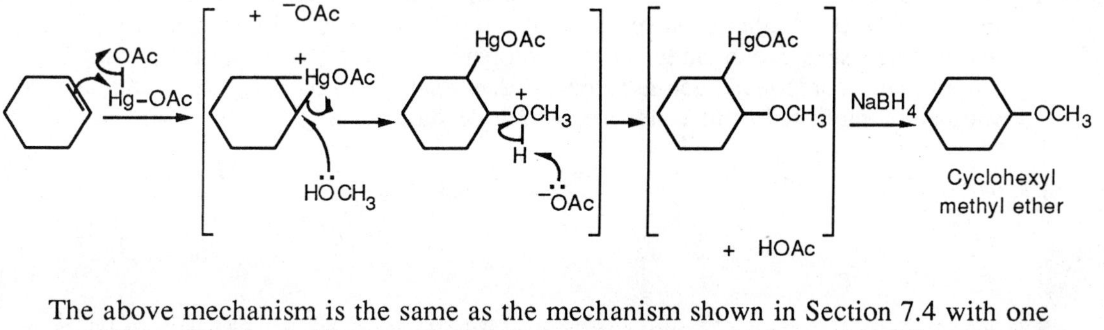

The above mechanism is the same as the mechanism shown in Section 7.4 with one exception: In this problem, methanol, rather than water, is the nucleophile, and an ether, rather than an alcohol, is the observed product.

7.43

Reaction of the double bond with Br_2 forms a cyclic bromonium ion.

The bromonium ion can be attacked by an electron pair from the nucleophilic –OH group to form the cyclic bromo ether

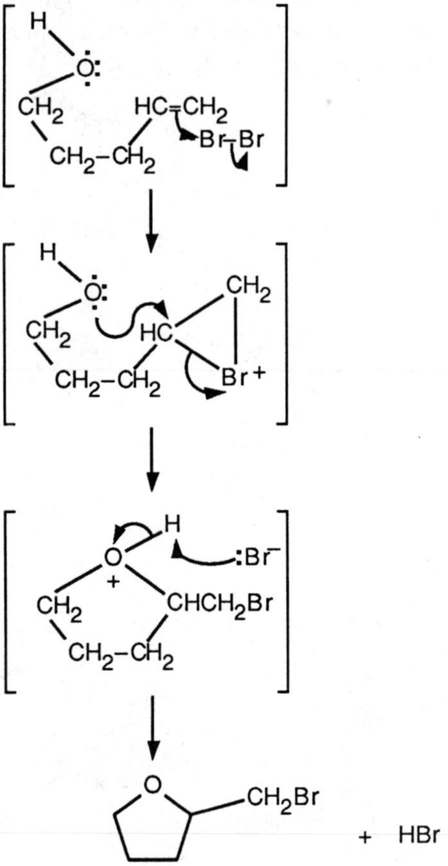

+ HBr

2–(Bromomethyl)tetrahydrofuran

The above mechanism is the same as that for halohydrin formation, shown in Section 7.3. In this case, the nucleophile is the hydroxyl group of 4–penten–1–ol.

7.44 a) Bromine dissolved in CCl_4 has a reddish-brown color. When an alkene is dissolved in Br_2/CCl_4, the double bond reacts with bromine, and the color disappears. This test distinguishes cyclopentene from cyclopentane, which does not react with Br_2.

b) An aromatic compound such as benzene is unreactive to the Br_2/CCl_4 reagent and can be distinguished from 2–hexene, which decolorizes Br_2/CCl_4.

7.45

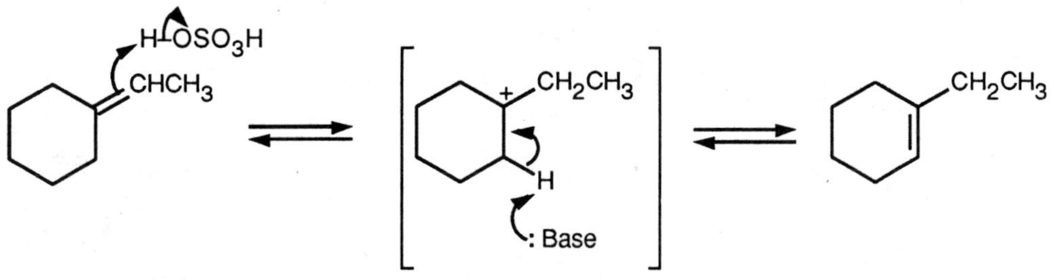

most stable carbocation

Protonation occurs to produce the most stable cation, which can then lose a proton to form either of two alkenes. Because 1–ethylcyclohexene is the major product of this equilibrium, it must be the more stable product.

7.46

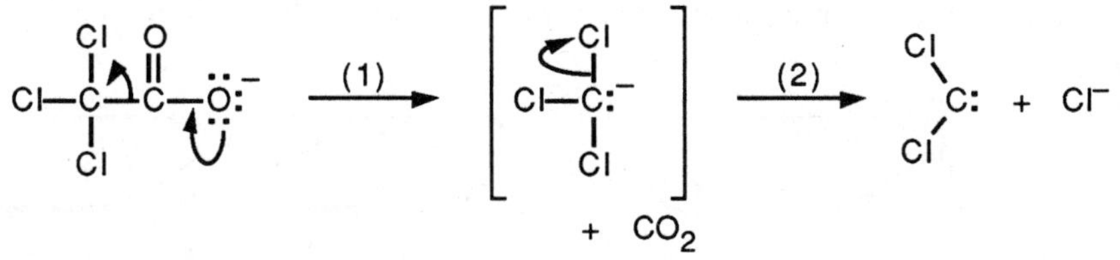

In step 1, carbon dioxide is lost from the trichloroacetate anion. In step 2, elimination of chloride anion produces dichlorocarbene. Step 2 is the same for both the above reaction and the base-induced elimination of HCl from chloroform.

7.47 α–Terpinene, $C_{10}H_{16}$, has three degrees of unsaturation -- two double bonds and one ring.

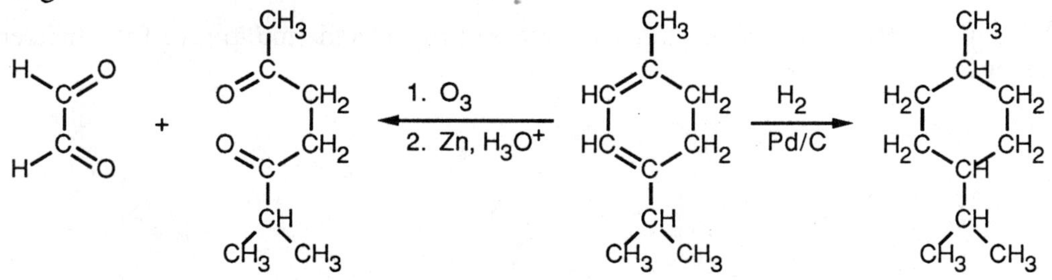

6–Methylheptane–   α–Terpinene
2,5–dione

7.48 Make models of the *cis* and *trans* diols. Notice that it is much easier to form a five-membered cyclic periodate from the *cis* diol than from the *trans* diol. We therefore predict that the *cis* periodate intermediate will be of lower energy than the *trans* periodate intermediate because of the lack of strain in the *cis* periodate ring.

Remembering that any factor that lowers the energy of a transition state or intermediate should also lower $\Delta G^{\ddagger}$ and increase the rate of reaction, we predict that diol cleavage should proceed much more rapidly for *cis* diols than for *trans* diols.

7.49

The intermediate carbocation can be attacked by the electrons of the adjacent bromine to form a cyclic bromonium ion, which is attacked from the opposite side to yield anti product.

Here, the intermediate carbocations are attacked by bromide ion to give four different products.

7.50

Study Guide for Chapter 7

After studying this chapter, you should be able to:

(1) Predict the products of reactions of alkenes (7.3, 7.4, 7.5, 7.6, 7.8, 7.10, 7.12, 7.14, 7.16, 7.17, 7.19, 7.20, 7.21, 7.23, 7.24, 7.33, 7.35, 7.38).

(2) Choose the correct alkene starting material to yield a given product (7.9, 7.11, 7.15, 7.18, 7.22, 7.36, 7.37).

(3) Deduce the structure of an alkene from its molecular formula and products of cleavage (7.25, 7.26, 7.27, 7.28, 7.29, 7.30, 7.31, 7.34, 7.39, 7.40, 7.47).

(4) Formulate mechanisms for reactions of alkenes (7.7, 7.32, 7.41, 7.42, 7.43, 7.45, 7.46, 7.48, 7.49, 7.50).

Chapter 8 – Alkynes

8.1

a) 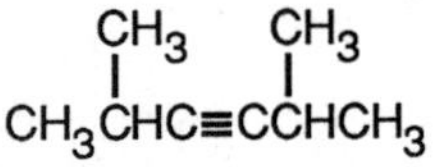
$$CH_3CHC\equiv CCHCH_3$$
with CH₃ and CH₃ substituents

2,5–Dimethyl–3–hexyne

b) 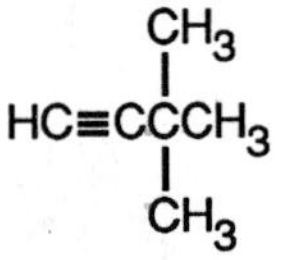
$$HC\equiv CCCH_3$$
with CH₃ groups

3,3–Dimethyl–1–butyne

c)
$$CH_3CH=CHCH=CHC\equiv CCH_3$$

2,4–Octadiene–6–yne
(not 4,6–Octadien–2–yne)

d) 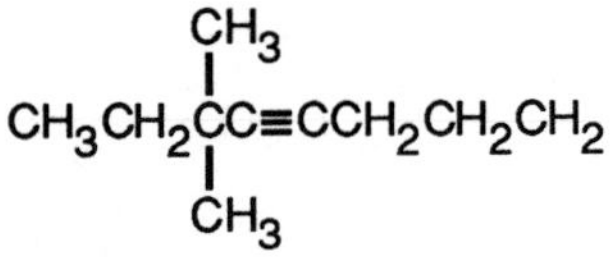
$$CH_3CH_2CC\equiv CCH_2CH_2CH_2$$
with CH₃ groups

3,3–Dimethyl–4–octyne

e) 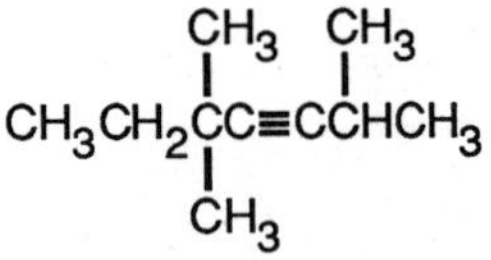
$$CH_3CH_2CC\equiv CCHCH_3$$
with CH₃, CH₃, CH₃ groups

2,5,5–Trimethyl–3–heptyne

f)

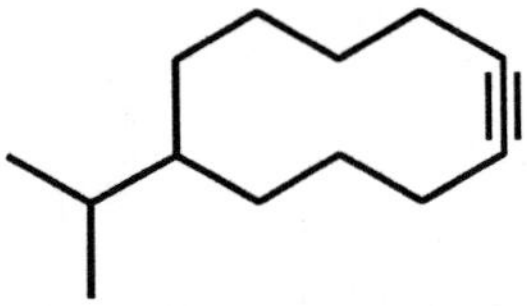

6–Isopropylcyclodecyne

8.2

$$CH_3CH_2CH_2CH_2C\equiv CH$$

1–Hexyne

$$CH_3CH_2CH_2C\equiv CCH_3$$

2–Hexyne

$$CH_3CH_2C\equiv CCH_2CH_3$$

3–Hexyne

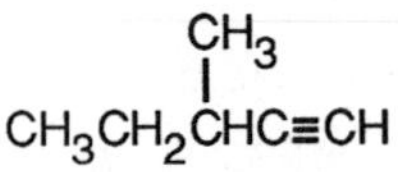

$$CH_3CH_2CHC\equiv CH$$
with CH₃

3–Methyl–1–pentyne

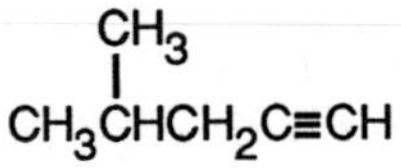

$$CH_3CHCH_2C\equiv CH$$
with CH₃

4–Methyl–1–pentyne

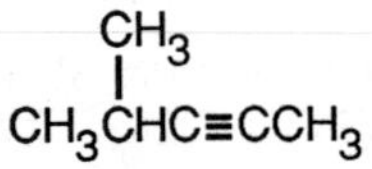

$$CH_3CHC\equiv CCH_3$$
with CH₃

4–Methyl–2–pentyne

$$CH_3CC\equiv CH$$
with CH₃, CH₃

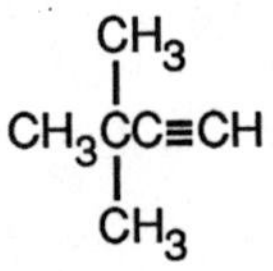

3,3–Dimethyl–1–butyne

8.3

a)
$$CH_3CH_2CH_2C\equiv CH \ + \ 2\,Cl_2 \ \longrightarrow \ CH_3CH_2CH_2CCl_2CHCl_2$$

b)

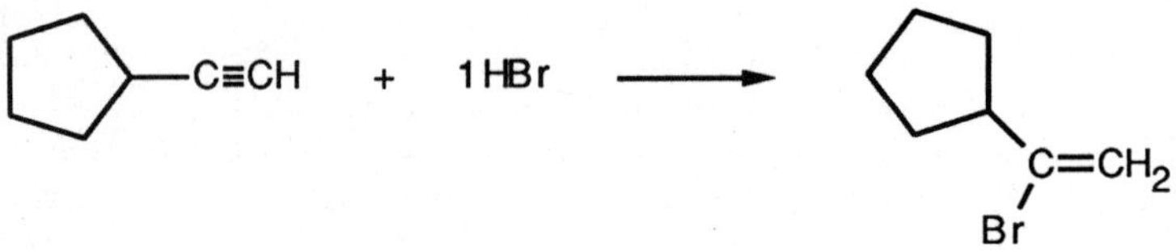

$$\text{cyclopentyl}-C\equiv CH \ + \ 1\,HBr \ \longrightarrow \ \text{cyclopentyl}-C=CH_2$$
with Br

c) $CH_3CH_2CH_2CH_2C{\equiv}CCH_3$ +   1 HBr $\longrightarrow$

$$CH_3CH_2CH_2CH_2 \underset{Br}{\overset{H}{C{=}C}} CH_3$$

+

$$CH_3CH_2CH_2CH_2 \underset{H}{\overset{Br}{C{=}C}} CH_3$$

8.4

$$CH_3CH_2CH_2C{\equiv}CCH_2CH_2CH_3 \xrightarrow[HgSO_4]{H_3O^+} CH_3CH_2CH_2CH_2\overset{O}{\overset{\|}{C}}CH_2CH_2CH_3$$

This symmetrical alkyne yields only one product.

$$CH_3CH_2CH_2C{\equiv}CCH_2\overset{CH_3}{\overset{|}{C}}HCH_3 \xrightarrow[HgSO_4]{H_3O^+}$$

$$CH_3CH_2CH_2CH_2\overset{O}{\overset{\|}{C}}CH_2\overset{CH_3}{\overset{|}{C}}HCH_3$$

+

$$CH_3CH_2CH_2\overset{O}{\overset{\|}{C}}CH_2CH_2\overset{CH_3}{\overset{|}{C}}HCH_3$$

Two ketone products result from hydration of 2–methyl–4–octyne.

8.5

a)

$$CH_3CH_2CH_2C{\equiv}CH \xrightarrow[HgSO_4]{H_3O^+} \left[CH_3CH_2CH_2\overset{OH}{\overset{|}{C}}{=}CH_2 \right] \longrightarrow CH_3CH_2CH_2\overset{O}{\overset{\|}{C}}CH_3$$

b)

$$CH_3CH_2C{\equiv}CCH_3 \xrightarrow[HgSO_4]{H_3O^+} CH_3CH_2\overset{O}{\overset{\|}{C}}CH_2CH_3 + CH_3CH_2CH_2\overset{O}{\overset{\|}{C}}CH_3$$

8.6

a)

$-C{\equiv}CH \xrightarrow[\text{2. } H_2O_2, \text{ }^-OH]{\text{1. } R_2BH,THF}$ $-CH_2CHO$

b) $(CH_3)_2CHC{\equiv}CCH(CH_3)_2 \xrightarrow[\text{2. } H_2O_2, \text{ }^-OH]{\text{1. } R_2BH,THF} (CH_3)_2CHCH_2\overset{O}{\overset{\|}{C}}CH(CH_3)_2$

8.7   Boron becomes attached to the less substituted carbon when BH_3 reacts with an alkene. Of the two alkenes that might be used to form disiamylborane (($CH_3)_2C{=}CHCH_3$ or $(CH_3)_2CHCH{=}CH_2$), only the first yields the desired borane. Reaction stops after BH_3 adds to two equivalents of alkene because addition to a third equivalent is sterically hindered.

$$2\ (CH_3)_2C=CHCH_3 \xrightarrow{\ BH_3,\ THF\ } \left[H{-}B{-}CHCH(CH_3)_2 \right]_2 \begin{array}{c}CH_3\\|\end{array}$$

2-Methyl-2-butene   Disiamylborane

8.8 Disiamylborane is prepared by addition of BH_3 to two equivalents of 2-methyl-2-butene (problem 8.7). When disiamylborane adds to an alkyne, an alkenylborane is produced. Oxidation of this alkenylborane with $H_2O_2/^-OH$ then yields a carbonyl compound, along with two equivalents of 3-methyl-2-butanol.

$$\left[(CH_3)_2CHCH{-}BH \right]_2 \ +\ R{-}C{\equiv}CH \xrightarrow{\ THF\ } \left[(CH_3)_2CHCH{-}B{-}CH{=}CHR \right]_2$$

$$\downarrow H_2O_2/^-OH$$

$$R{-}CH_2CHO \ +\ 2\ (CH_3)_2CHCHCH_3$$

3-Methyl-2-butanol

8.9

a) $CH_3CH_2CH_2CH_2CH_2C{\equiv}CCH_3 \xrightarrow[\ 2.\ H_2O\ ]{\ 1.\ Li/NH_3\ }$ (structure of trans-2-Octene)

2-Octyne   trans-2-Octene

b) $CH_3CH_2CH_2C{\equiv}CCH_2CH_3 \xrightarrow[\ Lindlar\ ]{\ H_2\ }$ (structure of cis-3-Heptene)

3-Heptyne   cis-3-Heptene

c) $CH_3CH_2CHC{\equiv}CH$ (with CH_3 substituent) $\xrightarrow[\text{or } H_2, \text{Lindlar}]{\ 1.\ Li/NH_3\ \ 2.\ H_2O\ }$ $CH_3CH_2CHCH{=}CH_2$ (with CH_3 substituent)

3-Methyl-1-pentyne   3-Methyl-1-pentene

8.10 A base that is strong enough to deprotonate acetone must be the conjugate base of an acid weaker than acetone. In this problem, only $Na^+\ ^-C{\equiv}CH$ is a base strong enough to deprotonate acetone.

8.11

Alkyne	R'X (X=Br or I)	Product
a) $CH_3CH_2CH_2C{\equiv}CH$	CH_3X	$CH_3CH_2CH_2C{\equiv}CCH_3$
or		2-Hexyne
$HC{\equiv}CCH_3$	$CH_3CH_2CH_2X$	

b) $(CH_3)_2CHC\equiv CH$ CH_3CH_2X $(CH_3)_2CHC\equiv CCH_2CH_3$

2–Methyl–3–hexyne

c)   [structure: cyclohexane with $C\equiv CH$] CH_3X [structure: cyclohexane with $C\equiv CCH_3$]

d) $(CH_3)_2CHCH_2C\equiv CH$ CH_3X $(CH_3)_2CHCH_2C\equiv CCH_3$

or   5–Methyl–2–hexyne

$HC\equiv CCH_3$ $(CH_3)_2CHCH_2X$

e) $HC\equiv CC(CH_3)_3$ CH_3CH_2X $CH_3CH_2C\equiv CC(CH_3)_3$

2,2–Dimethyl–3–hexyne

Products (b), (c), and (e) can be synthesized by only one route because only primary halides can be used for acetylide alkylations.

8.12

$CH_3C\equiv CH$ $\xrightarrow[\text{2. } CH_3Br,\ THF]{\text{1. } NaNH_2,\ NH_3}$ $CH_3C\equiv CCH_3$ $\xrightarrow[\substack{\text{Lindlar}\\\text{catalyst}}]{H_2}$ [structure: cis alkene with H, H on top and CH_3, CH_3 on bottom]

cis–2–Butene

8.13

a)   [structure: benzene ring with $C\equiv CH$] $\xrightarrow[H_2O]{KMnO_4}$ [structure: benzene ring with $COOH$] $+$ CO_2

b) $CH_3(CH_2)_7C\equiv C(CH_2)_7C\equiv C(CH_2)_7CH_3$

$\Big\downarrow \begin{array}{l} KMnO_4 \\ H_2O \end{array}$

$2\ CH_3(CH_2)_7COOH$ + $HOOC(CH_2)_7COOH$

8.14 The starting material is $CH_3CH_2CH_2C\equiv CCH_2CH_2CH_3$.

a) Either O_3 or $KMnO_4$ cleaves 4–octyne into two four-carbon fragments.

$CH_3CH_2CH_2C\equiv CCH_2CH_2CH_3$ $\xrightarrow[H_2O]{KMnO_4}$ $2\ CH_3CH_2CH_2COOH$

Butanoic acid

b) To reduce a triple bond to a double bond with *cis* stereochemistry use H_2/Pd with Lindlar catalyst.

$$CH_3CH_2CH_2C{\equiv}CCH_2CH_2CH_3 \quad \xrightarrow[\text{Lindlar}]{H_2}$$

$$\underset{\substack{\\ cis-4-\text{Octene}}}{\underset{H \qquad\qquad H}{\overset{\substack{CH_3CH_2CH_2 \qquad CH_2CH_2CH_3}}{C{=}C}}}$$

c) Addition of HBr to *cis*–4–octene (part b) yields 4–bromooctane.

$$\underset{H \qquad\quad H}{\overset{\substack{CH_3CH_2CH_2 \qquad CH_2CH_2CH_3}}{C{=}C}} \quad \xrightarrow{\text{HBr}} \quad \underset{\text{4–Bromooctane}}{CH_3CH_2CH_2CHBrCH_2CH_2CH_2CH_3}$$

Alternatively, lithium/ammonia reduction of 4–octyne, followed by addition of HBr, gives 4–bromooctane.

d) Hydration of *cis*–4–octene (part b) yields 4–hydroxyoctane (4–octanol).

$$\underset{H \qquad\quad H}{\overset{\substack{CH_3CH_2CH_2 \qquad CH_2CH_2CH_3}}{C{=}C}} \quad \xrightarrow[\text{2. NaBH}_4]{\text{1. Hg(OAc)}_2,\ H_2O} \quad \underset{\text{4–Hydroxyoctane}}{\overset{\overset{\displaystyle OH}{|}}{CH_3CH_2CH_2CHCH_2CH_2CH_2CH_3}}$$

e) Addition of Cl_2 to 4–octene (part b) yields 4,5–dichlorooctane.

$$\underset{H \qquad\quad H}{\overset{\substack{CH_3CH_2CH_2 \qquad CH_2CH_2CH_3}}{C{=}C}} \quad \xrightarrow{\text{Cl}_2} \quad \underset{\text{4,5–Dichlorooctane}}{CH_3CH_2CH_2CHClCHClCH_2CH_2CH_3}$$

8.15 The following syntheses are explained in detail in order to illustrate "retrosynthetic" logic -- the system of planning syntheses by working backwards.

a) 1. $CH_3CH_2CH_2CH_2CH_2CH_2CH_2CH_2CH_2CH_3$. An immediate precursor might be an alkene or alkyne. Try n–$C_8H_{17}C{\equiv}CH$, which can be reduced to decane by H_2/Pd.

   2. The alkyne n–$C_8H_{17}C{\equiv}CH$ can be formed by alkylation of $HC{\equiv}C{:}^-Na^+$ by n–$C_8H_{17}Br$.

   3. $HC{\equiv}C{:}^-Na^+$ can be formed by treatment of $HC{\equiv}CH$ with $NaNH_2$, NH_3.

The complete sequence:

$$HC{\equiv}CH \quad \xrightarrow[\text{NH}_3]{\text{NaNH}_2} \quad HC{\equiv}C{:}^-Na^+ \quad \xrightarrow[\text{THF}]{n-\text{C}_8\text{H}_{17}\text{Br}} \quad n-C_8H_{17}C{\equiv}CH \quad \xrightarrow{\text{H}_2/\text{Pd}} \quad n-C_{10}H_{22}$$

b) 1. An immediate precursor to $CH_3CH_2CH_2CH_2C(CH_3)_3$ might be $HC{\equiv}CCH_2CH_2C(CH_3)_3$, which, when hydrogenated, yields 2,2–dimethylhexane.

   2. $HC{\equiv}CCH_2CH_2C(CH_3)_3$ can be formed by alkylation of $HC{\equiv}C{:}^-Na^+$ with $BrCH_2CH_2C(CH_3)_3$.

The complete sequence:

$$HC\equiv CH \xrightarrow[\text{NH}_3]{\text{NaNH}_2} HC\equiv C:^- Na^+$$

$$HC\equiv C:^- Na^+ + BrCH_2CH_2C(CH_3)_3 \longrightarrow HC\equiv CCH_2CH_2C(CH_3)_3$$

$$\downarrow \begin{array}{l} 2H_2 \\ Pd \end{array}$$

$$CH_3CH_2CH_2CH_2C(CH_3)_3$$

c) 1. $CH_3CH_2CH_2CH_2CH_2CHO$ can be made by treating $CH_3CH_2CH_2CH_2C\equiv CH$ with disiamylborane followed by H_2O_2, ^-OH.

2. $CH_3CH_2CH_2CH_2C\equiv CH$ can be synthesized from $CH_3CH_2CH_2CH_2Br$ and $HC\equiv C:^- Na^+$.

The complete sequence:

$$HC\equiv CH \xrightarrow[\text{NH}_3]{\text{NaNH}_2} HC\equiv C:^- Na^+ \xrightarrow[\text{THF}]{CH_3CH_2CH_2CH_2Br} CH_3CH_2CH_2CH_2C\equiv CH$$

$$\downarrow \begin{array}{l} 1.\ H\text{--}BR_2 \\ 2.\ H_2O_2/^-OH \end{array}$$

$$CH_3CH_2CH_2CH_2CH_2\overset{\overset{\displaystyle O}{\|}}{C}H$$

d) 1. $CH_3CH_2CH_2CH_2CH_2\overset{\overset{\displaystyle O}{\|}}{C}CH_3$. This ketone is the product of hydration of
$CH_3CH_2CH_2CH_2CH_2C\equiv CH$ with $H_2SO_4, H_2O, HgSO_4$.

2. $CH_3CH_2CH_2CH_2CH_2Br$ + $HC\equiv C:^- Na^+$ $\xrightarrow{\text{THF}}$ $CH_3CH_2CH_2CH_2CH_2C\equiv CH$

The complete sequence:

$$HC\equiv CH \xrightarrow[\text{NH}_3]{\text{NaNH}_2} HC\equiv C:^- Na^+ \xrightarrow[\text{THF}]{CH_3CH_2CH_2CH_2CH_2Br} CH_3CH_2CH_2CH_2CH_2C\equiv CH$$

$$\downarrow \begin{array}{l} H_2SO_4, H_2O \\ Hg^{2+} \end{array}$$

$$CH_3CH_2CH_2CH_2CH_2\overset{\overset{\displaystyle O}{\|}}{C}CH_3$$

8.16 a) 2,2–Dimethyl–3–hexyne   b) 2,5–Octadiyne
c) 3,6–Dimethyl–2–hepten–4–yne   d) 3,3–Dimethyl–1,5–hexadiyne
e) 1,3–Hexadien–5–yne   f) 3,6–Diethyl–2–methyl–4–octyne

8.17

a) $CH_3CH_2CH_2C\equiv C\overset{\overset{\displaystyle CH_3}{|}}{\underset{\underset{\displaystyle CH_3}{|}}{C}}CH_2CH_3$

b) $CH_3C\equiv CC\equiv C\overset{\overset{\displaystyle CH_3}{|}}{C}HCH_2\overset{\underset{\displaystyle CH_2CH_3}{|}}{C}HC\equiv CH$

c) $(CH_3)_3CC\equiv CC(CH_3)_3$

d)

e) $HC\equiv CCH=CHCH=CHCH_3$

f) $CH_3CH_2C\equiv CCH_2C(CH_3)_2CHClCH=CH_2$

g) $CH_3CH_2CH_2CH_2\overset{\overset{\displaystyle CH_3CHCH_2CH_3}{|}}{C}C\equiv CH$

h) $CH_3CH_2CH_2\overset{\overset{\displaystyle C(CH_3)_3}{|}}{C}C\equiv CCH(CH_3)_2$

8.18 a) $(CH_3)_3CCH_2CH_2C\equiv CCH_2CH_3$. <u>correct name</u>; 7,7–Dimethyl–3–octyne
This compound is an *octyne*.

b) $HC\equiv CC(CH_3)_2CH_2CH_2CH(CH_3)_2$. <u>correct name</u>; 3,3,6–Trimethyl–1–heptyne
Start numbering from the opposite end.

c) $CH_3CH=CHCH_2CH(CH_3)C\equiv CH$. <u>correct name</u>; 3–Methyl–5–hepten–1–yne
Try not to break up the name of the compound more than necessary.

d) $HC\equiv CCH_2\overset{\overset{\displaystyle CH_3}{|}}{C}HCH_2CH_2\overset{\overset{\overset{\displaystyle CH_3}{|}}{\displaystyle CHCH_3}}{|}$... <u>correct name</u> 4,7,8–Trimethyl–1–nonyne
Choose the longest chain, and number from the opposite end.

e) $CH_3CH_2CH=CHC\equiv CH$. <u>correct name</u>; 3–Hexen–1–yne

f)

<u>correct name</u> 1–Ethynyl–3–methylcyclohexane
The ring positions are numbered incorrectly.

8.19 a) $CH_3CH=CHC\equiv CC\equiv CCH=CHCH=CHCH=CH_2$.
1,3,5,11–Tridecatetraen-7,9–diyne

Using *E–Z* notation: (3E,5E,11E)–1,3,5,11–Tridecatetraen–7,9–diyne
The parent alkane of this hydrocarbon is tridecane.

b) $CH_3C\equiv CC\equiv CC\equiv CC\equiv CCH=CH_2$. 1–Tridecen–3,5,7,9,11–pentayne
This hydrocarbon is also of the tridecane family.

8.20

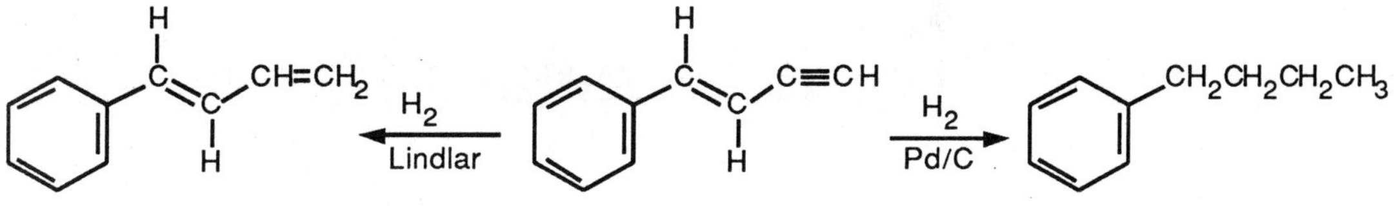

B

H₂ / Lindlar

H₂ / Pd/C

A

8.21 a) An acyclic alkane with eight carbons has the formula C_8H_{18}. C_8H_{10} has eight fewer hydrogens, or four fewer pairs of hydrogens, than C_8H_{18}. Thus, C_8H_{10} contains four degrees of unsaturation.

b) Because only one equivalent of H_2 is absorbed over the Lindlar catalyst, *one* triple bond is present.

c) Three equivalents of H_2 are absorbed when reduction is done over a palladium catalyst; two of them hydrogenate the triple bond already found to be present. Therefore, one *double* bond must also be present.

d) C_8H_{10} must contain one ring.

Many structures for C_8H_{10} are possible.

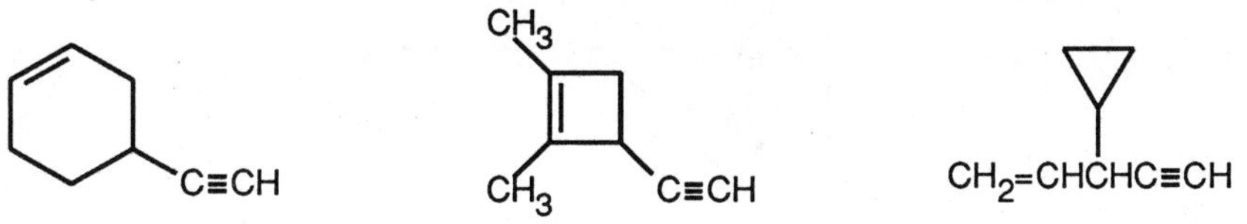

8.22

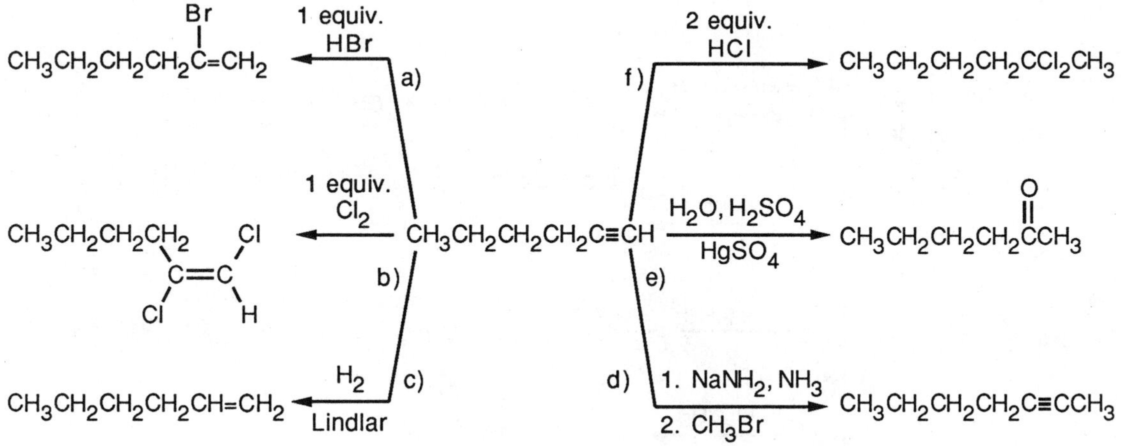

8.23

CH$_3$(CH$_2$)$_3$C≡C(CH$_2$)$_3$CH$_3$

a) $\xrightarrow[\text{Lindlar}]{\text{H}_2}$

CH$_3$(CH$_2$)$_3$ C=C (CH$_2$)$_3$CH$_3$ (cis, both H on same side)

b) $\xrightarrow{\text{Li, NH}_3}$

CH$_3$(CH$_2$)$_3$ C=C H / H C=C (CH$_2$)$_3$CH$_3$ (trans)

c) $\xrightarrow{\text{1 equiv Br}_2}$

CH$_3$(CH$_2$)$_3$ C=C Br / Br C=C (CH$_2$)$_3$CH$_3$

d) $\xrightarrow[\text{2. H}_2\text{O}_2,\ ^-\text{OH}]{\text{1. BH}_3,\ \text{THF}}$

e) $\xrightarrow[\text{HgSO}_4]{\text{H}_2\text{O, H}_2\text{SO}_4}$

CH$_3$(CH$_2$)$_3$C(=O)(CH$_2$)$_4$CH$_3$

f) $\xrightarrow[\text{Pd/C}]{\text{xs H}_2}$ CH$_3$(CH$_2$)$_8$CH$_3$

8.24

CH$_3$CH$_2$CH$_2$C≡CCH$_3$

a) $\xrightarrow{\text{2 equiv Br}_2}$ CH$_3$CH$_2$CH$_2$CBr$_2$CBr$_2$CH$_3$

b) $\xrightarrow{\text{1 equiv HBr}}$

CH$_3$CH$_2$CH$_2$ C=C H / Br C=C CH$_3$ +   CH$_3$CH$_2$CH$_2$ C=C Br / H C=C CH$_3$

c) $\xrightarrow{\text{xs HBr}}$ CH$_3$CH$_2$CH$_2$CBr$_2$CH$_2$CH$_3$ +   CH$_3$CH$_2$CH$_2$CH$_2$CBr$_2$CH$_3$

d) $\xrightarrow{\text{Li, NH}_3}$

CH$_3$CH$_2$CH$_2$ C=C H / H C=C CH$_3$

e) $\xrightarrow[\text{HgSO}_4]{\text{H}_2\text{O, H}_2\text{SO}_4}$ CH$_3$CH$_2$CH$_2$C(=O)CH$_2$CH$_3$ +   CH$_3$CH$_2$CH$_2$CH$_2$C(=O)CH$_3$

8.25

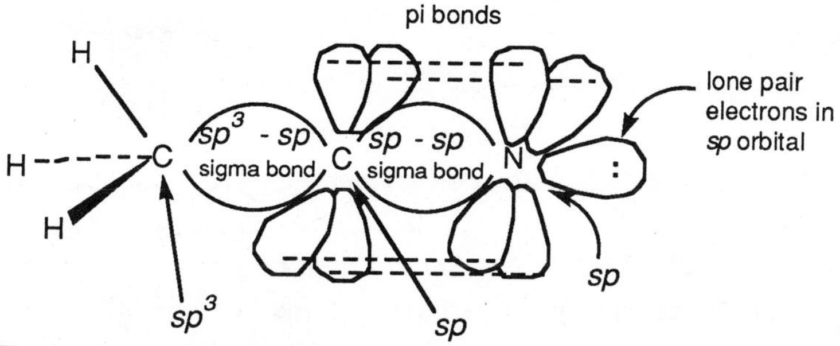

Each of the two pi bonds between carbon and nitrogen is formed by overlap of one *p* orbital of carbon with one *p* orbital of nitrogen.

8.26

a) $CH_3CH_2C{\equiv}CH$ $\xrightarrow[\text{HgSO}_4]{\text{H}_2\text{O, H}_2\text{SO}_4}$ $CH_3CH_2\overset{\overset{\displaystyle O}{\|}}{C}CH_3$

b) $CH_3CH_2C{\equiv}CH$ $\xrightarrow[\text{2. H}_2\text{O}_2,\,^-\text{OH}]{\text{1. Disiamylborane}}$ $CH_3CH_2CH_2CHO$

c) $\xrightarrow[\text{2. CH}_3\text{I, THF}]{\text{1. NaNH}_2,\,\text{NH}_3}$

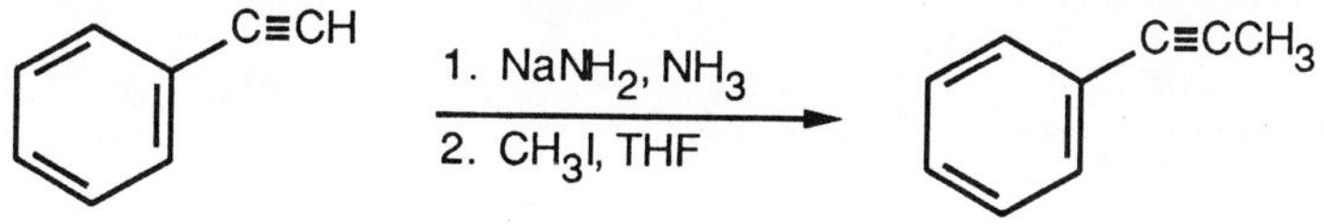

d) $\xrightarrow[\text{Lindlar}]{\text{H}_2}$

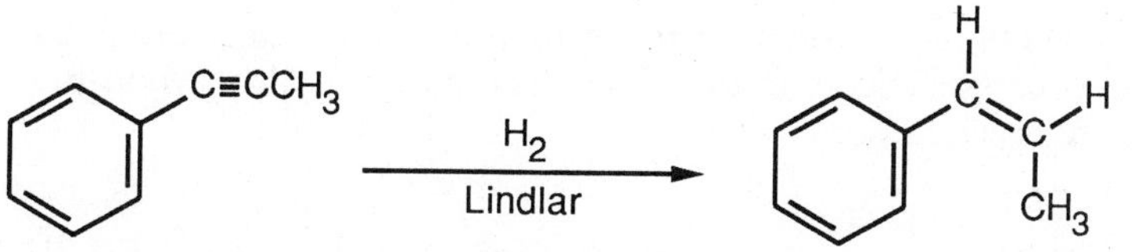

e) $CH_3CH_2C{\equiv}CH$ $\xrightarrow[\text{H}_2\text{O}]{\text{KMnO}_4}$ $CH_3CH_2COOH \ + \ CO_2$

f) $CH_3CH_2CH_2CH_2CH{=}CH_2$ $\xrightarrow[\text{CCl}_4]{\text{Br}_2}$ $CH_3CH_2CH_2CH_2CHBrCH_2Br$

$\downarrow$ 1. 2 NaNH$_2$
2. H$_3$O$^+$

$CH_3CH_2CH_2CH_2C{\equiv}CH$

8.27

a)

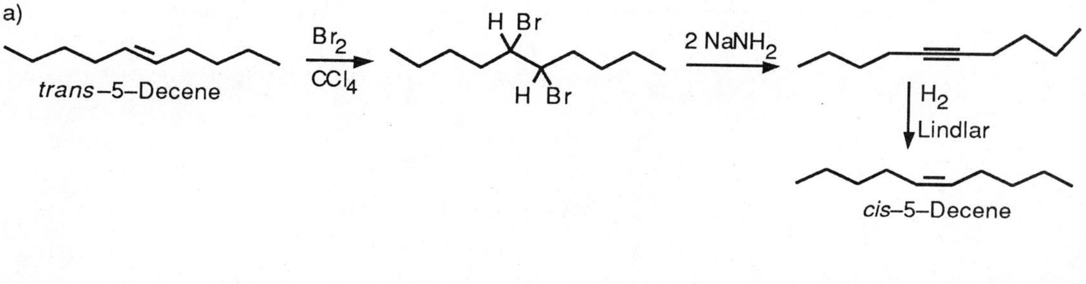

trans–5–Decene $\xrightarrow[\text{CCl}_4]{\text{Br}_2}$ $\xrightarrow{\text{2 NaNH}_2}$

$\downarrow$ $\underset{\text{Lindlar}}{\text{H}_2}$

cis–5–Decene

b)

cis-5-Decene $\xrightarrow[\text{CCl}_4]{\text{Br}_2}$ (H Br / H Br intermediate) $\xrightarrow{\text{2 NaNH}_2}$ (internal alkyne)

$\downarrow$ Li/NH$_3$

trans-5-Decene

8.28 Both KMnO$_4$ and O$_3$ oxidation of alkynes yield carboxylic acids; terminal alkynes give CO$_2$ also.

a) $CH_3(CH_2)_5C\equiv CH$ $\xrightarrow[\text{H}_2\text{O}]{\text{KMnO}_4}$ $CH_3(CH_2)_5COOH$ $+$ CO_2

b)

(phenyl)$C\equiv CCH_3$ $\xrightarrow[\text{H}_2\text{O}]{\text{KMnO}_4}$ (phenyl)COOH $+$ CH_3COOH

c) In this case, the existence of only one product, a diacid, shows that the original hydrocarbon contains a ring.

$\xrightarrow[\text{H}_2\text{O}]{\text{KMnO}_4}$ $HOOC(CH_2)_8COOH$

d) Notice that the products of this ozonolysis contain aldehyde and ketone functional groups, as well as a carboxylic acid and CO$_2$. The parent hydrocarbon must thus contain a double and a triple bond.

$CH_3CH=\overset{\overset{\displaystyle CH_3}{|}}{C}CH_2CH_2C\equiv CH$ $\xrightarrow[\text{2. Zn, H}_3\text{O}^+]{\text{1. O}_3}$ CH_3CHO $+$ $CH_3\overset{\overset{\displaystyle O}{||}}{C}CH_2CH_2COOH$ $+$ CO_2

e)

$\xrightarrow[\text{2. Zn, H}_3\text{O}^+]{\text{1. O}_3}$ $OHCCH_2CH_2CH_2CH_2\overset{\overset{\displaystyle O}{||}}{C}COOH$ $+$ CO_2

8.29

a) $CH_3CH_2CH_2C\equiv CH$ $\xrightarrow[\text{Lindlar}]{\text{H}_2}$ $CH_3CH_2CH_2CH=CH_2$ $\xrightarrow[\text{2. Zn/H}_3\text{O}^+]{\text{1. O}_3}$ $CH_3CH_2CH_2CHO$ $+$ $CH_2=O$

b) $(CH_3)_2CHCH_2C\equiv CH$ $\xrightarrow[\text{2. CH}_3\text{CH}_2\text{Br}]{\text{1. NaNH}_2, \text{NH}_3}$ $(CH_3)_2CHCH_2C\equiv CCH_2CH_3$

$\downarrow$ Li/NH$_3$

$\underset{(CH_3)_2CHCH_2}{\overset{H}{\diagdown}}C=C\underset{H}{\overset{CH_2CH_3}{\diagup}}$

8.30

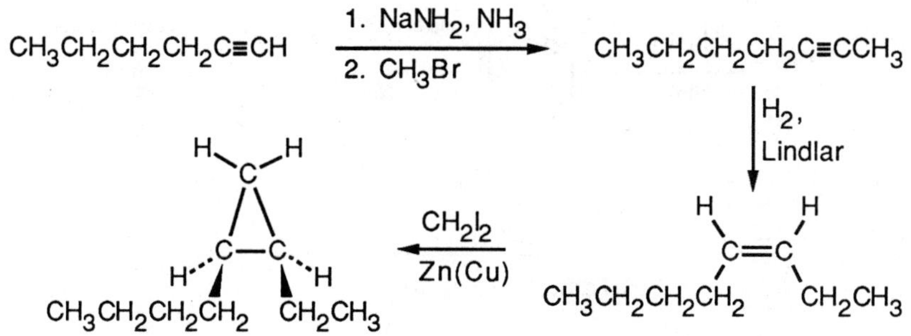

CH3CH2CH2CH2C≡CH
1. NaNH2, NH3
2. CH3Br
CH3CH2CH2CH2C≡CCH3
H2,
Lindlar
CH3CH2CH2CH2
CH2CH3
CH2I2
Zn(Cu)

8.31

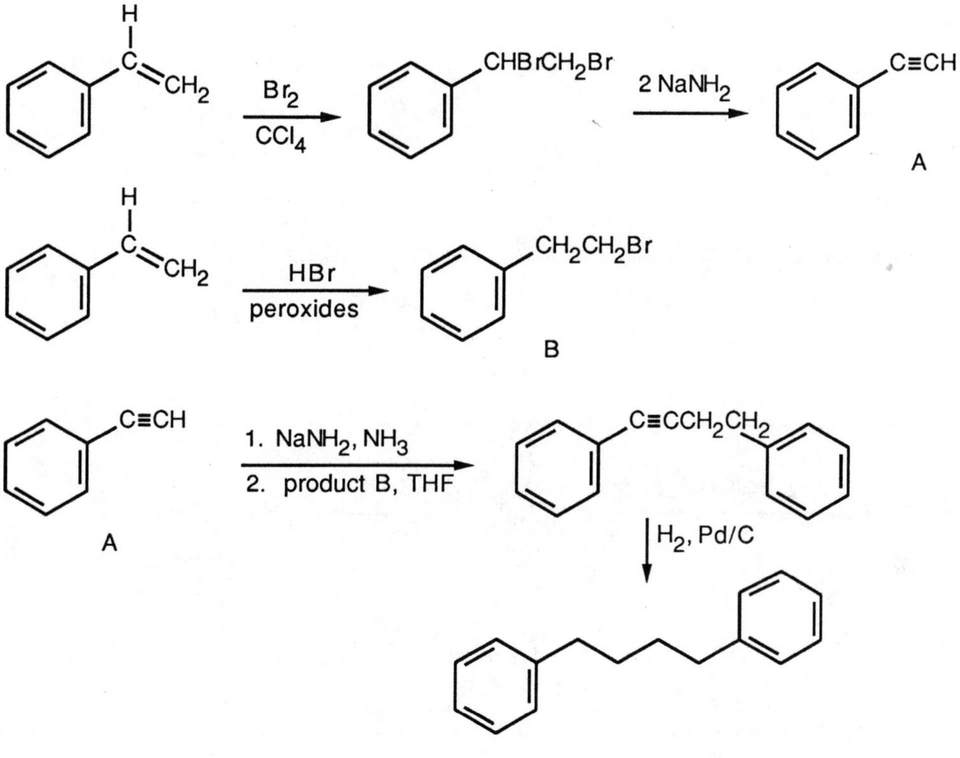

CH2
Br2
CCl4
CHBrCH2Br
2 NaNH2
C≡CH
A
HBr
peroxides
CH2CH2Br
B
1. NaNH2, NH3
2. product B, THF
C≡CCH2CH2
H2, Pd/C

8.32

a) $$CH_3(CH_2)_4C{\equiv}CH \xrightarrow[2.\ H_2O_2,\ ^{-}OH]{1.\ \text{Disiamylborane}} CH_3(CH_2)_5CHO$$

b)

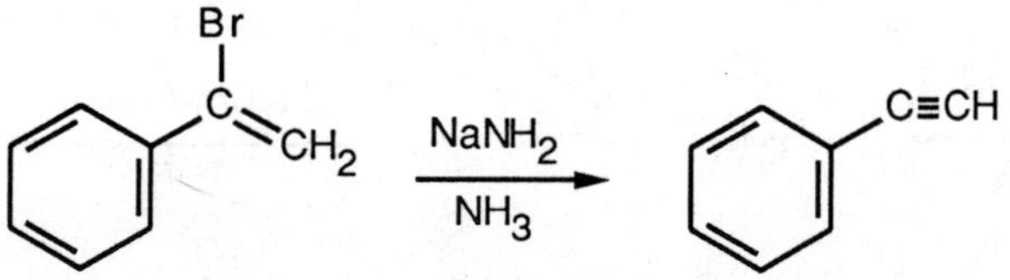

Br
CH2
NaNH2
NH3
C≡CH

c)

8.33

a) $CH_3CH_2C{\equiv}CH \xrightarrow{2\ Cl_2} CH_3CH_2CCl_2CHCl_2$

1,1,2,2-Tetrachlorobutane

b) $CH_3CH_2C{\equiv}CH \xrightarrow[\text{Lindlar}]{H_2} CH_3CH_2CH{=}CH_2 \xrightarrow[\text{peroxides}]{HBr} CH_3CH_2CH_2CH_2Br$

$CH_3CH_2C{\equiv}CH \xrightarrow[\text{2. } CH_3CH_2CH_2CH_2Br,\ THF]{\text{1. } NaNH_2,\ NH_3} CH_3CH_2C{\equiv}CCH_2CH_2CH_2CH_3$

$\downarrow \begin{array}{l}H_2\\ Pd/C\end{array}$

Octane

c) $CH_3CH_2C{\equiv}CH \xrightarrow[\text{2. } H_2O_2,\ ^-OH]{\text{1. Disiamylborane}} CH_3CH_2CH_2CHO$

Butanal

8.34

a) $HC{\equiv}CH \xrightarrow[NH_3]{NaNH_2} HC{\equiv}C{:}^-Na^+ \xrightarrow[THF]{CH_3CH_2CH_2Br} CH_3CH_2CH_2C{\equiv}CH$

b) $HC{\equiv}CH \xrightarrow[NH_3]{NaNH_2} HC{\equiv}C{:}^-Na^+ \xrightarrow[THF]{CH_3CH_2Br} CH_3CH_2C{\equiv}CH$

$\downarrow \begin{array}{l}NaNH_2\\ NH_3\end{array}$

$CH_3CH_2C{\equiv}CCH_2CH_3 \xleftarrow[THF]{CH_3CH_2Br} [CH_3CH_2C{\equiv}C{:}^-Na^+]$

c) $HC{\equiv}CH \xrightarrow[NH_3]{NaNH_2} HC{\equiv}C{:}^-Na^+ \xrightarrow{(CH_3)_2CHCH_2Br} (CH_3)_2CHCH_2C{\equiv}CH$

$\downarrow \begin{array}{l}H_2 \quad \text{or Li/NH}_3\\ \text{Lindlar}\end{array}$

$(CH_3)_2CHCH_2CH{=}CH_2$

d)   Product from a) $\xrightarrow[NH_3]{NaNH_2} [CH_3CH_2CH_2C{\equiv}C{:}^-Na^+] \xrightarrow[THF]{CH_3CH_2CH_2Br}$

$CH_3CH_2CH_2C{\equiv}CCH_2CH_2CH_3 \xrightarrow[Hg^{2+}]{H_2SO_4,\ H_2O} CH_3CH_2CH_2\overset{\displaystyle O}{\overset{\|}{C}}CH_2CH_2CH_2CH_3$

e) 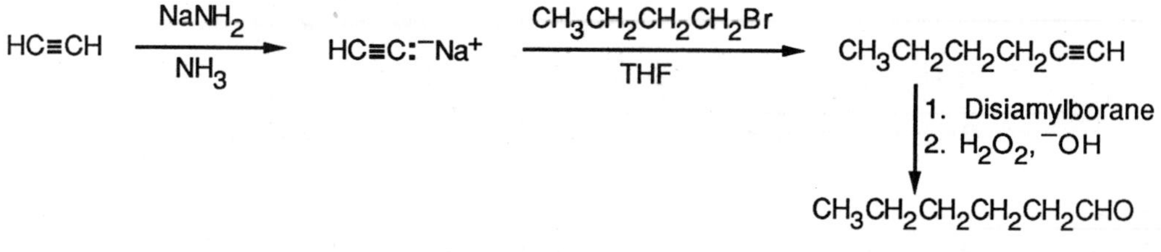 $HC \equiv CH$ $\xrightarrow[NH_3]{NaNH_2}$ $HC \equiv C:^- Na^+$ $\xrightarrow[THF]{CH_3CH_2CH_2CH_2Br}$ $CH_3CH_2CH_2CH_2C \equiv CH$

$\downarrow$ 1. Disiamylborane
2. $H_2O_2, ^-OH$

$CH_3CH_2CH_2CH_2CH_2CHO$

8.35

a) 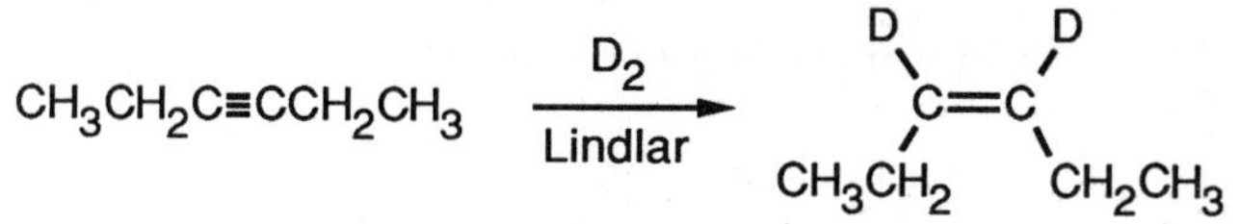 $CH_3CH_2C \equiv CCH_2CH_3$ $\xrightarrow[Lindlar]{D_2}$

D   D
\\   /
C=C
/   \\
CH_3CH_2 CH_2CH_3

b) 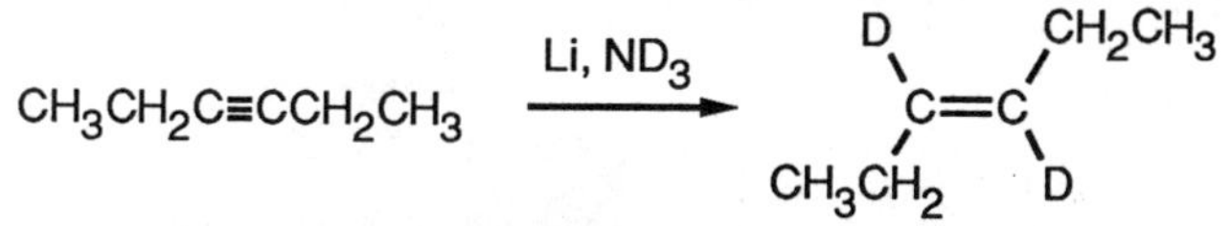 $CH_3CH_2C \equiv CCH_2CH_3$ $\xrightarrow{Li, ND_3}$

D CH_2CH_3
\\   /
C=C
/   \\
CH_3CH_2 D

c) $CH_3CH_2CH_2C \equiv CH$ $\xrightarrow[NH_3]{NaNH_2}$ $[CH_3CH_2CH_2C \equiv C:^- Na^+]$ $\xrightarrow{D_3O^+}$ $CH_3CH_2CH_2C \equiv CD$

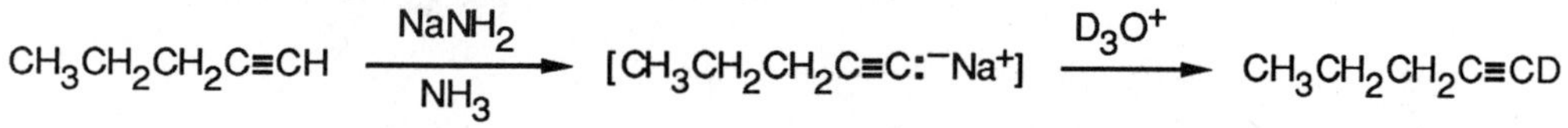

d)

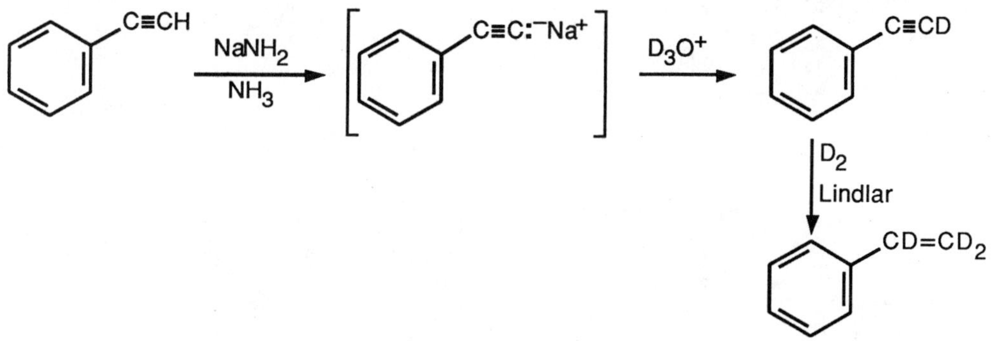

(phenyl)−$C \equiv CH$ $\xrightarrow[NH_3]{NaNH_2}$ $\left[\text{(phenyl)}-C \equiv C:^- Na^+ \right]$ $\xrightarrow{D_3O^+}$ (phenyl)−$C \equiv CD$

$\downarrow$ D_2 / Lindlar

(phenyl)−$CD=CD_2$

8.36

pi bonds

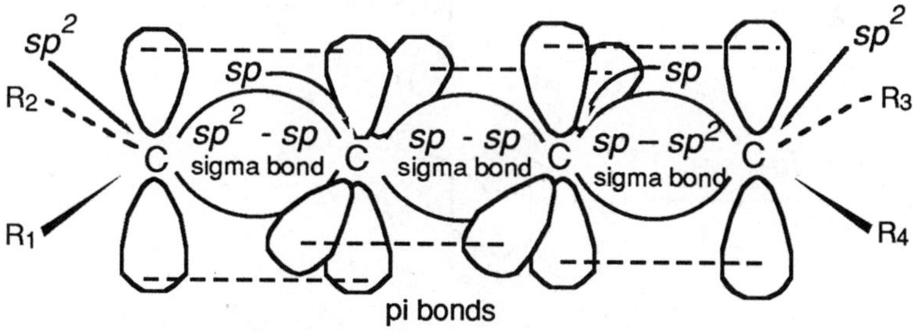

This simplest cumulene is pictured above. The carbons at the end of the cumulated double bonds are sp^2 hybridized and form one pi bond to the "interior" carbons. The interior carbons are sp hybridized; each carbon forms two pi bonds – one to an "exterior" carbon and one to the other interior carbon. If you build a model of this cumulene, you can see that the substituents all lie in the same plane. This cumulene can thus exhibit *cis–trans* isomerism, just as simple alkenes can.

In general, the substituents of any cumulene with an odd number of adjacent double bonds lie in a plane; these cumulenes can exhibit *cis–trans* isomerism. The relationship of substituents at the ends of any even-numbered cumulene will be explained later.

8.37

$$HC\equiv CH \xrightarrow[\text{2. } BrCH_2(CH_2)_6CH_2Br]{\text{1. } NaNH_2, NH_3}$$

8.38 Muscalure is a C_{23} alkene. The only functional group present is the double bond between C_9 and C_{10}. Since our synthesis begins with acetylene, we can assume that the double bond can be produced by hydrogenation of a triple bond.

$$HC\equiv CH \xrightarrow[\text{NH}_3]{\text{NaNH}_2} HC\equiv C:^- Na^+ \xrightarrow[\text{THF}]{CH_3(CH_2)_6CH_2Br} CH_3(CH_2)_7C\equiv CH$$

$$\downarrow NaNH_2, NH_3$$

$$CH_3(CH_2)_7C\equiv C(CH_2)_{12}CH_3 \xleftarrow[\text{THF}]{CH_3(CH_2)_{11}CH_2Br} [CH_3(CH_2)_7C\equiv C:^- Na^+]$$

$$\downarrow H_2, \text{ Lindlar}$$

CH_3(CH_2)_6CH_2   CH_2(CH_2)_{11}CH_3
$$\text{C}=\text{C}$$
H   H

(Z)-9-Tricosene

8.39

$$\xrightarrow{3 H_2/Pd}$$ B (cyclohexane with CH_2CH_2CH_3)

A (cyclohexylidene with H and C≡CH)
$$\xrightarrow[\text{2. Zn, H}_3O^+]{\text{1. O}_3}$$ cyclohexanone (=O)   plus other fragments

$$\xrightarrow[\text{2. CH}_3I]{\text{1. NaNH}_2, NH_3}$$ C (cyclohexylidene with H and C≡CCH_3)

8.40

$$\xleftarrow[\text{Pa}]{H_2}$$

CH_2−C≡C−C≡C−CH_2
| |
CH_2−C≡C−C≡C−CH_2

A

$$\xrightarrow[\text{2. Zn, H}_3O^+]{\text{1. O}_3}$$

2 HOOCCH_2CH_2COOH
+
2 HOOC−COOH

8.41

$$CH_3\overset{\overset{\displaystyle O}{\|}}{C}CH_3 \xrightarrow[\text{2. } H_3O^+]{\text{1. } HC\equiv C\!:^-Na^+} CH_3\underset{\underset{\displaystyle C\equiv CH}{|}}{\overset{\overset{\displaystyle OH}{|}}{C}}CH_3 \xrightarrow{H_3O^+} CH_2\!=\!\overset{\overset{\displaystyle CH_3}{|}}{C}C\equiv CH \xrightarrow[\text{Lindlar}]{H_2} CH_2\!=\!\overset{\overset{\displaystyle CH_3}{|}}{C}CH\!=\!CH_2$$

2–Methyl–
1,3–butadiene

8.42 1) Erythrogenic acid contains six degrees of unsaturation (see Sec. 6.1 for method of calculating unsaturation equivalents for compounds containing elements other than C and H).

2) One of these double bonds is contained in the carboxylic acid functional group –COOH; thus, five other degrees of unsaturation are present.

3) Because five equivalents of H_2 are absorbed on catalytic hydrogenation, erythrogenic acid contains no rings.

4) The presence of both aldehyde and carboxylic acid products of ozonolysis indicates that both double and triple bonds are present in erythrogenic acid.

5) Only two ozonolysis products contain aldehyde functional groups; these fragments must have been double-bonded to each other in erythrogenic acid. $H_2C=CH(CH_2)_4C\equiv$

6) The other ozonolysis products result from cleavage of triple bonds. However, not enough information is available to tell in which order the fragments were attached. The two possible structures:

<u>A</u> $H_2C=CH(CH_2)_4C\equiv C-C\equiv C(CH_2)_7COOH$
<u>B</u> $H_2C=CH(CH_2)_4C\equiv C(CH_2)_7C\equiv CCOOH$

One method of distinguishing between the two possible structures is to treat erythrogenic acid with two equivalents of H_2, using Lindlar catalyst. The resulting trialkene can then be ozonized. The fragment that originally contained the carboxylic acid will still have it intact and can be identified as the terminal carboxylic acid fragment.

Study Guide for Chapter 8

After studying this chapter, you should be able to:

(1) Name and draw alkynes, according to IUPAC conventions (8.1, 8.2, 8.16, 8.17, 8.18, 8.19).

(2) Predict the products of reactions of alkynes (8.3, 8.4, 8.6, 8.9, 8.12, 8.14, 8.20, 8.22, 8.23, 8.24, 8.26, 8.29, 8.30, 8.32).

(3) Choose the correct alkyne starting material to yield a given product (8.5, 8.8, 8.11).

(4) Use cleavage and/or hydrogenation products to determine the structure of an unknown alkyne (8.13, 8.21, 8.28, 8.39, 8.40, 8.42).

(5) Explain the hybridization of the triple bond, and account for the weak acidity of alkynes (8.10, 8.25, 8.36).

(6) Carry out transformation and syntheses involving alkynes (8.15, 8.27, 8.31, 8.33, 8.34, 8.36, 8.37, 8.38, 8.41).

Chapter 9 – Stereochemistry

9.1 Chiral: screw, beanstalk, shoe.
Achiral: screwdriver, hammer.

9.2 Use the following rules to locate centers that are *not* stereogenic.

1. All $-CH_3$ and $-CX_3$ carbons are nonstereogenic.
2. All $-CH_2$ and $-CX_2-$ carbons are nonstereogenic.
3. All $-C{=}C-$ and $-C{\equiv}C-$ carbons are nonstereogenic.

By rule 3, all benzene-ring carbons are nonstereogenic

a) 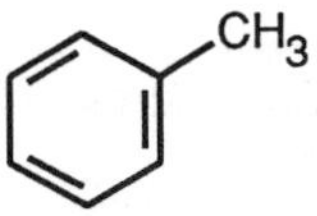

Toluene
achiral

b) 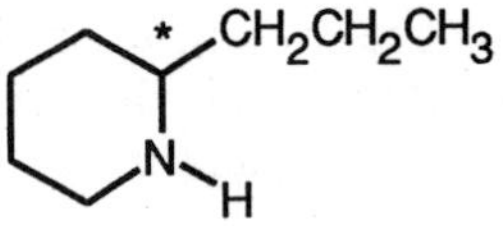

Coniine
chiral

c) 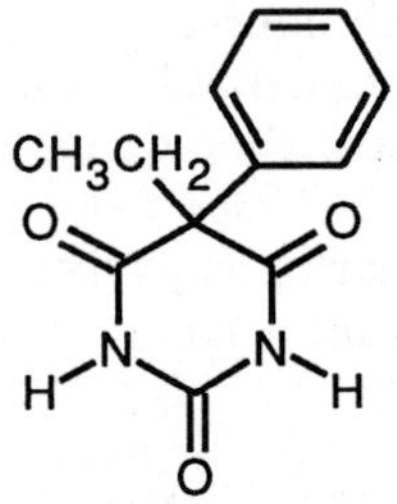

Phenobarbital
achiral

9.3

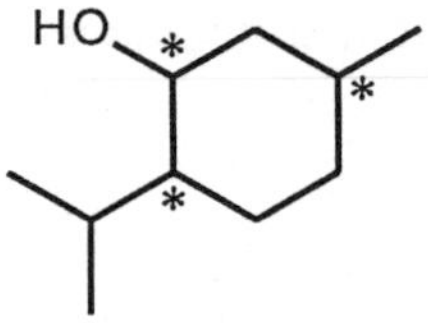

Menthol

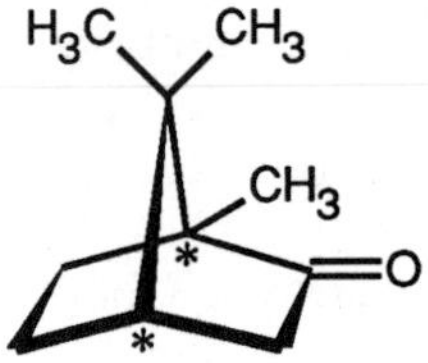

Camphor

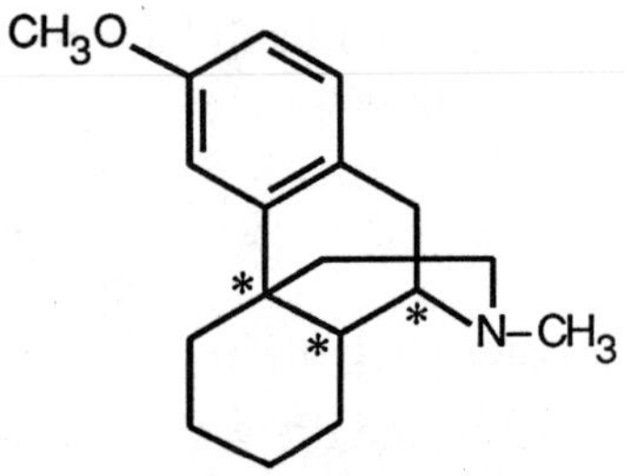

Dextromethorphan

9.4

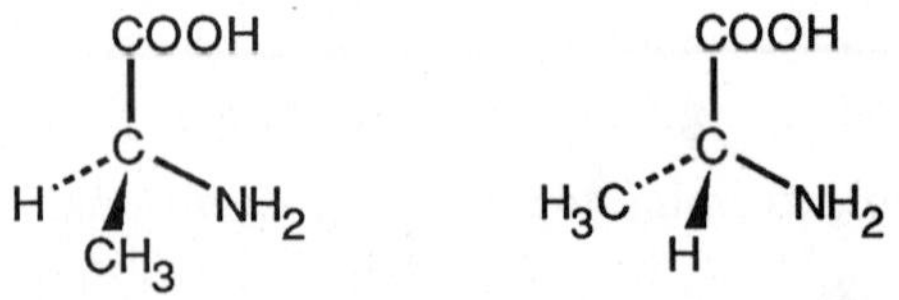

Alanine

9.5 We know that observed rotation is directly proportional to concentration. Thus, if the concentration of a sample is halved, the observed rotation is also halved. In this problem, halving the concentration of a solution with observed rotation of +90° would give a solution with a rotation of +45°. If, instead, the actual rotation were –270°, dilution would produce a solution with a rotation of –135°. The value of observed rotation after dilution allows us to know the original sign of rotation.

9.6

Use the formula $[\alpha]_D = \dfrac{\alpha}{l \times C}$ where

$[\alpha]_D$ = specific rotation

α = observed rotation

l = path length of cell (in dm)

C = concentration (in g/mL)

In this problem:

$\alpha = +1.2°$

$l = 5\,cm = 0.5\,dm$

$C = 1.5\,g/10mL = 0.15\,g/mL$

$$[\alpha]_D = \frac{+1.2°}{0.5\,dm \times 0.15\,g/mL} = +16°$$

9.7 Use the sequence rules in section 9.6.

a) By rule 1, –H is of lowest priority, and –Br is of highest priority. By rule 2, –CH$_2$CH$_2$OH is of higher priority than –CH$_2$CH$_3$.

Highest ⟶ Lowest

–Br, –CH$_2$CH$_2$OH, –CH$_2$CH$_3$, –H

b) By rule 3, –COOH can be considered as $-\overset{\displaystyle O-}{\underset{\displaystyle O-}{C}}-OH$. Since three "oxygens" are attached to a

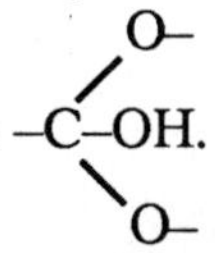

–COOH carbon and only one oxygen is attached to –CH$_2$OH, –COOH is of higher priority than –CH$_2$OH. –CO$_2$CH$_3$ is of higher priority than –COOH by rule 2, and –OH is of highest priority by rule 1.

Highest ⟶ Lowest

–OH, –CO$_2$CH$_3$, –COOH, –CH$_2$OH

c)   –NH$_2$, –CN, –CH$_2$NHCH$_3$, –CH$_2$NH$_2$

d)   –Br, –Cl, –CH$_2$Br, –CH$_2$Cl

9.8 The following scheme may be used to assign *R,S* configurations to stereogenic centers:

Step 1. For each stereogenic center, rank substituents by the Cahn–Ingold–Prelog system; give the number 4 to the lowest priority substituent. For part (a)

Substituent	Priority
–Br	1
–COOH	2
–CH$_3$	3
–H	4

Step2. Manipulate the molecule so that the lowest priority group is oriented toward the rear. To avoid errors, use a molecular model of the compound.

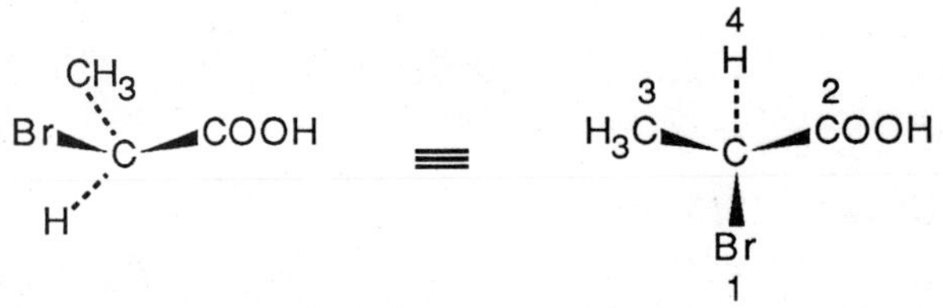

Step 3. Find the direction of rotation of the arrows that go from group 1 to group 2 to group 3. If the arrows have a clockwise rotation, the configuration is *R*; if the arrows have a counterclockwise rotation, the configuration is *S*. Here, the configuration is *S*.

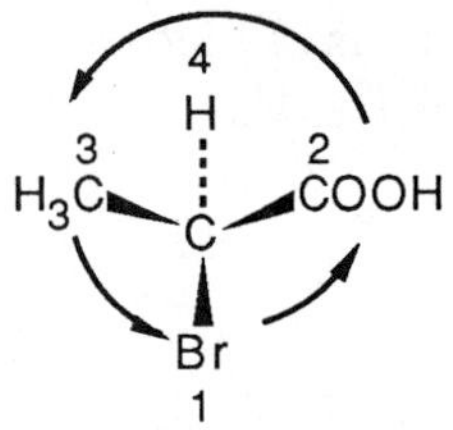

b)

c)

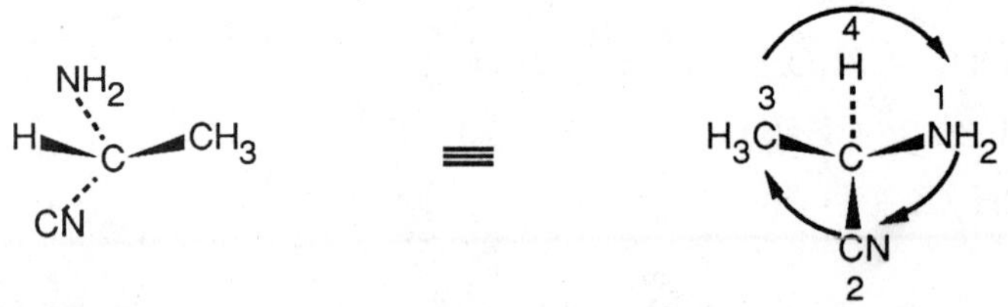

9.9 *R,S* assignments for more complicated molecules can be made by using a slight modification of the rules in problem 9.8. It is especially important to use molecular models when a compound has more than one stereogenic center.

Step 1. Assign priorities to groups on the top stereogenic center.

Substituent	Priority
–Br	1
–CH(OH)CH$_3$	2
–CH$_3$	3
–H	4

Step 2. Orient the model so that –H bonded to the first stereogenic center points back.

Step 3. Find the direction of the arrows that travel from 1 to 2 to 3 and assign an *R,S* configuration to the first stereogenic center.

Step 4. Repeat steps 1-3 for the next stereogenic center.

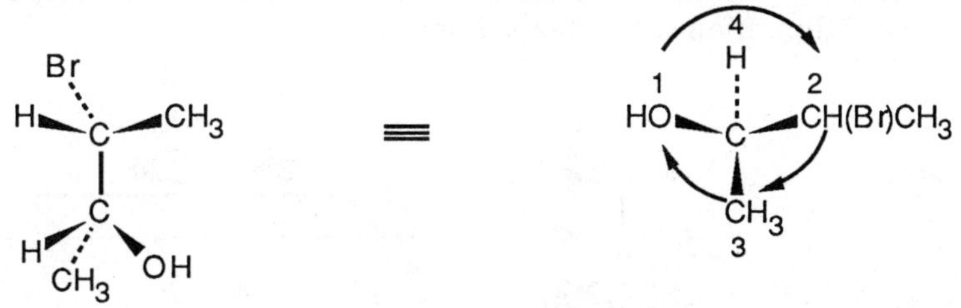

b)   *S,R*
c)   *R,S*
d)   *S,S*

a, d are enantiomers and are diastereomeric to b, c.
b, c are enantiomers and are diastereomeric to a, d.

9.10

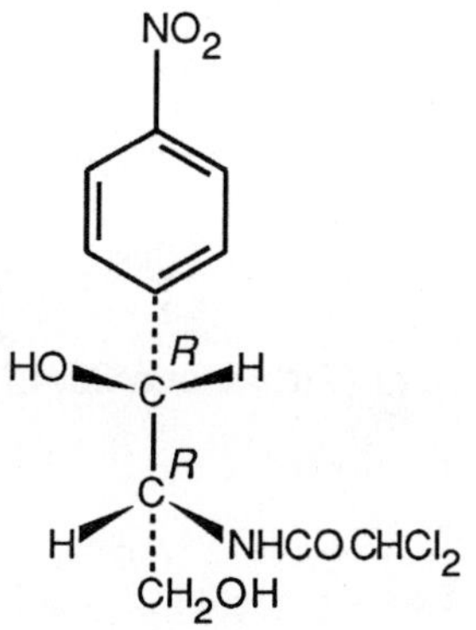

Chloramphenicol

9.11 To decide if a structure represents a *meso* compound, try to locate a plane of symmetry that divides the molecule into two halves that are mirror images. Molecular models may be helpful.

a)

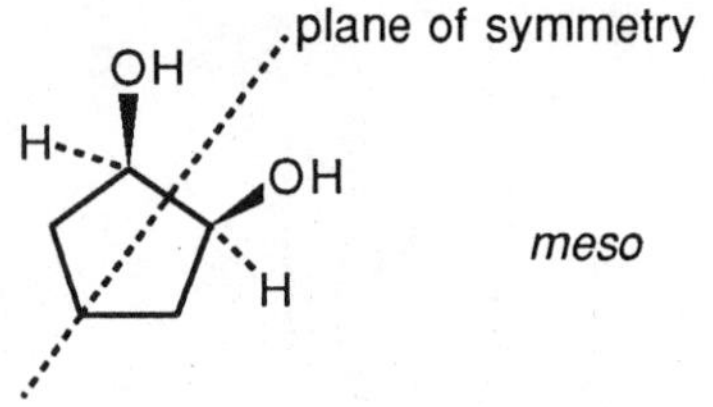

meso

b) and c) are not *meso* structures.

d)

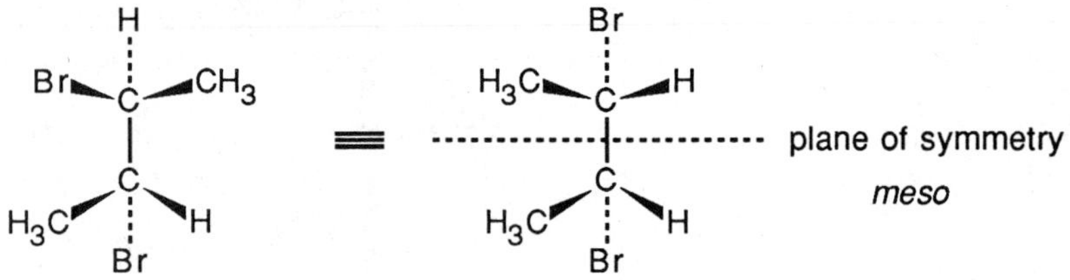

plane of symmetry

meso

9.12 In order for a molecule to exist as a meso form, it must possess a plane of symmetry. 2,3–Dibromobutane can exist as a pair of enantiomers *or* as a meso compound, depending on the configurations at carbons 2 and 3.

a)

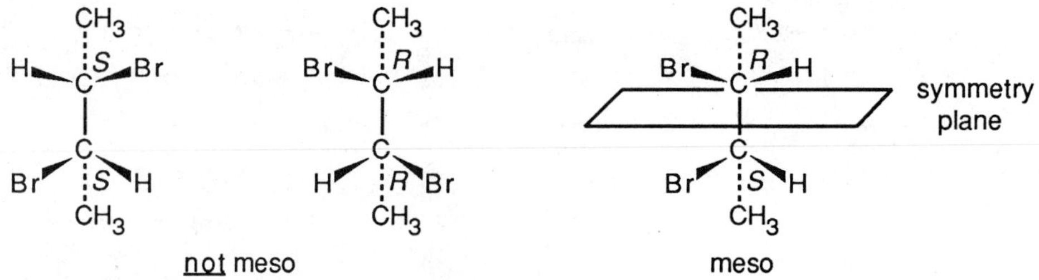

not meso   meso

b)   2,3–Dibromopentane has no symmetry plane and thus can't exist in a meso form.
c)   2,4–Dibromopentane can exist in a meso form.

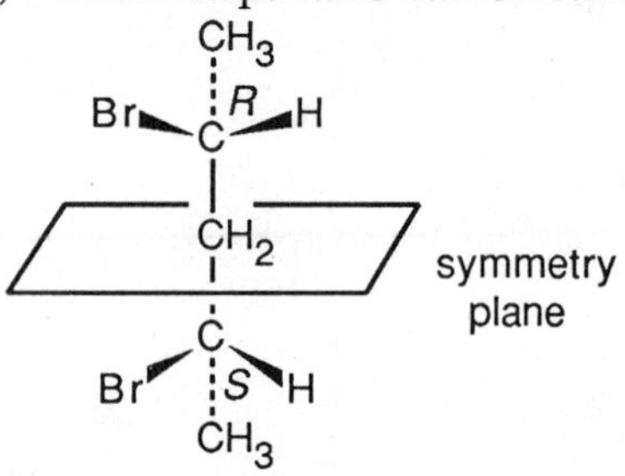

symmetry plane

2,4–Dibromopentane can also exist as a pair of enantiomers (2*R*,4*R* and 2*S*,4*S*) that are not meso compounds.

9.13

Morphine has five stereogenic centers and, in principle, can have $2^5 = 32$ stereoisomers. Many of these stereoisomers are too strained to exist.

9.14 Two manipulations of Fischer projections are allowable.
1.   A Fischer projection may be rotated on the page by 180°, but not by 90° or 270°.
2.   Holding one group of a Fischer projection steady, we may rotate the other three groups either clockwise or counterclockwise.
If we hold the –COOH group of projection A steady and rotate the other three groups by 120°, we arrive at a projection identical with projection B. Thus A and B are identical.

If projection C is rotated 180°, the resulting projection has two groups superimposable with projection B. However, no manipulation can make all four groups superimposable; C is thus an enantiomer of A and B.

Likewise, projection D is not superimposable on projection B but is identical to projection C. Thus, A and B are identical and enantiomeric with C and D, which are also identical.

9.15 a)   Manipulate two groups in the first structure to the positions they occupy in the second structure.

For example:

Now, holding –CH₃ steady, rotate the groups clockwise.

$$\begin{array}{c} CHO \\ H\!-\!\!\!\!\!\rotatebox{0}{}\!\!\!-Cl \\ \boxed{CH_3}\ \text{steady} \end{array} \quad = \quad \begin{array}{c} H \\ Cl\!-\!\!\!-CHO \\ CH_3 \end{array}$$

Although this projection resembles the second one in placement of –CH₃ and –H, –Cl and –CHO are interchanged. The two projections thus represent enantiomers.

b)   As above, rotate the first projection by 180°.

$$\begin{array}{c} CH_2OH \\ H\!-\!\!\!-OH \\ CHO \end{array} \quad = \quad \begin{array}{c} CHO \\ HO\!-\!\!\!-H \\ CH_2OH \end{array}$$

If we now compare this structure to the second structure of the pair, two of the groups occupy the same relative position.

$$\begin{array}{c} CHO \\ HO\!-\!\!\!-H \\ CH_2OH \end{array} \quad vs. \quad \begin{array}{c} CHO \\ HO\!-\!\!\!-CH_2OH \\ H \end{array}$$

There is, however, no rotation that can make these two structures superimposable; they are enantiomers.

9.16 One of the hardest spatial problems in organic chemistry is to visualize two-dimensional drawings as three-dimensional chemical structures. This difficulty becomes particularly troublesome when it is necessary to assign *R,S* configurations to structures, especially if they are drawn as Fischer projections. Working out these assignments is easier when models are used, but it is still possible to determine a configuration from a two-dimensional drawing.

    The following system may be used for assigning *R,S* configurations. Problem (a) is used as an example.

Step 1. Rank substituents in priority order.

Substituents	Priority	
–Br	1	high
–COOH	2	
–CH₃	3	
–H	4	low

Step2. Use either of the allowable manipulations of Fischer projections to bring the group of lowest priority to the top of the projection. In this case, hold –Br steady and rotate the other three groups clockwise.

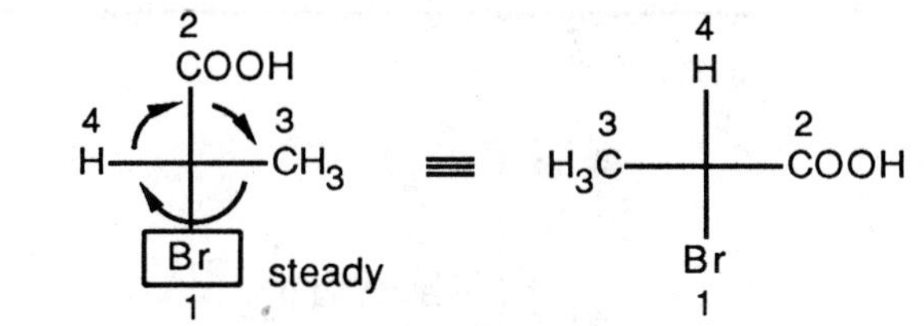

Step 3. Indicate on the rotated projection the direction of the arrows that proceed from group 1 to group 3. If the direction of the arrows is clockwise, the configuration is *R*; if the direction is counterclockwise, the configuration is *S*. Here the configuration is *S*.

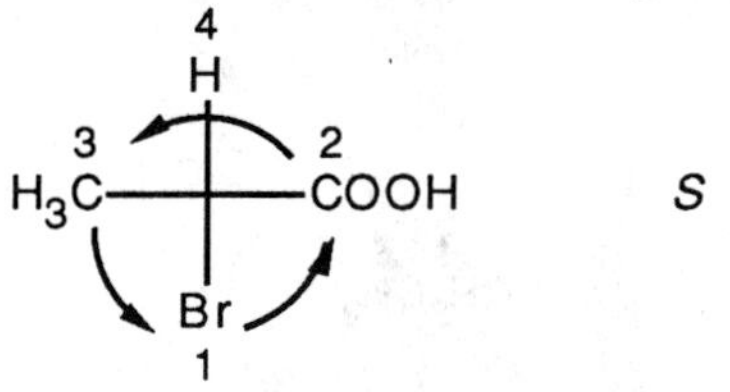

b)

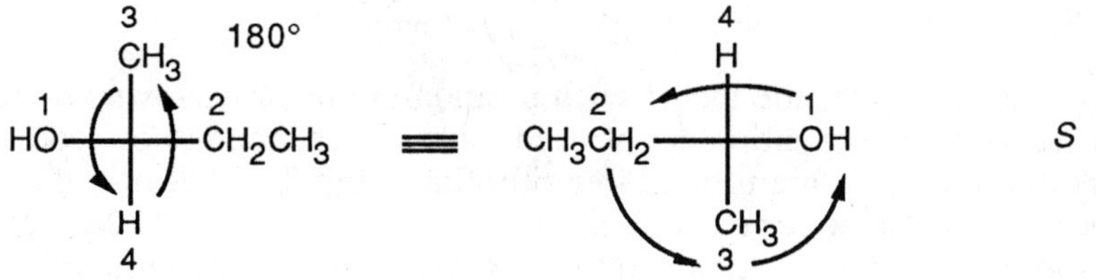

c)

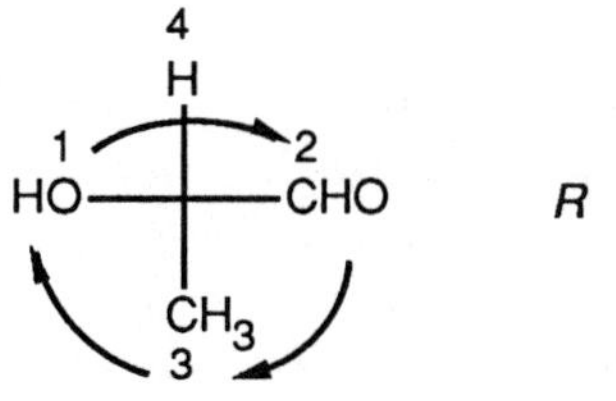

9.17 For simplicity, consider only top-side attack of bromine.

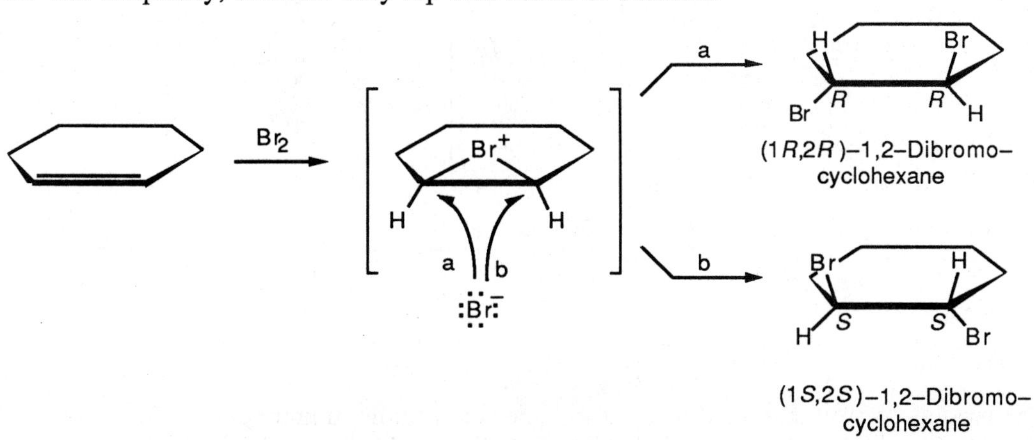

(1*R*,2*R*)–1,2–Dibromo–cyclohexane

(1*S*,2*S*)–1,2–Dibromo–cyclohexane

The product of addition of Br_2 to cyclohexene is a racemic mixture of the (1*R*,2*R*) and (1*S*,2*S*) enantiomers. The products of bottom-side attack are identical to those of top-side attack.

9.18 Possible bromonium ion intermediates are shown below. In this problem, assume that attack of :Br⁻ is somewhat more likely at carbon 2.

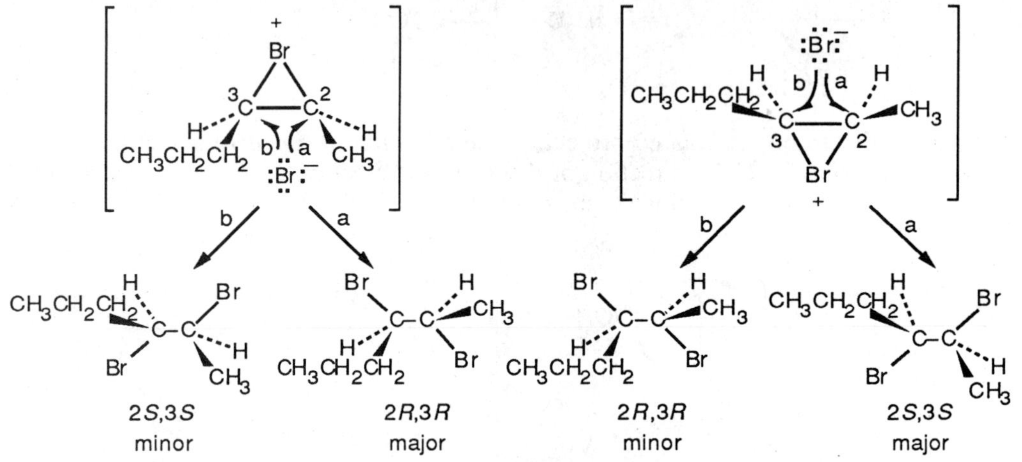

2S,3S   2R,3R   2R,3R   2S,3S
minor   major   minor   major

The products of attack of bromide ion on each bromonium ion are shown above. Notice that the major products are enantiomers of each other, as are the minor products. Because the bromonium ions are formed in a 50:50 mixture, and because the percent attack at carbon 2 is the same for each bromonium ion, the amount of (2R,3R) and (2S,3S)–dibromohexanes will be equal, and the product will be a racemic mixture.

9.19 Remember from Problem 9.18 that attack at carbon 2 is assumed to be more likely.

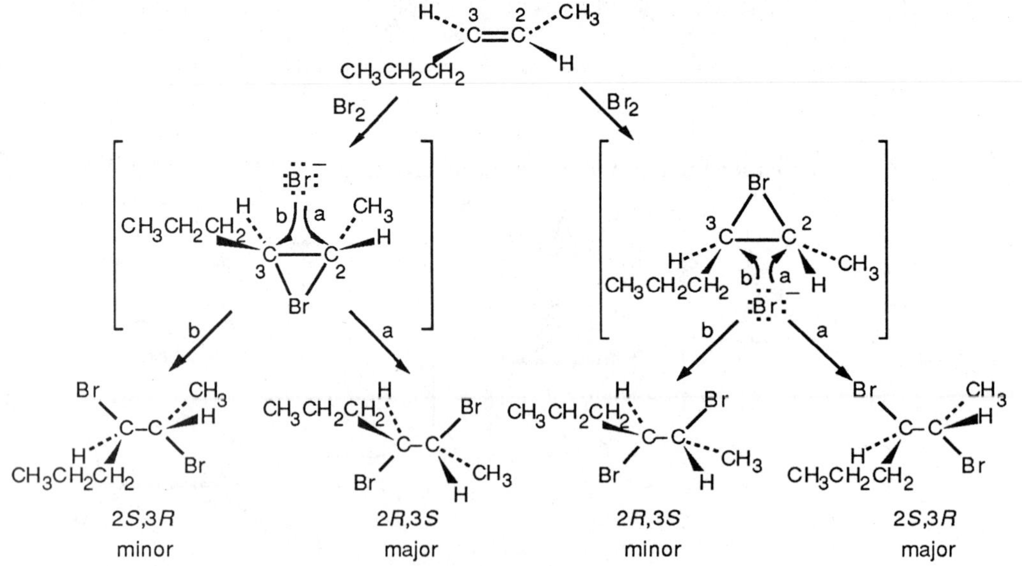

2S,3R   2R,3S   2R,3S   2S,3R
minor   major   minor   major

The product of bromination of *trans*–2–hexene is a racemic mixture of (2S,3R)–2,3–dibromohexane and (2R,3S)–2,3–dibromohexane. The reasoning is explained in Problem 9.18.

9.20 Look back to Figure 9.21, which shows the reaction of *R*–4–methyl–1–hexene with HBr. In a similar way, we can write a reaction mechanism for the reaction of HBr with *S*–4–methyl–1–hexene.

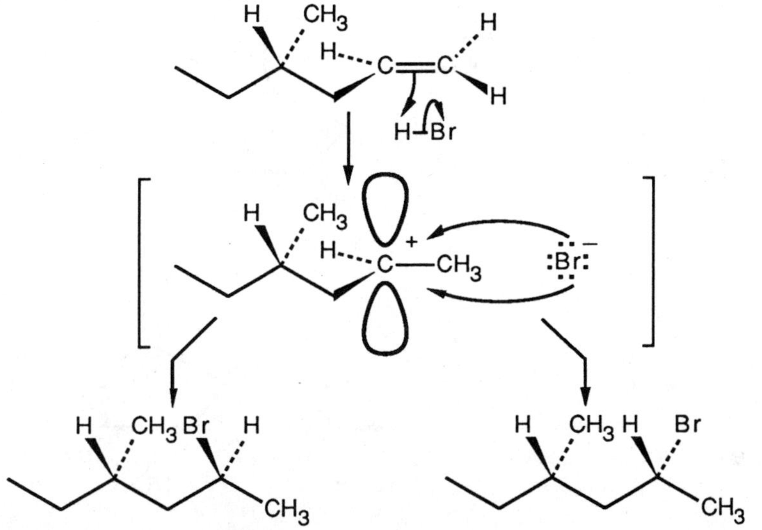

(2*S*,4*S*)–2–Bromo–4–methylhexane   (2*R*,4*S*)–2–Bromo–4–methylhexane

The 2*S*,4*S* stereoisomer is the enantiomer of the 2*R*,4*R* isomer, and the transition states leading to the formation of these two isomers are enantiomeric and of equal energy. Thus, the 2*S*,4*S* and 2*R*,4*R* enantiomers are formed in equal amounts. A similar argument can be used to show that the 2*R*,4*S* and 2*S*,4*R* isomers are formed in equal amounts. The product of the reaction of HBr with racemic starting material is a racemic mixture of the four possible stereoisomers; the product mixture is optically inactive.

Note that the ratio of (2*R*,4*R* + 2*S*,4*S*):(2*R*,4*S* + 2*S*,4*R*) is not 50:50. Nevertheless, the product mixture is optically inactive because the enantiomers are formed in equal amounts.

9.21

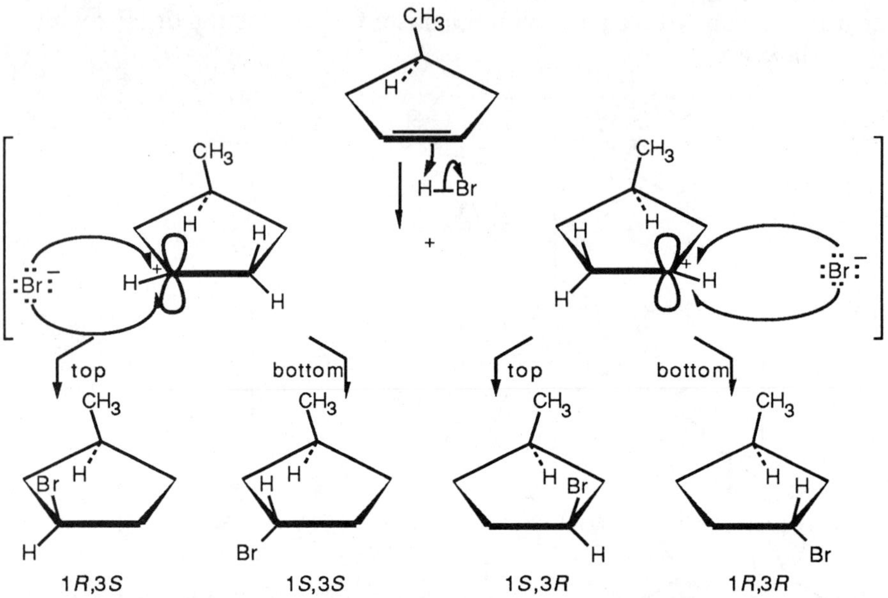

$1R,3S$ $1S,3S$ $1S,3R$ $1R,3R$

Two enantiomeric carbocations are formed. Each carbocation can be attacked by bromide from either the top or the bottom to yield four stereoisomers of 1–bromo–3–methylcyclopentane. The same argument used in Problem 9.20 can be used to show that the $1S,3R$ and $1R,3S$ enantiomers are formed in equal amounts, and the $1S,3S$ and $1R,3R$ isomers are formed in equal amounts. The product mixture is optically inactive.

9.22

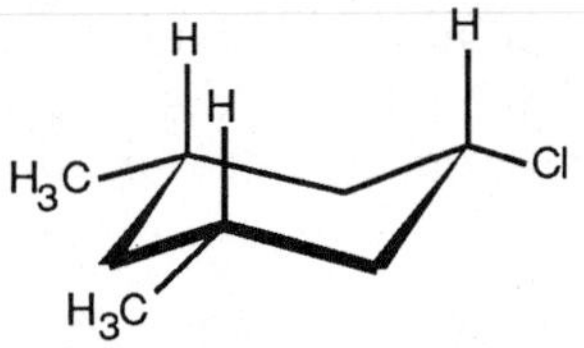

There are four stereoisomers of 1–chloro–3,5–dimethylcyclohexane. It is possible for (1) both methyl groups to be *cis* to chlorine; (2) both methyl groups to be *trans* to chlorine; (3) one methyl group to be *cis* and the other to be *trans* to chlorine (two isomers). The geometric isomer in which all three groups are *cis* is the most stable because the three groups are equatorial.

9.23

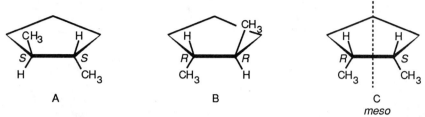

A and B are enantiomers and are chiral. Compound C is their diastereomer and is a *meso* compound.

9.24

for Cholic acid; $[\alpha]_D = \dfrac{+2.22°}{0.1 \text{ dm x } 0.6 \text{ g/mL}} = \dfrac{+2.22°}{0.06} = +37.0°$

9.25

for Ecdysone; $[\alpha]_D = \dfrac{+0.087°}{0.2 \text{ dm x } 0.007 \text{ g/mL}} = +62°$

9.26 a) *Chirality* is the property of "handedness" -- the property of an object that causes it to be non-superimposable on its mirror image.

  b) A *stereogenic center* of a molecule is an atom that is bonded to four different atoms or groups of atoms.

  c) *Optical activity* is the property of a sample that causes it to rotate the plane of polarization of plane-polarized light.

  d) A *diastereomer* is a stereoisomer that is not the mirror image of another stereoisomer.

  e) An *enantiomer* is one of a pair of stereoisomers that have a mirror image relationship.

  f) A *racemate* is a 50:50 mixture of (+) and (–) enantiomers that behaves as if it were a pure compound and that is optically inactive.

9.27

a) $CH_3CH_2CH_2\overset{*}{C}H(CH_3)CH_2CH(CH_3)_2$. 2,4–Dimethylheptane has one
                                              stereogenic center.

b) $CH_3CH_2C(CH_3)_2CH_2CH(CH_2CH_3)_2$. 3–Ethyl–5,5–dimethylheptane is achiral.

c)
  Cl—⬡—Cl   *cis*–1,4–Dichlorocyclohexane is *achiral*. Notice the plane of symmetry that passes through the –Cl groups.

d) $CH_3C\equiv C\overset{*}{C}H(CH_3)\overset{*}{C}H(CH_3)C\equiv CCH_3$. 4,5–Dimethyl–2,6–octadiyne has two
                                                        stereogenic centers.

The chirality of the compound depends upon the configuration at each of the stereogenic centers. The *R,R* and *S,S* isomers are chiral enantiomers; the *R,S* isomer is an achiral *meso* compound.

9.28

a) $CH_3CH_2CH_2\overset{*}{C}HClCH_3$

2–Chloropentane

b) $CH_3CH_2CH_2CH_2\overset{*}{C}HOHCH_3$

2–Hexanol

c) $CH_3CH_2\overset{\text{CH}_3}{\underset{*}{\overset{|}{C}}}HCH=CH_2$

3–Methyl–1–pentene

d) $CH_3CH_2CH_2CH_2\overset{\text{CH}_3}{\underset{*}{\overset{|}{C}}}HCH_2CH_3$

3–Methylheptane

9.29

$CH_3CH_2CH_2CH_2CH_2OH$

achiral

$CH_3CH_2CH_2\overset{*}{C}H(OH)CH_3$

chiral

$CH_3CH_2CH(OH)CH_2CH_3$

achiral

$CH_3CH_2\overset{*}{\underset{\text{CH}_2OH}{\overset{\text{CH}_3}{\overset{|}{C}}}}H$

chiral

$CH_3CH_2\overset{\text{CH}_3}{\underset{\text{CH}_3}{\overset{|}{C}}}OH$

achiral

$CH_3\overset{*}{C}H(OH)\overset{\text{CH}_3}{\underset{\text{CH}_3}{\overset{|}{C}}}H$

chiral

$HOCH_2CH_2\overset{\text{CH}_3}{\underset{\text{CH}_3}{\overset{|}{C}}}H$

achiral

$(CH_3)_3CCH_2OH$

achiral

9.30 Draw the five C_6H_{14} hexanes.

$CH_3CH_2CH_2CH_2CH_2CH_3$

3 kinds of –H

$CH_3CH_2CH_2CH(CH_3)_2$

5 kinds of –H

$CH_3CH_2CH(CH_3)CH_2CH_3$

4 kinds of –H

$(CH_3)_2CHCH(CH_3)_2$

2 kinds of –H

$CH_3CH_2C(CH_3)_3$

3 kinds of –H

17 monobromohexanes can be formed from the hexane isomers. You may need to draw all the bromohexanes to find those that are chiral.
The nine chiral bromohexanes:

$CH_3CH_2CH_2CH_2\overset{*}{C}HBrCH_3$

$CH_3CH_2CH_2\overset{*}{C}HBrCH_2CH_3$

$CH_3CH_2CH_2\overset{*}{C}H(CH_3)CH_2Br$

$CH_3CH_2\overset{*}{C}HBrCH(CH_3)_2$

$CH_3\overset{*}{C}HBrCH_2CH(CH_3)_2$

$CH_3CH_2\overset{*}{C}H(CH_3)CH_2CH_2Br$

$CH_3CH_2\overset{*}{C}H(CH_3)\overset{*}{C}HBrCH_3$

$(CH_3)_2CH\overset{*}{C}H(CH_3)CH_2Br$

$CH_3\overset{*}{C}HBrC(CH_3)_3$

9.31

a) $CH_3CH_2\overset{*}{C}H(OH)CH_3$

b) $CH_3CH_2\overset{*}{C}H(COOH)CH_3$

This carboxylic acid has no rings or carbon-carbon double or triple bonds. (The formula $C_5H_{10}O_2$ indicates that there is one double bond present, but it is in the carboxylic acid functional group.)

c) $CH_3\overset{*}{C}HBr\overset{*}{C}H(OH)CH_3$

d) $CH_3\overset{*}{C}HBrCHO$

9.32 Chiral: golf club, monkey wrench.

Achiral: basketball, fork, wine glass, snowflake.

9.33

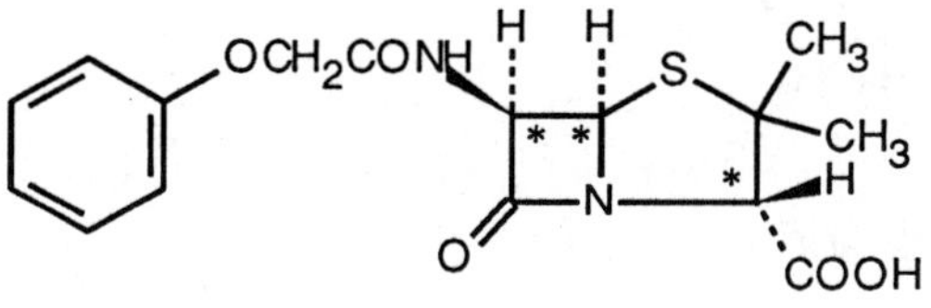

Pencillin V
three stereogenic carbons

9.34

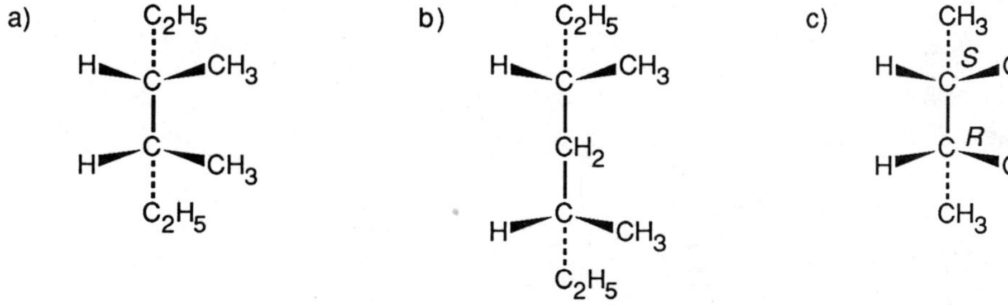

9.35 The specific rotation of (2R,3R)–2,3–dichloropentane is equal in magnitude and opposite in sign to the specific rotation of (2S,3S)–2,3–dichloropentane because the compounds are enantiomers. There is no predictable relationship between the specific rotations of the 2R,3S and 2R,3R diastereomers.

9.36–9.37

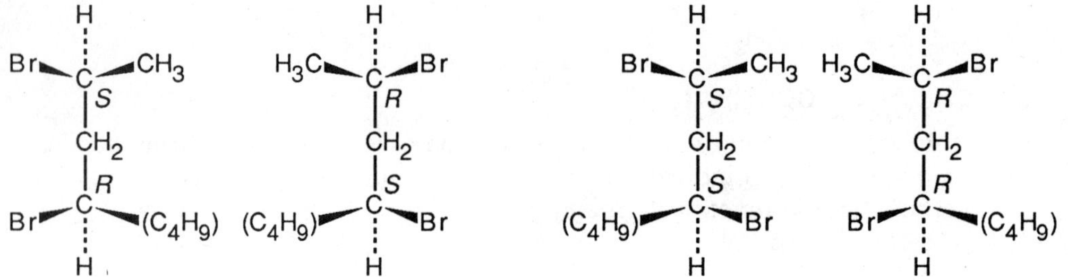

2S,4R   2R,4S   2S,4S   2R,4R

enantiomers   enantiomers

The 2R,4S stereoisomer is the enantiomer of the 2S,4R stereoisomer.
The 2S,4S and 2R,4R stereoisomers are diastereoisomers of the 2S,4R stereoisomer.

9.38

Highest priority ⟶ Lowest priority

a) –C(CH₃)₃, –CH=CH₂, –CH(CH₃)₂, –CH₂CH₃

b) , –C≡CH, –C(CH₃)₃, –CH=CH₂

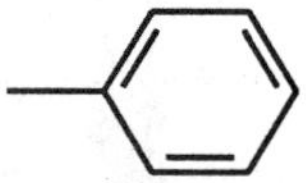

c) –COOCH₃, –COCH₃, –CH₂OCH₃, –CH₂CH₃

d) –Br, –CH₂Br, –CN, –CH₂CH₂Br

9.39

a)   b)   c)

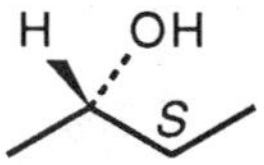

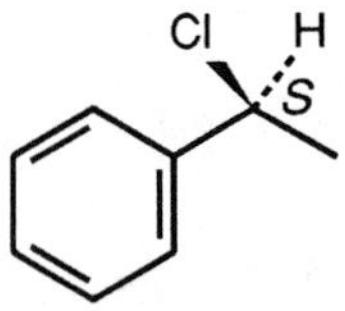

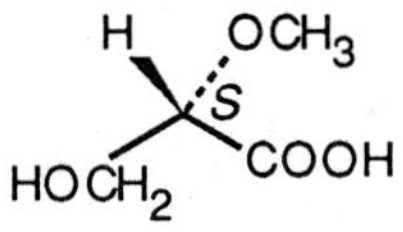

9.40

a)   b)

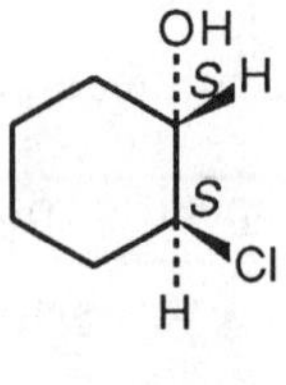

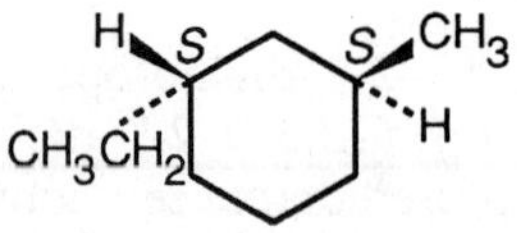

9.41

a)

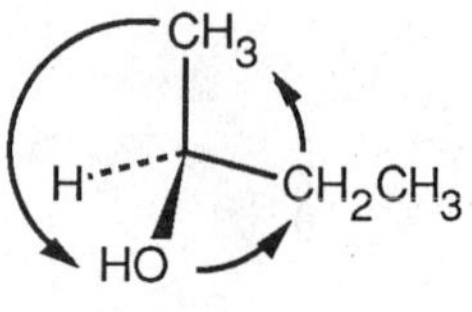

(S)–2–Butanol

b)

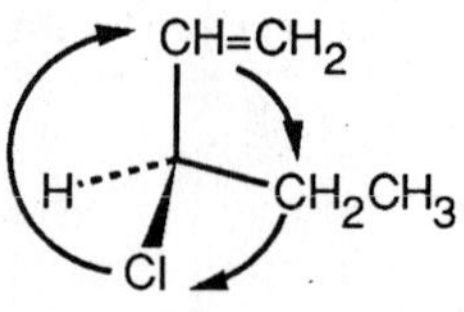

(R)–3–Chloro–1–pentene

9.42

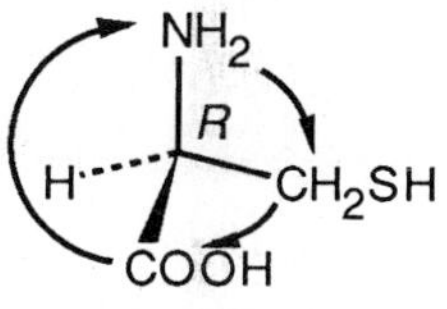

(R)–Cysteine

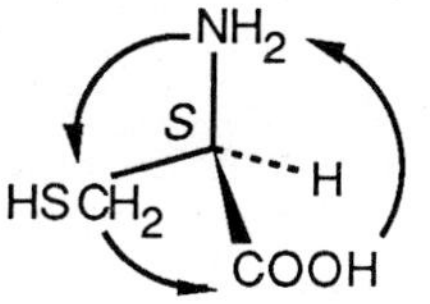

(S)–Cysteine

9.43 Identical molecules: b, c, d.
  Pair of enantiomers: a.

9.44

a)

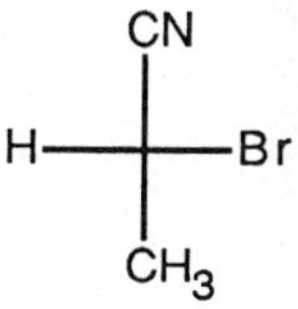

R

b)

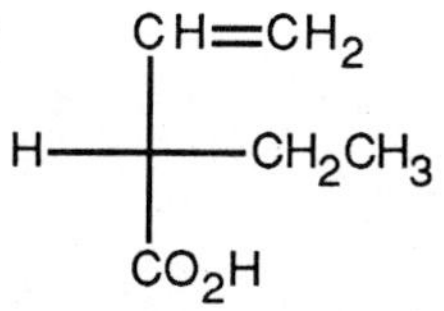

S

c)

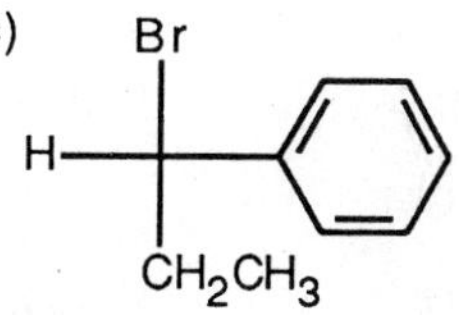

S

9.45

a)

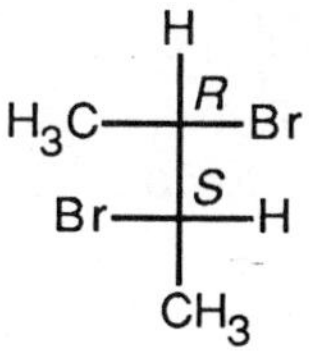

b)

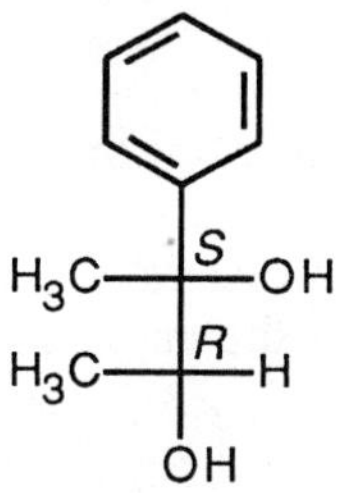

c)

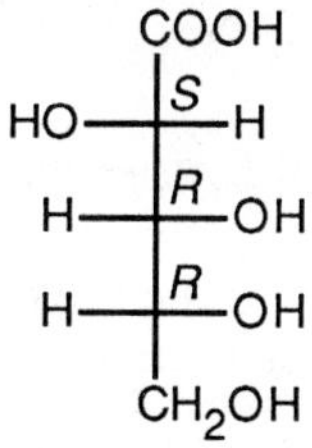

d)

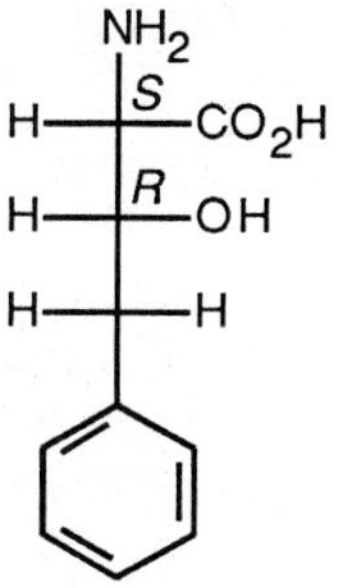

9.46

a)

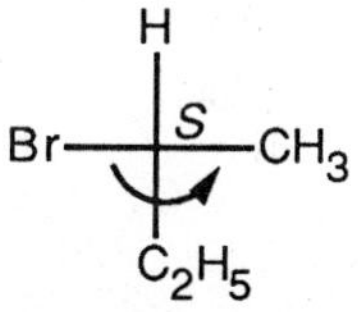

(S)–2–Bromobutane
$CH_3CH_2CHBrCH_3$

b)

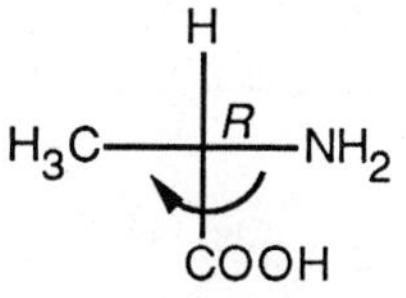

(R)–Alanine
$CH_3CH(NH_2)CO_2H$

c)

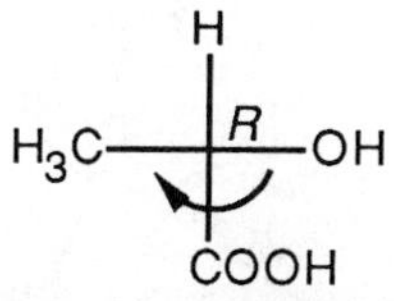

(R)–2–Hydroxypropanoic acid
$CH_3CH(OH)COOH$

d)

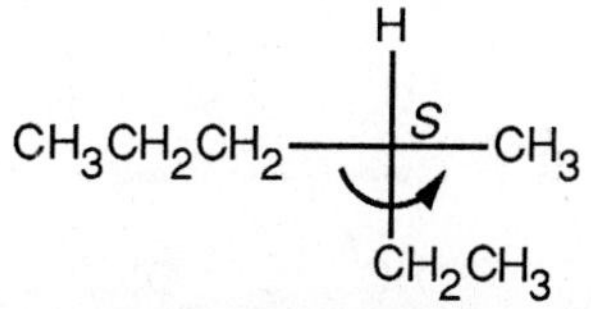

(S)–3–Methylhexane
$CH_3CH_2CH_2CH(CH_3)CH_2CH_3$

9.47

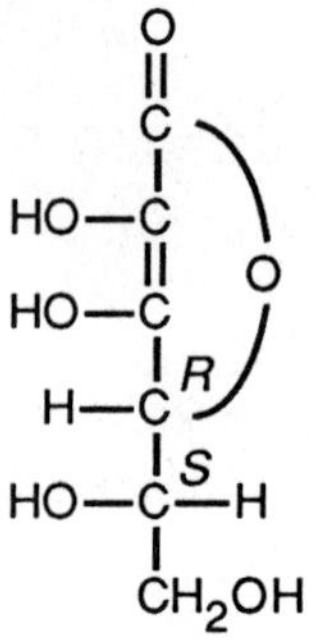

9.48

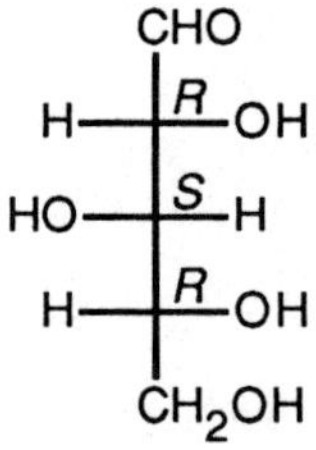

(+)-Xylose

9.49–9.50 The initial product of hydroxylation of a double bond is a cyclic *osmate*. Drawings of the osmates for *cis*–2–butene and *trans*–2–butene are shown below. No carbon-oxygen bonds are broken in the cleavage step; cleavage occurs at osmium-oxygen bonds. The final stereochemistry, therefore, is the same as that of the initial adduct.

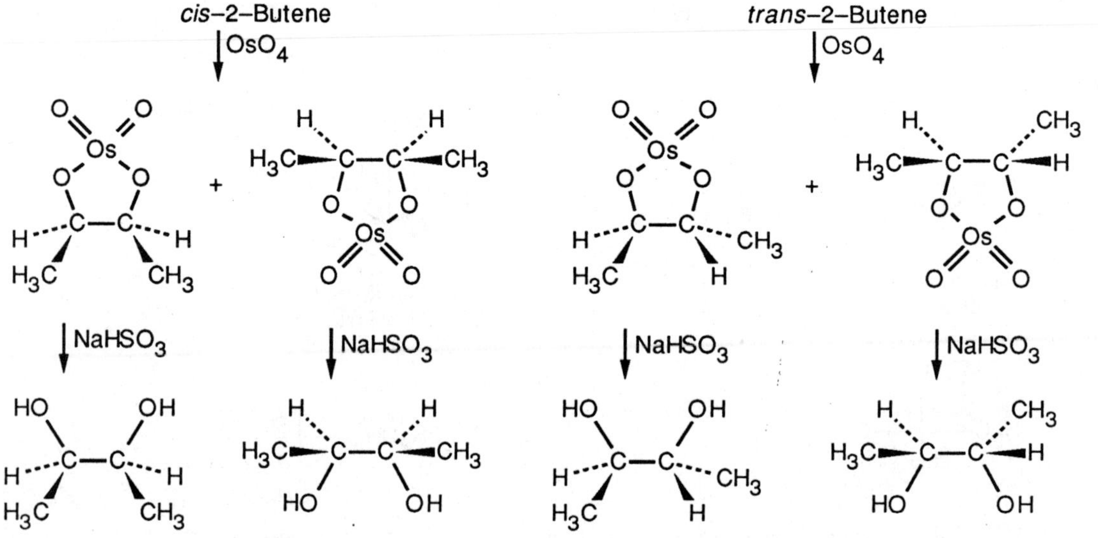

cis–2–Butene

trans–2–Butene

meso–2,3–Butanediol

(2R,3R)–2,3–Butanediol   (2S,3S)–2,3–Butanediol

racemic

9.51

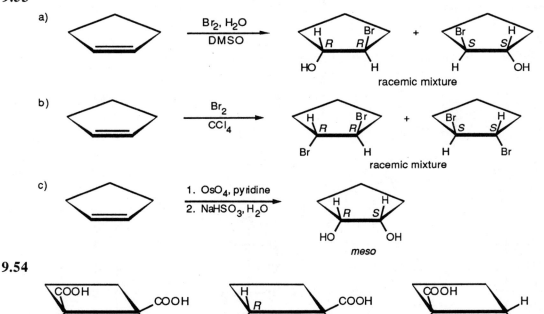

Peroxycarboxylic acids can attack either the "top" side or the "bottom" side of a double bond. The epoxide resulting from "top-side" attack on *cis*–4–octene has two stereogenic centers, but because it has a plane of symmetry, it is a *meso* compound. The two epoxides are identical.

9.52

The epoxide formed by "top-side" attack of a peroxyacid on *trans*–4–octene is pictured. This epoxide has two stereogenic centers of *R* configuration. The epoxide formed by "bottom-side" attack has *S,S* configuration. The two epoxide enantiomers are formed in equal amounts and constitute a racemic mixture.

9.53

a) Br_2, H_2O / DMSO

racemic mixture

b) Br_2 / CCl_4

racemic mixture

c) 1. OsO_4, pyridine
   2. $NaHSO_3, H_2O$

meso

9.54

A   B   C

B and C are enantiomers and are optically active. Compound A is their diastereomer and is a *meso* compound.

The two isomeric cyclobutane–1,3–dicarboxylic acids are diastereomers and are both *meso* compounds.

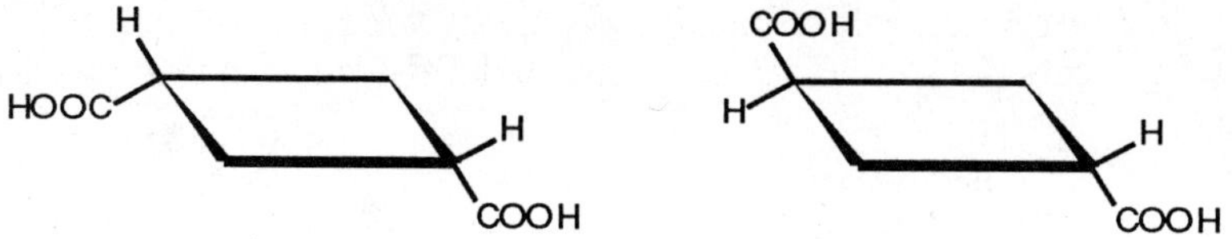

9.55

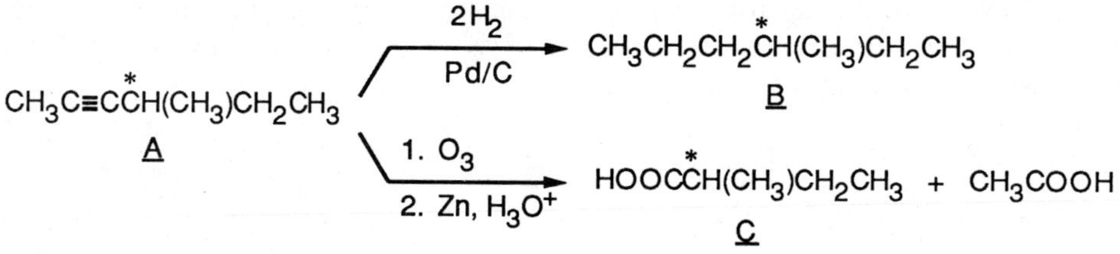

$$CH_3C\equiv CCH(CH_3)CH_2CH_3 \quad \text{(A)}$$

2 H$_2$ / Pd/C → $CH_3CH_2CH_2\overset{*}{C}H(CH_3)CH_2CH_3$ (B)

1. O$_3$; 2. Zn, H$_3$O$^+$ → $HOOC\overset{*}{C}H(CH_3)CH_2CH_3$ + CH_3COOH (C)

9.56 A has four multiple bonds/rings.

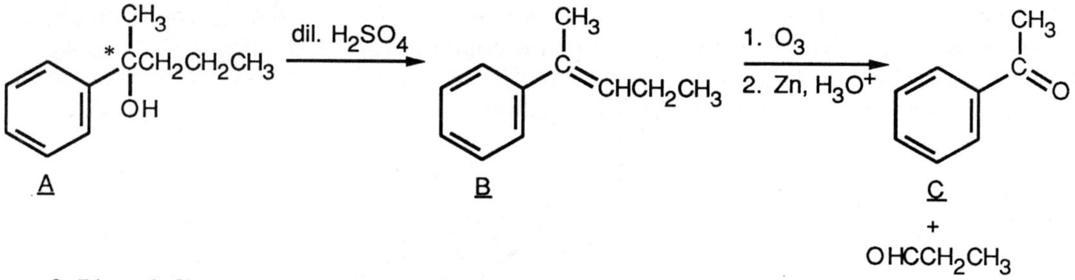

2–Phenyl–3–pentanol is also a satisfactory answer.

9.57

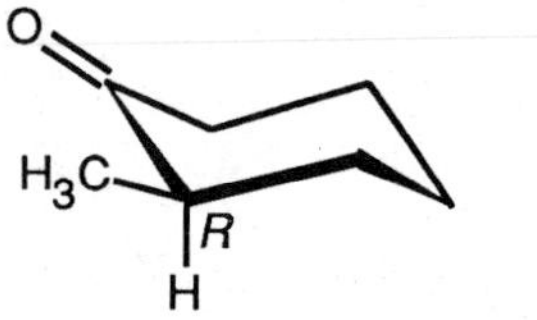

(R)–2–Methylcyclohexanone

9.58

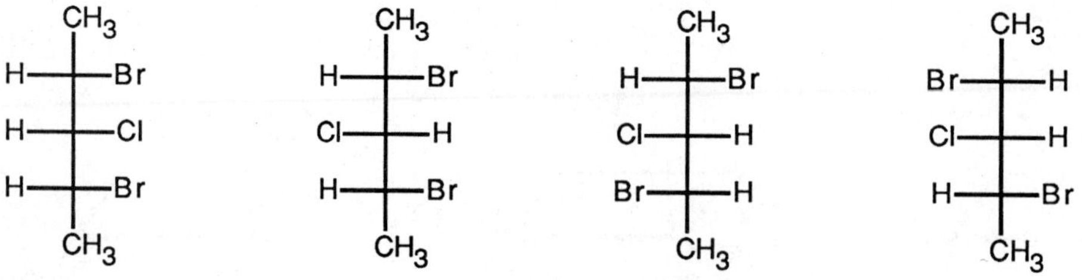

There are four stereoisomers of 2,4–dibromo–3–chloropentane. C and D are enantiomers that should be optically active. A and B are actively inactive, meso compounds and are diastereomers.

9.59 A tetrahedrane can be chiral. Notice that the orientations of the four substituents in space are the same as the orientations of the four substituents of a tetrasubstituted carbon atom. If you were able to make a model of a tetrasubstituted tetrahedrane (without having your models fall apart), you would also be able to make a model of its mirror image.

9.60 If we analyze all carbon atoms for chirality according to the rules of Problem 9.2, we find that no chiral carbon atoms are present. Yet mycomycin is chiral. How can this be?

Make a model of mycomycin. For simplicity, call –CH=CHCH=CHCH$_2$COOH "A" and –C≡CC≡CH "B". Remember from Chapter 6 that the carbon atoms of an allene are linear and that the pi bonds formed are perpendicular to each other. Attach substituents at the sp^2 carbons.

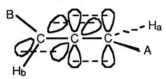

Notice that the substituents $\underline{A}$, H$_a$, and all carbon atoms lie in a plane that is perpendicular to the plane that contains B, H$_b$, and all carbon atoms.

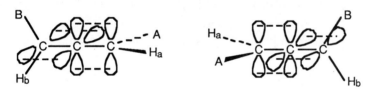

Now, make another model identical to the first, except for an exchange of $\underline{A}$ and H$_a$. This new allene is not superimposable on the original allene; in fact, the two allenes are mirror images. The two allenes are enantiomers and are chiral for the same reason that tetrasubstituted carbon atoms are chiral -- they possess no plane of symmetry.

9.61 4–Methylcyclohexylideneacetic acid is chiral for the same reason as an allene; it possesses no plane of symmetry and is not superimposable on its mirror image. As in the case of allenes, the two functional groups at one end lie in a plane perpendicular to the plane that contains the two functional groups at the other end.

9.62

a)

S-1-Chloro-
2-methylbutane

S-1,4-Dichloro-
2-methylbutane

R-1,2-Dichloro-
2-methylbutane

S-1,2-Dichloro-
2-methylbutane

50:50 mixture

b)   Chlorination at carbon 4 yields optically active product; chlorination at carbon 2 yields optically inactive product.

c)   Radical chlorination reactions taking place at a stereogenic center occur with racemization; radical chlorination reactions at a site other than the stereogenic center do not affect the stereochemistry of the stereogenic center.

9.63

50%   50%

a)   Reaction of a Grignard reagent with an achiral starting material, such as propanal, yields racemic product.

b)   The product consists of a 50:50 mixture of R-2-butanol and its enantiomer, S-2-butanol.

9.64

(2S,3R)-3-Phenyl-
2-butanol

(2R,3R)-3-Phenyl-
2-butanol

a)   Reaction of a Grignard reagent with a chiral starting material yields chiral products; the product mixture is usually optically active.

b)   The two products are a mixture of the 2S,3R and 2R,3R diastereomers of 3-phenyl-2-butanol. The product ratio can't be predicted, but it is not 50:50.

Study Guide for Chapter 9

After studying this chapter, you should be able to:

(1) Calculate the specific rotation of an optically active compound (9.5, 9.6, 9.25, 9.26).

(2) Determine if an object or a molecule is chiral (9.1, 9.2).

(3) Locate stereogenic centers in molecules (9.3, 9.13, 9.33).

(4) Assign priorities to substituents around a stereogenic carbon (9.7, 9.38).

(5) Assign *R,S* designations to stereogenic centers (9.8, 9.9, 9.10, 9.23, 9.39, 9.40).

(6) Given a stereoisomer, draw its enantiomer and/or diastereomers (9.36, 9.37, 9.54, 9.58).

(7) Decide if a stereoisomer is a meso compound and locate its plane of symmetry (9.11, 9.12).

(8) Manipulate Fischer projections to see if they are identical (9.14, 9.15, 9.43).

(9) Assign *R,S* configurations to Fischer projections (9.16, 9.44, 9.45, 9.47, 9.48).

(10) Draw Fischer projections of chiral compounds (9.46).

(11) Draw chiral molecules corresponding to a given formula (9.28, 9.29, 9.30, 9.31, 9.34, 9.41, 9.42, 9.55, 9.56, 9.57).

(12) Predict the stereochemistry of reaction products (9.17, 9.18, 9.19, 9.20, 9.21, 9.49, 9.50, 9.51, 9.52, 9.53, 9.62, 9.63, 9.64).

(13) Understand how a compound without a stereogenic carbon may be chiral (9.60, 9.61).

(14) Define the important terms in this chapter (9.26).

10.1

a) $CH_3CH_2CH_2CH_2I$ 1–Iodobutane

b) $(CH_3)_2CHCH_2CH_2Cl$ 1–Chloro–3–methylbutane

c) $BrCH_2CH_2CH_2C(CH_3)_2CH_2Br$ 1,5–Dibromo–2,2–dimethylpentane

d) $(CH_3)_2CCICH_2CH_2Cl$ 1,3–Dichloro–3–methylbutane

e) $CH_3CHICH(CH_2CH_2Cl)CH_2CH_3$ 1–Chloro–3–ethyl–4–iodopentane

f) $CH_3CHBrCH_2CH_2CHCICH_3$ 2–Bromo–5–chlorohexane

10.2

a) $CH_3CH_2CH_2C(CH_3)_2CHCICH_3$ 2–Chloro–3,3–dimethylhexane

b) $CH_3CH_2CH_2CCl_2CH(CH_3)_2$ 3,3–Dichloro–2–methylhexane

c) $CH_3CH_2CBr(CH_2CH_3)_2$ 3–Bromo–3–ethylpentane

d)   1,1–Dibromo–4–isopropylcyclohexane

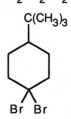

$CH_3CHCH_2CH_3$

e) $CH_3CH_2CH_2CH_2CH_2CHCH_2CHCICH_3$ 4–*sec*–Butyl–2–chlorononane

f) $C(CH_3)_3$ 1,1–Dibromo–4–*tert*–butylcyclohexane

10.3

	Product	Site of Chlorination
	$CH_3CH_2CH_2CH(CH_3)CH_2Cl$ 1–Chloro–2–methylpentane	a
	$CH_3CH_2CH_2C(CH_3)_2Cl$ 2–Chloro–2–methylpentane	b
$CH_3CH_2CH_2CH(CH_3)_2$ e d c b a 2–Methylpentane $\xrightarrow[hv]{Cl_2}$	$CH_3CH_2CHClCH(CH_3)_2$ 3–Chloro–2–methylpentane	c
	$CH_3CHClCH_2CH(CH_3)_2$ 2–Chloro–4–methylpentane	d
	$CH_2ClCH_2CH_2CH(CH_3)_2$ 1–Chloro–4–methylpentane	e

Chlorination at sites b and e yields achiral products. The products of chlorination at sites a, c and d are chiral; each product is formed as a racemic mixture of enantiomers.

10.4

$$CH_3-CH_2-\overset{\overset{\displaystyle a}{\overset{\displaystyle CH_3}{|}}}{\underset{\underset{\displaystyle b}{\underset{\displaystyle H}{|}}}{C}}-CH_3$$
d c   a

2–Methylbutane

Type of H	a	b	c	d
Number of H of each type	6	1	2	3
Relative reactivity	1.0	5.0	3.5	1.0
Number times reactivity	6.0	5.0	7.0	3.0
Percent chlorination	29%	24%	33%	14%

$CH_3CH_2CH(CH_3)_2$ $\xrightarrow[hv]{Cl_2}$

$CH_3CH_2CH(CH_3)CH_2Cl$ + $CH_3CH_2CCl(CH_3)_2$
        29%   24%

+ $CH_3CHClCH(CH_3)_2$ + $CH_2ClCH_2CH(CH_3)_2$
        33%   14%

10.5 $(CH_3)_2CH–H$ + $Cl\cdot$ → $(CH_3)_2CH\cdot$ + $H–Cl$
$\Delta H° = 95$ kcal/mol $\Delta H° = 103$ kcal/mol

$\Delta H° = 95$ kcal/mol $- 103$ kcal/mol $= -8$ kcal/mol.

$(CH_3)_2CH–H$ + $Br\cdot$ → $(CH_3)_2CH\cdot$ + $H–Br$
$\Delta H° = 95$ kcal/mol $\Delta H° = 88$ kcal/mol

$\Delta H° = 95$ kcal/mol $-$ 88 kcal/mol $= +7$ kcal/mol.

The reaction of Br• with a secondary hydrogen atom is more selective. In the endothermic reaction of $(CH_3)_2CH_2$ with Br• the transition state resembles the isopropyl radical. Since the secondary isopropyl radical is much more stable than the primary radical, the reaction is much more likely to proceed by secondary hydrogen abstraction and thus to be more selective.

10.6

Abstraction of hydrogen by a bromine radical yields an allylic radical.

The allylic radical reacts with Br_2 to produce A and B.

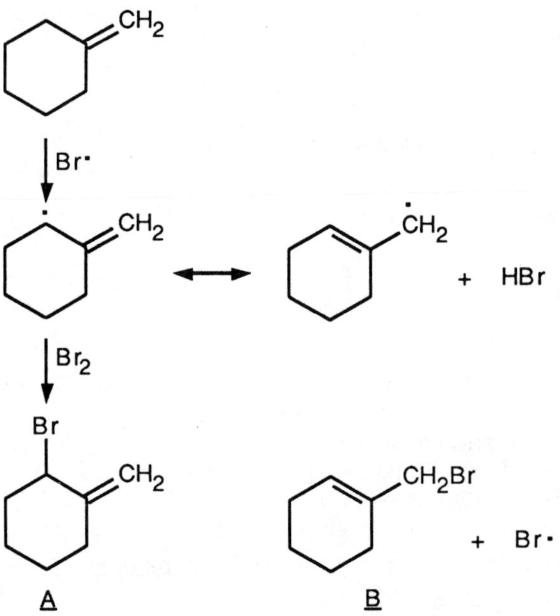

Product B, which has a trisubstituted double bond, forms in preference to product A, which has a disubstituted double bond.

10.7

a)

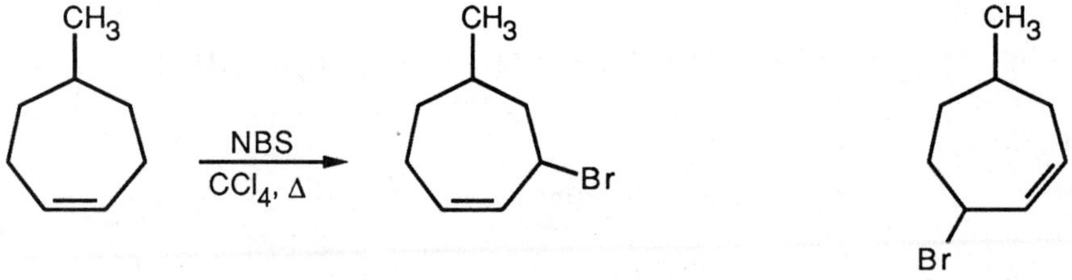

5–Methylcycloheptene   3–Bromo–5–methylcycloheptene   3–Bromo–6–methylcycloheptene

b)

$$CH_3\overset{\underset{|}{CH_3}}{\underset{\cdot}{C}}-CH=CHCH_2CH_3$$

$\updownarrow$

$$CH_3\overset{\underset{|}{CH_3}}{C}=CH-\overset{\cdot}{C}HCH_2CH_3$$

$$CH_3\overset{\underset{|}{CH_3}}{C}H-CH=CH\overset{\cdot}{C}HCH_3$$

$\updownarrow$

$$CH_3\overset{\underset{|}{CH_3}}{C}H-\overset{\cdot}{C}HCH=CHCH_3$$

$$CH_3\overset{\underset{|}{CH_3}}{C}HCH=CHCH_2CH_3 \quad \xrightarrow{\underset{radical}{Br\cdot}}$$

$$\xrightarrow{\underset{CCl_4,\ \Delta}{NBS}}$$

$$CH_3\overset{\underset{|}{CH_3}}{\underset{\underset{Br}{|}}{C}}-CH=CHCH_2CH_3$$

+

$$CH_3\overset{\underset{|}{CH_3}}{C}=CH\overset{\underset{Br}{|}}{C}HCH_2CH_3$$

+

$$CH_3\overset{\underset{|}{CH_3}}{C}H-CH=CH\overset{\underset{Br}{|}}{C}HCH_3$$

+

$$CH_3\overset{\underset{|}{CH_3}}{C}H-\overset{\underset{Br}{|}}{C}HCH=CHCH_3$$

Two different allylic radicals can form, and four different bromohexenes can be produced.

10.8 Table 5.4 shows that the bond dissociation energy for $C_6H_5CH_2-H$ is 85 kcal/mol. This value is even smaller than the bond dissociation energy for allylic hydrogens, and thus it is relatively easy to form the $C_6H_5CH_2\cdot$ radical. The high bond dissociation energy for formation of $C_6H_5\cdot$, 112 kcal/mol, indicates the bromination on the benzene ring will not occur; the only product of reaction with NBS is $C_6H_5CH_2Br$.

10.9

a)

b)

c)

10.10

a)

b)

c) 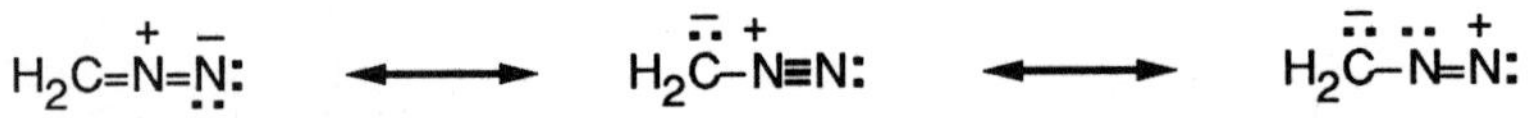 $H_2C=\overset{+}{N}=\overset{-}{\underset{\cdot\cdot}{N}}:$ ⟷ $H_2\overset{-}{C}-\overset{+}{N}\equiv N:$ ⟷ $H_2\overset{\cdot\cdot}{C}-\overset{\cdot\cdot}{\underset{\cdot\cdot}{N}}\overset{+}{\equiv}N:$

d) 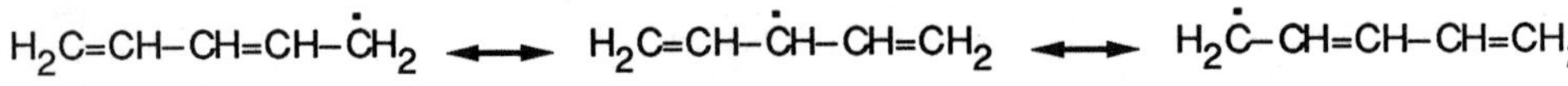 $H_2C=CH-CH=CH-\overset{\cdot}{C}H_2$ ⟷ $H_2C=CH-\overset{\cdot}{C}H-CH=CH_2$ ⟷ $H_2\overset{\cdot}{C}-CH=CH-CH=CH_2$

10.11

a) 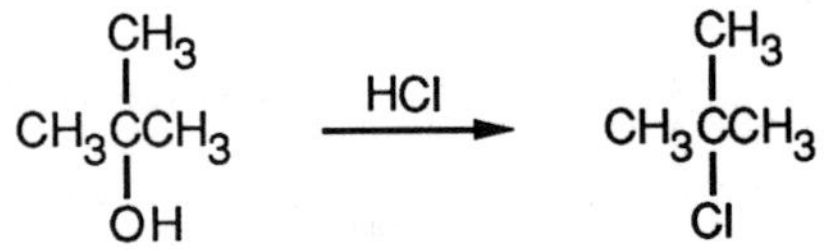 $\underset{\underset{OH}{|}}{\overset{\overset{CH_3}{|}}{CH_3CCH_3}}$ $\xrightarrow{\text{HCl}}$ $\underset{\underset{Cl}{|}}{\overset{\overset{CH_3}{|}}{CH_3CCH_3}}$

HCl is a good reagent for converting a tertiary alcohol to a tertiary chloride.

b) 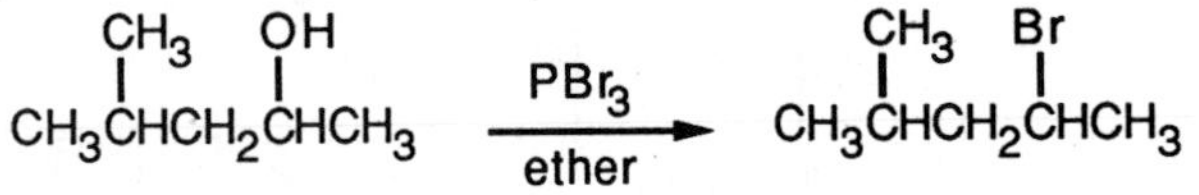 $\underset{}{\overset{\overset{CH_3\quad OH}{|\qquad|}}{CH_3CHCH_2CHCH_3}}$ $\xrightarrow[\text{ether}]{\text{PBr}_3}$ $\underset{}{\overset{\overset{CH_3\quad Br}{|\qquad|}}{CH_3CHCH_2CHCH_3}}$

Use PBr₃ for converting secondary alcohols to alkyl bromides.

c) $HOCH_2CH_2CH_2CH_2CH(CH_3)_2$ $\xrightarrow[\text{ether}]{\text{PBr}_3}$ $BrCH_2CH_2CH_2CH_2CH(CH_3)_2$

d) 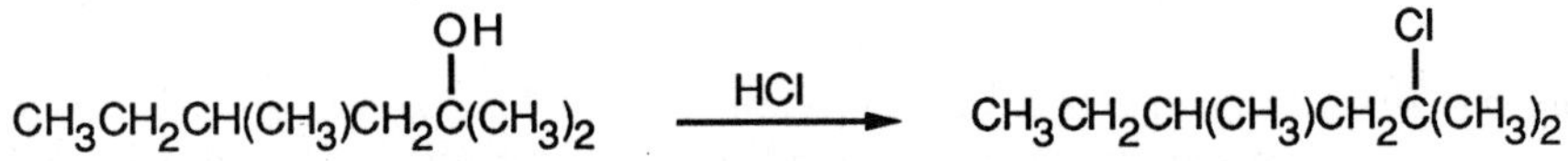 $CH_3CH_2CH(CH_3)CH_2\overset{\overset{OH}{|}}{C}(CH_3)_2$ $\xrightarrow{\text{HCl}}$ $CH_3CH_2CH(CH_3)CH_2\overset{\overset{Cl}{|}}{C}(CH_3)_2$

10.12 Table 8.2 shows that the pK_a of CH₃–H is 60. Since CH₄ is a very weak acid, $:\overset{-}{C}H_3$ is a very strong base. Alkyl Grignard reagents have pK_as close to that of $:\overset{-}{C}H_3$; alkenyl Grignard reagents are somewhat weaker bases. Both reactions (a) and (b) occur as written because of the extreme base strength of $:\overset{-}{C}H_3$.

a) CH_3MgBr + $H-C\equiv C-H$ $\longrightarrow$ CH_4 + $H-C\equiv C-MgBr$

       stronger   stronger   weaker   weaker
         base   acid   acid   base

b) CH_3MgBr + NH_3 $\longrightarrow$ CH_4 + $H_2N-MgBr$

       stronger   stronger   weaker   weaker
         base   acid   acid   base

10.13 If Grignard reagents react with proton donors to convert R–MgX into R–H, they will also react with *deuterium* donors to convert R–MgX into R–D. In this case:

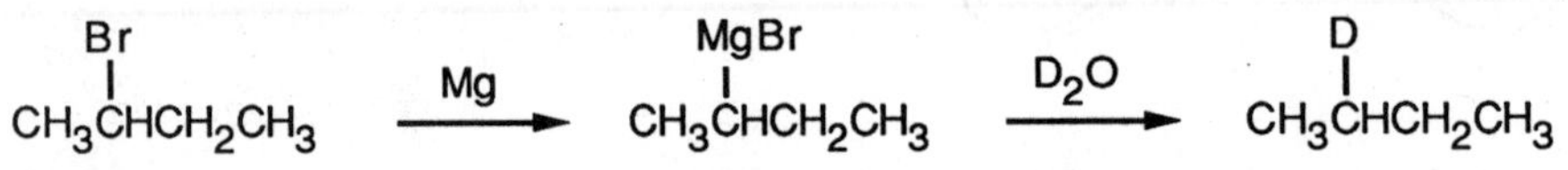

$\underset{}{\overset{\overset{Br}{|}}{CH_3CHCH_2CH_3}}$ $\xrightarrow{\text{Mg}}$ $\underset{}{\overset{\overset{MgBr}{|}}{CH_3CHCH_2CH_3}}$ $\xrightarrow{\text{D}_2\text{O}}$ $\underset{}{\overset{\overset{D}{|}}{CH_3CHCH_2CH_3}}$

10.14 A Grignard reagent can't be prepared from a compound containing a functional group that is a good proton donor. As the Grignard reagent starts to form, it is immediately quenched by the proton source. The –COOH, –OH, and –NH₂ functional groups are too acidic to be used for preparation of a Grignard reagent. Terminal alkynes are also too acidic; for example $BrCH_2C\equiv CH$ does not form a Grignard reagent.

10.15

a)

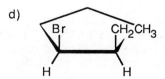

Br → NBS / CCl₄ → (CH₃)₂CuLi → CH₃

3–Methylcyclohexene

b) $2\ CH_3CH_2CH_2CH_2Br \xrightarrow[\text{pentane}]{4\ Li} 2\ CH_3CH_2CH_2CH_2Li$

$+\ 2\ LiBr \ \Big\downarrow \begin{array}{l}CuI\\ \text{ether}\end{array}$

$CH_3CH_2CH_2CH_2CH_2CH_2CH_2CH_3 \xleftarrow[\text{ether}]{CH_3CH_2CH_2CH_2Br} (CH_3CH_2CH_2CH_2)_2CuLi$
Octane

c) $CH_3CH_2CH_2CH=CH_2 \xrightarrow[\text{peroxides}]{HBr} CH_3CH_2CH_2CH_2CH_2Br$

$2\ CH_3CH_2CH_2CH_2CH_2Br \xrightarrow[\text{pentane}]{4\ Li} 2\ CH_3(CH_2)_3CH_2Li\ +\ 2\ LiBr$

$\Big\downarrow \begin{array}{l}CuI\\ \text{ether}\end{array}$

$CH_3(CH_2)_8CH_3 \xleftarrow[\text{ether}]{CH_3(CH_2)_3CH_2Br} [CH_3(CH_2)_3CH_2]_2CuLi$
Decane

10.16

a) $(CH_3)_2CHCHBrCHBrCH_2CH(CH_3)_2$ 3,4–Dibromo–2,6–dimethylheptane

b) $CH_3CH=CHCH_2CHICH_3$ 5–Iodo–2–hexene

c) $(CH_3)_2CBrCH_2CHClCH(CH_3)_2$ 2–Bromo–4–chloro–2,5–dimethylhexane

d) $CH_3CH_2CH(CH_2Br)CH_2CH_2CH_3$ 3–(Bromomethyl)hexane

e) $ClCH_2CH_2CH_2C\equiv CCH_2Br$ 1–Bromo–6–chloro–2–hexyne

10.17

a) $CH_3CH_2CH(CH_3)CHClCHClCH_3$ 2,3–Dichloro–4–methylhexane

b) $(CH_3CH_2)_2CBrCH_2CH(CH_3)_2$ 4–Bromo–4–ethyl–2–methylhexane

c) $(CH_3)_3CCHIC(CH_3)_3$ 3–Iodo–2,2,4,4–tetramethylpentane

d)   *cis*–1–Bromo–2–ethylcyclopentane

Br   CH₂CH₃

H   H

10.18 Abstraction of hydrogen by Br· can produce either of two allylic radicals. The first radical, resulting from abstraction of a secondary hydrogen, is more likely to be formed.

$$CH_3\overset{\cdot}{C}HCH{=}CHCH_3 \quad\longleftrightarrow\quad CH_3CH{=}CH\overset{\cdot}{C}HCH_3$$

and

(identical resonance forms)

$$CH_3CH_2CH{=}CH\overset{\cdot}{C}H_2 \quad\longleftrightarrow\quad CH_3CH_2\overset{\cdot}{C}HCH{=}CH_2$$

Reaction of the radical intermediates with a bromine source leads to a mixture of products:

$CH_3CH_2CHBrCH{=}CH_2$ 3–Bromo–1–pentene

and

$CH_3CHBrCH{=}CHCH_3$ cis– and trans–4–Bromo–2–pentene

and

$CH_3CH_2CH{=}CHCH_2Br$ cis– and trans–1–Bromo–2–pentene

The major product is 4–bromo–2–pentene, instead of the desired product, 1–bromo–2–pentene.

10.19 Three different allylic radical intermediates can be formed. Bromination of these intermediates can yield as many as five bromoalkenes. This is definitely not a good reaction to use in a synthesis.

(allylic; secondary hydrogen abstracted)   3–Bromo–2–methylcyclohexene

(allylic; secondary hydrogen abstracted)   3–Bromo–1–methylcyclohexene

3–Bromo–3–methylcyclohexene

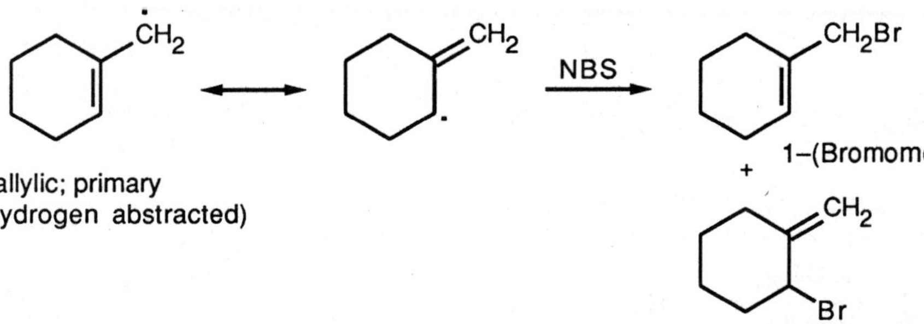

(allylic; primary hydrogen abstracted)

1–(Bromomethyl)cyclohexene

2–Bromomethylenecyclohexane

10.20

a)

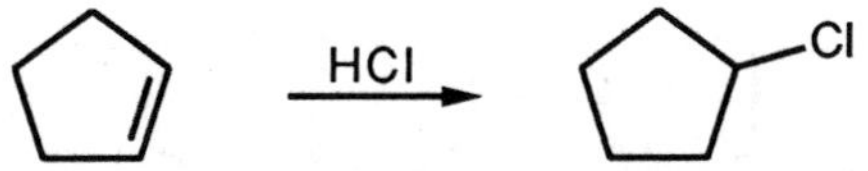

HCl

Chlorocyclopentane

b)

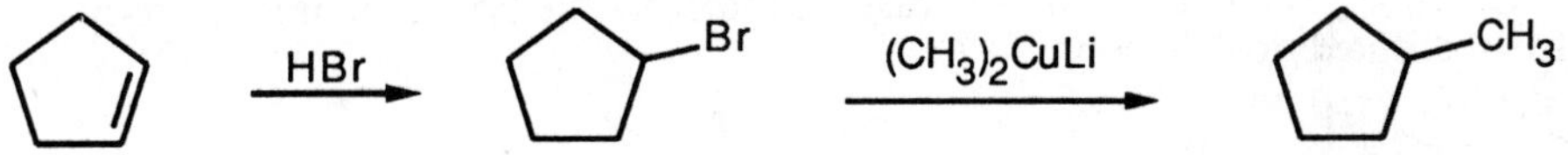

HBr

$(CH_3)_2CuLi$

Methylcyclopentane

c)

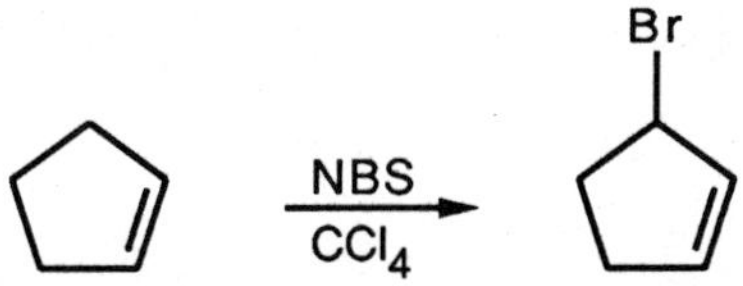

NBS / CCl_4

3–Bromocyclopentene

d)

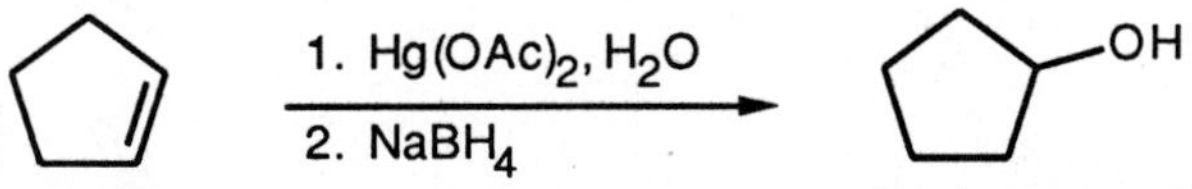

1. $Hg(OAc)_2$, H_2O
2. $NaBH_4$

Cyclopentanol

e)

HBr

2 Li / Pentane

2

CuI / ether

ether

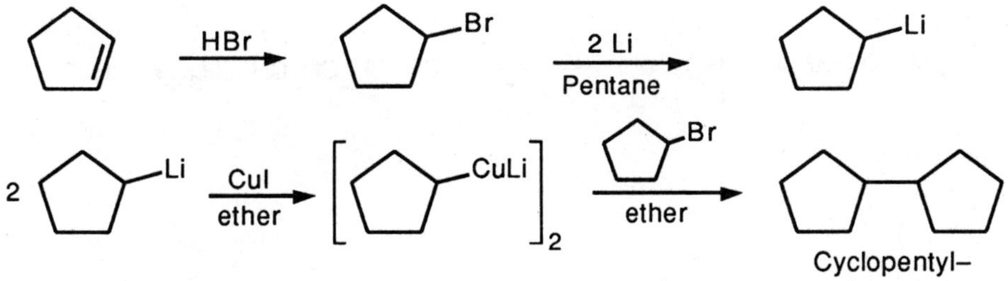

Cyclopentyl–cyclopentane

f)

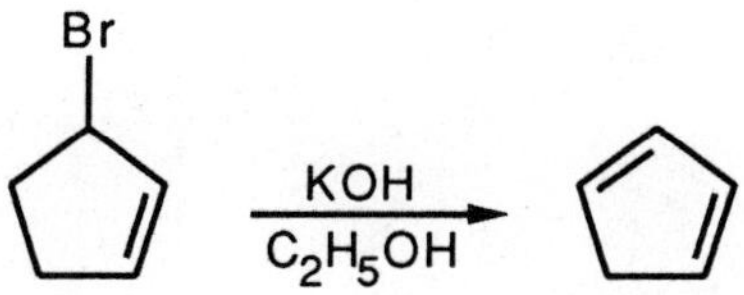

KOH / C_2H_5OH

(from c)

1,3–Cyclopentadiene

10.21

a)

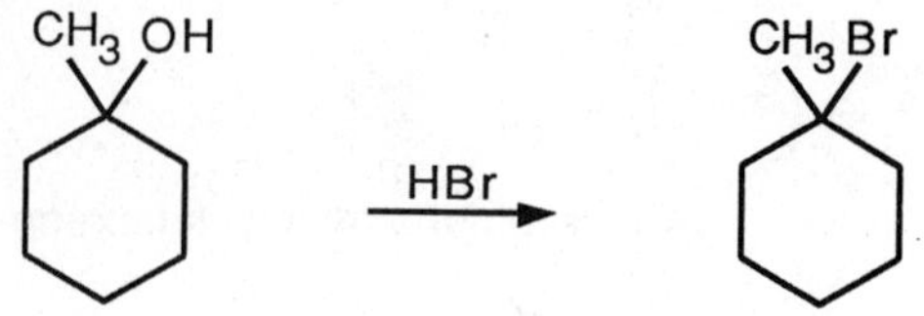

HBr

This is a good method for converting a tertiary alcohol to a bromide.

b) $CH_3CH_2CH_2CH_2OH$ $\xrightarrow[\text{pyridine}]{\text{SOCl}_2}$ $CH_3CH_2CH_2CH_2Cl$

c)

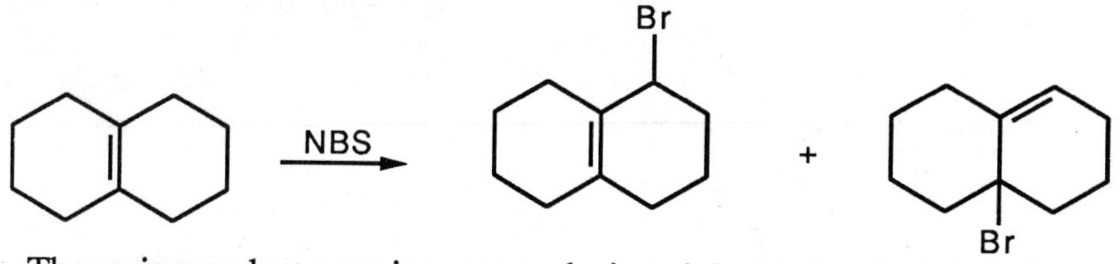

NBS

+

The major product contains a tetrasubstituted double bond; the minor product contains a trisubstituted double bond.

d)

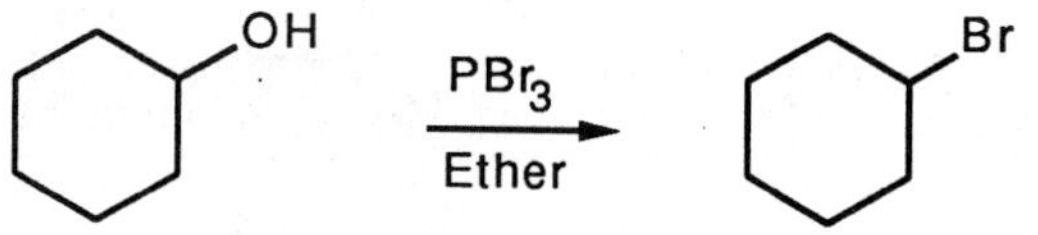

$\xrightarrow[\text{Ether}]{\text{PBr}_3}$

This is a good method for converting a primary or secondary alcohol to a bromide.

e) $CH_3CH_2CHBrCH_3$ $\xrightarrow[\text{Ether}]{\text{Mg}}$ $\underset{A}{CH_3CH_2\overset{\displaystyle MgBr}{\underset{|}{C}}HCH_3}$ $\xrightarrow{H_2O}$ $\underset{B}{CH_3CH_2CH_2CH_3}$

f) $2\ CH_3CH_2CH_2CH_2Br$ $\xrightarrow[\text{Pentane}]{\text{4 Li}}$ $\underset{A}{2\ CH_3CH_2CH_2CH_2Li}$ $\xrightarrow{\text{CuI}}$ $\underset{B}{(CH_3CH_2CH_2CH_2)_2CuLi}$

g) $CH_3CH_2CH_2CH_2Br$ + $(CH_3)_2CuLi$ $\xrightarrow{\text{Ether}}$ $CH_3CH_2CH_2CH_2CH_3$ + CH_3Cu + $LiBr$

10.22

$$CH_3CH_2CH_2\overset{\overset{CH_3}{|}}{C}HCH_3 \xrightarrow[h\nu]{Cl_2}$$

$$CH_3CH_2CH_2\overset{\overset{CH_3}{|}}{\underset{*}{C}}HCH_2Cl \quad + \quad CH_3CH_2CH_2\overset{\overset{CH_3}{|}}{\underset{\underset{Cl}{|}}{C}}CH_3$$

1–Chloro–2–methylpentane   2–Chloro–2–methylpentane

+

$$CH_3CH_2\overset{\overset{CH_3}{|}}{\underset{*}{C}}H\overset{\overset{}{}}{\underset{\underset{Cl}{|}}{C}}HCH_3 \quad + \quad CH_3\overset{\overset{Cl}{|}}{\underset{*}{C}}HCH_2\overset{\overset{CH_3}{|}}{C}HCH_3$$

3–Chloro–2–methylpentane   2–Chloro–4–methylpentane

+

$$ClCH_2CH_2CH_2\overset{\overset{CH_3}{|}}{C}HCH_3$$

1–Chloro–4–methylpentane

Three of the above products are chiral (stereogenic centers are starred). None of the products are optically active; each chiral product is a racemic mixture.

10.23

Abstraction of a proton at the chiral center of S–3–methyl–hexane produces an achiral radical intermediate, which reacts with bromine to form a 1:1 mixture of R and S enantiomeric, chiral bromoalkanes. The product mixture is optically inactive.

10.24

+ other products

Abstraction of a proton from carbon 4 yields a chiral radical intermediate. Reaction of this intermediate with chlorine does not occur with equal probability from each side, and the two diastereomeric products are not formed in 1:1 ratio. The first product is optically active, and the second product is a meso compound.

10.25

for Cl (1) Cl• + CH₃–H → CH₃• + H–Cl

104 kcal/mol   103 kcal/mol

ΔH° = 104 kcal/mol – 103 kcal/mol = +1 kcal/mol

(2) $CH_3\cdot$ + Cl–Cl → Cl· + CH_3–Cl
       58 kcal/mol   84 kcal/mol

$\Delta H° = 58\,\text{kcal/mol} - 84\,\text{kcal/mol} = -26\,\text{kcal/mol}$
$\Delta H°_{total} = +1\,\text{kcal/mol} - 26\,\text{kcal/mol} = -25\,\text{kcal/mol}$

for Br· (1) Br· + CH_3–H → $CH_3\cdot$ + H–Br
            104 kcal/mol   88 kcal/mol

$\Delta H° = 104\,\text{kcal/mol} - 88\,\text{kcal/mol} = +16\,\text{kcal/mol}$

(2) $CH_3\cdot$ + Br–Br → Br· + CH_3–Br
       46 kcal/mol   70 kcal/mol

$\Delta H° = 46\,\text{kcal/mol} - 70\,\text{kcal/mol} = -24\,\text{kcal/mol}$
$\Delta H°_{total} = +16\,\text{kcal/mol} - 24\,\text{kcal/mol} = -8\,\text{kcal/mol}$

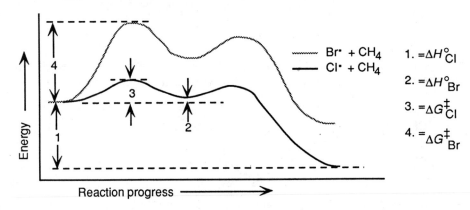

Br· + CH_4 1. $= \Delta H°_{Cl}$
Cl· + CH_4 2. $= \Delta H°_{Br}$
                3. $= \Delta G^{\ddagger}_{Cl}$
                4. $= \Delta G^{\ddagger}_{Br}$

Reaction between Cl· and CH_4 is likely to be faster. ΔH_o for formation of the $CH_3\cdot$ intermediate is lower for chlorination than for bromination, and thus $\Delta G^{\ddagger}$ is likely to be lower also, according to the Hammond Postulate.

10.26

10.27 Two allylic radicals can form:

and

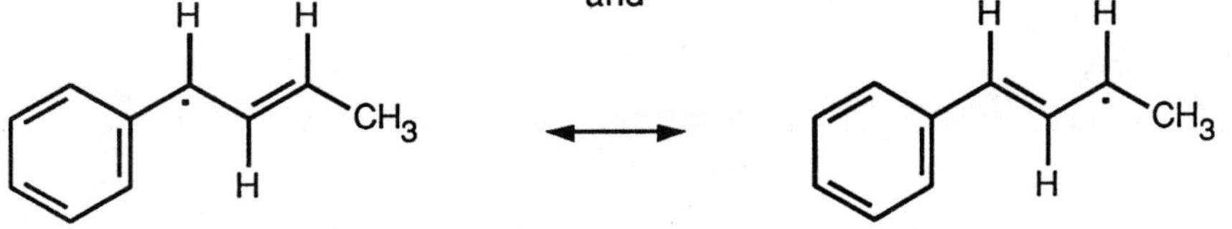

The second radical is much more likely to form because it is both allylic and benzylic, and it yields the following products:

+

10.28 Resonance forms do not differ in the position of nuclei. The two structures in (a) are not resonance forms; the carbon and hydrogen atoms outside the ring occupy different positions in each structure.

not resonance structures

The pairs of structures in parts (b), (c), and (d) are resonance forms.

10.29

a)

b)

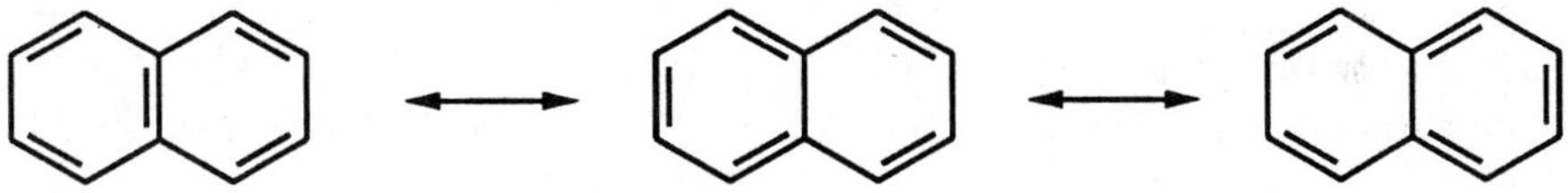

c)

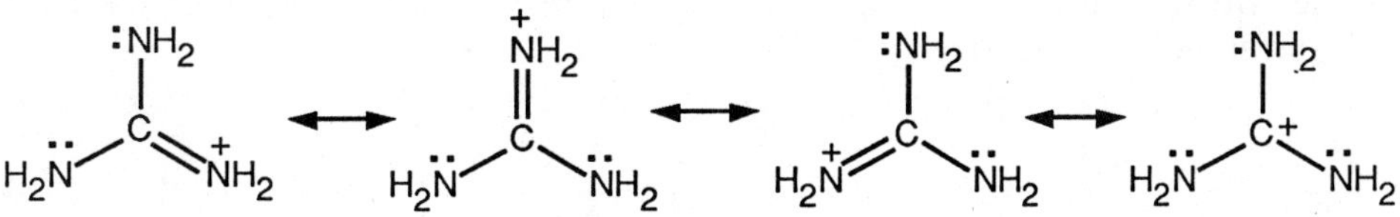

d) $H_3C-\overset{..}{\underset{..}{S}}-\overset{+}{C}H_2 \longleftrightarrow H_3C-\overset{+}{\underset{..}{S}}=CH_2$

e) $H_2C=CH-\overset{+}{C}H_2 \longleftrightarrow \overset{+}{H_2C}-CH=CH_2$

f) $H_2C=CH-CH=CH-\overset{+}{C}H-CH_3 \longleftrightarrow H_2C=CH-\overset{+}{C}H-CH=CH-CH_3$

$$\updownarrow$$

$$\overset{+}{H_2C}-CH=CH-CH=CH-CH_3$$

10.30

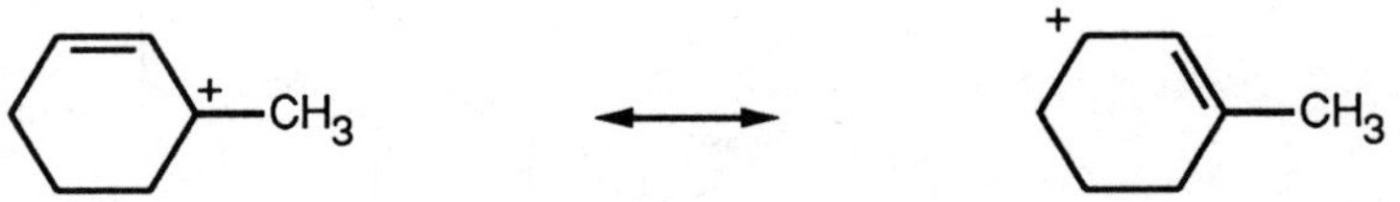

A	B
tertiary carbocation	secondary carbocation
disubstituted double bond	trisubstituted double bond

In general, tertiary carbocations (as in A) are more stable than secondary carbocations (as in B). However, trisubstituted double bonds (as in B) are more stable than disubstituted double bonds (as in A). Since these two facts contradict each other, it's not possible to predict which of the two resonance forms predominates unless additional information is given. (Actually, A predominates.)

10.31 All of these reactions involve addition of a dialkylcopper reagent $((CH_3CH_2CH_2CH_2)_2CuLi)$ to an alkyl halide. The dialkylcopper is prepared by treating 1--bromobutane with lithium, followed by addition of CuI:

$$2\ CH_3CH_2CH_2CH_2Br \xrightarrow[\text{Pentane}]{2\ Li} 2\ CH_3CH_2CH_2CH_2Li \xrightarrow[\text{Ether}]{CuI} (CH_3CH_2CH_2CH_2)_2CuLi$$

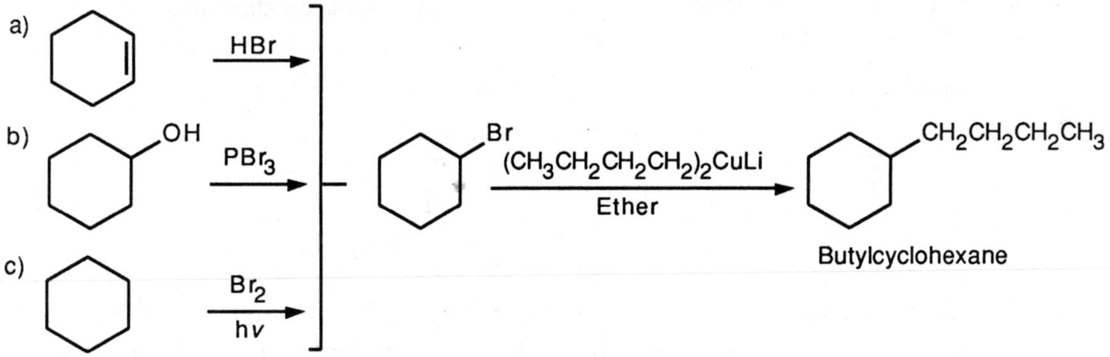

a) HBr

b) OH PBr₃

c) Br₂ hv

(CH₃CH₂CH₂CH₂)₂CuLi Ether

Butylcyclohexane

10.32 The two structures are not resonance forms because the position of the carbon atoms is different in each form.

10.33 a) Fluoroalkanes do not normally form Grignard reagents.

b) Two allylic radicals can be produced.

(1)   (2)

(3)   (4)

Instead of a single product, as many as four bromide products may result.

c) Dialkylcopper reagents do not react with fluoroalkanes.

10.34 Pairs (a) and (d) represent resonance structures; pairs (b) and (c) do not. For (b) and (c), a proton differs in position in each structure of the pair. The structures in (b) and (c) represent constitutional isomers, rather than resonance structures.

Study Guide for Chapter 10

After studying this chapter, you should be able to:

(1)   Draw and name alkyl halides (10.1, 10.2, 10.16, 10.19).

(2)   Understand the mechanism of radical halogenation and the stability order of radicals (10.3, 10.4, 10.5, 10.6, 10.7, 10.8, 10.18, 10.19, 10.22, 10.23, 10.24, 10.25, 10.27).

(3)   Draw resonance structures (10.9, 10.10, 10.26, 10.28, 10.29, 10.30, 10.32, 10.34).

(4)   Prepare alkyl halides (10.11, 10.20, 10.21).

(5)   Prepare Grignard reagents and dialkylcopper reagents and use them in syntheses (10.12, 10.13, 10.14, 10.15, 10.31, 10.33).

11.1

11.2

11.3 If back-side attack were necessary for S_N2 reaction, a molecule with a hindered "back-side" would not be able to react by an S_N2 mechanism. In this problem approach by the hydroxide ion from the back side of the bromoalkane is blocked by the rigid ring system, and displacement can't occur.

11.4

a) $CH_3CH_2CH_2CH_2Br$ + NaI $\longrightarrow$ $CH_3CH_2CH_2CH_2I$

b) $CH_3CH_2CH_2CH_2Br$ + KOH $\longrightarrow$ $CH_3CH_2CH_2CH_2OH$

c) $CH_3CH_2CH_2CH_2Br$ + HC≡C⁻ Li⁺ $\longrightarrow$ $CH_3CH_2CH_2CH_2C≡CH$

d) $CH_3CH_2CH_2CH_2Br$ + NH_3 $\longrightarrow$ $CH_3CH_2CH_2CH_2\overset{+}{N}H_3Br^-$

11.5 Since the pK_b's of triethylamine (10.75) and quinuclidine (10.95) are similar, the difference in reaction rate must be due to a factor other than basicity.

Build molecular models of triethylamine and quinuclidine. A model of the most stable conformation of triethylamine shows that the ethyl groups interfere with approach of the nitrogen lone pair electrons to iodomethane. In quinuclidine, however, the hydrocarbon framework is rigidly held back from the nitrogen lone pair. It is sterically easier for quinuclidine to approach methyl iodide, and reaction occurs at a faster rate.

11.6 a) $(CH_3)_2N:^-$ is more nucleophilic. A negatively charged reagent is more nucleophilic than its conjugate acid.

b) $(CH_3)_3N$ is more nucleophilic than $(CH_3)_3B$. $(CH_3)_3B$ is non-nucleophilic because it has no lone electron pair.

c) Since nucleophilicity increases in going down a column of the periodic table, H_2S is more nucleophilic than H_2O.

11.7 In this problem, we are comparing two effects -- the effect of the substrate and the effect of the leaving group.

Most reactive $\longrightarrow$ Least reactive

$CH_3OTos > CH_3Br > (CH_3)_2CHCl >> (CH_3)_3CCl$

11.8

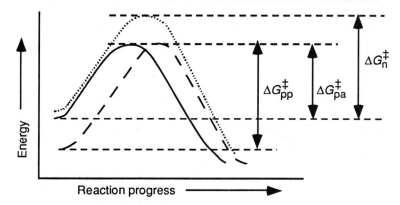

Polar protic solvents (dashed line) increase $\Delta G^{\ddagger}$ for S_N2 reactions by hydrogen-bonding to the nucleophile and lowering its energy level.

Polar aprotic solvents (solid line) do not hydrogen-bond to the nucleophile. Since the energy level of the nucleophile is higher, $\Delta G^{\ddagger}$ is smaller.

Non-polar solvents (dotted line) don't lower the energy level of the nucleophile. In addition, non-polar solvents can't lower the energy level of the charged transition state. Because of these two factors, $\Delta G^{\ddagger}$ for non-polar solvents is higher, and the reaction rate is slower.

11.9

Attack by acetate can occur on either side of the planar, achiral carbocation intermediate. The resulting acetate product is a racemic mixture.

11.10 If reaction had proceeded with complete inversion, the product would have had a specific rotation of +53.6°. If complete racemization had occurred, $[\alpha]_D$ would have been zero.

The observed rotation was +5.3°. Since $\dfrac{+5.3°}{+53.6°} = 0.099$, 9.9% of the original tosylate was inverted. The remaining 90.1% of the product must have been racemized.

11.11 S_N1 reactivity is related to carbocation stability. Thus, substrates that form the most stable carbocations are the most reactive in S_N1 reactions.

Most Reactive ──────────────⟶ Least Reactive

$H_2C=CHCHBrCH_3$ > $CH_3CHBrCH_3$ > CH_3CH_2Br > $H_2C=CHBr$

      allylic   secondary   primary   vinylic

11.12

$CH_3CHBrCH=CH_2 \rightleftharpoons \left[CH_3\overset{+}{C}HCH=CH_2 \longleftrightarrow CH_3CH=CH\overset{+}{C}H_2 \right] \rightleftharpoons CH_3CH=CHCH_2Br$

The two bromobutenes undergo S_N1 reaction at the same rate because they both form the same allylic carbocation.

11.13

1–Chloro–1,2–diphenylethane

Nucleophilic substitution of 1–chloro–1,2–diphenylethane proceeds via an S_N1 mechanism because of stabilization of the carbocation intermediate by the phenyl group at C-1. In an S_N1 reaction, the rate-limiting step is carbocation formation; all subsequent steps occur at a faster rate and do not affect the rate of reaction. After carbocation formation, the rate does not depend on the identity of the nucleophile that combines with the carbocation. In this example, $F:^-$ and $(CH_3CH_2)_3N:$ react at the same rate.

11.14 The first step in an S_N1 displacement is ionization of the substrate to form a planar, sp^2–hybridized carbocation and a leaving group. The carbocation that would form from ionization of this haloalkane can't become planar because of the rigid structure of the rest of the molecule. Because it's not possible to form the necessary carbocation, an S_N1 reaction can't occur.

11.15 The major elimination product in each case has the most substituted double bond.

a) $CH_3CH_2\overset{Br}{\overset{|}{C}H}\overset{CH_3}{\overset{|}{C}H}CH_3 \longrightarrow CH_3CH_2CH=\overset{CH_3}{\overset{|}{C}}CH_3 + CH_3CH=CH\overset{CH_3}{\overset{|}{C}H}CH_3$

                                              major   minor

                                (trisubstituted double bond) (disubstituted double bond)

b)

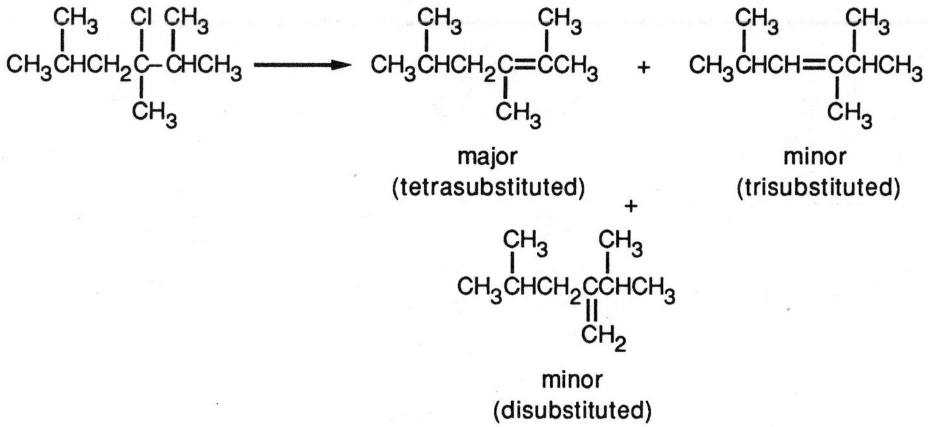

CH$_3$CHCH$_2$C—CHCH$_3$ ⟶ CH$_3$CHCH$_2$C=CCH$_3$ + CH$_3$CHCH=CCHCH$_3$

major
(tetrasubstituted)

minor
(trisubstituted)

+

CH$_3$CHCH$_2$CCHCH$_3$

minor
(disubstituted)

c)

major
(trisubstituted)

minor
(monosubstituted)

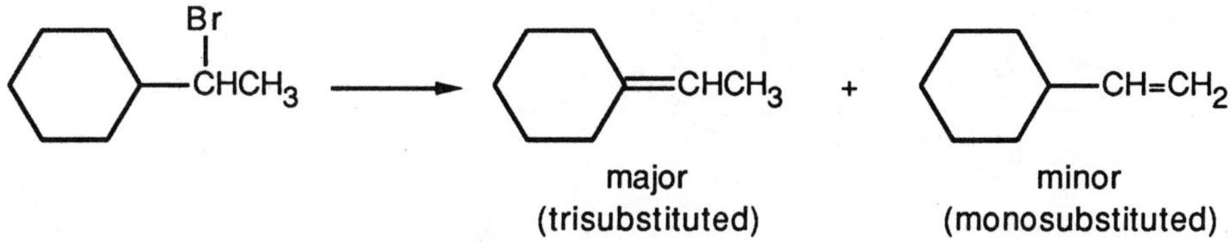

11.16

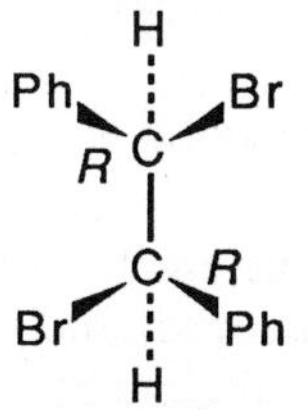

(1R,2R)–1,2–Dibromo–1,2–diphenylethane

Convert this drawing into a Newman projection and draw the conformation having *anti-periplanar* geometry for –H and –Br.

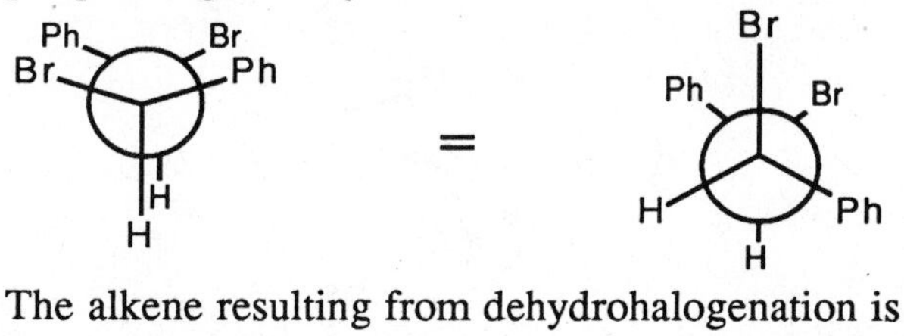

=

The alkene resulting from dehydrohalogenation is (Z)–1–bromo–1,2–diphenylethylene.

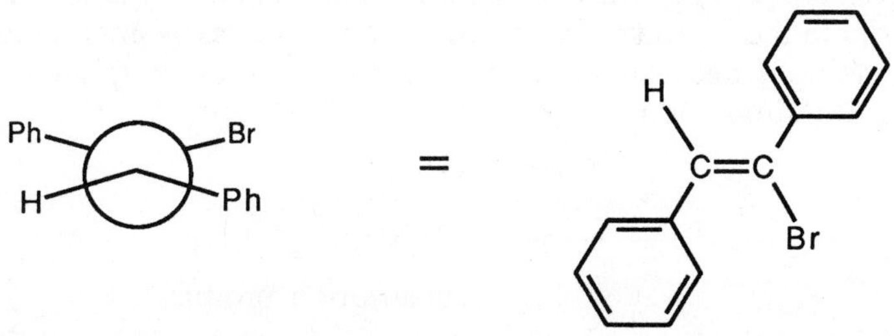

=

11.17

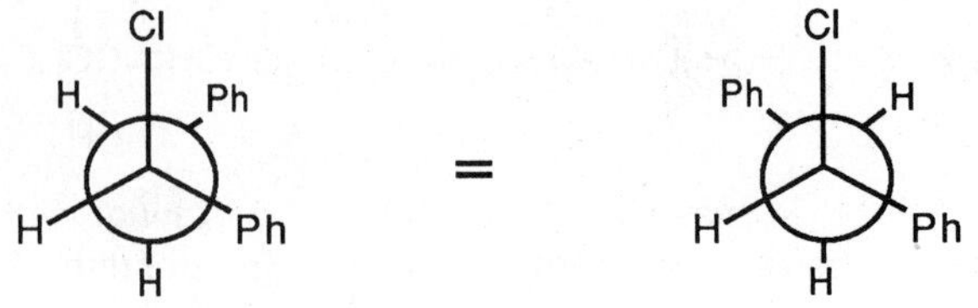

A   B

These two Newman projections place –H and –Cl in the correct *anti*-periplanar geometry for E2 elimination.

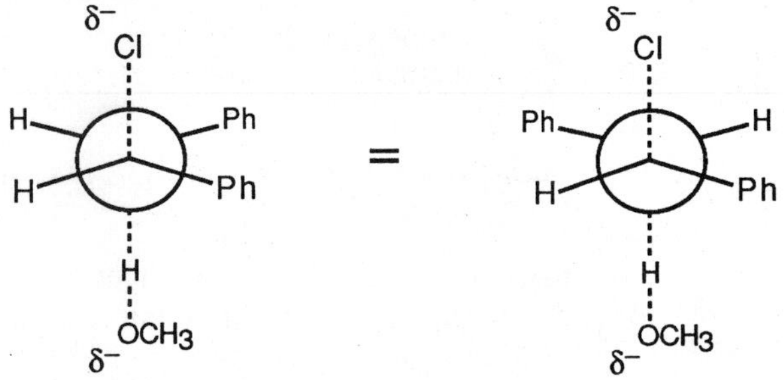

T.S. A‡   T.S. B‡

Either transition state $\underline{A}^{\ddagger}$ or $\underline{B}^{\ddagger}$ can form when 1–chloro–1,2–diphenylethane undergoes E2 elimination. Steric interactions of the two phenyl groups in T.S. $\underline{A}^{\ddagger}$ make this transition state (and the product resulting from it) of higher energy than transition state $\underline{B}^{\ddagger}$. Formation of the product from $\underline{B}^{\ddagger}$ is therefore favored, and *trans*–1,2–diphenylethylene is the major product.

11.18

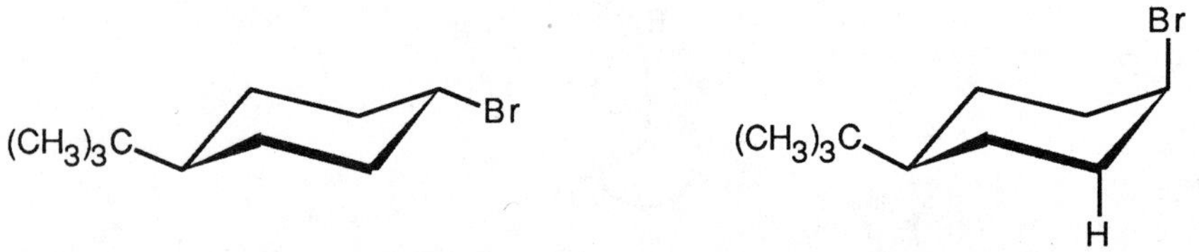

trans   cis

The more stable conformations of the two isomers are pictured above; the *tert*–butyl group is always equatorial in the more stable conformation. The *cis* isomer should react faster under E2 conditions because –Br and –H are in the *anti*-periplanar arrangement that favors E2 elimination.

11.19

a) $CH_3CH_2CH_2CH_2Br$ + NaN_3 $\longrightarrow$ $CH_3CH_2CH_2CH_2N_3$

      primary   substitution product

Since the halide is primary and a substitution product is formed, this is a S_N2 reaction.

b)

$$CH_3CH_2\overset{\overset{\displaystyle Cl}{|}}{C}HCH_2CH_3 \quad + \quad KOH \quad \longrightarrow \quad CH_3CH_2CH=CHCH_3$$

secondary   strong base   elimination product

This is a E2 reaction since a secondary halide reacts with a strong base to yield elimination product.

c)

tertiary   substitution product

This is a S_N1 reaction. Tertiary halides form substitution products only by the S_N1 route.

11.20 a) CH_3I reacts faster than CH_3Br because $:I^-$ is a better leaving group than $:Br^-$.

b) CH_3CH_2I reacts faster with $:OH^-$ in dimethylsulfoxide (DMSO) than in ethanol. Ethanol, a protic solvent, hydrogen-bonds with hydroxide ion and decreases its reactivity.

c) Under the S_N2 conditions of this reaction, CH_3Cl reacts faster than $(CH_3)_3CCl$. Approach of the nucleophile to the bulky $(CH_3)_3CCl$ molecule is hindered.

d) $H_2C=CHCH_2Br$ reacts faster because vinylic halides such as $H_2C=CHBr$ are unreactive to displacement reactions.

11.21

$$CH_3CH_2\overset{\overset{\displaystyle CH_3}{|}}{C}HCH_2I \quad + \quad ^-{:}CN \quad \longrightarrow \quad CH_3CH_2\overset{\overset{\displaystyle CH_3}{|}}{C}HCH_2CN \quad + \quad :\overset{..}{\underset{..}{I}}:^-$$

(primary halide)

This is a S_N2 reaction, in which reaction rate depends on the concentration of both alkyl halide and of nucleophile.

a) Halving the concentration of cyanide and doubling the concentration of alkyl halide will not change the reaction rate.

b) Tripling the concentrations of both cyanide and alkyl halide will cause a ninefold increase in reaction rate.

11.22

$$CH_3CH_2\overset{\overset{\displaystyle CH_3}{|}}{\underset{\underset{\displaystyle I}{|}}{C}}CH_3 \quad + \quad CH_3CH_2OH \quad \longrightarrow \quad CH_3CH_2\overset{\overset{\displaystyle CH_3}{|}}{\underset{\underset{\displaystyle OCH_2CH_3}{|}}{C}}CH_3 \quad + \quad HI$$

(tertiary halide)

This is an S_N1 reaction, whose reaction rate depends only on the concentration of 2–iodo–2–methyl-butane.

Tripling the concentrations of alkyl halide will triple the rate of reaction.

11.23 In this problem you have been provided with the reagents necessary to synthesize the desired compounds, using just one substitution reaction.

a) CH_3Br + $Na^+ {}^-C{\equiv}CCH(CH_3)_2$ $\longrightarrow$ $CH_3C{\equiv}CCH(CH_3)_2$

NOT $CH_3C{\equiv}C^-Na^+$ + $BrCH(CH_3)_2$. The strong base $CH_3C{\equiv}\bar{C}{:}$ brings about E2 elimination producing $CH_3C{\equiv}CH$ and $H_2C{=}CHCH_3$.

b) $CH_3CH_2CH_2CH_2Br$ + NaCN $\longrightarrow$ $CH_3CH_2CH_2CH_2CN$

c) $H_3C{-}Br$ + $\overset{..}{:}\overset{..}{O}C(CH_3)_3$ $\longrightarrow$ $H_3C{-}OC(CH_3)_3$. The reaction of $CH_3\overset{..}{\underset{..}{O}}{:}^-$ with $Br{-}C(CH_3)_3$ results in elimination, not substitution.

d) $CH_3CH_2CH_2Br$ + NH_3 $\longrightarrow$ $CH_3CH_2CH_2NH_2$ + HBr

e) $\left[\langle\bigcirc\rangle {-}P{:} \right]_3$ + CH_3Br $\longrightarrow$ $\left[\langle\bigcirc\rangle \overset{+}{{-}P}CH_3 \right]_3$ Br^-

f)

11.24 a) The difference in this pair of reactions is in the *leaving group*. Since $^-$OTos is a better leaving group than $^-$Cl (see Section 11.5), S_N2 displacement by iodide on CH_3–OTos proceeds faster.

b) The *substrates* in these two reactions are different. Bromoethane is a primary bromoalkane and bromocyclohexane is a secondary bromoalkane. Since S_N2 reactions proceed faster at primary, rather than secondary, carbon atoms, S_N2 displacement on bromoethane is a faster reaction.

c) Ethoxide ion and cyanide ion are different *nucleophiles*. Since $:CN^-$ is more reactive than $CH_3CH_2O^-$ in S_N2 reactions, S_N2 displacement on 2-bromopropane by CN^- proceeds at a faster rate.

d) The *solvent* in each reaction is different. S_N2 reactions run in hexamethylphosphoramide (HMPA) proceed faster than those run in other solvents. Thus, S_N2 displacement by acetylide ion on bromomethane proceeds faster in HMPA than in benzene.

11.25 Because 1–bromopropane is a primary haloalkane, the mode of reaction is either S_N2 or E2, depending on the basicity and the amount of steric hindrance in the nucleophile.

a) $CH_3CH_2CH_2Br$ + $NaNH_2$ $\longrightarrow$ $CH_3CH_2CH_2NH_2$

b) $CH_3CH_2CH_2Br$ + $\overset{+}{K}{:}\overset{-}{O}C(CH_3)_3$ $\longrightarrow$ $CH_3CH{=}CH_2$

K^+ $^-$O-t-Butyl is a hindered, strong base that causes elimination, not substitution.

c) $CH_3CH_2CH_2Br$ + NaI $\longrightarrow$ $CH_3CH_2CH_2I$

d) $CH_3CH_2CH_2Br$ + NaCN $\longrightarrow$ $CH_3CH_2CH_2CN$

e) $CH_3CH_2CH_2Br$ + $Na^+ {}^-C{\equiv}CH$ $\longrightarrow$ $CH_3CH_2CH_2C{\equiv}CH$ + $CH_3CH{=}CH_2$

f) $CH_3CH_2CH_2Br$ + Mg $\longrightarrow$ $CH_3CH_2CH_2MgBr$ $\overset{H_2O}{\longrightarrow}$ $CH_3CH_2CH_3$

11.26 Remember two rules used to predict nucleophilicity:

1) In comparing nucleophiles that have the same attacking atom, nucleophilicity parallels basicity. (In other words, a more basic nucleophile is a more effective nucleophile.)

2) Nucleophilicity increases in going down a column of the periodic table. Use Table 11.1 if necessary.

	More Nucleophilic	Less Nucleophilic	Reason
a)	$^-\ddot{N}H_2$	$:NH_3$	Rule 1
b)	$CH_3CO\ddot{O}:^-$	$H_2\ddot{O}:$	Rule 1
c)	$:\ddot{F}:^-$	BF_3	BF_3 is not a nucleophile
d)	$(CH_3)_3P:$	$(CH_3)_3N:$	Rule 2
e)	$:\ddot{I}:^-$	$:\ddot{C}l:^-$	Rule 2
f)	$^-:C\equiv N$	$^-:\ddot{O}CH_3$	Table 11.1

11.27 An alcohol is converted to an ether by two different routes in this series of reactions. The two resulting ethers have identical structural formulas but differ in sign of specific rotation. Therefore, at some step or steps in these reaction sequences, inversion of configuration at the chiral carbon must have occurred. Let's study each step of the Phillips and Kenyon series to find where inversion is occurring.

In *step 1,* the relatively acidic hydroxyl proton reacts with potassium metal to produce a potassium alkoxide. Since the bond between carbon and oxygen has not been broken, no inversion has occurred in this step.

The potassium alkoxide acts as a nucleophile in the S_N2 displacement on CH_3CH_2Br in *step 2*. It is the C–Br bond of bromoethane, however, not the C–O bond of the alkoxide, that is broken; no inversion at the chiral carbon occurs in step 2.

The starting alcohol reacts with tosyl chloride in *step 3*. Again, since the O–H, and not the C–O, bond of the alcohol is broken, no inversion can occur at this step.

Inversion does occur at *step 4*. The ⁻OTos group is displaced by CH_3CH_2OH. The C–O bond of the tosylate (OTos) is broken, and a new C–O bond is formed.

Notice the specific rotations of the two enantiomeric products. The product of steps 1 and 2 should be enantiomerically pure because neither reaction has affected the C–O bond. Reaction 4 proceeds with some racemization at the chiral carbon to give a smaller value of $[\alpha]_D$.

11.28 a) Substitution does not take place with secondary alkyl halides when a strong, bulky base is used. Elimination occurs instead, and produces $H_2C=CHCH_2CH_3$ and $CH_3CH=CHCH_3$.

b) Reaction of this secondary fluoroalkane with hydroxide yields both elimination and substitution products.

c) $SOCl_2$ in pyridine converts primary and secondary alcohols to chlorides by an S_N2 mechanism. 1–Methyl–1–cyclohexanol is a tertiary alcohol, and does not undergo S_N2 substitution. Instead, E2 elimination occurs to give 1–methylcyclohexane.

11.29 S_N1 reactivity:

Most reactive ——————————————————————————→ Least reactive

a)

(most stable carbocation)

b) $(CH_3)_3C–Br$ > $(CH_3)_3C–F$ > $(CH_3)_3C–OH$

(best leaving group)

c)

(most stable carbocation)

11.30 S_N2 reactivity:

Most reactive ——————————————————————————→ Least reactive

a) $CH_3CH_2CH_2Cl$ > $CH_3CH_2CHClCH_3$ > $(CH_3)_3CCl$

(primary carbon atom)

b) $(CH_3)_2CHCH_2Br$ > $(CH_3)_2CHCHCH_3$ (with Br) > $(CH_3)_3CCH_2Br$

(least sterically hindered carbon atom)

c) $CH_3CH_2CH_2OTos$ > $CH_3CH_2CH_2Br$ > $CH_3CH_2CH_2OCH_3$

(best leaving group)

11.31

R–2–Bromooctane

(R)–2–Bromooctane is a secondary bromoalkane, which undergoes S_N2 substitution. Since S_N2 reactions proceed with inversion of configuration, the configuration at the chiral carbon atom is inverted. (This does not necessarily mean that all R isomers become S isomers after an S_N2 reaction; the R–S designation refers to the priorities of groups, and priorities may change when the nucleophile is varied.)

Nucleophile	Product
a) ⁻:CN	S
b) CH₃CO:⁻	S
c) CH₃S:⁻	S
d) :Br:⁻	S + R

2-Bromooctane is 100% racemized after 50% of the original R-2-bromooctane has reacted with Br⁻.

11.32 a) The rates of both S_N1 and S_N2 reactions are affected by the use of polar solvents. S_N1 reactions are accelerated because polar solvents stabilize developing charges in the transition state. Most S_N2 reactions, however, are slowed down by polar *protic* solvents because these solvents hydrogen-bond to the nucleophile and decrease its reactivity. Polar aprotic solvents solvate nucleophiles without hydrogen bonding and increase nucleophile reactivity in S_N2 reactions.

b) Good leaving groups (weak bases whose negative charge can be dispersed) increase the rates of S_N1 and S_N2 reactions.

c) A good attacking nucleophile accelerates the rate of an S_N2 reaction. Since the nucleophile is involved in the rate-limiting step of an S_N2 reaction, a good attacking nucleophile lowers the energy of the transition state and increases the rate of reaction. Choice of nucleophile has no effect on the rate of a S_N1 reaction because attack of the nucleophile occurs after the rate-limiting step.

d) Because the rate-limiting step in an S_N2 reaction involves attack of the nucleophile on the substrate, any factor that makes approach of the nucleophile more difficult slows down the rate of reaction. Especially important is the degree of crowding at the reacting carbon atom. Tertiary carbon atoms are too crowded to allow S_N2 substitution to occur. Even steric hindrance one carbon atom away from the reacting site causes a drastic slowdown in rate of reaction.

The rate-limiting step in an S_N1 reaction involves formation of a carbocation. Any structural factor in the substrate that stabilizes carbocations will increase the rate of reaction. Substrates that are tertiary, allylic, or benzylic react the fastest.

11.33

This is an excellent method of ether preparation since iodomethane is very reactive in S_N2 displacements.

Reaction of a secondary haloalkane with a basic nucleophile yields both substitution and elimination products. This is obviously a less satisfactory method of ether preparation.

11.34

Methoxide removes a proton from the hydroxyl group of 4–bromo–1–butanol.

S_N2 displacement of :Br⁻ by the alkoxide oxygen yields the cyclic ether tetrahydrofurna.

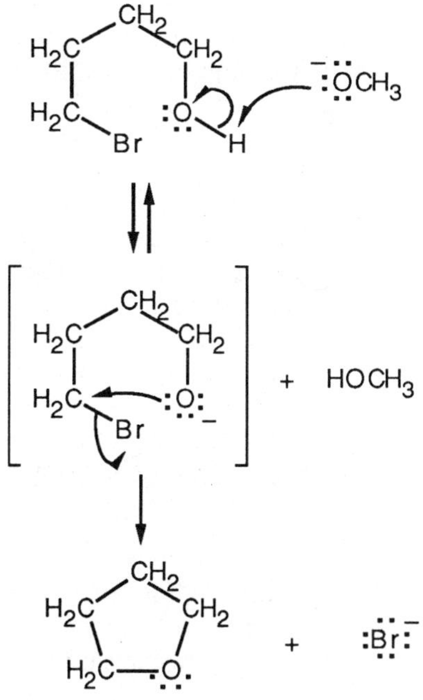

Tetrahydrofuran

$CH_3OCH_2CH_2CH_2CH_2OH$ is also produced.

11.35

$BrCH_2CH_2Br$ + 2 NaOH ⟶ $HOCH_2CH_2OH$

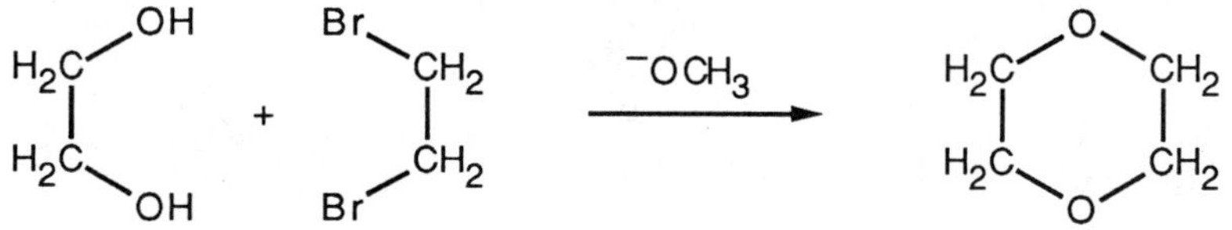

11.36

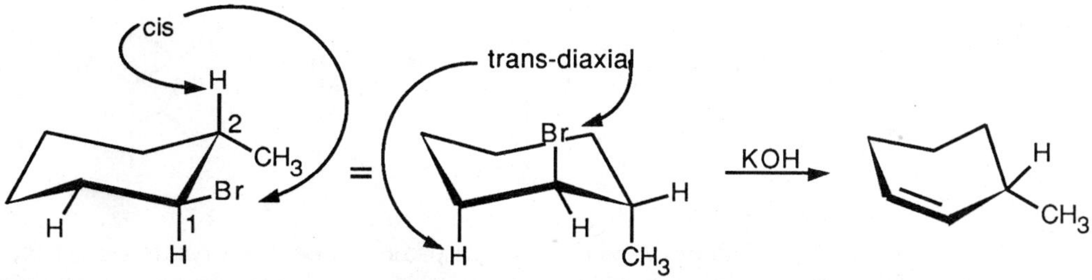

E2 reactions require that the two atoms to be eliminated have a trans-diaxial relationship. Since it's impossible for bromine and the hydrogen at C2 to be trans-diaxial, elimination occurs in the opposite direction to yield 3–methylcyclohexene, the non-Zaitsev product.

11.37

a)

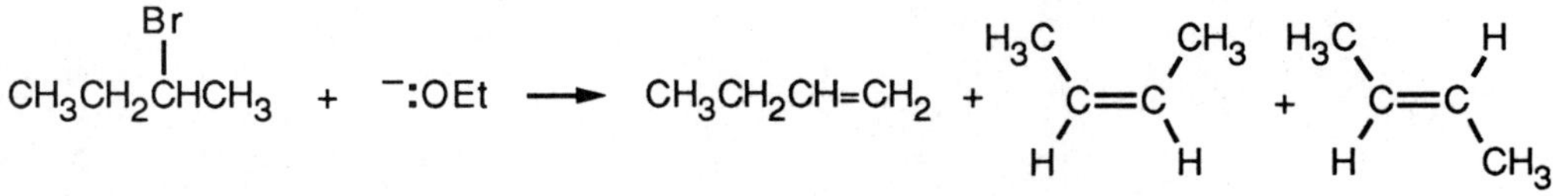

Br
|
CH₃CH₂CHCH₃ + ⁻:OEt ⟶ CH₃CH₂CH=CH₂ + (cis/trans 2-butene structures)

b)

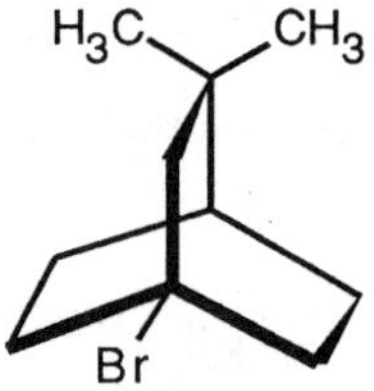

does not undergo nucleophilic substitution
(see Problems 11.3 and 11.14)

c)

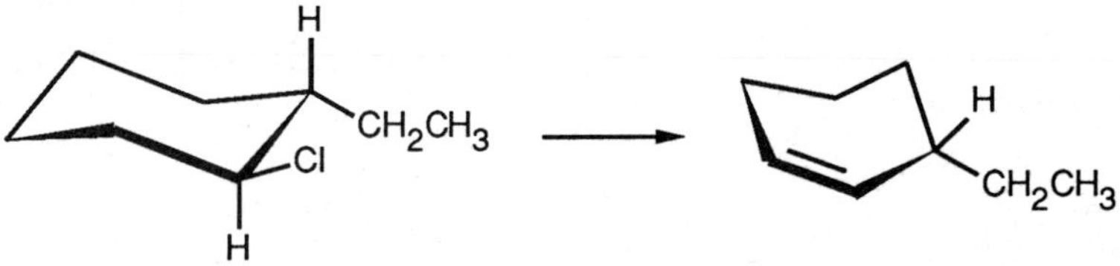

This alkyl halide gives the less substituted cycloalkene (non-Zaitsev product).
Elimination to form Zaitsev product is not likely to occur because the –Cl and –H
involved cannot assume the *anti*-periplanar geometry preferred for E2 elimination.

d)   (CH₃)₃C–OH + HCl $\xrightarrow{0°}$ (CH₃)₃C–Cl

11.38

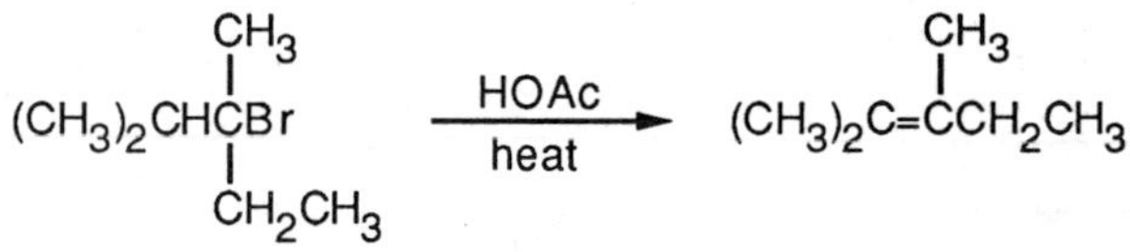

$$\underset{\underset{CH_2CH_3}{|}}{\overset{\overset{CH_3}{|}}{(CH_3)_2CHCBr}} \xrightarrow[\text{heat}]{\text{HOAc}} \underset{}{\overset{\overset{CH_3}{|}}{(CH_3)_2C=CCH_2CH_3}}$$

This alkene has the most substituted double bond.

11.39

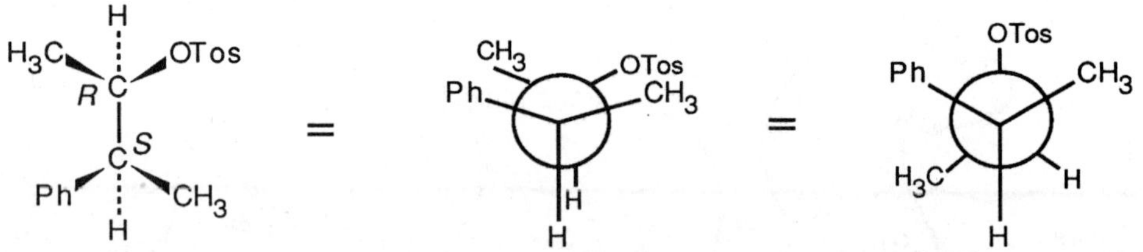

Draw the Newman projection that corresponds to the tosylate of
(2R,3S)–3–phenyl–2–butanol. The Newman projection can be rotated until the –OTos
and the –H on the adjoining carbon atom are *anti*–periplanar. Even though this
conformation has several *gauche* interactions, it is the only conformation in which
–OTos and –H are 180° apart.

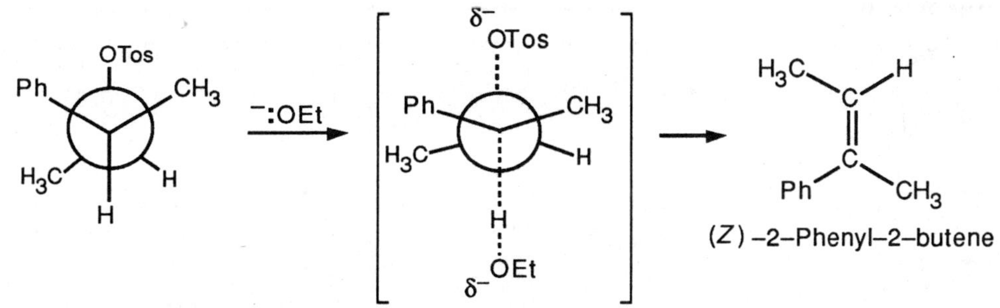

(Z)–2–Phenyl–2–butene

Elimination yields the Z isomer of 2–phenyl–2–butene. Refer to Chapter 6 for the method of assigning E, Z designation.

11.40 By the same argument used in Problem 11.39, you can show that elimination from (2R,3R)–3– phenyl–2–butyl tosylate give the E–alkene.

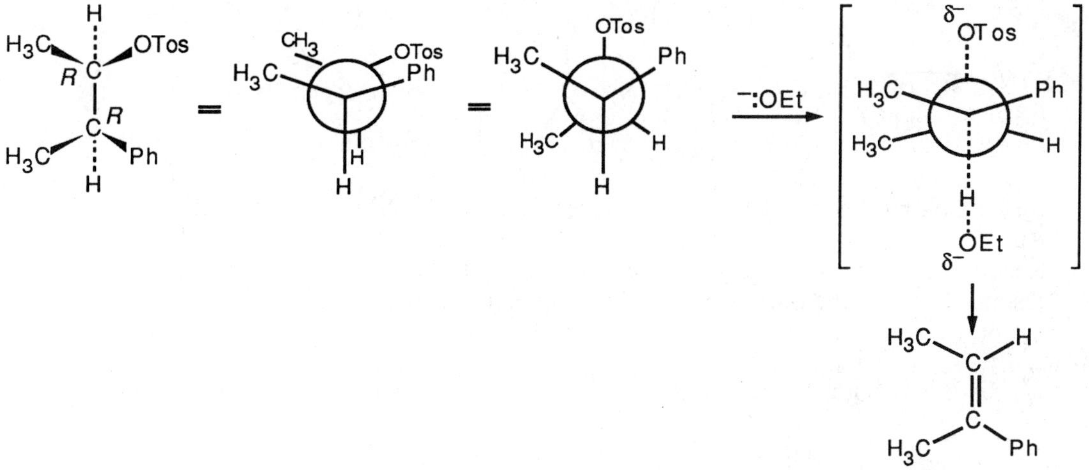

(E)–2–Phenyl–2–butene

The 2S,3S isomer also forms the E–alkene; the 2S,3R isomer yields Z–alkene.

11.41

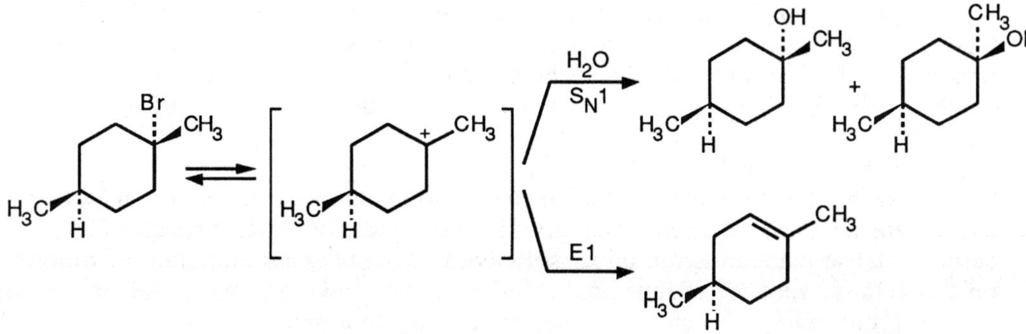

This tertiary bromoalkane reacts by S_N1 and E1 routes to yield alcohol and alkene products.

11.42 S_N2 reactivity:

Most reactive ⟶ Least reactive

$CH_3CH_2CH_2CH_2Br$ > $CH_3\overset{\overset{\displaystyle CH_3}{|}}{C}HCH_2Br$ > $CH_3CH_2\overset{\overset{\displaystyle Br}{|}}{C}H_2CH_3$ > $CH_3\overset{\overset{\displaystyle Br}{|}}{\underset{\underset{\displaystyle CH_3}{|}}{C}}CH_3$

1–Bromobutane   1–Bromo–2–methyl–   2–Bromobutane   2–Bromo–2–methyl–
                          propane   propane

11.43

S_N2 attack by the lone pair electrons associated with carbon gives the nitrile product. Attack by the lone pair electrons associated with nitrogen yields isonitrile product. Reaction to form nitrile is more likely because carbon has a formal charge of –1 in this resonance form, the only form that has complete octets for both atoms.

11.44

(*E*)–2–Chloro–2–butene–1,4–dioic acid   (*Z*)–2–Chloro–2–butene–1,4–dioic acid

Hydrogen and chlorine are *anti* to each other in the Z isomer and are *syn* in the E isomer. Since the Z isomer reacts fifty times faster than the E isomer, elimination must proceed more favorably when the substituents to be eliminated are *anti* to one another. This is the same stereochemical result as occurs in E2 eliminations of alkyl halides.

11.45 Since 2–butanol is a secondary alcohol, substitution can occur by either an S_N1 or S_N2 route, depending on reaction conditions. Two factors favor an S_N1 mechanism in this case. (1) The reaction is run under solvolysis (solvent as nucleophile) conditions in a polar, protic solvent. (2) Dilute acid converts a poor leaving group ($^-$OH) into a good leaving group (OH_2), which dissociates from 2–butanol more easily.

Protonation of oxygen . . .

. . . is followed by loss of water to form a planar carbocation.

Attack of water from either side of the planar cation yields racemic product.

11.46 The chiral tertiary alcohol (R)–3–methyl–3–hexanol reacts with HBr by an S_N1 pathway. HBr protonates the hydroxyl group, which dissociates to yield a planar, achiral carbocation. Attack by the nucleophilic bromide anion can occur from either side of the carbocation to produce ($\pm$)3–bromo–3– methylhexane.

11.47 Since carbon-deuterium bonds are slightly stronger than carbon-hydrogen bonds, more energy is required to break a C–D bond than to break a C–H bond. In a reaction where either a carbon-deuterium or a carbon-hydrogen bond is broken in the rate-limiting step, a higher percentage of C–H bond-breaking will occur because the energy of activation for C–H breakage is lower.

   In E2 reaction of this problem, the transition state involves breaking either a C–H or C–D bond.

Transition state $\underline{A}^{\ddagger}$ is of higher energy than transition state $\underline{B}^{\ddagger}$ because more energy is required to break the C–D bond. The product that results from transition state $\underline{B}^{\ddagger}$ is thus formed in greater abundance.

11.48 One of the steric requirements of E2 elimination is the need for periplanar geometry, which optimizes orbital overlap in the transition state leading to alkene product. Two types of periplanar arrangements of substituents -- *syn* and *anti* are possible.

A model of the deuterated bromo compound shows that the deuterium, bromine, and the two carbon atoms that will constitute the double bond all lie in a plane. This arrangement of atoms leads to *syn* elimination. Even though *anti* elimination is usually preferred, it does not occur for this compound. Models show that *anti*-periplanar arrangement of bromine, hydrogen, and the two carbons can't occur because of the rigidity of the molecule.

11.49

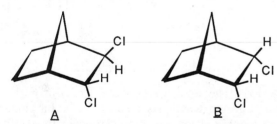

We concluded in Problem 11.48 that E2 elimination in compounds of this bicyclic structure occurs with *syn*-periplanar geometry. In compound $\underline{A}$, –H and –Cl can be eliminated via the *syn*-periplanar route. Since neither *syn* nor *anti*-periplanar elimination is possible for $\underline{B}$, elimination occurs by a slower E1 route.

11.50

Diastereomer *8* reacts much more slowly in an E2 reaction. No pair of hydrogen and chlorine atoms can assume the *anti*-periplanar orientation preferred for E2 elimination.

11.51 The two pieces of evidence indicate that the reaction proceeds by an S_N2 mechanism. S_N2 reactions proceed much faster in polar aprotic solvents such as DMF, and methyl esters react faster than ethyl esters. This reaction is an S_N2 displacement on a methyl ester by iodide.

Other experiments can provide additional evidence for an S_N2 mechanism. We can determine if the reaction is second-order by varying the concentration of LiI. We can also vary the type of nucleophile to distinguish the mechanism from an S_N1 mechanism, which does not depend on the type of nucleophile.

11.52 Cl:⁻ is a relatively poor leaving group and acetate is a relatively poor nucleophile; a displacement involving these two groups proceeds at a very slow rate. I:⁻, however, is both a good nucleophile and a good leaving group. 1–Chlorooctane therefore reacts preferentially with iodide to form 1–iodooctane. Only a small amount of 1–iodooctane is formed (because of the low concentration of iodide ion), but 1–iodooctane is more reactive than 1–chlorooctane toward substitution by acetate. Reaction with acetate produces 1–octyl acetate and regenerates iodide ion. The whole process can now be repeated with another molecule of 1–chlorooctane. The net result is production of 1–octyl acetate; no iodide is consumed.

11.53 Two optically inactive structures are possible for compound X. Any other structure consistent with the series of reactions is optically active.

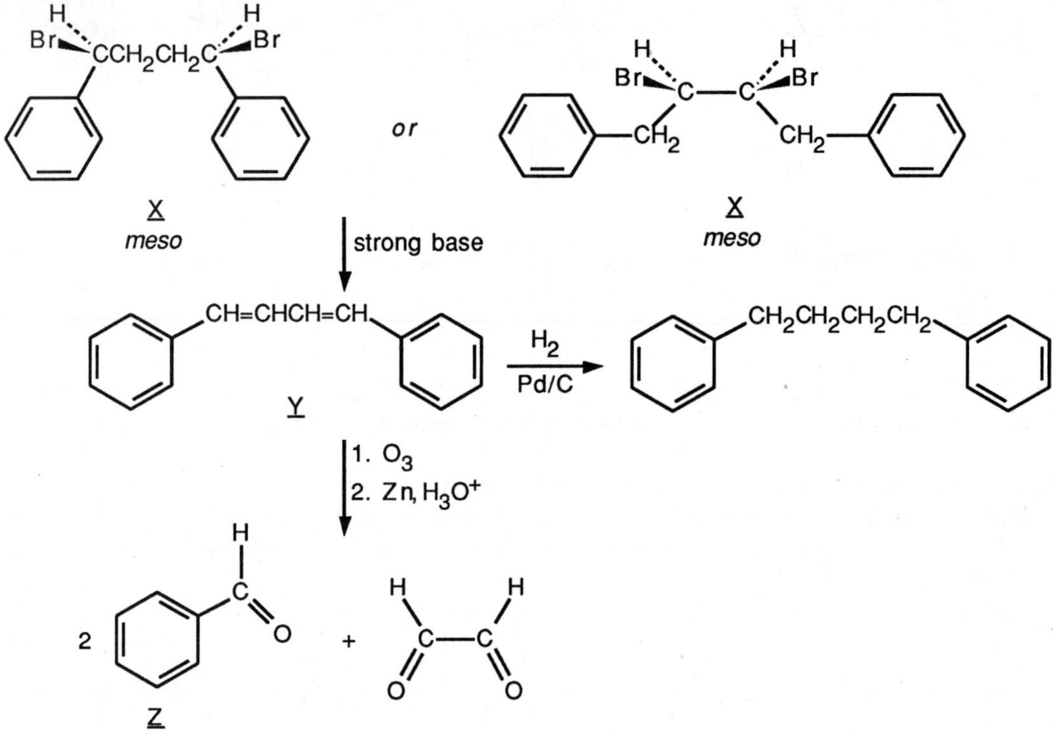

X
meso
or
X
meso
strong base
CH=CHCH=CH
Y
H2
Pd/C
CH2CH2CH2CH2
1. O3
2. Zn, H3O+
2
Z
+

11.54

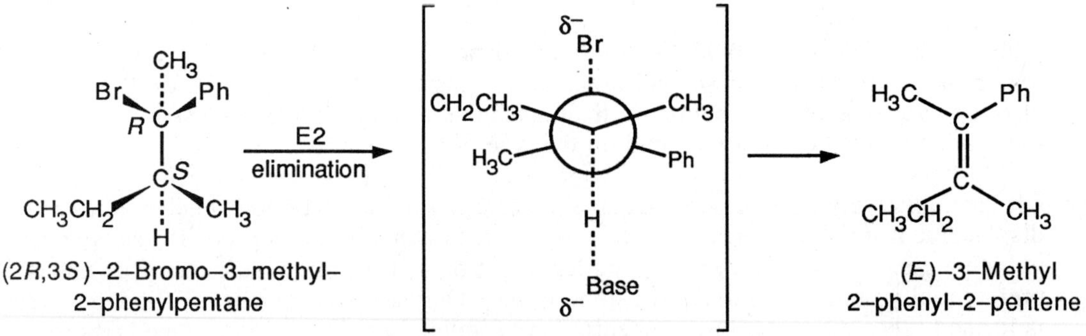

E2
elimination
Base
(2R,3S)–2–Bromo–3–methyl–2–phenylpentane
(E)–3–Methyl 2–phenyl–2–pentene

The 2*S*,3*R* isomer also yields *E* product.

11.55

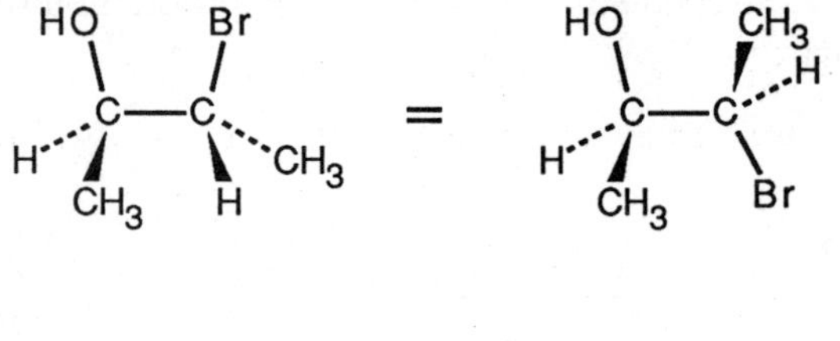

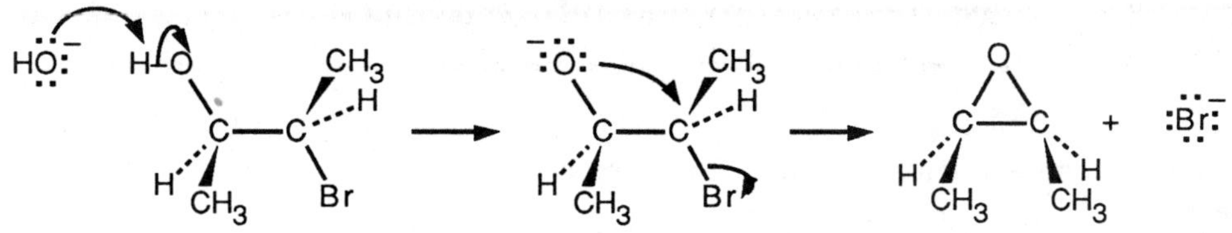

This reaction is an intramolecular S_N2 displacement.

11.56

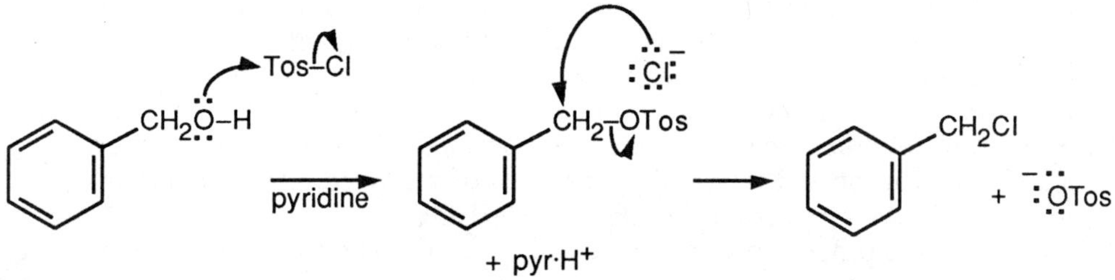

+ pyr·H$^+$

11.57

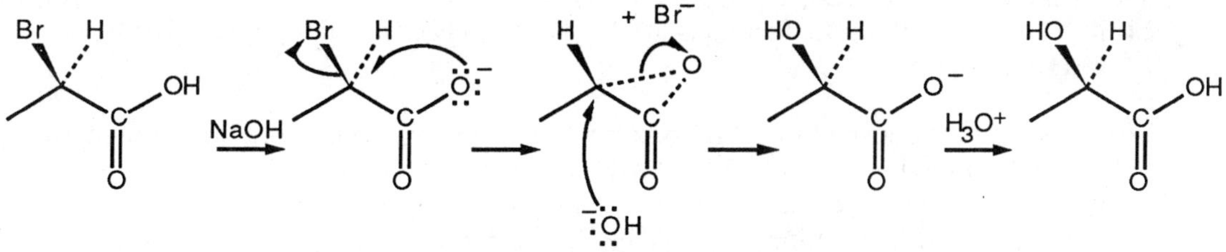

Study Guide for Chapter 11

After studying this chapter, you should be able to:

(1)   Formulate the mechanism of:
(a) S_N2 reactions (11.1, 11.2, 11.3, 11.27, 11.43, 11.51, 11.55, 11.56, 11.57).
(b) S_N1 reactions (11.9, 11.10, 11.12, 11.45, 11.46).
(c) Elimination reactions (11.16, 11.17, 11.18, 11.39, 11.40, 11.44, 11.47).

(2)   Predict the effect of substrate, nucleophile or base:
(a) S_N2 reactions (11.8, 11.20, 11.21, 11.24, 11.26, 11.30, 11.32, 11.42).
(b) S_N1 reactions (11.11, 11.13, 11.14, 11.22, 11.29, 11.32).
(c) Elimination reactions (11.48, 11.49, 11.50).

(3)   Predict the products of
(a) Substitution reactions (11.4, 11.23, 11.25, 11.331, 11.33).
(b) Elimination reactions (11.15, 11.36, 11.38, 11.41, 11.53, 11.54).

(4)   Classify reactions as S_N1, S_N2, E1 or E2 (11.19, 11.28).

(5)   Use substitution and elimination reactions in synthetic sequences (11.34, 11.35).

12.1 The following systematic approach may be helpful.

   a) $M^+\cdot = 86$
   1. Find the compound of molecular weight 86 that contains only –C and –H. Remember that a hydrocarbon with n carbon atoms can contain no more than $2n + 2$ hydrogen atoms. Here, C_6H_{14} is the correct formula.
   2. Find the formula corresponding to $M^+\cdot = 86$ that contains carbon, hydrogen, and one oxygen atom. If one oxygen atom (atomic weight = 16) is added to the base formula from step one, one carbon atom (atomic weight 12) and four hydrogen atoms (atomic weight 4) must be removed. This formula is $C_5H_{10}O$.
   3. Proceed to find the remaining molecular formulas. Each time one oxygen is added, one carbon and four hydrogens must be removed. The remaining formulas for $M^+\cdot = 86$ are $C_4H_6O_2$ and $C_3H_3O_3$.

   b) $M^+\cdot = 128$. The procedure is the same as in part a). The hydrocarbon having $M^+\cdot = 128$ is C_9H_{20}. The formula containing one oxygen is $C_8H_{16}O$. The remaining formulas are $C_7H_{12}O_2$, $C_6H_8O_3$, $C_5H_4O_4$.

   c) $M^+\cdot = 156$. Possible formulas are $C_{11}H_{24}$, $C_{10}H_{20}O$, $C_9H_{16}O_2$, $C_8H_{12}O_3$, $C_7H_8O_4$, $C_6H_4O_5$.

12.2 Use the method described in 12.1a to solve this problem. The hydrocarbon (containing only C and H) having $M^+\cdot = 218$ is $C_{16}H_{26}$. Since nootkatone also contains oxygen, we must consider only those formulas that include oxygen. Using the previous procedure, we can determine that $C_{15}H_{22}O$, $C_{14}H_{18}O_2$, $C_{13}H_{14}O_3$, $C_{12}H_{10}O_4$, $C_{11}H_6O_5$ are possible formulas for nootkatone. The actual formula of nootkatone is $C_{15}H_{22}O$.

12.3 Each carbon atom has a 1.11% probability of being ^{13}C and a 98.89% probability of being ^{12}C. The ratio of the height of the ^{13}C peak to the height of the ^{12}C peak for a one-carbon compound is (1.11/98.9) x 100% = 1.12%. For a six-carbon compound, the contribution to $(M+1)^+\cdot$ from ^{13}C is 6 x (1.11/98.9) x 100% = 6.72%. For benzene, the relative height of $(M+1)^+\cdot$ is 6.72% of the height of $M\cdot^+$.

   A similar line of reasoning can be used to calculate the contribution to $(M+1)^+\cdot$ from ^2H. The natural abundance of ^2H is 0.015%; the ratio of a ^2H peak to a ^1H peak for a one-hydrogen compound is 0.015%. For a six-hydrogen compound the contribution to $(M+1)^+\cdot$ from ^2H is 6 x 0.015% = 0.09%.

   For benzene, $(M+1)^+\cdot$ is 6.81% of $M^+\cdot$. Notice that ^2H contributes very little to the size of $(M+1)^+\cdot$.

12.4 Carbon is a tetravalent element; nitrogen is a trivalent element. If the structural unit –CH_2– (formula weight 14) is replaced by the structural unit –NH– (formula weight 15), the molecular weight of the resulting compound increases by one. Since all neutral hydrocarbons have even molecular weights (C_nH_{2n+2}, C_nH_{2n}, etc.) the resulting nitrogen-containing compound has an odd-numbered molecular weight and molecular ion. If two –CH_2– units are replaced by two –NH– units, the molecular weight of the resulting compound increases by two and remains an even number. You can continue this argument to prove that an odd number of –NH– groups results in an odd-numbered molecular weight and molecular ion. You can also prove that an odd number

of –NH$_2$ or –N– groups also results in an odd-numbered molecular ion.

12.5   Because M$^+$• is an odd number, pyridine contains an odd number of nitrogen atoms. If pyridine contained one nitrogen atom (atomic weight 14) the remaining atoms would have a formula weight of 65, corresponding to –C$_5$H$_5$. C$_5$H$_5$N is, in fact, the molecular formula of pyridine.

12.6   The structural formula of 2,2–dimethylpropane and of its molecular ion are given below.

$$
\begin{array}{c}
CH_3 \\
| \\
CH_3CCH_3 \\
| \\
CH_3
\end{array}
\xrightarrow{e^-}
\left[
\begin{array}{c}
CH_3 \\
| \\
CH_3CCH_3 \\
| \\
CH_3
\end{array}
\right]^{+\cdot}
\;+\; e^-
$$

When the molecular ion fragments, neutral and positively charged species are produced. The fragment of $m/z = 57$ corresponds to C$_4$H$_9$$^+$. The base peak usually represents the cation best able to stabilize positive charge. Since tertiary carbocations are relatively stable, C$_4$H$_9$$^+$ is most likely to be the *tert*–butyl cation.

12.7

$$
CH_3CH_2CH=C
\begin{array}{c}
\diagup CH_3 \\
\diagdown CH_3
\end{array}
\qquad\qquad
CH_3CH_2CH_2CH=CHCH_3
$$

2–Methyl–2–pentene   2–Hexene

Fragmentation occurs to a greater extent at the weakest carbon-carbon bonds; the positive charge remains with the fragment that is more able to stabilize it. A table of bond dissociation energies (Table 5.4) shows that allylic bonds have lower bond dissociation energies than the other bonds in these two compounds. Thus, the principal fragmentations of these compounds yield allylic cations.

$$
^+CH_2CH=C
\begin{array}{c}
\diagup CH_3 \\
\diagdown CH_3
\end{array}
\qquad\qquad
^+CH_2CH=CHCH_3
$$

$m/z = 69$ $m/z = 55$

Spectrum (b), which has $m/z = 55$ as its base peak, corresponds to 2–hexene. Although spectrum (a) has $m/z = 41$ as its base peak, the peak at $m/z = 69$ is almost as abundant. Spectrum (a) corresponds to 2–methyl–2–pentene.

12.8

$$E = h\upsilon = \frac{hc}{\lambda} \; ; \; h = 6.62 \times 10^{-34} \text{ J sec}, c = 3 \times 10^{10} \text{ cm/sec}$$

for $\lambda = 10^{-4}$ cm (infrared radiation)

$$E = \frac{6.62 \times 10^{-34} \text{ J sec} \times 3 \times 10^{10} \text{ cm/sec}}{10^{-4} \text{ cm}} = \frac{2 \times 10^{-23}}{10^{-4}} = 2 \times 10^{-19} \text{ J}$$

for $\lambda = 3 \times 10^{7}$ cm (x-radiation)

$$E = \frac{2 \times 10^{-23} \text{ J}}{3 \times 10^{-7}} = 7 \times 10^{-17} \text{ J}$$

Thus, an x-ray is of higher energy than infrared radiation.

12.9   First, convert radiation in cm to radiation in Hz by the equation:

$$\upsilon = \frac{c}{\lambda} \; ; \; \upsilon = \frac{3 \times 10^{10} \text{ cm/sec}}{9 \times 10^{-4} \text{ cm}} = 3 \times 10^{13} \text{ Hz}$$

The equation $E = h\upsilon$ says that the greater the value of υ, the greater the energy. Thus, radiation with $\upsilon = 3 \times 10^{13}$ Hz ($\lambda = 9 \times 10^{-4}$ cm) is higher in energy than radiation with $\upsilon = 4 \times 10^{9}$ Hz.

12.10

a) $$E = \frac{2.86 \times 10^{-3} \text{ kcal/mol}}{\lambda \text{ (in cm)}} \; ; \; \text{in this part, } \lambda = 5 \times 10^{9} \text{ cm}$$

$$= \frac{2.86 \times 10^{-3} \text{ kcal/mol}}{5 \times 10^{-9}}$$

$$= 0.57 \times 10^{6} \text{ kcal/mol}$$

$$= 5.7 \times 10^{5} \text{ kcal/mol for gamma rays}$$

b) $E = 9.5 \times 10^{3}$ kcal/mol for X-rays

c) $$\upsilon = \frac{c}{\lambda} \; ; \; \lambda = \frac{c}{\upsilon} = \frac{3 \times 10^{10} \text{ cm/sec}}{6 \times 10^{15} \text{ Hz}} = 5 \times 10^{-6} \text{ cm}$$

$$E = \frac{2.86 \times 10^{-3} \text{ kcal/mol}}{5 \times 10^{-6}} = 5.7 \times 10^{2} \text{ kcal/mol for ultraviolet light}$$

d) $E = 67$ kcal/mol for visible light

e) $E = 1.4$ kcal/mol for infrared radiation

f) $E = 9.5 \times 10^{-3}$ kcal/mol for microwave radiation

g) 102.5 MHz $= 1.025 \times 10^{8}$ Hz $= \upsilon$

$$\lambda = \frac{c}{\upsilon} = \frac{3 \times 10^{10} \text{ cm/sec}}{1.025 \times 10^{8} \text{ Hz}} = 2.93 \times 10^{2} \text{ cm}$$

$$E = \frac{2.86 \times 10^{-3} \text{ kcal/mol}}{2.93 \times 10^{2}} = 0.976 \times 10^{-5} \text{ kcal/mol} = 9.76 \times 10^{-6} \text{ kcal/mol for KFAT}$$

12.11

Wavenumber = $\dfrac{1}{\text{wavelength}}$;   wavenumber has units of cm^{-1}. 1 μm = 10^{-4} cm.

a) 3.10 μm = 3.10 x 10^{-4} cm; $\dfrac{1}{3.1 \times 10^{-4} \, cm}$ = 3225 cm^{-1}

b) 5.85 μm; 1710 cm^{-1}

c) $\dfrac{1}{2250 \, cm^{-1}}$ = 4.44 x 10^{-4} cm = 4.44 μm

d) 970 cm^{-1}; 10.3 μm

12.12 a) A compound with a strong absorption at 1710 cm^{-1} contains a carbonyl group and is either a ketone or aldehyde.
b) A nitro compound has a strong absorption at 1540 cm^{-1}.
c) A compound showing both carbonyl (1720 cm^{-1}) and –OH (2500-3000 cm^{-1} broad) absorptions is a carboxylic acid.

12.13 To use IR spectroscopy to distinguish between isomers, find a strong IR absorption present in one isomer that is absent in the other isomer.

a) CH_3CH_2OH

Strong hydroxyl band at 3400–3640 cm^{-1}.

CH_3OCH_3

No band in the region 3400–3640 cm^{-1}.

b) $CH_3CH_2CH_2CH_2CH=CH_2$

Alkene bands at 3020–3100 cm^{-1} and at 1650–1670 cm^{-1}.

No bands in alkene region

c) CH_3CH_2COOH

Strong, broad band at 2500–3100 cm^{-1}.

$HOCH_2CH_2CHO$

Strong band at 3400–3640 cm^{-1}.

12.14 Based on what we know at present, we can identify three absorptions in this spectrum.

a) The absorption at 2100 cm^{-1} is due to a –C≡C– stretch.

b) The absorption at 2900 cm^{-1} is due to a $-\overset{|}{\underset{|}{C}}-H$ stretch

c) The absorption at 3300 cm^{-1} is due to a ≡C–H stretch.

12.15 a) An ester next to a double bond absorbs at 1715 cm^{-1}. The alkene double bond absorbs at 1650-1670 cm^{-1}.

b)

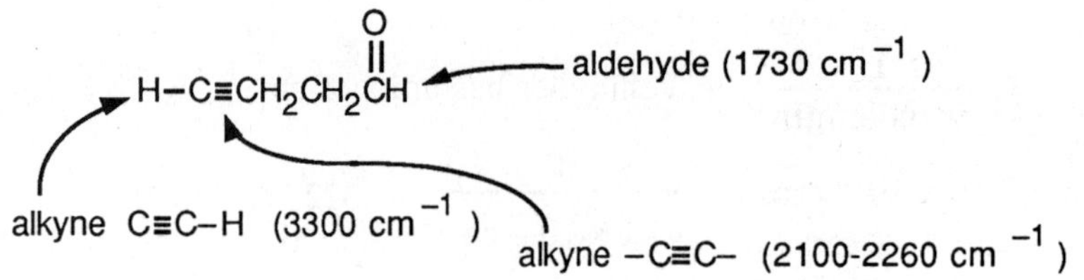

H$-$C$\equiv$CH$_2$CH$_2$CH ←—— aldehyde (1730 cm^{-1})

(with C=O shown above CH)

alkyne C$\equiv$C$-$H (3300 cm^{-1})

alkyne $-$C$\equiv$C$-$ (2100-2260 cm^{-1})

12.16 In this problem, all formulas must represent hydrocarbons.

a) For $M^{+\cdot}$ = 64, the only possible molecular formula is C_5H_4.

b) For $M^{+\cdot}$ = 186, possible formulas are $C_{14}H_{18}$ and $C_{15}H_6$.

c) For $M^{+\cdot}$ = 158, the only reasonable formula is $C_{12}H_{14}$.

d) Three formulas for $M^{+\cdot}$ = 220 are possible: $C_{16}H_{28}$, $C_{17}H_{16}$, and $C_{18}H_4$.

12.17

	$M^{+\cdot}$	Molecular Formula	Degree of Unsaturation
a)	86	C_6H_{14}	0
b)	110	C_8H_{14}	2
c)	146	$C_{11}H_{14}$	5
d)	190	$C_{14}H_{22}$	4
		$C_{15}H_{10}$	11

12.18

	$M^{+\cdot}$	Molecular Formula	Degree of Unsaturation	Possible Structure
a)	132	$C_{10}H_{12}$	5	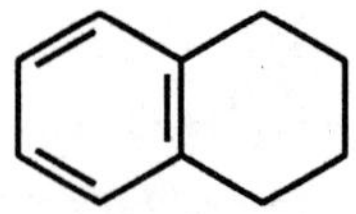
b)	166	$C_{13}H_{10}$	9	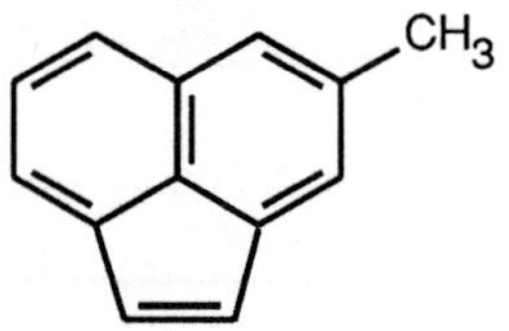
		$C_{12}H_{22}$	2	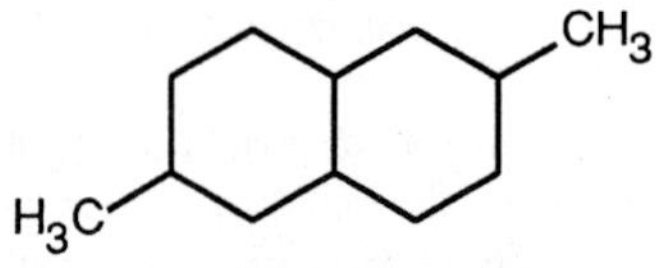
c)	84	C_6H_{12}	1	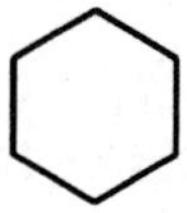

12.19 Remember that compounds in this problem may contain carbon, hydrogen, oxygen, and nitrogen. In addition, the molecular ions of many compounds may have the same value of $M^{+\cdot}$. Some of the less likely molecular formulas -- those with few carbon or hydrogen atoms -- have been omitted.

a) $M^{+\cdot} = 74$. Any nitrogen-containing compound that shows a molecular ion at $M^{+\cdot} = 74$ must have an even number of nitrogen atoms.

Compounds containing:

C, H;	C_6H_2.
C, H, O;	$C_4H_{10}O$, $C_3H_6O_2$, $C_2H_2O_3$.
C, H, N;	$C_3H_{10}N_2$, CH_6N_4.
C, H, N, O;	$C_2H_6N_2O$, $CH_2N_2O_2$.

b) $M^{+\cdot} = 131$ has an odd number of nitrogen atoms; no hydrocarbons correspond to this molecular ion.

C, H, N;	C_9H_9N, $C_7H_5N_3$, $C_6H_{17}N_3$, $C_4H_{13}N_5$.
C, H, N, O;	$C_7H_{17}NO$, C_8H_5NO, $C_6H_{13}NO_2$, $C_5H_9NO_3$, $C_4H_5NO_4$, $C_5H_{13}N_3O$, $C_4H_9N_3O_2$, $C_3H_5N_3O_3$, $C_3H_9N_5O$.

12.20 Reasonable molecular formulas for camphor are $C_{10}H_{16}O$, $C_9H_{12}O_2$, $C_8H_8O_3$. The actual formula, $C_{10}H_{16}O$, corresponds to three degrees of unsaturation. The ketone functional group accounts for one of these. Since camphor is a saturated compound, the other two degrees of unsaturation are due to two rings.

Camphor

12.21 The molecular formula of nicotine is $C_{10}H_{14}N_2$. To find the base formula, subtract the number of nitrogens from the number of hydrogens. The base formula of nicotine, $C_{10}H_{12}$, indicates the presence of five degrees of unsaturation; two of these are due to the two rings, and the other three are due to three double bonds.

Nicotine

12.22 In order to simplify this problem, neglect the ^{13}C and ^2H isotopes in determining the molecular ions of these compounds.

a) The formula weight of $-CH_3$ is 15; the atomic masses of the two bromine isotopes are 79 and 81. The two molecular ions of bromoethane occur at $M^{+\cdot} = 96$ (49.5%) and $M^{+\cdot} = 94$ (50.5%).

b) The formula weight of $-C_6H_{13}$ is 85; atomic masses of the two chlorine isotopes are 35 and 37. The two molecular ions of chlorohexane occur at $M^+\cdot = 122$ (24.5%) and $M^+\cdot = 120$ (75.5%).

12.23 Again, neglect ^{13}C and ^{2}H in these calculations.

a) Finding the molecular ions of chloroform is a probability exercise.
1) The probability that all three chlorine atoms are ^{37}Cl is $(0.245)^3 = 0.015$.
2) The probability that two chlorine atoms are ^{37}C and one is ^{35}Cl is $3(0.245)(0.245)(0.755) = 3(0.045) = 0.135$. The factor 3 enters the calculations because three permutations of two ^{37}Cl's and one ^{35}Cl are possible.
3) The probability that one chlorine atom is ^{37}Cl and two are ^{35}Cl is $3(0.245)(0.755)(0.755) = 3(0.140) = 0.420$.
4) The probability that all chlorine atoms are ^{35}Cl is $(0.755)^3 = 0.430$.
5) The mass of: $CH^{37}Cl^{37}Cl^{37}Cl = 124$
$$CH^{37}Cl^{37}Cl^{35}Cl = 122$$
$$CH^{37}Cl^{35}Cl^{35}Cl = 120$$
$$CH^{35}Cl^{35}Cl^{35}Cl = 118.$$
6) There, molecular ions for chloroform occur at:

$M^+\cdot$	124	122	120	118
Abundance	1.5%	13.5%	42.0%	43.0%

b) The molecular ions for Freon 12:

$M^+\cdot$	124	122	120
Abundance	6.0%	37.0%	57.0%

12.24

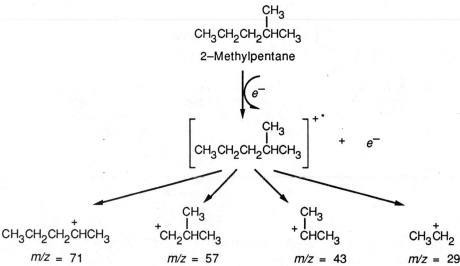

The molecular ion, at $m/z = 86$, is present in very low abundance. The base peak, at $m/z = 43$, represents a stable secondary carbocation.

12.25 Before doing the hydrogenation, familiarize yourself with the mass spectra of cyclohexene and cyclohexane. Note that $M^{+\cdot}$ is different for each compound.

After the reaction is underway, inject a sample from the reaction mixture onto the GC/MS. There should be two obvious peaks that can be unambiguously identified by their mass spectra. As the reaction proceeds, one of these peaks will increase in size and one will decrease; by this point you should be able to identify these peaks without running additional mass spectra. When the reaction is complete the gas chromatogram should show one peak whose mass spectrum is superimposable with that of cyclohexane.

12.26 See Problem 12.11 for the method of solution.
   a) 3355 cm^{-1} b) 1720 cm^{-1} c) 2030 cm^{-1}

12.27 a) 5.70 μm b) 3.08 μm c) 5.80 μm d) 5.62 μm

12.28 $CH_3CH_2C \equiv CH$ shows absorptions at $2100\text{-}2260 \text{ cm}^{-1}$ and at 3300 cm^{-1} that are due to the terminal alkyne bond.
   $CH_2 = CHCH = CH_2$ has absorptions in the regions $1650\text{-}1670 \text{ cm}^{-1}$ and $3020\text{-}3100$ that are due to the double bonds. No absorptions occur in the alkyne region.
   $CH_3C \equiv CCH_3$. For reasons we won't discuss, symmetrically substituted alkynes such as 2–butyne do not show a $C \equiv C$ bond in the IR. This alkyne is distinguished from the other isomers in that it shows no absorptions in either the alkyne or alkene regions.

12.29 Two enantiomers have identical physical properties (other than the sign of specific rotation). Thus, their IR spectra are also identical.

12.30 Diastereomers have different physical properties and chemical behavior. Hence the IR spectra of two diastereomers are also expected to be different.

12.31 a) Absorptions at 3300 cm^{-1} and 2150 cm^{-1} are due to a terminal triple bond. Possible structures:

$$CH_3CH_2CH_2C \equiv CH \qquad (CH_3)_2CHC \equiv CH$$

   b) IR absorption at 3400 cm^{-1} is due to a hydroxyl group. Since no double bond absorption is present, the compound must be a cyclic alcohol.

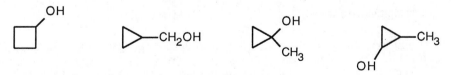

   c) Absorption at 1715 cm^{-1} is due to a ketone. The only possible structure is $CH_3CH_2COCH_3$.

d) Absorptions at 1600 cm^{-1} and 1500 cm^{-1} are due to an aromatic ring. Possible
structures:

12.32 a) HC≡CCH$_2$NH$_2$

Alkyne absorptions at
3300 cm^{-1}, 2100-2260 cm^{-1}
Amine absorption at
3310-3500 cm^{-1}.

CH$_3$CH$_2$C≡N

Nitrile absorption at
2210-2260 cm^{-1}

b) CH$_3$COCH$_3$

Strong ketone absorption
at 1715 cm^{-1}

CH$_3$CH$_2$CHO

Strong aldehyde absorption
at 1730 cm^{-1}

12.33 Spectrum (b) differs from spectrum (a) in several respects. Note in particular the
absorbances at 715 cm^{-1} (s), 1140 cm^{-1} (s), 1650 cm^{-1} (m), and 3000 cm^{-1} (m) in
spectrum (b). The absorbances at 1650 cm^{-1} (C=C stretch) and 3000 cm^{-1} (=C–H stretch)
can be found in Table 12.1. They allow us to assign spectrum (b) to cyclohexene and
spectrum (a) to cyclohexane.

12.34 a) CH$_3$C≡CCH$_3$ does not exhibit a terminal ≡C–H stretching vibration at 3300 cm^{-1}, as
CH$_3$CH$_2$C≡CH does.
b) CH$_3$COCH=CHCH$_3$, a conjugated unsaturated ketone, shows a strong ketone
absorbance at 1690 cm^{-1}; CH$_3$COCH$_2$CH=CH$_2$, a nonconjugated ketone, shows a
ketone absorption at 1710 cm^{-1}.
c) CH$_3$CH$_2$CHO exhibits an aldehyde band at 1725 cm^{-1}; H$_2$C=CHOCH$_3$ shows
characteristic alkene absorbances, as well as a C–O stretch near 1200 cm^{-1}.

12.35

1–Methylcyclohexanol   1–Methylcyclohexene

The infrared spectrum of the starting alcohol shows a broad absorption at 3400-3640 cm^{-1},
due to an O–H stretch, and another strong absorption at 1050-1100 cm^{-1}, due to a C–O
stretch. The alkene product exhibits medium intensity absorbances at 1645-1670 cm^{-1} and
at 3000-3100 cm^{-1}. Monitoring the *disappearance* of one of the alcohol absorptions allows
one to decide when the alcohol is totally dehydrated. It is also possible to monitor the
appearance of one of the alkene absorbances.

12.36

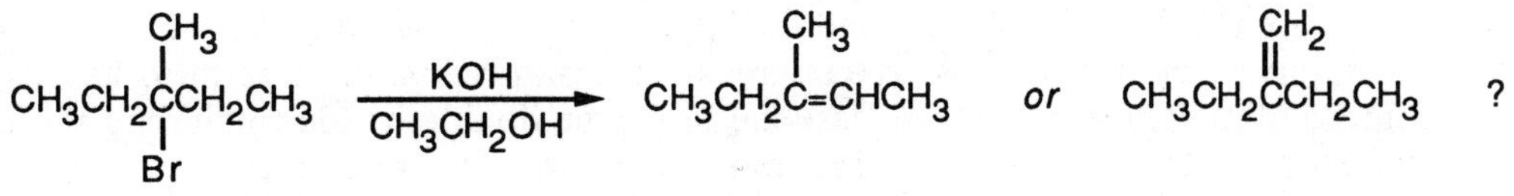

$$CH_3CH_2\underset{\underset{Br}{|}}{\overset{\overset{CH_3}{|}}{C}}CH_2CH_3 \xrightarrow[CH_3CH_2OH]{KOH} CH_3CH_2\overset{\overset{CH_3}{|}}{C}=CHCH_3 \quad or \quad CH_3CH_2\overset{\overset{CH_2}{||}}{C}CH_2CH_3 \quad ?$$

3–Bromo–3–methylpentane   3–Methyl–2–pentene   3–Ethyl–1–butene

The IR spectra of both products show the characteristic absorptions of alkenes in the regions 3020-3100 cm^{-1} and 1650 cm^{-1}. However, in the region 700-1000 cm^{-1}, 2–ethyl–1–butene shows a strong absorption at 890 cm^{-1} that is typical of 2,2–disubstituted $R_2C=CH_2$ alkenes. The presence or absence of this peak should help to identify the product. (3–Methyl–2–pentene is the major product of the dehydrobromination reaction.)

12.37

Compound	Distinguishing Absorption	Due to
a) $CH_3CH_2\overset{\overset{O}{||}}{C}CH_3$	1715 cm^{-1}	$\diagdown_{\diagup}C=O$ (ketone)
b) $(CH_3)_2CHCH_2C{\equiv}CH$	2140 cm^{-1} 3300 cm^{-1}	$-C{\equiv}C-$ $-C{\equiv}C-H$
c) $(CH_3)_2CHCH_2CH=CH_2$	910 cm^{-1} , 990 cm^{-1} 1650-1670 cm^{-1} 3020-3100 cm^{-1}	$R-CH=CH_2$ $-\overset{|}{C}=\overset{|}{C}-$ $=\overset{|}{C}-H$
d) $CH_3CH_2CH_2\overset{\overset{O}{||}}{C}OCH_3$	1735 cm^{-1}	$R-\overset{\overset{O}{||}}{C}-OR$ (ester)
e) ⬡$-\overset{\overset{O}{||}}{C}CH_3$	1690 cm^{-1}	$\diagdown_{\diagup}C=O$ (ketone next to aromatic ring)

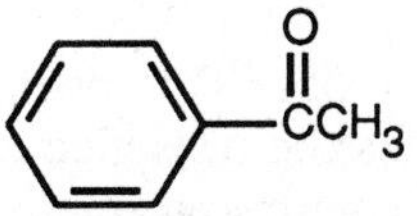

12.38 By manipulating symbols, we can arrive at the following expressions:

$E = h\nu = hc/\lambda = hc\,\tilde{\upsilon}$ where $\tilde{\upsilon}$ is the wavenumber. The last expression shows that, as $\tilde{\upsilon}$ increases, the energy needed to cause IR absorption increases, indicating greater bond strength. Thus, an ester C=O bond ($\tilde{\upsilon}$ = 1735 cm^{-1}) is stronger than a ketone C=O bond ($\tilde{\upsilon}$ = 1715 cm^{-1}).

12.39 Possible molecular formulas containing carbon, hydrogen, and oxygen and having $M^{+}\cdot$ = 150 are $C_{10}H_{14}O$, $C_9H_{10}O_2$, and $C_8H_6O_3$. The first formula has four degrees of unsaturation, the second has five degrees of unsaturation, and the third has six degrees of unsaturation. Since carvone has three double bonds (including the ketone) and one ring, $C_{10}H_{14}O$ is the correct molecular formula for carvone.

Carvone

12.40 The intense absorption at 1690 cm^{-1} is due to a carbonyl group next to a double bond.

12.41 The peak of maximum intensity (base peak) in the mass spectrum occurs at m/z = 67. This peak does *not* represent the molecular ion, since $M^{+}\cdot$ of a hydrocarbon must be an even number. Careful inspection reveals the molecular ion peak at m/z = 68. $M^{+}\cdot$ = 68 corresponds to a hydrocarbon of molecular formula C_5H_8 with a degree of unsaturation of two.

Fairly intense peaks in the mass spectrum occur at m/z = 67, 53, 40, 39, and 27. The peak at m/z = 67 corresponds to loss of one hydrogen atom, and the peak at m/z = 53 represents loss of a methyl group. The unknown hydrocarbon thus contains a methyl group.

Significant IR absorptions occur at 2130 cm^{-1} (–C≡C– stretch) and at 3320 cm^{-1} (≡C–H stretch). These bands indicate that the unknown hydrocarbon is a terminal alkyne. Possible structures for C_5H_8 are $CH_3CH_2CH_2C≡H$ and $(CH_3)_2CHC≡CH$. [In fact, 1–pentyne is correct.]

12.42 The molecular ion, $M^{+}\cdot$ = 70, corresponds to the molecular formula C_5H_{10}. This compound has one double bond or ring.

The base peak in the mass spectrum occurs at m/z = 55. This peak represents loss of a methyl group from the molecular ion and indicates the presence of a methyl group in the unknown hydrocarbon. All other peaks occur with low intensity.

In the IR spectrum, it is possible to distinguish absorbances at 1660 cm^{-1} and at 3000 cm^{-1}; the 2960 cm^{-1} absorption is rather hard to detect because it occurs as a shoulder on the alkane C–H stretch at 2850-2960 cm^{-1}. These two absorptions are due to a double bond.

Since no absorptions occur in the region 890 cm^{-1} – 990 cm^{-1}, we can exclude terminal alkenes as possible structures. The remaining possibilities for C_5H_{10} are $CH_3CH_2CH=CHCH_3$ and $(CH_3)_2C=CHCH_3$. [2–Methyl–2–butene is correct.]

Study Guide for Chapter 12

After studying this chapter, you should be able to:

(1)   Write molecular formulas corresponding to a given molecular ion (12.1, 12.2, 12.4, 12.5, 12.16, 12.17, 12.18, 12.19, 12.20, 12.21, 12.39).

(2)   Use the natural abundance of isotopes to calculate molecular ions (12.3, 12.22, 12.23).

(3)   Use mass spectra to determine molecular weights and base peaks, and to distinguish between hydrocarbons (12.6, 12.7, 12.24, 12.41, 12.42).

(4)   Calculate the energy of electromagnetic radiation (12.8, 12.9, 12.10, 12.38).

(5)   Convert from wavelength to wavenumber, and *vice versa* (12.11, 12.26, 12.27).

(6)   Identify functional groups by their infrared absorptions (12.12, 12.13, 12.14, 12.15, 12.28, 12.29, 12.30, 12.32, 12.33, 12.34, 12.37, 12.40, 12.41, 12.42).

(7)   Use IR and MS information to monitor reaction progress (12.25, 12.36, 12.37).

13.1

$$E = \frac{2.86 \times 10^{-3}\,\text{kcal/mol}}{\lambda \text{ (in cm)}}$$

$$\lambda = \frac{c}{\upsilon} = \frac{3 \times 10^{10}\,\text{cm/sec}}{\upsilon}$$

here υ = 56 MHz, or 5.6×10^7 Hz

so $\lambda = \dfrac{3 \times 10^{10}\,\text{cm/sec}}{5.6 \times 10^7\,\text{Hz}} = 0.54 \times 10^3\,\text{cm}$

$$E = \frac{2.86 \times 10^{-3}\,\text{kcal/mol}}{0.54 \times 10^3} = 5.3 \times 10^{-6}\,\text{kcal/mol}$$

Compare this value with $E = 5.7 \times 10^{-6}$ kcal/mol for ^{1}H. It takes less energy to spin-flip a ^{19}F nucleus than to spin-flip a ^{1}H nucleus.

13.2

$$\lambda = \frac{c}{\upsilon} = \frac{3 \times 10^{10}\,\text{cm/sec}}{\upsilon}$$

here υ = 100 MHz = 100×10^6 Hz, or 10^8 Hz

so $\lambda = \dfrac{3 \times 10^{10}\,\text{cm/sec}}{10^8\,\text{Hz}} = 3 \times 10^2\,\text{cm}$

$$E = \frac{2.86 \times 10^{-3}\,\text{kcal/mol}}{3 \times 10^2} = 9.5 \times 10^{-6}\,\text{kcal/mol}$$

Increasing the spectrometer frequency increases the amount of energy needed for resonance.

13.3

a)

This alkene has two different types of carbon and shows two signals in its ^{13}C NMR. Since all protons are equivalent, only one ^{1}H NMR signal appears.

b)

All carbon atoms are equivalent, as are all hydrogen atoms. Consequently the ^{13}C NMR spectrum and the ^1H NMR spectrum of cyclohexane each show one signal.

c)

$$CH_3\overset{\displaystyle O}{\overset{\displaystyle \|}{C}}CH_3$$

Acetone shows two ^{13}C NMR signals and one ^1H NMR signal.

d)

$$(CH_3)_3\overset{\displaystyle O}{\overset{\displaystyle \|}{C}}COCH_3$$

a   bc d

Four signals appear in the ^{13}C NMR spectrum of this ester because four different kinds of carbon atoms are present. The ^1H NMR shows two signals.

e)

^{13}C: Three signals
^1H: Two signals

f)

^{13}C: Three signals
^1H: Two signals

13.4

b H CH_3 a

c H   Cl

2–Chloropropene has three kinds of protons. Protons b and c differ because one is *cis* to the chlorine and the other is *trans*.

13.5

a)

$$\delta = \frac{\text{Observed chemical shift (\# Hz away from TMS)}}{\text{Spectrometer frequency (MHz)}}$$

δ = Parts per million. Here, δ = 2.1 ppm

$$2.1 \text{ ppm} = \frac{\text{Observed chemical shift}}{60 \text{ (MHz)}}$$

$$126 \text{ Hz} = \text{Observed chemical shift}$$

b) If the ^1H NMR spectrum of acetone were recorded at 100 MHZ, the position of absorption would still be 2.1 δ because measurements given in ppm or δ units are independent of the operating frequency of the NMR spectrometer.

c) $2.1 \delta = \dfrac{\text{Observed chemical shift}}{100 \text{ MHz}}$; observed chemical shift = 210 Hz

13.6

$$\delta = \frac{\text{Observed chemical shift (in Hz)}}{60 \text{ MHz}}$$

a) $\delta = \dfrac{436 \text{ Hz}}{60 \text{ MHz}} = 7.27$ ppm for $CH\,Cl_3$

b) $\delta = \dfrac{183 \text{ Hz}}{60 \text{ MHz}} = 3.05$ ppm for CH_3Cl

c) $\delta = \dfrac{208 \text{ Hz}}{60 \text{ MHz}} = 3.47$ ppm for CH_3OH

d) $\delta = \dfrac{318 \text{ Hz}}{60 \text{ MHz}} = 5.30$ ppm for CH_2Cl_2

13.7

$$\overset{4}{C}H_3\overset{3}{C}H_2\overset{2}{C}O_2\overset{1}{C}H_3$$

δ(ppm)	Assignment
9.3	4
27.6	3
51.4	1
174.6	2

13.8

a)

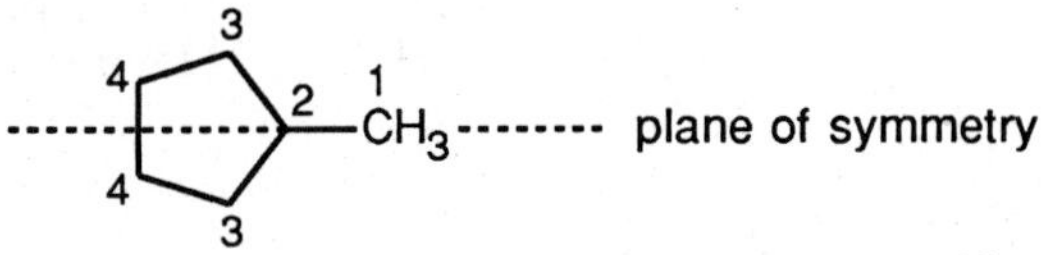

Four resonance lines are observed in the ^{13}C NMR spectrum of methylcyclopentane.

b)

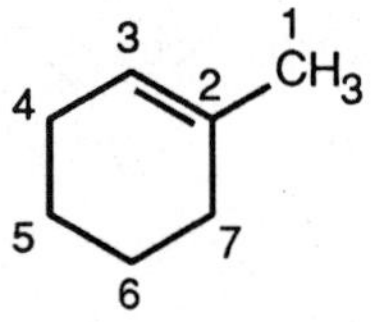

Seven resonance lines are seen. No two carbon atoms in 1–methylcyclohexene are equivalent because no plane of symmetry is present.

c)

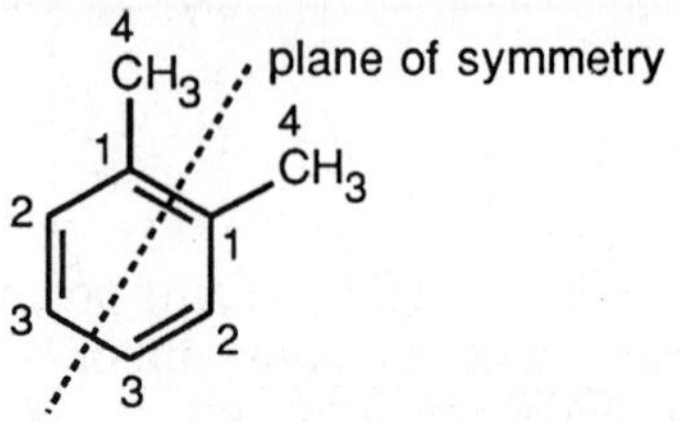

Four resonance lines are observed. A plane of symmetry causes one half of the carbon atoms to be equivalent to the other half.

d)

$$\underset{H}{\overset{\underset{5}{H_3C}}{}} \underset{CH_3}{\overset{1}{CH_3}}$$

Five resonance lines are observed. Carbons 1 and 2 are non-equivalent because of the double bond stereochemistry.

13.9   Each part of this problem has several correct answers.
  a) 1–Methylcyclohexene (see Problem 13.8b) and 1,3–dimethylcyclopentene show seven resonance lines.

b) $\underset{H_3C}{\overset{H_3C}{}} CHCH_2CH_2CH_3$ Two of the six carbons are equivalent

c) $\underset{H_3C}{\overset{H_3C}{}} CHCH_2Cl$ The two methyl groups are equivalent.

13.10

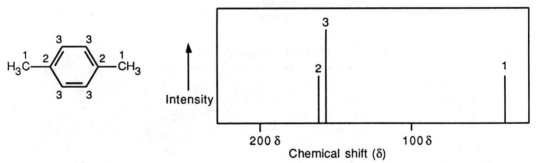

The ^{13}C spectrum of p–dimethylbenzene shows three lines. The ratio of peak areas is 1:1:2.

13.11

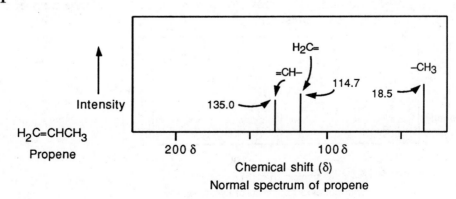

Normal spectrum of propene

a) Propene has three different carbon atoms and shows three carbon resonances.

b) In a spin-coupled spectrum:

The *methyl* carbon resonance splits into four peaks having a 1:3:3:1 ratio of intensity.

The $H_2C=$ carbon resonance splits into three peaks having a 1:2:1 ratio of intensity.

The $=CH-$ carbon resonance splits into two peaks having a 1:1 ratio of intensity.

13.12

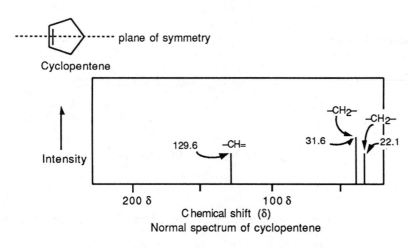

Cyclopentene

Normal spectrum of cyclopentene

a) The ^{13}C spectrum of cyclopentene shows three peaks.

b)

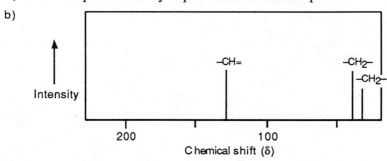

Chemical shift (δ)

In an integrated spectrum, peak areas are proportional to the relative number of carbon atoms each peak represents.

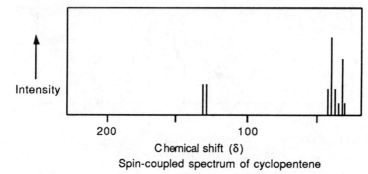

Chemical shift (δ)

Spin-coupled spectrum of cyclopentene

13.13 Either of two products may result from radical addition of HBr to 2–methylpropene: $(CH_3)_2CHCH_2Br$ (non-Markovnikov product) and $(CH_3)_3CBr$ (Markovnikov product). ^{13}C NMR can easily distinguish between them.

	$(CH_3)_3CBr$	$(CH_3)_3CHCH_2Br$
Number of non-equivalent carbon atoms	2	3
Ratio of peak area	3:1	2:1:1
Spin-spin splitting	one quartet one singlet	one quartet one triplet one doublet
Approximate position or resonance	8–30 δ (q) 25–65 δ (s)	8–30 δ (q), (d) 25–65 δ (t)

The product of HBr addition to 2–methylpropene in the presence of peroxides has a ^{13}C NMR spectrum identical to that of $(CH_3)_2CHCH_2Br$.

13.14

Compound	Kinds of non–equivalent protons
a) $\overset{1}{C}H_3\overset{2}{C}H_2Br$	Two
b) $\overset{1}{C}H_3O\overset{2}{C}H_2\overset{3}{C}H$ with $\overset{4}{C}H_3$ groups	Four
c) $\overset{1}{C}H_3\overset{2}{C}H_2\overset{3}{C}H_2NO_2$	Three
d) aromatic ring with CH3	Four
e) substituted alkene	Five

The two protons attached to the double bond are non-equivalent.

f)

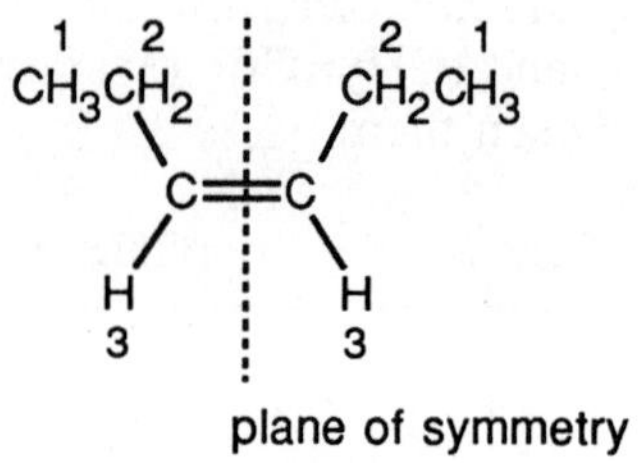

$$\underset{3}{\overset{1\ 2}{CH_3CH_2}} \quad \underset{3}{\overset{2\ 1}{CH_2CH_3}}$$

C=C

H   H
3   3

plane of symmetry

Three

13.15

Compound	δ	Kind of proton
a) Cyclohexane	1.43	secondary alkyl
b) CH_3COCH_3	2.17	methyl ketone
c) C_6H_6	7.37	aromatic
d) Glyoxal	9.70	aldehyde
e) CH_2Cl_2	5.30	protons adjacent to two halogens
f) $(CH_3)_3N$	2.12	methyl protons adjacent to nitrogen

13.16

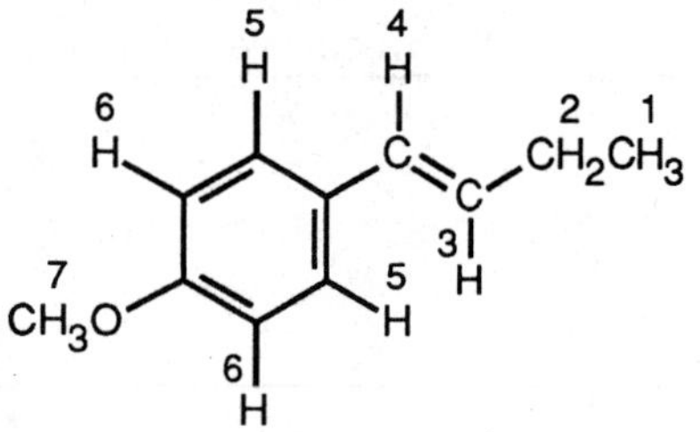

5   4
H   H
6
H.   C CH_2CH_3
           C
          3
7   5 H
CH_3O H
   6
   H

Seven different kinds of protons are present in the above structure.

Proton	δ	Kind of proton
1	1.0	primary alkyl
2	1.8	allylic
3	6.1	vinylic
4	6.2	vinylic
	(different from proton 3)	
5	7.2	aromatic
6	6.7	aromatic
7	3.8	ether

Note: The two "5" protons are equivalent to each other, as are the two "6" protons, because of free rotation around the bond joining the aromatic ring and the alkenyl side chain.

13.17

Compound	Proton	Number of Adjacent Protons	Splitting
a) $\overset{1}{C}HBr_2\overset{2}{C}H_3$	1	3	quartet
	2	1	doublet
b) $\overset{1}{C}H_3O\overset{2}{C}H_2\overset{3}{C}H_2Br$	1	0	singlet
	2	2	triplet
	3	2	triplet
c) $Cl\overset{1}{C}H_2\overset{2}{C}H_2\overset{1}{C}H_2Cl$	1	2	triplet
	2	4	quintet
d) $\overset{1}{C}H_3\overset{2}{C}H_2O\overset{3}{C}CH(CH_3)_2$ (with $\overset{O}{\parallel}$ on C3)	1	2	triplet
	2	3	quartet
	3	6	septet
	4	1	doublet

13.18

a) CH_3OCH_3

b) $CH_3CHClCH_3$

c) $ClCH_2CH_2OCH_2CH_2Cl$

d) $CH_3CH_2\overset{O}{\overset{\parallel}{C}}OCH_3$

13.19 The ^1H NMR spectrum shows two signals, corresponding to two types of hydrogens. These signals are in the ratio 9:14, or 2:3. Since the unknown contains 10 hydrogens, four protons are of one type and six are of the other type.

The upfield signal at 1.10 δ is due to saturated primary protons. The downfield signal at 3.50 δ is due to protons on carbon adjacent to an electronegative atom -- in this case, oxygen.

The signal at 1.1 δ is a *triplet*, and indicates two neighboring protons. The signal at 3.5 δ is a *quartet*, and indicates three neighboring protons. The unknown compound is diethyl ether, $CH_3CH_2OCH_2CH_3$.

13.20

3–Bromo–1–phenyl–1–propene

Coupling of the C2 proton to the C1 vinylic proton occurs with $J = 16$ Hz and causes the signal of the C2 proton to be split into a doublet. The C2 proton is also coupled to the two C3 protons with $J = 8$ Hz. This splitting causes each leg of the C2 proton doublet to be split into a triplet, producing six lines in all. Because of the size of the coupling constants, two of the lines coincide, and a quintet is observed.

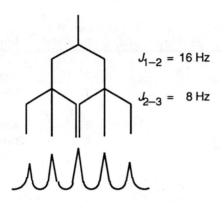

J_{1-2} = 16 Hz

J_{2-3} = 8 Hz

13.21

Compound	Number of Absorptions	Integrated peak area		Chemical Shift δ	Splitting
1 2 3 4 $CH_3CH_2CH_2OH$	4	1	3	0.92	triplet
		2	2	1.57	sextet
		3	2	3.58	triplet
		4	1	2–5	singlet
1 2 3 1 $CH_3CH(OH)CH_3$	3	1	6	1.20	doublet
		2	1	4.00	septet
		3	1	2–5	singlet

Since the number of absorptions, peak areas and splitting patterns differ, these two possible products can be easily distinguished by ^{1}H NMR. The product of hydroboration is $CH_3CH_2CH_2OH$.

13.22 See Problem 13.6 for the method of solution.
　　a) 2.18 δ　　b) 4.78 δ　　c) 7.52 δ　　d) 9.05 δ

13.23 See Problem 13.5 for the method of solution.
　　a) 168 Hz　　b) 276 Hz　　c) 504 Hz　　d) 616 Hz

13.24 a) Since the symbol "δ" indicates ppm downfield from TMS, chloroform absorbs at 7.3 ppm.

　　b) 　　$$\delta = \frac{\text{Observed chemical shift (\# Hz from TMS)}}{\text{Spectrometer frequency (in MHz)}}$$

　　　7.3 ppm $= \dfrac{\text{Chemical shift}}{360 \text{ MHz}}$; 7.3 ppm x 360 MHz = chemical shift

　　Chemical shift = 2600 Hz

　　c) δ is still 7.3 because the chemical shift measured in δ is independent of the operating frequency of the spectrometer.

13.25–13.26

Compound	Number of ^{13}C Absorptions	Splitting in spin-coupled spectrum			
		Singlets	Doublets	Triplets	Quartets
a)	5	1 (carbon 2)	0	3 (carbons 3, 4, 5)	1 (carbon 1)
b) $\overset{1}{C}H_3\overset{2}{C}H_2O\overset{3}{C}H_3$	3	0	0	1 (carbon 2)	2 (carbons 1, 3)
c)	6	1 (carbon 2)	1 (carbon 3)	3 (carbons 4, 5, 6)	1 (carbon 1)
d)	6	1 (carbon 2)	2 (carbons 1, 3)	1 (carbon 5)	2 (carbons 4, 6)
e)	4	0		1 (carbon 2)	2 (carbons 3, 4)	1 (carbon 1)
f)	4	1 (carbon 1)	0	3 (carbons 2, 3, 4)	0

13.27 ^{13}C NMR absorptions occur over a range of 250 ppm; ^{1}H NMR absorptions generally occur over a range of 10 ppm. The spread of peaks in ^{13}C NMR is much greater than accidental overlap is less likely. In addition, normal ^{13}C NMR spectra are uncomplicated by spin-splitting. The total number of lines is smaller and, again, overlap is less common.

13.28 a) The *chemical shift* is the exact position at which a nucleus absorbs rf energy in an NMR spectrum.

b) *Spin-spin splitting* is the splitting of a single NMR resonance into multiple lines. Spin-spin splitting occurs when the effective magnetic field felt by a nucleus is influenced by the small magnetic moments of adjacent nuclei. In ^{13}C NMR, the signal of a carbon bonded to n protons is split into $n + 1$ peaks. In ^{1}H NMR, the signal of a proton with n equivalent neighboring protons is split into $n + 1$ peaks. The magnitude of spin-spin splitting is given by the coupling constant J.

c) The *applied magnetic field* is the magnetic field that is externally applied to a sample by an NMR spectrometer.

d) The *spectrometer operating frequency* is the frequency of applied rf energy used by the spectrometer to bring a magnetic nucleus into resonance. The rf energy required depends on the magnetic field strength and on the nature of the nucleus being observed.

e) If the NMR signal of nucleus A is split by the spin of adjacent nucleus B, there is reciprocal splitting of the signal of nucleus B by the spin of nucleus A. The spins of the two nuclei are said to be coupled. The distance between two individual peaks within the multiplet of A is the same as the distance between two individual peaks within the multiplet of B. This distance, measured in Hz, is called the *coupling constant*.

f) The right side of an NMR spectrum is the *upfield* side; the left side is the *downfield* side. Nuclei that absorb upfield are more shielded and absorb at a higher applied magnetic field. Nuclei that absorb downfield are less shielded and absorb at a lower applied magnetic field.

13.29

Compound	Non-equivalent protons
a)	4
b) CH$_3$CH$_2$CH$_2$OCH$_3$ (1 2 3 4)	4
c)	2
d)	6

e)

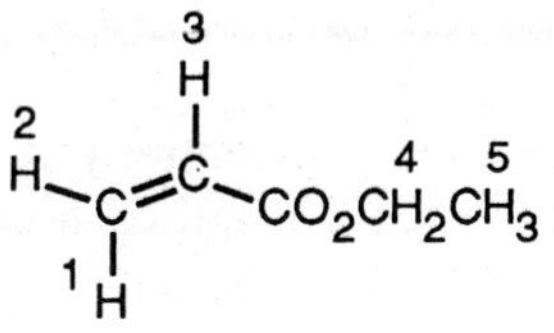

5

13.30

Lowest Chemical Shift ⟶ Highest Chemical Shift

CH_4 < Cyclohexane < CH_3COCH_3 < CH_2Cl_2, $H_2C=CH_2$ < benzene

0.23   1.43   2.17   5.30   5.33   7.37

13.31

^1H NMR:

Compound	Number of peaks	Peak Assignment	Splitting Pattern	
a) $(CH_3)_3\overset{2}{C}H$ (positions 1, 2)	2	1	doublet	(9H)
		2	multiplet (dectet)	(1H)
b) $\overset{1}{C}H_3\overset{2}{C}H_2\overset{3}{C}OOCH_3$	3	1	triplet	(3H)
		2	quartet	(2H)
		3	singlet	(3H)
c) (alkene structure)	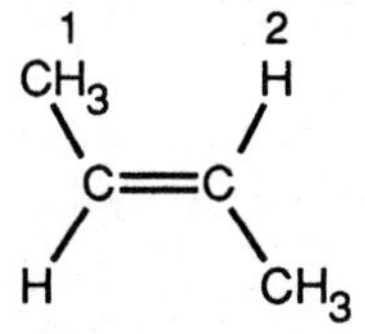 2	1	doublet	(6H)
		2	quartet	(2H)

^{13}C NMR:

Compound	Number of peaks	Peak Assignment	Splitting Pattern
a) $(CH_3)_3\overset{2}{C}H$ (positions 1, 2)	2	1	quartet
		2	doublet
b) $\overset{1}{C}H_3\overset{2}{C}H_2\overset{3}{C}O\overset{4}{O}CH_3$	4	1	quartet
		2	triplet
		3	singlet
		4	quartet
c) (alkene structure)	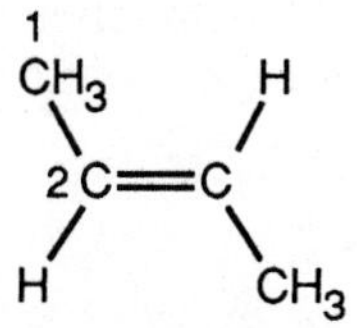 2	1	quartet
		2	doublet

13.32

^{1}H: CH$_3$CH$_2$COOCH(CH$_3$)$_2$
    1   2   3   4

^{13}C: CH$_3$CH$_2$COOCH(CH$_3$)$_2$
       1   2   3   4   5

Peak Assignment	Splitting Pattern	
1	triplet	(3H)
2	quartet	(2H)
3	septet	(1H)
4	doublet	(6H)
1	quartet	
2	triplet	
3	singlet	
4	doublet	
5	quartet	

13.33 Use of ^{13}C NMR to distinguish between the two isomers has been described in the text. ^{1}H NMR can also be useful.

A   B

Isomer A has only four kinds of protons because of symmetry. Its vinylic proton absorption (4.5 – 6.5 δ) represents two hydrogens. Isomer B contains six different kinds of protons. Its ^{1}H NMR shows an unsplit methyl group signal and one vinylic proton signal of relative area 1. These differences make it possible to distinguish between A and B.

13.34 First, examine each isomer for structural differences that are obviously recognizable in the NMR spectrum. If it is not possible to pick out distinguishing features immediately, it may be necessary to sketch an approximate spectrum of each isomer for comparison.

a) CH$_3$CH=CHCH$_2$CH$_3$ has two vinylic protons with chemical shifts at 5.4 – 5.5 δ. Because ethylcyclopropane shows no signal in this region, it should be easy to distinguish one isomer from the other.

b) CH$_3$CH$_2$OCH$_2$CH$_3$ has two kinds of protons; its ^{1}H NMR spectrum consists of two resonances -- a triplet and a quartet. CH$_3$OCH$_2$CH$_2$CH$_3$ has four different types of protons, and its spectrum is more complex. In particular, the methyl group bonded to oxygen shows an unsplit singlet absorption.

c) Because these isomers have the same number and types of resonances, it is necessary to examine the ^{1}H NMR spectrum of each isomer for identification.

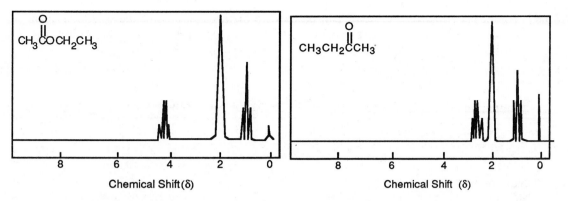

The absorptions at either 2.1 – 2.4 δ (protons adjacent to a carbonyl group) or at 3.3 – 4.0 δ (protons adjacent to oxygen) can be used to identify the isomers.

d) Each isomer contains four different kinds of protons -- two kinds of methyl protons and two kinds of vinylic protons. For the first isomer, the methyl resonances are both singlets, whereas for the second isomer, one resonance is a singlet and one is a doublet.

13.35

a) $(CH_3)_4C$ b)   c)

13.36 a) C_3H_6O contains one double bond or ring.
b) Possible structures for C_3H_6O include:

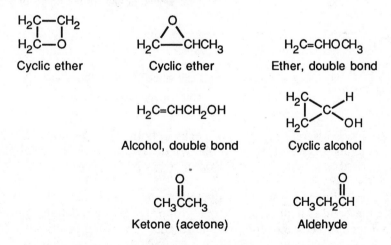

c) Carbonyl functional groups absorb at 1715 cm^{-1} in the infrared. Only the last two compounds above show an infrared absorption in this region.
d) Because the aldehyde from part b) contains three different kinds of protons, its ^1H NMR spectrum exhibits three resonances. The ketone shows only one resonance. Since the unknown compound of this problem shows only one ^1H NMR absorption (in the methyl ketone region), it must be acetone.

13.37 Either ^{1}H NMR or ^{13}C NMR can be used to distinguish among these isomers. In either case, it is first necessary to find the number of different kinds of protons or carbon atoms.

Compound	H$_2$C—CH$_2$ $\vert$ $\vert$ H$_2$C—CH$_2$	H$_2$C=CHCH$_2$CH$_3$	CH$_3$CH=CHCH$_3$	(CH$_3$)$_2$C=CH$_2$
Kinds of protons	1	5	2	2
Kinds of carbon atoms	1	4	2	3
Number of ^{1}H NMR peaks	1	5	2	2
Number of ^{13}C NMR peaks	1	4	2	3

^{13}C NMR is the preferred method for identifying these compounds; each isomer differs in the number of absorptions in its ^{13}C NMR spectrum.

^{1}H NMR can also be used to distinguish among the isomers. The two isomers that show two ^{1}H NMR peaks differ in their splitting patterns.

13.38

	Number of peaks	Distinguishing Absorptions
^{1}H	5	Unsplit vinylic peak, relative area 1
^{13}C	7	Two vinylic peaks (one singlet, one doublet)
^{1}H	4	Split vinylic peak, relative area 2
^{13}C	5	One vinylic peak (doublet)

The two isomers have different numbers of peaks in both ^{1}H NMR and ^{13}C NMR. In addition, the distinguishing absorptions in the vinylic region of both the ^{1}H and ^{13}C spectra make it possible to identify each isomer by its NMR spectrum.

13.39 The ketone IR absorption of 3–methyl–2–cyclohexenone occurs near 1690 cm^{-1} because the double bond is next to the ketone group. The ketone IR absorption of 4–cyclopentenyl methyl ketone occurs near 1715 cm^{-1}, the usual position for ketone absorption.

13.40 BrCH$_2$CH$_2$CH$_2$Br.

13.41

a) $(CH_3)_2CHCCH_3$ with a carbonyl group (C=O) on the carbon

b) A structure showing:

$$\begin{array}{c} CH_3 \\ \\ Br \end{array} C=C \begin{array}{c} H \\ \\ H \end{array}$$

13.42 Possible structures for $C_4H_7ClO_2$ are $CH_3CH_2CO_2CH_2Cl$ and $ClCH_2CO_2CH_2CH_3$. Chemical-shift data can distinguish between these two possible structures.

$$CH_3CH_2\overset{\overset{\displaystyle O}{\|}}{C}OCH_2Cl \qquad\qquad ClCH_2\overset{\overset{\displaystyle O}{\|}}{C}OCH_2CH_3$$

I   II

In I, the protons attached to the carbon bonded to both oxygen and chlorine ($-OCH_2Cl$) absorb far downfield ($5.0 - 6.0\ \delta$). Because no signal is present in this region of the ^1H NMR spectrum given, the unknown must be II.

13.43

a)
$$\begin{array}{c} H_3C \\ \\ Cl \end{array} C=C \begin{array}{c} H \\ \\ CH_2Cl \end{array} \quad Z$$

The *E* isomer is also a correct answer.

b) $C(CH_3)_3$ on a benzene ring

c) $BrCH_2CH_2\overset{\overset{\displaystyle O}{\|}}{C}CH_3$

d) $CH_2CH_2CH_2Br$ on a benzene ring

13.44 The molecular formula of the isomers, $C_{12}H_{16}$, corresponds to five multiple bonds and/or rings. We will try to differentiate between the isomers by using both ^1H NMR and ^{13}C NMR.

A   B

^1H *NMR*. Isomer **A** has two kinds of protons -- those in four-membered rings and those in the eight-membered ring. Each group of eight protons absorbs in the allylic region of the spectrum (C=C–C–H; $1.5 - 2.5\ \delta$). Two kinds of protons are also present in isomer **B**, but you will have to build a model to see the difference between them. One group of eight protons points toward a double bond; the other group of eight protons points away from a double bond. Both groups of protons are also allylic and absorb in the region of $1.5 - 2.5\ \delta$. ^1H NMR thus cannot be used to distinguish between **A** and **B**.

^{13}C *NMR.* Isomer $\underline{A}$ has three different kinds of carbon atoms:

Carbon atom		Quantity	Chemical shift
=C̶		four	100–150 δ
̶CH_2̶	(in 4-membered rings)	four	15–55 δ
̶CH_2̶	(in 8-membered rings)	four	15–55 δ

Isomer $\underline{B}$ contains only two different kinds of carbon atoms:

	Quantity	Chemical shift
=C̶	four	100–150 δ
̶CH_2̶	eight	15–55 δ

Three resonances should appear in the ^{13}C NMR spectrum of $\underline{A}$; only two resonances should appear in the ^{13}C NMR spectrum of $\underline{B}$. Although ^1H NMR cannot distinguish between $\underline{A}$ and $\underline{B}$, ^{13}C NMR will solve the problem.

13.45

a) $(CH_3)_2CHCH_2Br$

b) $CH_3CHCH_2CH_2Cl$
$|$
Cl

13.46

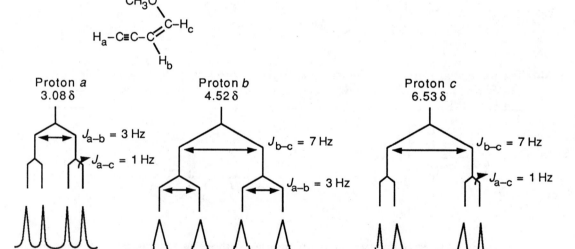

The absorption in the ^1H NMR spectrum can be identified by comparison with the tree diagrams. H_a absorbs at 3.08 δ, H_b absorbs at 4.52 δ, and H_c absorbs at 6.35 δ.

13.47

Carbon	δ (ppm)
1	14
2	61
3	166
4	
5	
6	127–133 (four absorptions)
7	

Ethyl benzoate

13.48 Compound A (4 multiple bonds and/or rings) must be symmetrical because it exhibits only six peaks in its ^{13}C NMR spectrum. Saturated carbons account for two of these peaks (δ = 15, 28 ppm), and unsaturated carbons account for the other four (δ = 119, 129, 131, 143 ppm).

^{1}H NMR shows a triplet (3H at 1.1 δ), and a quartet (2H at 2.5 δ), indicating the presence of an ethyl group. The other signals (4H at 6.9 – 7.3 δ are due to aromatic protons.

Compound A

13.49

a) $(CH_3O)_2CHCH_3$

b)

c)

13.50 The peak in the mass spectrum at m/z = 84 is probably the molecular ion of the unknown compound and corresponds to a molecular weight of 84 (C_6H_{12} -- one double bond or ring).

^{13}C NMR shows three different kinds of carbons and indicates a symmetrical hydrocarbon. The absorption at 132 δ is due to a vinylic carbon atom. A reasonable structure for the unknown is:

$$CH_3CH_2CH=CHCH_2CH_3$$

3–Hexene

Study Guide for Chapter 13

After studying this chapter, you should be able to:

(1)   Understand the principle of NMR and be able to define important terms (13.1, 13.2, 13.27, 13.28).

(2)   Calculate the relationship between delta value, chemical shift and spectrometer operating frequency (13.5, 13.6, 13.22, 13.23, 13.24).

(3)   Identify non-equivalent carbons and hydrogens, and predict the number of signals appearing in the ^{1}H NMR and ^{13}C NMR spectra of compounds (13.3, 13.4, 13.8, 13.14, 13.25, 13.29).

(4)   Assign resonances to specific carbons or hydrogens of a given structure (13.7, 13.30, 13.47, 13.48).

(5)   Propose structures for compounds, given their NMR spectrum (13.9, 13.19, 13.35, 13.40, 13.41, 13.42, 13.43, 13.45, 13.48, 13.49).

(6)   Describe or sketch spectra corresponding to a given compound (13.10, 13.11, 13.12, 13.15, 13.16, 13.18).

(7)   Predict ^{1}H splitting patterns, using tree diagrams if necessary (13.17, 13.20, 13.31, 13.32, 13.46).

(8)   Use NMR to distinguish between isomers (13.34, 13.36, 13.37, 13.38).

(9)   Use NMR to identify reaction products (13.13, 13.21, 13.33, 13.44).

14.1

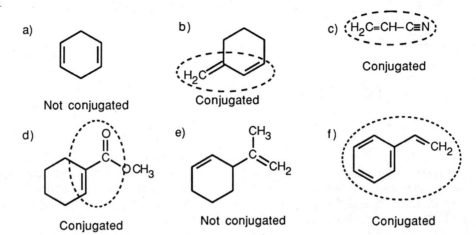

a) Not conjugated

b) Conjugated

c) Conjugated

d) Conjugated

e) Not conjugated

f) Conjugated

14.2

We would expect $\Delta H_{hydrog} = 30.3 + 30.3 = 60.6$ kcal/mol for allene if the heat of hydrogenation for each double bond were the same as that for an isolated double bond. The observed $\Delta H_{hydrog} = 71.3$ kcal/mol is 10.7 kcal/mol *higher* than the expected value. Allene is higher in energy (less stable) than a non-conjugated diene, which in turn is less stable than a conjugated diene.

14.3

$CH_3CH=CHCH=CH_2$ 1,3–Pentadiene

Product	Name	Results from:
(1) $CH_3CH=CHCHClCH_3$	4–Chloro–2–pentene	1,2 Addition 1,4 Addition
(2) $CH_3CH_2CHClCH=CH_2$	3–Chloro–1–pentene	1,2 Addition
(3) $CH_3CH_2CH=CHCH_2Cl$	1–Chloro–2–pentene	1,4 addition

14.4

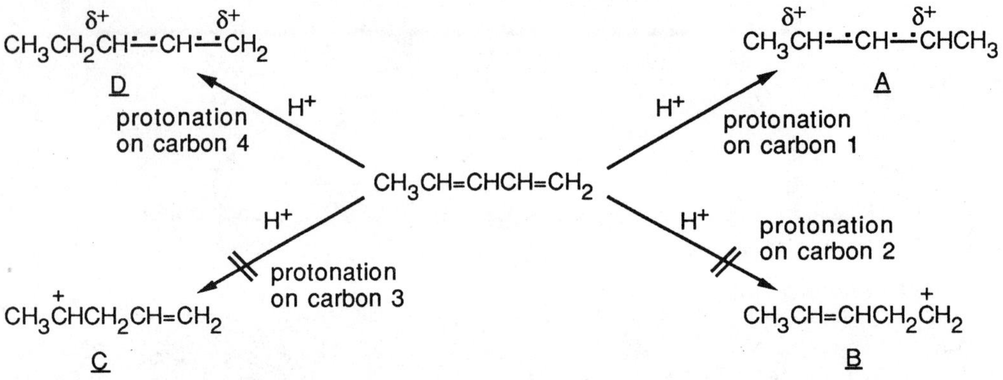

$$CH_3CH_2CH \overset{\delta+}{\cdots} CH \overset{\delta+}{\cdots} CH_2$$

D

protonation
on carbon 4 H^+

$$CH_3CH \overset{\delta+}{\cdots} CH \overset{\delta+}{\cdots} CHCH_3$$

A

H^+ protonation
on carbon 1

$$CH_3CH=CHCH=CH_2$$

H^+ protonation
on carbon 2

H^+

protonation
on carbon 3

$$CH_3\overset{+}{C}HCH_2CH=CH_2$$

C

$$CH_3CH=CHCH_2\overset{+}{C}H_2$$

B

The positive charge of allylic carbocation **A** is delocalized over three secondary carbons; the positive charge of carbocation **D** is delocalized over secondary carbons and one primary carbon. We therefore predict that carbocation **A** is the major intermediate formed, and that product (1) in problem 14.3 predominates. Note that product (1) results from both 1,2– and 1,4–addition.

14.5

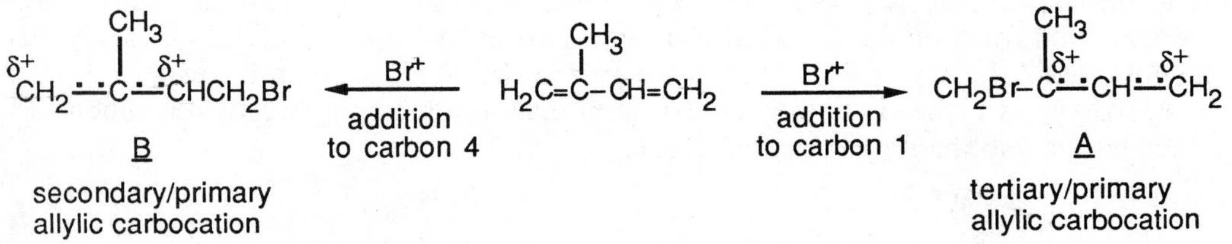

$$\overset{CH_3}{\underset{\text{}}{\overset{|}{\underset{}{CH_2 \overset{\delta+}{\cdots} C \overset{\delta+}{\cdots} CHCH_2Br}}}}$$

B

secondary/primary
allylic carbocation

$\xleftarrow{Br^+}$ addition to carbon 4

$$\overset{CH_3}{\underset{}{\overset{|}{\underset{}{H_2C=C-CH=CH_2}}}}$$

$\xrightarrow{Br^+}$ addition to carbon 1

$$\overset{CH_3}{\underset{}{\overset{|}{\underset{}{CH_2Br-C \overset{\delta+}{\cdots} CH \overset{\delta+}{\cdots} CH_2}}}}$$

A

tertiary/primary
allylic carbocation

Tertiary/primary allylic carbocation **A** is more stable than secondary/primary allylic carbocation **B**. Since the products formed from the more stable intermediate predominate, 3,4–dibromo–3–methyl– 1–butene is the major product of 1,2 addition of bromine to isoprene.

14.6   Figure 14.4 shows the three pi molecular orbitals of an allylic pi system. An allyl radical has three pi electrons. Two of them occupy the bonding molecular orbital, and the third electron occupies the non-bonding orbital.

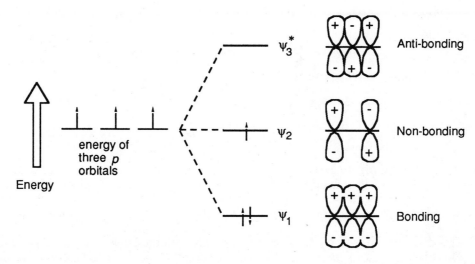

14.7

CH₃CHBrCH=CH₂ $\underset{:\ddot{B}\ddot{r}:}{\overset{-:\ddot{B}\ddot{r}:}{\rightleftharpoons}}$ $[\overset{+}{CH_3CHCH=CH_2}]$

3–Bromo–1–butene

CH₃CH=CHCH₂Br $\underset{:\ddot{B}\ddot{r}:}{\overset{-:\ddot{B}\ddot{r}:}{\rightleftharpoons}}$ $[CH_3CH=CH\overset{+}{CH_2}]$

1–Bromo–2–butene

Allylic halides can undergo slow ionization to form stabilized carbocations. Both 3–bromo–1–butene and 1–bromo–2–butene form the same allylic carbocation, pictured above, upon ionization. Addition of bromide ion to the allylic carbocation then occurs to form a mixture of bromobutenes. Since the reaction is run under equilibrium conditions, the thermodynamically more stable 1–bromo–2–butene predominates.

14.8

H₂C=CHCH=CH₂ + X₂

1,2 → $H_2\overset{X}{\underset{|}{C}}CHCH=CH_2$ monosubstituted double bond
         $\underset{|}{X}$

1,4 → $H_2\overset{X}{\underset{|}{C}}CH=CHCH_2$ disubstituted double bond
                          $\underset{|}{X}$

1,4–adducts are more stable than 1,2–adducts because disubstituted double bonds are more stable than monosubstituted double bonds (see Chapter 6).

14.9

Good Dienophiles: (a)$H_2C=CHCCl$ (with C=O), (d) [structure]

Poor Dienophiles: (b)$H_2C=CHCH_2CH_2COOCH_3$, (c) [structure], (e) [structure]

Compound (a) and (d) are good dienophiles because they have electron-withdrawing groups conjugated with a double bond. Alkene (c) is a poor dienophile because it has no electron-withdrawing functional group. Compounds (b) and (e) are poor dienophiles because their electron-withdrawing groups are not conjugated with the double bond.

14.10 a) This diene has an *s-cis* conformation and should undergo Diels-Alder cycloaddition.
   b) This diene has an *s-trans* conformation. Because the double bonds are in a fused ring system, it is not possible for them to rotate to an *s-cis* conformation.
   c) Rotation can also occur about the single bond of this *s-trans* diene. The resulting *s-cis* conformation, however, has an unfavorable steric interaction of the interior methyl group with a hydrogen at carbon 1. Rotation to the *s-cis* conformation is therefore not likely.

s–trans
(more stable)

s–cis
(less stable)

14.11 The difference in reactivity of the three cyclic dienes is due to steric factors. As the "non-diene" part of the molecule becomes larger, the carbon atoms at the end of the diene portion of the ring are forced farther apart. Overlap with the pi system of the dienophile in the pericyclic transition state is poorer, and reaction is slower.

14.12

$$200 \text{ nm} = 200 \times 10^{-7} \text{ cm} = 2 \times 10^{-5} \text{ cm}$$
$$400 \text{ nm} = 400 \times 10^{-7} \text{ cm} = 4 \times 10^{-5} \text{ cm}$$

$$E = \frac{2.86 \times 10^{-3} \text{ kcal/mol}}{\lambda \text{ (in cm)}}$$

for $\lambda = 2 \times 10^{-5}$ cm

$$E = \frac{2.86 \times 10^{-3} \text{ kcal/mol}}{2 \times 10^{-5}} = 1.4 \times 10^{2} \text{ kcal/mol}$$

for $\lambda = 4 \times 10^{-5}$ cm

$$E = \frac{2.86 \times 10^{-3} \text{ kcal/mol}}{4 \times 10^{-5}} = 0.72 \times 10^{2} \text{ kcal/mol} = 72 \text{ kcal/mol}$$

The energy of electromagnetic radiation occurs over the range of 72–140 kcal/mol.

14.13

Energy (in kcal/mol)	UV	IR	^{1}H NMR (at 60 MHz)
	72 – 140	1.1 – 11	5.7×10^{-6}

The energy required for UV transitions is greater than the energy required for IR or ^{1}H NMR transitions.

14.14

$$\varepsilon = \frac{A}{C \cdot l}$$

Where ε = molar absorptivity
A = absorbance
l = sample path length (in cm)
C = concentration (in M)

In this problem

$\varepsilon = 50{,}100 = 5.01 \times 10^{4}$
$l = 1.0$ cm
$A = 0.735$

$$C = \frac{A}{\varepsilon \times l} = \frac{0.735}{5.01 \times 10^{4} \times 1.0} = 1.47 \times 10^{-5} \text{ M}$$

14.15 All compounds having *alternating* single and multiple bonds should show ultraviolet absorption in the range 200–400 nm. Only compound (a) is not UV-active. All of the compounds pictured below are UV active.

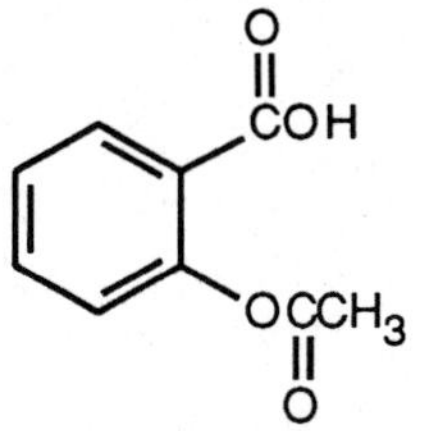

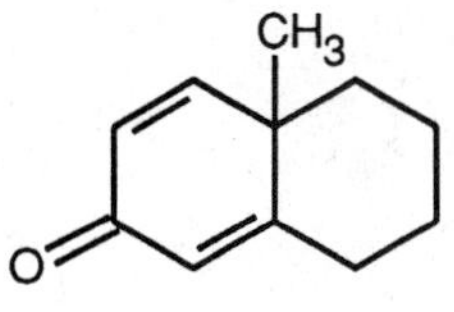

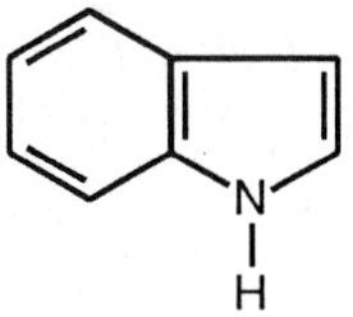

14.16 a) 3–Methyl–2,4–hexadiene
c) 2,3,5–Heptatriene
b) 1,3,5–Heptatriene
d) 3–Propyl–1,3–pentadiene

14.17

a) 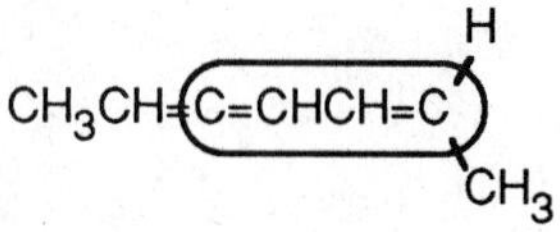 $CH_3CH=C=CHCH=C$ ⟨H, CH_3⟩

b)

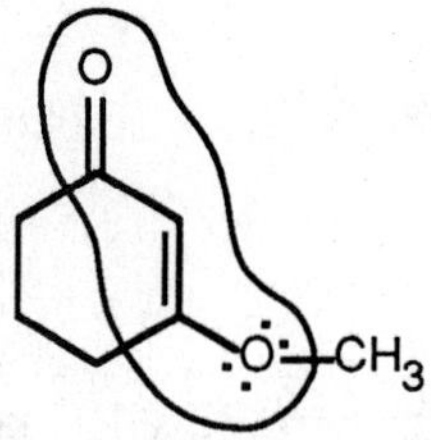

c)

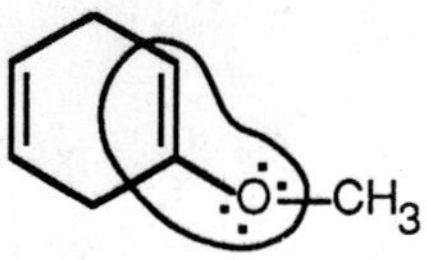

d)

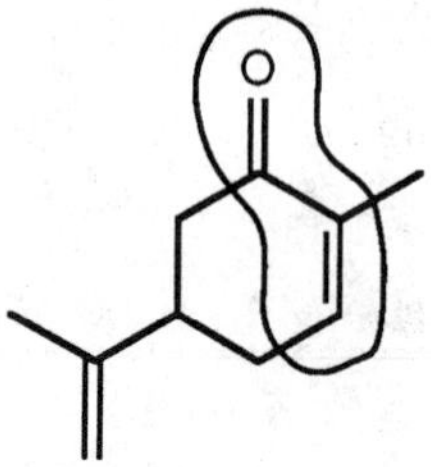

e)

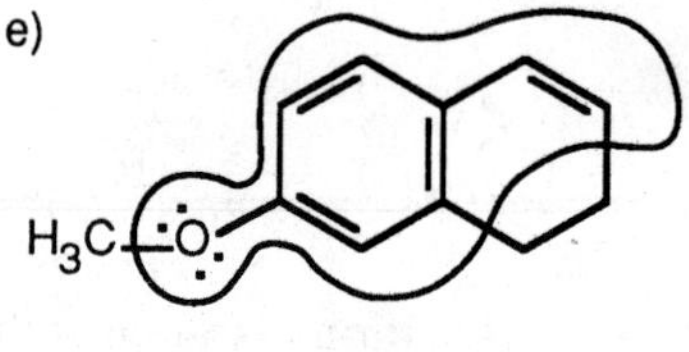

f)

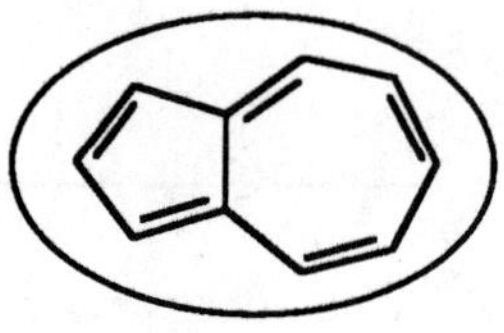

14.18

a)

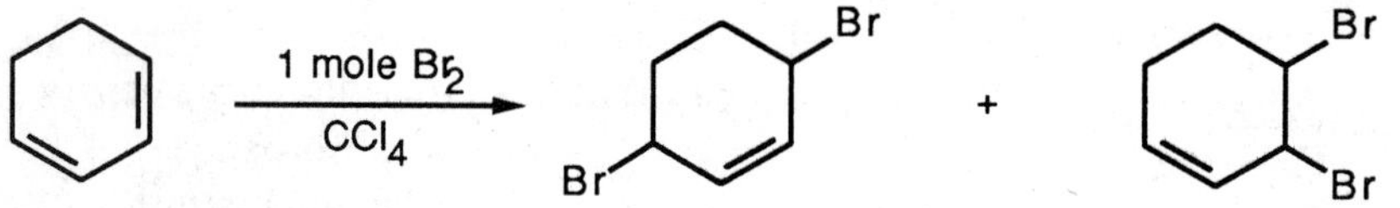

b)

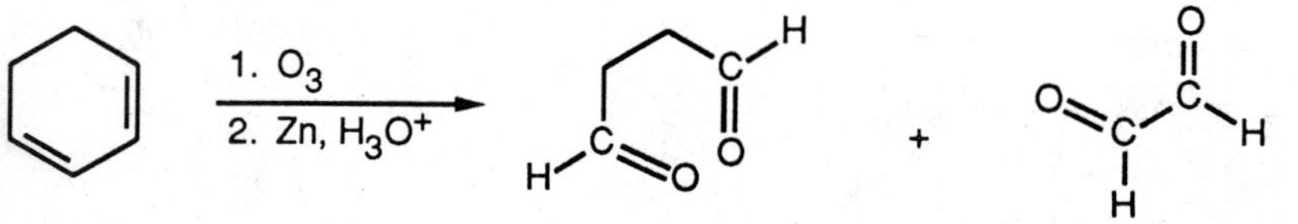

c)

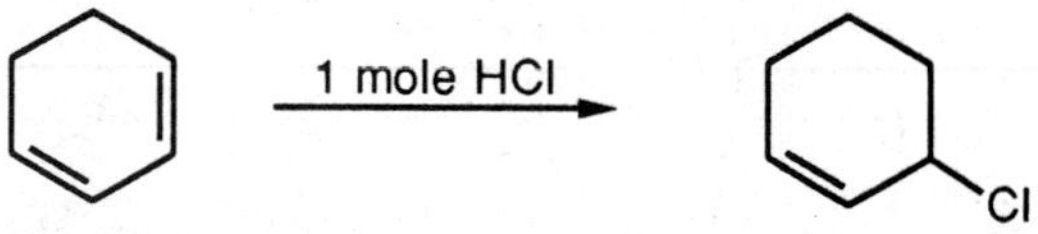

d)

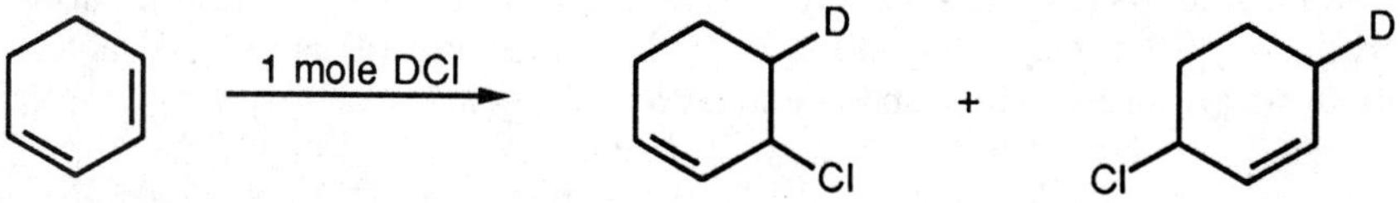

e)

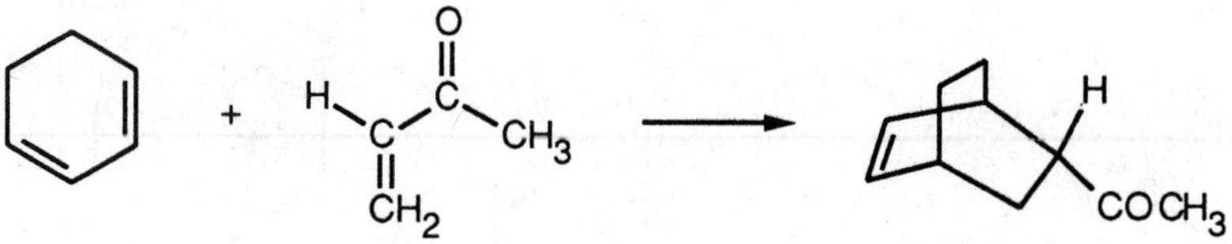

f)

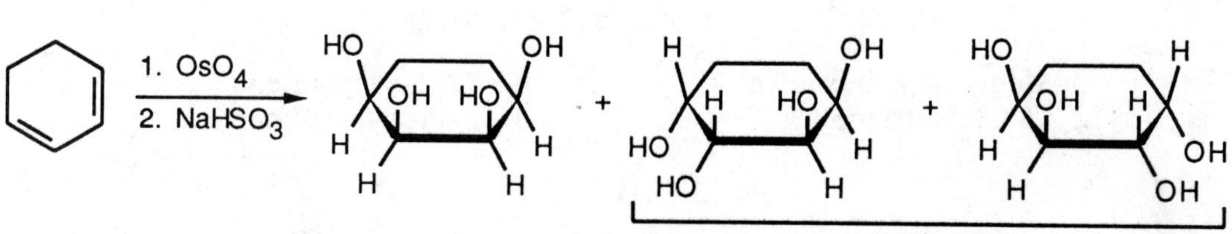

14.19

Conjugated Dienes:	$CH_3CH=CHCH=CH_2$ 1,3–Pentadiene	$\overset{\displaystyle CH_3}{\underset{\displaystyle }{H_2C=CHC=CH_2}}$ 2–Methyl–1,3–butadiene
Cumulated Dienes:	$CH_3CH_2CH=C=CH_2$ 1,2–Pentadiene	$CH_3CH=C=CHCH_3$ 2,3–Pentadiene
	$H_2C=C=C(CH_3)_2$ 3–Methyl–1,2–butadiene	
Non-conjugated Diene:	$H_2C=CHCH_2CH=CH_2$ 1,4–Pentadiene	

14.20

	$CH_3CH_2C\equiv CCH_2CH_3$ 3–Hexyne	$CH_3CH=CHCH=CHCH_3$ 2,4–Hexadiene	$CH_3CH_2CH=C=CHCH_3$ 2,3–Hexadiene
^1H:	2 peaks triplet, quartet below 2.0 δ	3 peaks two in region 4.5–6.5 δ	5 peaks two in region 4.5–6.5 δ
^{13}C:	3 peaks 8–55 δ : 2 65–85 δ : 1	3 peaks 8–30 δ : 1 100–150 δ : 2	6 peaks 8–55 δ : 3 100–150 δ : 2 ~ 200 δ : 1 (sp carbon)
UV Absorption?	No	Yes	No

2,4–Hexadiene can easily be distinguished from the other two isomers; it is the only isomer that absorbs in the UV region. The other two isomers show significant differences in their ^1H and ^{13}C NMR spectra and can be identified by either technique.

14.21 In order to absorb in the 200–400 nm range, an alkene must be conjugated. Since the double bonds of allene are not conjugated, allene does not absorb light in the UV region. Allene absorbs light at a wavelength lower than 200 nm, in the range where unconjugated double bonds slow absorption.

14.22

a)

b)

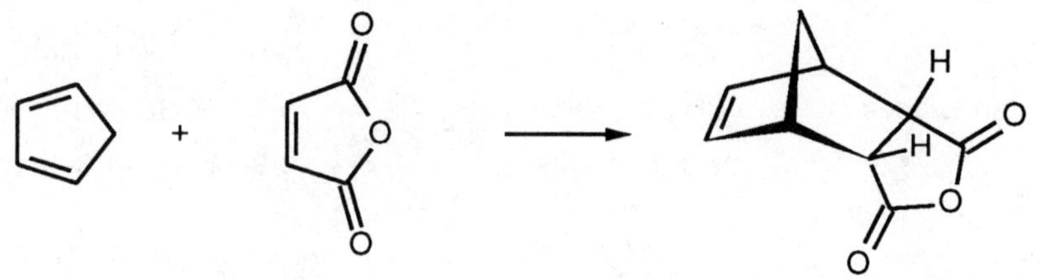

c)

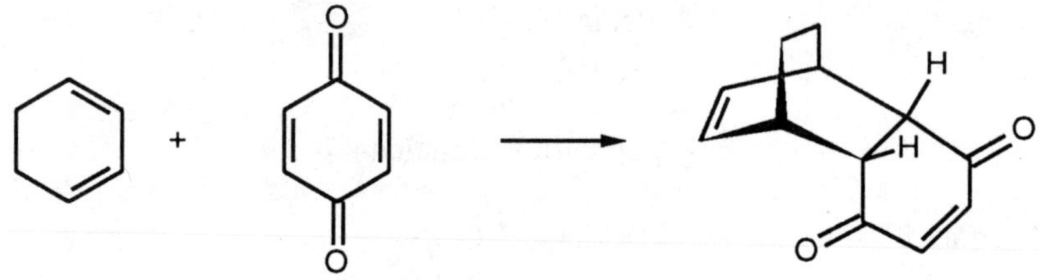

If two moles of cyclohexadiene are present for each mole of dienophile, you can also obtain a second product:

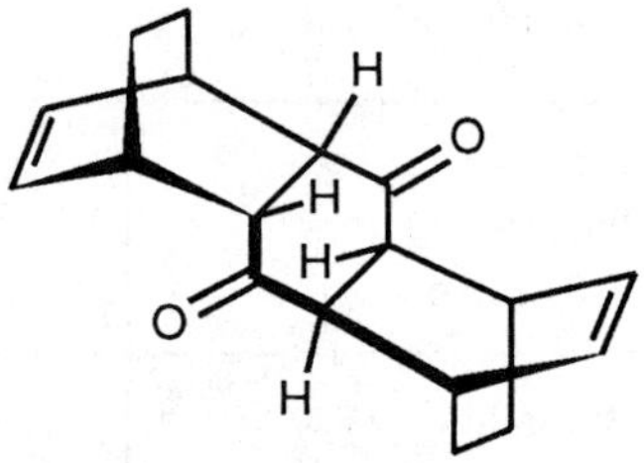

d)

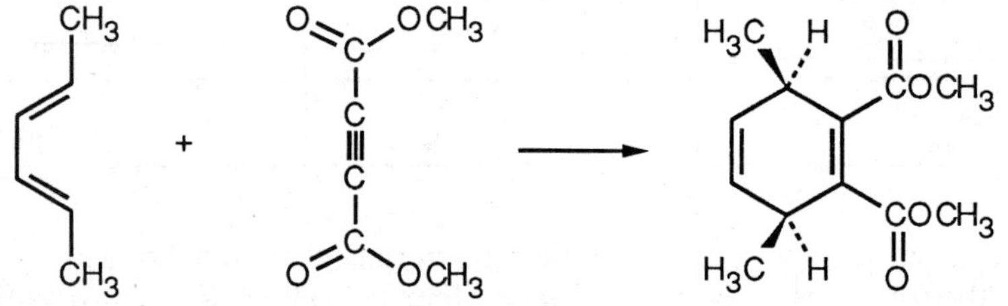

14.23

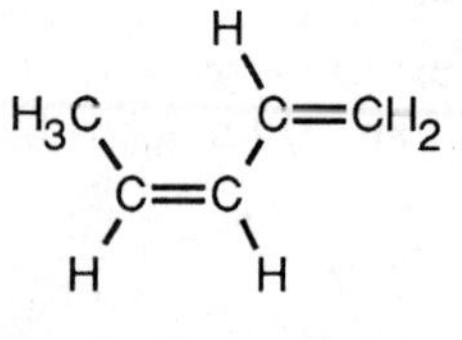

cis–1,3–Pentadiene 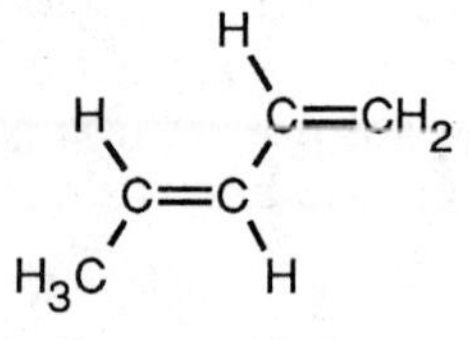   trans–1,3–Pentadiene

Both pentadienes are more stable in *s–trans* conformations. To undergo Diels-Alder reactions, they must rotate about the single bond between the double bonds to assume *s–cis* conformations.

cis–1,3–Pentadiene   trans–1,3–Pentadiene

It's not difficult for the *trans* isomer to assume the *s–cis* conformation. When *cis*–1,3–pentadiene rotates to the *s–cis* conformation, steric interaction occurs between the methyl-group protons and a proton on carbon 1. Since it's more difficult for *cis*–1,3–pentadiene to assume the *s–cis* conformation, it is less reactive in the Diels-Alder reaction.

14.24 Only compounds having alternating multiple bonds show $\pi \rightarrow \pi^*$ ultraviolet absorptions in the 200–400 nm range. Of the compounds shown, only pyridine (b) absorbs in this range.

14.25 $HC \equiv CC \equiv CH$ can't be used as a Diels-Alder diene because it is linear. The end carbons are too far apart to be able to react with a dienophile in a cyclic transition state.

14.26 Among the possible structures:

14.27 Protonation on carbon 1:

Protonation on carbon 2:

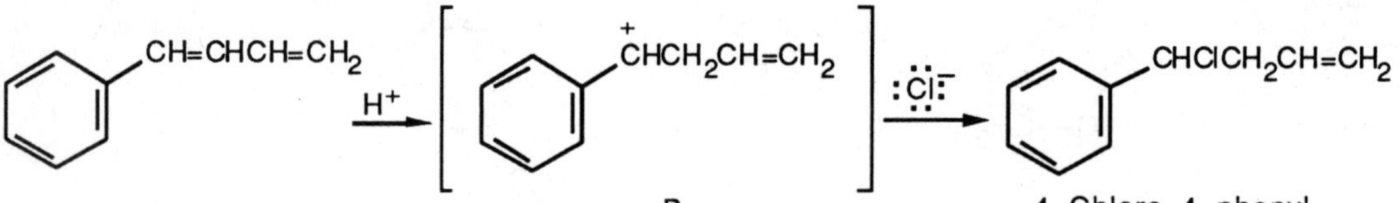

$$\text{C}_6\text{H}_5\text{CH=CHCH=CH}_2 \xrightarrow{\text{H}^+} \left[\text{C}_6\text{H}_5\overset{+}{\text{C}}\text{HCH}_2\text{CH=CH}_2 \right] \xrightarrow{:\overset{..}{\text{Cl}}:^-} \text{C}_6\text{H}_5\text{CHClCH}_2\text{CH=CH}_2$$

B

4–Chloro–4–phenyl–
1–butene

Protonation on carbon 3:

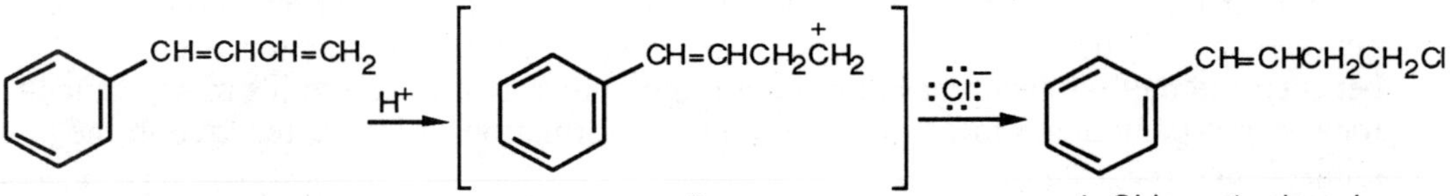

$$\text{C}_6\text{H}_5\text{CH=CHCH=CH}_2 \xrightarrow{\text{H}^+} \left[\text{C}_6\text{H}_5\text{CH=CHCH}_2\overset{+}{\text{C}}\text{H}_2 \right] \xrightarrow{:\overset{..}{\text{Cl}}:^-} \text{C}_6\text{H}_5\text{CH=CHCH}_2\text{CH}_2\text{Cl}$$

C

4–Chloro–1–phenyl–
1–butene

Protonation on carbon 4:

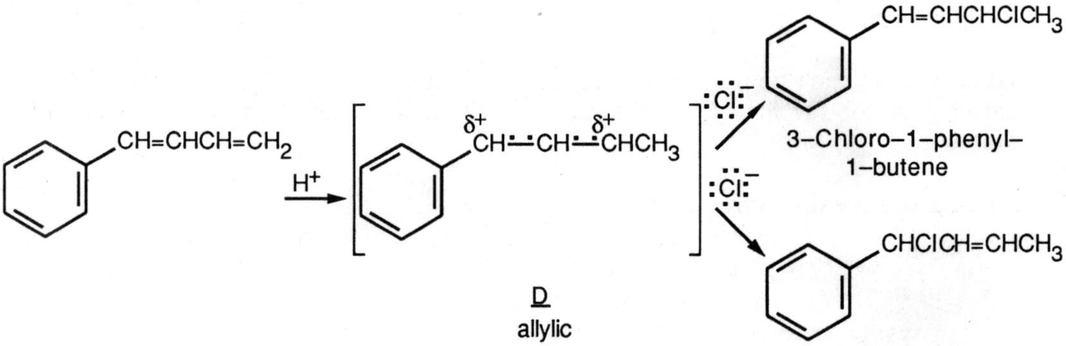

D
allylic

CH=CHCHClCH₃

3–Chloro–1–phenyl–
1–butene

CHClCH=CHCH₃

1–Chloro–1–phenyl–
2–butene

Carbocation D is most stable because it can use the pi systems of both the benzene ring and the side chain to further delocalize positive charge.
3–Chloro–1–phenyl–1–butene is the major product because it results from cation D and because its double bond can conjugate with the benzene ring to provide extra stability.

14.28

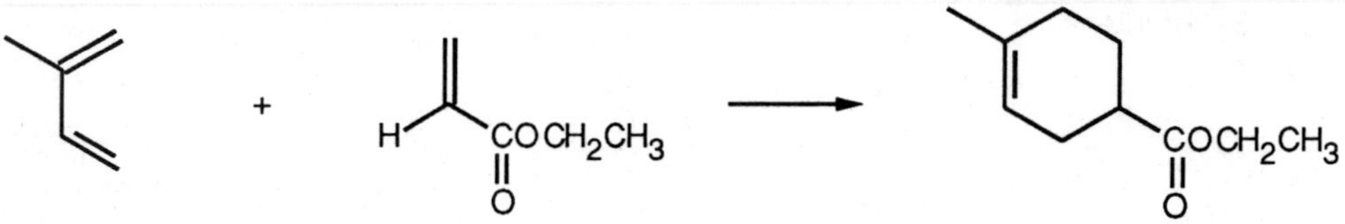

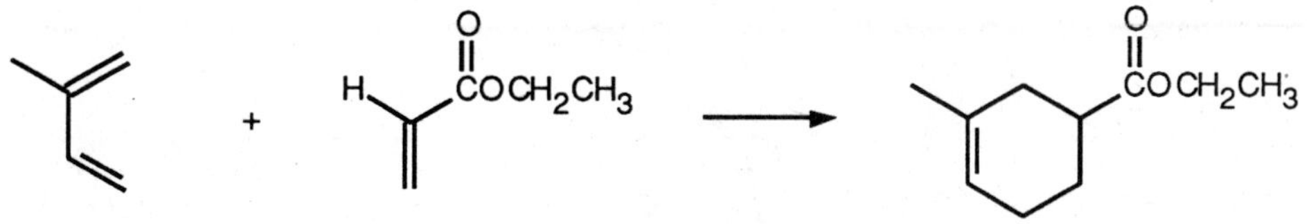

Two different orientations of the dienophile ester group are possible in the cyclic transition state, and two different products can form.

14.29 The most reactive dienophiles contain electron-withdrawing groups.

Most reactive ⟶ Least reactive

$(NC)_2C=C(CN)_2$ > $H_2C=CHCHO$ > $H_2C=CHCH_3$ > $(CH_3)_2C=C(CH_3)_2$

| Four electron-withdrawing groups | One electron-withdrawing group | | Four electron-*donating* groups |

14.30 First, find the six-membered ring formed by the Diels-Alder reaction. Locate the new bonds; you should then be able to identify the diene and the dienophile.

a)

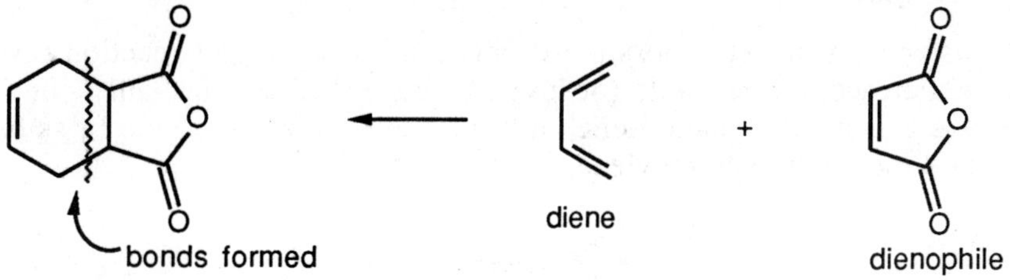

b)

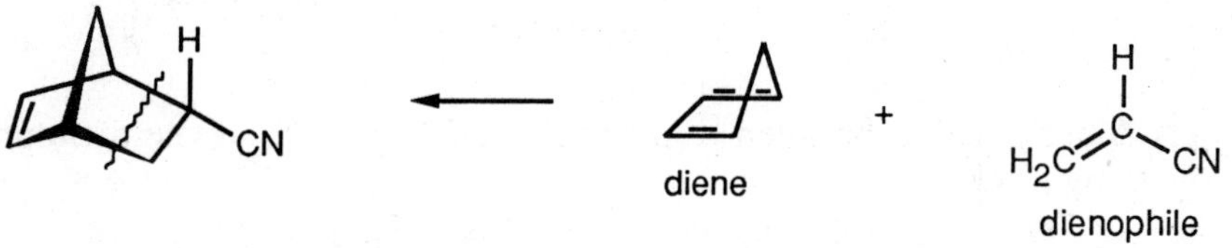

c)

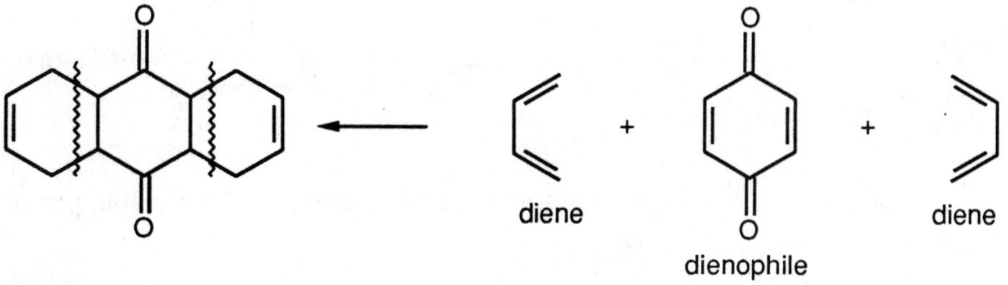

d)

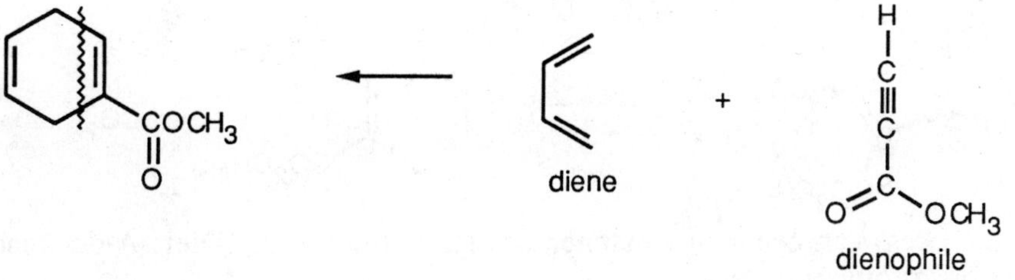

14.31

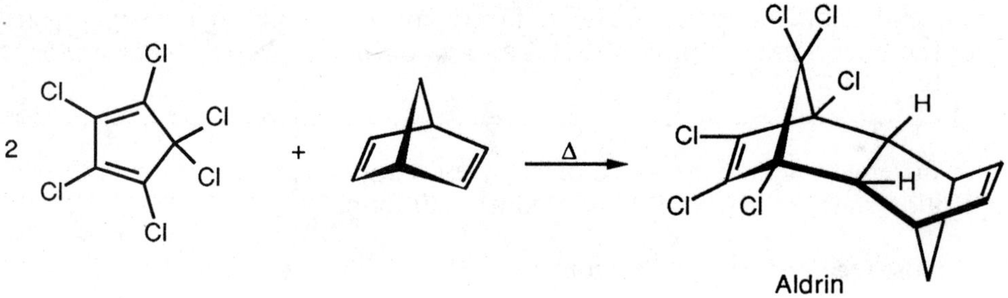

Aldrin

14.32

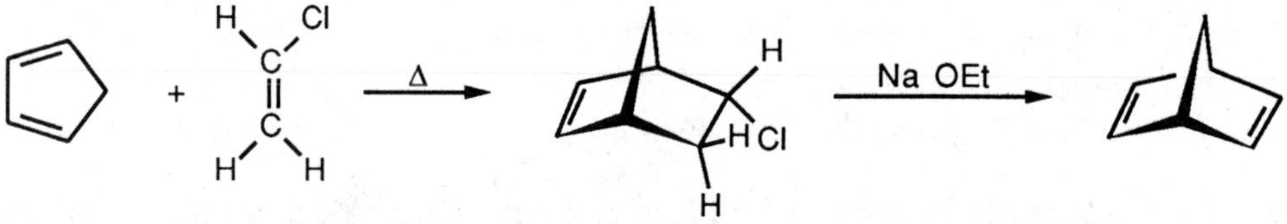

Diels-Alder reaction   E2 elimination

14.33 In the instances when it is possible to make the Diels-Alder reaction reversible, the products are much more stable (of lower energy) than the reactants. In this case, the reactant is a non-conjugated diene, and the products are benzene (a stable, conjugated molecule) and ethylene.

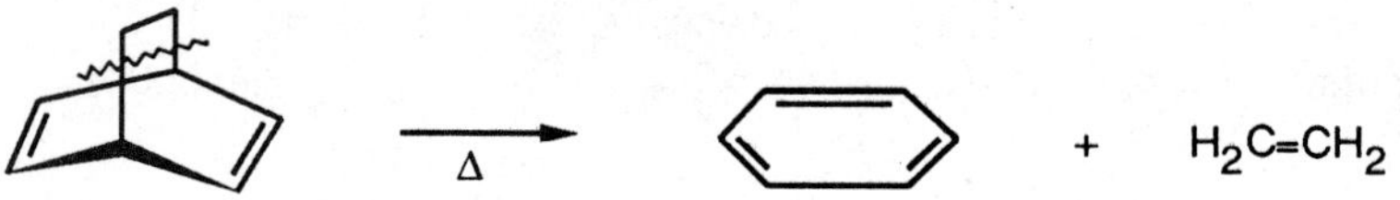

$+$ $H_2C=CH_2$

14.34 A Diels-Alder reaction between α–pyrone (diene) and the alkyne dienophile yields the following product.

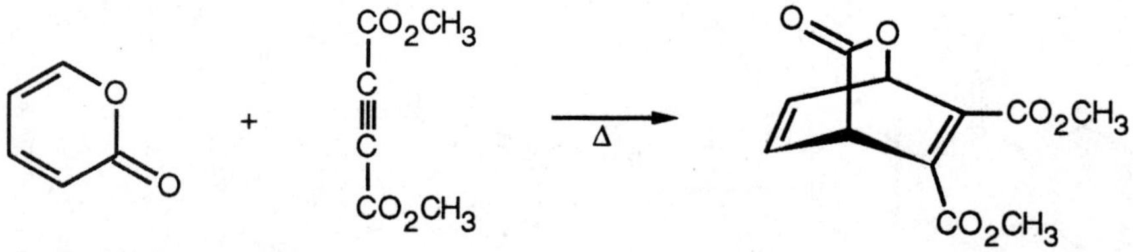

The double bonds in this product are not conjugated. A more stable product is formed by loss of CO_2.

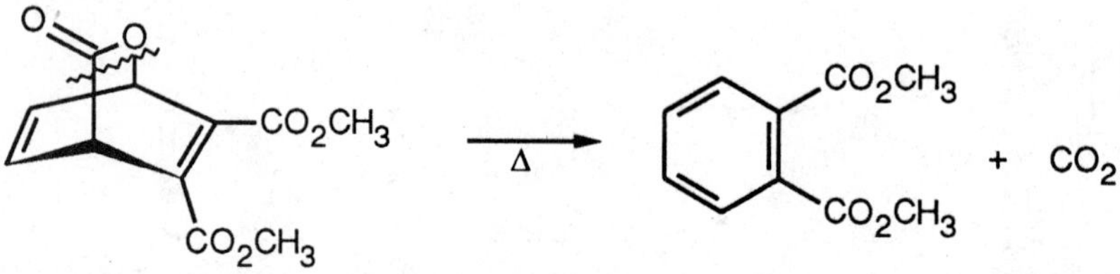

$+$ CO_2

This process can occur in a manner similar to the reverse Diels-Alder reaction of the previous problem.

14.35 The value of λ_{max} in the ultraviolet spectrum of dienes becomes larger with increasing methyl substitution. Since energy is inversely related to λ_{max}, the energy needed to produce ultraviolet absorption decreases with increasing methyl substitution.

Diene	# of–CH$_3$ groups	λ_{max}	$\lambda_{max} - \lambda_{max}$ (butadiene)
	0	217 nm	0
	1	220	3
	1	223	6
	2	226	9
	2	227	10
	3	232	15
	4	240	23

The average increase in λ_{max} is 5 nm per methyl group.

14.36

1,3,5–Hexatriene
λ_{max} = 258 nm

2,3–Diemthyl–1,3,5–hexatriene
λ_{max} ~ 268 nm

From Problem 14.35, we concluded that one methyl group increases λ_{max} of a conjugated diene by 5 nm. Since 2,3–dimethyl–1,3,5–hexatriene has two methyl substituents, its UV λ_{max} should be about 10 nm longer than the λ_{max} of 1,3,5–hexatriene.

14.37 a) ß–Ocimene, $C_{10}H_{16}$, has three degrees of unsaturation. Catalytic hydrogenation yields a hydrocarbon of formula $C_{10}H_{22}$. ß–Ocimene thus contains three double bonds, and no rings.

b) The ultraviolet absorption at 232 nm indicates that ß–ocimene is conjugated.

c) The carbon skeleton, as determined from hydrogenation, is:

$$CH_3CH_2\overset{\overset{\displaystyle CH_3}{|}}{C}HCH_2CH_2CH_2\overset{\overset{\displaystyle CH_3}{|}}{C}HCH_3$$

2,6–Dimethyloctane

Ozonolysis data is used to determine the location of the double bonds. The acetone fragment, which comes from carbon atoms 1 and 2 of 2,6–dimethyloctane, fixes the position of one double bond. Formaldehyde results from ozonolysis of a double bond at the other end of ß–ocimene. Placement of the other fragments to conform to the carbon skeleton yields the following structural formula for ß–ocimene.

$$H_2C=CH\overset{\overset{\displaystyle CH_3}{|}}{C}=CHCH_2CH=\overset{\overset{\displaystyle CH_3}{|}}{C}CH_3$$

β–Ocimene

d)

$$H_2C=CH\overset{\overset{\displaystyle CH_3}{|}}{C}=CHCH_2CH=\overset{\overset{\displaystyle CH_3}{|}}{C}CH_3$$

β–Ocimene

$\xrightarrow{\text{1. O}_3}{\text{2. Zn, H}_3\text{O}^+}$ $H_2C=O$ + $O=CH\overset{\overset{\displaystyle CH_3}{|}}{C}=O$ + $O=CHCH_2CH=O$ + $O=\overset{\overset{\displaystyle CH_3}{|}}{C}-CH_3$

Formal–dehyde   Pyruval–dehyde   Malonal–dehyde   Acetone

$\xrightarrow{\text{H}_2/\text{Pd}}$ $CH_3CH_2\overset{\overset{\displaystyle CH_3}{|}}{C}HCH_2CH_2CH_2\overset{\overset{\displaystyle CH_3}{|}}{C}HCH_3$

14.38 Much of what was proven for ß–ocimene is also true for myrcene, since both hydrocarbons have the same carbon skeleton and contain conjugated double bonds. The difference between the two isomers is in the placement of double bonds.

The ozonolysis fragments are 2–oxopentanedial (five carbon atoms), acetone (three carbon atoms), and two equivalents of formaldehyde (one carbon atom each). Putting these fragments together in a manner consistent with the data gives the following structural formula for myrcene:

$$H_2C=CH\overset{\overset{\displaystyle CH_2}{||}}{C}CH_2CH_2CH=\overset{\overset{\displaystyle CH_3}{|}}{C}CH_3$$

Myrcene

$\xrightarrow{\text{1. O}_3}{\text{2. Zn, H}_3\text{O}^+}$ $2H_2C=O$ + $O=CH\overset{\overset{\displaystyle O}{||}}{C}CH_2CH_2CH=O$ + $O=\overset{\overset{\displaystyle CH_3}{|}}{C}CH_3$

$\xrightarrow{\text{H}_2/\text{Pd}}$ $CH_3CH_2\overset{\overset{\displaystyle CH_3}{|}}{C}HCH_2CH_2CH_2\overset{\overset{\displaystyle CH_3}{|}}{C}HCH_3$

14.39

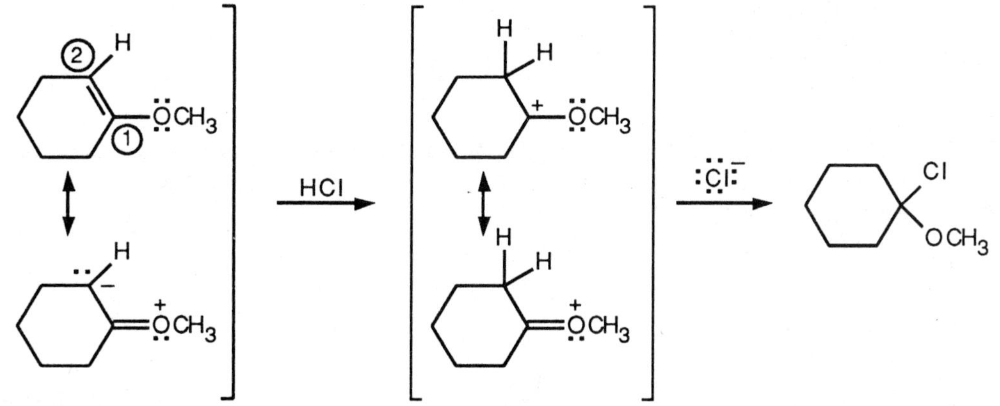

Conjugation with
the oxygen non-
bonded electrons
makes the double
bond more
nucleophilic (Fig. 14.11).

Reaction with HCl
yields a cation
intermediate that
can be stabilized by
the oxygen electrons.

Addition of
Cl⁻ leads to
the observed
product.

There are two reasons why the other regioisomer is not formed: (1) Carbon (1) (above) is less nucleophilic than carbon (2); (2) The cation intermediate that would result from protonation at carbon (1) can't be stabilized by the oxygen electrons.

14.40 Double bonds can be conjugated not only with other multiple bonds but also with the lone-pair electrons of atoms such as oxygen and nitrogen. p–Toluidine has the same number of double bonds as benzene, yet its λ_{max} is 31 nm greater. The electron pair of the nitrogen atom can conjugated with the pi electrons of the three double bonds or the ring, extending the pi system and increasing λ_{max}.

14.41 Hydrogen ion protonates the nitrogen atom of p–toluidine and prevents its lone pair of electrons from conjugating with the ring double bonds. λ_{max} is therefore lowered to a value very close to that of benzene.

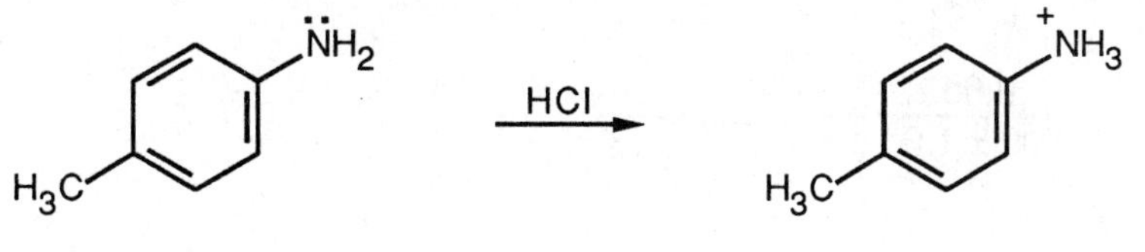

14.42 Dilute NaOH serves to remove the proton from the –OH group, leaving the phenoxide anion.

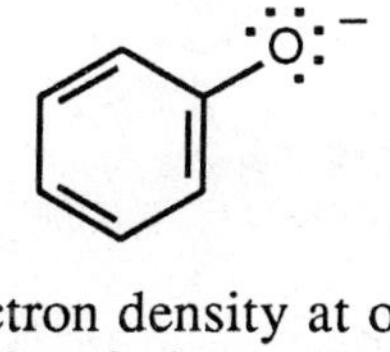

The increased electron density at oxygen makes it easier for an oxygen electron pair to conjugate with the pi electrons of the ring double bonds. The extended conjugation increases λ_{max} in a manner similar to p–toluidine (Problem 14.40).

14.43   a) Hydrocarbon $\underline{A}$ (4 multiple bonds and/or rings):

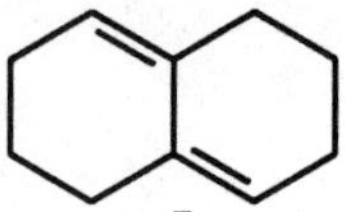

or

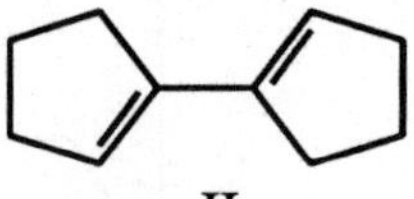

I   II

b) Rotation about the central single bond of II allows the double bond to assume the s–cis conformation necessary for a Diels-Alder reaction. Rotation is not possible for I.

c)

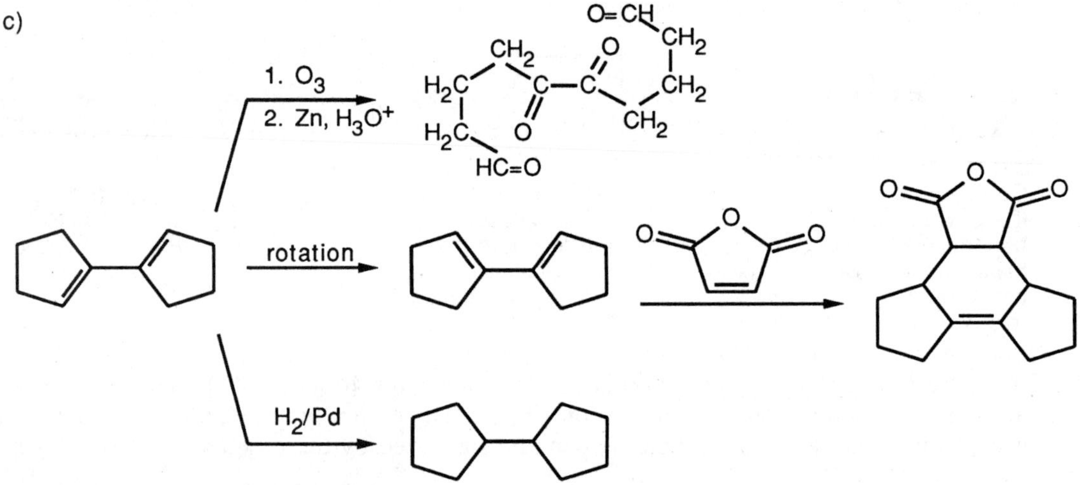

1. O_3
2. Zn, H_3O^+

rotation

H_2/Pd

14.44

$H_2C=CHCH=CH_2$ $\xrightarrow[\text{1,4–Addition}]{Br_2,\ CCl_4}$ $BrCH_2CH=CHCH_2Br$ $\xrightarrow[\text{Pd/C}]{H_2}$ $BrCH_2CH_2CH_2CH_2Br$

$\downarrow$ 2NaCN

$NCCH_2CH_2CH_2CH_2CN$

Adiponitrile

14.45

$$C = \frac{A}{\varepsilon \times l} = \frac{0.065}{11{,}900 \times 1.00\ \text{cm}} = \frac{6.5 \times 10^{-2}}{1.19 \times 10^4} = 5.5 \times 10^{-6}\,\text{M}$$

Study Guide for Chapter 14

After studying this chapter, you should be able to:

(1)   Locate conjugated portions of molecules (14.1, 14.17).

(2)   Draw and name conjugated dienes (14.16, 14.19).

(3)   Understand the reasons for the stability of conjugated molecules (14.2, 14.6).

(4)   Predict the products of electrophilic additions to conjugated molecules (14.3, 14.4, 14.5, 14.18, 14.26, 14.27, 14.39, 14.44).

(5)   Understand the concept of kinetic *vs* thermodynamic control of reactions (14.7, 14.8).

(6)   Predict the products of the Diels-Alder reaction. You should be able to identify compounds that are good dienes and dienophiles (14.9, 14.10, 14.11, 14.22, 14.23, 14.25, 14.28, 14.29, 14.30, 14.31, 14.32, 14.33, 14.34).

(7)   Calculate the energy required for ultraviolet absorption, and use molar absorptivity to calculate concentration (14.12, 14.13, 14.14, 14.45).

(8)   Predict if and where a compound absorbs radiation in the ultraviolet region (14.15, 14.21, 14.24, 14.35, 14.36, 14.40, 14.41, 14.42).

(9)   Deduce the structure of an unknown diene from structural and spectral data (14.20, 14.37, 14.38, 14.43).

Chapter 15 – Benzene and Aromaticity

15.1

a)

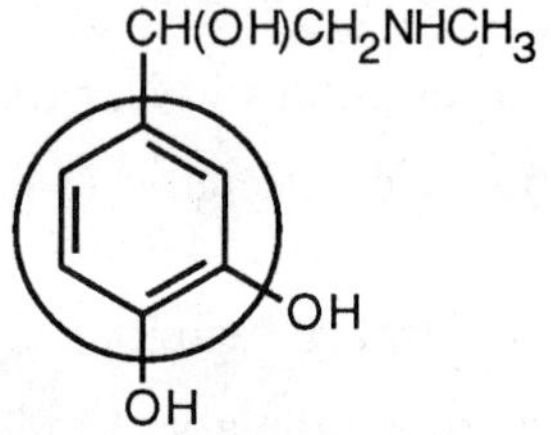

Adrenaline

b)

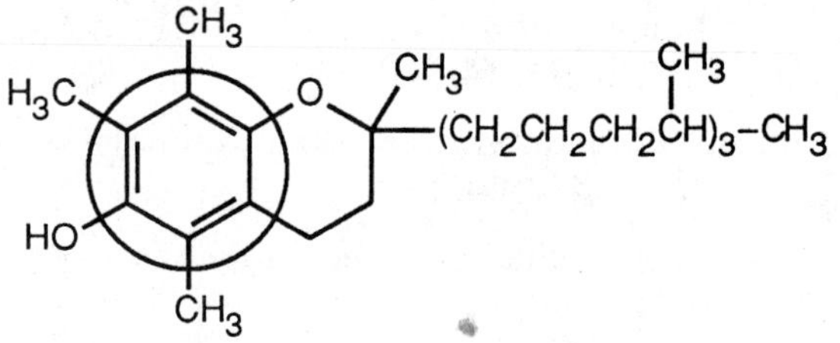

Vitamine E

c)

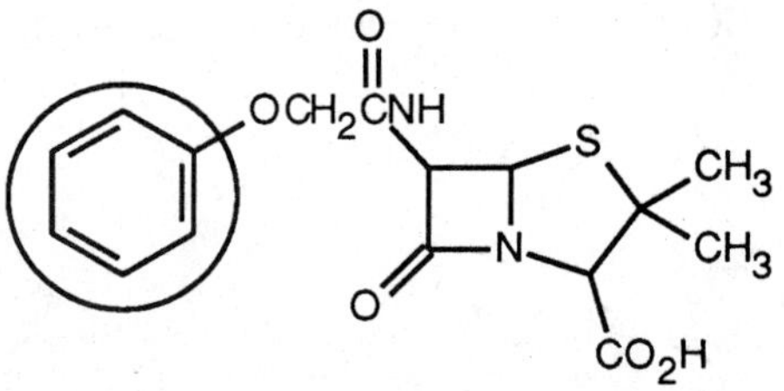

Penicillin V

15.2 An ortho disubstituted benzene has two substituents in a 1,2 relationship. A meta disubstituted benzene has its two substituents in a 1,3 relationship. A para disubstituted benzene has its two substituents in a 1,4 relationship.

a)

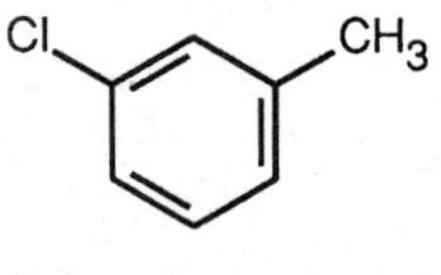

meta disubstituted

b)

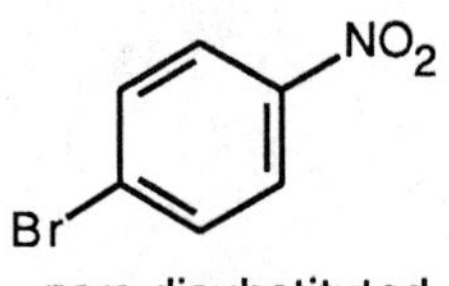

para disubstituted

c)

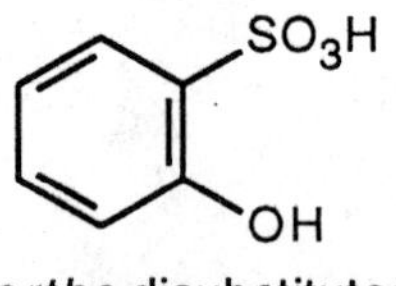

ortho disubstituted

15.3

a)

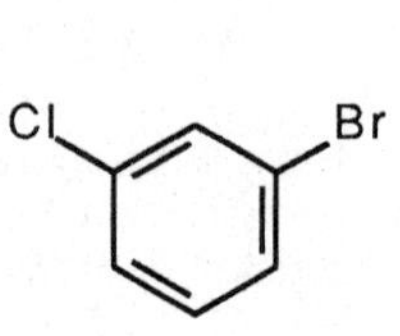

m–Bromochlorobenzene

b)

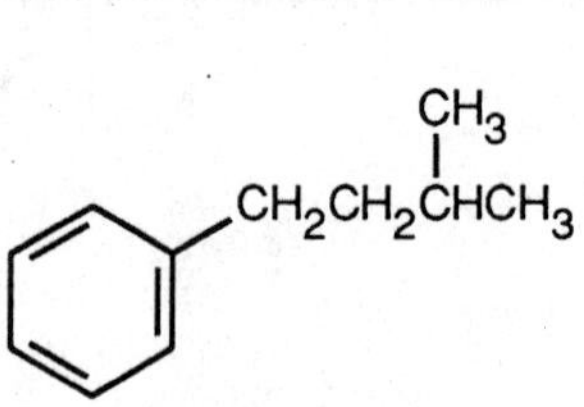

(3–Methylbutyl)benzene

c)

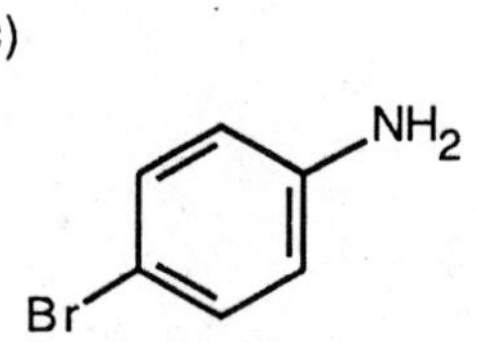

p–Bromoaniline

d)

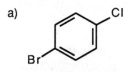

2,4–Dichlorotoluene

e)

O₂N

1–Ethyl–2,4–dinitro–benzene

f)

1,2,3,5–Tetra-methylbenzene

15.4

a)

Br

p–Bromochlorobenzene

b)

Br

p–Bromotoluene

c)

m–Chloroaniline

d)

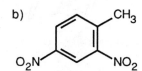

1–Chloro–3,5–dimethylbenzene

15.5

a)

The correct name is 1–Bromo–2–chlorobenzene. You must use the lowest possible combination of numbers. You may also use the *ortho* designation.

b)

O₂N

The correct name is 2,4–dinitrotoluene. This is the same mistake as in a).

c)

Br

The correct name is *p* -bromotoluene or 1-bromo-4-methylbenzene.

d)

H₃C

The correct name is 2–chloro–1,4–dimethylbenzene. Compounds with more than two substituents must be named by citing the number of each substituent on the ring.

15.6

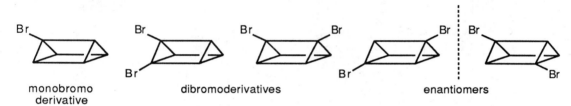

1,2,4–Tribromobenzene   1,2,3-Tribromobenzene   1,3,5–Tribromobenzene

According to Kekulé theory, four tribromobenzenes are possible. Kekulé would say that the two 1,2,4–tribromobenzenes rapidly interconvert, and that only three isomers can be isolated.

15.7   Since all carbon atoms are equivalent in Ladenburg "benzene", only one monobromo derivative is possible. Four dibromo derivatives, including a pair of enantiomers, are possible.

monobromo   dibromoderivatives   enantiomers
derivative

Dewar "benzene", which is a bent molecule, has two different kinds of carbons. Three monobromo derivatives, including a pair of enantiomers, are possible.

enantiomers

The dibromo derivatives include three pairs of enantiomers and three other dibromo Dewar "benzenes".

+   +   +
enantiomer enantiomer enantiomer

15.8

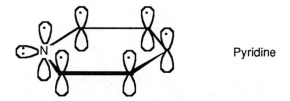

If *o*-xylene exists only as a structure **A**, ozonolysis would cause cleavage at the bonds indicated and would yield two moles of pyruvaldehyde and one mole of glyoxal for each mole of **A** consumed. If *o*-xylene exists only as structure **B**, ozonolysis would yield one mole of 2,3-butanedione and two moles of glyoxal. If *o*-xylene exists as a resonance hybrid of **A** and **B**, the ratio of ozonolysis products would be glyoxal : pyruvaldehyde : 2,3-butanedione = 3:2:1. Since this ratio is identical to the experimentally determined ratio, we know that **A** and **B** contribute equally to the structure of *o*-xylene.

15.9

Pyridine

The electronic descriptions of pyridine and benzene are very similar. The pyridine ring is formed by the sigma overlap of carbon and nitrogen sp^2 orbitals. In addition, six *p* orbitals, perpendicular to the plane of the ring, hold six electrons. These six *p* orbitals form six molecular orbitals that allow electrons to be delocalized over the pi system of the pyridine ring. The lone pair of nitrogen electrons occupies an sp^2 orbital that lies in the plane of the ring.

15.10

Cyclodecapentaene has $4n + 2$ pi electrons (n = 2), but it is not flat. If cyclodecapentaene were flat, the hydrogen atoms circled would crowd each other across the ring. To avoid this interaction, cyclodecapentaene twists so that it is neither planar nor aromatic.

15.11

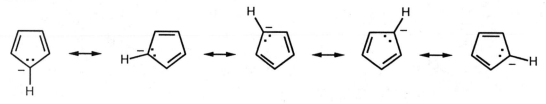

A compound that can be described by several resonance forms has a structure that can be represented by no one form. The structure of the cyclopentadienyl anion is a combination of all of the above structures and contains only one kind of carbon atom and one kind of hydrogen atom. All carbon-carbon bond lengths are equivalent; all carbon-hydrogen bonds lengths are equivalent. Both the ^{1}H NMR and ^{13}C NMR spectra show only one absorption.

15.12 When cyclooctatetraene accepts two electrons, it becomes a $(4n + 2)$ pi electron aromatic ion. Cyclooctatetraenyl dianion is planar with a carbon-carbon bond angle of 135° (a regular octagon).

15.13

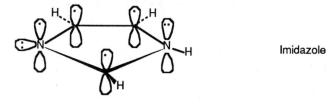

Imidazole

The aromatic heterocycle imidazole contains six pi electrons. Each carbon contributes one electron, the nitrogen bonded to hydrogen contributes two electrons, and the remaining nitrogen contributes one electron. Both nitrogens are sp^2 hybridized.

15.14

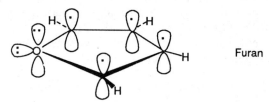

Furan

15.15

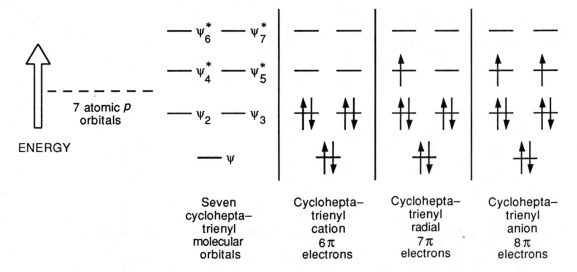

Seven cycloheptatrienyl molecular orbitals

Cycloheptatrienyl cation
6π electrons

Cycloheptatrienyl radial
7π electrons

Cycloheptatrienyl anion
8π electrons

The cycloheptatrienyl cation has six electrons (a Hückel number) and is aromatic.

15.16

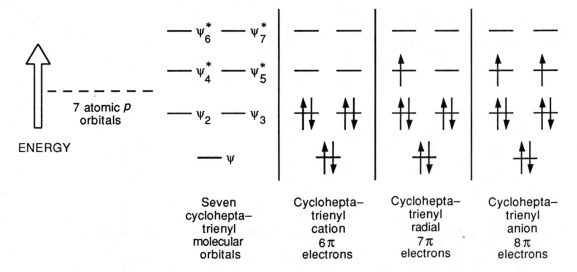

Notice that the bond between carbons 1 and 2 is represented as a double bond in two of the three resonance structures. The bond between carbons 2 and 3 is represented as a double bond in only one resonance structure. The C1–C2 bond has more double bond character in the resonance hybrid, and its carbon-carbon bond length is shorter than the C2–C3 bond length.

15.17 Naphthalene is a ten pi electron compound; the circle in each ring represents five electrons.

15.18 a) 2–Methyl–5–phenylhexane   b) *m*–Bromobenzoic acid
c) 1–Bromo–3,5–dimethylbenzene   d) *o*–Bromopropylbenzene
e) 1–Fluoro–2,4–dinitrobenzene   f) *p*–Chloroaniline

15.19

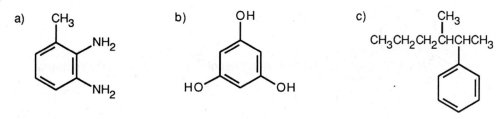

d)

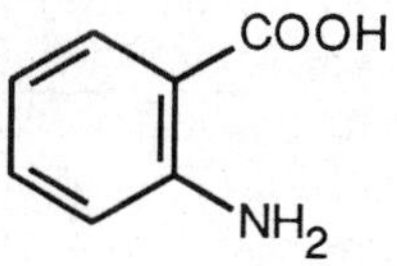

e)

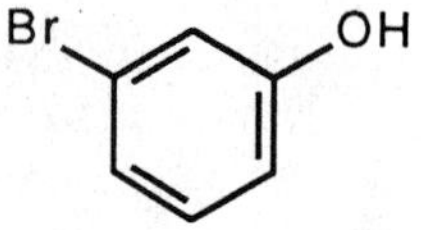

f)

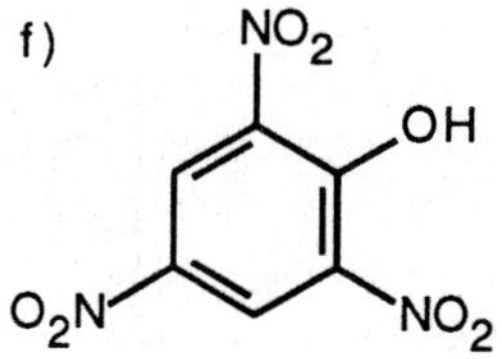

g)

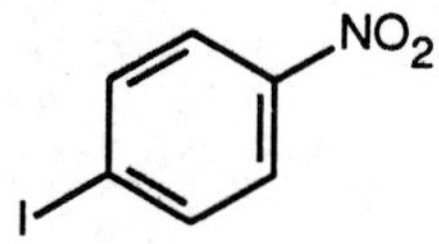

15.20

a)

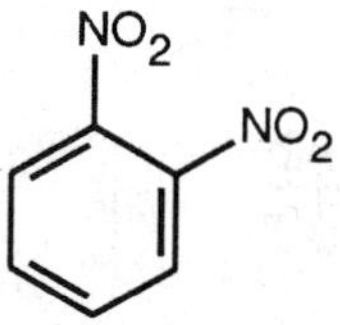

o-Dinitrobenzene

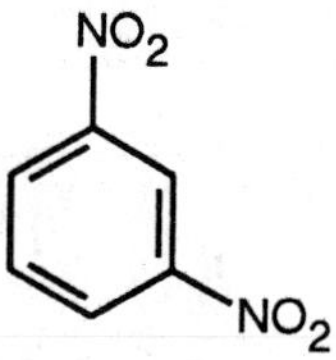

m-Dinitrobenzene

p-Dinitrobenzene

b)

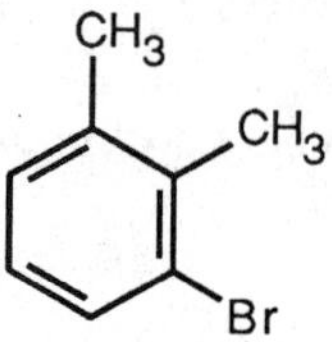

1–Bromo–2,3–dimethyl-
benzene

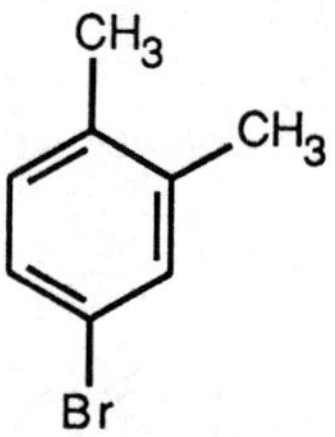

4–Bromo–1,2–dimethyl-
benzene

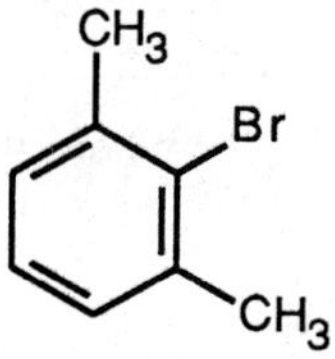

2–bromo–1,3–dimethyl-
benzene

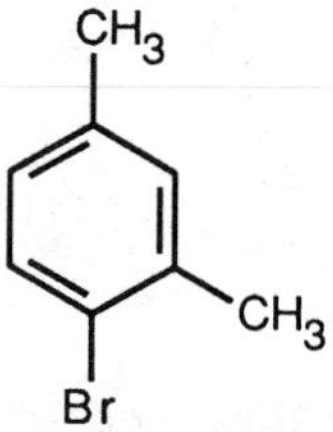

1–Bromo–2,4–dimethyl-
benzene

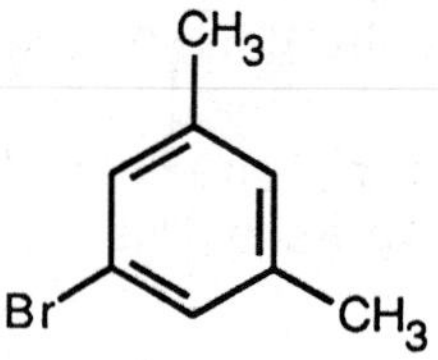

1–Bromo–3,5–dimethyl-
benzene

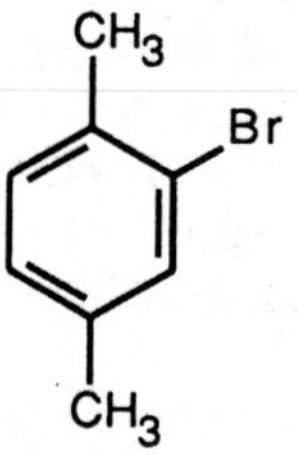

2–Bromo–1,4–dimethyl-
benzene

c)

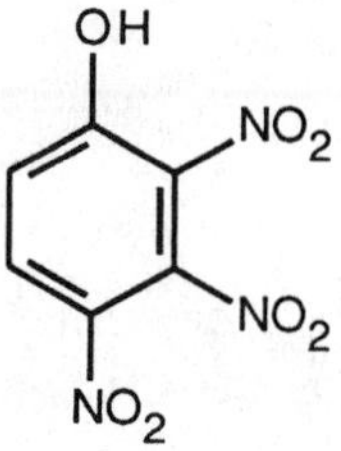

2,3,4–Trinitrophenol

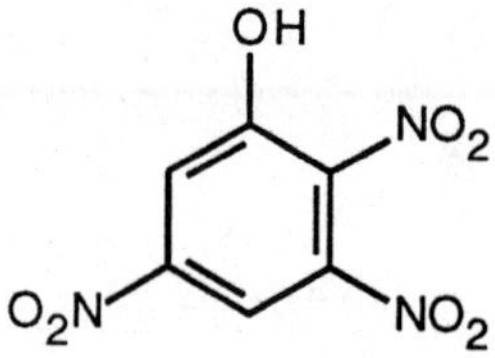

2,3,5–Trinitrophenol

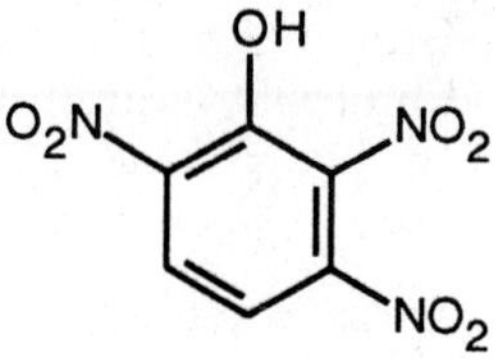

2,3,6–Trinitrophenol

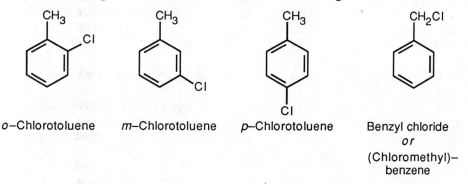

2,4,5–Trinitrophenol   2,4,6–Trinitrophenol   3,4,5–Trinitrophenol

15.21 All aromatic compounds of formula C_7H_7Cl have one ring and three double bonds.

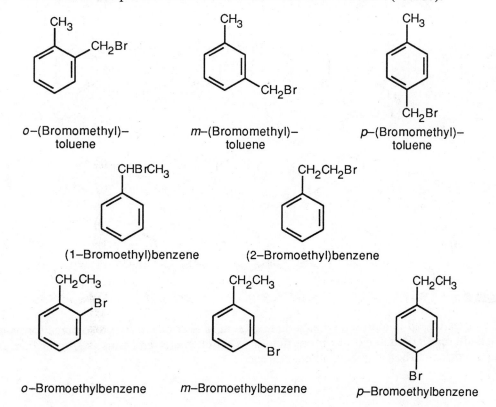

o–Chlorotoluene   *m*–Chlorotoluene   *p*–Chlorotoluene   Benzyl chloride
or
(Chloromethyl)–
benzene

15.22 Six of these compounds are illustrated and named in Problem (15.20b).

o–(Bromomethyl)–
toluene

m–(Bromomethyl)–
toluene

p–(Bromomethyl)–
toluene

(1–Bromoethyl)benzene

(2–Bromoethyl)benzene

o–Bromoethylbenzene   *m*–Bromoethylbenzene   *p*–Bromoethylbenzene

15.23 All compounds in this problem have four double bonds and/or rings and must be substituted benzenes, if they are to be aromatic. They may be substituted by methyl, ethyl, propyl, or butyl groups.

a)

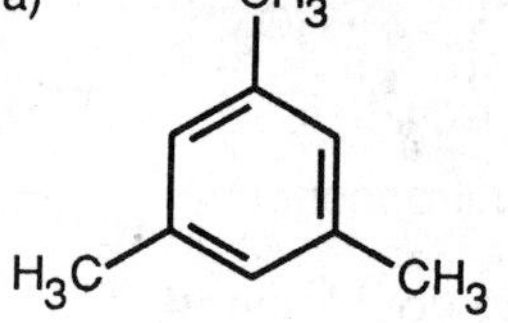

b)

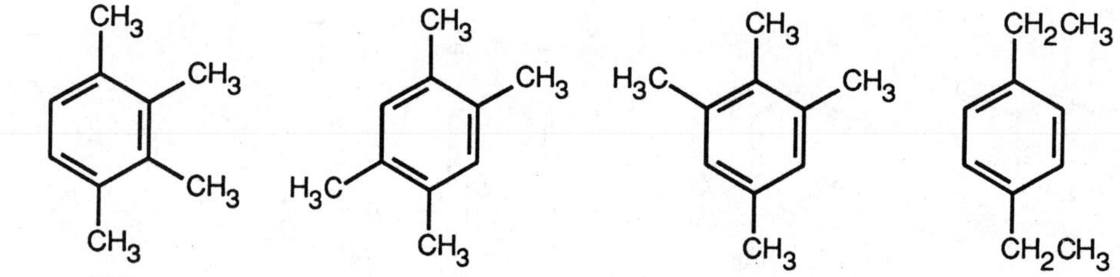

c)

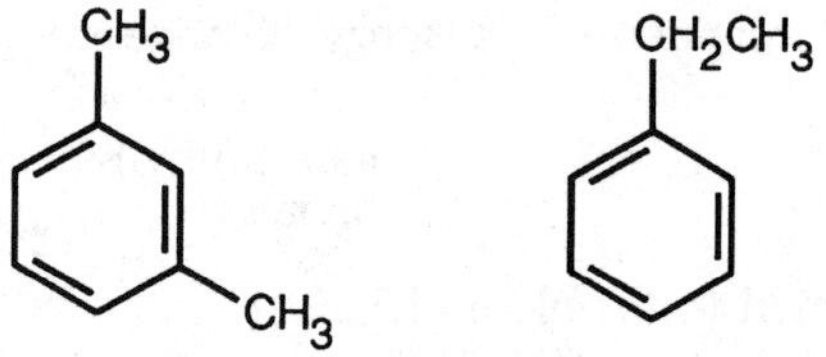

d)

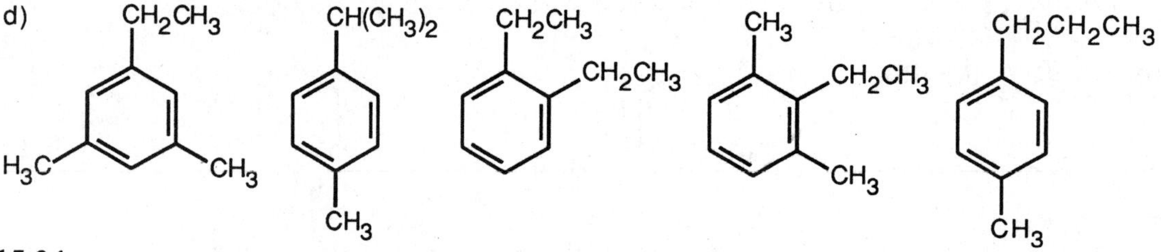

15.24

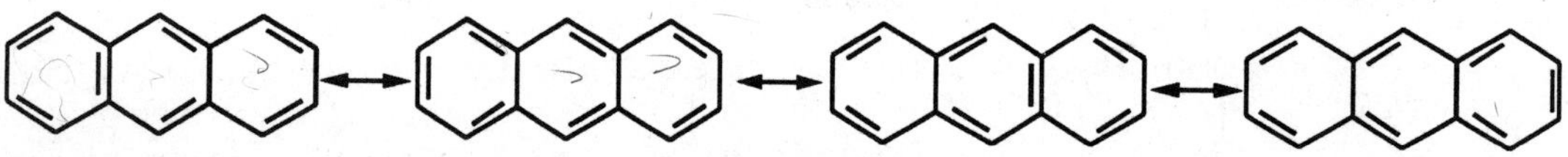

15.25–15.26

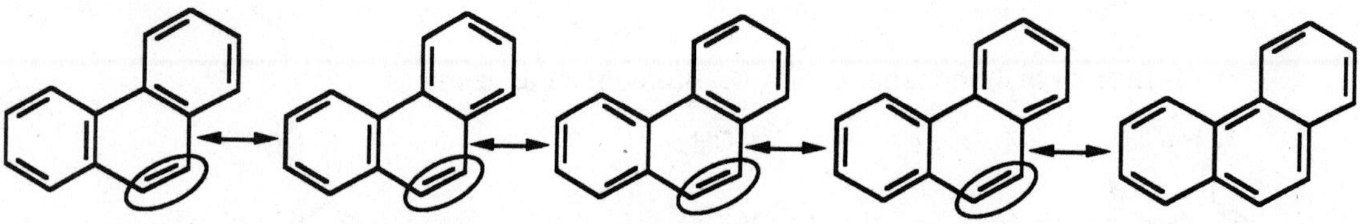

The circled bond is represented as a double bond in four of the five resonance forms of phenanthrene. This bond has more double-bond character and thus is shorter than the other carbon-carbon bonds of phenanthrene.

15.27 a) *Aromaticity* is a property of cyclic conjugated compounds having $(4n + 2)$ pi electrons. Aromatic compounds are usually stable and unreactive, and undergo substitution, rather than addition, reactions.

b) A system of alternating single and multiple bonds having overlapping p orbitals is said to be *conjugated*. Compounds having conjugated bonds are stabilized by their ability to delocalize pi electrons.

c) The *Hückel 4n + 2 rule* predicts that compounds having $(4n + 2)$ pi electrons (where $n = 0, 1, 2...$) in a cyclic conjugated system will be aromatic.

d) For some compounds it is possible to draw two or more Kekulé structures that differ only in the placement of electrons and not of atoms. These structures are called resonance forms. The actual structure of the compound can't be represented by any one of the resonance forms, but is a *resonance hybrid* of all of them.

15.28 The heat of hydrogenation is the amount of heat liberated when a compound reacts with hydrogen.

(1) Benzene + 3 H_2 → Cyclohexane $\qquad \Delta H_{hydrog} = 49.8$ kcal/mol

(2) 1,3–Cyclohexadiene + 2 H_2 → Cyclohexane $\qquad \Delta H_{hydrog} = 55.4$ kcal/mol

To find ΔH_{hydrog} for Benzene + H_2 → 1,3–Cyclohexadiene, subtract the ΔH_{hydrog} of reaction (2) from that of reaction (1):

49.8 kcal/mol – 55.4 kcal/mol = –5.6 kcal/mol

The heat of hydrogenation for this reaction is *negative*, and the reaction is endothermic.

15.29

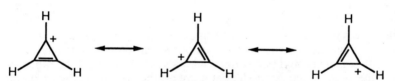

The product of the reaction of 3–chlorocyclopropene with $AgBF_4$ is the cyclopropenyl cation $C_3H_3^+$. The resonance structures of the cation indicate that all hydrogen atoms are equivalent; the ^1H NMR spectrum, which shows only one type of hydrogen atom, confirms this equivalence. The cyclopropenyl cation contains two pi electrons and should be aromatic, according to Hückel's rule. (Here, $n = 0$.)

15.30

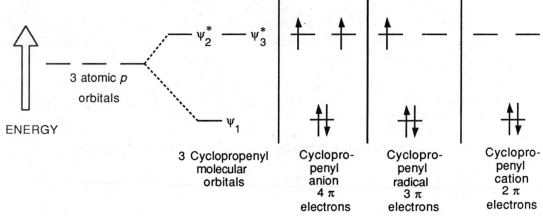

The cyclopropenyl cation is aromatic, according to Hückel's rule.

15.31 The circle in the cyclopropenyl cation represents two pi electrons.

15.32

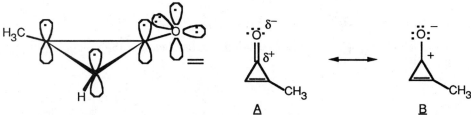

Methylcyclopropenone is a cyclic conjugated compound with a four electron pi system. The electronegative oxygen attracts the pi electrons of the carbon-oxygen pi bond. In resonance structure **B**, both carbonyl pi electrons are located on oxygen; the other two pi electrons remain in the ring. Since 2 is a Hückel number, the methylcyclopropenone ring fulfills the criteria of aromaticity and has the added stability of an aromatic ring.

15.33

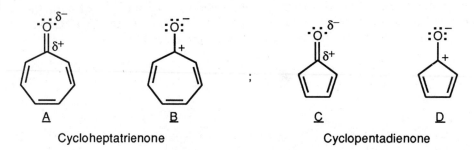

As in the previous problem, we can draw resonance forms in which both carbonyl pi electrons are located on oxygen. The cycloheptatrienone ring in **B** contains six pi electrons and should be stable, according to Hückel's rule. The cyclopentadienone ring in **D** contains four pi electrons and is not aromatic.

15.34 Check the number of electrons in the pi system of each compound. The species with a Hückel $(4n + 2)$ number of pi electrons is the most stable.

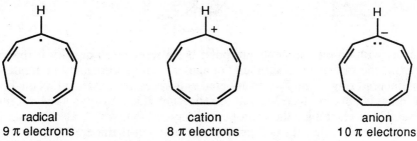

radical	cation	anion
9 π electrons	8 π electrons	10 π electrons

The ten pi electron anion is the most stable.

15.35

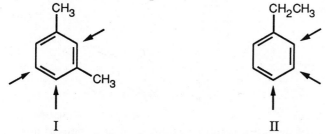

15.36 Compound $\underline{A}$ has four multiple bonds and/or rings. Possible structures that yield three monobromo substitution products are:

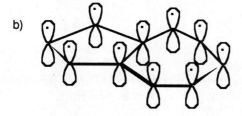

          I   II

Only structure I shows a six-proton singlet at 2.30 δ; I contains two identical methyl groups unsplit by other protons. The presence of four protons in the aromatic region of the ^{1}H NMR spectrum confirms that I is the correct structure.

15.37 a) Azulene is a planar, cyclic, conjugated hydrocarbon that has ten pi electrons. According to Hückel's rule it should be aromatic.

b)

c)

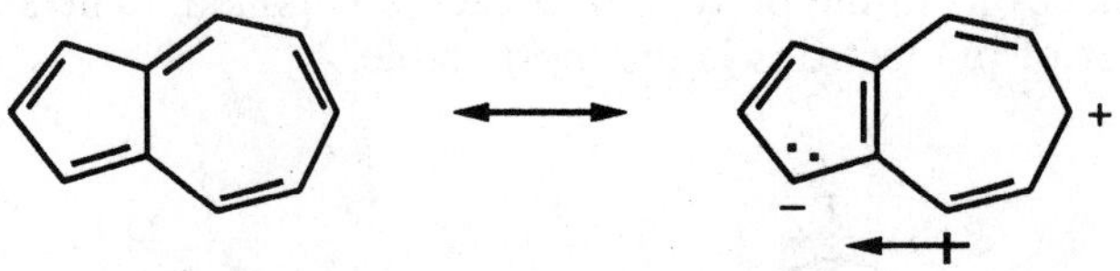
+
−

Molecules with dipole moments are polar because electron density is drawn from one part of the molecule to another. In azulene, electron density is greater in the five-membered ring. The five-membered ring resembles the cyclopentadienyl anion in having six pi electrons and in satisfying Hückel's rule. The seven-membered ring resembles the cycloheptatrienyl cation, which also has six pi electrons. Electron density is drawn from the seven-membered ring to the five-membered ring, satisfying Hückel's rule for both rings and producing a dipole moment.

15.38 The molecular weight of the hydrocarbon (120) corresponds to the structural formula C_9H_{12}, which indicates four double bonds and/or rings. The ^1H NMR singlet at 7.25 δ indicates the presence of five aromatic ring protons. The septet at 2.90 δ is due to a benzylic proton that has six neighboring protons.

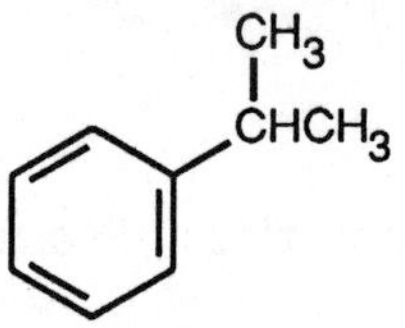
CH3
CHCH3

Isopropylbenzene

15.39

a)

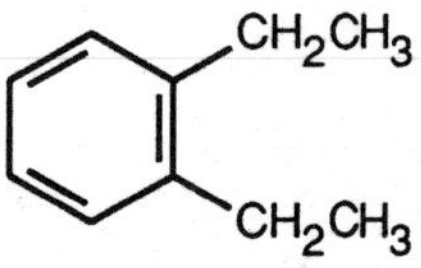
CH2CH3
CH2CH3

o–Diethylbenzene

b)

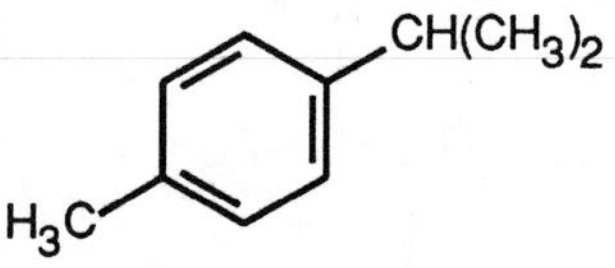
CH(CH3)2
H3C

p–Isopropyltoluene

The IR values are essential for deciding if the rings are *o, m* or *p*–substituted.

15.40

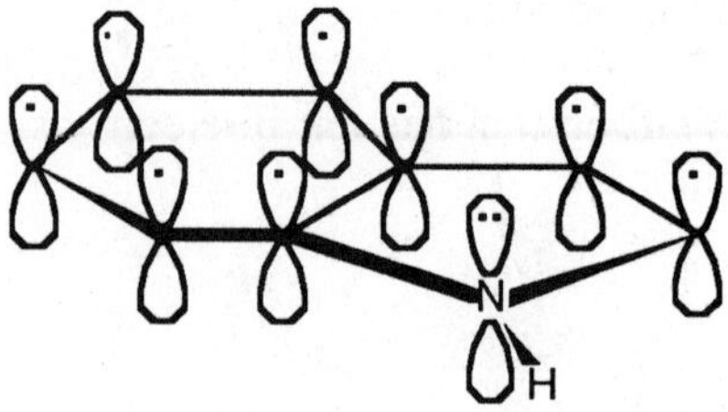
N
H

Indole has ten pi electrons and is aromatic. Two pi electrons come from the nitrogen atom. Indole and naphthalene both possess ten pi electrons in two rings.

15.41

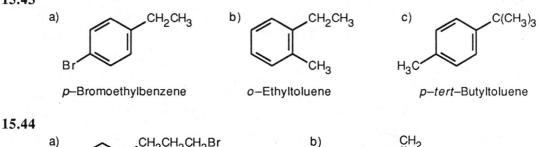

When 4–pyrone is protonated, structure **A** is produced. **B**, **C** and **D** are resonance forms of **A**. In **C** and **D**, a lone pair of electrons of the ring oxygen is delocalized into the ring to produce a six pi electron system, which should be stable, according to Hückel's rule.

15.42

The alkene double bond is protonated to yield an intermediate carbocation, which loses a proton to give a product in which the double bond is conjugated with the aromatic ring, as shown by the increased value of λ_{max}.

15.43

a) ![structure] CH$_2$CH$_3$, Br — *p*–Bromoethylbenzene

b) ![structure] CH$_2$CH$_3$, CH$_3$ — *o*–Ethyltoluene

c) ![structure] C(CH$_3$)$_3$, H$_3$C — *p*–*tert*–Butyltoluene

15.44

a) ![structure] CH$_2$CH$_2$CH$_2$Br

b) ![structure] CH$_2$=C(phenyl)$_2$

15.45 Pentalene has eight pi electrons; Hückel's rule predicts it to be unstable. Pentalene dianion, however, has ten pi electrons and should be stable.

15.46

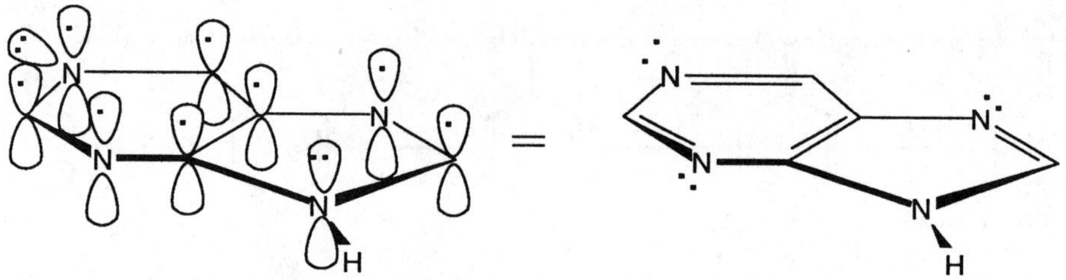

Purine is a ten-pi-electron aromatic molecule. The N–H nitrogen atom in the five-membered ring donates *both* electrons of its lone pair to the pi electron system, and each of the other three nitrogens donates one electron to the pi electron system.

Study Guide for Chapter 15

After studying this chapter, you should be able to:

(1)   Recognize aromatic compounds; name and draw substituted benzene compounds (15.1, 15.2, 15.3, 15.4, 15.15, 15.18, 15.19, 15.20, 15.21, 15.22, 15.23).

(2)   Draw resonance structures and molecular orbital diagrams for benzene and other cyclic conjugated molecules (15.6, 15.8, 15.9, 15.15, 15.16, 15.17, 15.24, 15.25, 15.26, 15.27, 15.30, 15.31).

(3)   Use Hückel's rule to predict aromaticity (15.10, 15.11, 15.12, 15.29, 15.32, 15.33, 15.34, 15.35, 15.37, 15.41, 15.45, 15.46).

(4)   Draw orbital pictures of cyclic conjugated molecules (15.13, 15.14, 15.40).

(5)   Use NMR and IR data to predict structures of aromatic compounds (15.36, 15.38, 15.39, 15.43, 15.44).

Chapter 16 – Chemistry of Benzene.
Electrophilic Aromatic Substitution

16.1

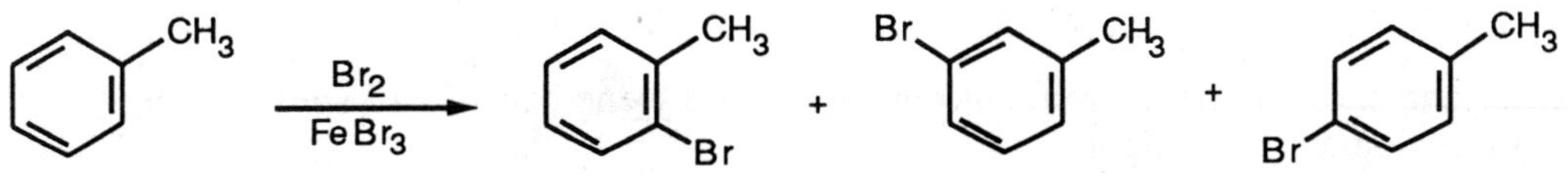

Toluene → o–Bromotoluene + m–Bromotoluene + p–Bromotoluene

16.2

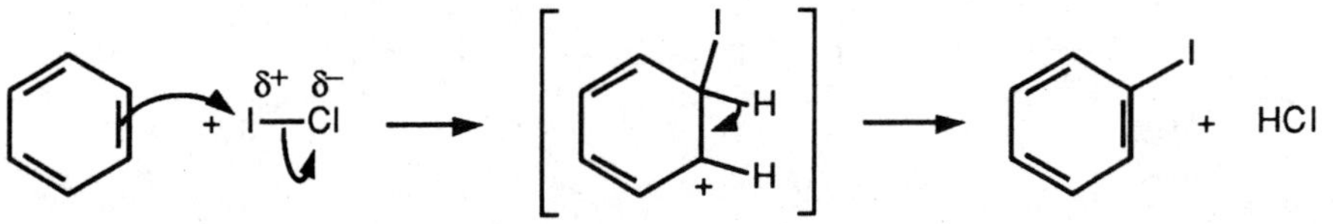

ICl can be represented as $\overset{\delta+}{I}\!\!-\!\!\overset{\delta-}{Cl}$ because chlorine is a more electronegative element than iodine. Iodine can act as an electrophile in electrophilic aromatic substitution.

16.3

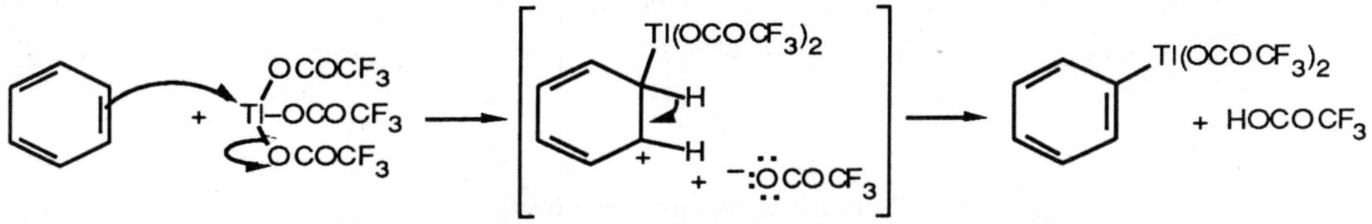

Thallium acts as an electrophile and substitutes on the aromatic ring.

16.4

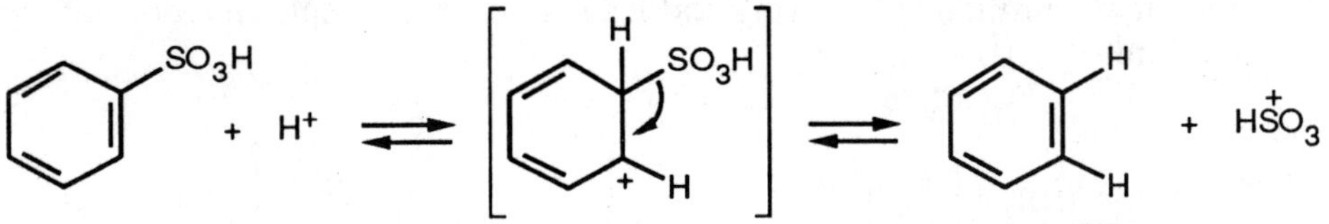

This mechanism is the reverse of the sulfonation mechanism illustrated in the text. H^+ is the electrophile in this reaction.

16.5

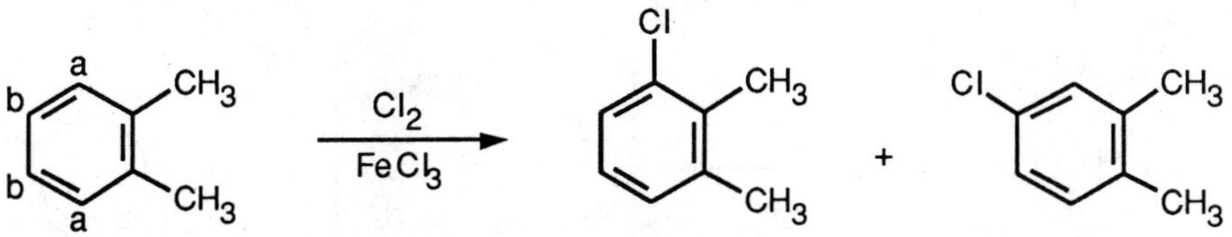

o–Xylene → A + B

Chlorination at position "a" of o–xylene yields product A; chlorination at position "b" yields product B.

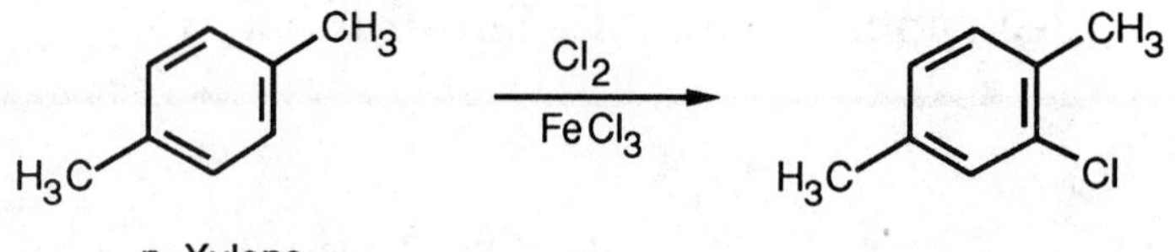

p–Xylene

Only one product results from chlorination of *p*–xylene because all sites for chlorination are equivalent.

16.6

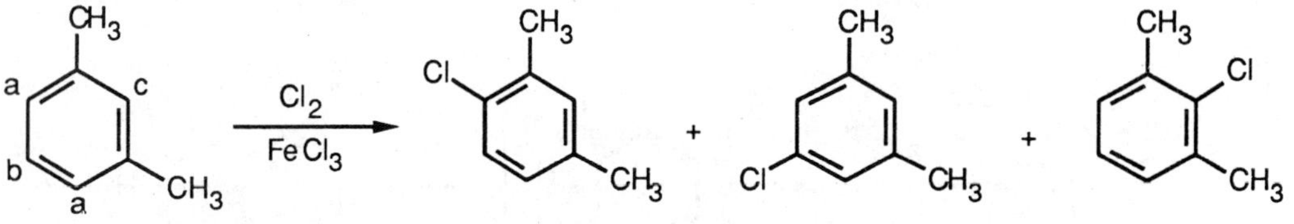

A B C

m–Xylene

Three products might be expected to form on chlorination of *m*–xylene. Product C is unlikely to form because substitution rarely occurs between two *meta* substituents. Product B is also unlikely to form for reasons to be explained later in the chapter.

16.7

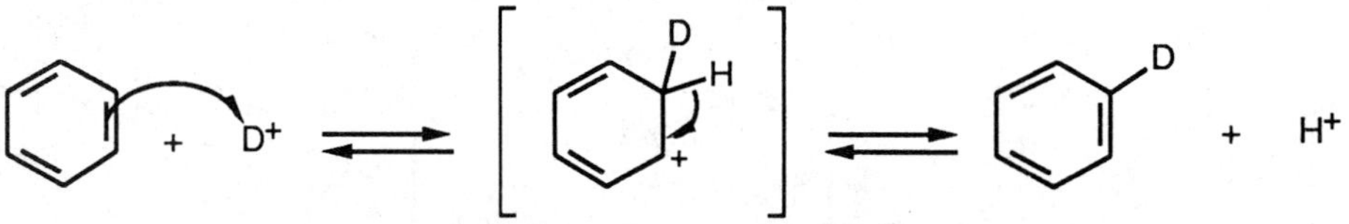

carbocation intermediate

Benzene can be protonated by strong acids. The resulting intermediate can lose either deuterium or hydrogen. If H^+ is lost, deuterated benzene is produced. Attack by D^+ can occur at all positions of the ring and leads to eventual replacement of all hydrogens by deuterium.

16.8

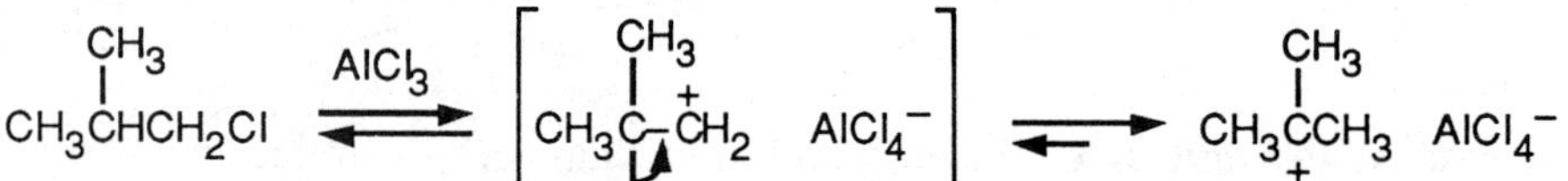

Isobutyl carbocation (primary)

tert–Butyl carbocation (tertiary)

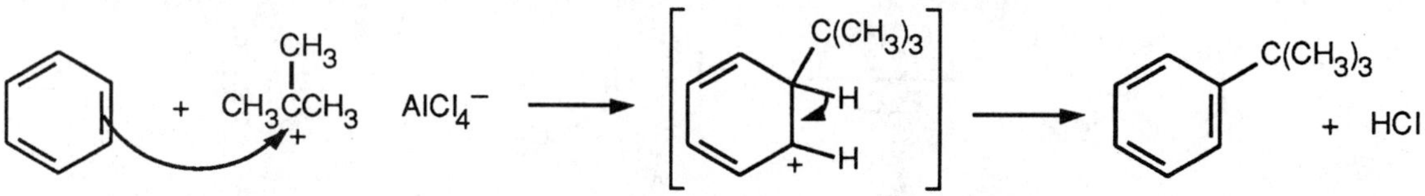

The isobutyl carbocation is initially formed when 1–chloro–2–methylpropane and $AlCl_3$ react. This carbocation rearranges, via a hydride shift, to the more stable *tert*–butyl carbocation, which can then alkylate benzene to form *tert*–butylbenzene.

16.9

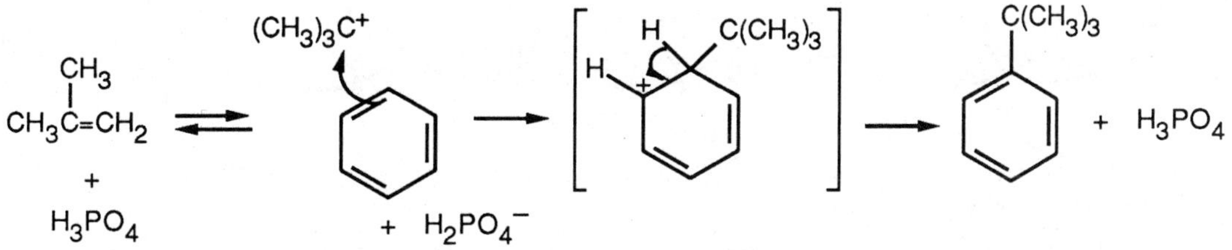

16.10 Carbocation rearrangements of alkyl halides occur: (1) if the initial carbocation is primary or secondary, and (2) if it is possible for the initial carbocation to rearrange to a more stable secondary or tertiary cation.

a) Although $CH_3\overset{+}{C}H_2$ is a primary carbocation, it can't rearrange to a more stable cation.

b) $CH_3CH_2CHClCH_3$ forms a secondary carbocation that doesn't rearrange.

c) $CH_3CH_2\overset{+}{C}H_2$ rearranges to the more stable $CH_3\overset{+}{C}HCH_3$.

d) $(CH_3)_3C\overset{+}{C}H_2$ (primary) undergoes an alkyl shift to yield $(CH_3)_2\overset{+}{C}CH_2CH_3$ (tertiary).

e) The cyclohexyl carbocation doesn't rearrange.

In summary:
No rearrangement: CH_3CH_2Cl, $CH_3CH_2CHClCH_3$, chlorocyclohexane.
Rearrangement: $CH_3CH_2CH_3Cl$, $(CH_3)_3CCH_2Cl$.

16.11 Use Figure 16.11 to find the activating and deactivating effects of groups.

Most Reactive ⟶ Least Reactive

a)   Phenol > toluene > benzene > nitrobenzene.
b)   Pehnol > benzene > chlorobenzene > benzoic acid.
c)   Aniline > benzene > bromobenzene > benzaldehyde.

16.12 An acyl substituent is deactivating. Once an aromatic ring has been acylated, it is much less reactive to further substitution. An alkyl substituent is activating. An alkyl-substituted ring is more reactive than an unsubstituted ring, and polysubstitution occurs readily.

16.13
a) *Para* attack:

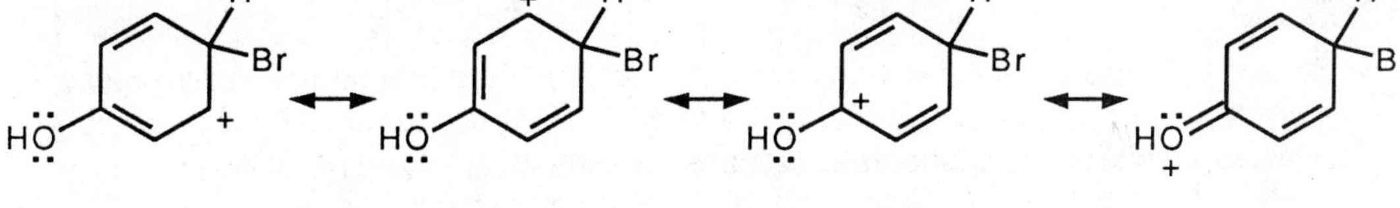

b) *Meta* attack:

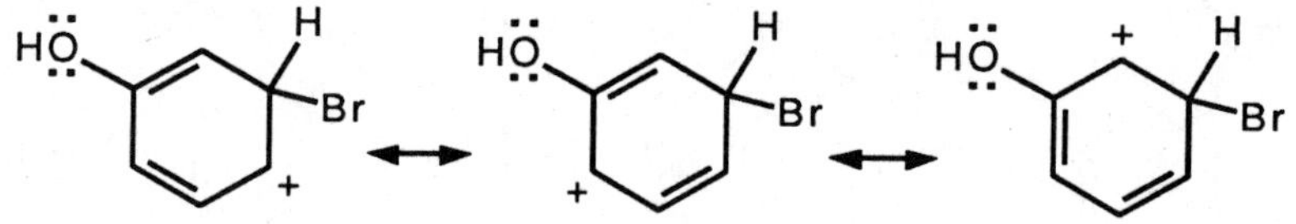

c) *Ortho* attack:

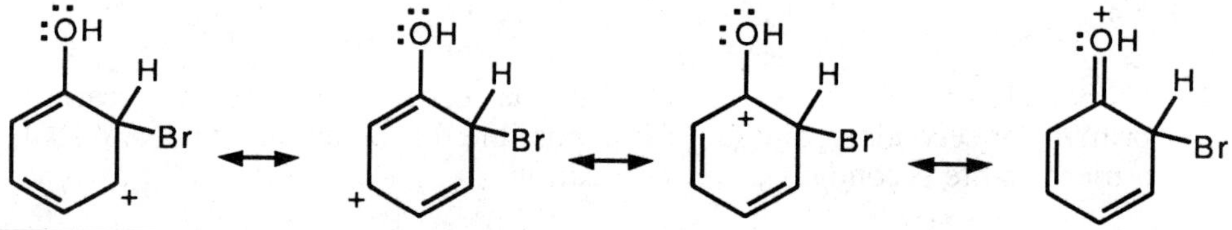

The more resonance forms that can be drawn, the greater is the extent of charge delocalization. Since the intermediates from *ortho* and *para* attack can be written as four resonance forms each, these intermediates are of lower energy. Reaction at the *ortho* and *para* positions is thus favored over reaction at the *meta* position.

16.14

Ortho attack:

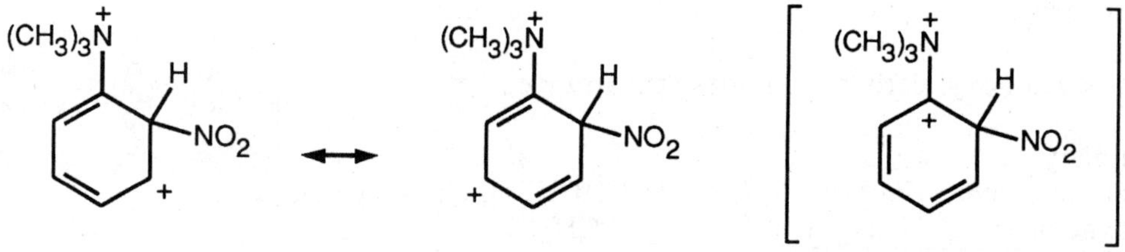

This form is not an important resonance form because two positive charges are next to each other.

Meta attack -- *MOST FAVORED*

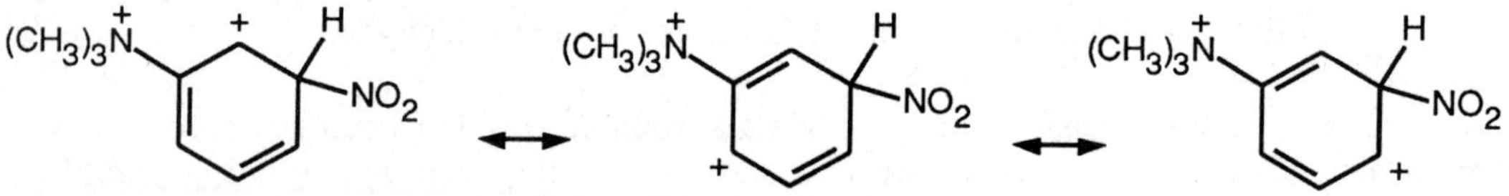

Para attack:

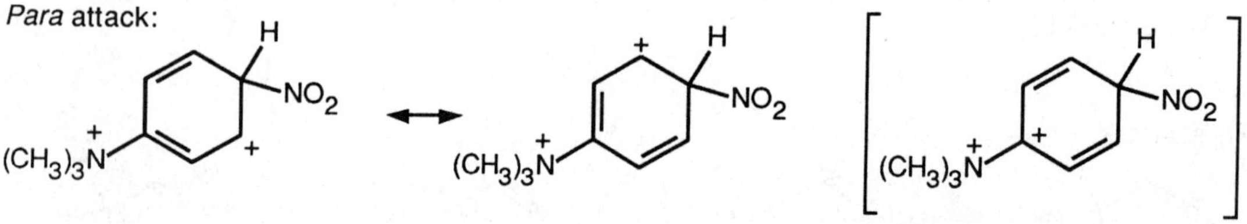

This form is not important.

Meta attack is most favorable because the cation intermediate is somewhat more stable than the intermediates from *ortho* and *para* attack.

16.15 Refer to Figure 16.14 in the text for the directing effects of substituents. You should memorize the effects of the most important groups.

a)

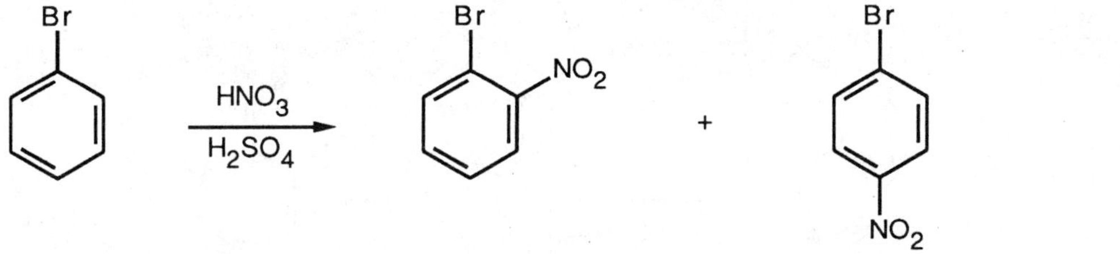

Even though bromine is a deactivator, it is an *ortho–para* director.

b)

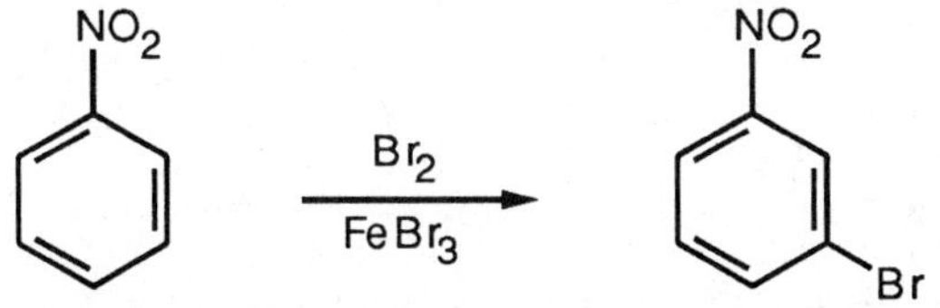

The $-NO_2$ group is a *m*–director.

c)

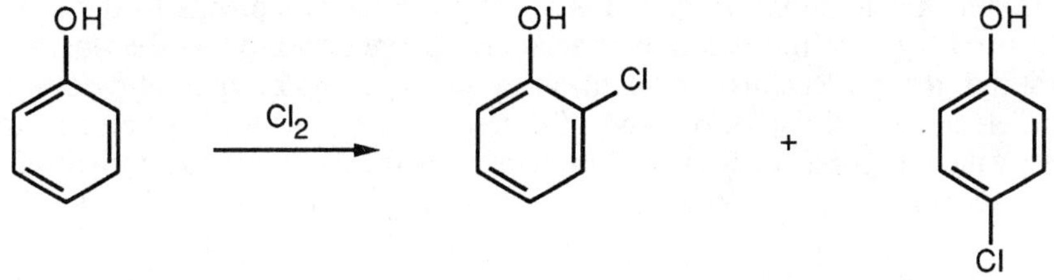

d)

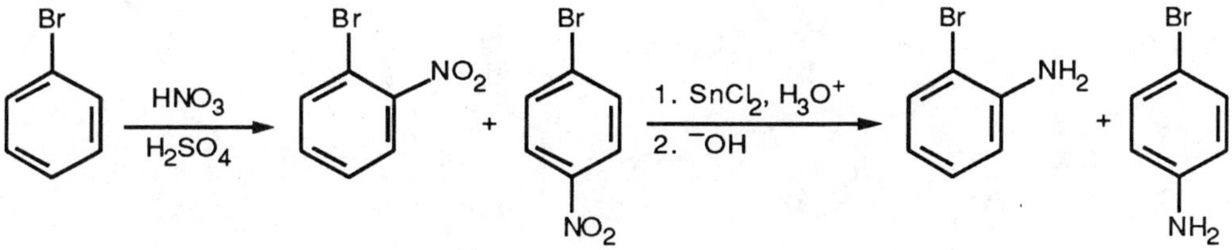

e)

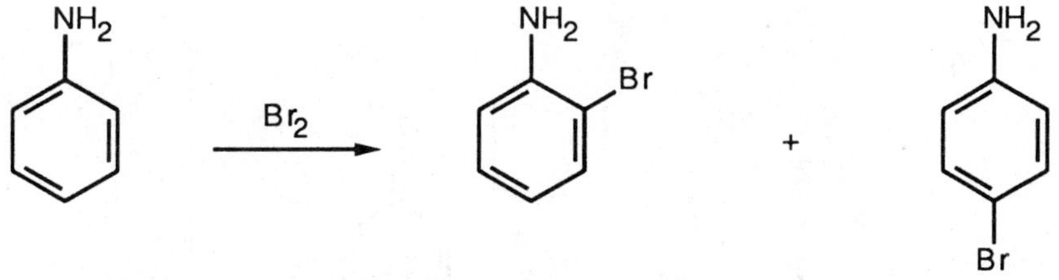

No catalyst is necessary since aniline is highly activated.

16.16 Toluene is more reactive toward electrophilic substitution than trifluoromethylbenzene. The electronegativity of the three fluorine atoms causes the trifluoromethyl group to be electron-withdrawing and deactivating toward electrophilic substitution.

16.17

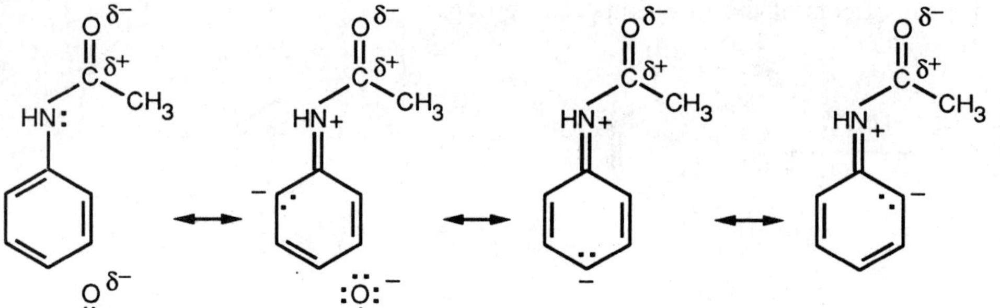

less
favored

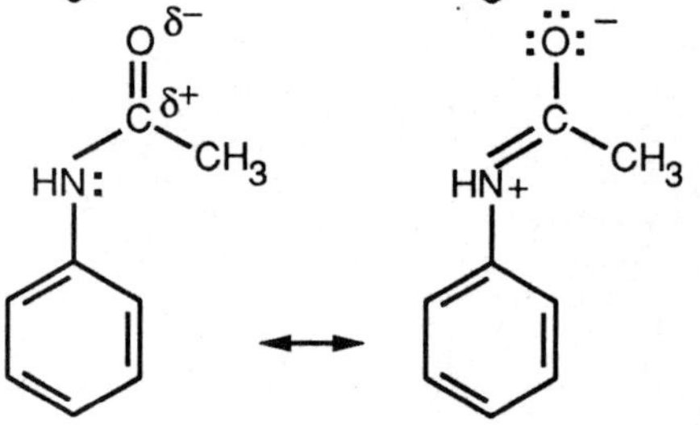

favored

For acetanilide, resonance donation of the nitrogen lone pair electrons to the aromatic ring is less favored because the positive charge on nitrogen is next to the positively polarized carbonyl group. Resonance donation to the carbonyl oxygen is favored because of the electronegativity of oxygen. Since the nitrogen lone pair electrons are less available to the ring, the reactivity of the ring toward electrophilic substitution is decreased, and acetanilide is less reactive than aniline toward electrophilic substitution.

16.18

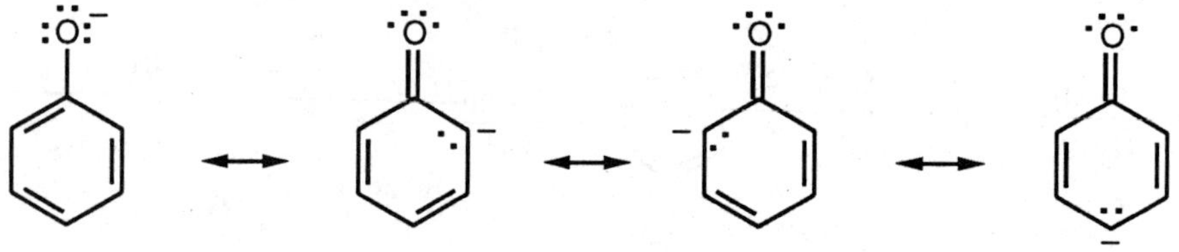

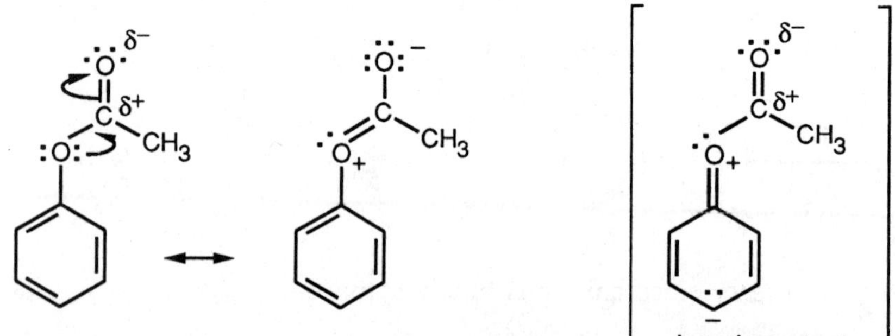

less important

Phenoxide ion is the most reactive and phenyl acetate is the least reactive towards electrophilic substitution. The full negative charge of the phenoxide anion can be delocalized into the ring, which becomes electron rich and strongly activated toward electrophilic aromatic substitution. For phenyl acetate, delocalization of an electron pair onto the electronegative carbonyl oxygen makes the ring less reactive.

16.19 *Ortho* attack:

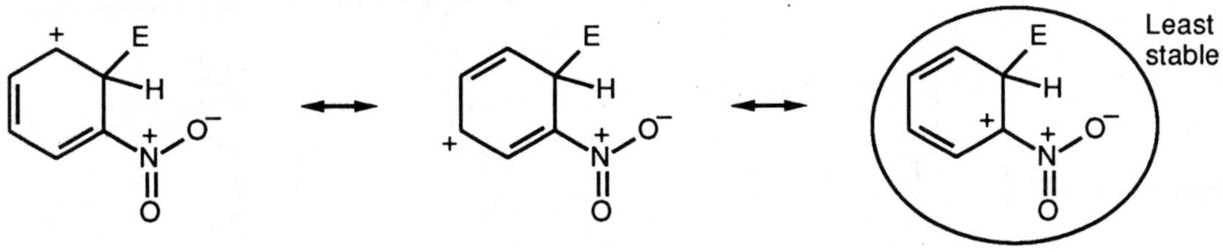

Meta attack:

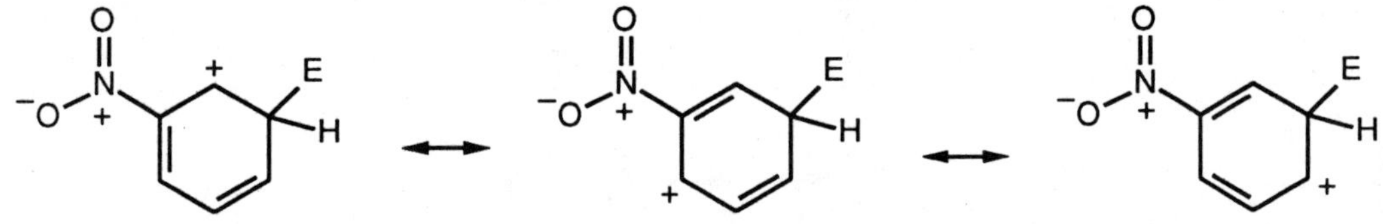

Para attack:

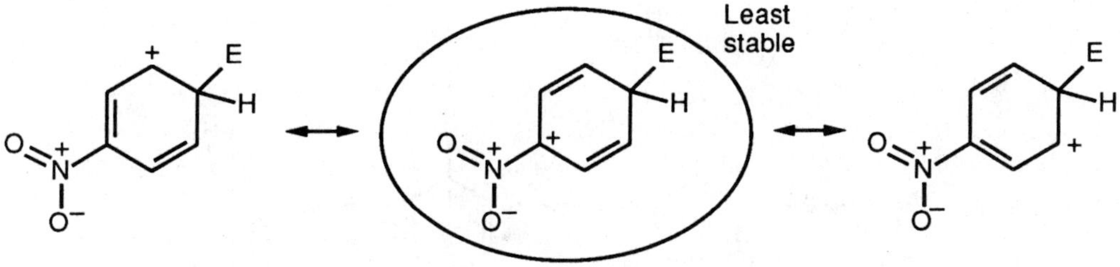

These intermediates are similar to those in Problem 16.14. The circled resonance forms are less stable, because they place two positive charges adjacent to each other. The intermediate from *meta* attack is thus favored.

16.20 The aromatic ring is deactivated toward electrophilic aromatic substitution by the combined electron-withdrawing inductive effect of electronegative nitrogen and oxygen. The lone pair of electrons of nitrogen can, however, stabilize by resonance the *ortho* and *para* substituted intermediates.

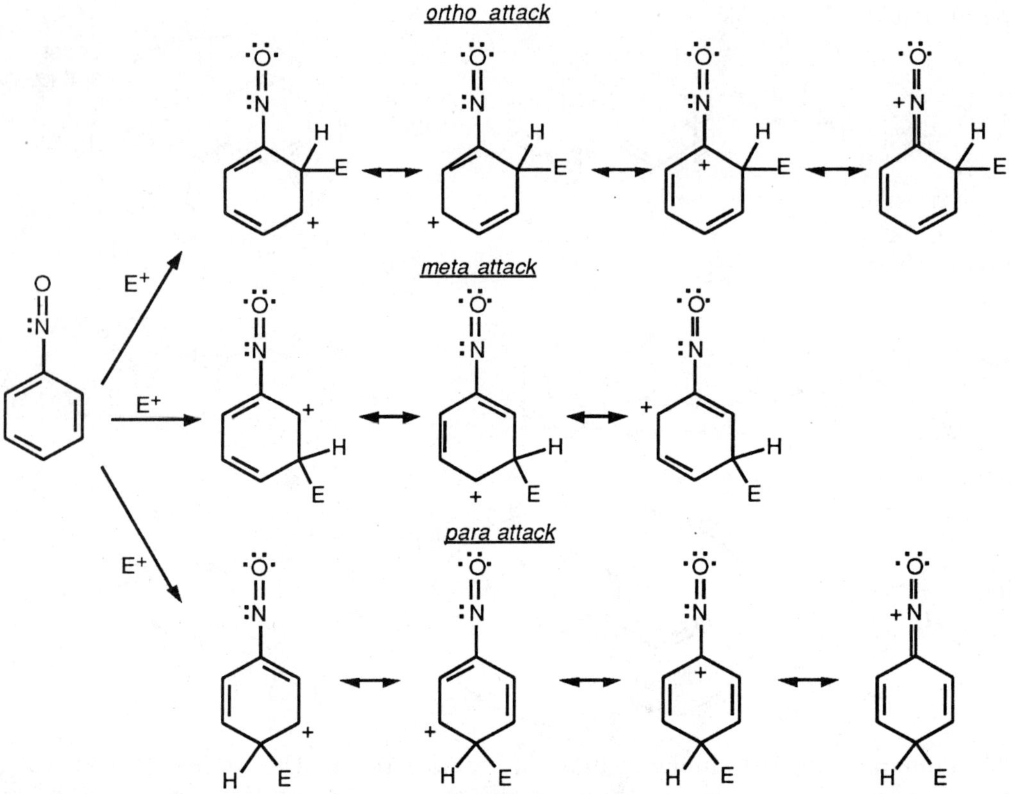

ortho attack

meta attack

para attack

This stabilization is more effective for intermediates of *ortho* and *para* than for *meta* attack.

16.21

a)

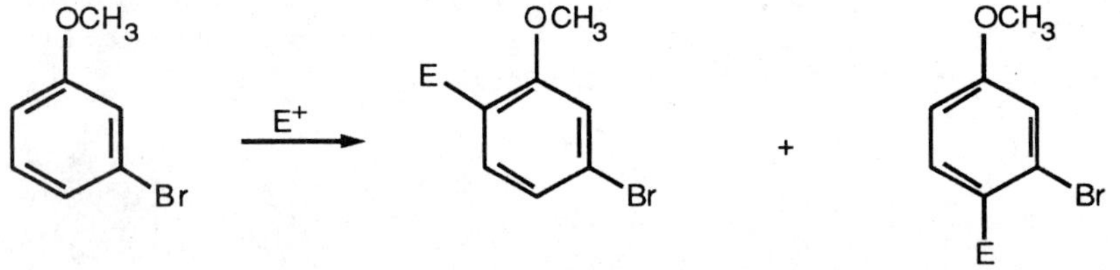

Both groups are *o–p* directors and direct substitution to the same positions. Attack does not occur between the two groups for steric reasons.

b)

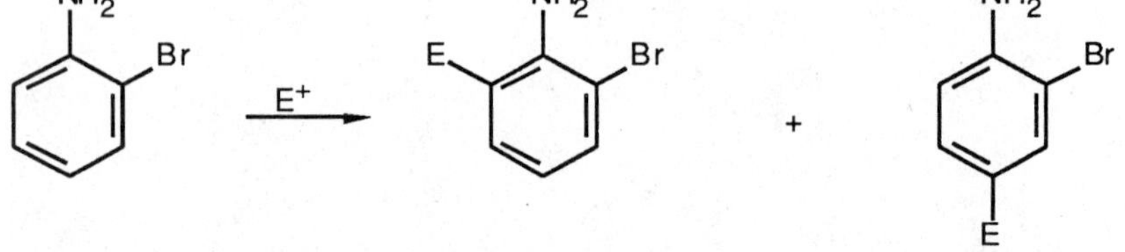

Both groups are *o–p* directors but direct substitution to different positions. Because –NH₂ group is a more powerful activator, substitution occurs *ortho* and *para* to it.

c)

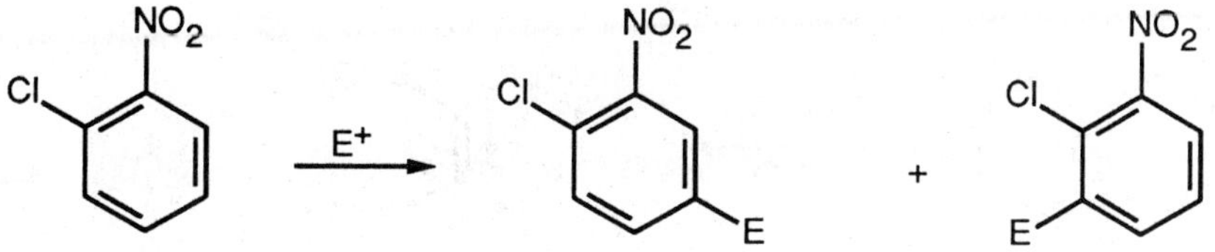

Both groups are deactivating, but they orient substitution toward the same positions.

16.22

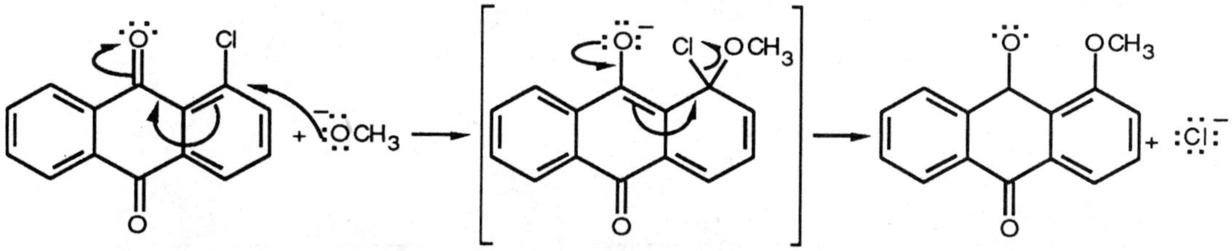

The carbonyl oxygens make the chlorine–containing ring electron-poor and vulnerable to attack by the nucleophile $^-$:OCH_3. They also stabilize the negatively charged Meisenheimer complex. This nucleophilic aromatic substitution occurs by an addition-elimination pathway.

16.23

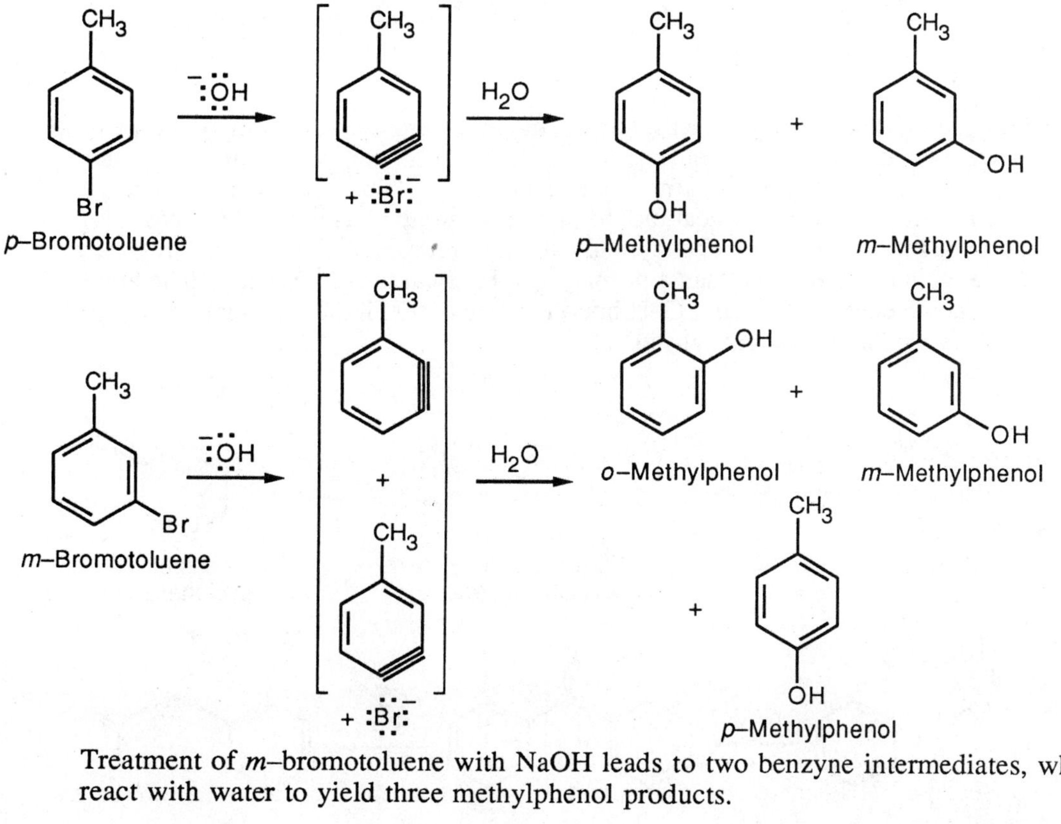

p–Bromotoluene

p–Methylphenol

m–Methylphenol

m–Bromotoluene

o–Methylphenol

m–Methylphenol

p–Methylphenol

Treatment of *m*–bromotoluene with NaOH leads to two benzyne intermediates, which react with water to yield three methylphenol products.

16.24

a)

$$\xrightarrow[\text{H}_2\text{O}]{\text{KMnO}_4}$$

(COOH, COOH product)

b)

CH(CH$_3$)$_2$

NO$_2$

$$\xrightarrow[\text{H}_2\text{O}]{\text{KMnO}_4}$$

COOH

NO$_2$

16.25

$$\xrightarrow[\text{AlCl}_3]{\text{CH}_3\text{CH}_2\text{Cl}}$$

CH$_2$CH$_3$

$$\xrightarrow[\text{(PhCO}_2)_2]{\text{NBS}}$$

CHBrCH$_3$

$$\xrightarrow[\text{CH}_3\text{CH}_2\text{OH}]{\text{KOH}}$$

CH=CH$_2$

16.26

Bond	CH$_3$CH$_2$–H	CH$_2$–H (benzyl)	H$_2$C=CHCH$_2$–H
Bond dissociation energy	98 kcal/mol	85 kcal/mol	87 kcal/mol

Bond dissociation energies measure the amount of energy that must be supplied to cleave a bond into two radical fragments. A radical is thus higher in energy and less stable than the compound it came from. Since the C–H bond dissociation energy is 98 kcal/mol for ethane and 85 kcal/mol for a methyl group C–H bond of toluene, less energy is required to form a benzyl radical than to form an ethyl radical. A benzyl radical is thus more stable than a primary alkyl radical by 13 kcal/mol. The bond dissociation energy of an allyl C–H bond is 87 kcal/mol, indicating that an allyl radical is nearly as stable as a benzyl radical.

16.27

H$_3$C, CH$_3$ (on ring)

$$\xrightarrow[\text{ethanol}]{\text{Li, NH}_3}$$

H$_3$C, CH$_3$ +   H$_3$C, CH$_3$

1,5–Dimethyl–
1,4–cyclohexadiene

1,3–Dimethyl–
1,4–cyclohexadiene

16.28

+

$$\xrightarrow{\text{AlCl}_3}$$

$$\xrightarrow[\text{Pd/C}]{\text{H}_2}$$

16.29

a)

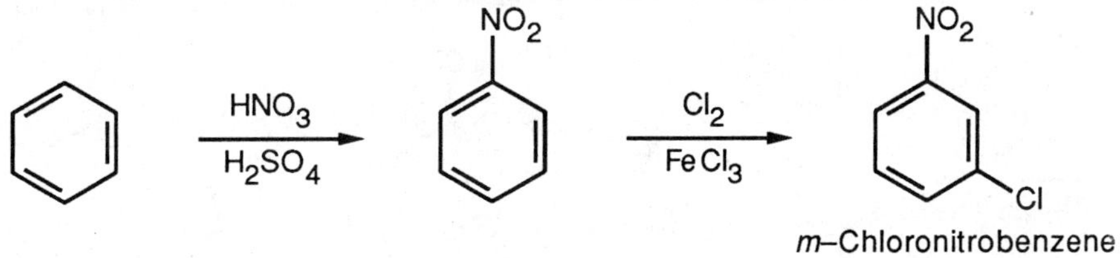

m–Chloronitrobenzene

b)

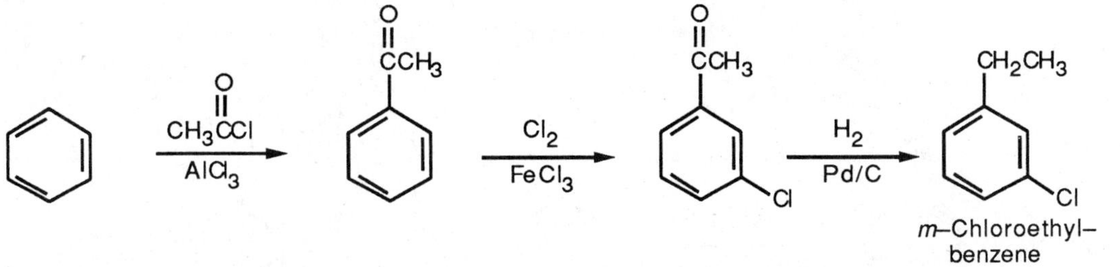

m–Chloroethyl-
benzene

c)   Two routes are possible:

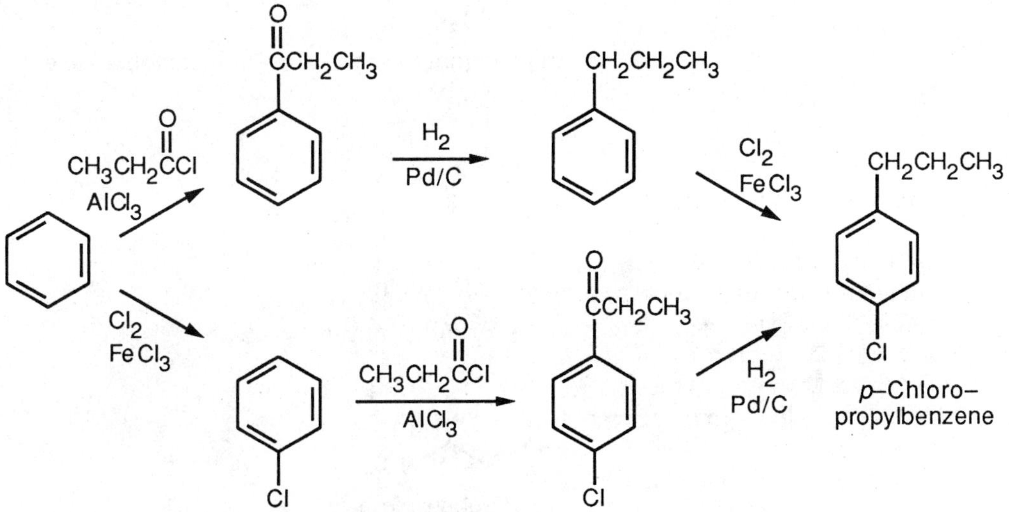

p–Chloro–
propylbenzene

16.30 a) Friedel-Crafts acylation, like Friedel-Crafts alkylation, does not occur at an aromatic ring carrying a meta-directing group. We will learn in a later chapter how to synthesize this compound by another route.

b) There are two problems with this synthesis as it is written:
   1. Rearrangement often occurs during Friedel-Crafts alkylations using primary halides.
   2. Even if *p*–chloropropylbenzene could be synthesized, introduction of the second –Cl group would occur *ortho* to the alkyl group.

A possible route to this compound:

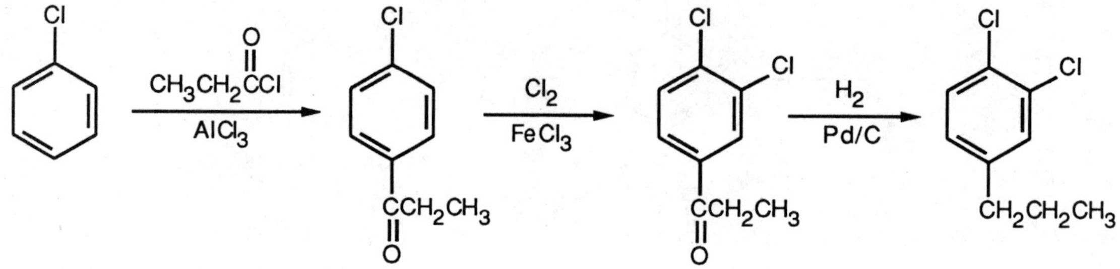

16.31

a)

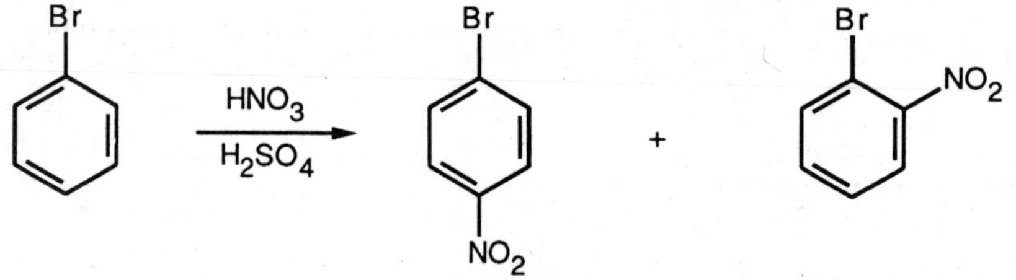

p–Bromonitrobenzene   o–Bromonitrobenzene

b)

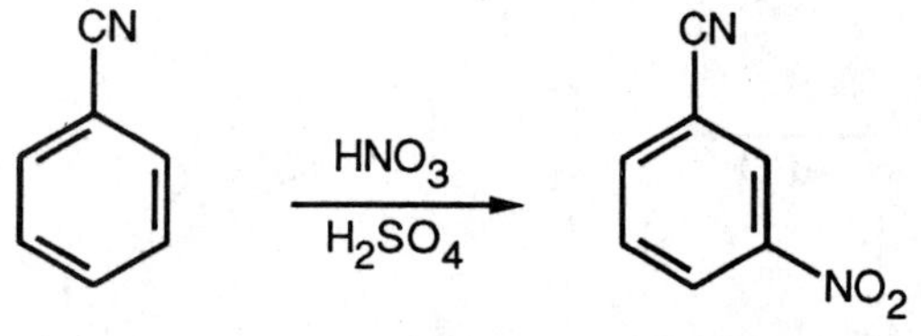

m–Nitrobenzonitrile

c)

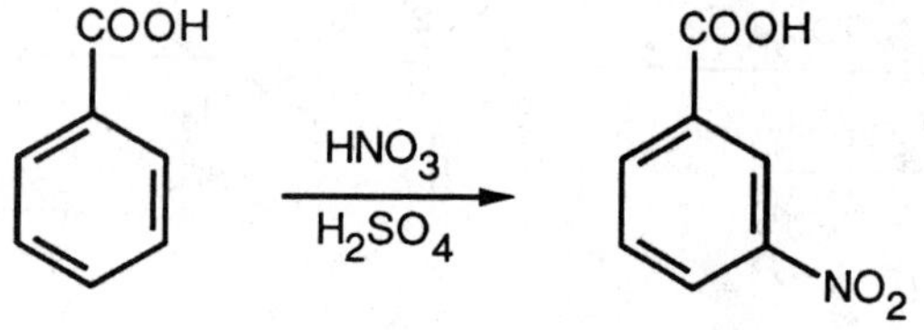

m–Nitrobenzoic acid

d)

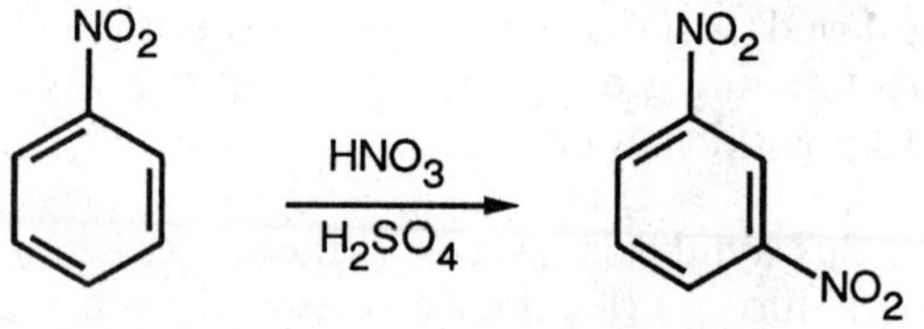

m–Dinitrobenzene

e)

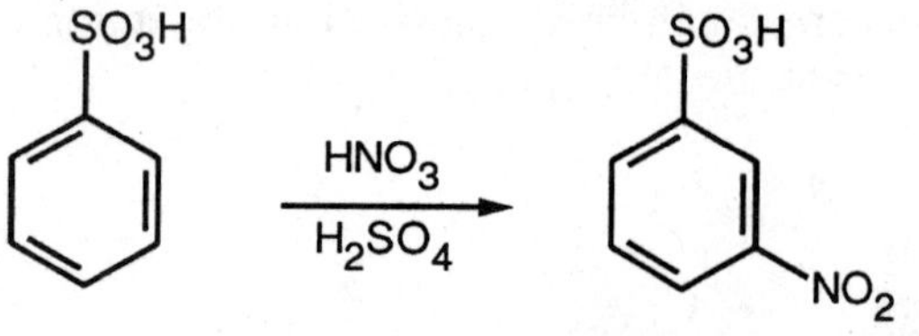

m–Nitrobenzenesulfonic acid

f)

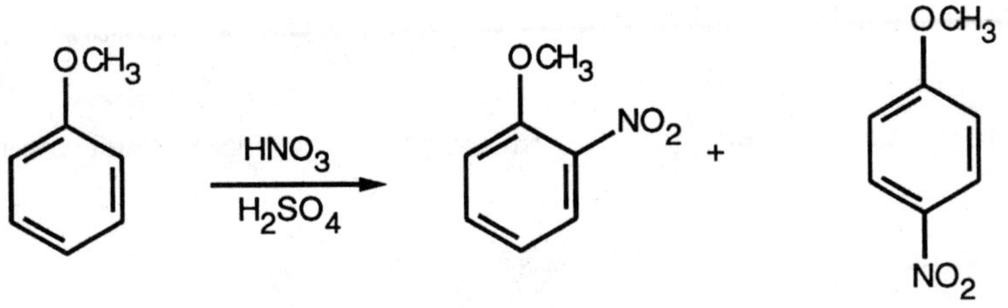

o–Methoxynitrobenzene   p–Methoxynitrobenzene

Only methoxybenzene reacts faster than benzene.

16.32   Most reactive ————————————————————> Least reactive

a) Benzene > Chlorobenzene > o-Dichlorobenzene
(b) Phenol > Nitrobenzene > p-Bromonitrobenzene
(c) o-Xylene > Fluorobenzene > Benzaldehyde
(d) p-Methoxybenzonitrile > p-Methylbenzonitrile > Benzonitrile

16.33

a)

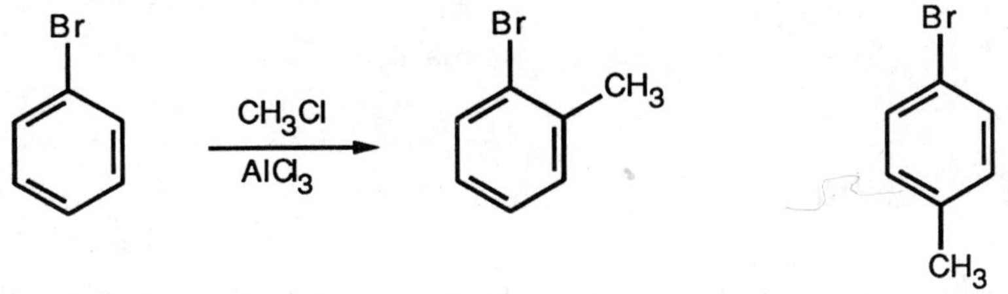

o–Bromotoluene   p–Bromotoluene

b)

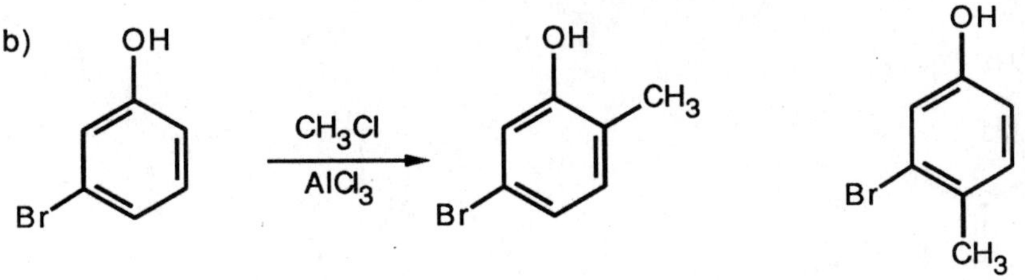

5–Bromo–2–methyl–   3–Bromo–4–methyl
phenol   phenol

c)

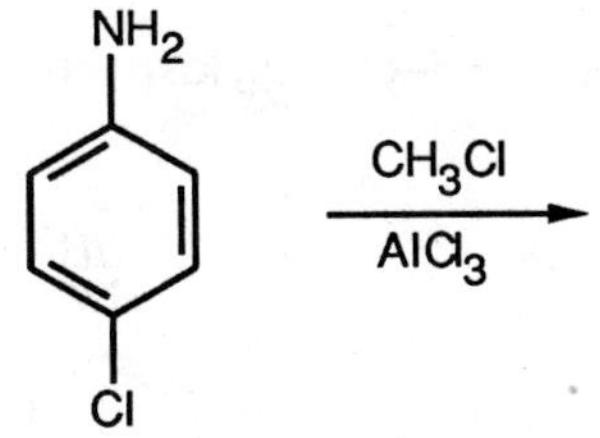

No reaction. AlCl₃ combines with–N̈H₂ to form a complex that deactivates the ring toward Friedel-Crafts alkylation.

d)

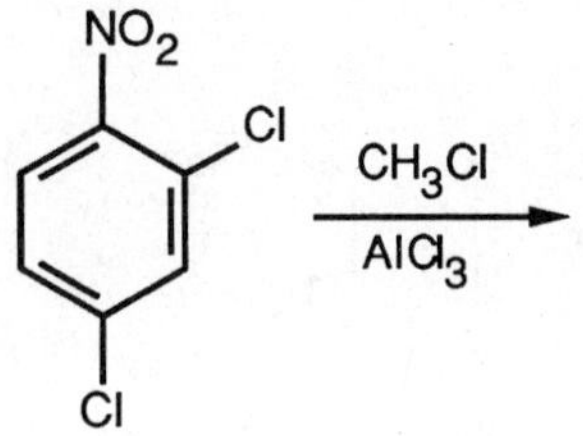

No reaction. The ring is deactivated

e)

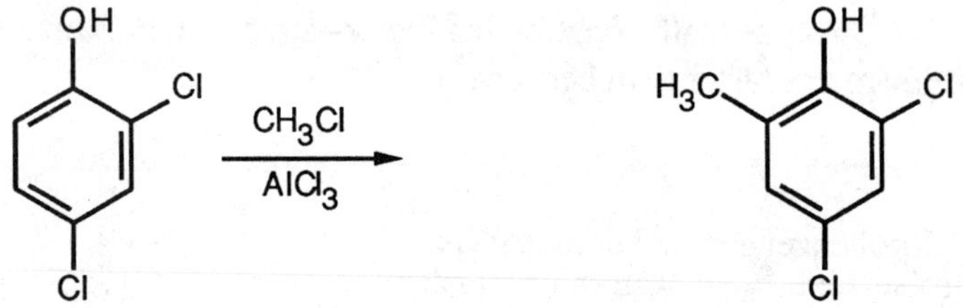

2,4–Dichloro–6–methylphenol

f)

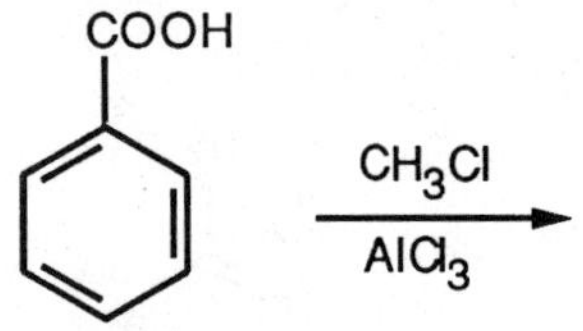

No reaction.

g)

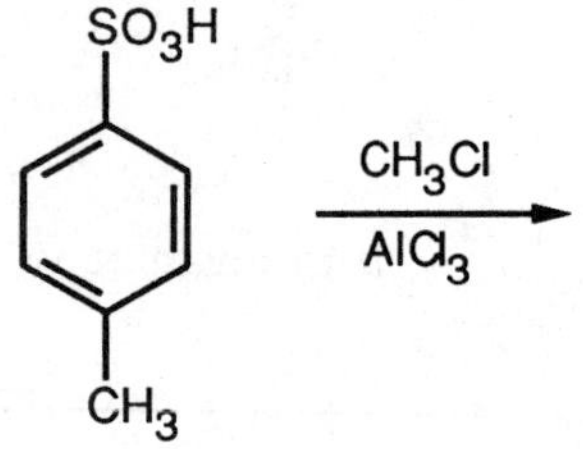

No reaction. The ring is deactivated

h)

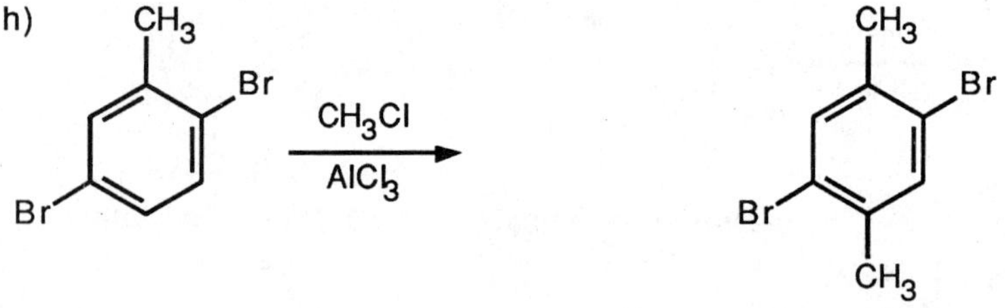

1,4–Dibromo–2,5–dimethylbenzene

16.34

a)

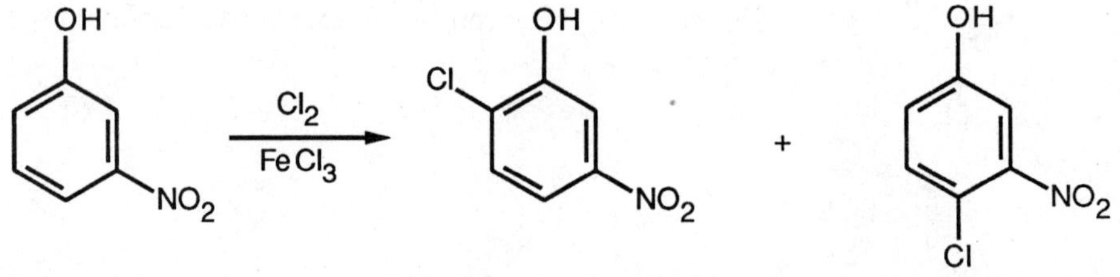

2–Chloro–5–nitrophenol   4–Chloro–3–nitrophenol

b)

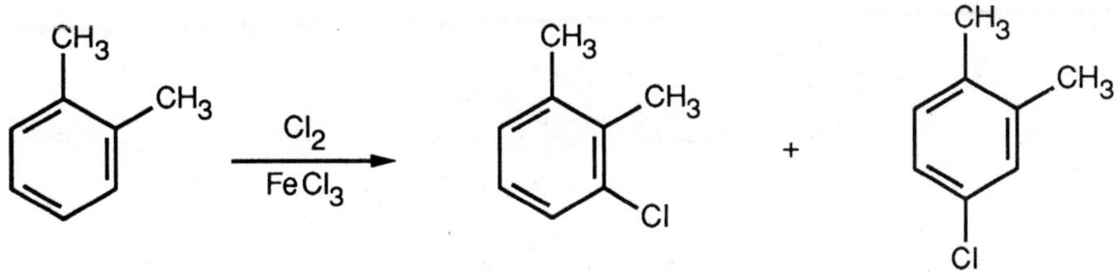

1–Chloro–2,3–
dimethylbenzene

+

4–Chloro–1,2–
dimethylbenzene

c)

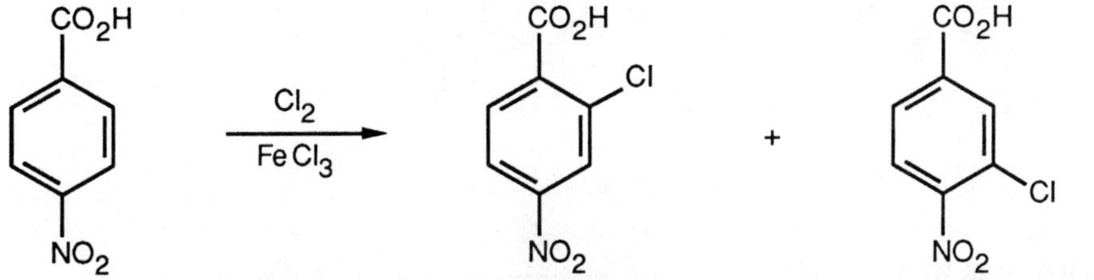

2–Chloro–4–nitro-
benzoic acid

+

3–Chloro–4–nitro-
benzoic acid

d)

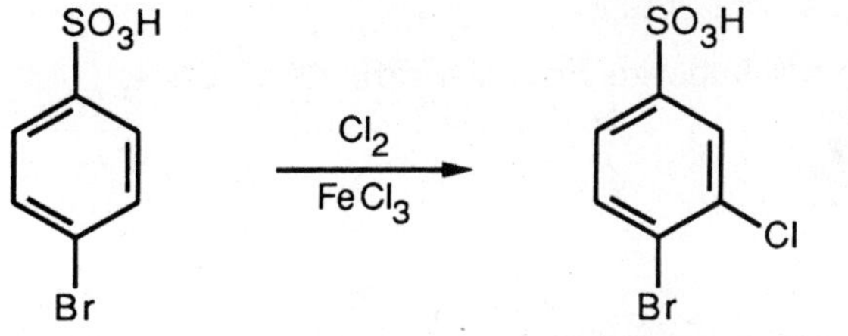

4–Bromo–3–chloro-
benzenesulfonic acid

16.35

a)

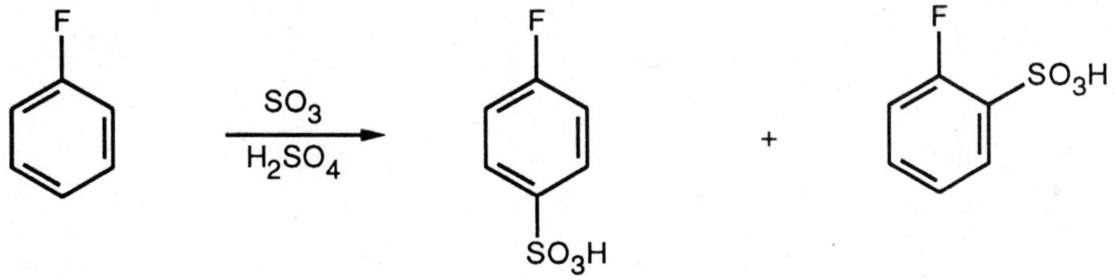

b)

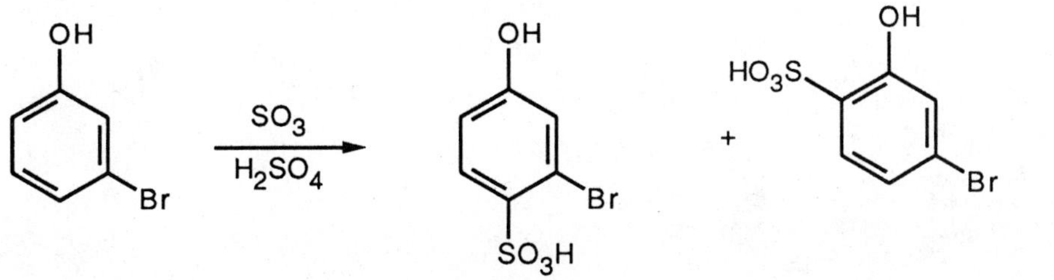

c)

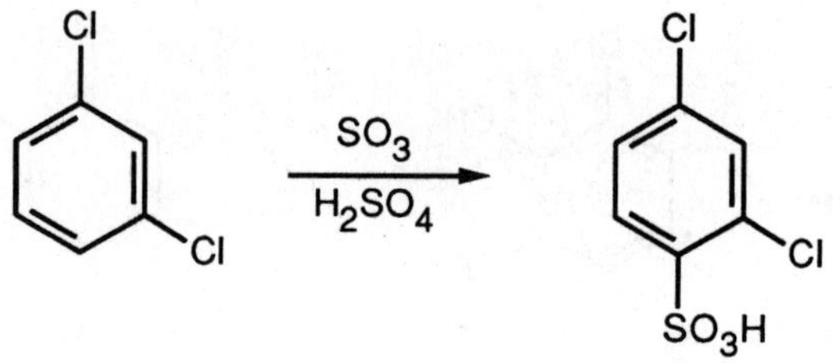

d)

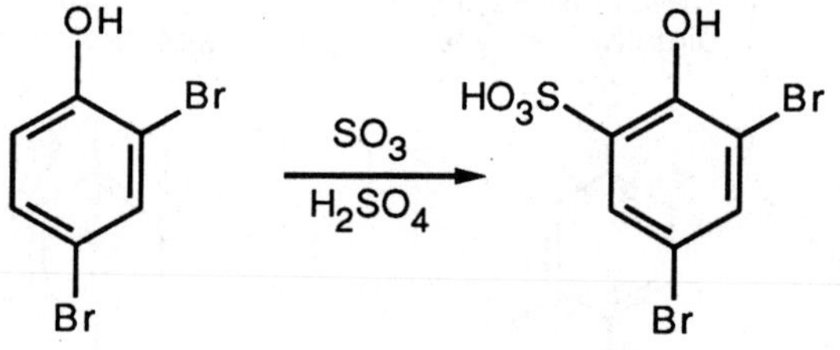

16.36 Most reactive ————————————————> Least reactive

Phenol > Toluene > p-Bromotoluene > Bromobenzene

Aniline and nitrobenzene do not undergo Friedel-Crafts alkylations.

16.37

Resonance structures show that bromination occurs in the *ortho* and *para* positions of the rings. The positively charged intermediate formed from *ortho* or *para* attack can be stabilized by resonance contributions from the second ring of biphenyl. This stabilization is not possible for *meta* attack.

16.38 Attack occurs on the unsubstituted ring because bromine is a deactivating group. Attack occurs at the *ortho* and *para* positions of the ring because the positively charged intermediate can be stabilized by resonance contributions from bromine and from the second ring (Problem 16.37).

16.39 The –CN group is a *meta*-directing deactivator for both inductive and resonance reasons. However, in 3–phenylpropanenitrile, the saturated side chain does not allow resonance interactions of –CN with the aromatic ring. In addition, the –CN group is too far from the ring for its inductive effect to be strongly felt. The side chain acts as an alkyl substituent, and *ortho–para* substitution is observed.

In 3–phenylpropenenitrile, the –CN group interacts with the ring through the pi electrons of the side chain. Drawings of resonance forms show that –CN deactivates the ring toward electrophilic substitution. Substitution occurs at the *meta* position.

16.40

Protonation of the double bond at carbon 2 of 1–phenylpropene leads to an intermediate that can be stabilized by resonance involving the phenyl ring.

16.41

$$CHCl_3 + AlCl_3 \rightleftharpoons {}^+CHCl_2\ {}^-AlCl_4$$

(Dichloromethyl)-
benzene

(Dichloromethyl)benzene can react with two additional equivalents of benzene to produce triphenylmethane.

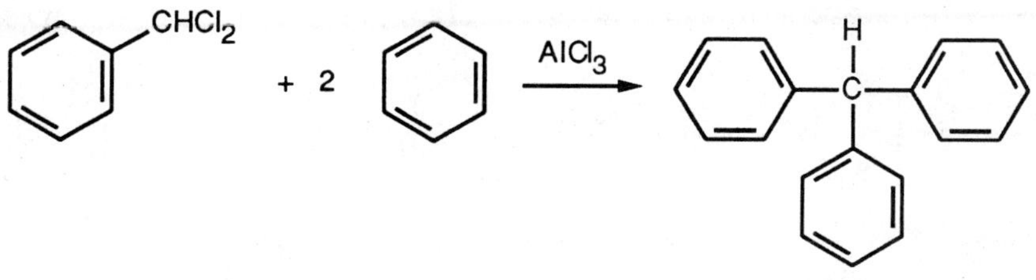

Triphenylmethane

16.42

a)

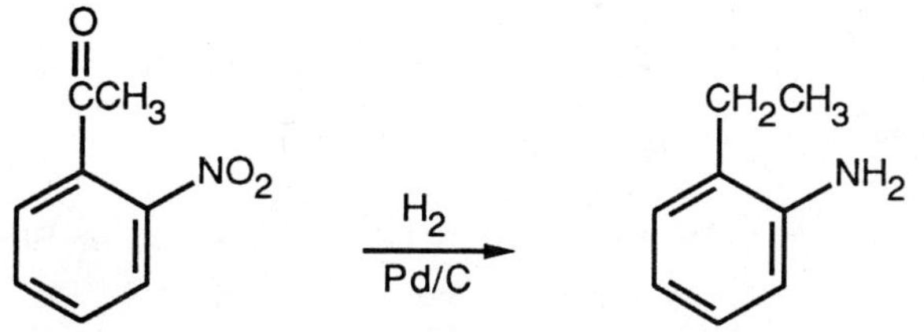

b) No reaction. Since this compound has no benzylic protons, it can't be oxidized by KMnO$_4$.

c)

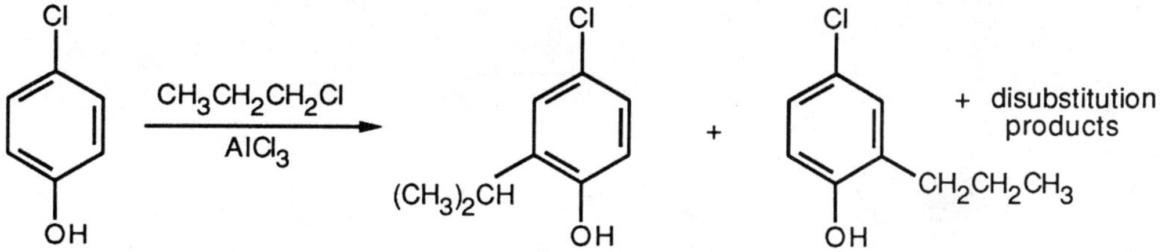

major product

16.43

a)

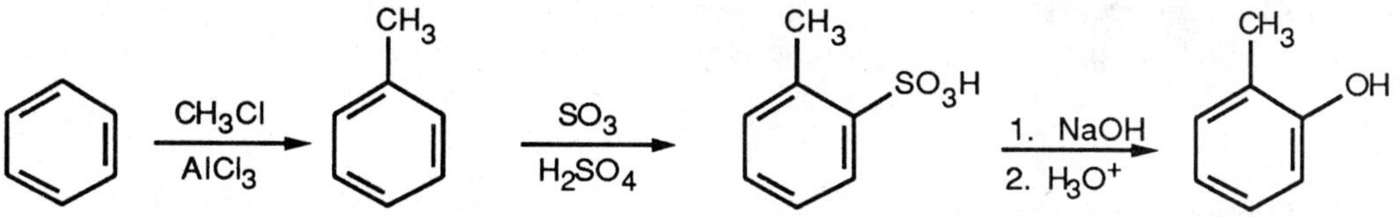

o-Methylphenol

b)

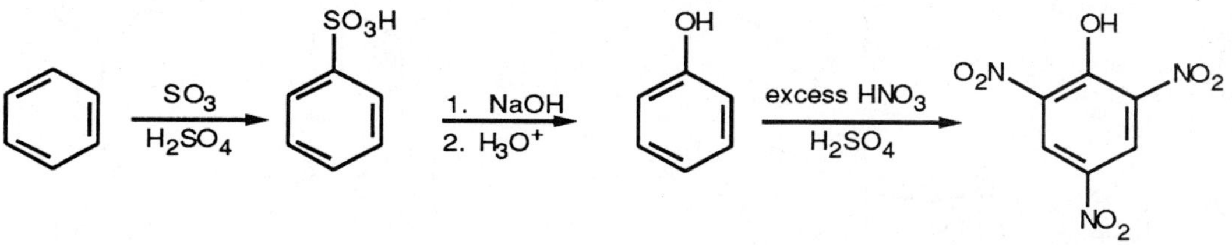

2,4,6-Trinitrophenol

c)

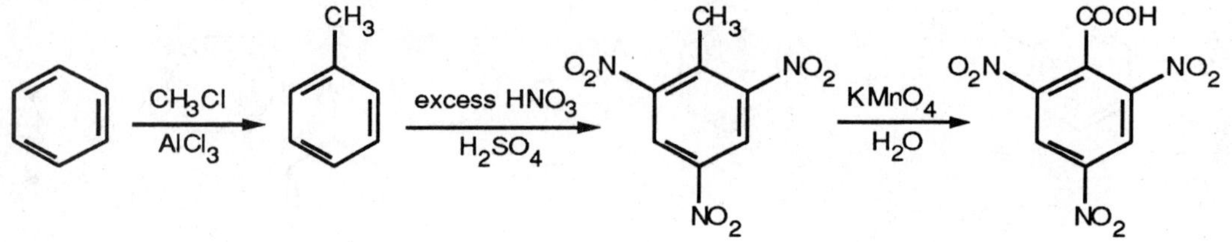

2,4,6–Trinitrobenzoic acid

d)

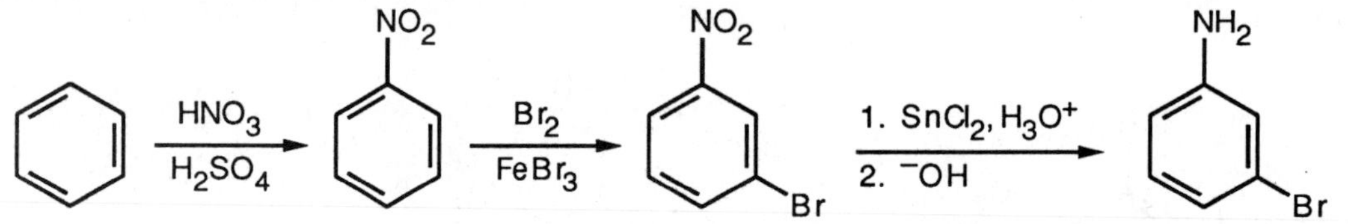

m–Bromoaniline

16.44

a)

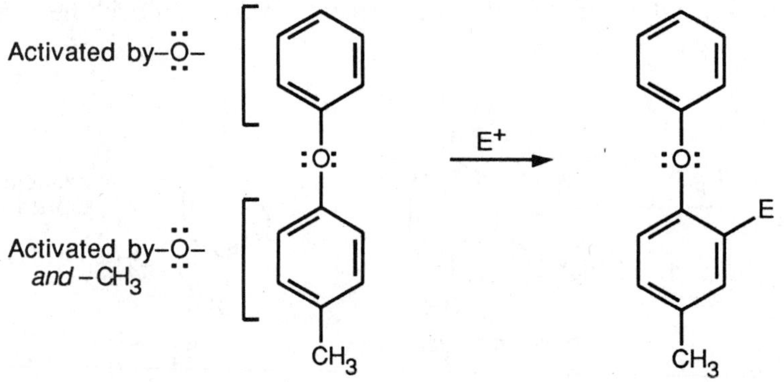

Activated by –Ö–

Activated by –Ö–
and –CH₃

Substitution occurs in the more activated ring. The position of substitution is determined by the more powerful activating group – in this case, the oxygen atom.

b)

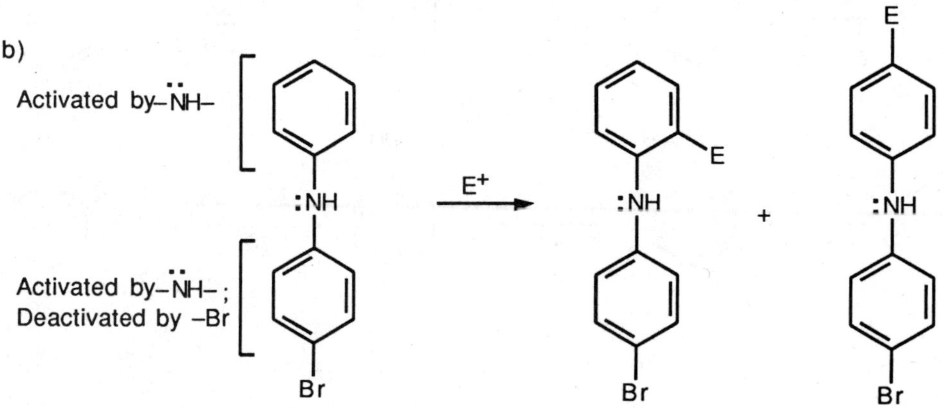

Activated by –N̈H–

Activated by –N̈H–;
Deactivated by –Br

The upper ring is more activated than the lower ring. –NH– is an *ortho-para* director.

c)

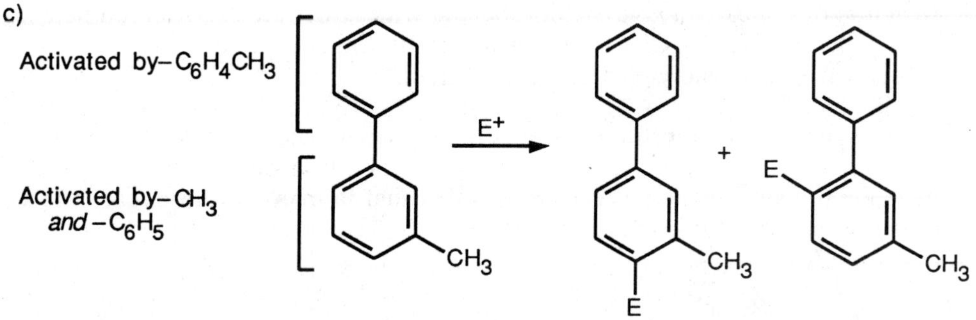

Activated by–$C_6H_4CH_3$

Activated by–CH_3
and –C_6H_5

Substitution occurs in the *ortho* and *para* positions of the more activated ring. Substitution does not occur between –C_6H_5 and –CH_3 for steric reasons.

16.45

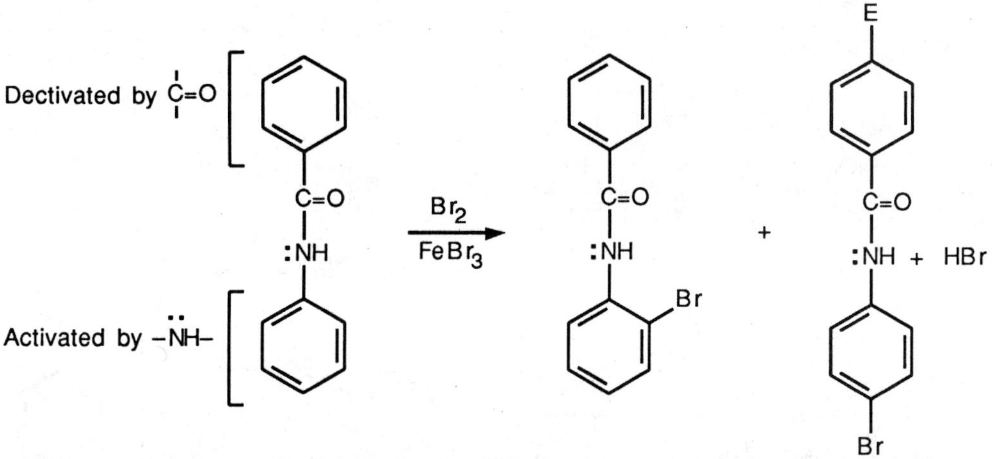

Dectivated by $\overset{|}{\underset{|}{C}}=O$

Activated by –$\ddot{N}H$–

Attack occurs in the activated ring and yields *ortho* and *para* bromination products. The intermediate is resonance-stabilized by overlap of the nitrogen lone pair electrons with the pi electrons of the substituted ring.

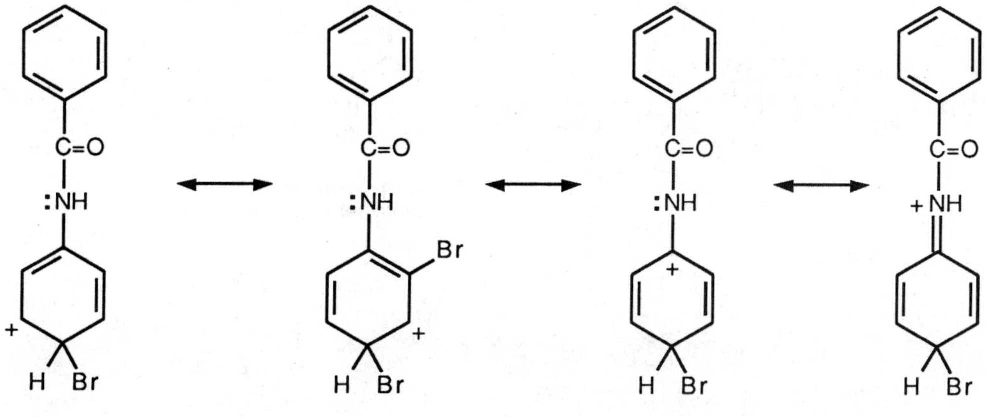

Similar drawings can be made of the resonance forms of the intermediate resulting from *ortho* attack.

16.46 Reaction of optically active 2-chlorobutane with $AlCl_3$ produces an ion pair $[CH_3{}^+CHCH_2CH_3\ {}^-AlCl_4]$. The planar, sp^2-hybridized carbocation is achiral, and its reaction with benzene yields racemic product.

16.47 When synthesizing substituted aromatic rings, it is necessary to introduce substituents in the proper order. A group that is introduced in the wrong order will not have the proper directing effect. Remember that in many of these reactions a mixture of *ortho* and *para* isomers may be formed.

a)

$$C_6H_6 \xrightarrow[\text{H}_2\text{SO}_4]{\text{SO}_3} \text{(SO}_3\text{H)} \xrightarrow[\text{2. H}_3\text{O}^+]{\text{1. NaOH}} \text{(OH)} \xrightarrow[\text{FeCl}_3]{\text{Cl}_2} \text{(OH, Cl)}$$

p–Chlorophenol

b)

$$C_6H_6 \xrightarrow[\text{H}_2\text{SO}_4]{\text{HNO}_3} \text{(NO}_2\text{)} \xrightarrow[\text{FeBr}_3]{\text{Br}_2} \text{(NO}_2\text{, Br)}$$

m–Bromonitrobenzene

c)

$$C_6H_6 \xrightarrow[\text{FeBr}_3]{\text{Br}_2} \text{(Br)} \xrightarrow[\text{H}_2\text{SO}_4]{\text{SO}_3} \text{(Br, SO}_3\text{H)} + \text{[Br, SO}_3\text{H]}$$

o–Bromobenzene-sulfonic acid

d)

$$C_6H_6 \xrightarrow[\text{H}_2\text{SO}_4]{\text{SO}_3} \text{(SO}_3\text{H)} \xrightarrow[\text{FeCl}_3]{\text{Cl}_2} \text{(SO}_3\text{H, Cl)}$$

m–Chlorobenzene-sulfonic acid

16.48

a)

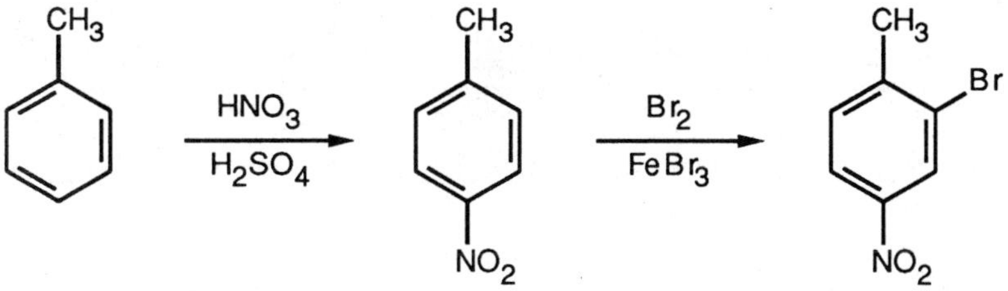

2–Bromo–4–nitrotoluene

b)

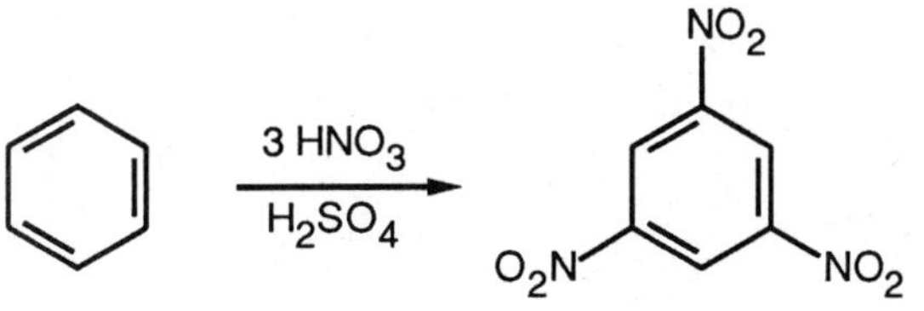

1,3,5–Trinitrobenzene

c)

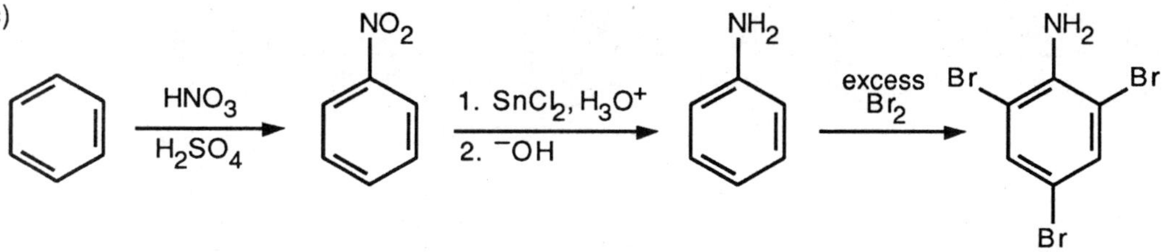

2,4,6–Tribromoaniline

No catalyst is needed for bromination because aniline is very activated toward substitution.

d)

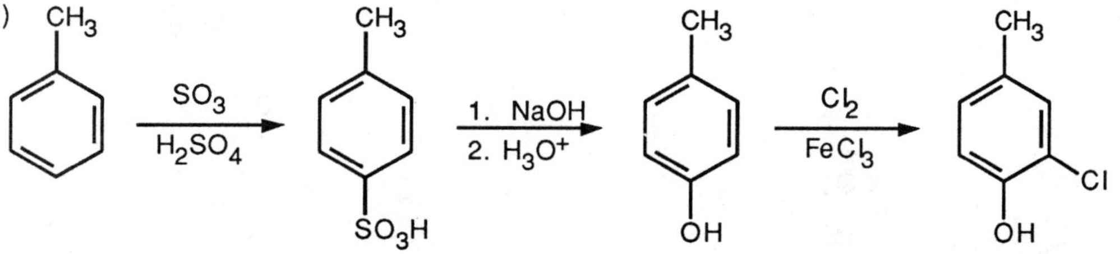

plus *ortho* isomer

2–Chloro–4–methylphenol

The –OH group is a stronger director than –CH$_3$.

16.49 a) Chlorination of toluene occurs at the *ortho* and *para* positions. To synthesize the given product, first oxidize toluene to benzoic acid and then chlorinate.

b) *p*–Nitrochlorobenzene is inert to Friedel-Crafts alkylation because the nitrochlorobenzene ring is deactivated.

c) The first two steps in the sequence are correct, but H$_2$/Pd reduces the nitro group as well as the ketone.

16.50

a)

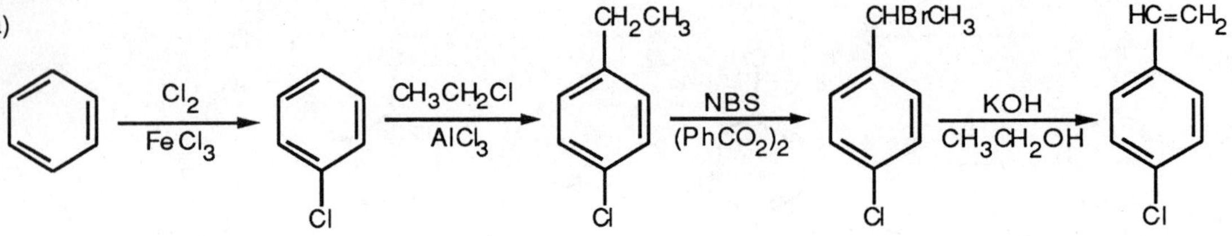

b)

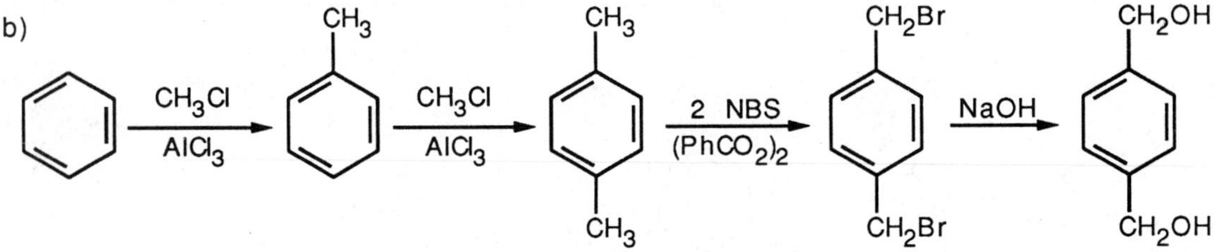

c)

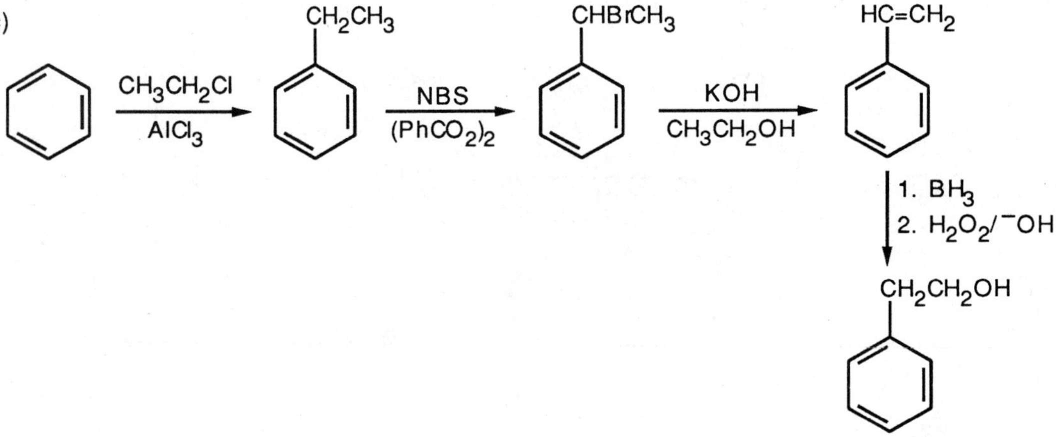

16.51

a)

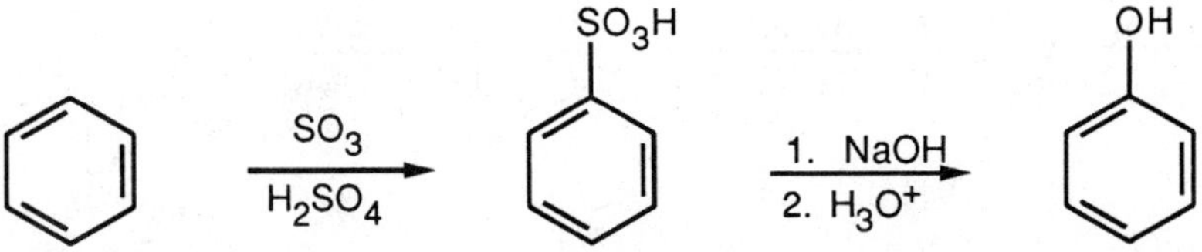

b)

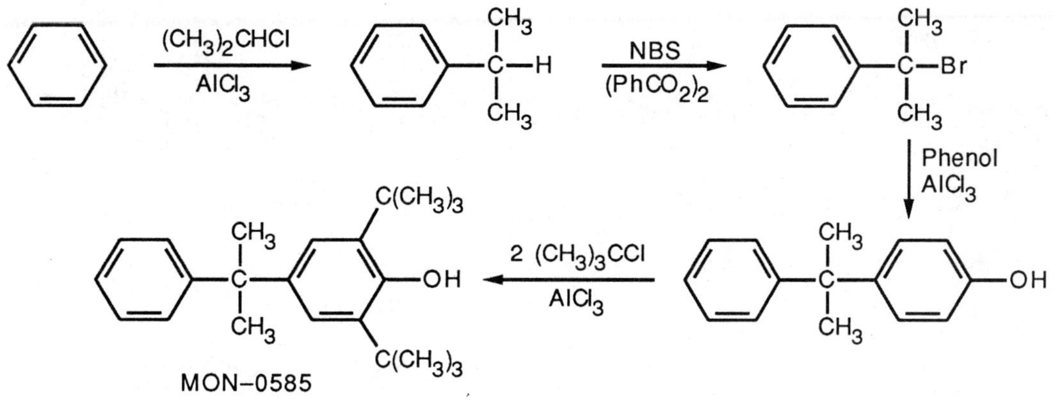

MON−0585

16.52

1)

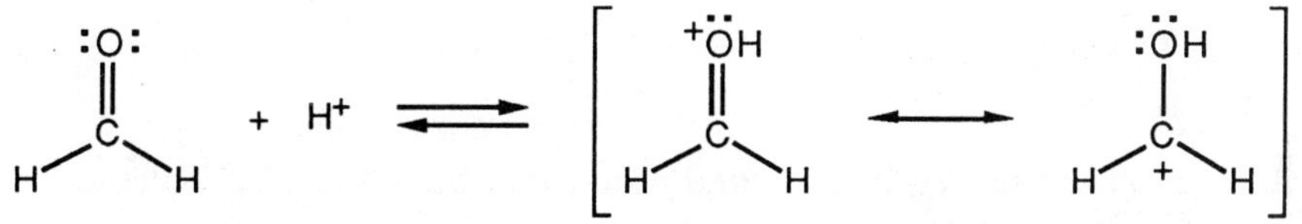

Formaldehyde is protonated to form a carbocation.

2)

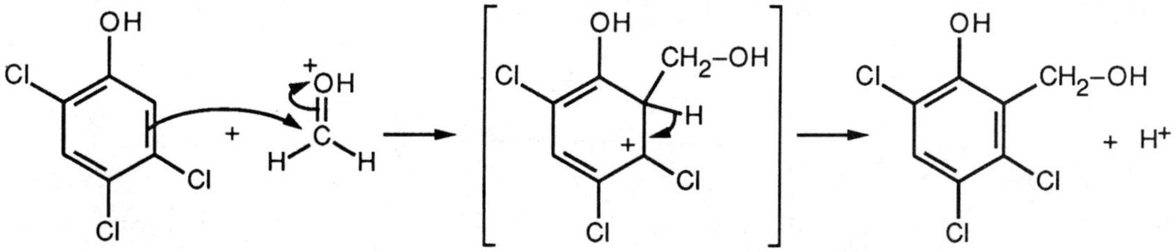

The formaldehyde cation acts as the electrophile in a substitution reaction at the 6 position of 2,4,5–trichlorophenol.

3)

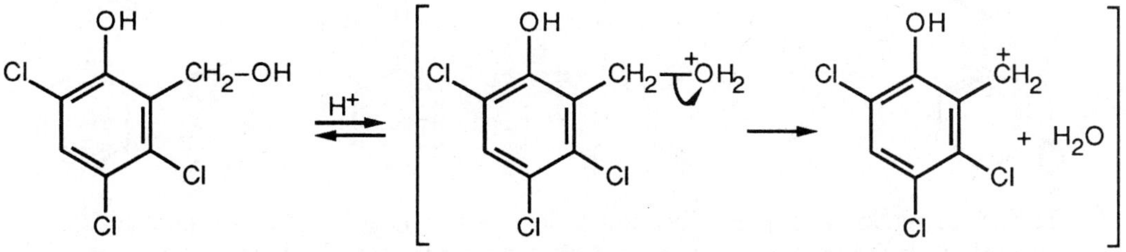

The product from step 2 is protonated by strong acid to produce a carbocation.

4)

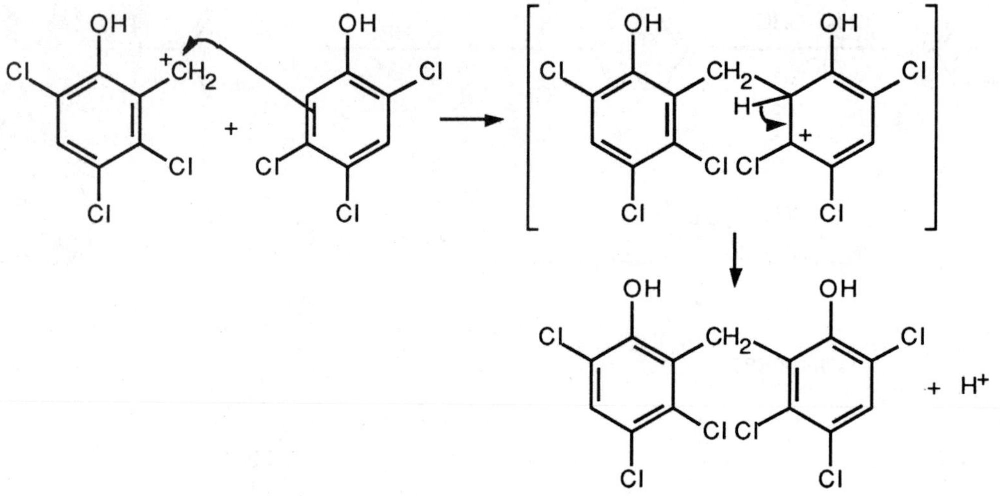

Hexachlorophene

This carbocation is attacked by a second molecule of 2,4,5–trichlorophenol to produce hexachlorophene.

16.53

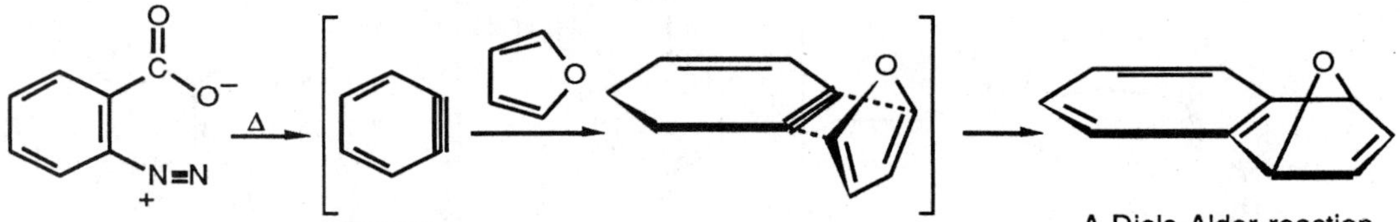

Benzyne
+ CO_2 + N_2

A Diels-Alder reaction

16.5

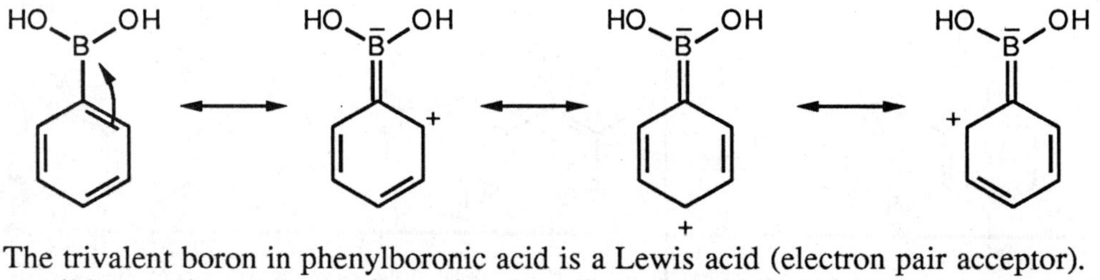

The trivalent boron in phenylboronic acid is a Lewis acid (electron pair acceptor). It is possible to write resonance forms for phenylboronic acid in which an electron pair from the phenyl ring is delocalized onto boron. In these resonance forms, the *ortho* and *para* positions of phenylboronic acid are the most electron-deficient; substitutions occur primarily at the *meta* position.

16.55 Resonance forms from the intermediate from attack at C–1:

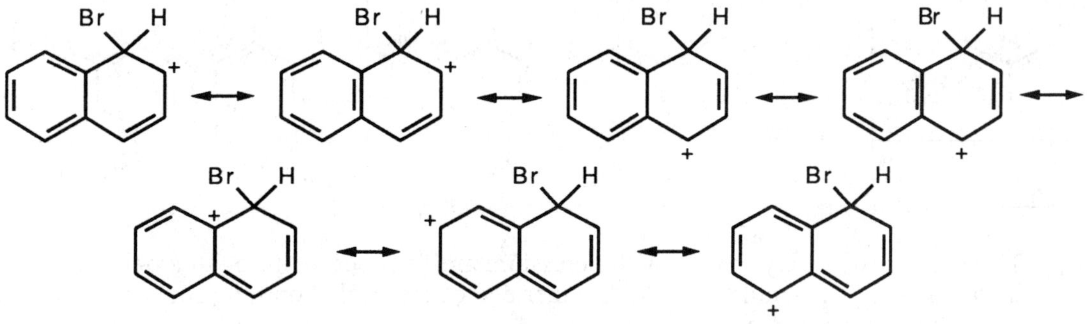

Resonance forms for the intermediate from attack at C2:

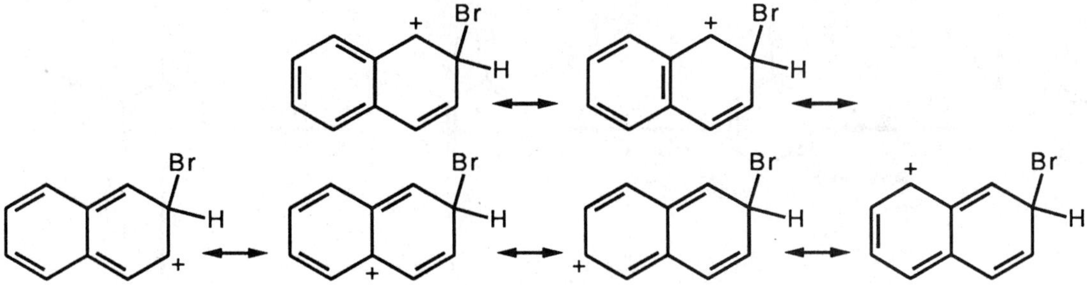

There are seven resonance forms for attack at C–1 and six for attack at C2. Look carefully at the forms, however. In the first four resonance structures for C–1 attack, the second ring is still fully aromatic. The positive charge has been delocalized into the second ring in the other three forms, destroying the ring's aromaticity. For C2 attack, only the first two resonance structures have a fully aromatic second ring. Since stabilization is lost when aromaticity is disrupted, the intermediate from C2 attack is less stable than the intermediate from C–1 attack, and C–1 attack is favored.

16.56

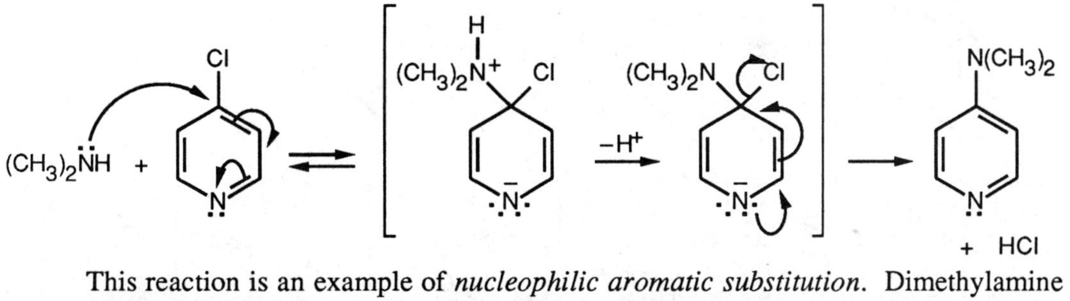

This reaction is an example of *nucleophilic aromatic substitution*. Dimethylamine is a nucleophile, and the nitrogen atom of the pyridine ring acts as an electron-withdrawing group that can stabilize the negatively-charged intermediate.

16.57

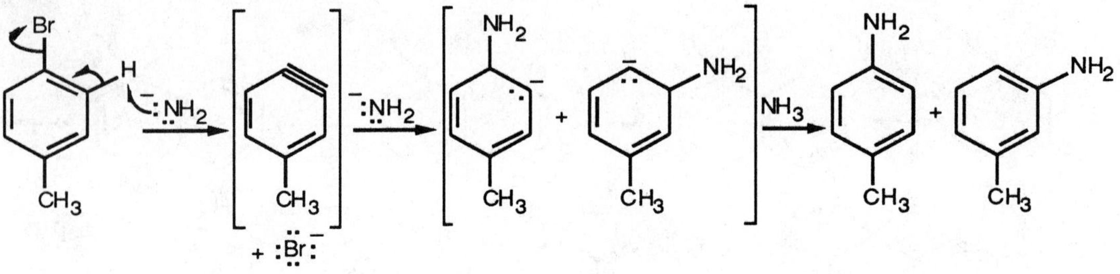

The reaction of an aryl halide with potassium amide proceeds through a *benzyne* intermediate. The amide can add to either end of the triple bond to produce the two methylanilines observed.

16.58

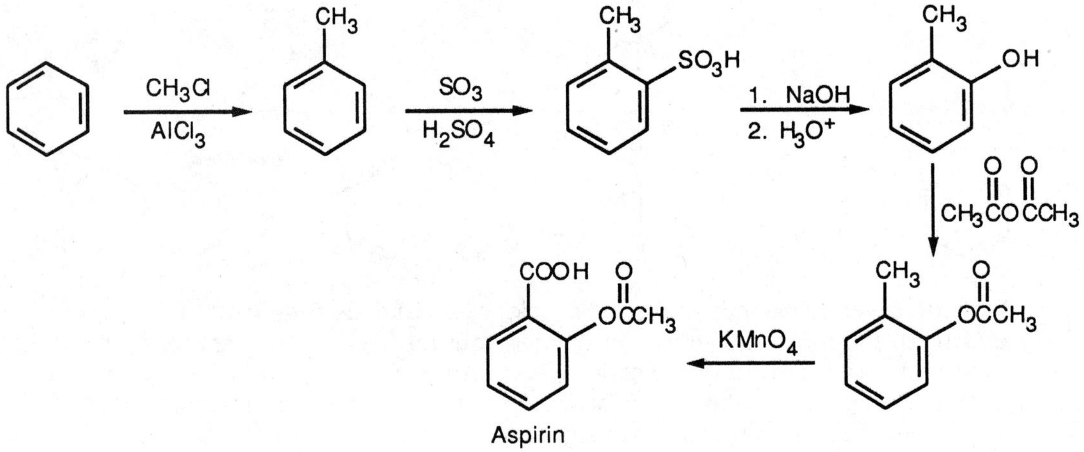

Aspirin

16.59

a)

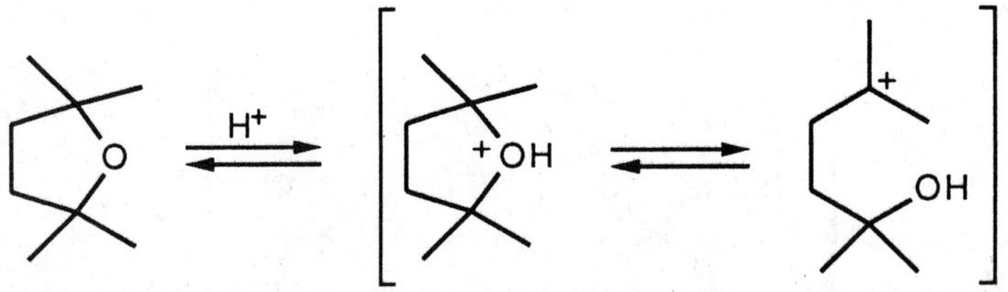

b)

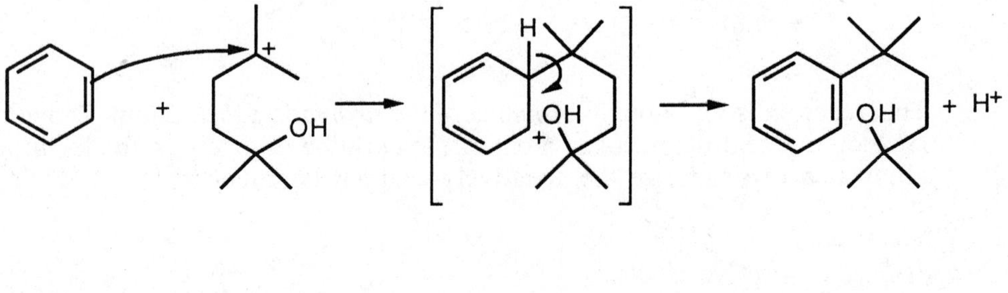

c)

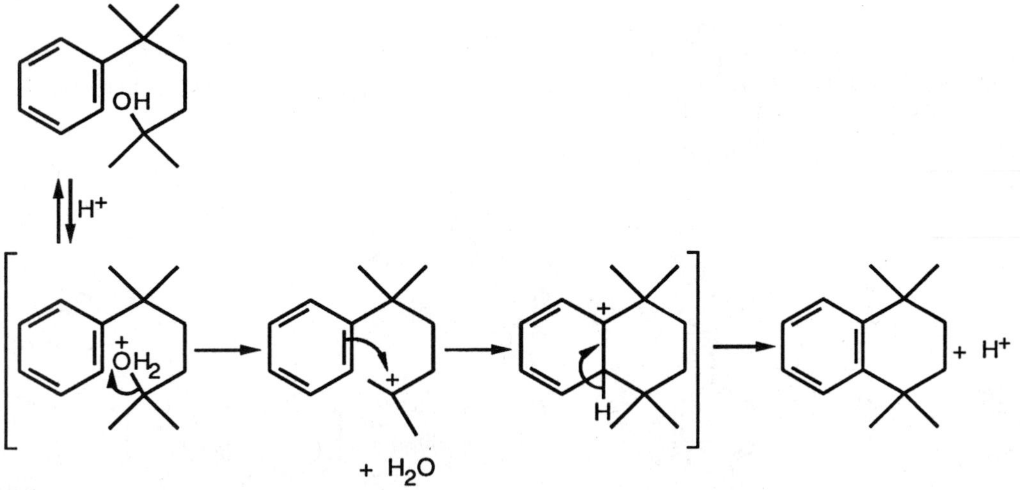
OH
H+
OH2
+ H+
+ H2O
H

16.60

1)

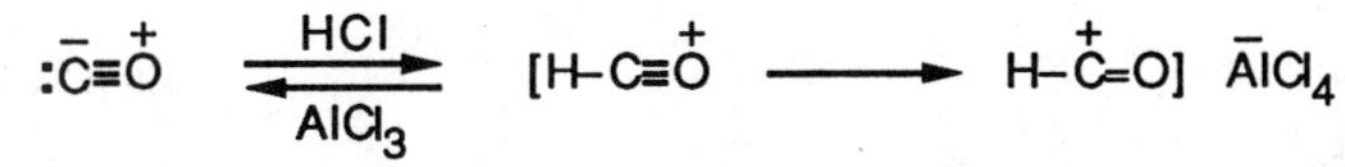
:C≡O
HCl
AlCl3
[H–C≡O
H–C=O] AlCl4

2)

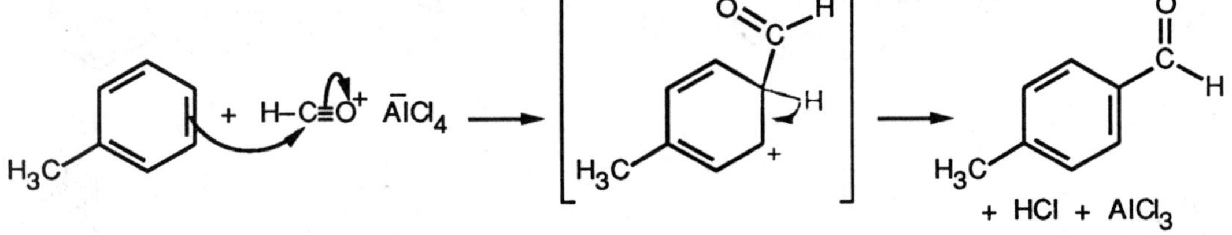
H3C
+ H–C≡O+ AlCl4
O
C
H
+ HCl + AlCl3

16.61

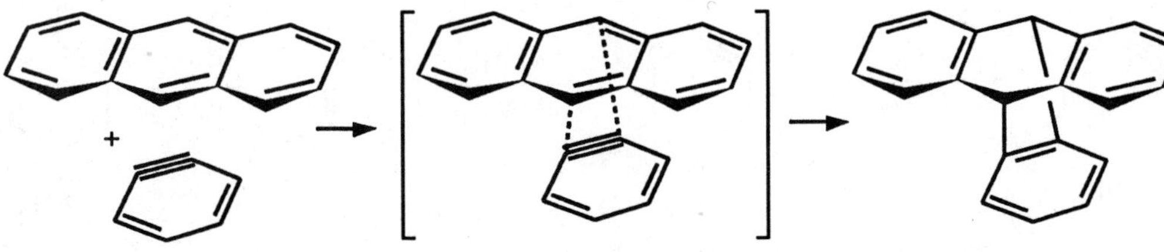

Triptycene

The reaction between benzyne and anthracene is a Diels-Alder reaction.

16.62

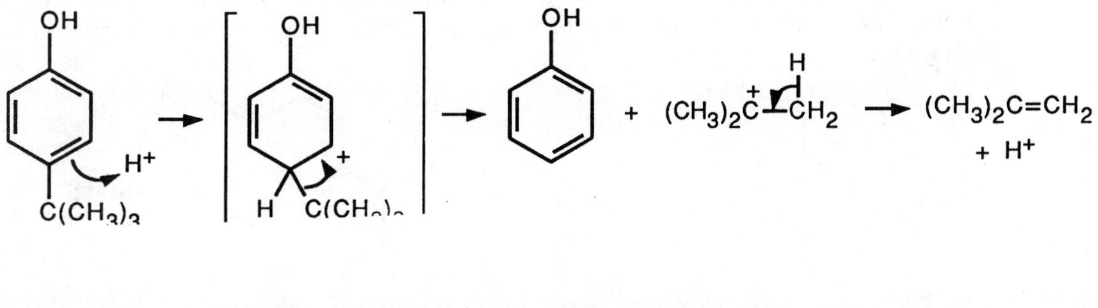
OH
H+
C(CH3)3
H
+ (CH3)2C–CH2
(CH3)2C=CH2
+ H+

16.63

a)

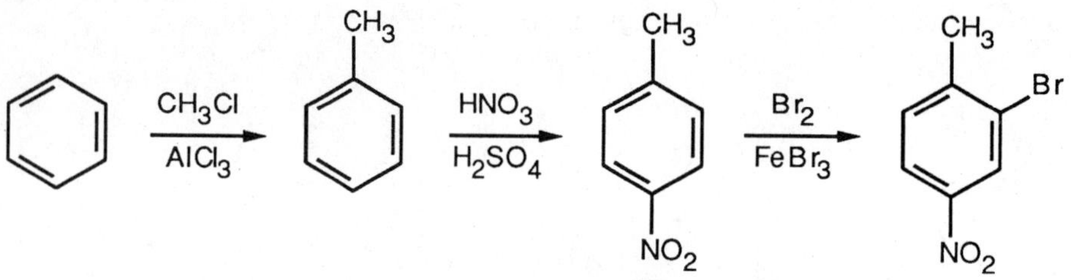

+ ortho isomer

b)

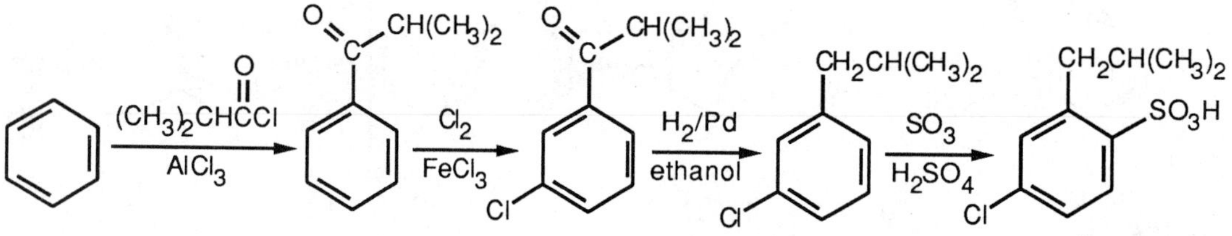

+ para isomer

c)

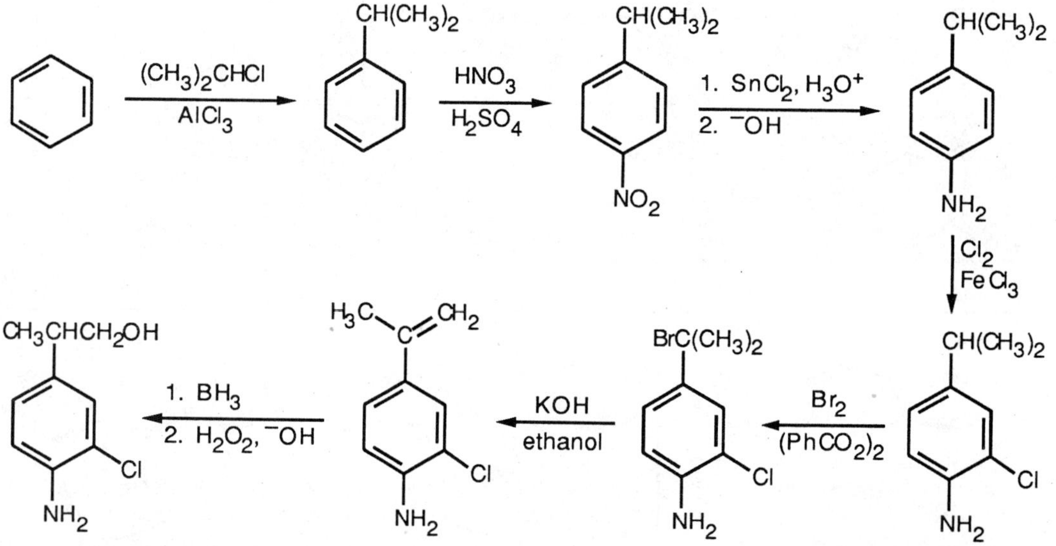

16.64 Problem 16.40 shows the mechanism of the addition of HBr to 1–phenylpropene and shows how the aromatic ring stabilizes the carbocation intermediate. Similar resonance forms can be drawn for the intermediates of reaction of the substituted styrenes with HBr. For the methoxy-substituted styrene, an additional form can be drawn. For the nitro-substituted styrene, no additional form is possible. In addition, one other resonance form is not important because it places two positive charges next to each other.

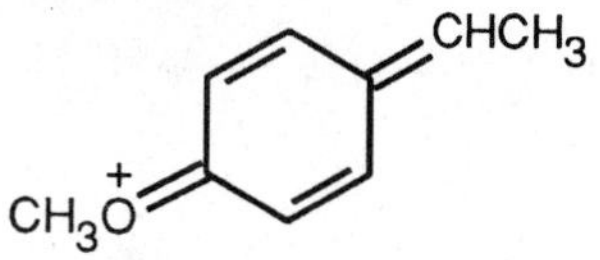

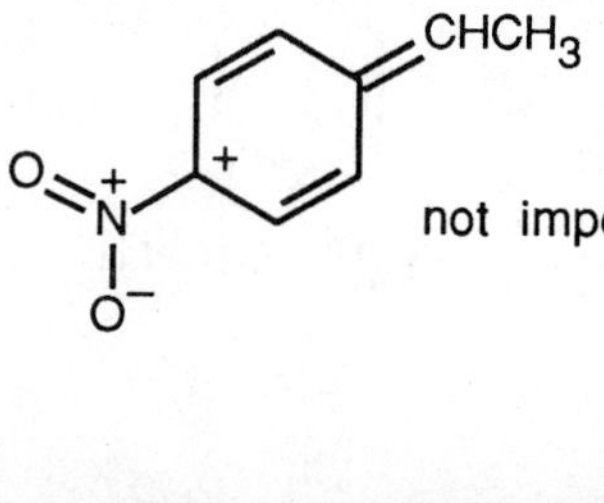

not important

Thus, the intermediate resulting from addition of HBr to the methoxy substituted styrene is more stable, and reaction of *p*–methoxystyrene should be faster.

16.65

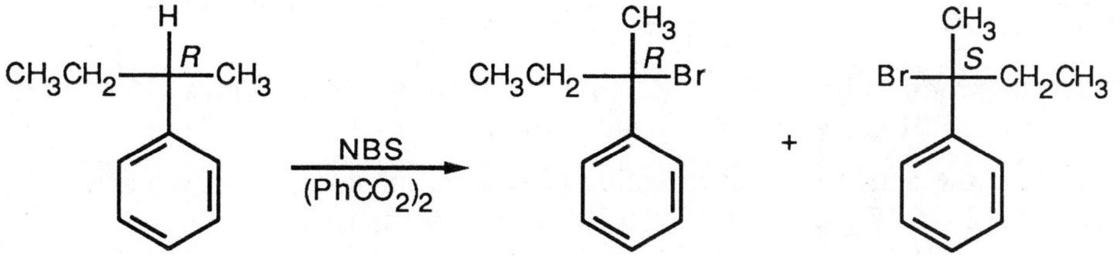

The intermediate benzyl radical is planar and achiral, and it reacts with bromine to produce a racemic mixture.

16.66

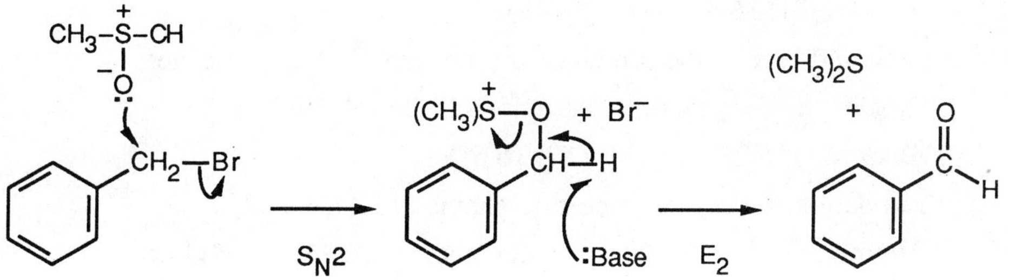

16.67

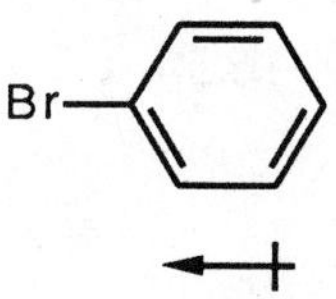

μ = 1.53 D

–Br has a strong electron-withdrawing inductive effect.

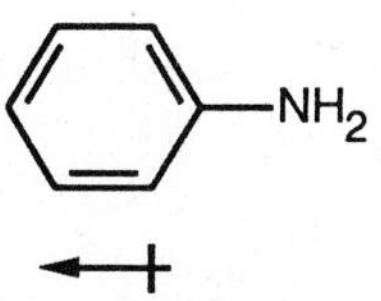

μ = 1.52 D

–NH$_2$ has a strong electron-donating resonance effect.

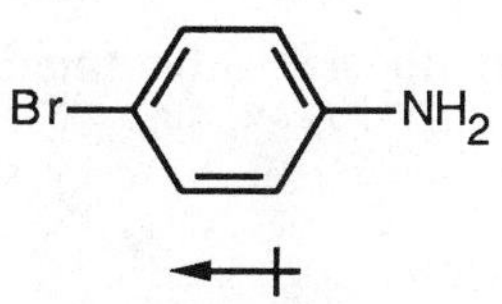

μ = 2.91 D

The polarities of the two groups are additive and produce a net dipole moment almost equal to the sum of the individual moments.

Study Guide for Chapter 16

After studying this chapter, you should be able to:

(1)   Formulate the mechanism of electrophilic aromatic substitution reactions (16.2, 16.3, 16.4, 16.7, 16.9, 16.41, 16.46, 16.52, 16.59, 16.60, 16.62).

(2)   Predict the products of electrophilic aromatic substitution reactions (16.1, 16.5, 16.6, 16.8, 16.10, 16.31, 16.33, 16.34, 16.35).

(3)   Understand the activating and directing effects of substituents on aromatic rings (16.11, 16.12, 16.17, 16.18, 16.19, 16.20, 16.54, 16.55, 16.67).

(4)   Use inductive and resonance arguments to predict orientation and reactivity in electrophilic substitution reactions (16.13, 16.14, 16.15, 16.16, 16.32, 16.36, 16.37, 16.39, 16.64).

(5)   Predict the position of electrophilic aromatic substitution of polysubstituted aromatic compounds (16.21, 16.38, 16.44, 16.45).

(6)   Understand and predict the products of other reactions of benzenes.

　　　(a) Nucleophilic aromatic substitution (16.22, 16.56).

　　　(b) Benzyne (16.23, 16.53, 16.57, 16.61).

　　　(c) Oxidation of alkylbenzene side-chains (16.24, 16.42).

　　　(d) Bromination of alkylbenzene side-chains (16.25, 16.26, 16.65).

　　　(e) Reduction of aromatic compounds (16.28, 16.42).

　　　(f) Birch reduction (16.27).

(7)   Synthesize substituted benzenes (16.29, 16.30, 16.43, 16.47, 16.48, 16.49, 16.50, 16.51, 16.58, 16.63).

Chapter 17 – Alcohols and Thiols

17.1

a)

OH OH
| |
$CH_3CHCH_2CHCH(CH_3)_2$

5–Methyl–2,4–hexanediol

b)

OH
|
$CH_2CH_2C(CH_3)_2$

2–Methyl–4–phenyl–2–butanol

c)

4,4–Dimethylcyclohexanol

d)

trans–2–Bromocyclopentanol

17.2

a)

CH_2CH_3
/
$CH_3CH=C$
\
CH_2OH

2–Ethyl–2–buten–1–ol

b)

3–Cyclohexen–1–ol

c)

trans–3–Chlorocycloheptanol

d)

OH
|
$CH_3CHCH_2CH_2CH_2OH$

1,4–Pentanediol

17.3 In general, the boiling points of a series of isomers decrease with branching. The more nearly spherical a compound becomes, the less surface area it has relative to a straight chain compound of the same molecular weight and functional group type. A smaller surface area allows fewer van der Waals interactions, the weak forces that attract covalent molecules to each other. In addition, branching makes it more difficult for hydroxyl groups to approach each other from all directions to hydrogen bond.

A given volume of 2–methyl–2–propanol contains fewer hydrogen bonds than the same volume of 1–butanol. Since it takes less energy to break the hydrogen bonds of 2–methyl–2–propanol, it is lower-boiling than 1–butanol or 2–butanol.

17.4   We learned in Chapter 16 that a nitro group is electron-withdrawing. Since electron-withdrawing groups stabilize alkoxide anions and lower pH, *p*–nitrobenzyl alcohol is more acidic than benzyl alcohol.

17.5

a)

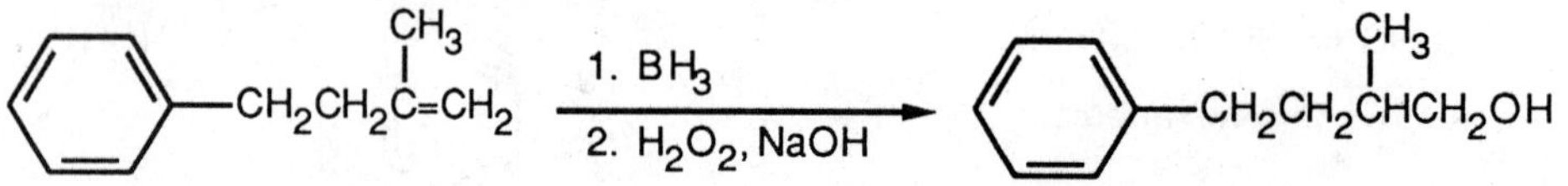

2-Methyl–4–phenyl–1–butanol

b)

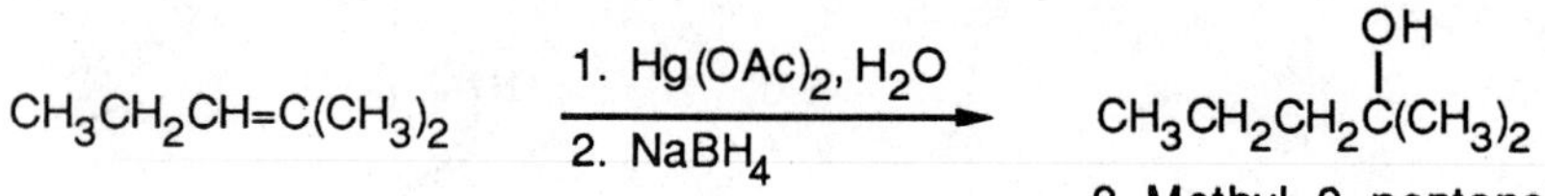

$$CH_3CH_2CH=C(CH_3)_2 \xrightarrow[\text{2. NaBH}_4]{\text{1. Hg(OAc)}_2, \text{H}_2\text{O}}$$

OH
|
$CH_3CH_2CH_2C(CH_3)_2$

2-Methyl–2–pentanol

c)

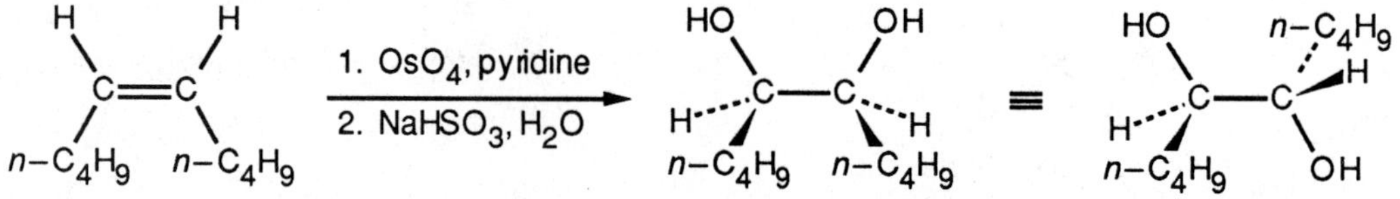

meso–5,6–Decanediol

17.6

Lowest oxidation state ⟶ Highest oxidation state

a)

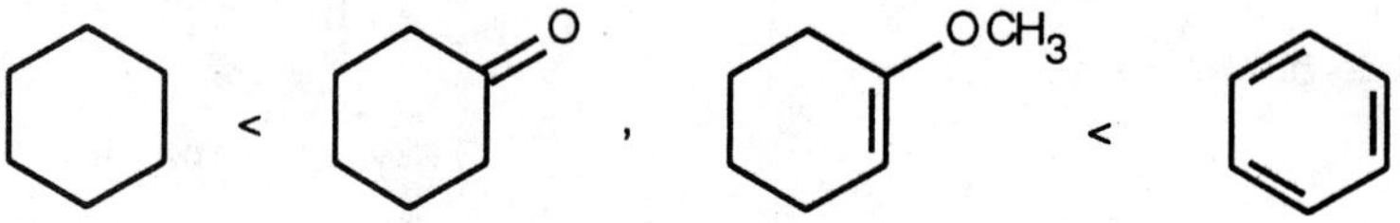

b) $CH_3CH_2NH_2$ < $H_2NCH_2CH_2NH_2$ < CH_3CN

17.7

a)

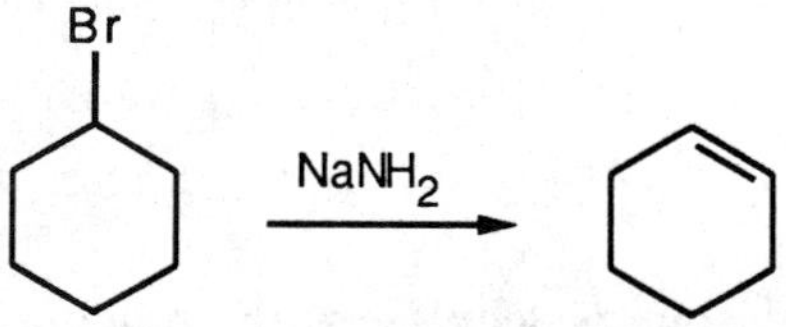

This reaction is neither an oxidation nor a reduction.

b)

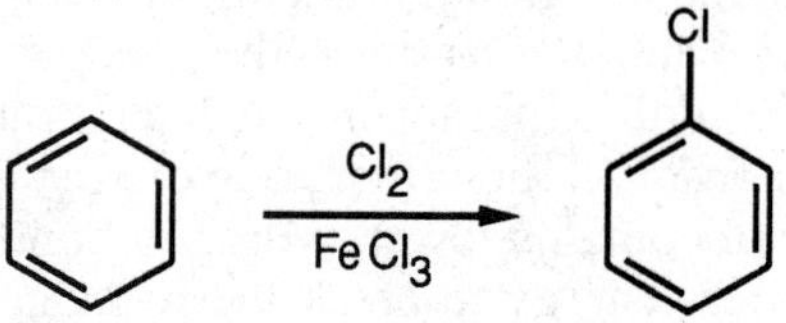

Because chlorine is substituted for hydrogen, this reaction is an oxidation.

c) $CH_3CH_2CH_2CH_2Br$ $\xrightarrow{\text{Mg}}$ $CH_3CH_2CH_2CH_2MgBr$ $\xrightarrow{H_3O^+}$ $CH_3CH_2CH_2CH_3$

This reaction is a reduction of 1–bromobutane; bromine is removed and hydrogen is added. Mg is oxidized.

d)

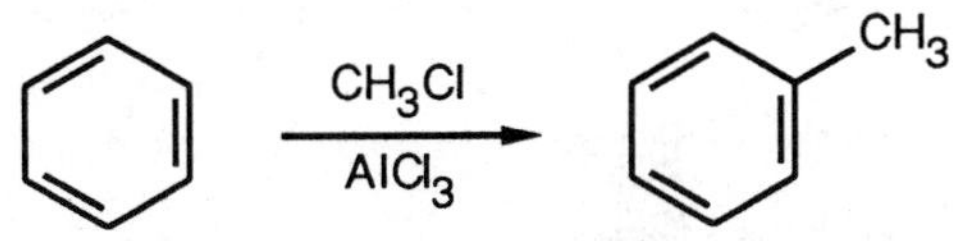

This reaction is neither an oxidation nor a reduction.

17.8

a) $\xrightarrow[\text{2. } H_3O^+]{\text{1. NaBH}_4}$

NaBH$_4$ reduces aldehydes and ketones without disturbing other functional groups.

b) $\xrightarrow[\text{2. } H_3O^+]{\text{1. LiAlH}_4}$

LiAlH$_4$ reduces both ketones and esters.

c)

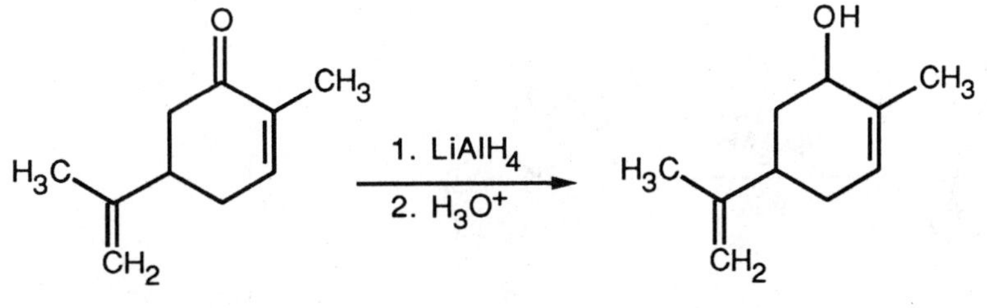

$\xrightarrow[\text{2. } H_3O^+]{\text{1. LiAlH}_4}$

LiAlH$_4$ reduces carbonyl functional groups without reducing double bonds.

d)

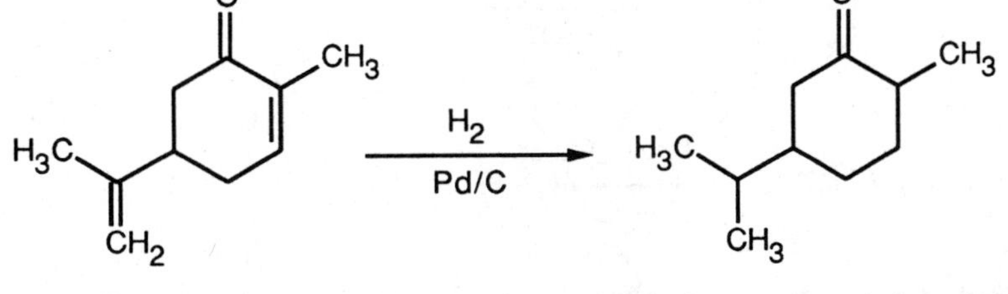

$\xrightarrow[\text{Pd/C}]{H_2}$

H$_2$, Pd/C reduces double bonds without reducing carbonyl functional groups.

17.9

a)

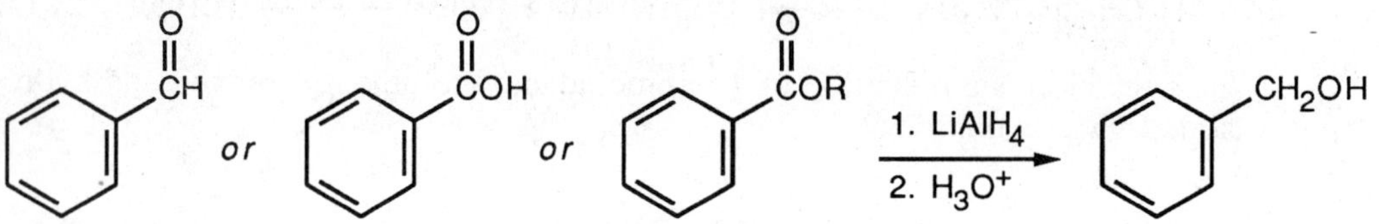

or

or

$\xrightarrow[\text{2. H}_3\text{O}^+]{\text{1. LiAlH}_4}$

b)

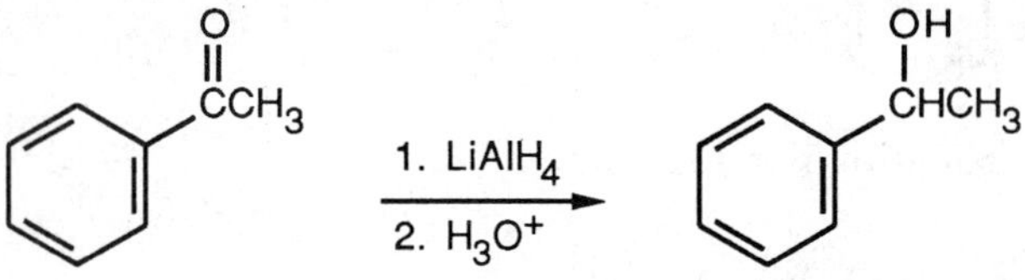

$\xrightarrow[\text{2. H}_3\text{O}^+]{\text{1. LiAlH}_4}$

c)

$\xrightarrow[\text{2. H}_3\text{O}^+]{\text{1. LiAlH}_4}$

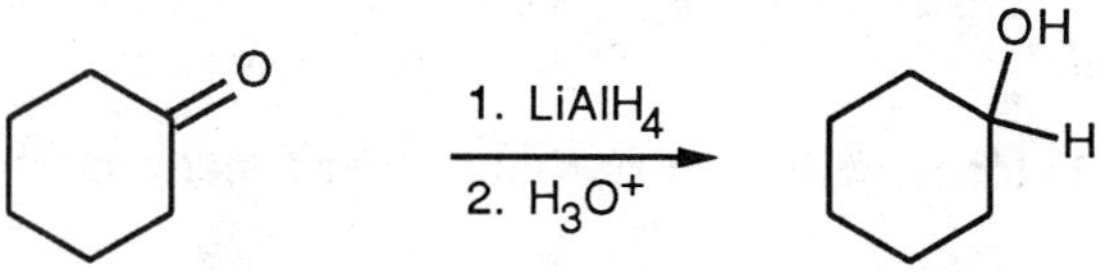

d) $(CH_3)_2CHCHO$ or $(CH_3)_2CHCOOH$ or $(CH_3)_2CHCOOR$

$\xrightarrow[\text{2. H}_3\text{O}^+]{\text{1. LiAlH}_4}$ $(CH_3)_2CHCH_2OH$

17.10

a)

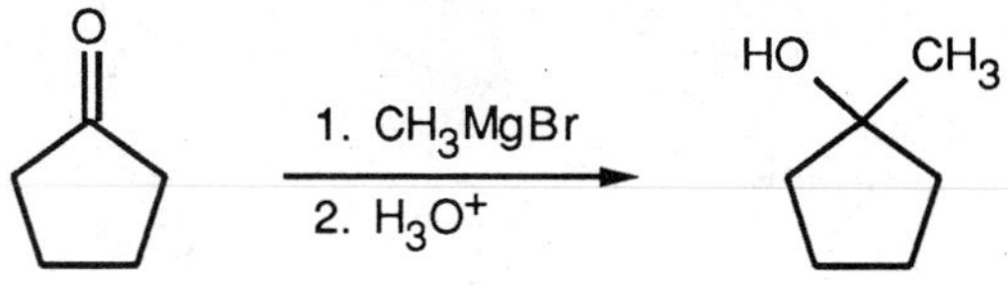

$\xrightarrow[\text{2. H}_3\text{O}^+]{\text{1. CH}_3\text{MgBr}}$

b)

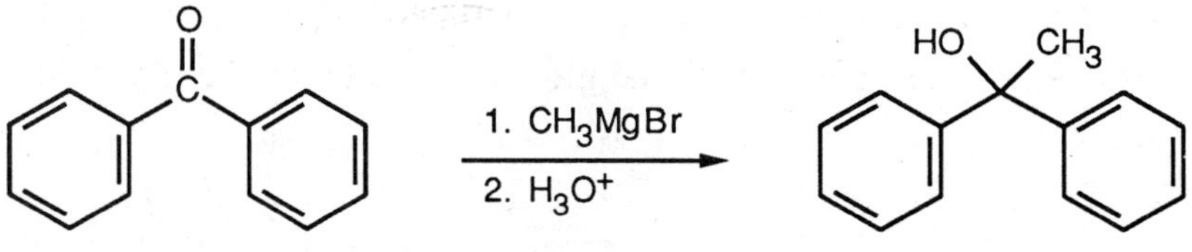

$\xrightarrow[\text{2. H}_3\text{O}^+]{\text{1. CH}_3\text{MgBr}}$

c) $CH_3CH_2CH_2CCH_2CH_3$ $\xrightarrow[\text{2. H}_3\text{O}^+]{\text{1. CH}_3\text{MgBr}}$ $CH_3CH_2CH_2\overset{\overset{\displaystyle OH}{|}}{\underset{\underset{\displaystyle CH_3}{|}}{C}}CH_2CH_3$

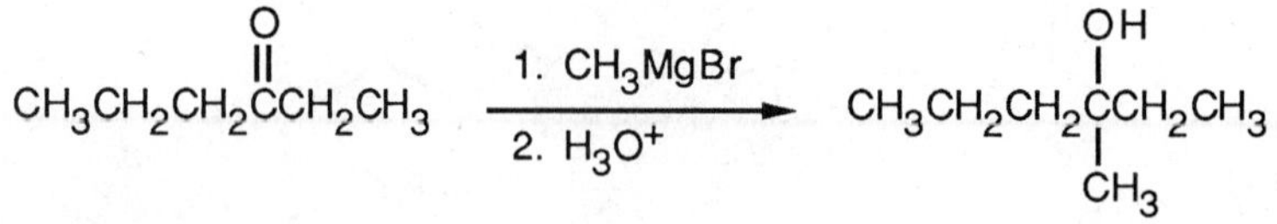

17.11 a) 2–Methyl–2–propanol is a tertiary alcohol. To synthesize a tertiary alcohol, start
with a ketone.

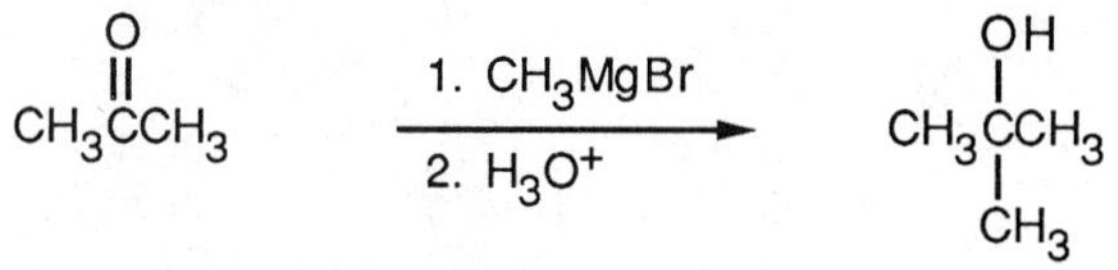

CH_3CCH_3 $\xrightarrow[\text{2. H}_3\text{O}^+]{\text{1. CH}_3\text{MgBr}}$ $CH_3\overset{\overset{\displaystyle OH}{|}}{\underset{\underset{\displaystyle CH_3}{|}}{C}}CH_3$

When two or more alkyl groups bonded to the carbon bearing the –OH group are the same, an alcohol can be synthesized from an ester and a Grignard reagent.

$$CH_3\overset{\overset{\displaystyle O}{\|}}{C}OR \quad \xrightarrow[\text{2. } H_3O^+]{\text{1. } 2\,CH_3MgBr} \quad CH_3\overset{\overset{\displaystyle OH}{|}}{\underset{\underset{\displaystyle CH_3}{|}}{C}}CH_3$$

2–Methyl–2–propanol

b) Since 1–methylcyclohexanol is a tertiary alcohol, start with a ketone.

1–Methylcyclohexanol

c) 3–Methyl–3–pentanol is a tertiary alcohol. Either a ketone or an ester can be used as a starting material.

$$CH_3CH_2\overset{\overset{\displaystyle O}{\|}}{C}CH_2CH_3 \quad \xrightarrow[\text{2. } H_3O^+]{\text{1. } CH_3MgBr}$$

or

$$CH_3CH_2\overset{\overset{\displaystyle O}{\|}}{C}CH_3 \quad \xrightarrow[\text{2. } H_3O^+]{\text{1. } CH_3CH_2MgBr}$$

or

$$CH_3\overset{\overset{\displaystyle O}{\|}}{C}OR \quad \xrightarrow[\text{2. } H_3O^+]{\text{1. } 2\,CH_3CH_2MgBr}$$

$$CH_3CH_2\overset{\overset{\displaystyle OH}{|}}{\underset{\underset{\displaystyle CH_3}{|}}{C}}CH_2CH_3$$

3–Methyl–3–pentanol

d) Three possible combinations of ketone plus Grignard reagent can be used to synthesize this tertiary alcohol.

$$CH_3CH_2\overset{\overset{\displaystyle O}{\|}}{C}CH_3 \quad \xrightarrow[\text{2. } H_3O^+]{\text{1. } C_6H_5MgBr}$$

or

$$CH_3CH_2\overset{\overset{\displaystyle O}{\|}}{C}- \qquad \xrightarrow[\text{2. } H_3O^+]{\text{1. } CH_3MgBr}$$

or

$$CH_3\overset{\overset{\displaystyle O}{\|}}{C}- \qquad \xrightarrow[\text{2. } H_3O^+]{\text{1. } CH_3CH_2MgBr}$$

$$CH_3CH_2\overset{\overset{\displaystyle OH}{|}}{C}CH_3$$

2–Phenyl–2–butanol

e) *Formaldehyde* must be used to synthesize this primary alcohol.

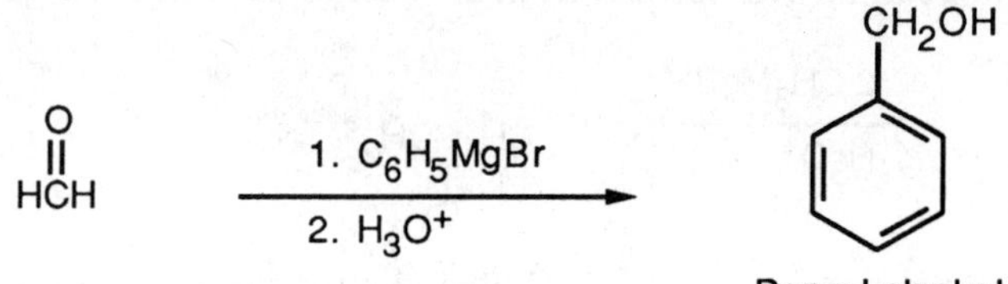
CH2OH
O
HCH
1. C6H5MgBr
2. H3O+

Benzyl alcohol

17.12 The first equivalent of CH_3MgBr reacts with the hydroxyl hydrogen of 4–hydroxycyclohexanone.

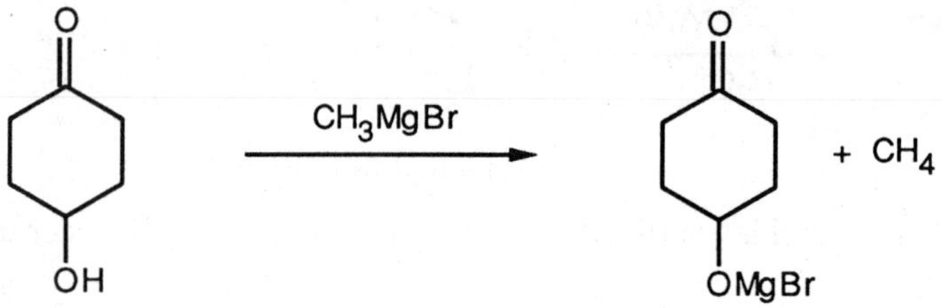
O
OH
CH3MgBr
O
+ CH4
OMgBr

The second equivalent of CH_3MgBr adds to the ketone in the expected manner to yield 1–methyl–1,4–cyclohexanediol.

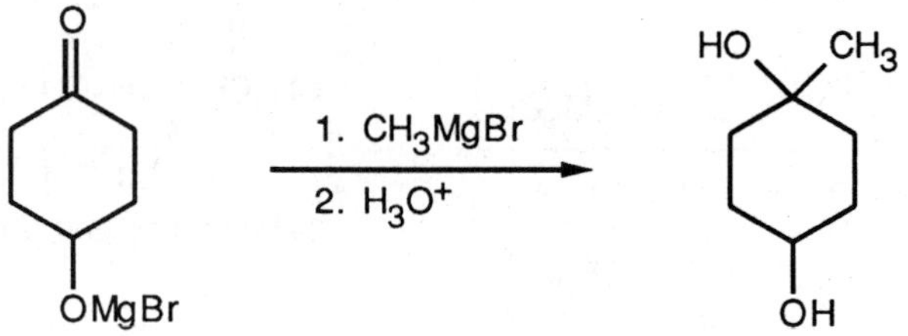
O
OMgBr
1. CH3MgBr
2. H3O+
HO
CH3
OH

17.13

a) 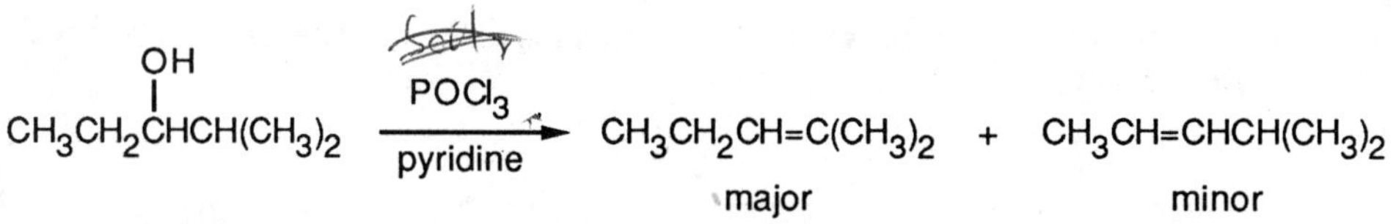
OH
CH3CH2CHCH(CH3)2
POCl3
pyridine
CH3CH2CH=C(CH3)2
+
CH3CH=CHCH(CH3)2
major
minor

b) 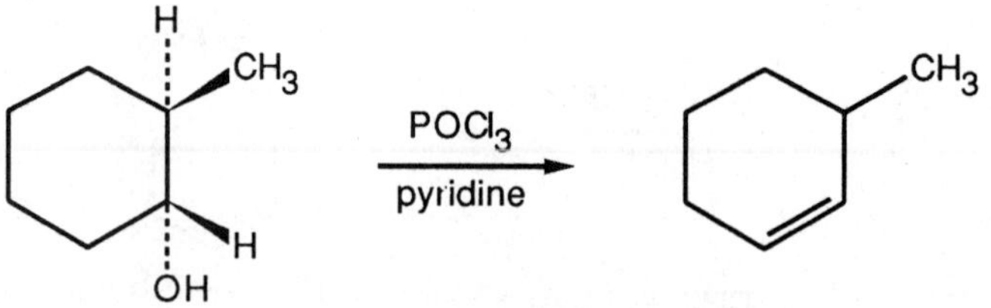
H
CH3
H
OH
POCl3
pyridine
CH3

In E2 elimination, dehydration proceeds most readily when the two groups to be eliminated have a *trans*–diaxial relationship. In this compound, the only hydrogen with the proper stereochemical relationship to the –OH group is at C6. Thus the non-Zaitsev product, 3–methylcyclohexene, is formed.

c)

$$\xrightarrow[\text{pyridine}]{\text{POCl}_3}$$

Here, the hydrogen at C2 is *trans* to the hydroxyl, and dehydration yields the Zaitsev product, 1–methylcyclohexene.

17.14

$$(CH_3)_3C-\underset{\underset{H}{|}}{\overset{\overset{OH}{|}}{C}}CH_3 \xrightleftharpoons[\substack{\text{protonation} \\ \text{of } -OH}]{H_2SO_4} \left[(CH_3)_3C-\underset{\underset{H}{|}}{\overset{\overset{+OH_2}{|}}{C}}CH_3 \xrightarrow{\text{loss of } H_2O} (CH_3)_2\overset{\overset{CH_3}{|}}{\underset{\underset{H}{|}}{C}}-\overset{+}{C}CH_3 + H_2O \right]$$

methyl shift

$$(CH_3)_2C=C(CH_3)_2 \xleftarrow[\text{loss of } H^+]{} \left[(CH_3)_2\overset{+}{C}-\overset{\overset{CH_3}{|}}{\underset{\underset{H}{|}}{C}}CH_3 \right]$$

17.15

a)

$$\xrightarrow[\text{reagent}]{\text{Jones}}$$

b) $CH_3\underset{\underset{CH_3}{|}}{\overset{}{C}}HCH_2OH \xrightarrow[\text{CH}_2\text{Cl}_2]{\text{PCC}} CH_3\underset{\underset{CH_3}{|}}{\overset{}{C}}HCHO$

c)

$$\xrightarrow[\text{reagent}]{\text{Jones}}$$

17.16

Starting Material	Jones	PCC
a) $CH_3CH_2CH_2CH_2CH_2CH_2OH$	$CH_3CH_2CH_2CH_2CH_2COOH$	$CH_3CH_2CH_2CH_2CH_2CHO$
b) $CH_3CH_2CH_2CH_2\underset{\underset{OH}{|}}{C}HCH_3$	$CH_3CH_2CH_2CH_2\overset{\overset{O}{\|}}{C}CH_3$	$CH_3CH_2CH_2CH_2\overset{\overset{O}{\|}}{C}CH_3$
c) $CH_3CH_2CH_2CH_2CH_2CHO$	$CH_3CH_2CH_2CH_2CH_2COOH$	no reaction

17.17

17.18 The infrared spectra of cholesterol and 5–cholesten–3–one each exhibit a unique absorption that makes it easy to distinguish between them. Cholesterol shows an –OH stretch at 3300-3600 cm^{-1}; 5–cholesten–3–one shows a C=O stretch at 1710 cm^{-1}. In the Jones oxidation of cholesterol to 5–cholesten–3–one, the –OH band will disappear and will be replaced by a C=O band. When oxidation is complete no –OH absorption should be visible.

17.19
a) No protons are adjacent to the hydroxyl group of 2–methyl–2–propanol; the –OH signal is *unsplit*.
b) Since one proton is adjacent to the hydroxyl proton of cyclohexanol, the –OH absorption is a *doublet*.
c) The –OH absorption of ethanol appears as a triplet because of splitting by two adjacent protons.
d) The –OH absorption is split into a *doublet*.
e) The signal for the hydroxyl proton of cholesterol is split into a *doublet*.
f) The –OH absorption of 1–methylcyclohexanol is *unsplit*.

17.20

a)
$$\underset{\text{2–Butanethiol}}{CH_3CH_2\overset{\overset{\displaystyle SH}{|}}{C}HCH_3}$$

b)
$$\underset{\text{2,2,6–Trimethyl–4–heptanethiol}}{(CH_3)_3CCH_2\overset{\overset{\displaystyle SH}{|}}{C}HCH_2CH(CH_3)_2}$$

c)
2–Cyclopentene–1–thiol

17.21

a) $CH_3CH=CHCO_2CH_3$ $\xrightarrow[\text{2. } H_3O^+]{\text{1. LiAlH}_4}$ $CH_3CH=CHCH_2OH$ $\xrightarrow{PBr_3}$ $CH_3CH=CHCH_2Br$

Methyl 2–butenoate

$\Big\downarrow \begin{array}{l} \text{1. } (H_2N)_2C=S \\ \text{2. } ^-OH,\, H_2O \end{array}$

b) $H_2C=CHCH=CH_2$ $\xrightarrow[\Delta]{HBr}$ $CH_3CH=CHCH_2Br$ $\xrightarrow[\text{2. } ^-OH,\, H_2O]{\text{1. } (H_2N)_2C=S}$ $CH_3CH=CHCH_2SH$

2–Butene–1–thiol

17.22
a) 2–Methyl–1,4–butanediol
b) 3–Ethyl–2–hexanol
c) *cis*–1,3–Cyclobutanediol
d) *cis*–2–Methyl–4–cyclohepten–1–ol
e) *cis*–3–Phenylcyclopentanol
f) 2,3–Dimethyl–3–hexanethiol

17.23 None of these alcohols has multiple bonds or rings.

CH$_3$CH$_2$CH$_2$CH$_2$CH$_2$OH

1–Pentanol

$$\underset{\text{2–Pentanol}}{\overset{\displaystyle OH}{CH_3CH_2CH_2\overset{|}{C}HCH_3}}$$

$$\underset{\text{3–Pentanol}}{\overset{\displaystyle OH}{CH_3CH_2\overset{|}{C}HCH_2CH_3}}$$

$$\underset{\text{2–Methyl–1–butanol}}{\overset{\displaystyle CH_3}{CH_3CH_2\overset{|}{C}HCH_2OH}}$$

$$\underset{\underset{\text{2–Methyl–2–butanol}}{\displaystyle OH}}{\overset{\displaystyle CH_3}{CH_3CH_2\overset{|}{\underset{|}{C}}CH_3}}$$

$$\underset{\underset{\text{3–Methyl–2–butanol}}{\displaystyle OH}}{\overset{\displaystyle CH_3}{CH_3\overset{|}{C}H\overset{|}{C}HCH_3}}$$

$$\underset{\text{3–Methyl–1–butanol}}{\overset{\displaystyle CH_3}{HOCH_2CH_2\overset{|}{C}HCH_3}}$$

$$\underset{\underset{\text{2,2–Dimethyl–1–propanol}}{\displaystyle CH_3}}{\overset{\displaystyle CH_3}{CH_3\overset{|}{\underset{|}{C}}CH_2OH}}$$

17.24 Primary alcohols react with Jones' reagent to form carboxylic acids, secondary alcohols yield ketones, and tertiary alcohols are unreactive to oxidation. Of the eight alcohols in the previous problem, only 2–methyl–2–butanol is unreactive to Jones oxidation.

CH$_3$CH$_2$CH$_2$CH$_2$CH$_2$OH $\xrightarrow{\text{Jones}}$ CH$_3$CH$_2$CH$_2$CH$_2$COOH

$$\overset{\displaystyle OH}{CH_3CH_2CH_2\overset{|}{C}HCH_3} \xrightarrow{\text{Jones}} \overset{\displaystyle O}{CH_3CH_2CH_2\overset{||}{C}CH_3}$$

$$\overset{\displaystyle OH}{CH_3CH_2\overset{|}{C}HCH_2CH_3} \xrightarrow{\text{Jones}} \overset{\displaystyle O}{CH_3CH_2\overset{||}{C}CH_2CH_3}$$

$$\overset{\displaystyle CH_3}{CH_3CH_2\overset{|}{C}HCH_2OH} \xrightarrow{\text{Jones}} \overset{\displaystyle CH_3}{CH_3CH_2\overset{|}{C}HCOOH}$$

$$\underset{\displaystyle OH}{\overset{\displaystyle CH_3}{CH_3\overset{|}{C}H\overset{|}{C}HCH_3}} \xrightarrow{\text{Jones}} \underset{\displaystyle O}{\overset{\displaystyle CH_3}{CH_3\overset{||}{\underset{|}{C}}\overset{|}{C}HCH_3}}$$

$$\overset{\displaystyle CH_3}{HOCH_2CH_2\overset{|}{C}HCH_3} \xrightarrow{\text{Jones}} \overset{\displaystyle CH_3}{HOOCCH_2\overset{|}{C}HCH_3}$$

$$\underset{\displaystyle CH_3}{\overset{\displaystyle CH_3}{CH_3\overset{|}{\underset{|}{C}}CH_2OH}} \xrightarrow{\text{Jones}} \underset{\displaystyle CH_3}{\overset{\displaystyle CH_3}{CH_3\overset{|}{\underset{|}{C}}COOH}}$$

17.25

a)

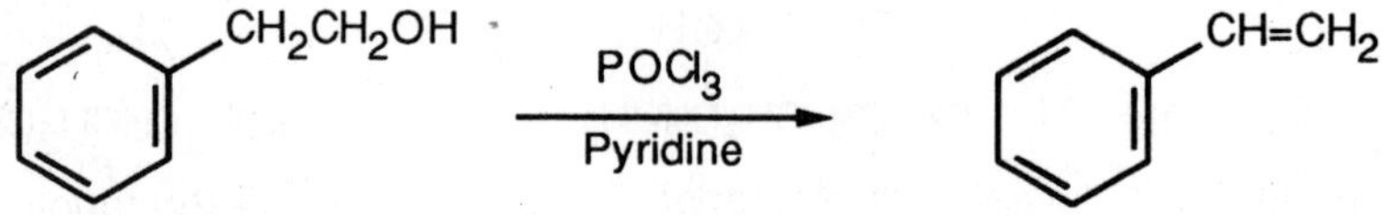

2-Phenylethanol →(POCl₃, Pyridine)→ Styrene

b)

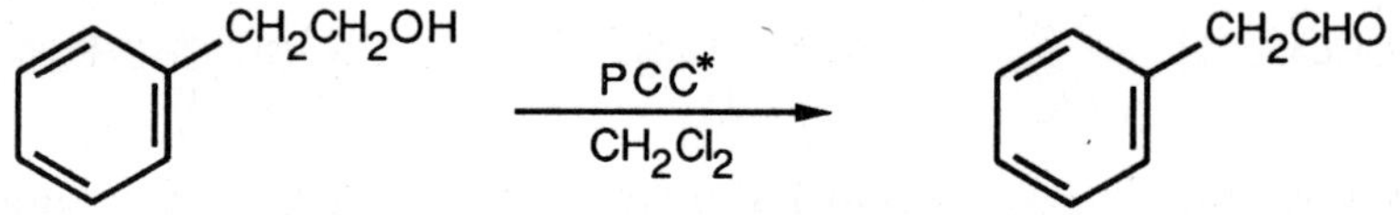

2-Phenylethanol →(PCC*, CH₂Cl₂)→ Phenylacetaldehyde

* $C_5H_6NCrO_3Cl$ (pyridinium chlorochromate)

c)

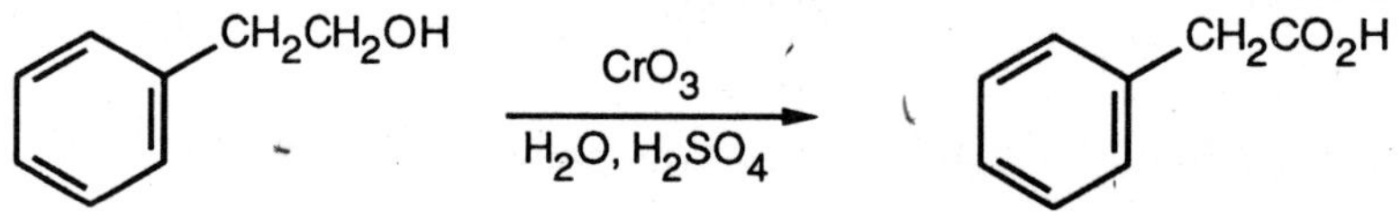

2-Phenylethanol →(CrO_3, H_2O, H_2SO_4)→ Phenylacetic acid

d)

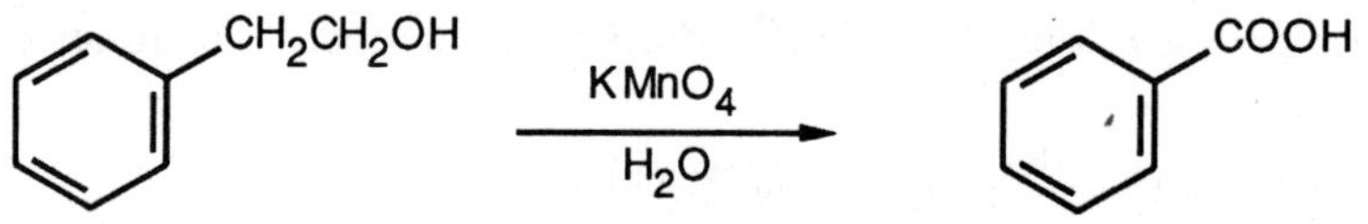

2-Phenylethanol →($KMnO_4$, H_2O)→ Benzoic acid

e)

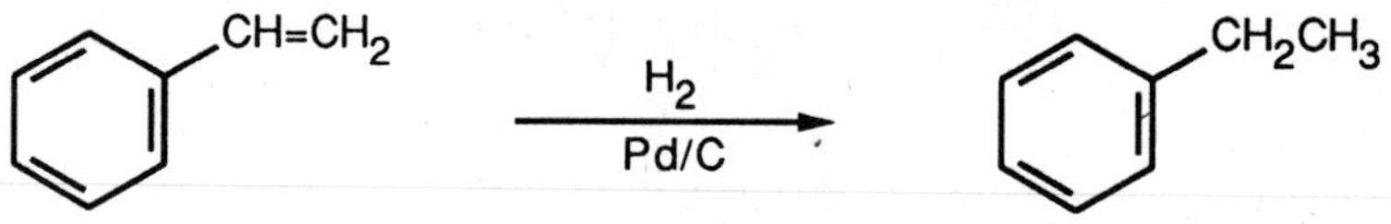

(from a) CH=CH₂ →(H_2, Pd/C)→ Ethylbenzene

f)

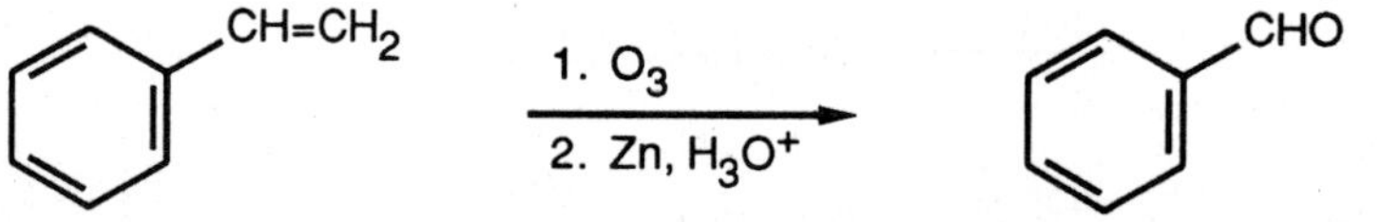

(from a) CH=CH₂ →(1. O_3 2. Zn, H_3O^+)→ Benzaldehyde

g)

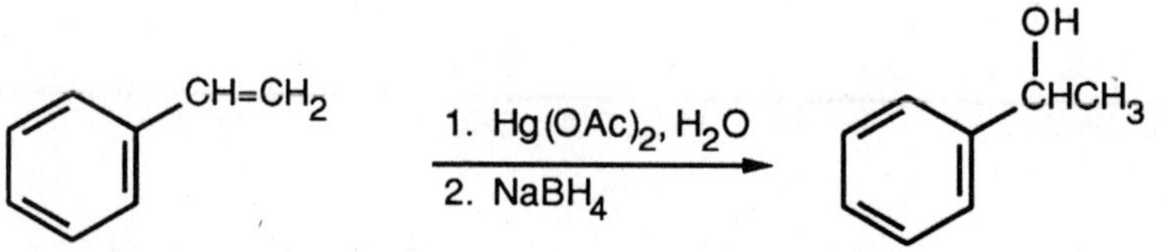

(from a) CH=CH₂ →(1. $Hg(OAc)_2$, H_2O 2. $NaBH_4$)→ 1-Phenylethanol

h)

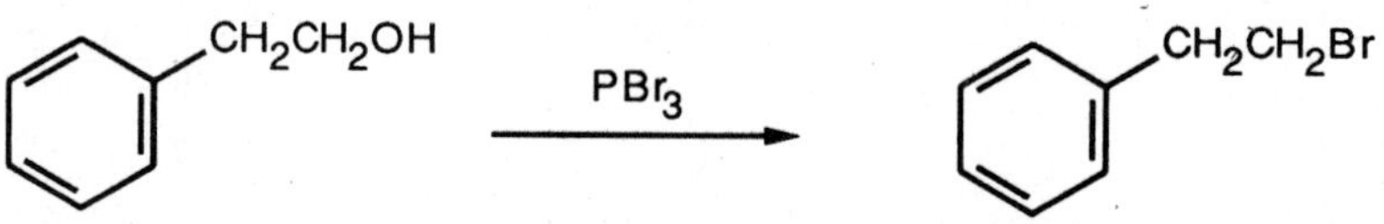

CH₂CH₂OH →(PBr_3)→ 2-Bromo-1-phenylethane

17.26

a)

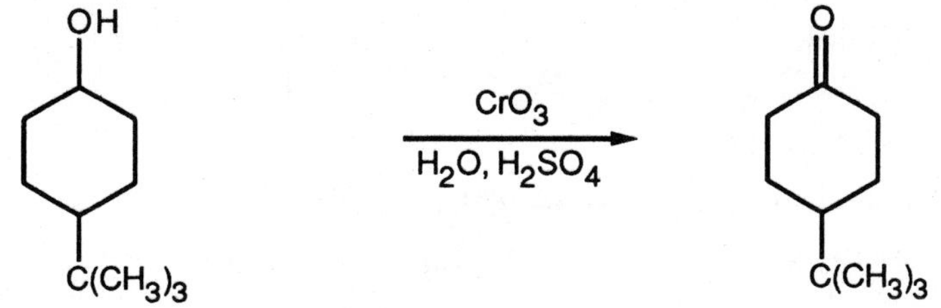

$$\xrightarrow[\text{H}_2\text{O, H}_2\text{SO}_4]{\text{CrO}_3}$$

b)

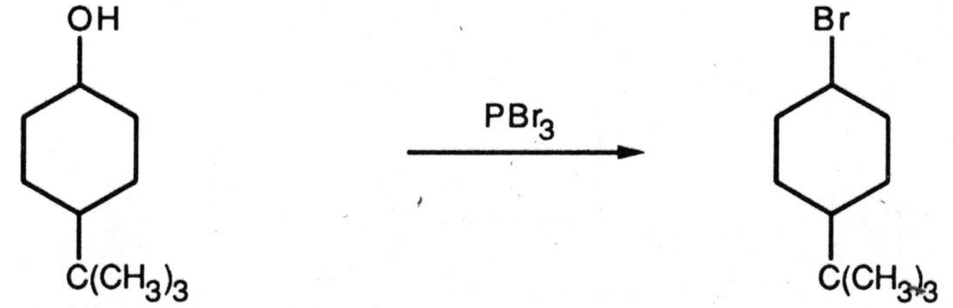

$$\xrightarrow{\text{PBr}_3}$$

c)

$$\xrightarrow[\text{Pyridine}]{\text{POCl}_3}$$

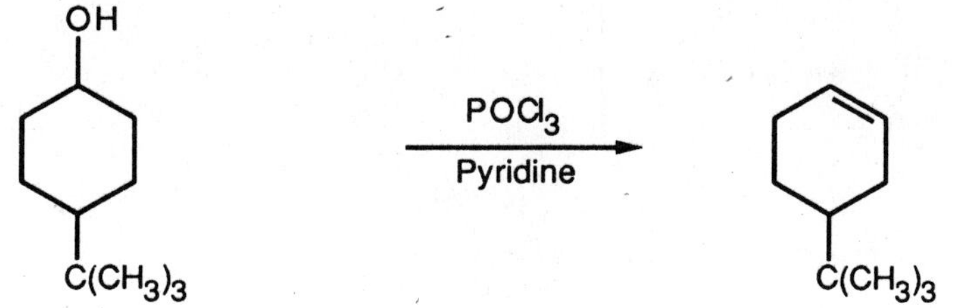

d) $CH_3CH_2CH_2OH$ $\xrightarrow[\text{CH}_2\text{Cl}_2]{\text{PCC}}$ CH_3CH_2CHO

e) $CH_3CH_2CH_2OH$ $\xrightarrow[\text{H}_2\text{O, H}_2\text{SO}_4]{\text{CrO}_3}$ CH_3CH_2COOH

f)

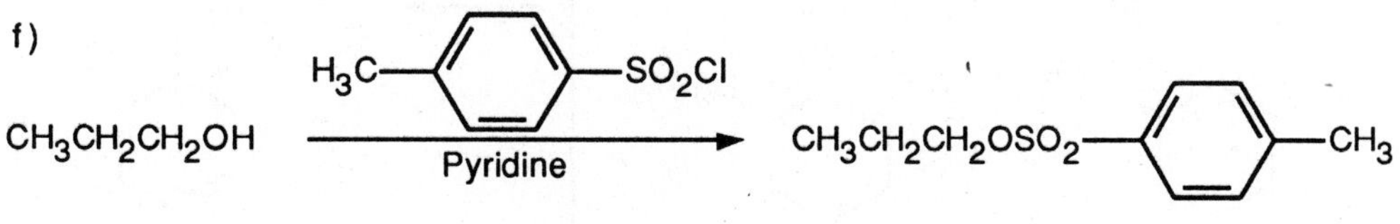

$CH_3CH_2CH_2OH$ $\xrightarrow{\text{Pyridine}}$ $CH_3CH_2CH_2OSO_2$—

g) $CH_3CH_2CH_2OH$ $\xrightarrow[\text{Pyridine}]{\text{SOCl}_2}$ $CH_3CH_2CH_2Cl$

h) $CH_3CH_2CH_2OH$ $\xrightarrow{\text{NaH}}$ $CH_3CH_2CH_2\ddot{\text{O}}{:}^-\text{Na}^+ \ + \ H_2$

17.27 In some of these problems, different combinations of Grignard reagent and carbonyl compound are possible. Remember that aqueous acid is added to the initial Grignard adduct to yield the alcohol.

a) CH_3MgBr + CH_3CH_2CHO

   or

   CH_3CHO + CH_3CH_2MgBr

$\longrightarrow$

   $\overset{\displaystyle OH}{\underset{\displaystyle |}{CH_3CHCH_2CH_3}}$

   2–Butanol

b)

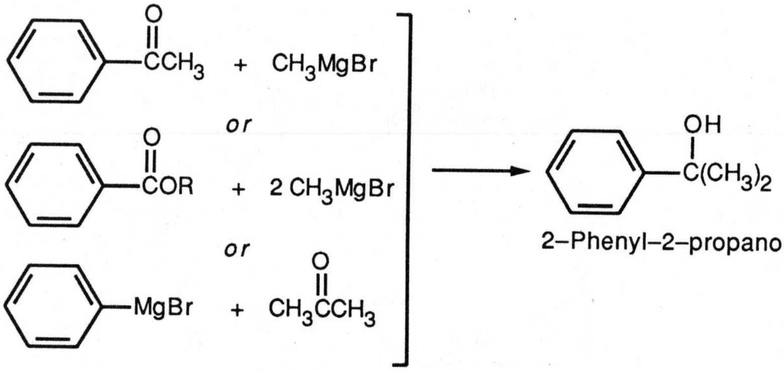

2–Phenyl–2–propanol

c) $\overset{\displaystyle CH_3}{\underset{\displaystyle |}{H_2C=C-MgBr}}$ + CH_2O $\longrightarrow$ $\overset{\displaystyle CH_3}{\underset{\displaystyle |}{H_2C=CCH_2OH}}$

   2–Methyl–2–propen–1–ol

d)

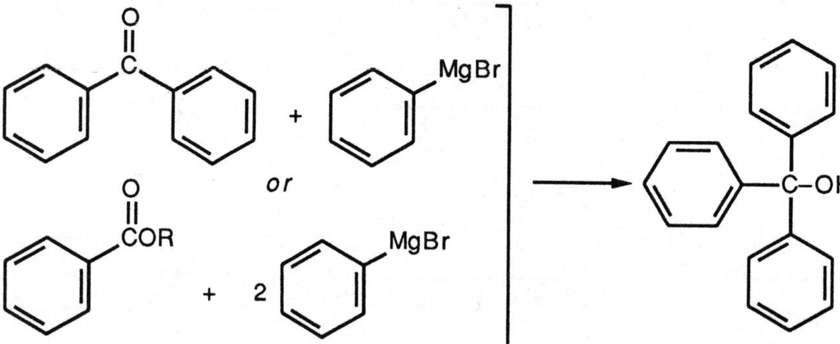

Triphenylmethanol

e) CH_3MgBr + $\overset{\displaystyle O}{\overset{\displaystyle ||}{HCCH_2CH_2CH_2Br}}$ $\longrightarrow$ $\overset{\displaystyle OH}{\underset{\displaystyle |}{CH_3CHCH_2CH_2CH_2Br}}$

   5–Bromo–2–pentanol

17.28

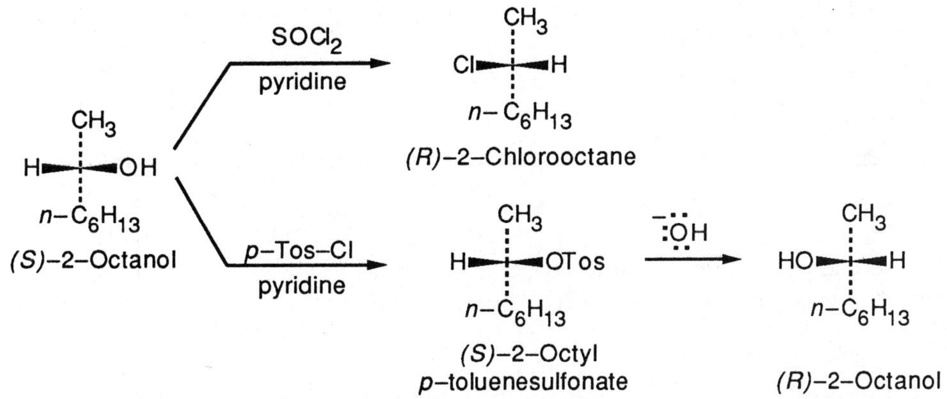

To form optically pure (R)–2–octanol, the poor hydroxide leaving group must be converted to the very good toluenesulfonate leaving group. Reaction of (S)–2–octyl–p–toluenesulfonate with hydroxide ion proceeds with inversion (S_N2 mechanism) to give (R)–2–octanol.

17.29

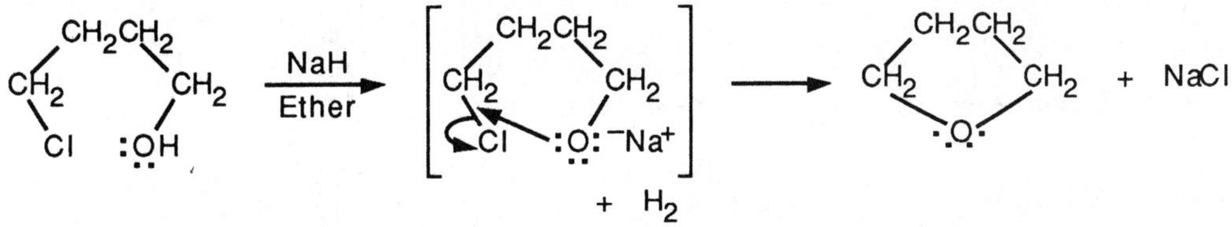

17.30

Compound	Carbonyl precursors

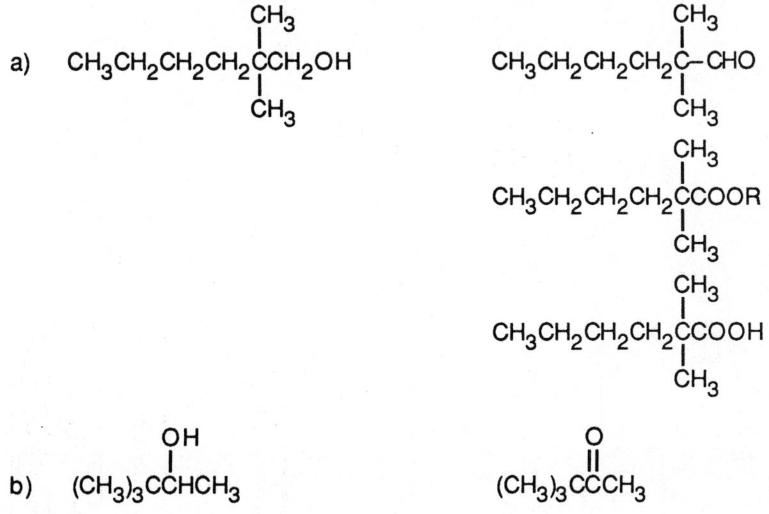

a) $CH_3CH_2CH_2CH_2CCH_2OH$ (with two CH_3 substituents)

$CH_3CH_2CH_2CH_2C-CHO$ (with two CH_3 substituents)

$CH_3CH_2CH_2CH_2CCOOR$ (with two CH_3 substituents)

$CH_3CH_2CH_2CH_2CCOOH$ (with two CH_3 substituents)

b) $(CH_3)_3CCHCH_3$ with OH

$(CH_3)_3CCCH_3$ with $\overset{O}{\overset{\|}{}}$

c)

d)

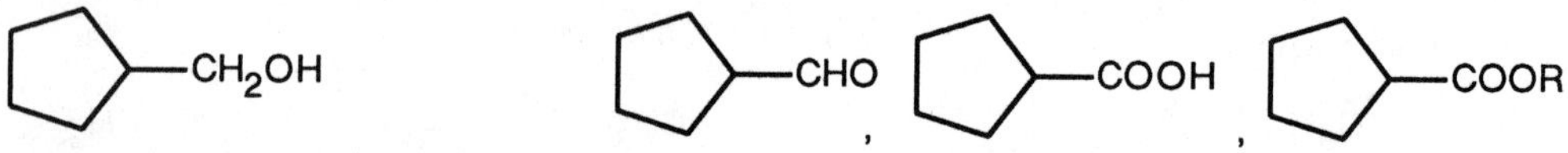

17.31 In these compounds you want to reduce some, but not all, of the functional groups present. To do this, you must select the correct reducing agent.

a)

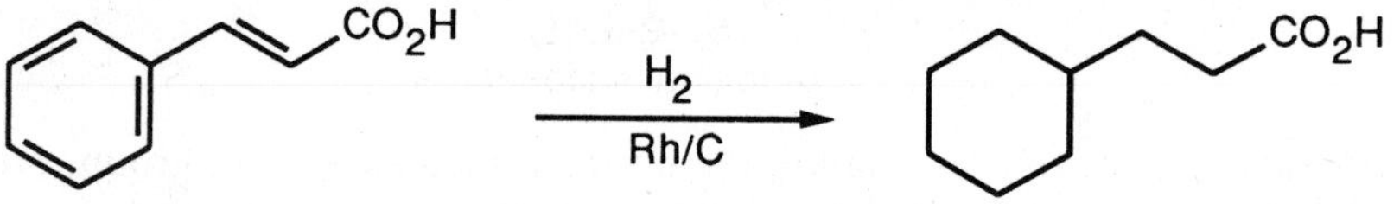

$$\xrightarrow[\text{Rh/C}]{\text{H}_2}$$

b)

$$\xrightarrow[\text{2. H}_3\text{O}^+]{\text{1. LiAlH}_4}$$

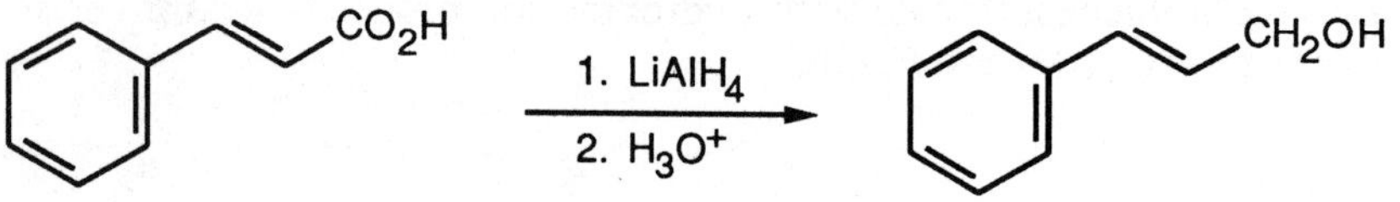

17.32

Grignard Reagent + Carbonyl Compound $\longrightarrow$ Product

a)   CH$_3$MgBr   +   CH$_3\overset{\overset{\displaystyle O}{\|}}{C}CH_3$

        or $\longrightarrow$ (CH$_3$)$_3$COH

    2 CH$_3$MgBr   +   CH$_3\overset{\overset{\displaystyle O}{\|}}{C}$OR

b) CH$_3$CH$_2$MgBr   +

c)   CH$_3$CH$_2$MgBr   +   CH$_3$CH$_2$$\overset{\overset{\displaystyle O}{\|}}{C}$-

        or

    2 CH$_3$CH$_2$MgBr   +   RO-$\overset{\overset{\displaystyle O}{\|}}{C}$-

        or

—MgBr   +   CH$_3$CH$_2\overset{\overset{\displaystyle O}{\|}}{C}CH_2CH_3$

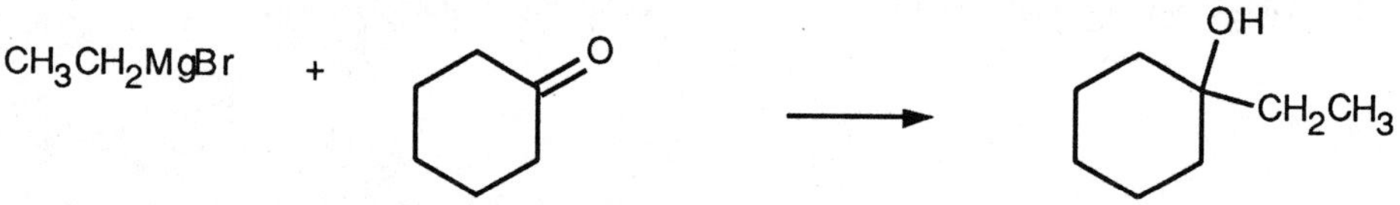

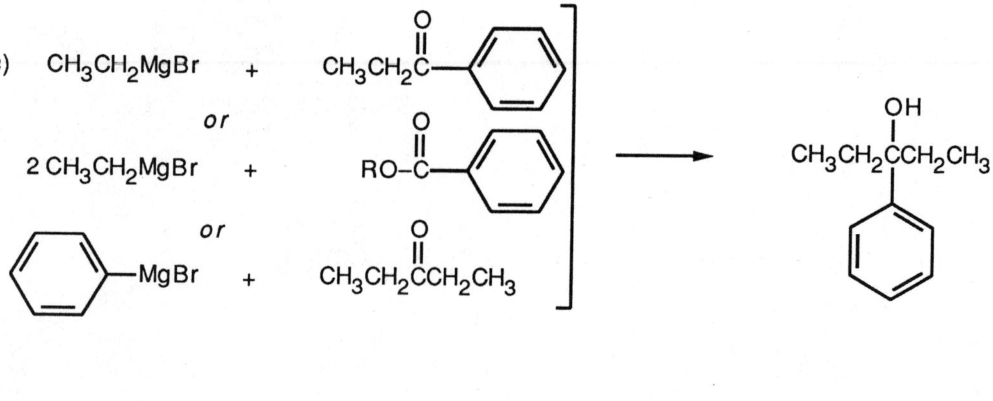

d)

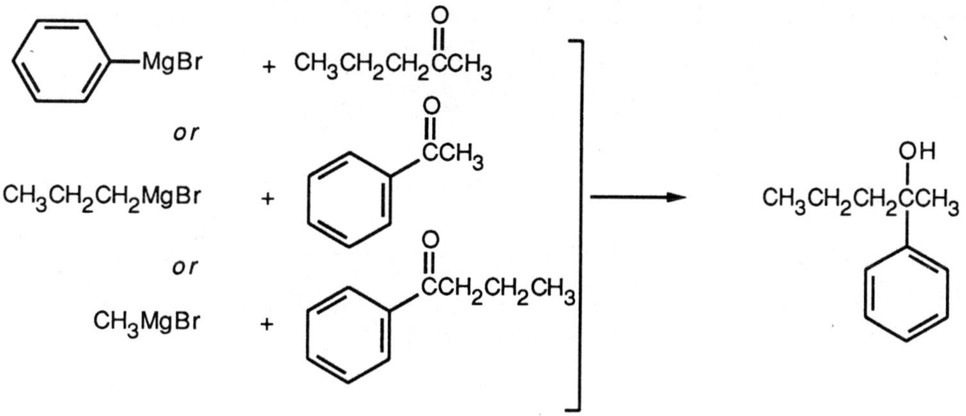

e)

f)

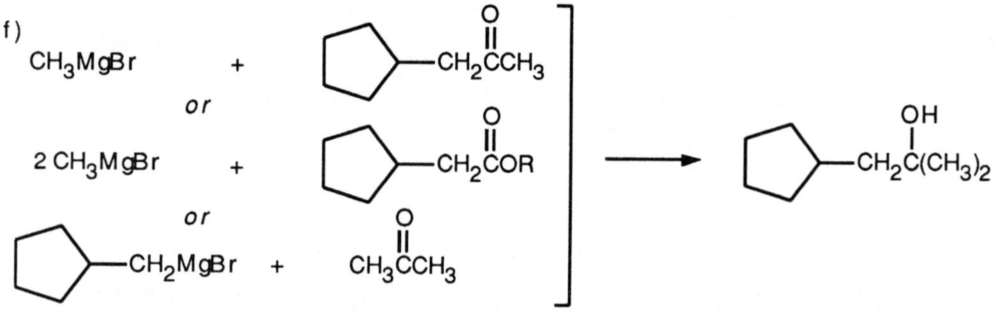

17.33

17.34

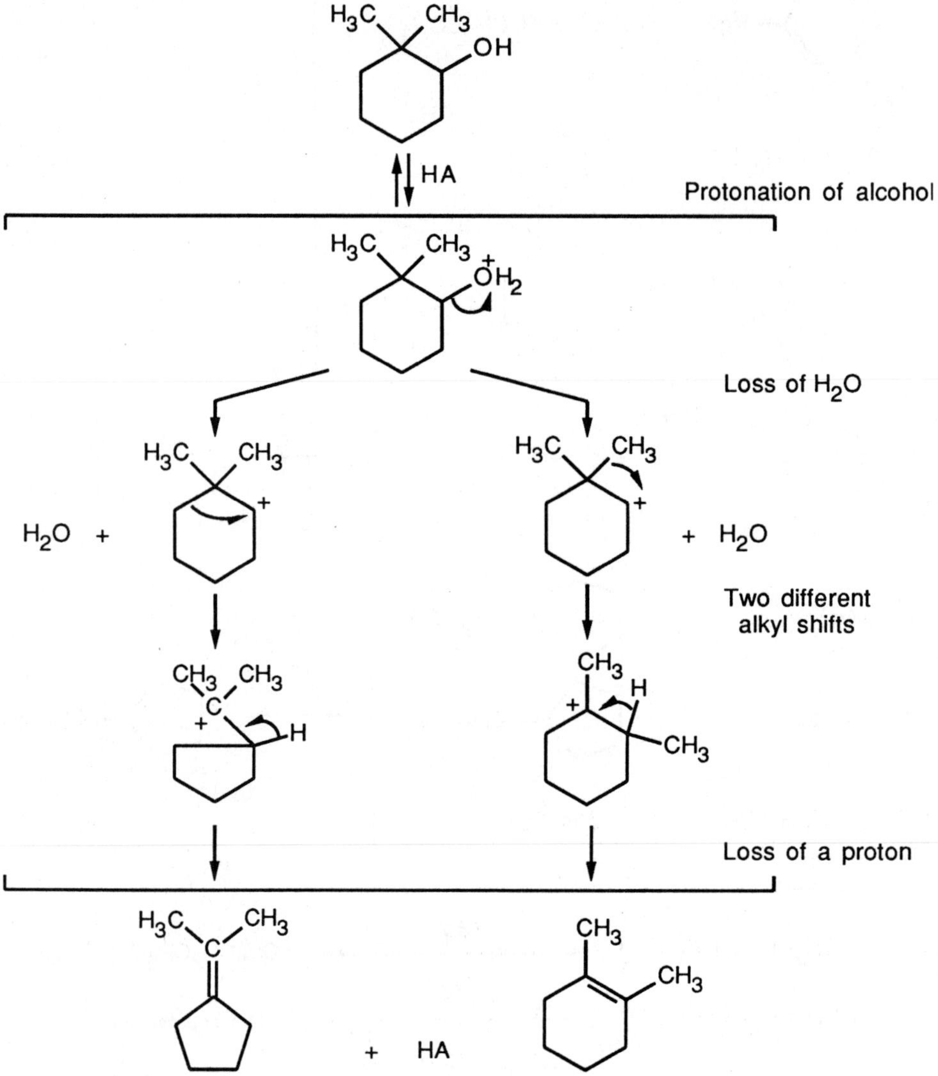

H3C
CH3
OH
HA
Protonation of alcohol
OH2
Loss of H2O
H2O +
+ H2O
Two different alkyl shifts
H
CH3
Loss of a proton
+ HA

17.35

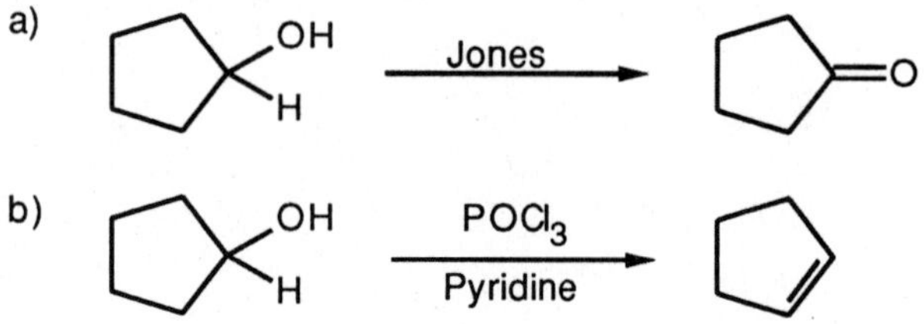

a)
OH
H
Jones
O
b)
OH
H
POCl3
Pyridine

c)

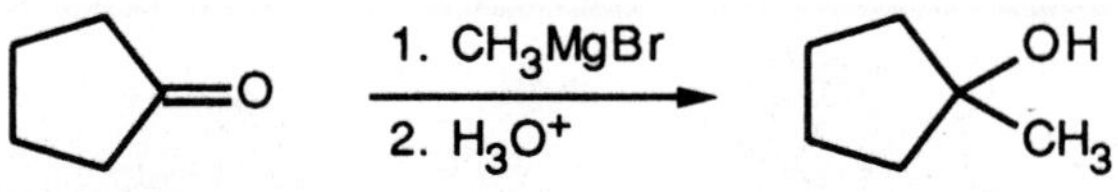

1. CH₃MgBr
2. H₃O⁺

(from a)

d)

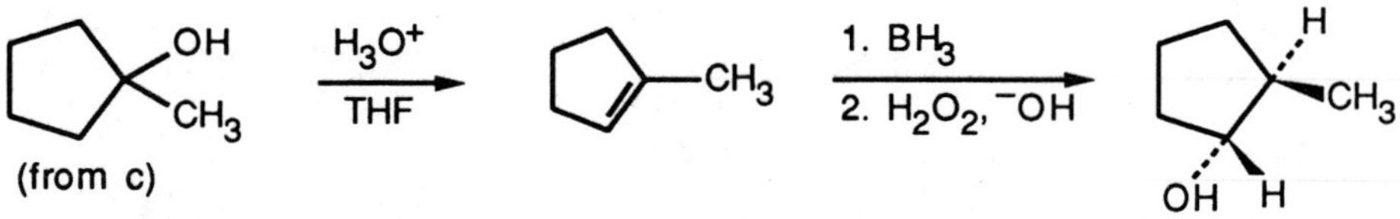

H₃O⁺
THF

1. BH₃
2. H₂O₂, ⁻OH

17.36

a)

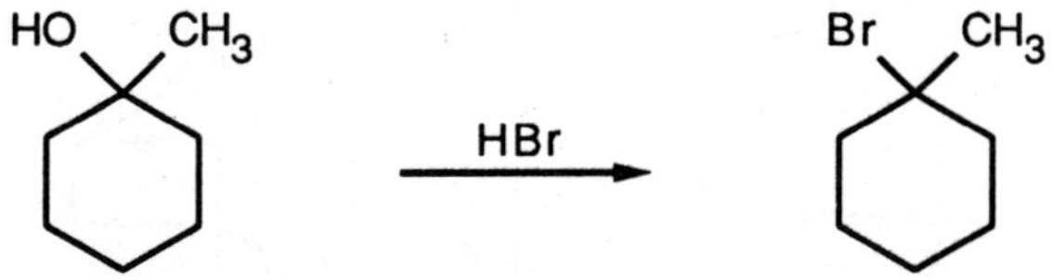

HBr

b)

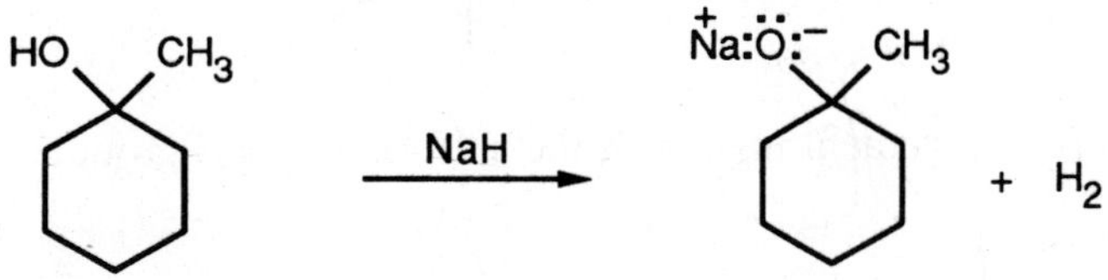

NaH

+ H₂

c)

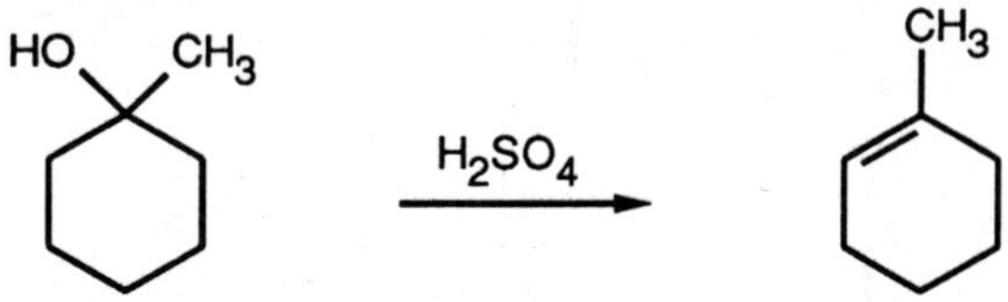

H₂SO₄

d)

Na₂Cr₂O₇

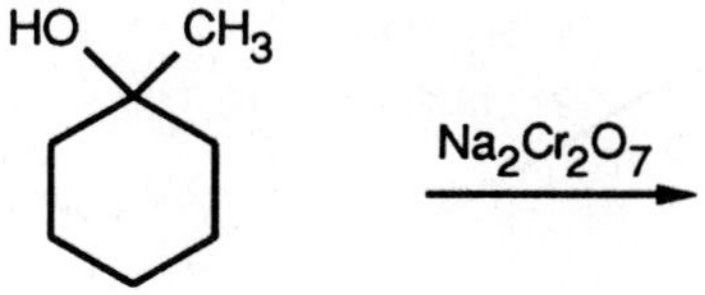

no reaction

17.37

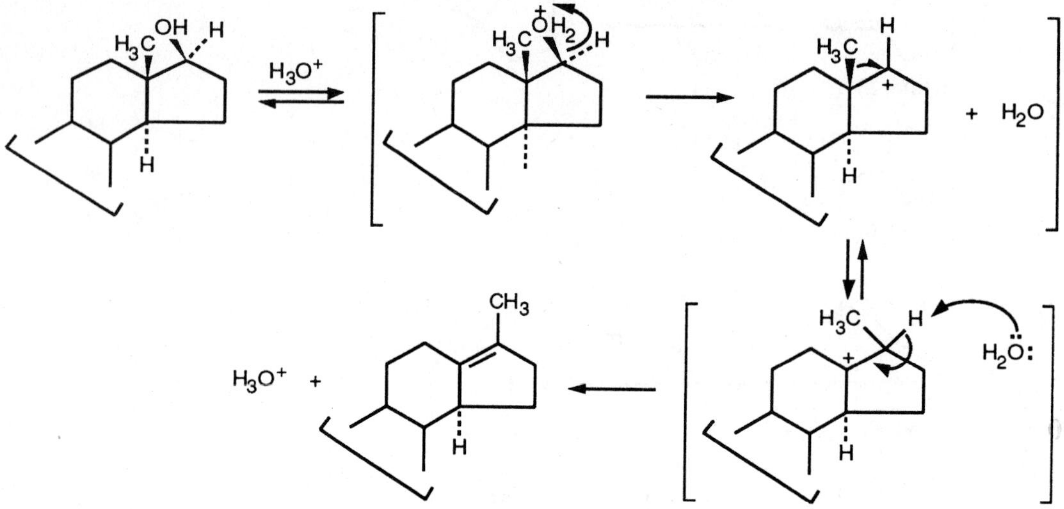

This is a carbocation rearrangement involving the shift of an alkyl group.

17.38

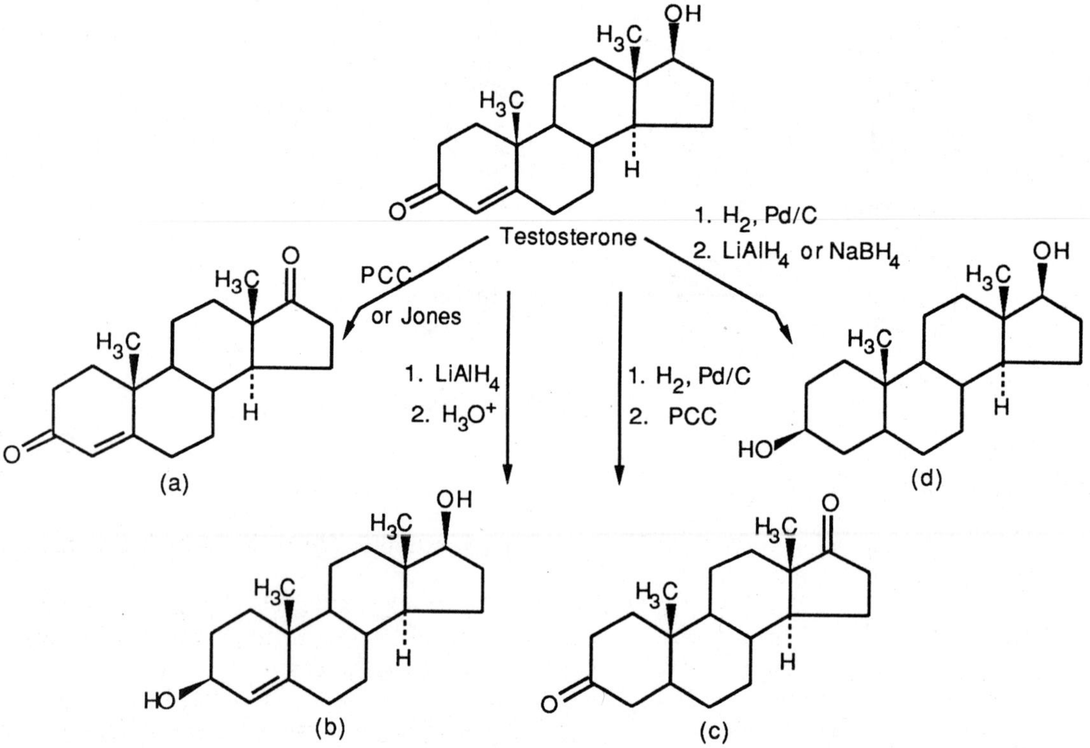

17.39

17.40

1-Methylcyclopentene   *trans*–2–Methylcyclo-   3–Methylcyclopentene
                        pentanol

The more stable product of dehydration of *trans*–2–methylcyclopentanol is
1–methylcyclopentene, which can be formed only via *syn*-elimination. The product of
anti-elimination is 3–methylcyclopentene. Since this product predominates, the
requirement of *anti*-geometry must be more important than formation of the more
stable product.

17.41

3-Methyl–3–buten–1–ol

The peak absorbing at 1.75 δ (3H) is due to the <u>d</u> protons. This peak, which occurs in
the allylic region of the spectrum, is unsplit.

    The peak absorbing at 2.25 δ (3H) is due to protons <u>c</u> and <u>a</u>. The peak is a triplet
because of splitting by the adjacent <u>b</u> protons. The absorption from proton <u>a</u> coincides
with the triplet and distorts it from its usual 1:2:1 ratio of intensity.

    The peak absorbing at 3.69 δ (2H) is due to the <u>b</u> protons. The adjacent
electronegative oxygen causes the peak to be downfield, and the adjacent –CH$_2$–
group splits the peak into a triplet.

    The peak at 4.75 δ (2H) is due to protons <u>e</u> and <u>f</u>. These two protons are so
similar that their absorptions overlap, and no splitting is observed.

17.42   1. $C_8H_{18}O_2$ has *no* double bonds or rings.
2. The IR band at 3350 cm^{-1} indicates the presence of a hydroxyl group.
3. The compound is probably symmetrical (simple NMR).
4. There is no splitting.

A structure that meets all of these criteria:

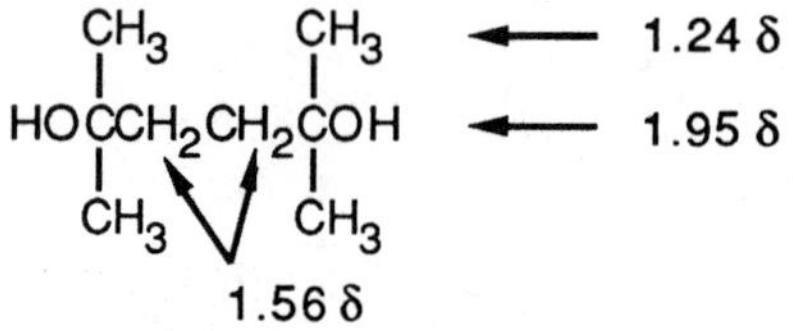

2,5–Dimethyl–2,5–hexanediol

17.43

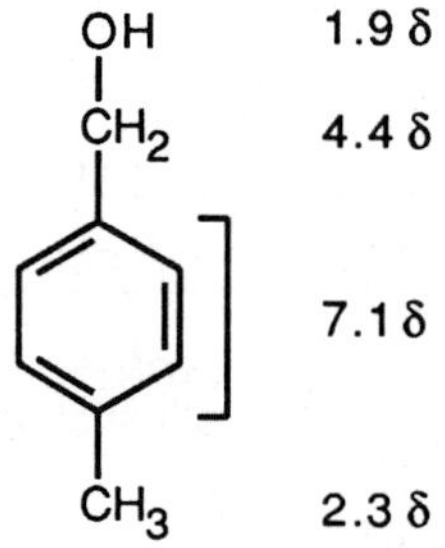

p–Methylbenzyl alcohol

17.44

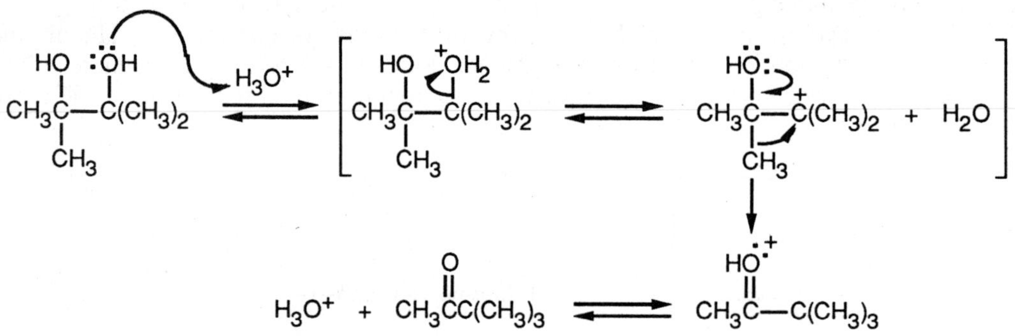

17.45   a) Compound <u>A</u> has one double bond or ring.
b) The infrared absorption at 3400 cm^{-1} indicates the presence of an alcohol. (The weak absorption at 1640 cm^{-1} is due to a C=C stretch.
c) (1) The absorptions at 1.63 δ and 1.70 δ are due to unsplit methyl protons. Because the absorptions are shifted slightly downfield, the protons are adjacent to an unsaturated center.
(2) The broad singlet at 3.83 δ is due to an alcohol proton.
(3) The doublet at 4.15 δ is due to two protons attached to a carbon bearing an electronegative atom (oxygen, in this case).
(4) The proton absorbing at 5.70 δ is a vinylic proton.

d)

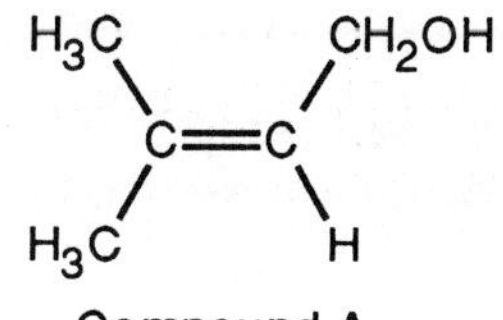

Compound A

3-Methyl-2-buten-1-ol

17.46

a) $CH_3CH_2CH_2OH$

b) 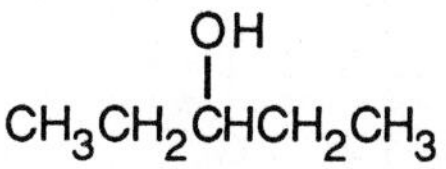 $CH_3CH_2\overset{\overset{\displaystyle OH}{|}}{C}HCH_2CH_3$

c)

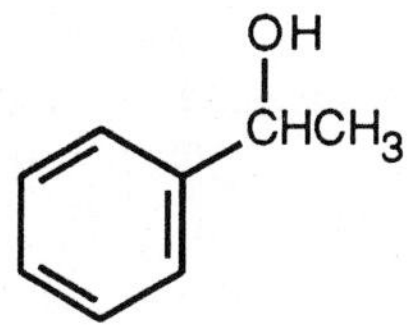

17.47

a)

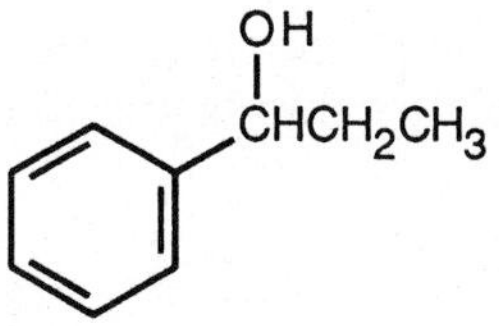

b)

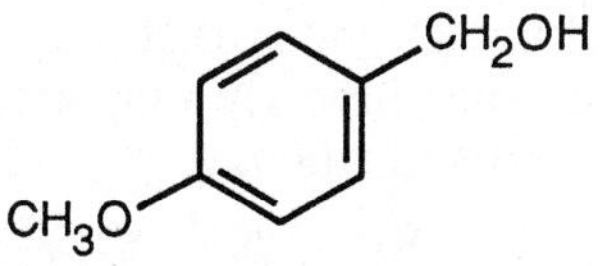

17.48

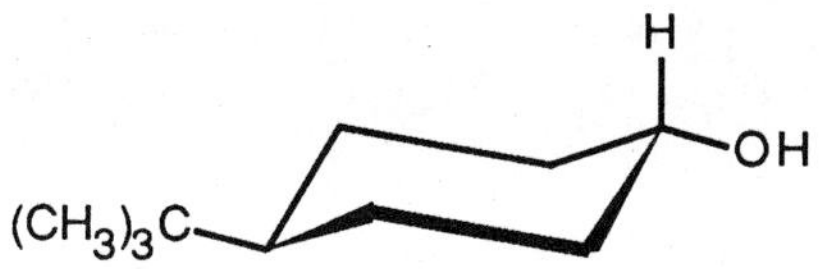

trans-4-tert-Butylcyclohexanol

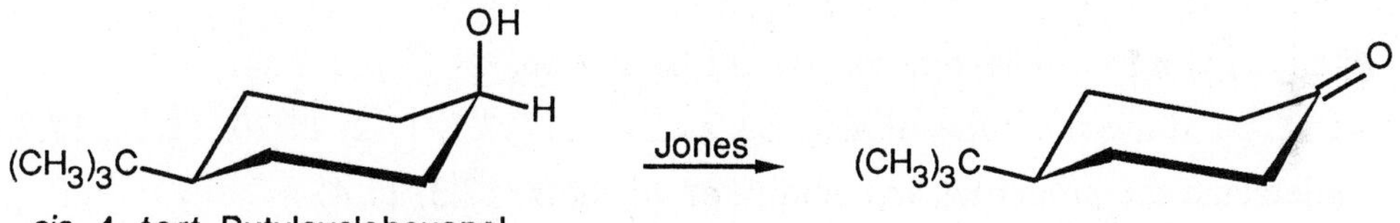

cis-4-tert-Butylcyclohexanol

Recall that the bulky *tert*-butyl group occupies the equatorial position. The *cis* isomer should oxidize faster than the *trans* isomer.

17.49

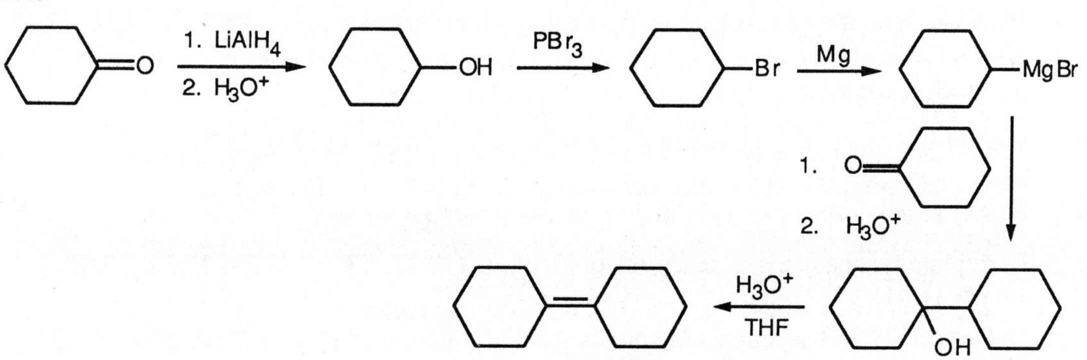

Bicyclohexylidene

17.50 CH₃SSCH₃

17.51 An alcohol adds to an aldehyde by a mechanism that we will study in a later chapter. The hydroxyl group of the intermediate undergoes oxidation (as shown in Section 17.9), and an ester is formed.

17.52 Phenol is a stronger acid than cyclohexanol because the phenoxide anion is stabilized by resonance involving the aromatic ring. No such relationship is possible for the cyclohexoxide anion.

Study Guide for Chapter 17

After studying this chapter, you should be able to:

(1)   Name and draw structures of alcohols and thiols (17.1, 17.2, 17.20, 17.22, 17.23).

(2)   Understand the properties and acidity of alcohols (17.3, 17.4).

(3)   Prepare alcohols:

    a) From alkenes (17.5).

    b) From reduction of carbonyl compounds (17.8, 17.9, 17.30, 17.31, 17.38).

    c) From reaction of carbonyl compounds with Grignard reagents (17.10, 17.11, 17.12, 17.27, 17.32).

    d) Prepare thiols (17.21).

(4)   Predict if a reaction is an oxidation, a reduction, or neither (17.6, 17.7).

(5)   Predict the products of reactions involving alcohols (17.13, 17.14, 17.15, 17.16, 17.24, 17.25, 17.26, 17.29, 17.33, 17.35, 17.36, 17.39, 17.48, 17.49).

(6)   Formulate mechanisms of reactions of alcohols (17.17, 17.29, 17.34, 17.37, 17.40, 17.44, 17.51).

(7)   Use chemical and spectroscopic data to identify alcohols (17.18, 17.19, 17.41, 17.42, 17.43, 17.45, 17.47, 17.50).

Chapter 18 – Ethers, Epoxides and Sulfides

18.1

a) $(CH_3)_2CHOCH(CH_3)_2$

2–Isopropoxypropane *or*
Diisopropyl ether

b) 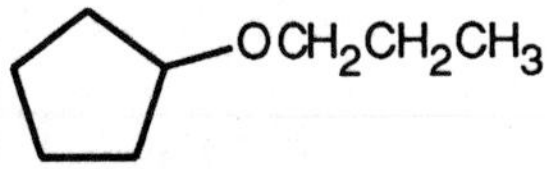

Propoxycyclopentane *or*
Cyclopentyl propyl ether

c) 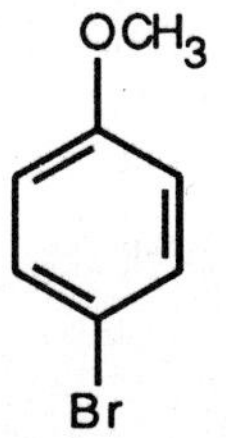

p–Bromoanisole *or*
4–Bromo–1–methoxybenzene

d) 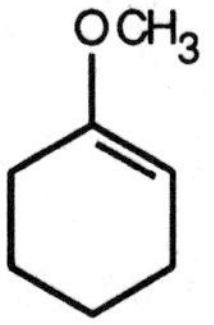

1–Methoxycyclohexene

e) $(CH_3)_2CHCH_2OCH_2CH_3$

1–Ethoxy–2–methylpropane *or*
Ethyl isobutyl ether

f) $H_2C{=}CHCH_2OCH{=}CH_2$

Allyl vinyl ether

18.2 The first step of the dehydration procedure is protonation of an alcohol; water is then displaced by another molecule of alcohol to form an ether. If two different alcohols are present either one can be protonated and either one can displace water, yielding a mixture of products.

If this procedure were used with ethanol and 1–propanol, the products would be diethyl ether, ethyl propyl ether, and dipropyl ether. If there were equimolar amounts of the alcohols, and if they were of equal reactivity, the product ratio would be diethyl ether : ethyl propyl ether : dipropyl = 1:2:1.

18.3

a) 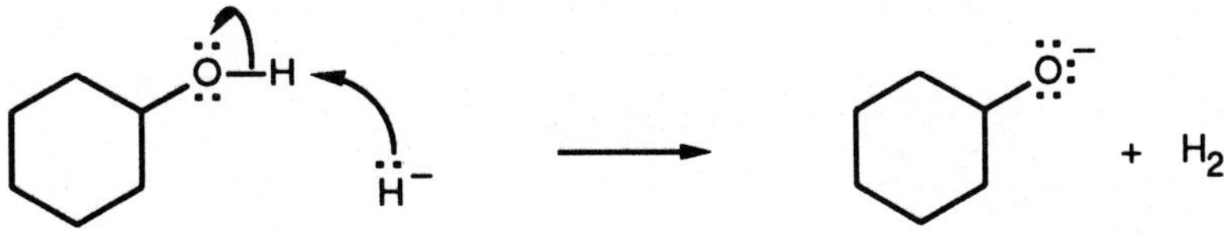

Formation of alkoxide ion.

b) 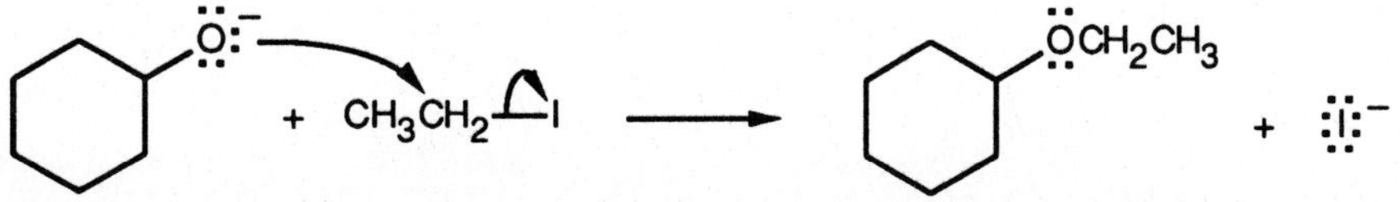

S_N2 displacement of iodide by alkoxide to yield an ether.

18.4

a)

$$CH_3CH_2CH_2\ddot{O}\!:^- Na^+ + CH_3Br$$

or

$$CH_3CH_2CH_2Br + CH_3\ddot{O}\!:^- Na^+$$

$\longrightarrow$ $CH_3CH_2CH_2OCH_3$ + NaBr

Methyl propyl ether

b)

:$\ddot{O}\!:^- Na^+$

[benzene ring] + CH_3Br $\longrightarrow$ [benzene ring with OCH_3] + NaBr

Anisole

c)

$$\begin{array}{c} CH_3 \\ | \\ CH_3\overset{|}{C}\ddot{O}\!:^- Na^+ \\ | \\ H \end{array} +$$

[benzene ring with CH_2Br] $\longrightarrow$ [benzene ring with $CH_2OCH(CH_3)_2$] + NaBr

Benzyl isopropyl ether

d)

$$\begin{array}{c} CH_3 \\ | \\ CH_3\overset{|}{C}CH_2\ddot{O}\!:^- Na^+ \\ | \\ CH_3 \end{array} + CH_3CH_2Br \longrightarrow \begin{array}{c} CH_3 \\ | \\ CH_3\overset{|}{C}CH_2OCH_2CH_3 \\ | \\ CH_3 \end{array} + NaBr$$

Ether 2,2–dimethyl
propyl ether

18.5 The compounds most reactive in the Williamson ether synthesis are also most reactive in any S_N2 process (review Chapter 11 if necessary).

Most reactive $\longrightarrow$ Least reactive

a) CH_3CH_2Br > $CH_3CHBrCH_3$ >> C_6H_5Br

primary   secondary   aromatic halide
(not reactive)

b) CH_3CH_2Br > CH_3CH_2Cl >> $ICH=CHCH_3$

better   poorer   vinylic
leaving group leaving group (not reactive)

18.6

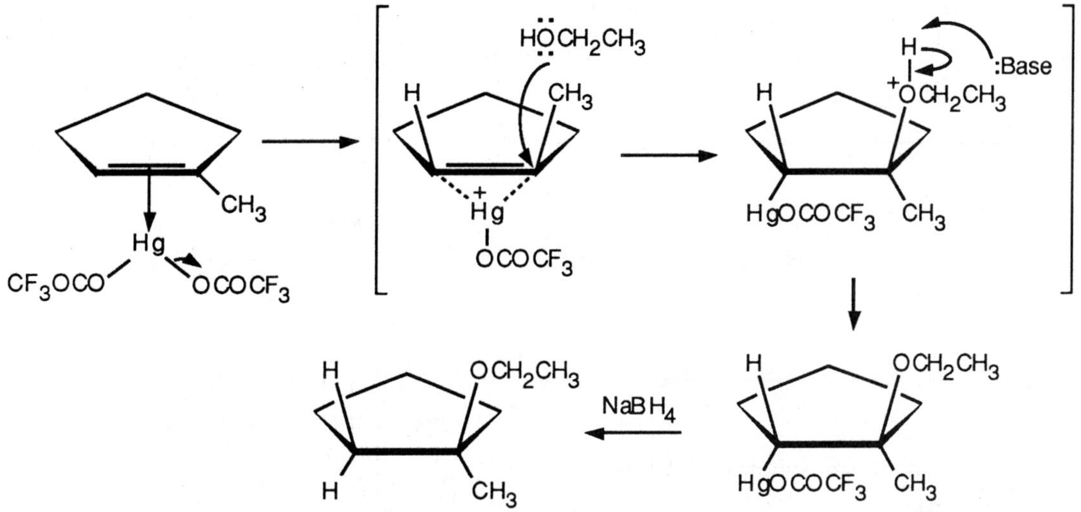

The reaction mechanism of alkoxymercuration/demercuration of an alkene is similar to other electrophilic additions we have studied. First, the cyclopentene pi electrons attack Hg^{2+} with formation of a mercurinium ion. Next, the nucleophilic alcohol displaces mercury. Markovnikov addition occurs because the carbon bearing the methyl group is better able to stabilize the partial positive charge arising from cleavage of the carbon-mercury bond. The ethoxyl and the mercuric groups are *trans* to each other. Finally, removal of mercury by $NaBH_4$ (the mechanism is not fully understood) results in the formation of 1–ethoxy–1–methylcyclopentane.

18.7   a) Either method of synthesis is appropriate.

1) Williamson

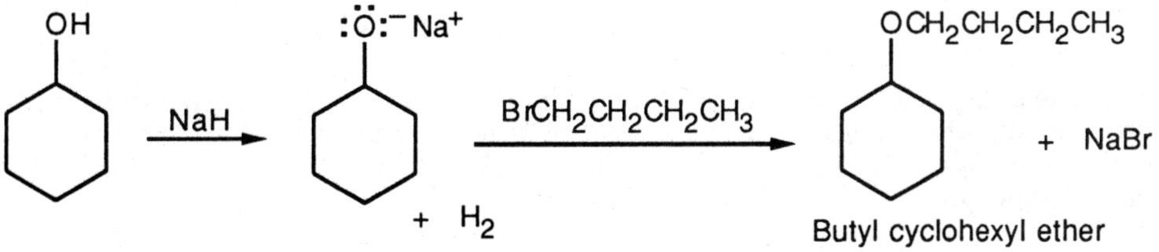

Butyl cyclohexyl ether

2) Alkoxymercuration

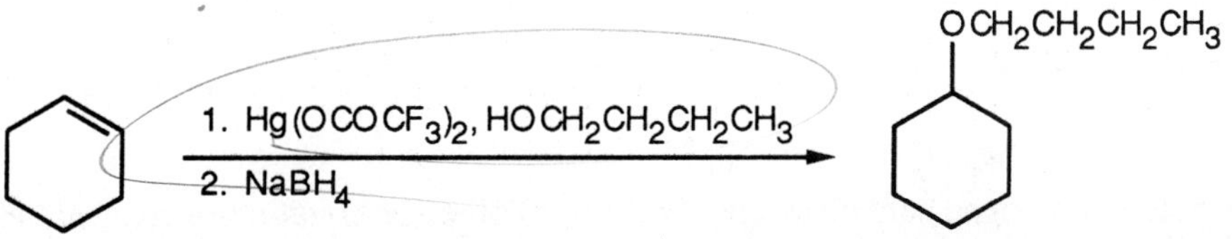

b) The Williamson synthesis must be used.

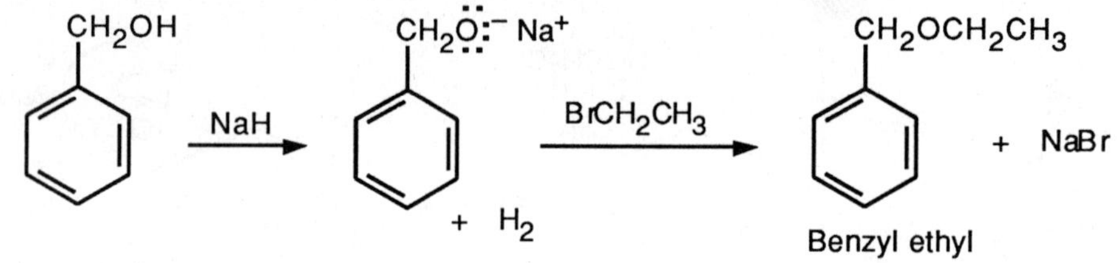

+ H₂

Benzyl ethyl
ether

c) Because both parts of the ether are somewhat hindered, use alkoxymercuration.

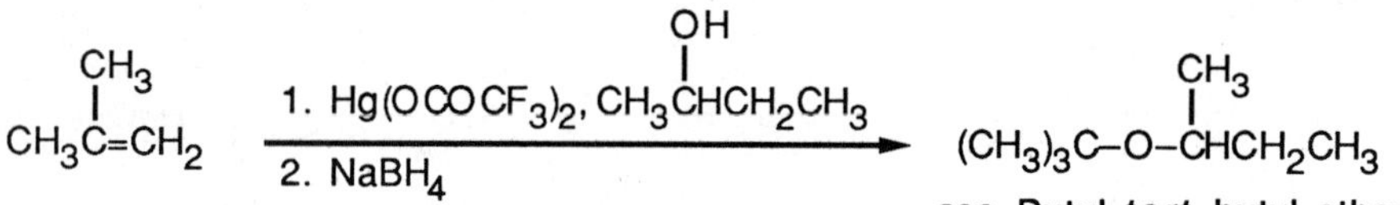

sec-Butyl tert-butyl ether

d) The Williamson synthesis is better.

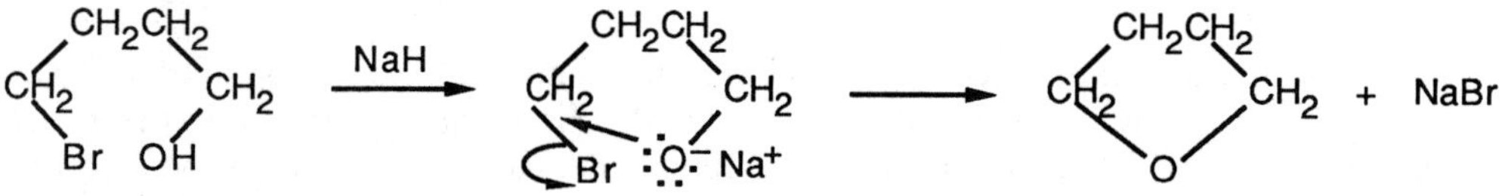

Tetrahydrofuran

18.8   Let R be any other alkyl function.

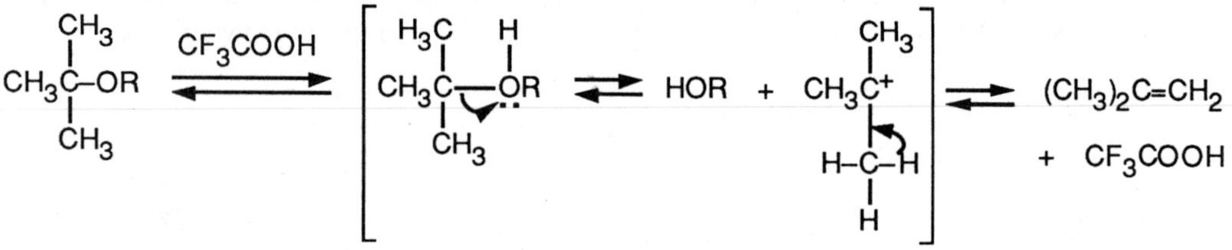

The first step of trifluoroacetic ether cleavage is protonation of the ether oxygen. The protonated intermediate collapses to form an alcohol and a tertiary carbocation. The carbocation loses a proton to form an alkene, isobutylene. This is an example of E_1 elimination.

18.9

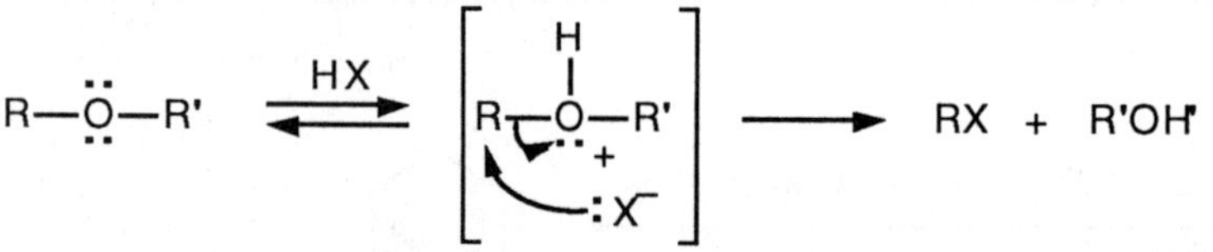

The first step in halogen acid cleavage of ethers involves protonation of oxygen. Halide then effects nucleophilic displacement to form an alcohol and an organic halide. The better the nucleophile, the more effective the displacement. Since :I⁻ and :Br⁻ are more nucleophilic than :Cl⁻, ether cleavage proceeds more smoothly with HI or HBr than with HCl.

18.10 Epoxidation by use of magnesium monoperoxyphthalate (MMPP) is a *syn*-addition of oxygen to a double bond; the original bond stereochemistry is retained.

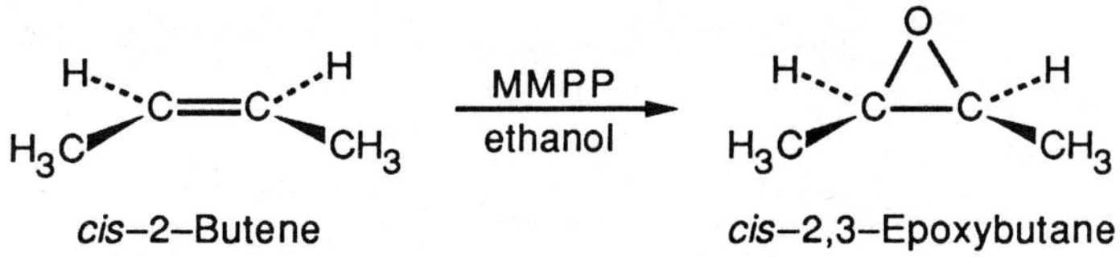

cis–2–Butene → *cis*–2,3–Epoxybutane

In the epoxide product, as in the alkene starting material, the methyl groups are *cis*.

18.11

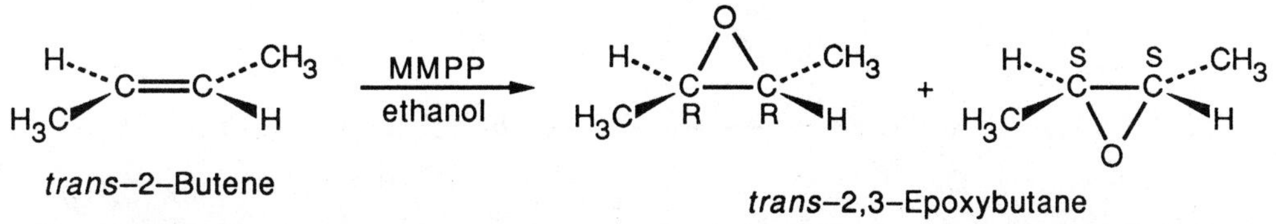

trans–2–Butene → *trans*–2,3–Epoxybutane

The argument in the previous problem can be used to show that reaction of *trans*–2–butene with MMPP yields *trans*–2,3–epoxybutane. A mixture of enantiomers is formed because MMPP can attack either the top or bottom of the double bond.

18.12

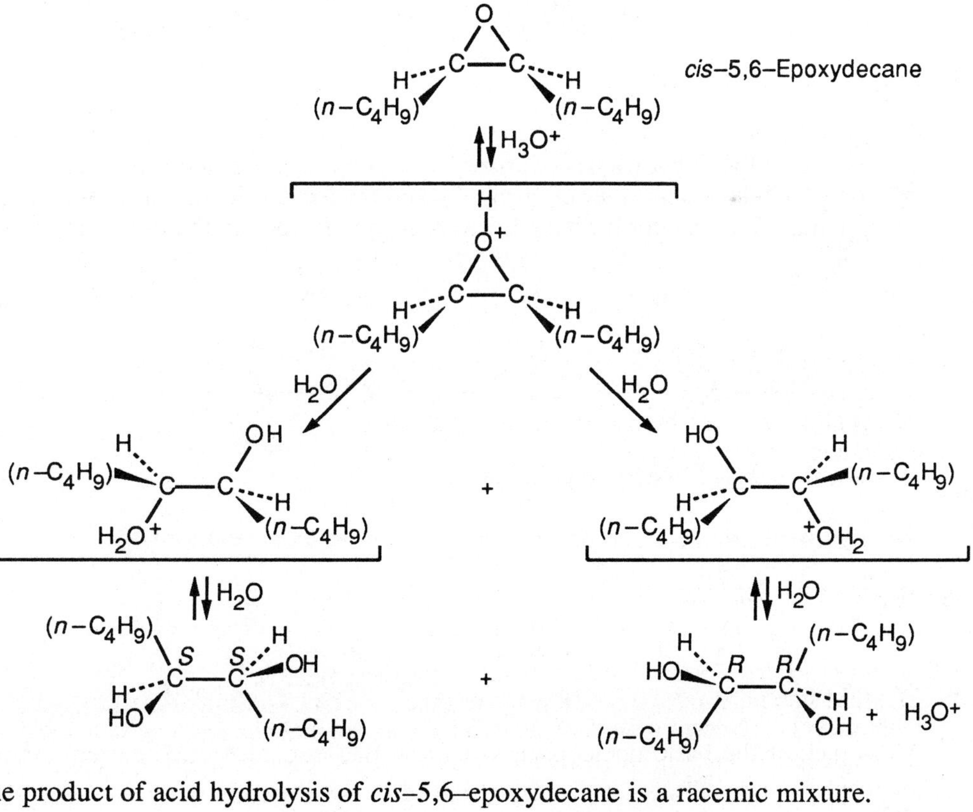

cis–5,6–Epoxydecane

The product of acid hydrolysis of *cis*–5,6–epoxydecane is a racemic mixture.

18.13

trans–5,6–Epoxydecane

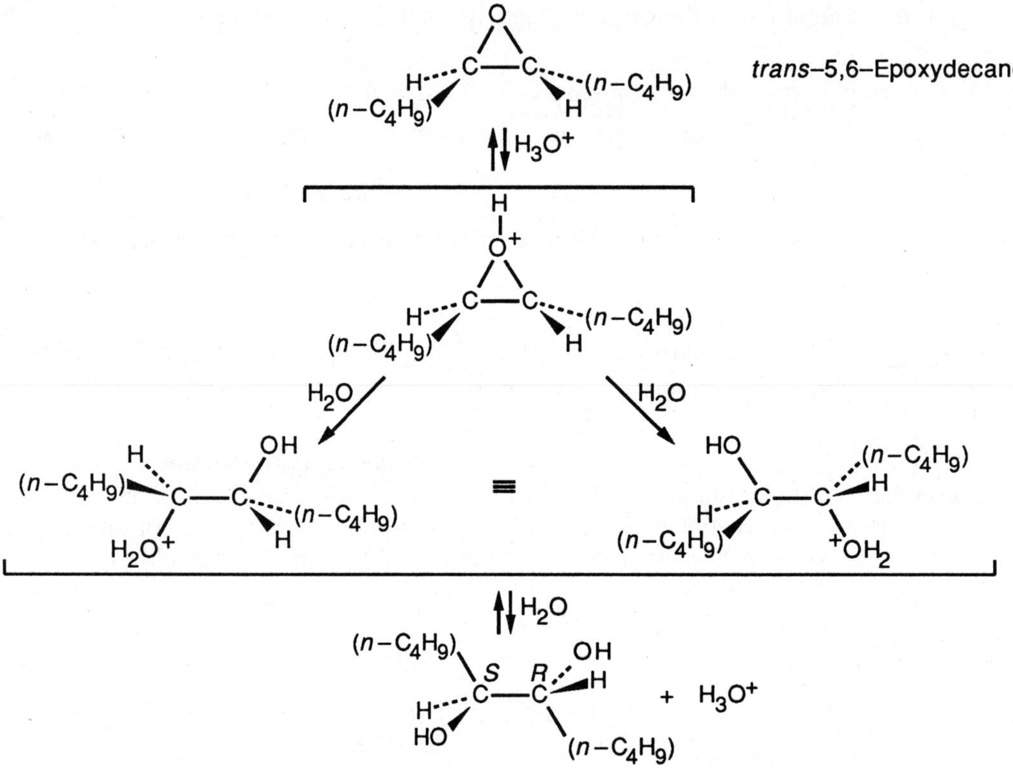

The product of acid hydrolysis of *trans*–5,6–epoxydecane is a *meso* compound. Acid hydrolysis of the enantiomer of *trans*–5,6–epoxydecane yields the same *meso* compound. This product is different from the one formed in the previous problem.

18.14

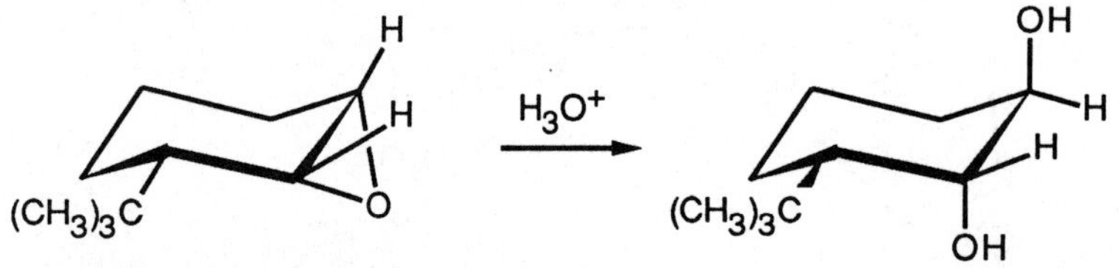

cis–3–*tert*–Butyl–1,2–epoxycyclohexane

The hydroxyl groups in the product have a *trans*–diaxial relationship.

18.15

a) 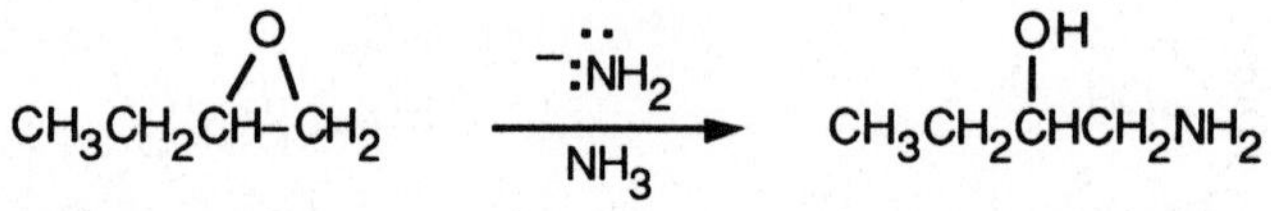 $CH_3CH_2CH-CH_2$ $\xrightarrow[NH_3]{^-:NH_2}$ $CH_3CH_2CHCH_2NH_2$
            \O/   OH

Attack of the basic nucleophile occurs at the less substituted carbon atom.

b)

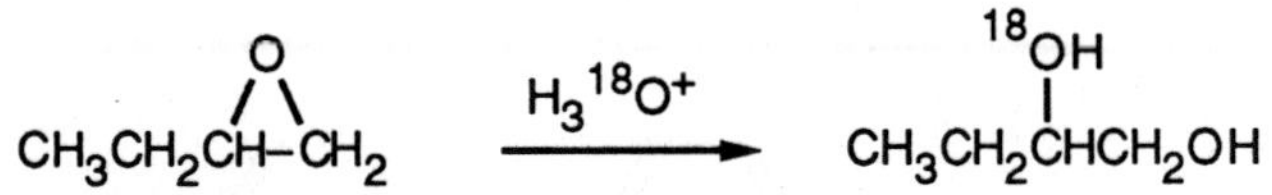

Attack of the nucleophile under acid conditions occurs at the more substituted carbon atom.

18.16

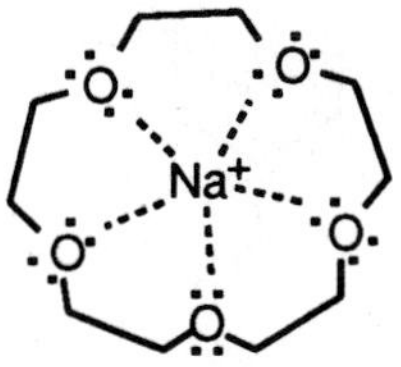

15–Crown–5 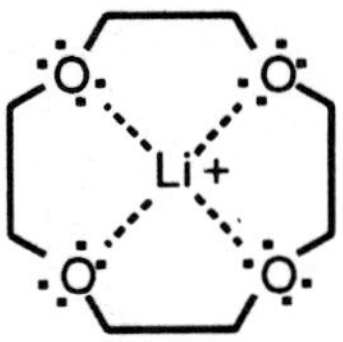   12–Crown–4

The ion-to-oxygen distance in 15-crown-5 is about 40% longer than the ion-to-oxygen distance in 12-crown-4.

18.17

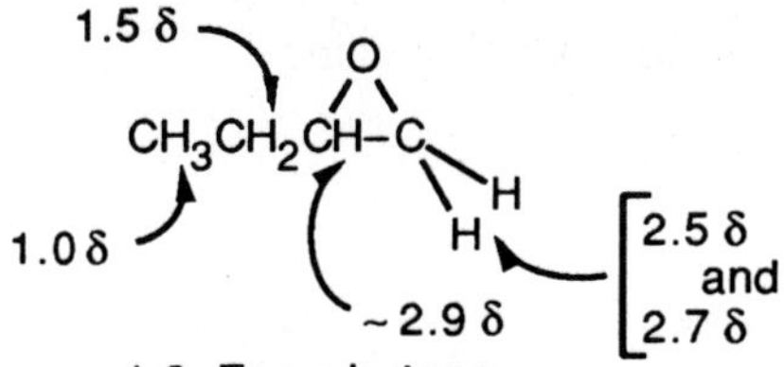

1,2–Epoxybutane

18.18

a) $CH_3CH_2SCH_3$

Ethyl methyl sulfide

b) $(CH_3)_3CSCH_2CH_3$

tert–Butyl ethyl sulfide

c)

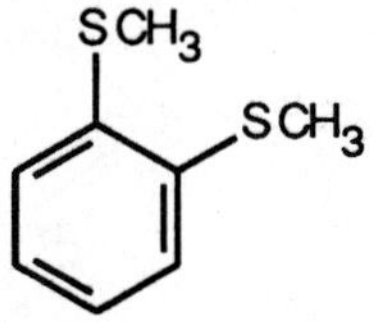

o–(Dimethylthio)benzene

d)

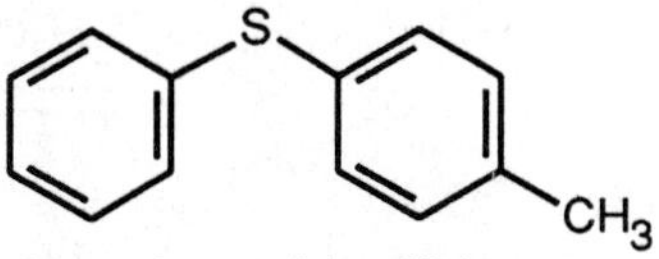

Phenyl p–tolyl sulfide or
p–(Phenylthio)toluene

18.19

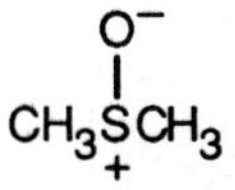

Dimethyl sulfoxide

CH_3SCH_3

Dimethyl sulfide

The boiling point of dimethyl sulfoxide is high because it is a dipolar compound. Dimethyl sulfoxide is miscible with water because it can hydrogen-bond with water.

18.20

a)

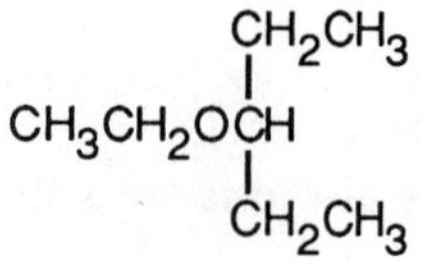

$$CH_3CH_2OCH\overset{\displaystyle CH_2CH_3}{\underset{\displaystyle CH_2CH_3}{|}}$$

b)

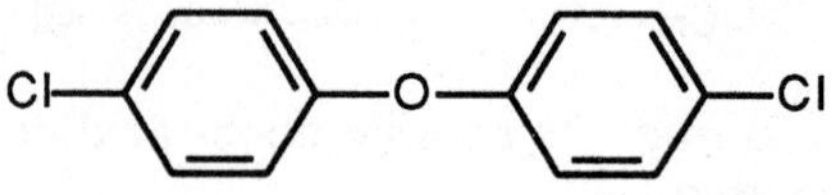

Cl—⬡—O—⬡—Cl

c)

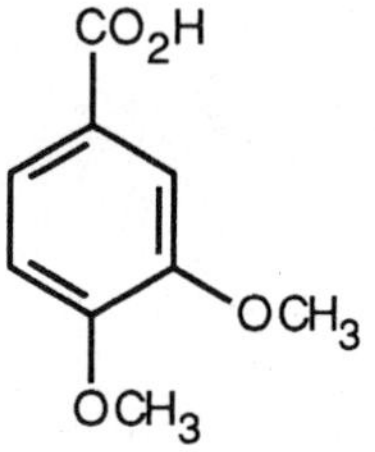

(structure: benzene ring with CO₂H at top, and OCH₃, OCH₃ at adjacent positions)

d)

(structure: cyclohexyl—O—cyclopentyl ether)

e)

(structure: phenol with OH, OCH₃ ortho, and CH₂CH=CH₂ at para)

18.21 a) *o*–Dimethoxybenzene

b) Cyclopropyl isopropyl ether *or*
    Isopropoxycyclopropane

c) 2–Methyltetrahydrofuran
e) 2,2–Dimethoxypropane
g) *p*–Nitroethoxybenzene *or*
    Ethyl *p*–nitrophenyl ether

d) Cyclohexyl cyclopropyl sulfide
f) 1,2–Epoxycyclopentane
h) 1,1–(Dimethylthio)cyclohexane

18.22

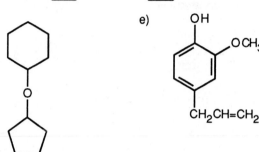

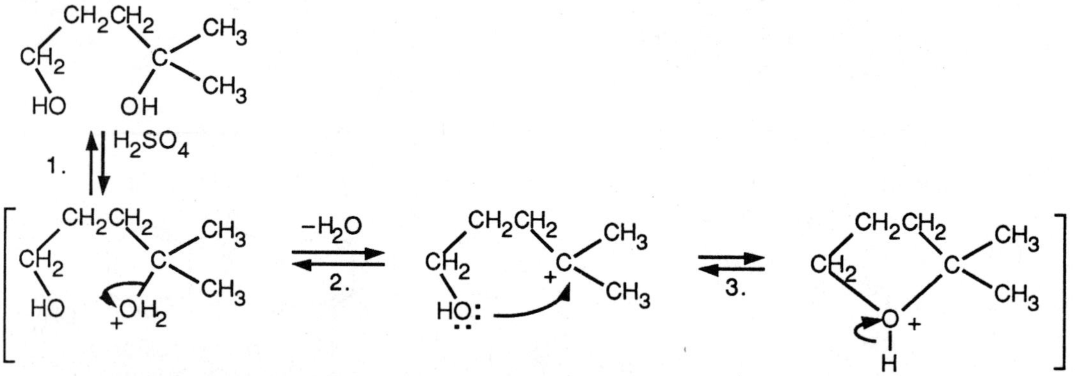

1. $\Updownarrow H_2SO_4$

(reaction scheme showing diol intermediate, −H₂O step 2., carbocation, step 3., oxonium intermediate)

The reaction involves: (1) protonation of the tertiary
hydroxyl group; (2) departure of water to form a tertiary
carbocation; (3) nucleophilic attack on the cation by
the other hydroxyl group. The tertiary hydroxyl group
is more likely to be eliminated because the resulting
carbocation is more stable.

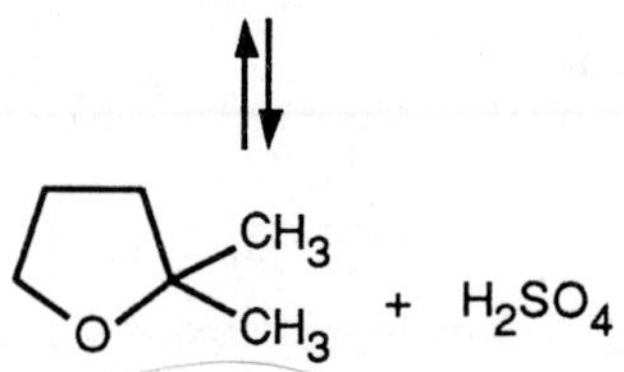

(product: 2,2-dimethyltetrahydrofuran) + H_2SO_4

18.23

a)

(cyclohexyl—O CH₂CH₃) $\xrightarrow[\text{H}_2\text{O}]{\text{HI}}$ (cyclohexyl—OH) + CH_3CH_2I

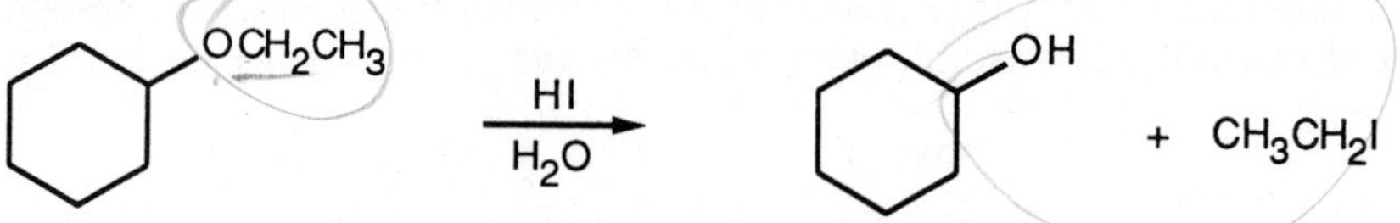

b)

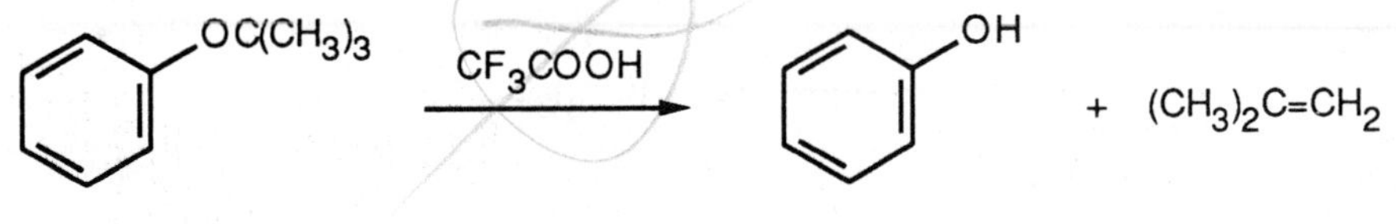

$$CF_3COOH$$

+ $(CH_3)_2C=CH_2$

c)

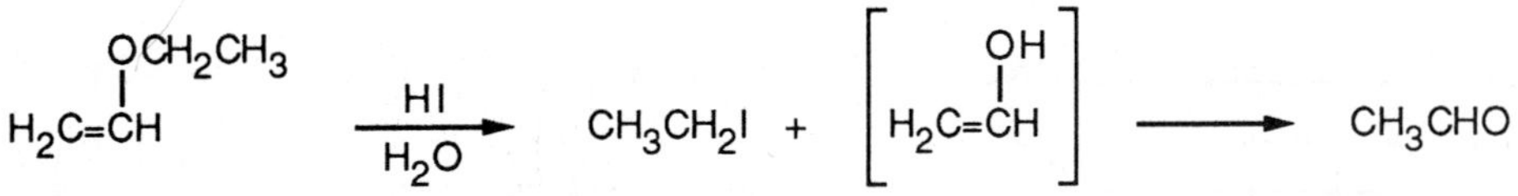

$$\xrightarrow[H_2O]{HI}$$ CH_3CH_2I + $\left[H_2C=\overset{\overset{\displaystyle OH}{|}}{CH} \right]$ ⟶ CH_3CHO

d) $(CH_3)_3CCH_2OCH_2CH_3$ $\xrightarrow[H_2O]{HI}$ $(CH_3)_3CCH_2OH$ + CH_3CH_2I

18.24

$$\frac{1.06 \text{ g vanillin}}{152 \text{ g/mol}} = 6.97 \times 10^{-3} \text{ mol vanillin}$$

$$\frac{1.60 \text{ g AgI}}{234.8 \text{ g/mol}} = 6.81 \times 10^{-3} \text{ mol AgI}$$

6.81×10^{-3} mol ⟶ 6.81×10^{-3} mol ⟶ 6.81×10^{-3} mol ⟶ 6.81×10^{-3} mol
AgI   :I⁻ CH_3I $-OCH_3$

Thus, 6.97×10^{-3} moles of vanillin contains 6.81×10^{-3} moles of methoxyl groups. Since the ratio of moles vanillin to moles methoxyl is approximately 1:1, each vanillin contains one methoxyl group.

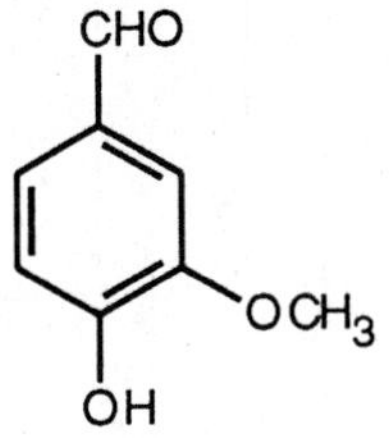

Vanillin

18.25

a)

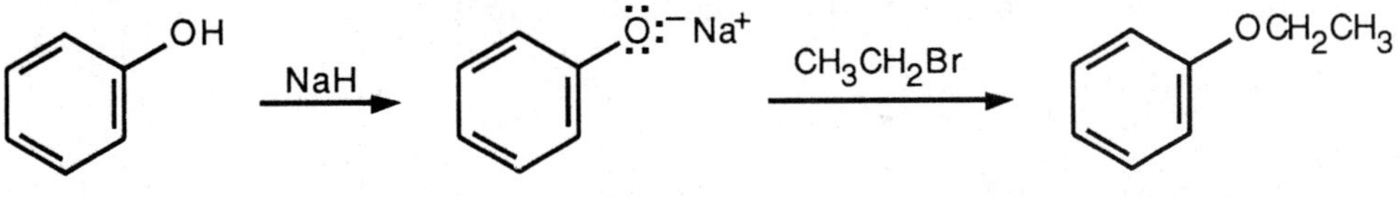

$$NaH$$

$$CH_3CH_2Br$$

b) 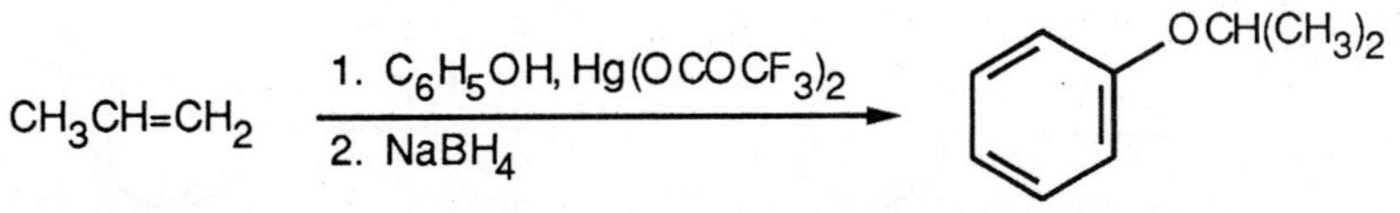 $CH_3CH=CH_2$ $\xrightarrow[\text{2. NaBH}_4]{\text{1. } C_6H_5OH, Hg(OCOCF_3)_2}$

c)

$$\xrightarrow[\text{ethanol}]{MMPP}$$

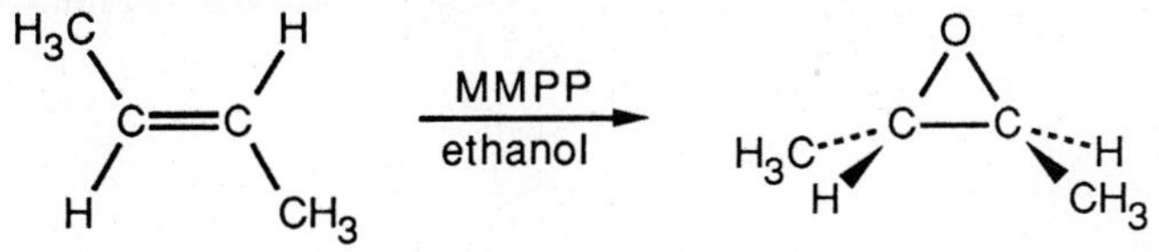

d)

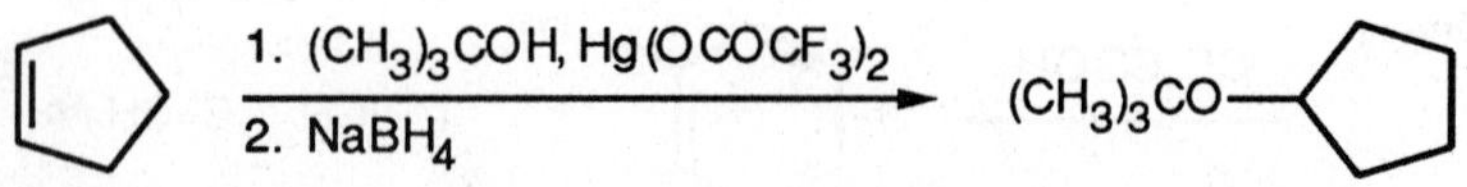

e)

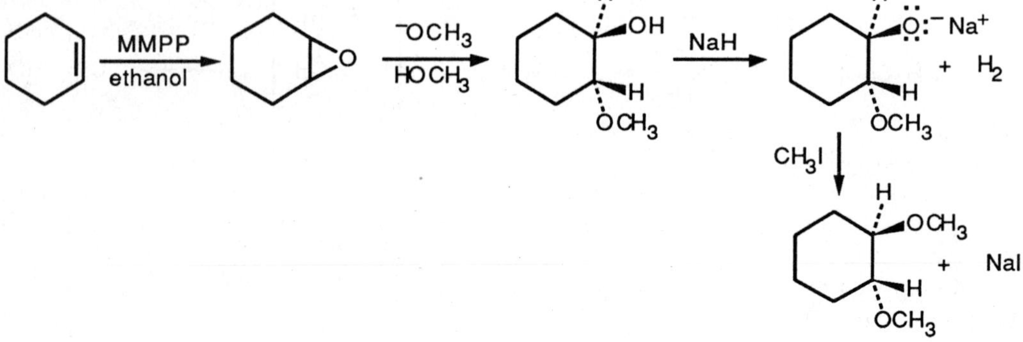

f)

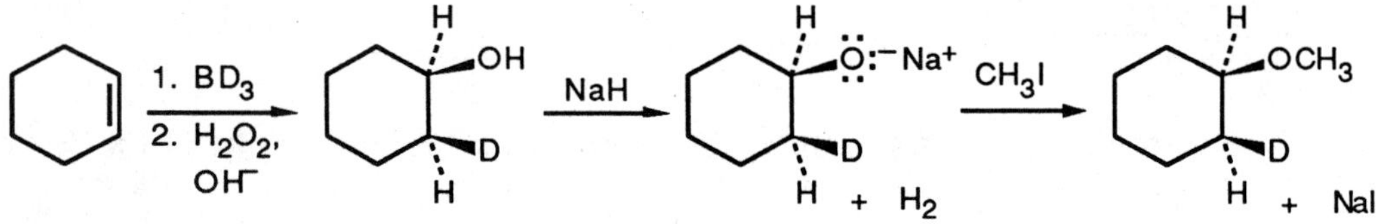

18.26

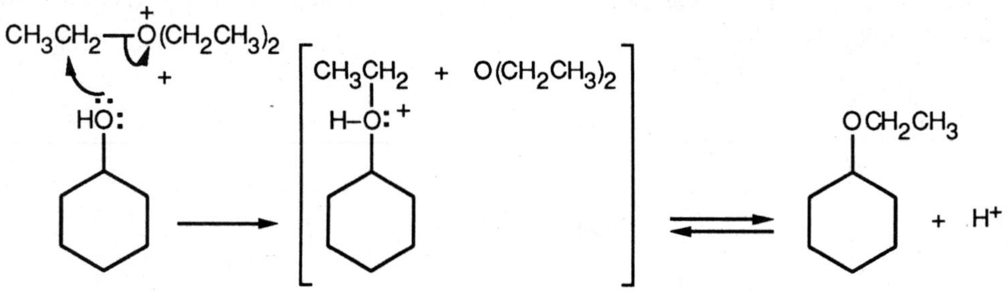

Trialkyloxonium salts are more reactive alkylating agents than alkyl iodides because a neutral ether is a better leaving group than an iodide.

18.27

a)

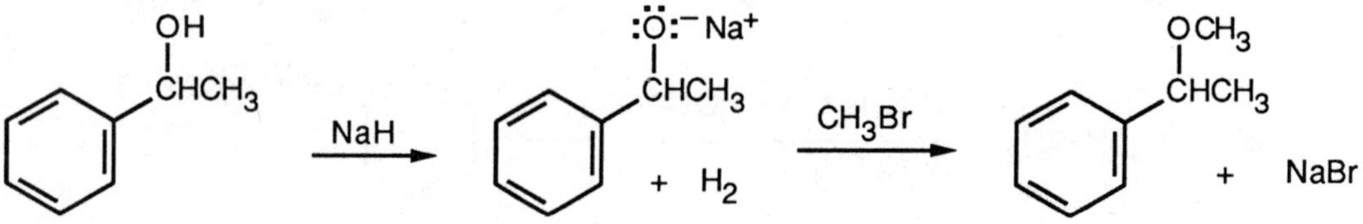

Methyl 1–phenylethyl ether

b)

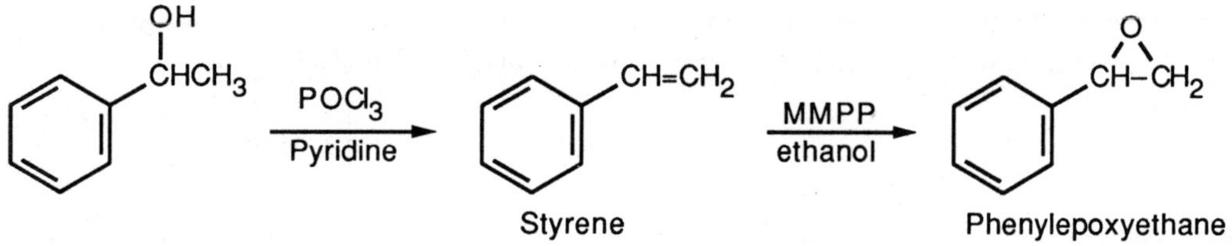
OH
CHCH$_3$
POCl$_3$
Pyridine
CH=CH$_2$
MMPP
ethanol
O
CH–CH$_2$
Styrene
Phenylepoxyethane

c)

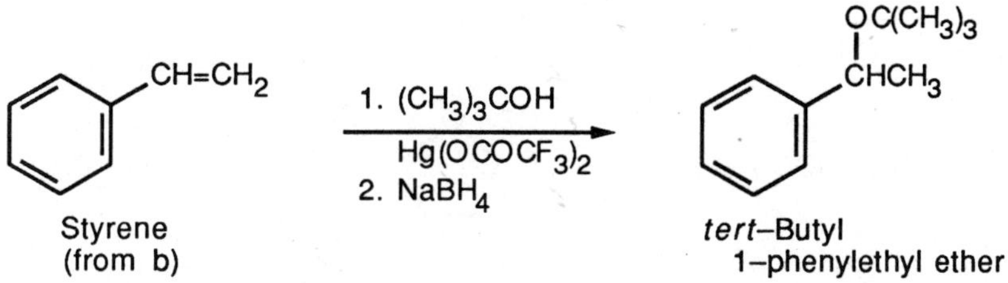
CH=CH$_2$
1. (CH$_3$)$_3$COH
Hg(OCOCF$_3$)$_2$
2. NaBH$_4$
OC(CH$_3$)$_3$
CHCH$_3$
Styrene
(from b)
tert–Butyl
1–phenylethyl ether

18.28

a)

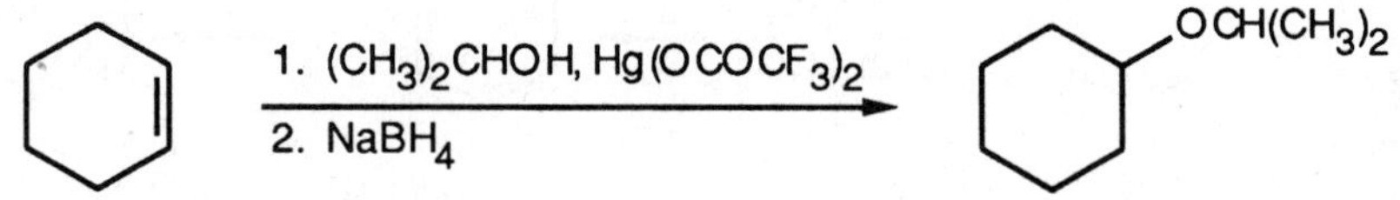
1. (CH$_3$)$_2$CHOH, Hg(OCOCF$_3$)$_2$
2. NaBH$_4$
OCH(CH$_3$)$_2$

b)

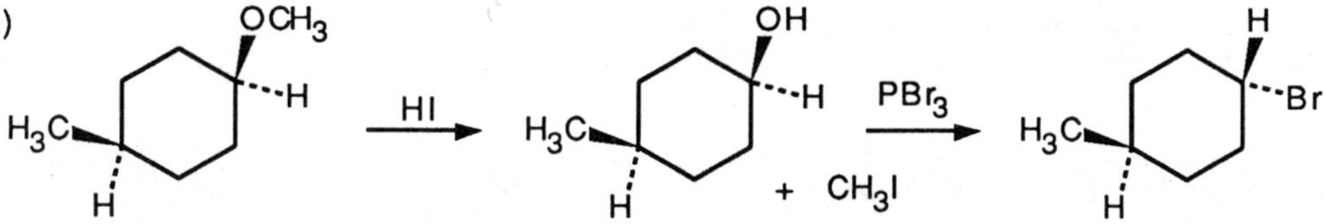
OCH$_3$
H
H$_3$C
H
HI
OH
H
H$_3$C
H
+ CH$_3$I
PBr$_3$
H
Br
H$_3$C
H

c)

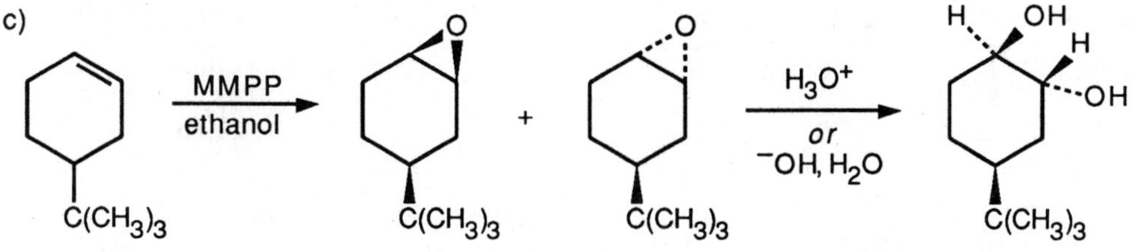
C(CH$_3$)$_3$
MMPP
ethanol
O
C(CH$_3$)$_3$
+
O
C(CH$_3$)$_3$
H$_3$O$^+$
or
$^-$OH, H$_2$O
H
OH
H
OH
C(CH$_3$)$_3$

d)

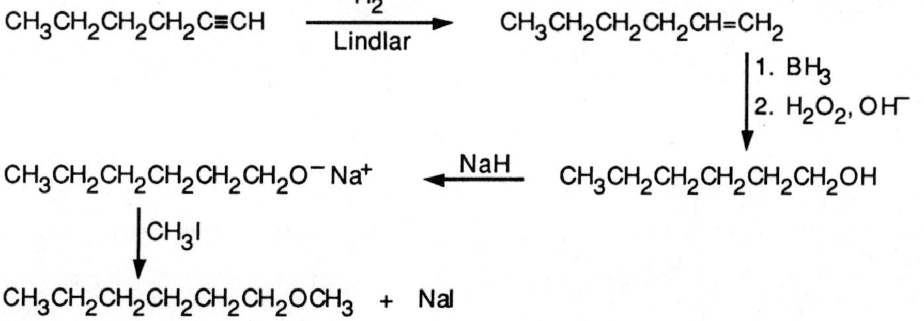
CH$_3$CH$_2$CH$_2$CH$_2$C≡CH
H$_2$
Lindlar
CH$_3$CH$_2$CH$_2$CH$_2$CH=CH$_2$
1. BH$_3$
2. H$_2$O$_2$, OH$^-$
CH$_3$CH$_2$CH$_2$CH$_2$CH$_2$CH$_2$OH
NaH
CH$_3$CH$_2$CH$_2$CH$_2$CH$_2$CH$_2$O$^-$ Na$^+$
CH$_3$I
CH$_3$CH$_2$CH$_2$CH$_2$CH$_2$CH$_2$OCH$_3$ + NaI

e) $CH_3CH_2CH_2CH_2C{\equiv}CH$ $\xrightarrow[\text{Lindlar}]{H_2}$ $CH_3CH_2CH_2CH_2CH{=}CH_2$

$\downarrow$ 1. $Hg(OCOCF_3)_2$, CH_3OH
   2. $NaBH_4$

$$CH_3CH_2CH_2CH_2\overset{\overset{\displaystyle OCH_3}{|}}{C}HCH_3$$

18.29

a)

$\xrightarrow{HI}$ $HOCH_2CH_2CH_2CH_2I$;

$\xrightarrow{2\ HI}$ $ICH_2CH_2CH_2CH_2I$

18.30

a)

$\xrightarrow[H_2SO_4]{SO_3}$ ... SO_3H $\xrightarrow[\text{2. }H_3O^+]{\text{1. NaOH, }\Delta}$ ... OH $\xrightarrow{NaH}$ ... $\ddot{O}{:}^- Na^+$ $+\ H_2$

b)

$\xrightarrow[AlCl_3]{CH_3Cl}$ ... CH_3 $\xrightarrow[CCl_4]{NBS}$ ... CH_2Br

c)

... $\ddot{O}{:}^- Na^+$ + ... CH_2Br $\longrightarrow$ ... CH_2O ...

from a   from b   Benzyl phenyl ether

18.31

$+$ $:\!\ddot{I}\!:^-$ $\longrightarrow$ ... $\ddot{O}{:}^-$ $+$ CH_3I

This reaction is an S_N2 displacement and can't occur at an aryl carbon.

18.32

$$CH_3\overset{\overset{\displaystyle CH_3}{|}}{C}{=}CH_2 \ \underset{}{\overset{H^+}{\rightleftharpoons}} \ \left[CH_3\overset{\overset{\displaystyle CH_3}{|}}{\underset{+}{C}}CH_3 \ + \ \overset{\overset{\displaystyle H}{|}}{:\!\ddot{O}}{-}R \ \rightleftharpoons \ CH_3\overset{\overset{\displaystyle CH_3}{|}}{\underset{\underset{\displaystyle H}{\overset{|}{\underset{+}{O}}-R}}{C}}CH_3 \right] \ \rightleftharpoons \ CH_3\overset{\overset{\displaystyle CH_3}{|}}{\underset{\overset{|}{OR}}{C}}CH_3 \ + \ H^+$$

Notice that this reaction is the reverse of acid-catalyzed cleavage of a tertiary ether.

18.33

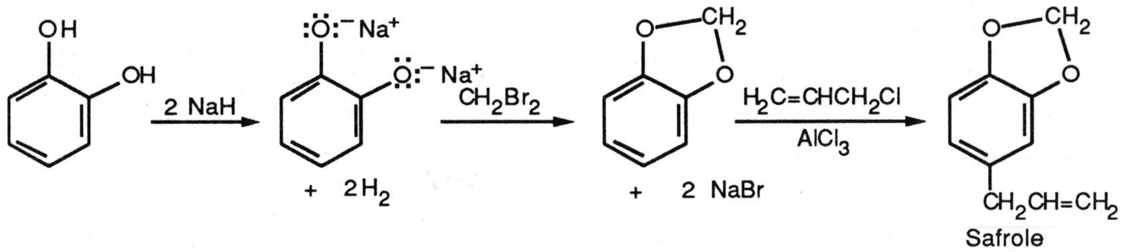

18.34

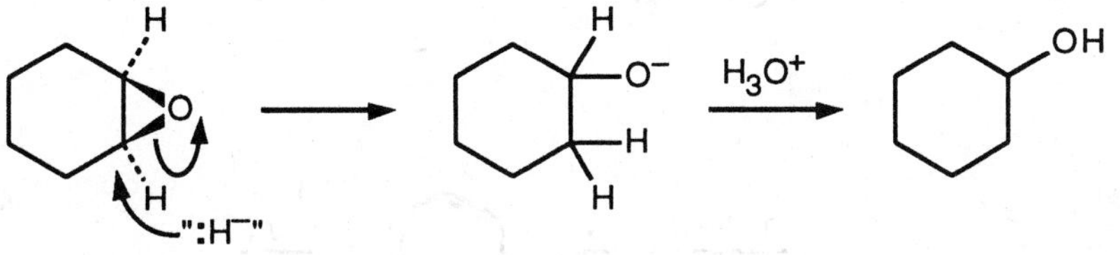

18.35

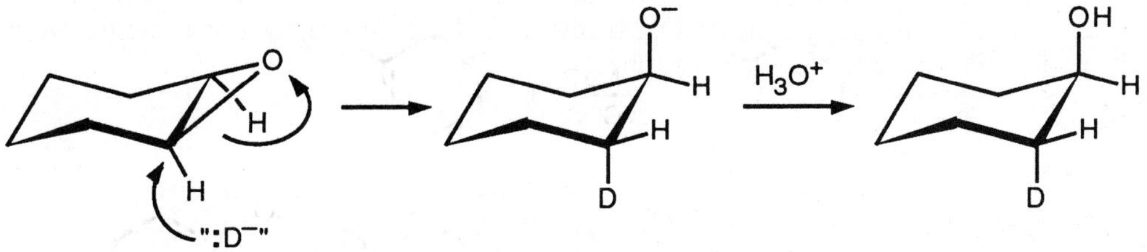

18.36 The mechanism of Grignard addition to oxetane is the same as the mechanism of Grignard addition to epoxides, described in Section 18.8. The reaction proceeds at a reduced rate because oxetane is less reactive than ethylene oxide. The four-membered ring oxetane is less strained, and therefore more stable, than the three-membered ethylene oxide ring.

18.37

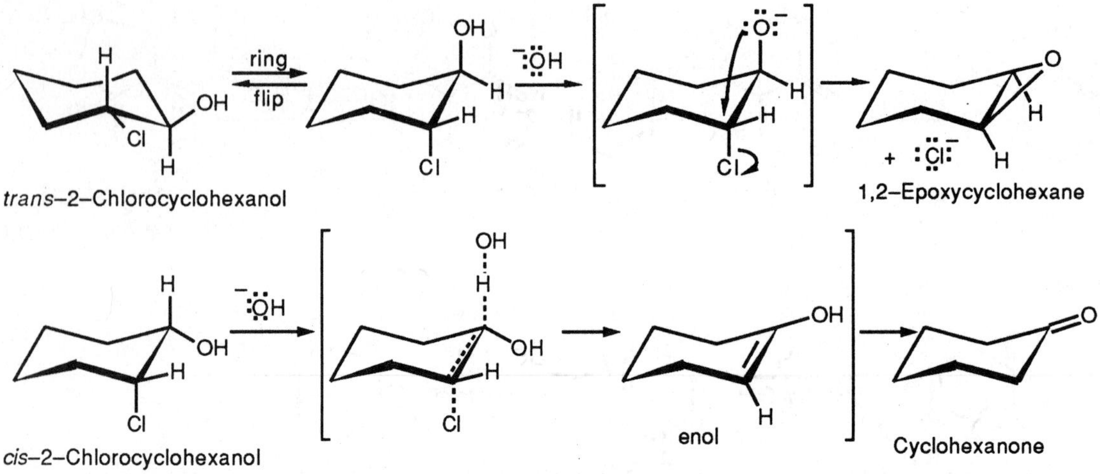

trans–2–Chlorocyclohexanol

1,2–Epoxycyclohexane

cis–2–Chlorocyclohexanol

enol

Cyclohexanone

In the *trans*–isomer, hydroxyl and chlorine are in the *trans* orientation that allows epoxide formation to occur as described in Section 18.7. Epoxidation can't occur for the *cis* isomer; instead, the base ⁻OH brings about E2 elimination, producing an enol, which rearranges to cyclohexanone.

18.38

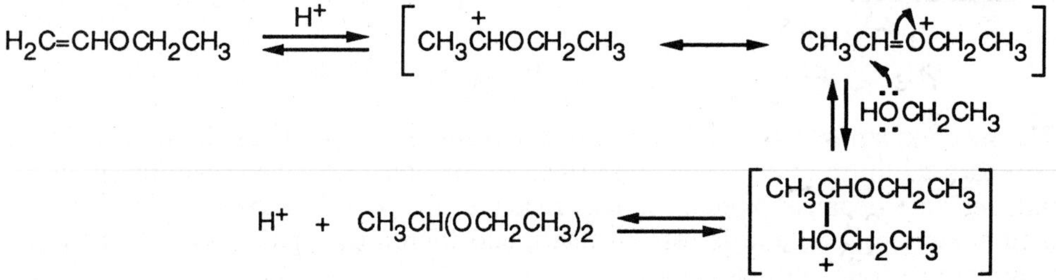

Addition of ethanol occurs with the observed regiochemistry because the initial carbocation is stabilized by resonance involving the ether oxygen.

18.39

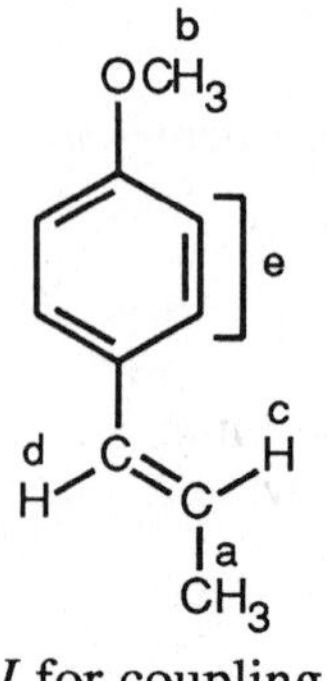

Anethole

Protons	δ	Multiplicity	Split by	J
a.	1.83	doublet	c	$J_{ac} = 8$
b.	3.75	singlet		
c.	6.08	quartet	a	$J_{ca} = 8$
d.	6.28	singlet		
e.	6.80, 7.23	multiplet		

J for coupling between protons c and d is very small.

18.40

18.41 $M^+ = 116$ corresponds to a sulfide of molecular formula $C_6H_{12}S$, indicating one degree of unsaturation.

18.42 A molecular model of bornene shows that approach to the upper face of the double bond is hindered by a methyl group. Reaction with RCO_3H occurs at the lower face of the double bond to produce epoxide **A**.

In the reaction of Br_2 and H_2O with bornene, the intermediate bromonium ion also forms at the lower face. Reaction with water yields a bromohydrin which, when treated with base, forms epoxide **B**.

18.43 Disparlure, $C_{19}H_{38}O$, contains one degree of unsaturation, which the ^1H NMR absorption at 2.8 δ identifies as an epoxide ring.

18.44

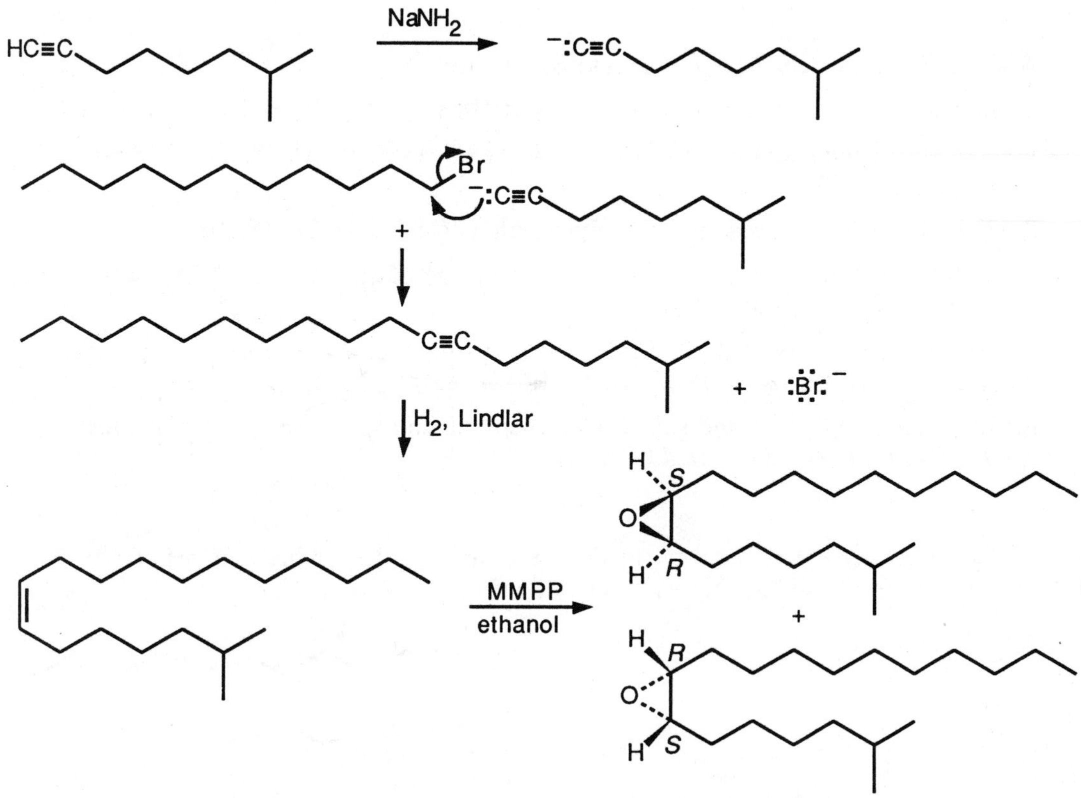

18.45

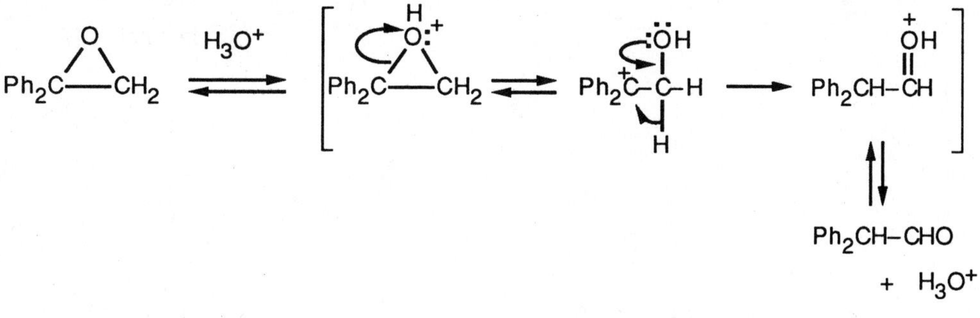

18.46

a)

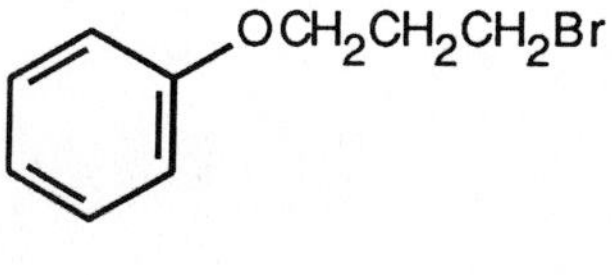

$OCH_2CH_2CH_2Br$

b) $CH_3CH(OCH_3)_2$

c)

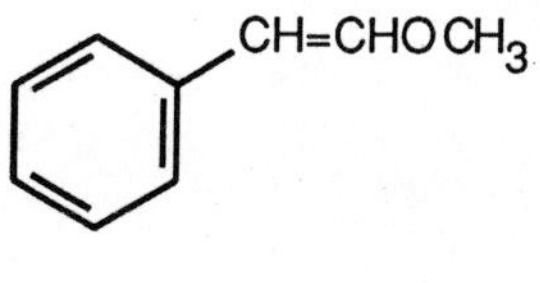

$CH=CHOCH_3$

Study Guide for Chapter 18

After studying this chapter, you should be able to:

(1)   Name and draw ethers, epoxides and sulfides (18.1, 18.16, 18.18, 18.20, 18.21).

(2)   Prepare ethers and epoxides (18.2, 18.3, 18.4, 18.5, 18.7, 18.10, 18.11, 18.25, 18.27, 18.30, 18.33, 18.40, 18.44).

(3)   Predict the products of reactions involving ethers (18.23, 18.28, 18.29).

(4)   Formulate mechanisms for preparation and cleavage of ethers (18.6, 18.8, 18.9, 18.22, 18.26, 18.31, 18.32, 18.36, 18.37, 18.38).

(5)   Formulate mechanisms and predict products of ring-opening reactions of epoxides (18.12, 18.13, 18.14, 18.15, 18.34, 18.35, 18.42, 18.45).

(6)   Identify ethers, epoxides and sulfides by chemical and spectroscopic techniques (18.17, 18.24, 18.39, 18.41, 18.43, 18.46).

Chapter 19 – Aldehydes and Ketones. Nucleophilic Addition Reactions

19.1

a)

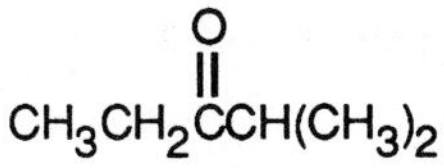

$CH_3CH_2CCH(CH_3)_2$ (with O double bond on the C)

2–Methyl–3–pentanone

b)

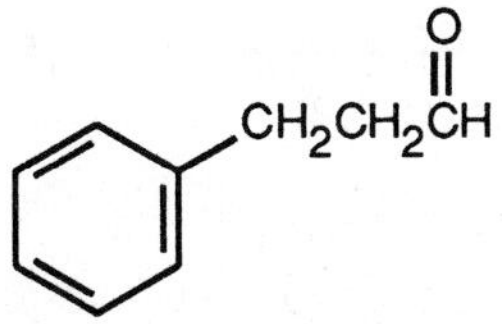

CH_2CH_2CH (with O double bond)

3–Phenylpropanal

c)

$CH_3CCH_2CH_2CH_2CCH_2CH_3$ (with two O double bonds)

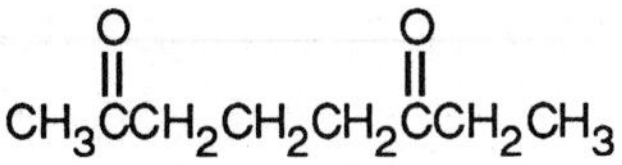

2,6–Octanedione

d)

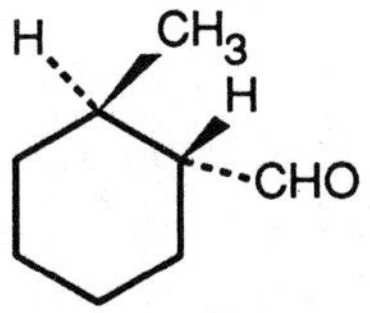

trans–2–Methylcyclohexane-
carbaldehyde

e)

$H\,CCH_2CH_2CH_2CH$ (with two O double bonds)

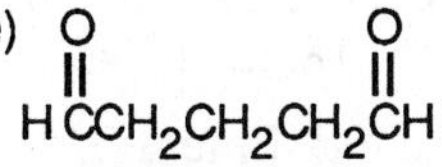

Pentanedial

f)

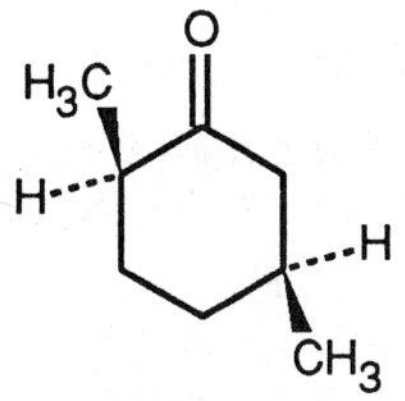

cis–2,5–Dimethylcyclohexanone

g)

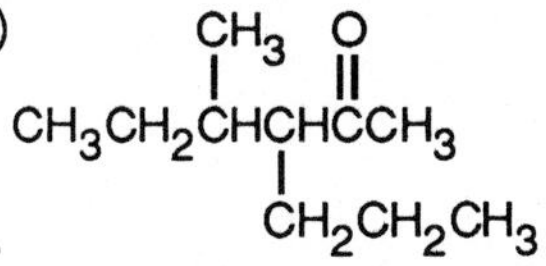

$CH_3CH_2CHCHCCH_3$ (with CH$_3$ and O above, CH$_2$CH$_2$CH$_3$ below)

4–Methyl–3–propyl–2–hexanone

h)

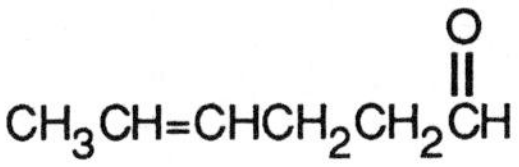

$CH_3CH=CHCH_2CH_2CH$ (with O double bond)

4–Hexenal

19.2

a)

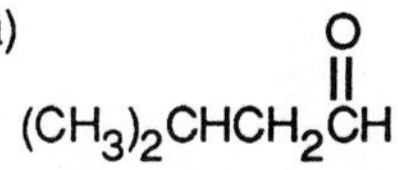

$(CH_3)_2CHCH_2CH$ (with O double bond)

3–Methylbutanal

b)

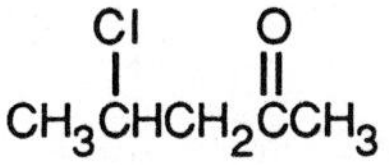

$CH_3CHCH_2CCH_3$ (with Cl above first CH, O above C)

4–Chloro–2–pentanone

c)

CH_2CH (with O double bond)

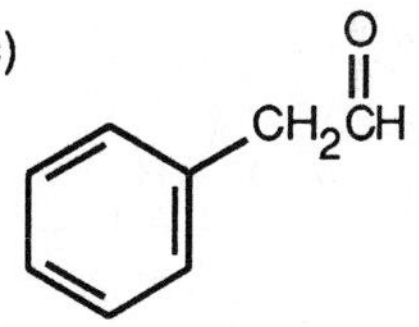

Phenylacetaldehyde

d)

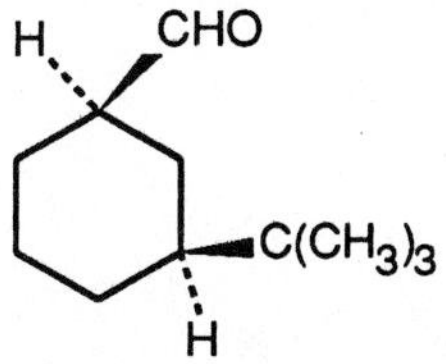

cis–3–tert–Butylcyclohexane-
carbaldehyde

e)

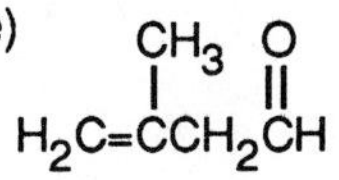

$H_2C=CCH_2CH$ (with CH$_3$ and O above)

3–Methyl–3–butenal

f)

$CH_3CH_2CHCH_2CH_2CHCH$ (with CH$_3$ above, O above, CHClCH$_3$ below)

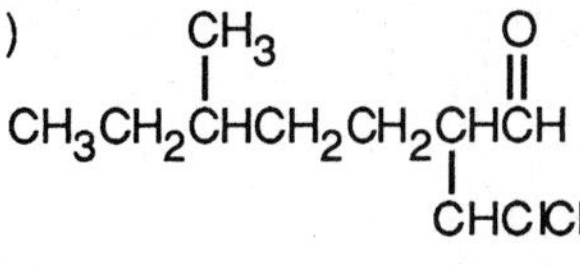

2–(1–Chloroethyl)–5–methylheptanal

19.3

a) $CH_3CH_2CH_2CH_2CH_2OH$ $\xrightarrow[CH_3Cl_2]{PCC}$ $CH_3CH_2CH_2CH_2CHO$

1-Pentanol   Pentanal

b) $CH_3CH_2CH_2CH_2CH=CH_2$ $\xrightarrow[\text{2. Zn, }H_3O^+]{\text{1. }O_3}$ $CH_3CH_2CH_2CH_2CHO$

1-Hexene

c) $CH_3CH_2CH_2CH_2COOCH_3$ $\xrightarrow[\text{2. }H_3O^+]{\text{1.   DIBAH}}$ $CH_3CH_2CH_2CH_2CHO$

19.4

a) 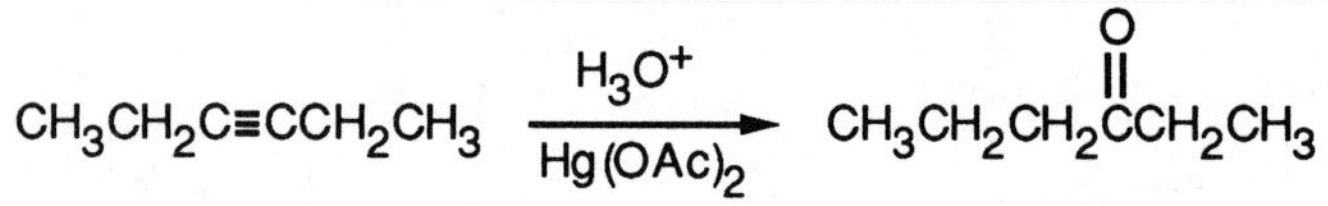 $CH_3CH_2C≡CCH_2CH_3$ $\xrightarrow[Hg(OAc)_2]{H_3O^+}$ $CH_3CH_2CH_2\overset{\displaystyle O}{\overset{\|}{C}}CH_2CH_3$

b)

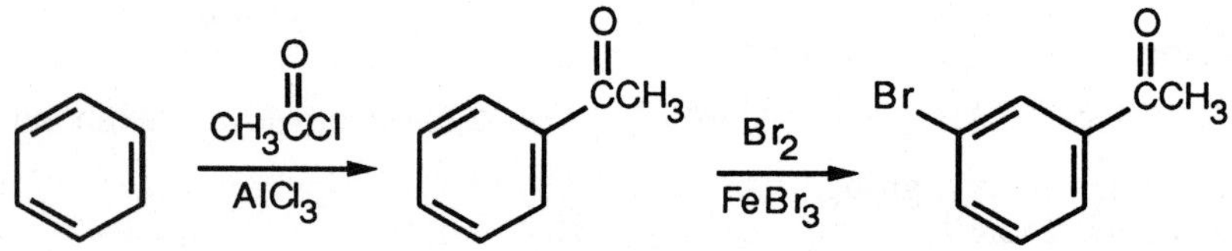

c)

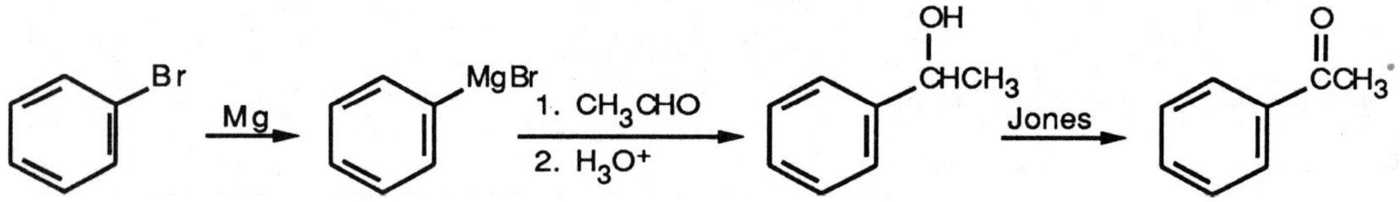

d)

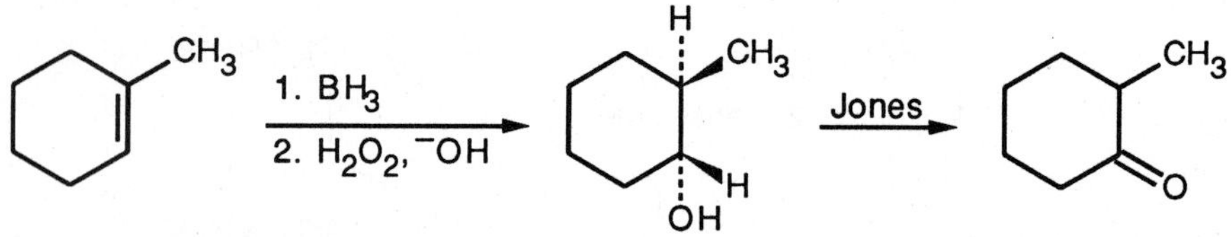

19.5

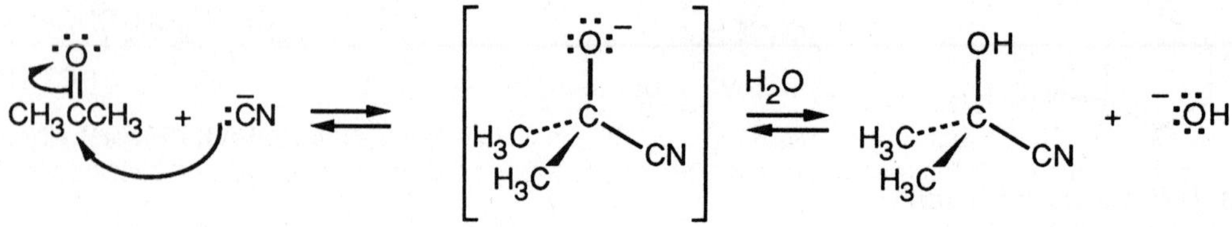

19.6 An aromatic aldehyde is less reactive than an aliphatic aldehyde toward nucleophilic addition. The partial positive charge of the carbonyl carbon can be delocalized into the aromatic ring.

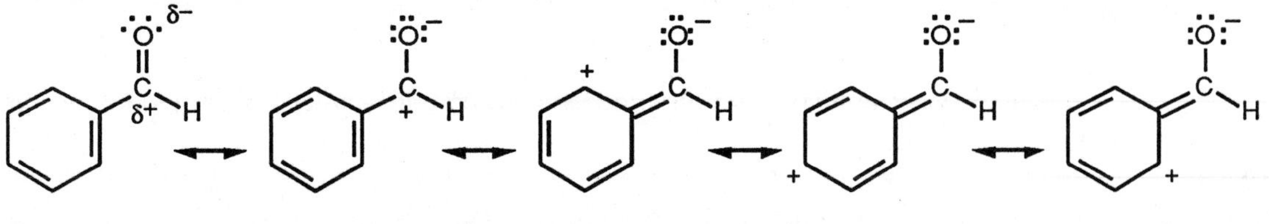

The carbonyl carbon is thus less electron-rich and less reactive toward nucleophiles. In addition, the resonance stabilization of an aromatic aldehyde is lost when a nucleophile adds to the aldehyde functional group.

19.7

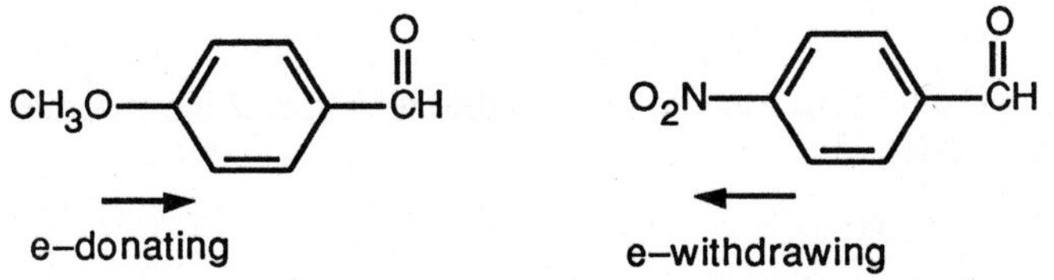

e–donating   e–withdrawing

The electron-donating methoxyl group makes the aldehyde carbon more electron-rich and less reactive toward nucleophiles than the aldehyde carbon of *p*–nitrobenzaldehyde.

19.8

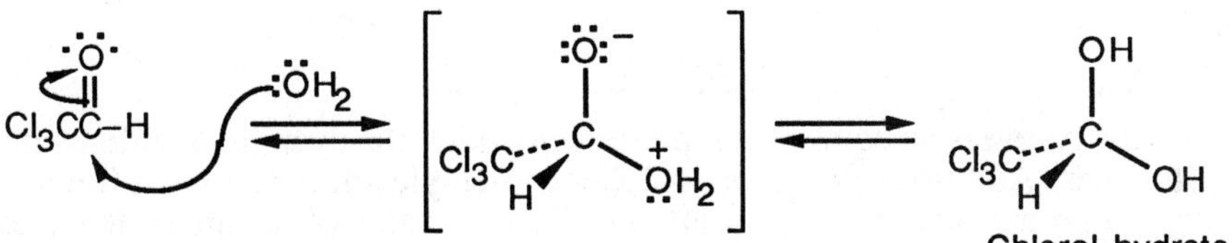

Chloral hydrate

19.9

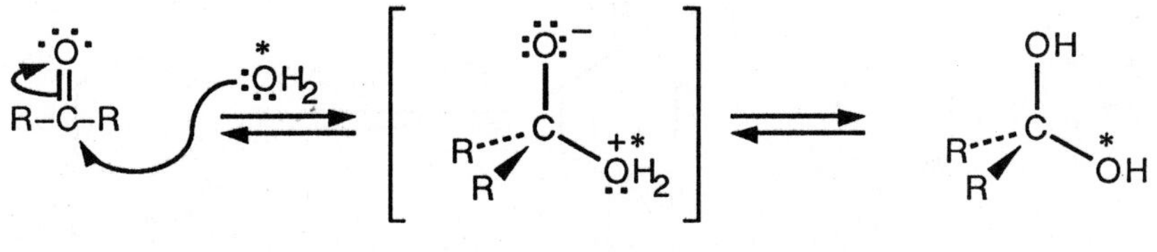

The above mechanism is similar to other nucleophilic addition mechanisms we have studied. Since all of these steps are reversible, we can write the above mechanism in reverse to show how labeled oxygen is incorporated into an aldehyde or ketone.

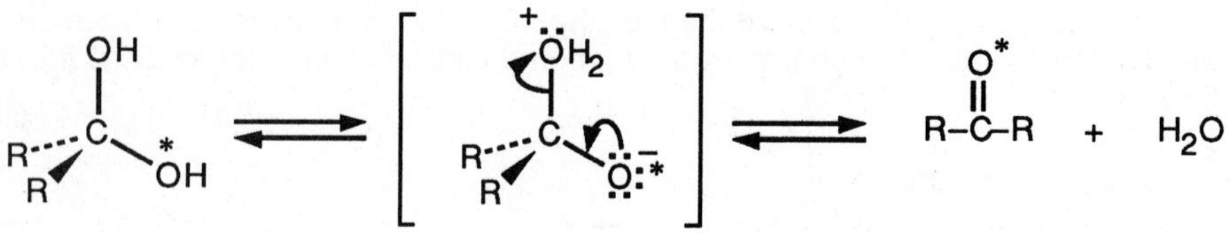

This exchange is very slow in water but proceeds more rapidly when either acid or base is present.

19.10

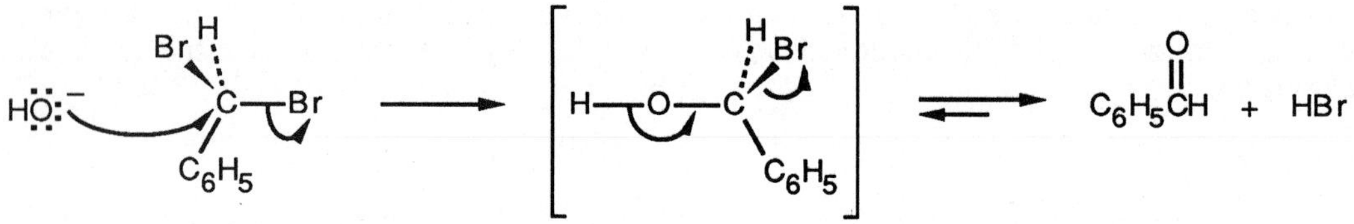

S_N2 addition of hydroxide to $C_6H_5CHBr_2$ yields an unstable tetrahedral intermediate, which loses HBr to yield benzaldehyde.

19.11

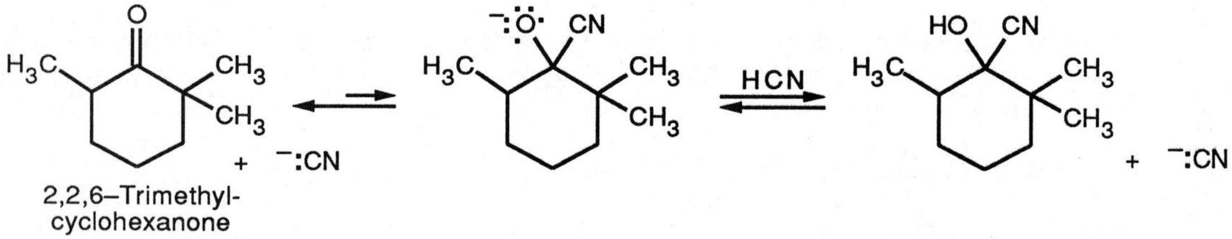

2,2,6–Trimethyl-cyclohexanone

Cyanohydrin formation is an equilibrium process in which the cyanide addition step is sterically sensitive. Addition of cyanide to 2,2,6–trimethylcyclohexanone is hindered by the three methyl groups, and the equilibrium lies far to the side of the unreacted ketone.

19.12

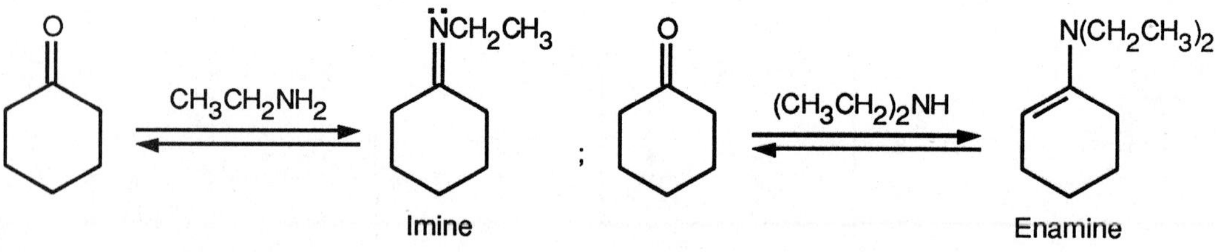

Imine   Enamine

19.13

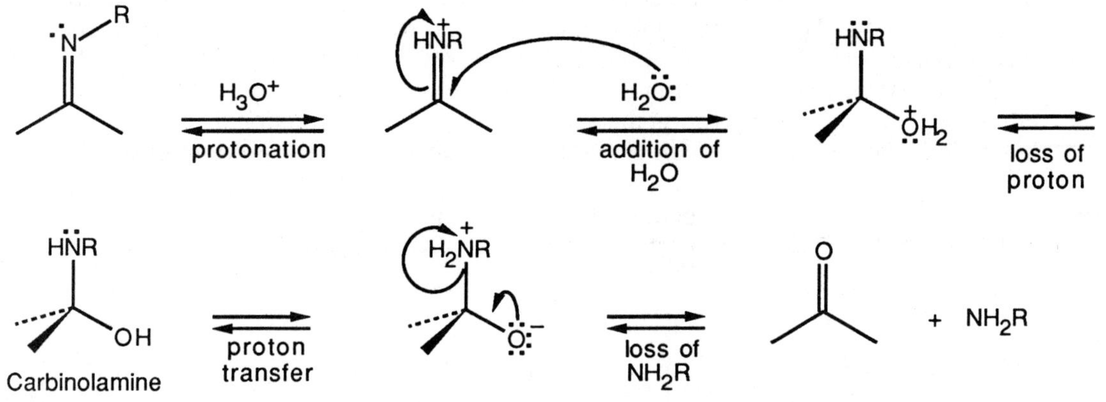

Carbinolamine

19.14

a)

Hemiacetal

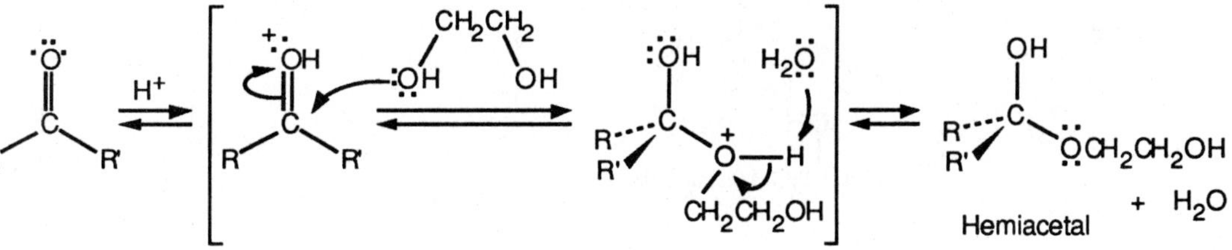

Formation of the hemiacetal is the first step.

b)

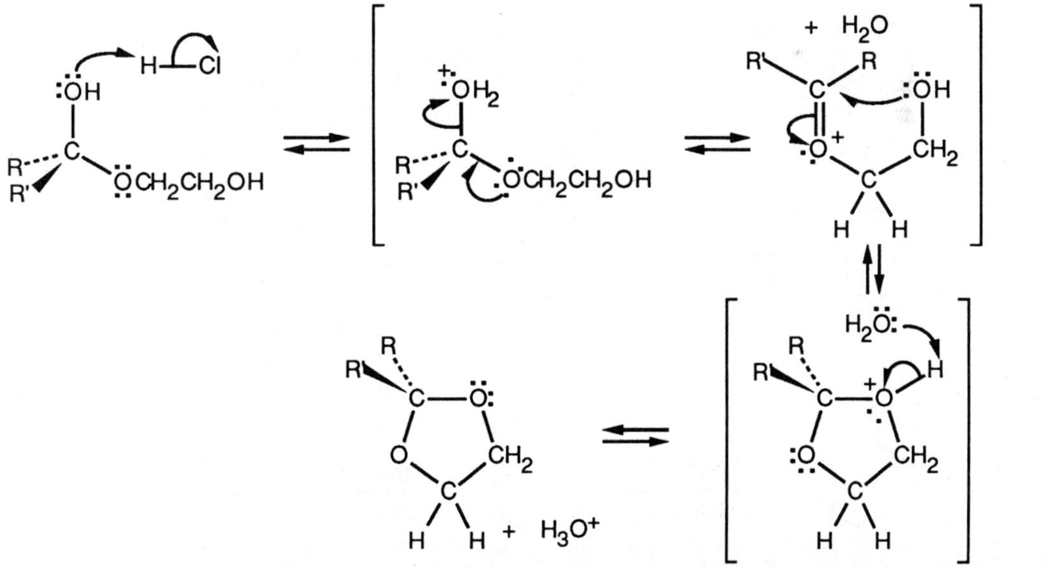

Protonation of the hemiacetal hydroxyl group is followed by loss of water. Attack by the second hydroxyl group of ethylene glycol forms the cyclic acetal ring.

19.15 In general, ketones are less reactive than aldehydes for both steric (excess crowding) and electronic reasons. If the keto-aldehyde in this problem were reduced with *one* equivalent of NaBH$_4$, the aldehyde functional group would be reduced in preference to the ketone.

For the same reason, reaction of the keto-aldehyde with *one* equivalent of ethylene glycol selectively forms the acetal of the aldehyde functional group. The ketone can then be reduced with NaBH$_4$ and the acetal protecting group cleaved.

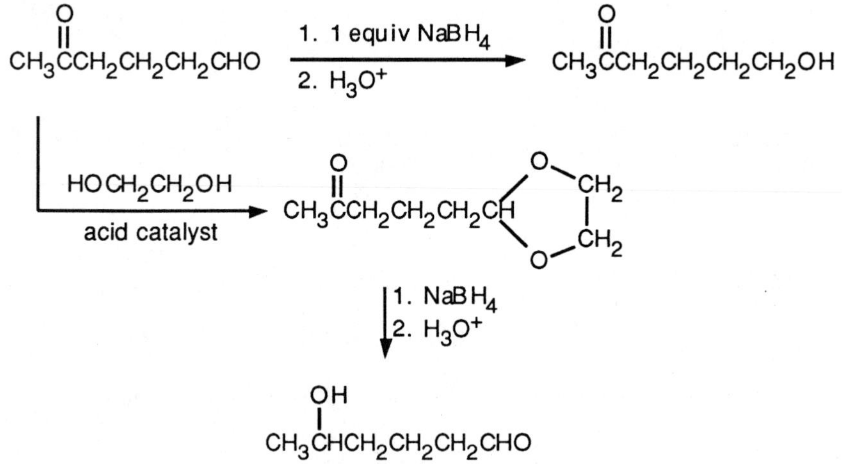

19.16

a)

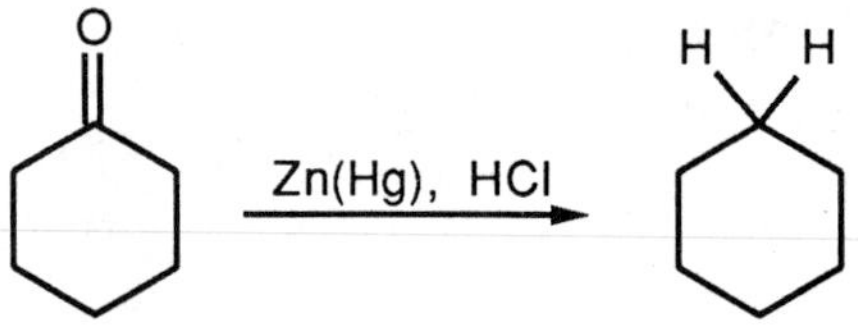

Clemmenson reduction

b)

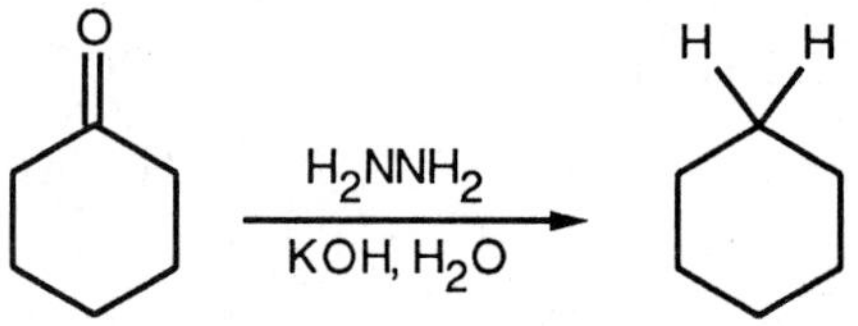

Wolff-Kishner reaction

c)

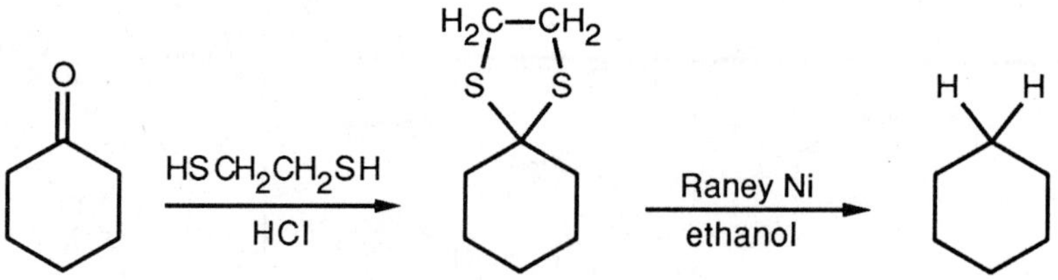

19.17 Because aldehydes are more reactive than ketones, the dithioacetal of the aldehyde is formed preferentially. Treatment with Raney nickel produces the desired ketone.

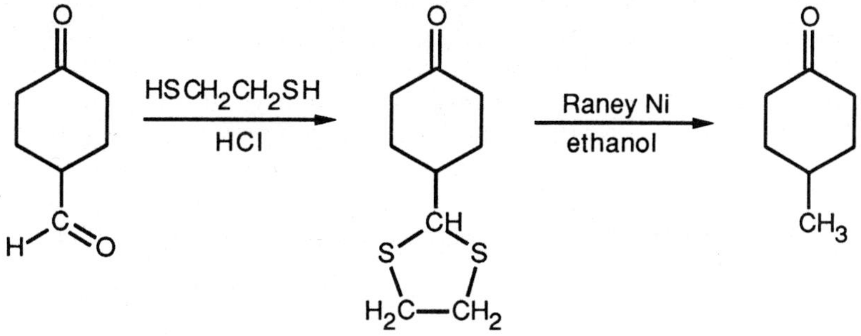

19.18

a)

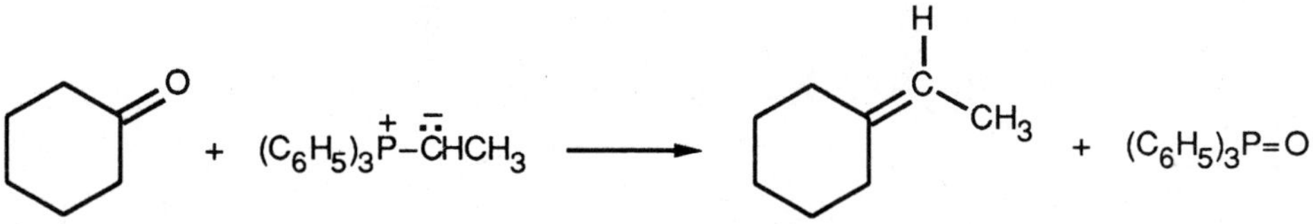

b)

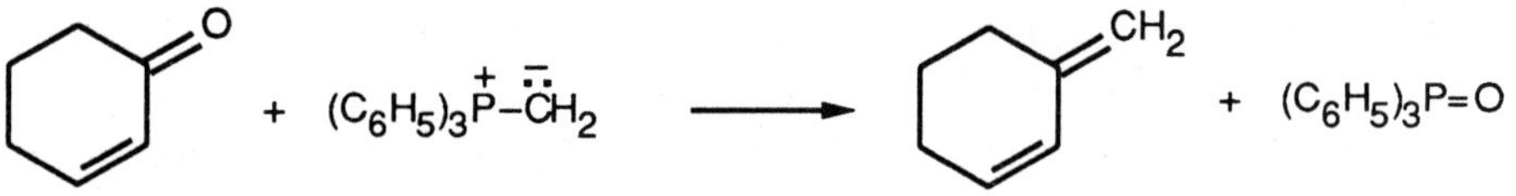

c)

$$CH_3\overset{O}{\overset{\|}{C}}CH_3 \ + \ (C_6H_5)_3\overset{+}{P}-\overset{..}{\overset{-}{C}}HCH_2CH_2CH_3 \longrightarrow (CH_3)_2C=CHCH_2CH_2CH_3 \ + \ (C_6H_5)_3P=O$$

d)

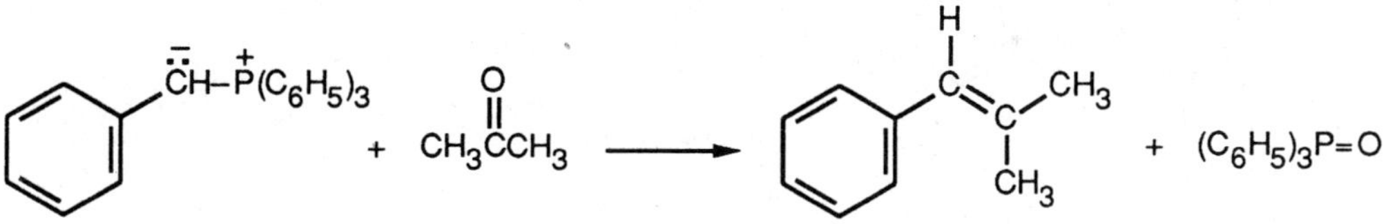

e)

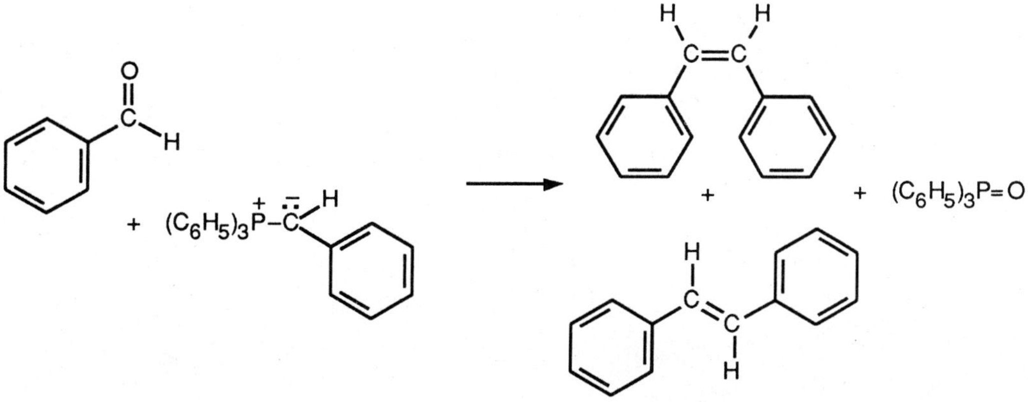

19.19 Suppose that *trimethyl*phosphine were to react with alkyl halide.

$$(CH_3)_3P\colon \; + \; RCH_2CH_2X \longrightarrow (CH_3)_3\overset{+}{P}CH_2CH_2R \; X^-$$

phosphonium salt

Treatment of the phosphonium salt with strong base would yield two different ylides.

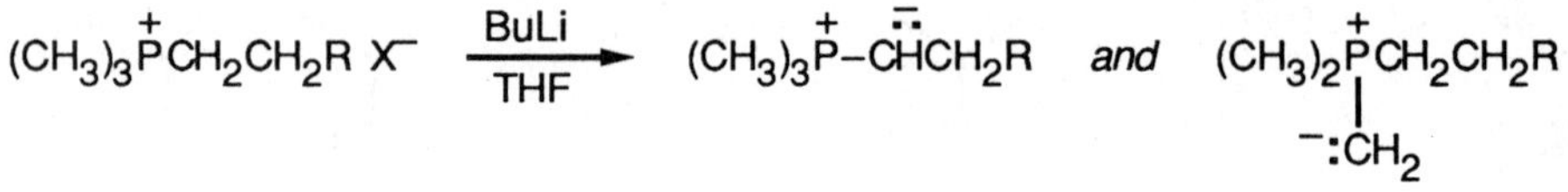

$$(CH_3)_3\overset{+}{P}CH_2CH_2R \; X^- \xrightarrow[\text{THF}]{\text{BuLi}} (CH_3)_3\overset{+}{P}-\overset{..}{\overset{-}{C}}HCH_2R \quad and \quad (CH_3)_2\overset{+}{P}CH_2CH_2R$$

$$^-\colon CH_2$$

Reaction of the ylides with a carbonyl compound would produce two different alkenes.

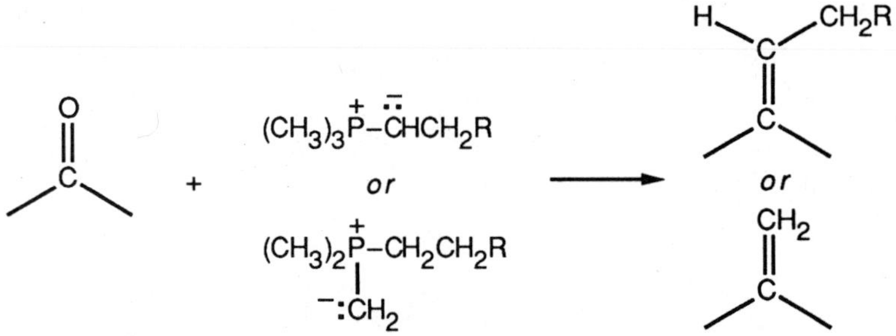

Another function of the phenyl groups of triphenylphosphine is stabilization of the ylide by delocalization of the positive charge on phosphorus.

19.20

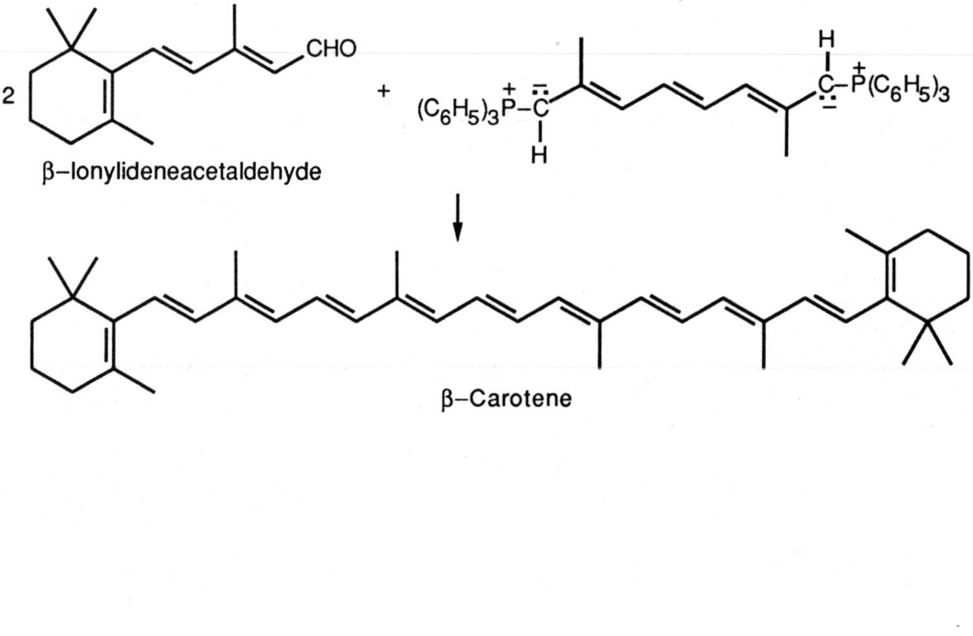

β–Ionylideneacetaldehyde

β–Carotene

19.21

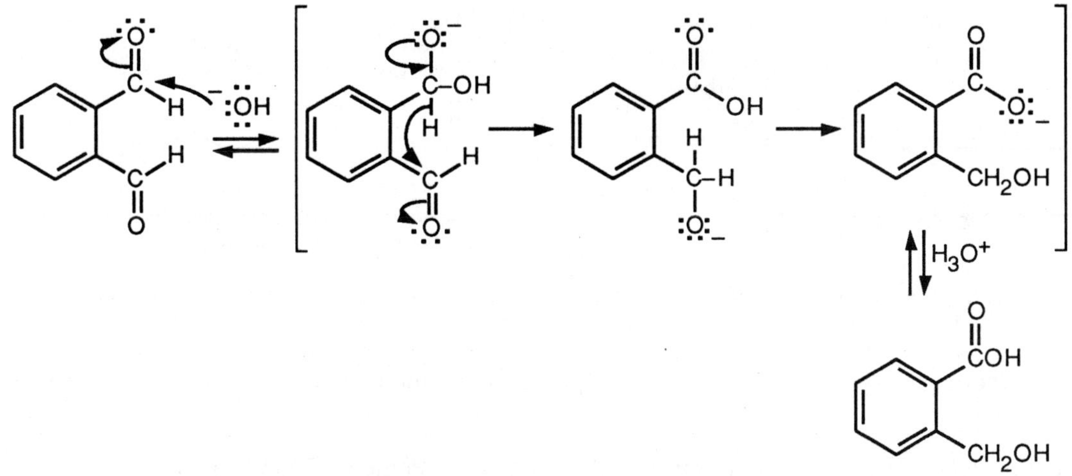
:OH
H
C–OH
OH
C–H
CH2OH
H3O+
COH

This is an internal Cannizzaro reaction.

19.22

a)

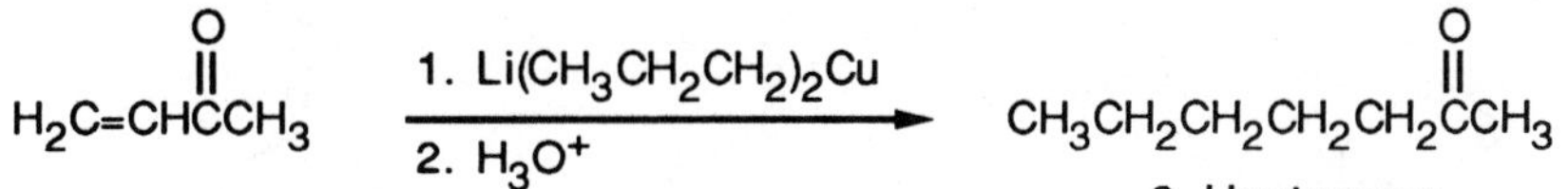
H2C=CHCCH3
1. Li(CH3CH2CH2)2Cu
2. H3O+
CH3CH2CH2CH2CH2CCH3

2–Heptanone

b)

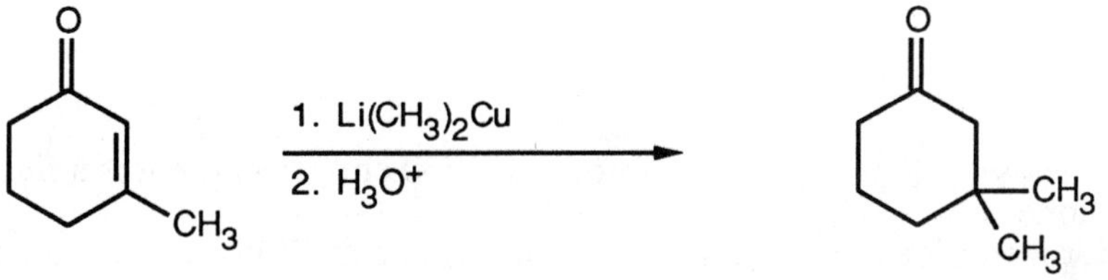
CH3
1. Li(CH3)2Cu
2. H3O+
CH3
CH3

3,3–Dimethylcyclohexanone

c)

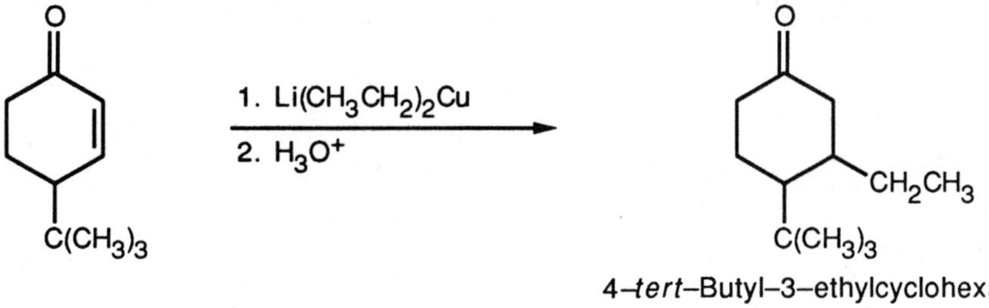
C(CH3)3
1. Li(CH3CH2)2Cu
2. H3O+
CH2CH3
C(CH3)3

4–*tert*–Butyl–3–ethylcyclohexanone

d)

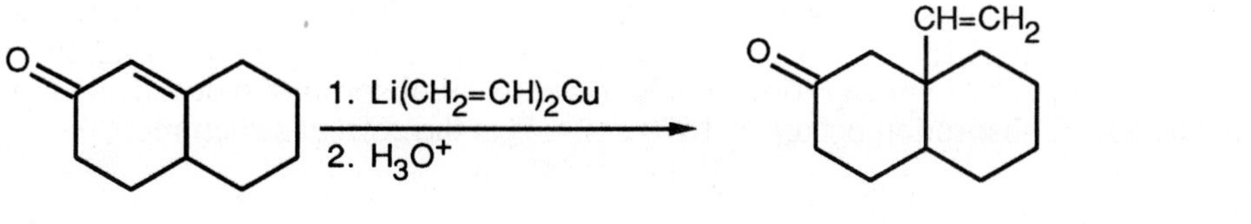
1. Li(CH2=CH)2Cu
2. H3O+
CH=CH2

19.23

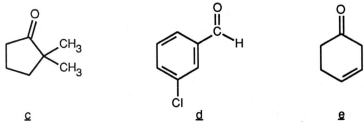

2–Cyclohexenone is a cyclic α,ß–unsaturated ketone whose carbonyl IR absorption occurs at 1685 cm^{-1}. If direct addition product <u>A</u> is formed, the carbonyl absorption will vanish and a hydroxyl absorption will appear at 3300 cm^{-1}. If conjugate addition produces <u>B</u>, the carbonyl absorption will shift to 1715 cm^{-1}, where 6-membered-ring saturated ketones absorb.

19.24 a) $H_2C=CHCH_2COCH_3$ absorbs at 1715 cm^{-1}. (4–Penten–2–one is not an
α,ß–unsaturated ketone.)

b) $CH_3CH=CHC3OCH_3$ is an α,ß–unsaturated ketone and absorbs at 1685 cm^{-1}.

<table>
<tr><td>

O

CH₃

CH₃

<u>c</u>
</td><td>

O
||
C–H

Cl

<u>d</u>
</td><td>

O

<u>e</u>
</td></tr>
</table>

c) 2,2–Dimethylcyclopentanone (<u>c</u>) absorbs at 1750 cm^{-1} because it is a five-membered
ring ketone.

d) *m*–Chlorobenzaldehyde shows a singlet absorption at 1705 cm^{-1} and a doublet at 2720
cm^{-1} and 2820 cm^{-1}.

e) 3–Cyclohexenone (<u>e</u>) absorbs at 1715 cm^{-1}.

f) $CH_3CH_2CH_2CH=CHCHO$ is an α,ß–unsaturated aldehyde that absorbs at 1705 cm^{-1}.

19.25

Compound <u>A</u> is a cyclic, non-conjugated keto-alkene whose carbonyl infrared absorption
should occur at 1715 cm^{-1}. Compound <u>B</u> is an α,ß–unsaturated, cyclic ketone; additional
conjugation with the phenyl ring should lower its IR absorption below 1685 cm^{-1} Because
the actual IR absorption occurs at 1670 cm^{-1}, <u>B</u> is the correct structure.

19.26

a)

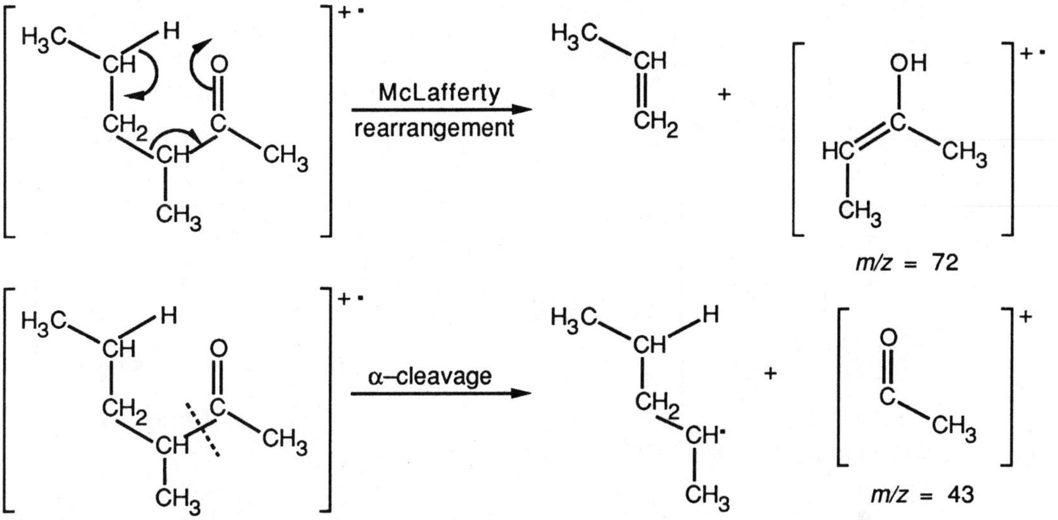

$m/z = 72$

$m/z = 114$
3-Methyl-2-hexanone

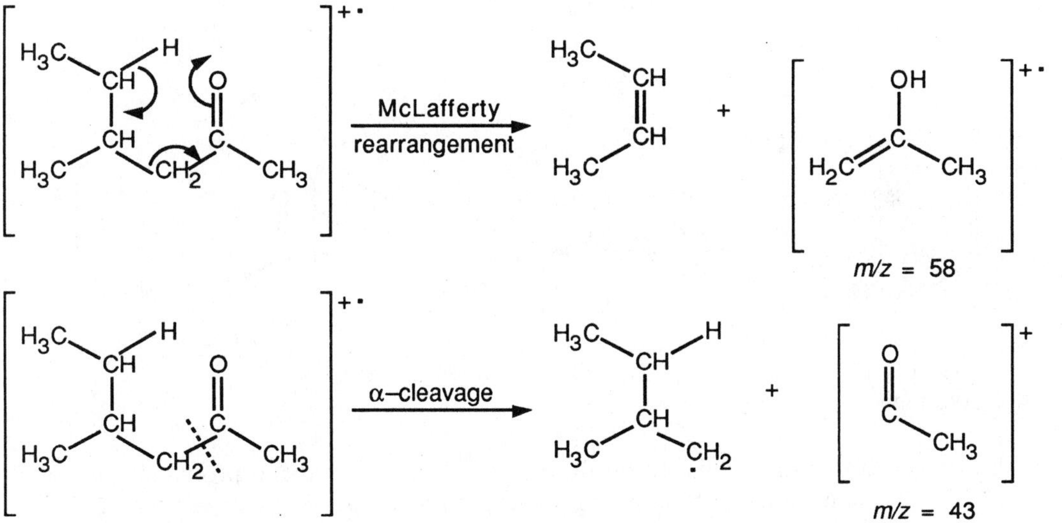

$m/z = 43$

$m/z = 58$

$m/z = 114$
4-Methyl-2-hexanone

$m/z = 43$

Both isomers exhibit peaks at $m/z = 43$ due to α–cleavage. The products of McLafferty rearrangement, however, occur at different values of m/z.

b)

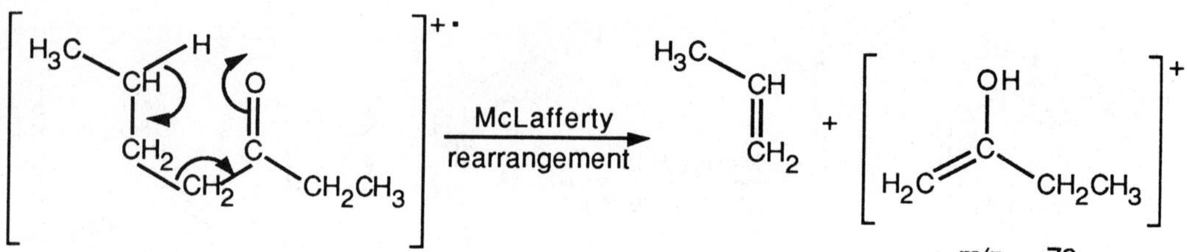

$m/z = 72$

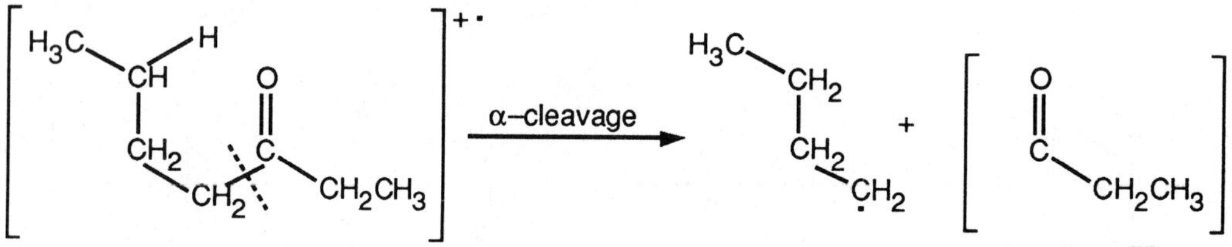

$m/z = 114$
3–Heptanone

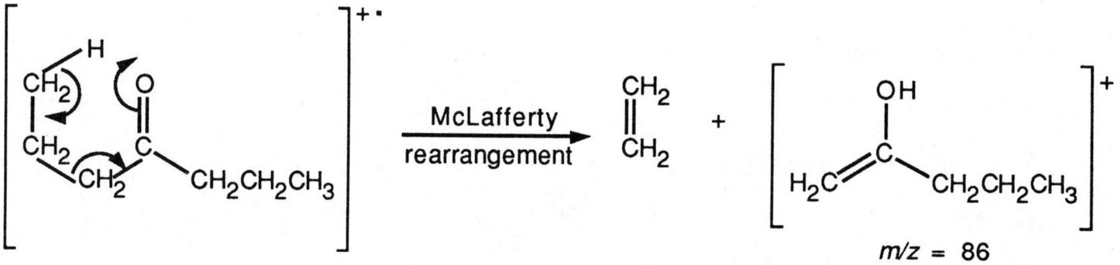

$m/z = 57$

$m/z = 86$

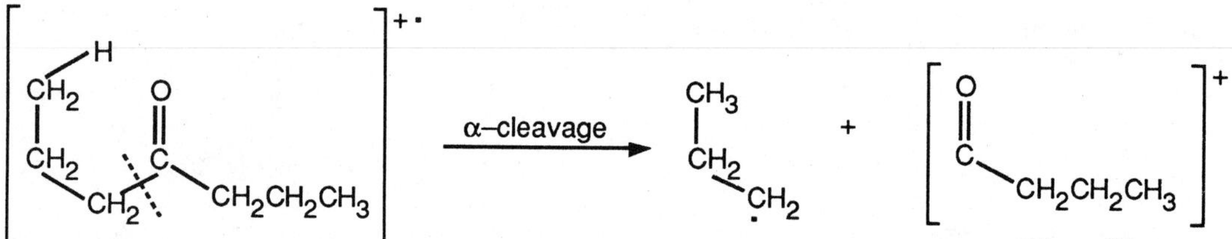

$m/z = 114$
4–Heptanone

$m/z = 71$

The isomers can be distinguished on the basis of both α–cleavage products ($m/z = 57$ vs $m/z = 71$) and McLafferty rearrangement products ($m/z = 72$ vs $m/z = 86$).

c)

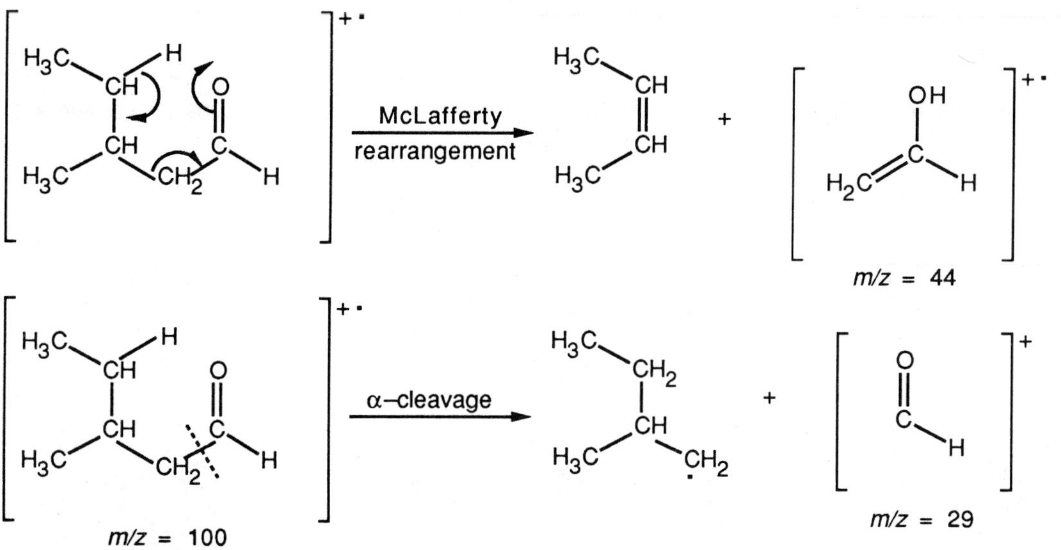

$m/z = 44$

$m/z = 100$
3–Methylpentanal

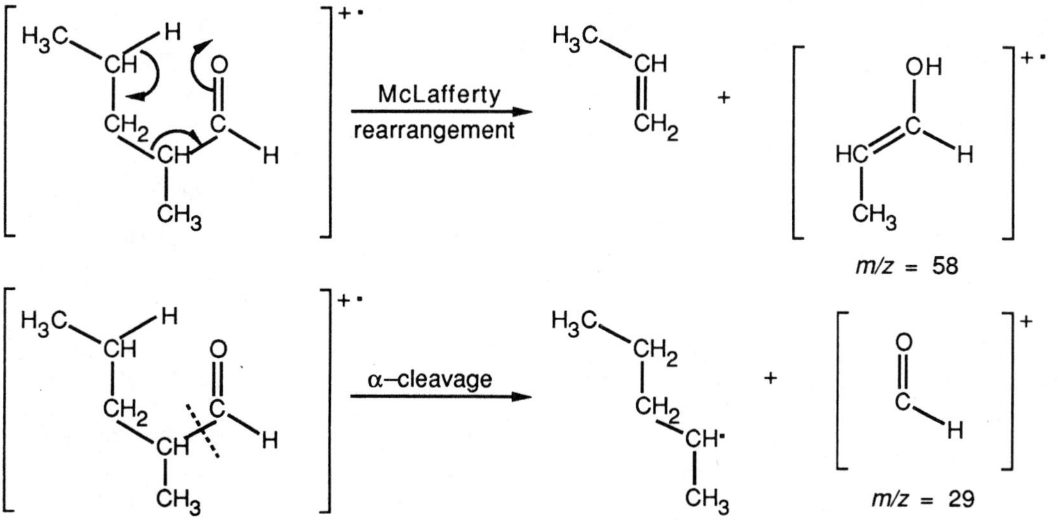

$m/z = 29$

$m/z = 58$

$m/z = 100$
2–Methylpentanal

$m/z = 29$

The fragments from McLafferty rearrangement, which occur at different values of m/z, serve to distinguish the two isomers.

19.27

a) 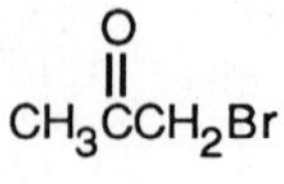

b) 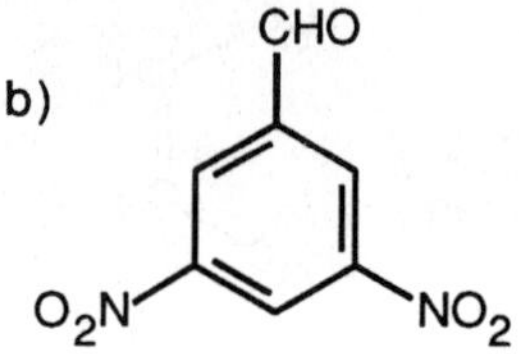

c) 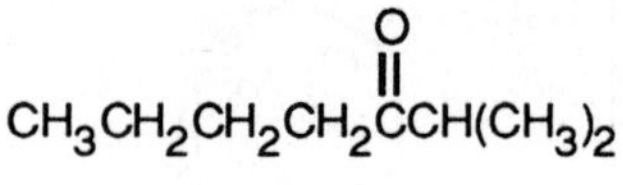

d) 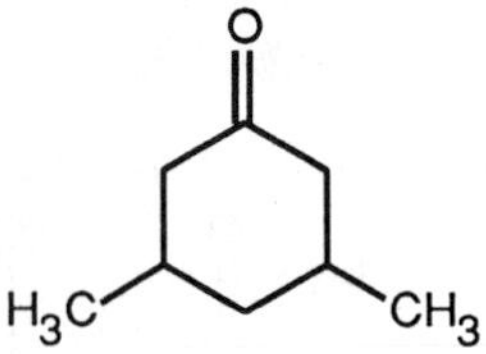

e) 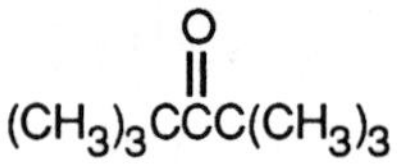

f) 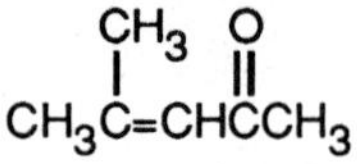

g) 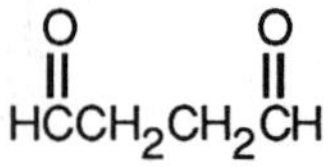

h) 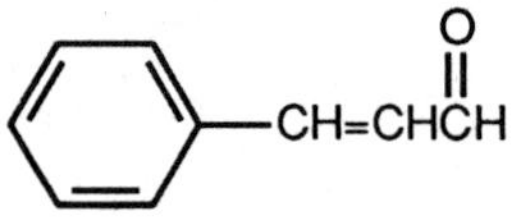

i) 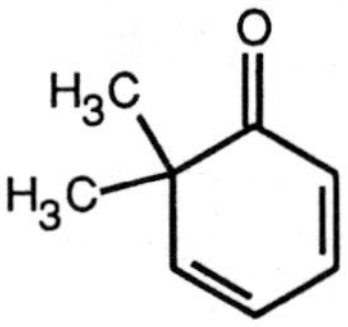

j) 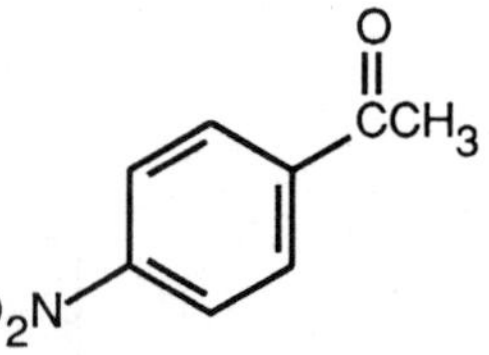

k) 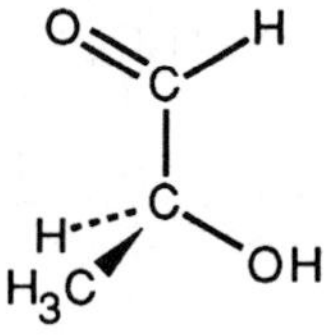

l) 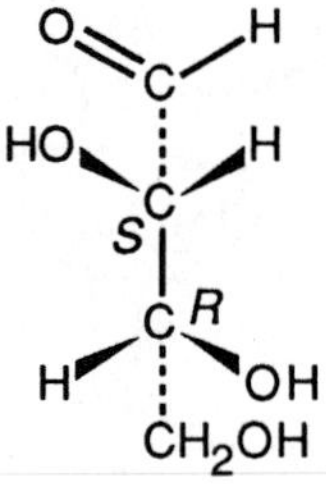

19.28

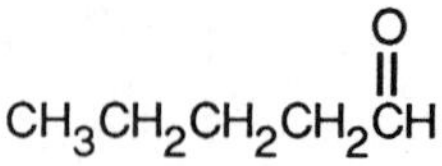

Pentanal

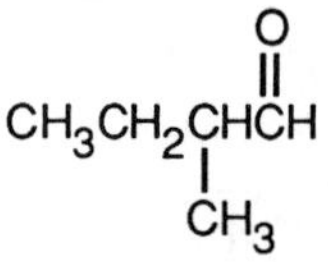

2–Methylbutanal

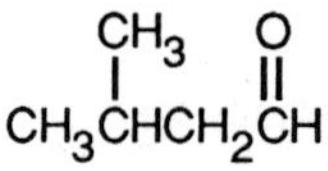

3–Methylbutanal

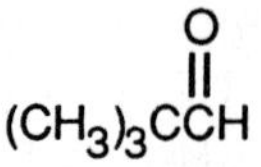

2,2–Dimethylpropanal

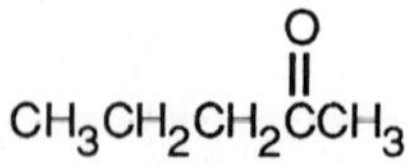

2–Pentanone

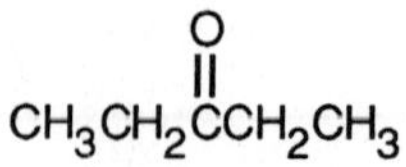

3–Pentanone

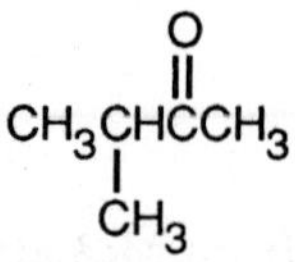

3–Methyl–2–butanone

19.29 a) 3–Methyl–3–cyclohexenone

b) (2*R*)–2, 3–Dihydroxypropanal (D–Glyceraldehyde)

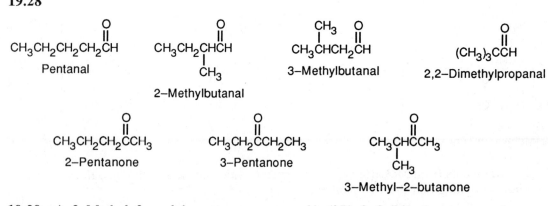

c) 5–Isopropyl–2–methyl–2–cyclohexenone   d) 2–Methyl–3–pentanone

e) 3–Hydroxybutanal   f) p–Benzenedicarbaldehyde

19.30

a)   The α,ß–unsaturated ketone C_6H_8O contains one ring. Possible structures:

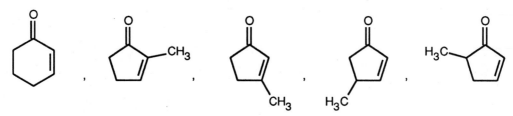

b) $CH_3\overset{O}{\overset{||}{C}}-\overset{O}{\overset{||}{C}}CH_3$ and many other structures.

c)

d)   and many other structures.

19.31

19.32

a) [structure: C₆H₅CH(OH)CH₃]

$$\text{OH}$$
$$\text{CHCH}_3$$

b) no reaction

c) [structure: C₆H₅C(=NOH)CH₃]

$$\overset{\cdot\cdot}{\text{N}}\!-\!\text{OH}$$
$$\text{CCH}_3$$

d) [structure]

$$\text{OH}$$
$$\text{C(CH}_3)_2$$

e) [structure]

$$\text{OCH}_3$$
$$\text{COCH}_3$$
$$\text{CH}_3$$

f) [structure]

$$\text{CH}_2\text{CH}_3$$

g) [structure]

$$\text{CH}_2$$
$$\text{CCH}_3$$

h) [structure]

$$\text{OH}$$
$$\text{CCH}_3$$
$$\text{CN}$$

19.33

a)

$$\xrightarrow[\text{KOH}]{\text{H}_2\text{NNH}_2}$$

b)

$$\xrightarrow[\text{2. H}_3\text{O}^+]{\text{1. Li(C}_6\text{H}_5)_2\text{Cu}}$$

$$\text{C}_6\text{H}_5$$

c)

$$\xrightarrow[\text{2. H}_3\text{O}^+]{\text{1. (C}_2\text{H}_5)_2\text{AlCN}}$$

$$\text{CN}$$

$$\xrightarrow{\text{H}_3\text{O}^+}$$

$$\text{COOH}$$

d)

$$\xrightarrow[\text{2. H}_3\text{O}^+]{\text{1. Li(CH}_3)_2\text{Cu}}$$

$$\text{CH}_3$$

$$\xrightarrow[\text{KOH}]{\text{H}_2\text{NNH}_2}$$

$$\text{CH}_3$$

or

$$\xrightarrow{\text{(C}_6\text{H}_5)_3\overset{+}{\text{P}}\!-\!\overset{\cdot\cdot}{\text{C}}\text{H}_2}$$

$$\text{CH}_2$$

$$\xrightarrow[\text{Pd/C}]{\text{H}_2}$$

$$\text{CH}_3$$

19.34 Glucose exists mainly as the cyclic hemiacetal, formed by the addition of the hydroxyl group of carbon atom 5 to the aldehyde group at carbon 1. A small amount of the open-chain aldehyde is in equilibrium with the cyclic hemiacetal. The open-chain aldehyde gives a positive Tollens' test, like any other aldehyde.

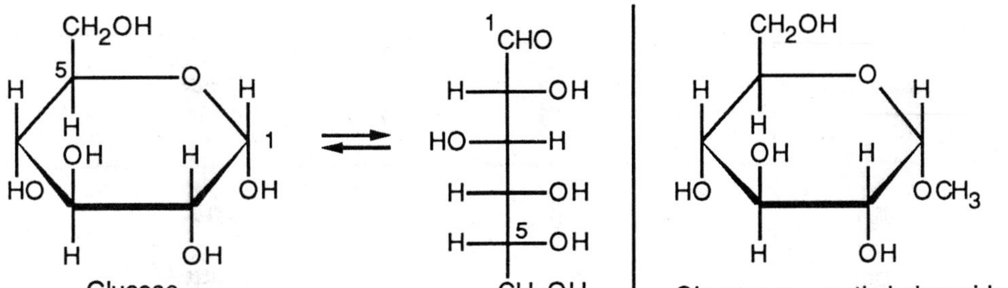

Glucose   Glucose α –methyl glycoside

Glucose α–methyl glycoside, an acetal, is not in equilibrium with an open-chain aldehyde and does not react with Tollens reagent.

19.35

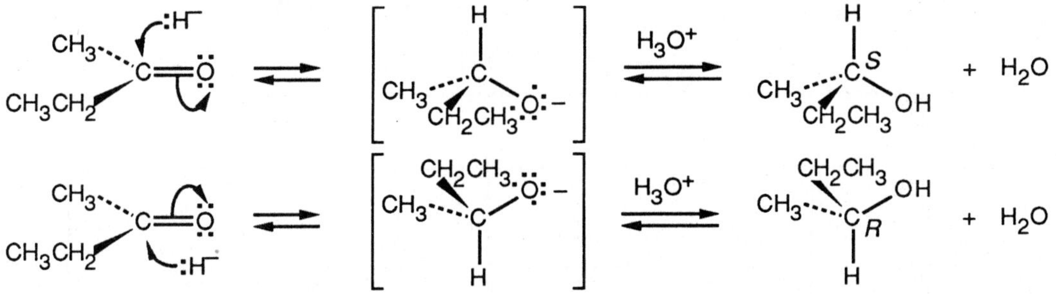

Attack can occur from either side of the planar carbonyl group to yield a racemic product mixture.

19.36 Remember:

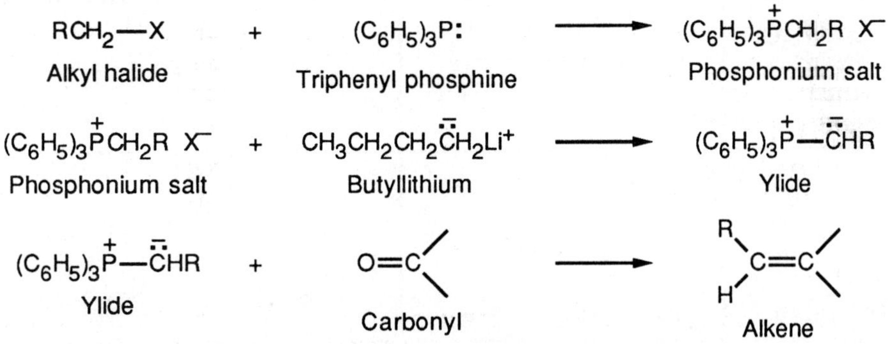

RCH$_2$—X   +   (C$_6$H$_5$)$_3$P: ⟶   (C$_6$H$_5$)$_3$$\overset{+}{P}CH_2$R X$^-$
Alkyl halide   Triphenyl phosphine   Phosphonium salt

(C$_6$H$_5$)$_3$$\overset{+}{P}CH_2$R X$^-$ +   CH$_3$CH$_2$CH$_2$$\overset{..}{C}H_2Li^+$ ⟶   (C$_6$H$_5$)$_3$$\overset{+}{P}$—$\overset{..}{C}$HR
Phosphonium salt   Butyllithium   Ylide

(C$_6$H$_5$)$_3$$\overset{+}{P}$—$\overset{..}{C}$HR   +   O=C$\diagup^{\diagdown}$ ⟶   alkene
Ylide   Carbonyl   Alkene

	Alkyl halide	Carbonyl	Product
a	C₆H₅CH₂Br	$H\overset{O}{\overset{\|}{C}}CH=CH$—C₆H₅	C₆H₅—CH=CHCH=CH—C₆H₅
b	C₆H₅CH₂Br	cyclohexanone	benzylidenecyclohexane
c	CH₃Br	2-cyclohexenone	3-methylenecyclohexene
d	CH₃Br	1-cyclohexene-1-carbaldehyde	1-vinylcyclohexene

19.37

	Carbonyl Compound	Grignard Reagent	Product
a	$CH_3\overset{O}{\overset{\|}{C}}H$	CH₃CH₂CH₂MgBr	$CH_3CH_2CH_2\overset{OH}{\overset{\|}{C}}HCH_3$
	$CH_3CH_2CH_2\overset{O}{\overset{\|}{C}}H$	CH₃MgBr	
b	CH₂O	CH₃CH₂CH₂MgBr	CH₃CH₂CH₂CH₂OH
c	cyclohexanone	C₆H₅MgBr	1-phenylcyclohexanol
d	$C_6H_5\overset{O}{\overset{\|}{C}}H$	C₆H₅MgBr	diphenylmethanol (C₆H₅)₂CHOH

19.38

a) $(C_6H_5)_3P: + Br-CH_2OCH_3 \longrightarrow (C_6H_5)_3\overset{+}{P}CH_2OCH_3 \; Br^-$

$\downarrow$ BuLi

$(C_6H_5)_3\overset{+}{P}-\overset{..}{C}HOCH_3 \; + \; LiBr$

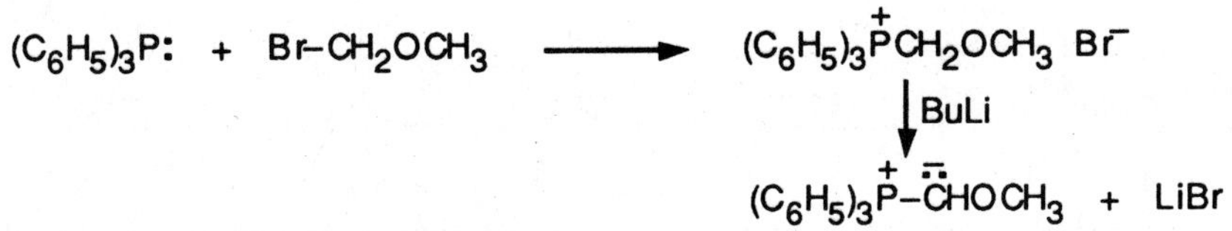

b)

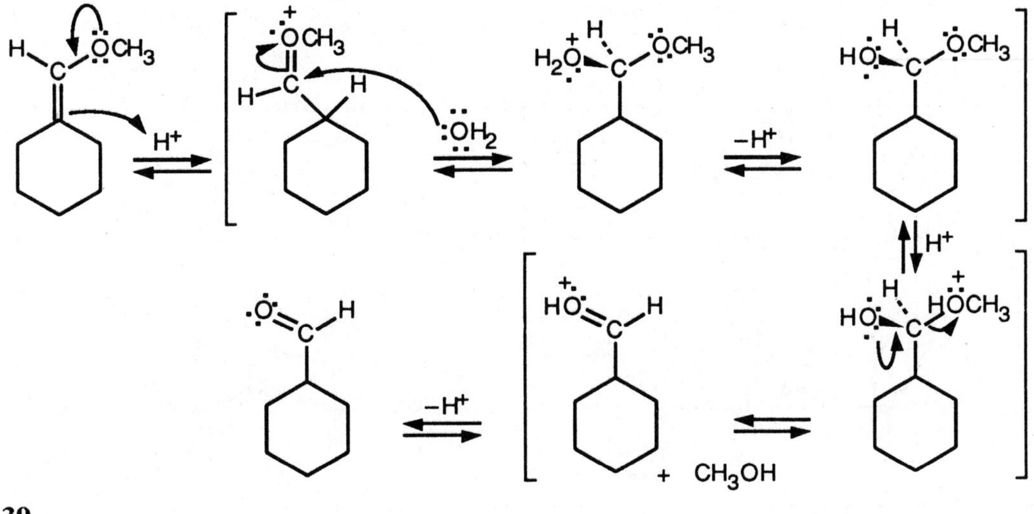

19.39

a)

Hemiacetal

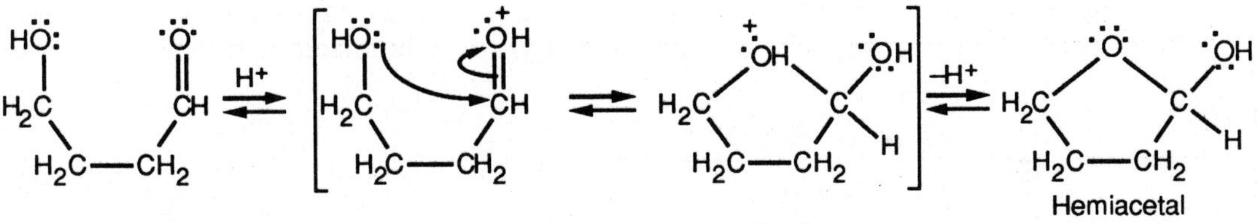

b)

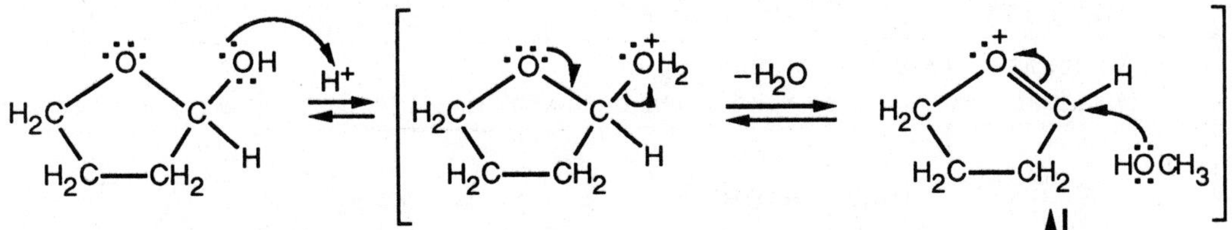

c)

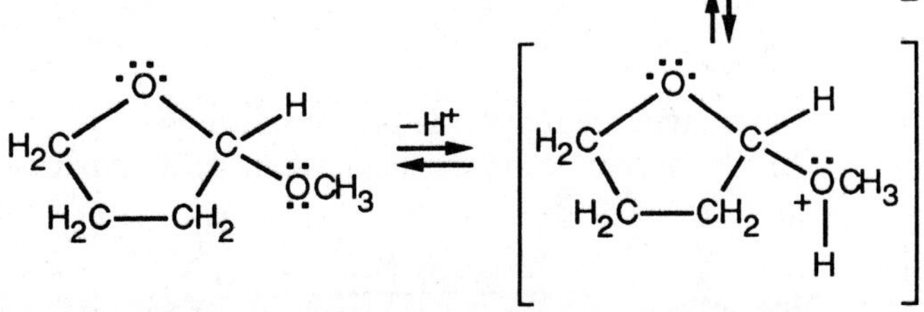

2–Methoxytetrahydropyran is a cyclic acetal. The hydroxyl oxygen of
4–hydroxybutanal reacts with the aldehyde to form the cyclic ether linkage.

19.40

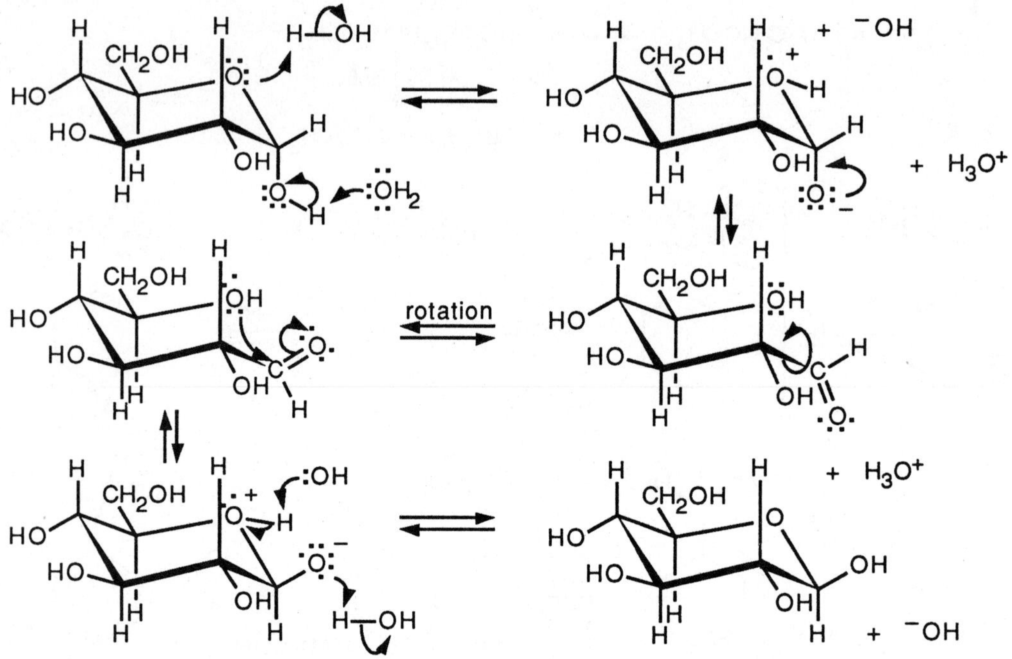

In this series of equilibrium steps, the hemiacetal ring of α–glucose opens to yield the free aldehyde. Rotation of the aldehyde group is followed by formation of the cyclic hemiacetal of ß–glucose. The reaction is catalyzed by both acid and base.

19.41

a)

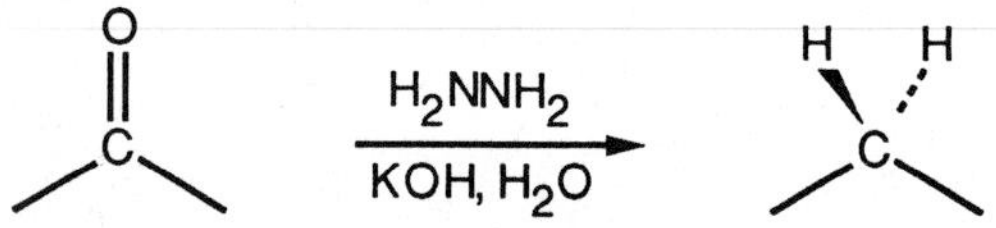

Advantage: reduction is one-step
Disadvantage: can't be used when base-sensitive functional groups are present

b)

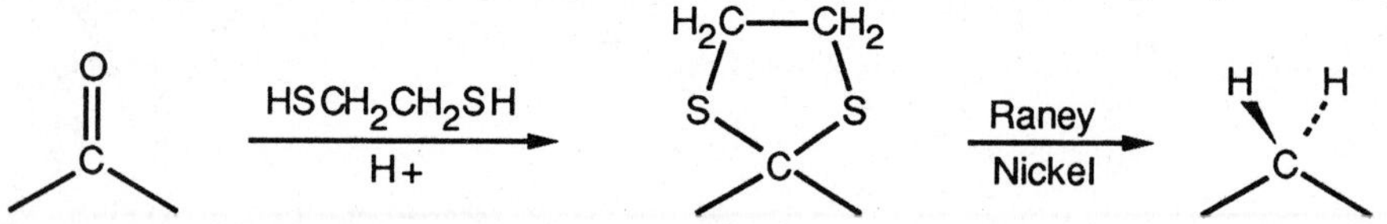

Advantage: reduction is only two steps
Disadvantage: can't be used when acid-sensitive functional groups are present

c)

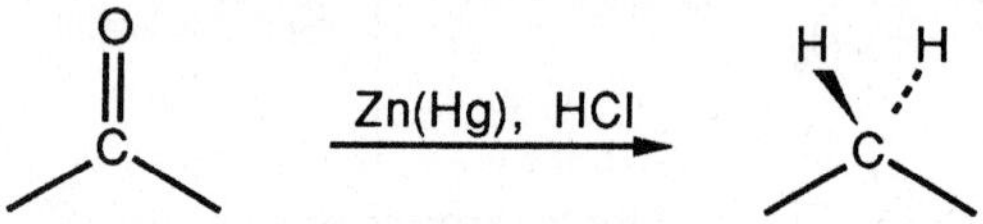

Advantage: reduction is one step
Disadvantage: can't be used when acid-sensitive functional groups are present

d)

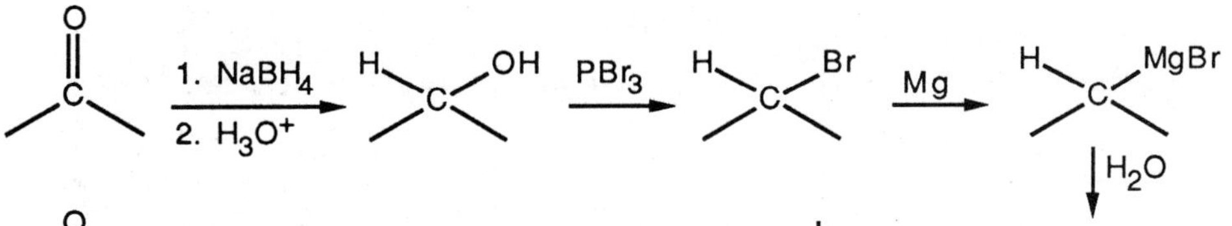

e)

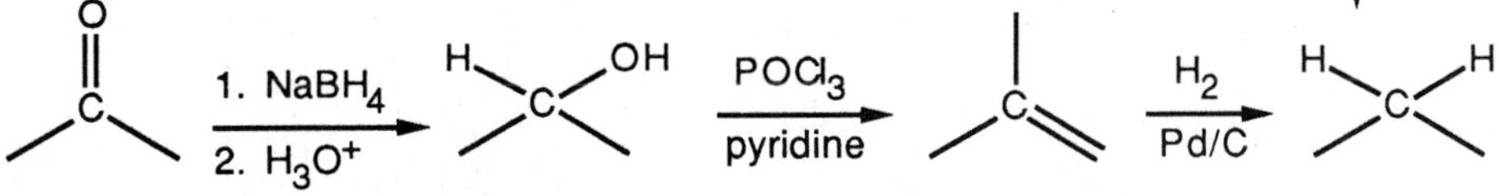

Disadvantage: these two methods require several steps

19.42

a)

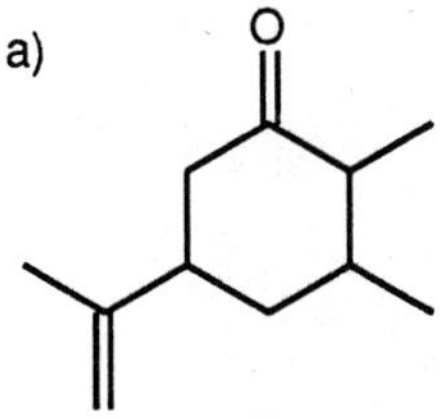

b)

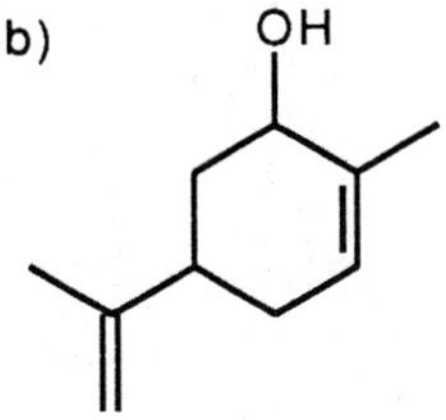

c)

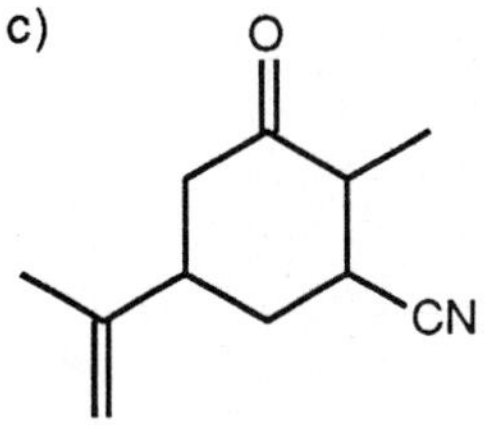

d)

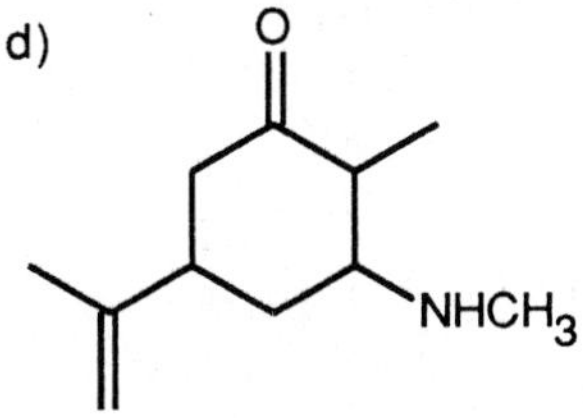

e)

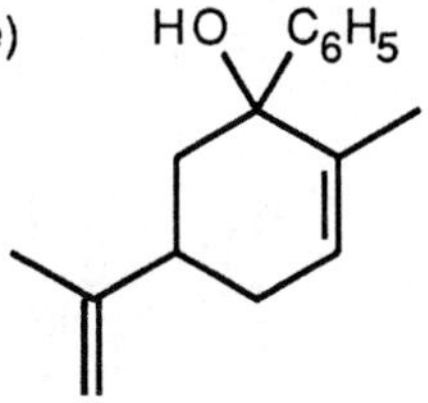

f)

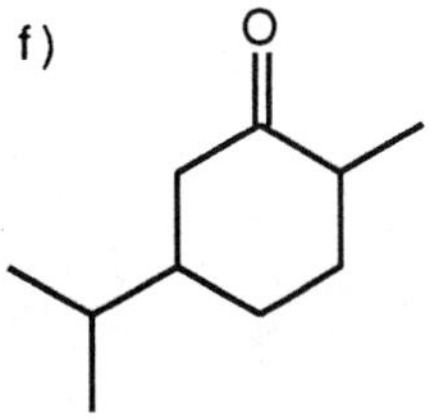

g) no reaction

h)

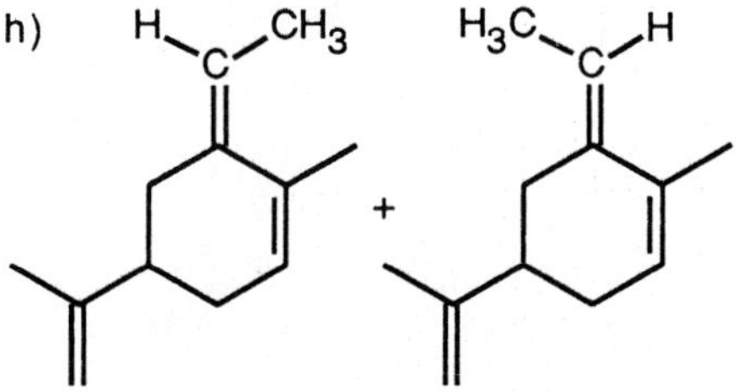

i)

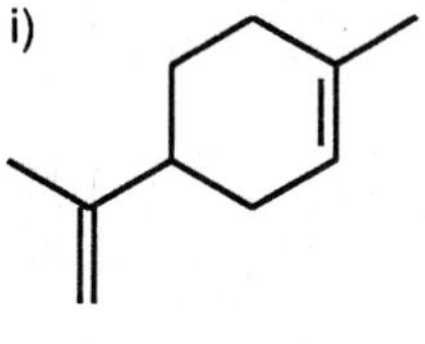

j)

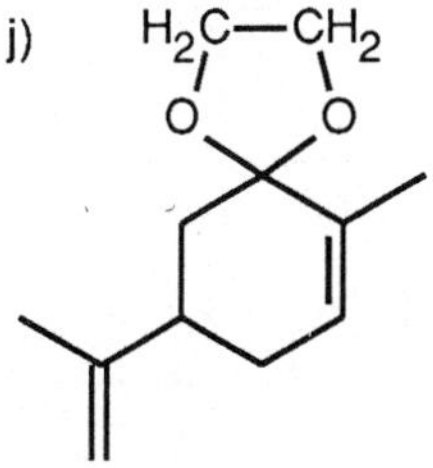

19.43

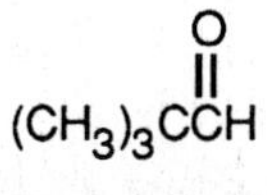

$(CH_3)_3CCH$

O
‖

2,2–Dimethylpropanal
Compound A

19.44

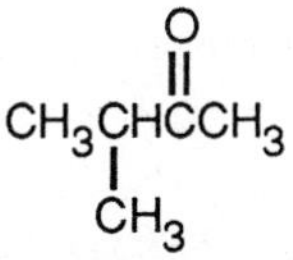

CH_3CHCCH_3

O
‖

CH_3

3–Methyl–2–butanone
Compound B

19.45

a)

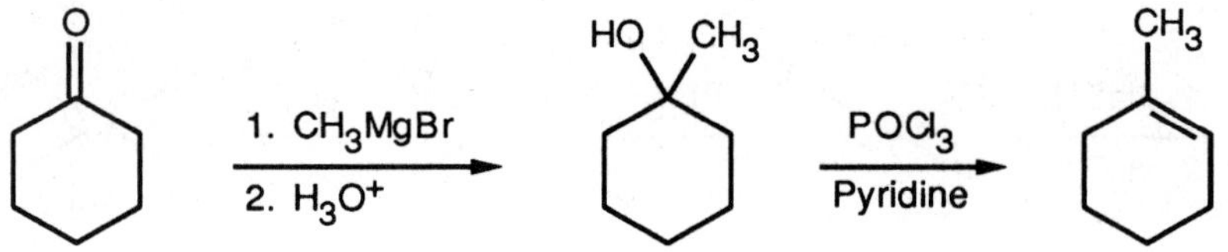

1-Methylcyclohexene

b)

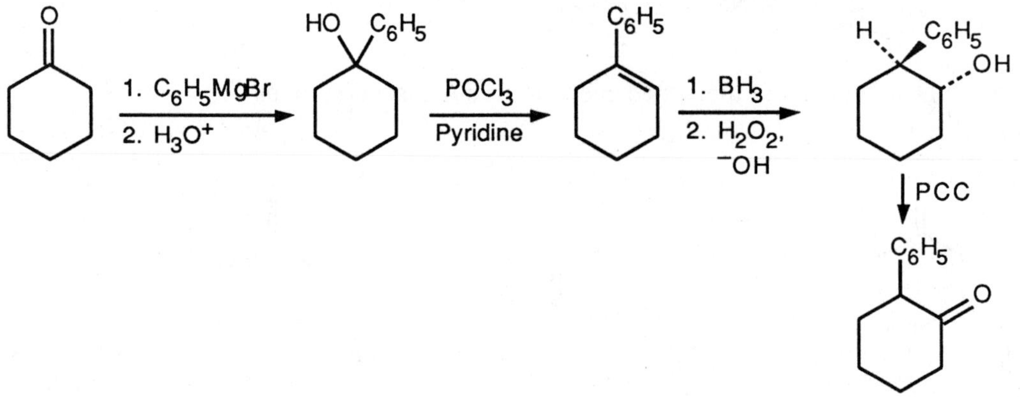

2-Phenylcyclohexanone

c)

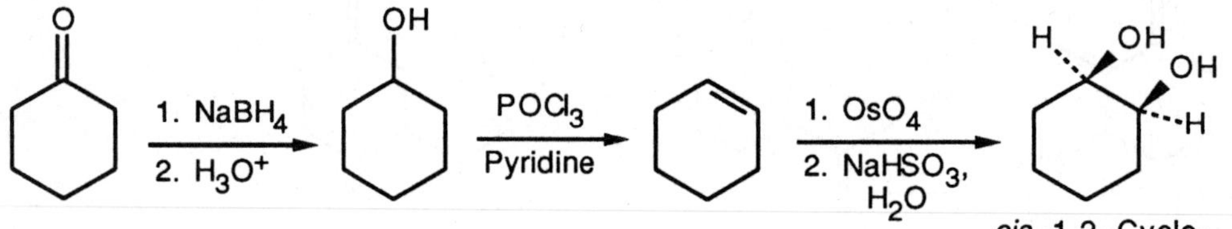

cis-1,2-Cyclo-
hexanediol

d)

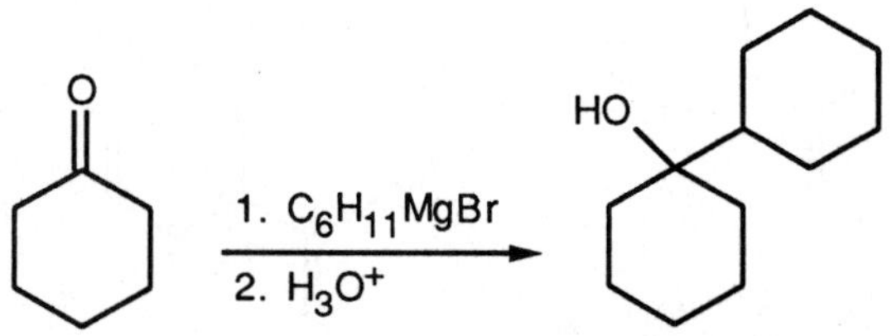

1-Cyclohexylcyclohexanol

19.46

Absorption:	Due to:
a) 1750 cm^{-1}	5-membered ring ketone
1685 cm^{-1}	α,β-unsaturated ketone

Absorption:	Due to:
b) 1710 cm^{-1}	5–membered ring *and* α,β–unsaturated ketone
c) 1750 cm^{-1}	5–membered ring ketone
d) 1705 cm^{-1}, 2720 cm^{-1}, 2820 cm^{-1} 1715 cm^{-1}	aromatic aldehyde aliphatic ketone

Compounds in parts b-d also show aromatic ring IR absorptions in the range 1450 cm^{-1} – 1600 cm^{-1} and in the range 690 – 900 cm^{-1}.

19.47 a) Basic silver ion does not oxidize secondary alcohols to ketones. Grignard addition to a conjugated ketone yields the 1,2 product, not the 1,4 product. The correct scheme:

b) The Jones oxidation converts primary alcohols to carboxylic acids, not to aldehydes. The correct scheme:

$$C_6H_5CH=CHCH_2OH \xrightarrow{\text{PCC}} C_6H_5CH=CHCHO$$

$$\downarrow \text{H}^+, \text{CH}_3\text{OH}$$

$$C_6H_5CH=CHCH(OCH_3)_2$$

c) Treatment of a cyanohydrin with H_3O^+ produces a carboxylic acid, not an amine. The correct scheme:

d) Raney Nickel is used on thioacetals, not hydrazones, to convert a carbonyl carbon to a saturated carbon. The correct scheme:

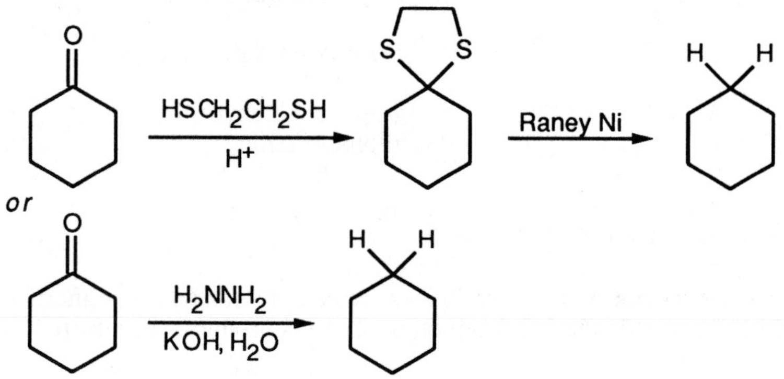

19.48

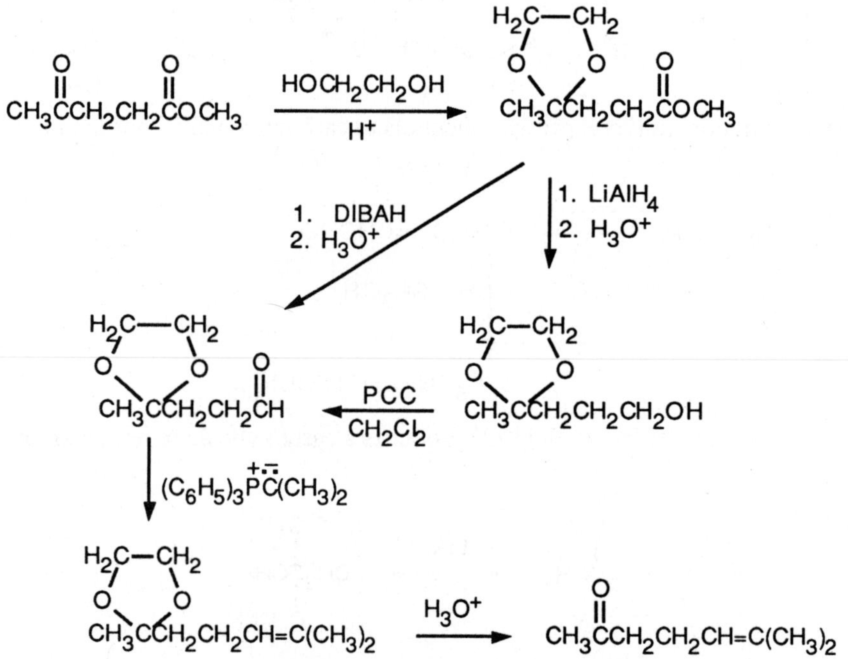

6–Methyl–5–hepten–2–one

19.49 Even though the product looks unusual, this reaction is made up of steps with which you are familiar.

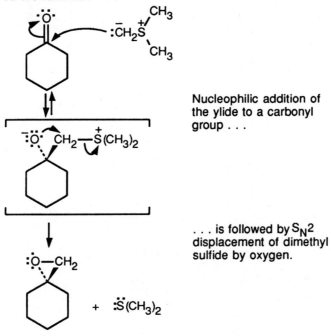

Nucleophilic addition of the ylide to a carbonyl group . . .

. . . is followed by S_N2 displacement of dimethyl sulfide by oxygen.

19.50

The above steps are the reverse of Problem 19.5.

This step is a nucleophilic addition of cyanide.

19.51

19.52

19.53

a) $(CH_3)_2CHCCH_2CH_3$ (with C=O on the central C)   b) $(CH_3)_2CHCH_2CH$ (with C=O on the terminal C)   c) 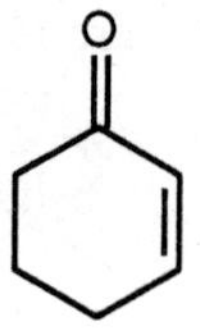

19.54

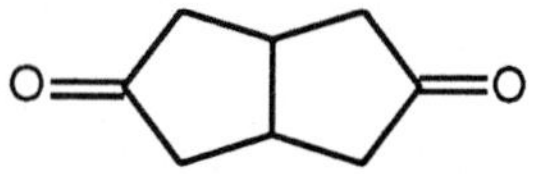

19.55 1) Aluminum, a Lewis Acid, complexes with the carbonyl oxygen.

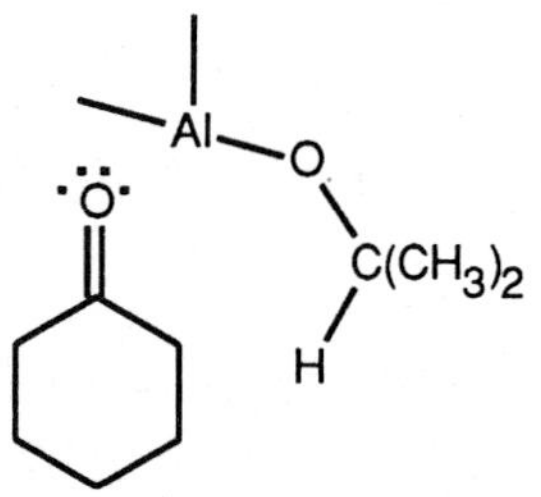

2) Complexation with aluminum makes the carbonyl functional group electrophilic and facilitates hydride transfer from isopropoxide.

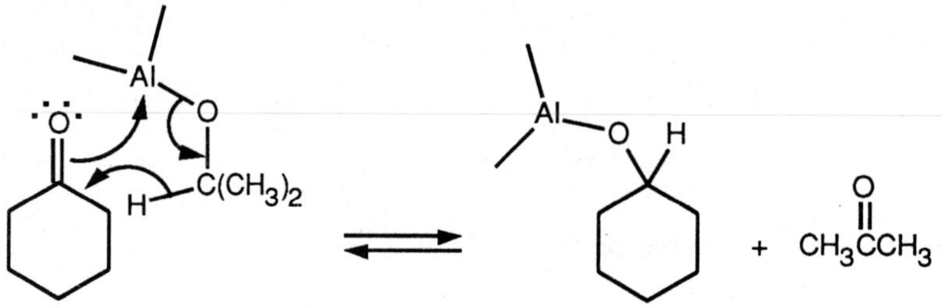

3) Treatment of the reaction mixture with aqueous acid cleaves the aluminum-oxygen complex and produces cyclohexanol

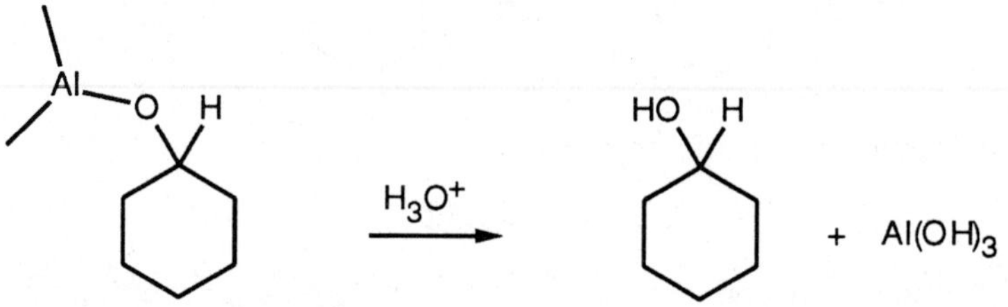

Both the MPV reaction and the Cannizzaro reaction are hydride transfers in which a carbonyl group is reduced by a second oxygen-containing functional group, which is oxidized.

19.56

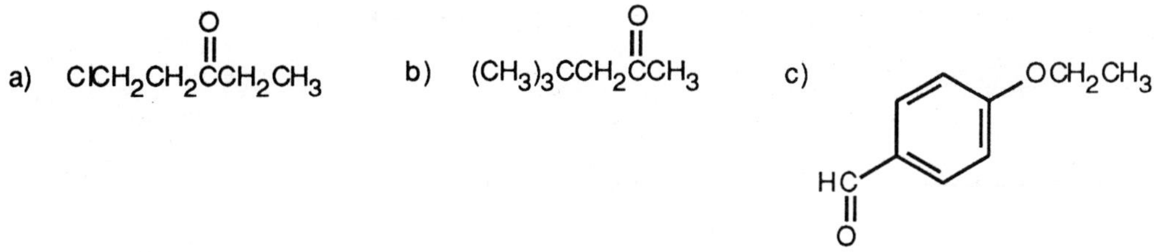

a) $ClCH_2CH_2\overset{\displaystyle O}{\overset{\|}{C}}CH_2CH_3$

b) $(CH_3)_3CCH_2\overset{\displaystyle O}{\overset{\|}{C}}CH_3$

c)

19.57

a)

b) $(CH_3O)_2CHCH_2\overset{\displaystyle O}{\overset{\|}{C}}CH_3$

c)

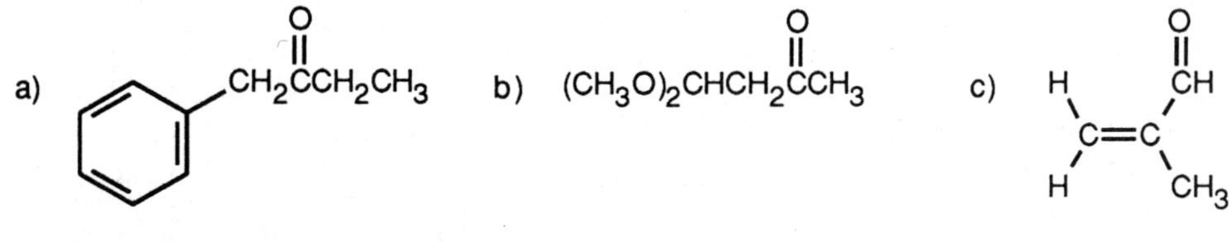

19.58 a) Nucleophilic addition of hydrazine to one of the carbonyl groups, followed by elimination of water, produces a hydrazone.

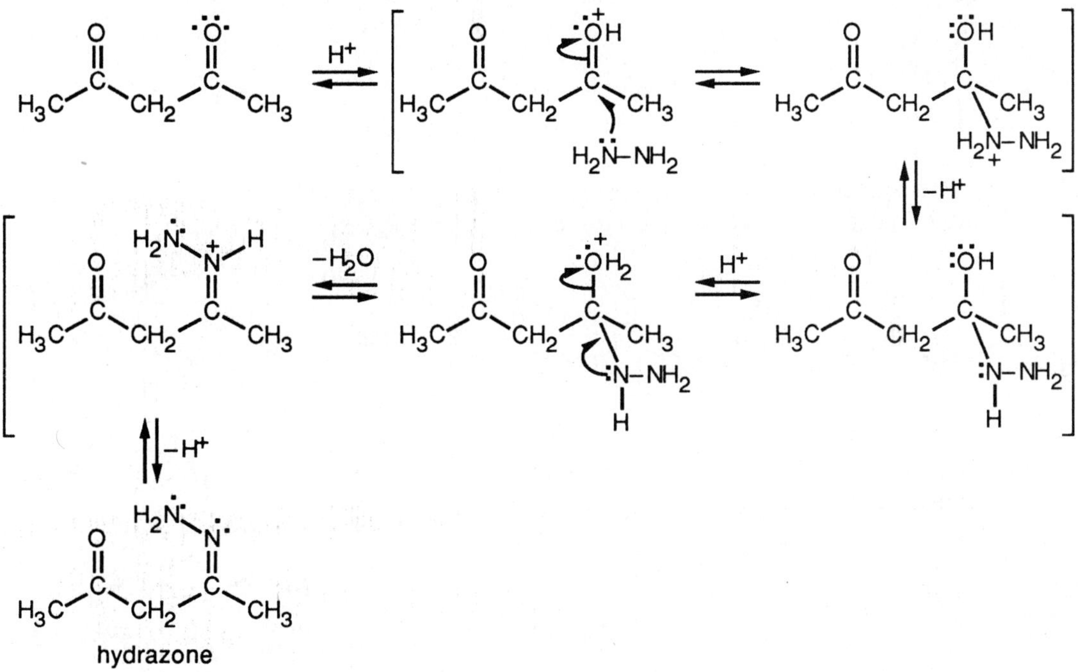

hydrazone

b) In a similar manner, the *other* nitrogen of hydrazine can add to the *other* carbonyl oxygen of 2,4–pentanedione to form the pyrazole.

19.59 The same sequence of steps used in the previous problem leads to the formation of 3,5–dimethylisoxazole when hydroxylamine is the reagent.

a)

b)

19.60

Study Guide for Chapter 19

After studying this chapter, you should be able to:

(1)   Name and draw aldehydes and ketones (19.1, 19.2, 19.27, 19.28, 19.29, 19.30).

(2)   Prepare aldehydes and ketones (19.3, 19.4).

(3)   Explain the difference in reactivity between aldehydes and ketones, and explain substituent effects on their reactivity (19.6, 19.7, 19.15).

(4)   Formulate mechanisms for nucleophilic addition reactions of aldehydes and ketones (19.5, 19.8, 19.9, 19.10, 19.11, 19.35, 19.38, 19.39, 19.40, 19.49, 19.50, 19.55, 19.58, 19.59, 19.60).

(5)   Predict the products of reactions of aldehydes and ketones with:

(a) Amines (19.2, 19.13).

(b) Alcohols (19.14, 19.34).

(c) Thiols (19.16, 19.17).

(d) Phosphorus ylides (19.18, 19.19, 19.20, 19.36).

(e) Strong base (19.21).

(f) Organocopper reagents (19.22).

(g) Grignard reagents (19.37).

(h) A combination of reagents (19.31, 19.32, 19.33, 19.41, 19.42).

(6)   Synthesize aldehydes and ketones (19.45, 19.47, 19.48).

(7)   Use spectroscopic information to determine the structure of aldehydes and ketones (19.23, 19.24, 19.25, 19.26, 19.43, 19.44, 19.46, 19.51, 19.52, 19.53, 19.54, 19.56, 19.57).

20.1

a) $(CH_3)_2CHCH_2COOH$

    3–Methylbutanoic acid

b) $CH_3CHBrCH_2CH_2COOH$

    4–Bromopentanoic acid

c) $CH_3CH=CHCH=CHCOOH$

    2,4–Hexadienoic acid

d) 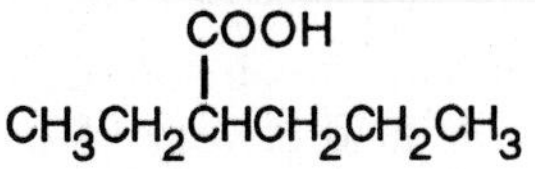
$$CH_3CH_2\overset{\overset{\displaystyle COOH}{|}}{C}HCH_2CH_2CH_3$$

    2–Ethylpentanoic acid

e)

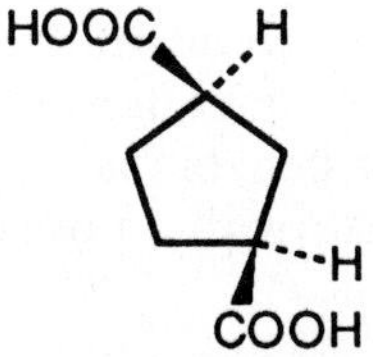

    cis–1,3–Cyclopentane-
    dicarboxylic acid

f)

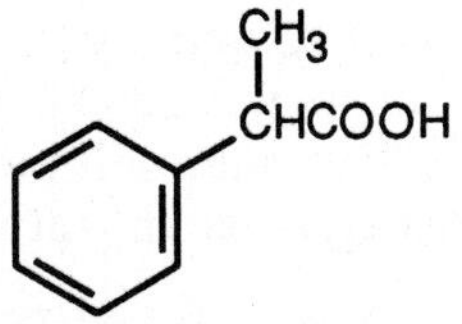

    2–Phenylpropanoic acid

20.2

a)

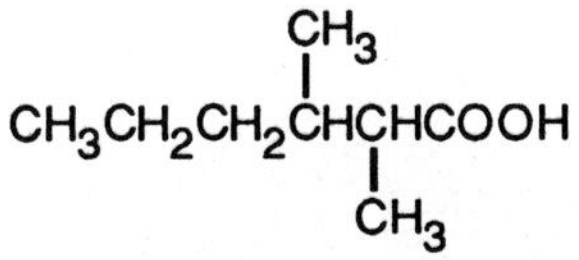

    2,3–Dimethylhexanoic acid

b) $(CH_3)_2CHCH_2CH_2COOH$

    4–Methylpentanoic acid

c)

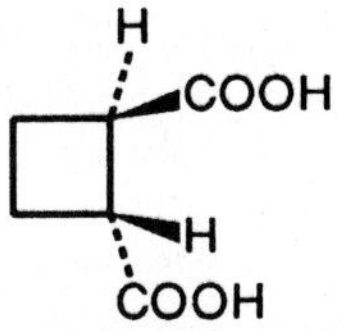

    trans–1,2–Cyclobutane-
    dicarboxylic acid

d)

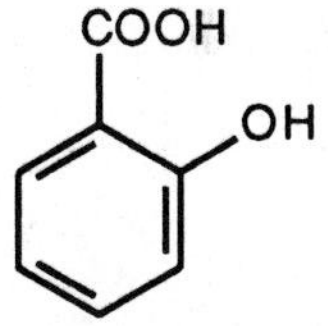

    o–Hydroxybenzoic acid

e)

    $(9Z, 12Z)$–Octadecadienoic acid

20.3 $\Delta G° = -RT \ln K_a = -2.303 \, RT \log K_a$. Here $R = 1.98$ cal/mol $\cdot$ K; $T = 300$ K

$\Delta G° = -2.303 \times 1.98$ cal/mol $\cdot$ K $\times 300$ K $\times \log K_a$

     $= -1.37 \times 10^3$ cal/mol $\times \log K_a$

     $= -1.37$ kcal/mol $\times \log K_a$

For Ethanol: $pK_a = 16$; $\log K_a = -16$
$$\Delta G^\circ = -1.37 \text{ kcal/mol} \times (-16) = +22 \text{ kcal/mol}$$

For Acetic acid: $pK_a = 4.72$; $\log K_a = -4.72$
$$\Delta G^\circ = -1.37 \text{ kcal/mol} \times (-4.72) = +6.47 \text{ kcal/mol}$$

Dissociation of acetic acid is favored. Since ΔG° for acetic acid is a smaller number, less energy is required for dissociation of acetic acid than for dissociation of ethanol.

20.4 Naphthalene is insoluble in water; benzoic acid is only slightly soluble. The *salt* of benzoic acid is very soluble in water, however, and we can take advantage of this solubility in separating naphthalene from benzoic acid.

Dissolve the mixture in an organic solvent, and extract with a dilute aqueous solution of sodium hydroxide or sodium bicarbonate, which will neutralize benzoic acid. Naphthalene will remain in the organic layer, and all benzoic acid, now converted to the benzoate salt, will be in the aqueous layer. To recover benzoic acid, remove the aqueous layer, acidify it with dilute mineral acid, and extract with an organic solvent.

20.5

$$Cl_2CHCOH + H_2O \underset{}{\overset{K_a}{\rightleftharpoons}} Cl_2CHCO^- + H_3O^+$$

$$K_a = \frac{[Cl_2CHCO_2^-][H_3O^+]}{[Cl_2CHCOOH]} = 5.5 \times 10^{-2}$$

	Initial molarity	Molarity after dissociation
Cl_2CHCOH	0.10	$0.10 - y$
Cl_2CHCO^-	0	y
H_3O^+	0	y

$$K_a = \frac{y \cdot y}{(0.10 - y)} = 5.5 \times 10^{-2}$$

Using the quadratic formula to solve for y, we find that $y = 0.052$.

$$\text{Percent dissociation} = \frac{0.052}{0.100} \times 100\% = 52\%.$$

20.6

Weaker acid ⟶ Stronger acid

a) CH_3CH_2COOH < $BrCH_2COOH$ < FCH_2COOH

Here, fluoride is the most electronegative group and can stabilize the carboxylate anion the best.

b)

The electron-withdrawing nitro group stabilizes the carboxylate anion. The methoxyl group, which is a resonance electron donor, destabilizes the carboxylate anion.

c) $CH_3CH_2NH_2$ < CH_3CH_2OH < CH_3CH_2COOH

20.7

pK_1 of oxalic acid is lower than that of a monocarboxylic acid because the carboxylate anion is stabilized both by resonance and by the inductive effect of the nearby second carboxylic acid group.

pK_2 of oxalic acid is higher than pK_1 for two reasons. (1) The first carboxylate group inductively destabilizes the negative charge resulting from dissociation of the second proton. (2) Electrostatic repulsion between the two adjacent negative charges destabilizes the dianion.

20.8   Inductive effects of functional groups are transmitted through bonds. As the length of the carbon chain increases, the effect of one functional group on another decreases. In this example, the influence of the second carboxyl group on the ionization of the first is barely felt by succinic and adipic acids.

20.9 A pK_a of 4.45 indicates that p–cyclopropylbenzoic acid is a weaker acid than benzoic acid. This, in turn, indicates that a cyclopropyl group must be electron-donating. Since electron-donating groups increase reactivity in electrophilic substitution reactions, p–cyclopropylbenzene should be more reactive than benzene toward electrophilic bromination.

20.10

Least acidic ──────────────────────────────────▶ Most acidic

a)

b)

20.11 Both of these methods are simple two-step transformations, and both methods can be used with compounds containing base-sensitive functional groups. Advantages of the cyanide hydrolysis method: (1) it can be used with acid-sensitive compounds; (2) it produces an optically active carboxylic acid from an optically active halide; (3) it can be used with compounds containing functional groups that interfere with Grignard reagent formation. The advantage of the Grignard method is that it can be used with tertiary halides and also with aryl and vinylic halides, which do not undergo S_N2 reactions.

20.12

a)

b) $(CH_3)_3CCl$ → $(CH_3)_3CCOOH$

1. Mg, ether
2. CO_2, ether
3. H_3O^+

c) $CH_3CH_2CH_2Br$ → $CH_3CH_2CH_2COOH$

1. Mg, ether
2. CO_2, ether *or* 1. NaCN
3. H_3O^+ 2. H_3O^+

20.13 a) 2,5–Dimethylhexanedioic acid   b) 2,2–Dimethylpropanoic acid
c) 3–Propylhexanoic acid   d) p–Nitrobenzoic acid
e) 1–Cyclodecenecarboxylic acid   f) 4,5–Dibromopentanoic acid

20.14

a)

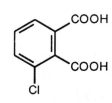

b)   HOOCCH$_2$CH$_2$CH$_2$CH$_2$CH$_2$COOH

c)   CH$_3$C≡CCH=CHCOOH

d)

CH$_2$CH$_3$
|
CH$_3$CH$_2$CH$_2$CH$_2$CHCH$_2$CHCOOH
|
CH$_2$CH$_2$CH$_3$

e)

COOH

COOH

Cl

f)   (C$_6$H$_5$)$_3$CCOOH

20.15 Acetic acid molecules are strongly associated because of hydrogen bonding. Molecules of the ethyl ester are much more weakly associated, and less heat is required to overcome the attractive forces between molecules of the ethyl ester. Even though the ethyl ester has a greater molecular weight, it boils at a lower temperature than the acid.

20.16

CH$_3$CH$_2$CH$_2$CH$_2$CH$_2$COOH
Hexanoic acid

CH$_3$
|
CH$_3$CH$_2$CH$_2$CHCOOH
2–Methylpentanoic acid

CH$_3$
|
CH$_3$CH$_2$CHCH$_2$COOH
3–Methylpentanoic acid

CH$_3$
|
CH$_3$CHCH$_2$CH$_2$COOH
4–Methylpentanoic acid

COOH
|
CH$_3$CH$_2$CHCH$_2$CH$_3$
2–Ethylbutanoic acid

CH$_3$
|
CH$_3$CH$_2$CCOOH
|
CH$_3$
2,2–Dimethylbutanoic acid

CH$_3$
|
CH$_3$CHCHCOOH
|
CH$_3$
2,3–Dimethylbutanoic acid

CH$_3$
|
CH$_3$CCH$_2$COOH
|
CH$_3$
3,3–Dimethylbutanoic acid

20.17

Least acidic ⟶ Most acidic

a)   CH$_3$COOH < HCOOH < HOOCCOOH
       Acetic acid   Formic acid   Oxalic acid

b)   p–Bromobenzoic acid < p–Nitrobenzoic acid < 2,4-Dinitrobenzoic acid

c) $C_6H_5CH_2CH_2COOH$ < $C_6H_5CH_2COOH$, $(C_6H_5)_2CHCOOH$

d) FCH_2CH_2COOH < ICH_2COOH < FCH_2COOH

20.18 Remember that the conjugate base of a weak acid is a strong base. In other words, the stronger the acid, the weaker the base derived from that acid.

Least basic ────────────────────────▶ Most basic

a)

$$\left[CH_3\overset{O}{\overset{\|}{C}}-\ddot{\underset{\cdot\cdot}{O}}:^- \right]_2 Mg^{2+} \; < \; Mg(OH)_2 \; < \; H_3C:^- \, Mg^+Br$$

b)

$<$ $<$ $HC{\equiv}C:^-\,Na^+$

c)

$$H\overset{O}{\overset{\|}{C}}-\ddot{\underset{\cdot\cdot}{O}}:^-\,Li^+ \; < \; H\ddot{\underset{\cdot\cdot}{O}}:^-\,Li^+ \; < \; CH_3CH_2\ddot{\underset{\cdot\cdot}{O}}:^-\,Li^+$$

20.19 Two factors are responsible for the difference in pK_2 values between these benzenedicarboxylic acids. (1) The mono-anion of phthalic acid is stabilized by hydrogen bonding of the remaining –COOH proton with the adjacent carboxylate group. This type of hydrogen bonding, which is not possible for terephthalic acid, stabilizes the phthalate monoanion. (2) The dianion resulting from the dissociation of

the second proton of phthalic acid has two negative charges close to one another. The resulting electrostatic repulsion destabilizes the phthalate dianion. The energy difference between mono- and di- anion is greater for phthalic acid, and pK_2 of phthalic acid is greater than pK_2 of terephthalic acid.

20.20

a) $CH_3CH_2CH_2\overset{O}{\overset{\|}{C}}OH$ $\xrightarrow[\text{2. } H_3O^+]{\text{1. } BH_3}$ $CH_3CH_2CH_2CH_2OH$

1–Butanol

b) $CH_3CH_2CH_2CH_2OH$ $\xrightarrow{PBr_3}$ $CH_3CH_2CH_2CH_2Br$

from a   1–Bromobutane

c) $CH_3CH_2CH_2CH_2Br$ $\xrightarrow{NaCN}$ $CH_3CH_2CH_2CH_2C{\equiv}N$

from b

$\downarrow H_3O^+$

$CH_3CH_2CH_2CH_2COOH$

Pentanoic acid

d) $CH_3CH_2CH_2CH_2Br$ $\xrightarrow{(CH_3)_3CO^-K^+}$ $CH_3CH_2CH{=}CH_2$

from b   1–Butene

e) $CH_3CH_2CH_2\overset{\overset{\displaystyle O}{\|}}{C}OH$ $\xrightarrow[CCl_4]{HgO,\ Br_2}$ $CH_3CH_2CH_2Br$

1–Bromopropane

f) $2\,CH_3CH_2CH_2CH_2Br$ $\xrightarrow{2Li}$ $2\,CH_3CH_2CH_2\ddot{C}H_2{}^- Li^+$

from b

$\downarrow CuI$

$(CH_3CH_2CH_2CH_2)_2Cu^- Li^+$

$(CH_3CH_2CH_2CH_2)_2CuLi$ + $CH_3CH_2CH_2CH_2Br$

$\downarrow$

$CH_3CH_2CH_2CH_2CH_2CH_2CH_2CH_3$

Octane

20.21

a) $CH_3CH_2CH_2CH_2OH$ $\xrightarrow[\text{reagent}]{\text{Jones}}$ $CH_3CH_2CH_2COOH$

b) $CH_3CH_2CH_2CH_2Br$ $\xrightarrow{NaOH}$ $CH_3CH_2CH_2CH_2OH$ $\xrightarrow[\text{reagent}]{\text{Jones}}$ $CH_3CH_2CH_2COOH$

c) $CH_3CH_2CH{=}CH_2$ $\xrightarrow[\text{2. } H_2O_2,\ ^-OH]{\text{1. } BH_3}$ $CH_3CH_2CH_2CH_2OH$ $\xrightarrow[\text{reagent}]{\text{Jones}}$ $CH_3CH_2CH_2COOH$

d) $CH_3CH_2CH_2Br$

1. NaCN
2. $^-$OH, H_2O
3. H_3O^+

$CH_3CH_2CH_2COOH$

or

1. Mg
2. CO_2
3. H_3O^+

e) $CH_3CH_2CH_2CH=CHCH_2CH_2CH_3$ $\xrightarrow[H_2O, K_2CO_3]{KMnO_4}$ $2\ CH_3CH_2CH_2COOH$

f) $CH_3CH_2CH_2CH_2COOH$ $\xrightarrow[CCl_4]{HgO,\ Br_2}$ $CH_3CH_2CH_2CH_2Br$; then proceed as in part b.

20.22

a)

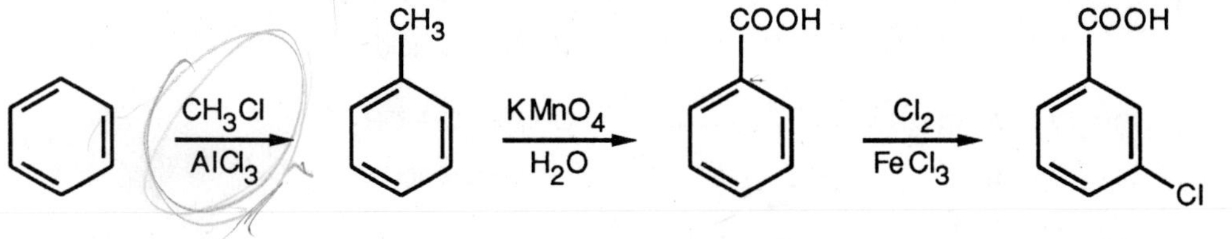

b)

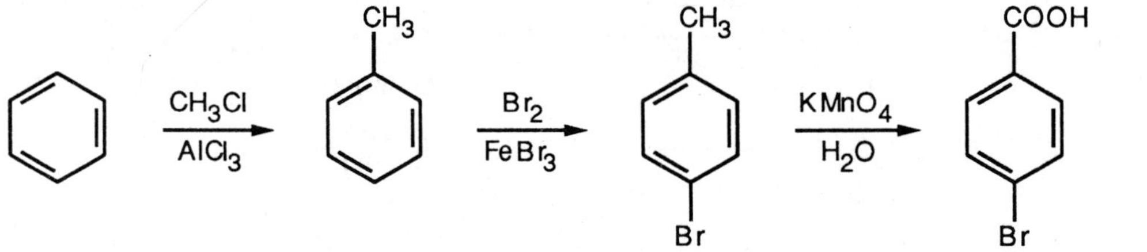

c)

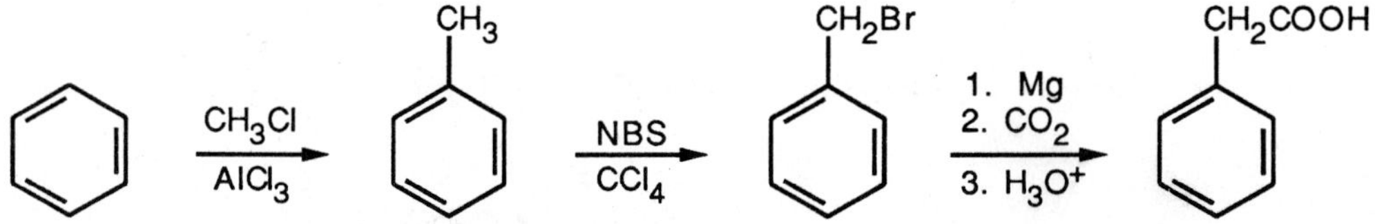

Alternatively, benzyl bromide can be treated with cyanide, and the resulting nitrile can be hydrolyzed.

20.23 a) $K_a = 8.4 \times 10^{-4}$ for lactic acid.
$pK_a = -\log(8.4 \times 10^{-4}) = -(-4 + \log 8.4) = -(-4 + 0.92) = -(-3.08) = 3.08$

b) $K_a = 5.6 \times 10^{-6}$ for acrylic acid.
$pK_a = -\log(5.6 \times 10^{-6}) = -(-6 + \log 5.6) = -(-6 + 0.75) = -(-5.25) = 5.25.$

20.24 a) $pK_a = 3.14$ for citric acid.
$K_a = 10^{-3.14} = 10^{-4} \cdot 10^{0.86} = 7.24 \times 10^{-4}$

b) $pK_a = 2.98$ for tartaric acid.
$K_a = 10^{-2.98} = 10^{-3} \cdot 10^{0.02} = 1.05 \times 10^{-3}.$

20.25

a. 1. BH$_3$
    2. H$_3$O$^+$

→ CH$_3$—⟨benzene ring⟩—CH$_2$OH

b. NBS
    CCl$_4$

→ BrCH$_2$—⟨benzene ring⟩—COOH

c. HgO, I$_2$
    CCl$_4$

→ CH$_3$—⟨benzene ring⟩—I

d. 1. CH$_3$MgBr
    2. H$_3$O$^+$

→ CH$_3$—⟨benzene ring⟩—COOH + CH$_4$

e. KMnO$_4$
    H$_3$O$^+$

→ HOOC—⟨benzene ring⟩—COOH

f. 1. LiAlH$_4$
    2. H$_3$O$^+$

→ CH$_3$—⟨benzene ring⟩—CH$_2$OH

Starting material: CH$_3$—⟨benzene ring⟩—COOH

20.26

a) CH$_3$CH$_2$Br $\xrightarrow{\text{Mg}}$ CH$_3$CH$_2$MgBr $\xrightarrow[\text{2. H}_3\text{O}^+]{\text{1. }^{13}\text{CO}_2}$ CH$_3$CH$_2^{13}$COOH

b) CH$_3$Br $\xrightarrow{\text{Mg}}$ CH$_3$MgBr $\xrightarrow[\text{2. H}_3\text{O}^+]{\text{1. }^{13}\text{CO}_2}$ CH$_3{}^{13}\overset{\displaystyle O}{\overset{\|}{C}}$OH $\xrightarrow[\text{2. H}_3\text{O}^+]{\text{1. BH}_3}$ CH$_3{}^{13}$CH$_2$OH

$\downarrow$ PBr$_3$

CH$_3{}^{13}$CH$_2\overset{\displaystyle O}{\overset{\|}{C}}$OH $\xleftarrow[\text{2. H}_3\text{O}^+]{\text{1. CO}_2}$ CH$_3{}^{13}$CH$_2$MgBr $\xleftarrow{\text{Mg}}$ CH$_3{}^{13}$CH$_2$Br

20.27

(CH$_3$)$_3$CCH$_2$COOH

3,3–Dimethylbutanoic acid

20.28

a)

b)

20.29 Either ^{13}C NMR or ^{1}H NMR can be used to distinguish among these three isomeric carboxylic acids.

Compound	Number of ^{13}C NMR signals	Number of ^{1}H NMR signals	splitting of ^{1}H NMR signals
$CH_3(CH_2)_3COOH$	5	5	1 triplet, peak area 3, 1.0 δ 1 triplet, peak area 2, 2.4 δ 2 multiplets, peak area 4, 1.5 δ 1 singlet, peak area 1, 12.0 δ
$(CH_3)_2CHCH_2COOH$	4	4	1 doublet, peak area 6, 1.0 δ 1 doublet, peak area 2, 2.4 δ 1 multiplet, peak area 1, 1.6 δ 1 singlet, peak area 1, 12.0 δ
$(CH_3)_3CCOOH$	3	2	1 singlet, peak area 9, 1.3 δ 1 singlet, peak area 1, 12.1 δ

20.30 a) Grignard carboxylation can't be used to prepare the carboxylic acid because of the acidic hydroxyl group. Use *nitrile hydrolysis*.

b) Either method produces the carboxylic acid in suitable yield. *Grignard carboxylation* is a better reaction for preparing a carboxylic acid from a secondary bromide. *Nitrile hydrolysis* produces a optically active carboxylic acid from an optically active bromide.

c) Neither method of acid synthesis yields the desired product. Any Grignard reagent formed will react with the carbonyl functional group present in the starting material. Reaction with cyanide occurs at the carbonyl functional group, producing a cyanohydrin, as well as at halogen. However, if the ketone is first protected by forming an acetal, *either method* can be used for producing a carboxylic acid.

d) Since the hydroxyl proton interferes with formation of the Grignard reagent, *nitrile hydrolysis* must be used to form the carboxylic acid.

20.31 2–Chloro–2–methylpentane is a tertiary alkyl halide and ⁻CN is a base. Instead of the desired S_N2 reaction of cyanide with a halide, E2 elimination occurs and yields 2–methyl–2–pentene.

20.32 a) BH_3 is a reducing agent, not an oxidizing agent. To obtain benzoic acid from toluene, use $KMnO_4$.

b) Use CO_2, instead of NaCN, to form the carboxylic acid, or eliminate Mg from this reaction scheme and form the acid by nitrile hydrolysis.

c) Reduction of a carboxylic acid with $LiAlH_4$ yields an alcohol, not an alkyl group.

d) Acid hydrolysis of the nitrile will also dehydrate the tertiary alcohol. Use basic hydrolysis to form the carboxylic acid.

20.33

Notice that the order of the reactions is very important. If toluene is oxidized first, the nitro group will be introduced in the *meta* position. If the nitro group is reduced first, oxidation to the carboxylic acid will reoxidize the –NH₂ group.

20.34 Before starting this type of problem, identify the functional groups present in the starting material. Lithocholic acid contains only alcohol and carboxylic acid functional groups. The given reagents can react with one, both, or neither functional group. Remember to keep track of stereochemistry.

20.35

Other routes to this compound are possible. Notice that the aldehyde functional group and the cyclohexyl group both serve to direct the aromatic chlorine to the correct position of the ring. Also, reaction of the hydroxy-acid with $SOCl_2$ converts –OH to –Cl and –COOH to –COCl. Treatment with H_2O regenerates the carboxylic acid.

20.36 a), b)   Use either ^1H NMR or ^{13}C NMR to distinguish between the isomers.

	Compound	# of ^{13}C NMR Absorptions	# of ^1H NMR Absorptions
a)	COOH (benzene with meta COOH groups)	5	4
	COOH (benzene with para COOH groups)	3	2
b)	$HOOCCH_2CH_2COOH$	2	2
	$CH_3CH(COOH)_2$	3	3

c) Use ^{1}H NMR to distinguish between these two compounds. The carboxylic acid proton of the first compound absorbs near 12 δ; the aldehyde proton of the second compound absorbs near 10 δ and is split into a triplet.

d) The cyclic acid shows four absorptions in both its ^{1}H NMR and ^{13}C NMR spectra. The unsaturated acid shows six absorptions in its ^{13}C NMR and five in its ^{1}H NMR spectrum; one of the ^{1}H NMR signals occurs in the vinyl (4.5 – 6.5 δ) region of the spectrum.

20.37

Substituent	pK_a	Acidity	*E.A.S. Reactivity
$-PCl_2$	3.59	Most acidic	Least reactive (most deactivating
$-OSO_2CH_3$	3.84		
$-CH=CHCN$	4.03		
$-HgCH_3$	4.10		
$-H$	4.20		
$-Si(CH_3)_3$	4.27	Least acidic	Most reactive (least deactivating)

* Electrophilic aromatic substitution

Recall from Section 20.5 that substituents that increase acidity also decrease reactivity in electrophilic aromatic substitution reactions. Of the above substituents, only $-Si(CH_3)_3$ is an activator.

20.38

$$CH_3CH_2OCH_2\overset{\overset{\textstyle O}{\|}}{C}OH$$

20.39 Both of these compounds contain four different kinds of protons (remember that the $H_2C=$ protons are non-equivalent). The carboxylic acid proton absorptions are easy to identify; the other three absorptions in each spectrum are more complex.

It is possible to assign the spectra correctly by studying the methyl group absorptions. The methyl group peak of crotonic acid is split into a doublet ($J = 10$) by the geminal ($CH_3CH=$) proton. (A small "second-order" splitting by the *trans* vinylic proton is also visible; on close inspection the doublet is seen to be a doublet of doublets.) The methyl group absorption of methacrylic acid appears to be a singlet, but second-order splitting by the $H_2C=$ protons shows it to be a quartet having very small J values. The first spectrum is that of crotonic acid and the second spectrum is that of methacrylic acid.

20.40

a)

b)

20.41

$$CH_3CH_2CH_2\overset{\overset{\displaystyle O}{\|}}{C}CH_3 \xrightarrow[\text{THF}]{CH_2=CHMgBr} CH_3CH_2CH_2\overset{\overset{\displaystyle CH_3}{|}}{\underset{\underset{\displaystyle OH}{|}}{C}}CH=CH_2 \xrightarrow[\text{2. } H_2O_2,\ ^-OH]{\text{1. } BH_3,\ THF} CH_3CH_2CH_2\overset{\overset{\displaystyle CH_3}{|}}{\underset{\underset{\displaystyle OH}{|}}{C}}CH_2CH_2OH$$

$$\Big\downarrow \text{Jones}$$

$$CH_3CH_2CH_2\overset{\overset{\displaystyle CH_3}{|}}{C}=CHCOOH \xleftarrow[\text{THF}]{H_3O^+} CH_3CH_2CH_2\overset{\overset{\displaystyle CH_3}{|}}{\underset{\underset{\displaystyle OH}{|}}{C}}CH_2COOH$$

3–Methyl–2–hexenoic acid

Dehydration should occur in the indicated direction to produce a double bond conjugated with the carboxylic acid carbonyl group.

Study Guide for Chapter 20

After studying this chapter, you should be able to:

(1)   Name and draw carboxylic acids (20.1, 20.2, 20.13, 20.14, 20.16).

(2)   Calculate dissociation constants, pH, $\Delta G°$ and percent dissociation for carboxylic acids (20.3, 20.5, 20.23, 20.24).

(3)   Explain the physical properties of carboxylic acids (20.4, 20.15).

(4)   Predict the effects of substituents on carboxylic acid acidity (20.6, 20.7, 20.8, 20.9, 20.10, 20.17, 20.18, 20.19, 20.37).

(5)   Prepare carboxylic acids (20.11, 20.12, 20.21, 20.22, 20.26, 20.28, 20.30, 20.31, 20.32, 20.33, 20.35, 20.41).

(6)   Recognize the types of reactions carboxylic acids undergo, and predict the products of these reactions (20.20, 20.25, 20.34).

(7)   Use spectroscopic techniques to identify carboxylic acids (20.27, 20.29, 20.36, 20.38, 20.39, 20.40).

21.1

a)

$$(CH_3)_2CHCH_2CH_2\overset{\overset{\displaystyle O}{\|}}{C}Cl$$

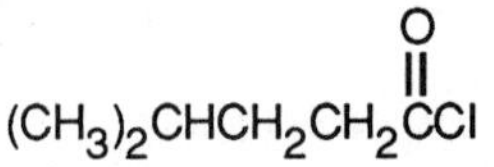

4–Methylpentanoyl chloride

b)

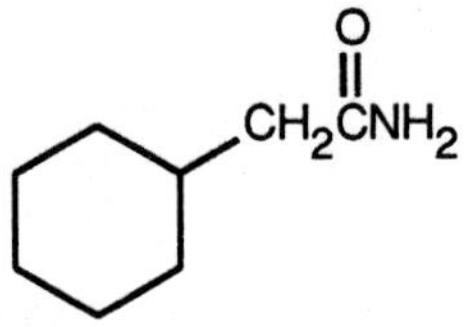

Cyclohexylacetamide

c) $CH_3CH_2CH(CH_3)CN$

2–Methylbutanenitrile

d)

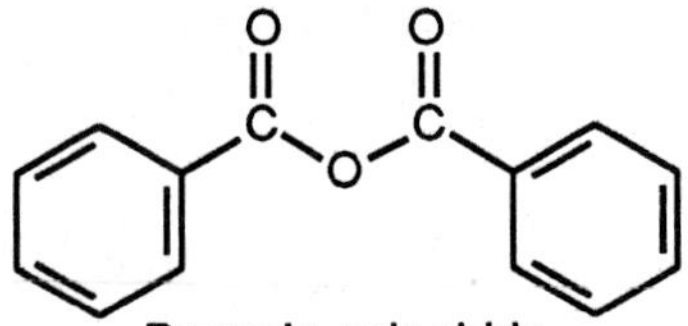

Benzoic anhydride

e)

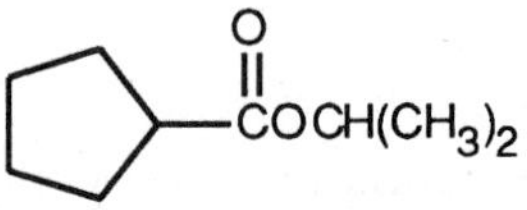

Isopropyl
cyclopentanecarboxylate

f)

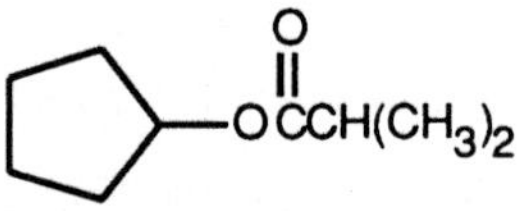

Cyclopentyl
2–methylpropanoate

g)

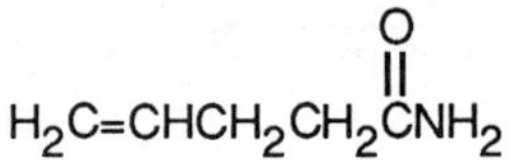

$$H_2C=CHCH_2CH_2\overset{\overset{\displaystyle O}{\|}}{C}NH_2$$

4–Pentenamide

h)

$$CH_3CH_2\overset{\overset{\displaystyle CN}{|}}{C}HCH_2CH_3$$

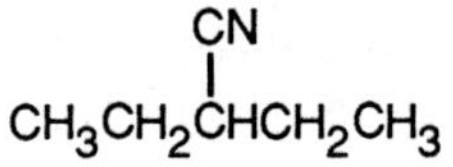

2–Ethylbutanenitrile

i)

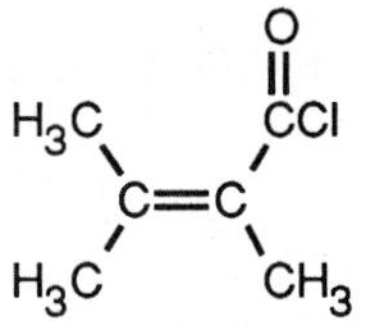

2,3–Dimethyl–2–butenoyl
chloride

j)

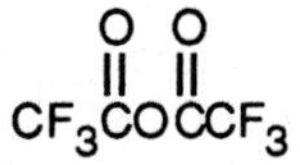

$$CF_3\overset{\overset{\displaystyle O}{\|}}{C}O\overset{\overset{\displaystyle O}{\|}}{C}CF_3$$

Bis(trifluoroacetic)anhydride

21.2

a) $CH_3CH_2CH=CHCN$ Correct name: 2–Pentenenitrile

   The nitrile carbon is at position 1.

b) $CH_3CH_2CH_2CONHCH_3$ Correct name: N–Methylbutanamide

   You must specify that the methyl group is bonded to nitrogen.

c)

$$CH_3$$
$$(CH_3)_2CHCH_2CHCOCl$$ Correct name: 2,4–Dimethylpentanoyl chloride

The prefix "di" must be put before "methyl".

d)

Correct name: Methyl 1–methylcyclohexanecarboxylate

The methyl group on the cyclohexane ring is at position 1. Cyclohexanecarboxylate" is one word.

21.3

Most reactive ⟶ Least reactive

a)

$$\underset{CH_3CCl}{\overset{O}{\|}} > \underset{CH_3COCH_3}{\overset{O}{\|}} > \underset{CH_3CNH_2}{\overset{O}{\|}}$$

b)

$$\underset{CH_3COCH(CF_3)_2}{\overset{O}{\|}} > \underset{CH_3COCH_2CCl_3}{\overset{O}{\|}} > \underset{CH_3COCH_3}{\overset{O}{\|}}$$

The most reactive acyl derivatives contain strongly electron-withdrawing groups in the alkyl portion of the structure.

21.4

$$\underset{\delta-\ \ \delta+\ \ \delta-}{F_3C-\overset{\overset{\displaystyle O^{\delta-}}{\|}}{C}-OCH_3} \qquad \underset{\delta+\ \ \delta-}{H_3C-\overset{\overset{\displaystyle O^{\delta-}}{\|}}{C}-OCH_3}$$

The strongly electron-withdrawing trifluoromethyl group makes the carbonyl carbon more electron-poor and more reactive toward nucleophiles than the methyl acetate carbonyl group. Methyl trifluoroacetate is thus more reactive than methyl acetate in nucleophilic acyl substitution reactions.

21.5

a)

$$\underset{\text{Acetic acid}}{CH_3\overset{\overset{\displaystyle O}{\|}}{C}-OH} + \underset{\text{Butanol}}{H-OCH_2CH_2CH_2CH_3} \underset{}{\overset{HCl}{\rightleftharpoons}} \underset{\text{Butyl acetate}}{CH_3\overset{\overset{\displaystyle O}{\|}}{C}-OCH_2CH_2CH_2CH_3} + H_2O$$

b)

$$\underset{\text{Butanoic acid}}{CH_3CH_2CH_2\overset{\overset{\displaystyle O}{\|}}{C}-OH} + \underset{\text{Methanol}}{H-OCH_3} \underset{}{\overset{HCl}{\rightleftharpoons}} \underset{\text{Methyl butanoate}}{CH_3CH_2CH_2\overset{\overset{\displaystyle O}{\|}}{C}-OCH_3} + H_2O$$

21.6

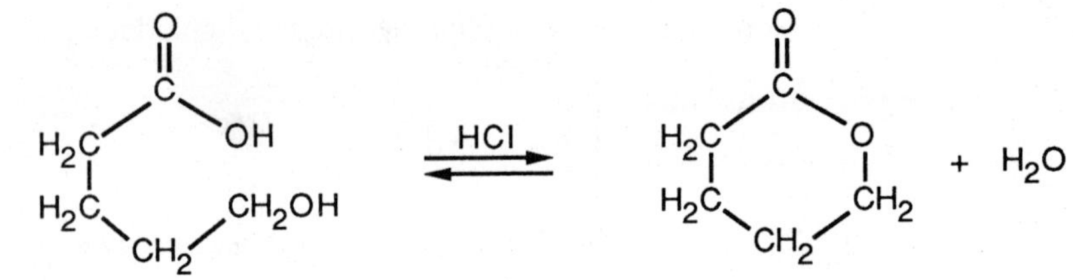

5-Hydroxypentanoic acid   a lactone

21.7

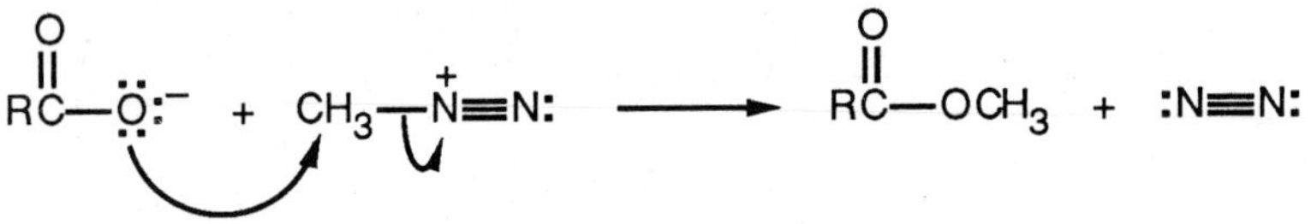

This reaction proceeds by a S_N2 mechanism, with N≡N as leaving group.

21.8

a)

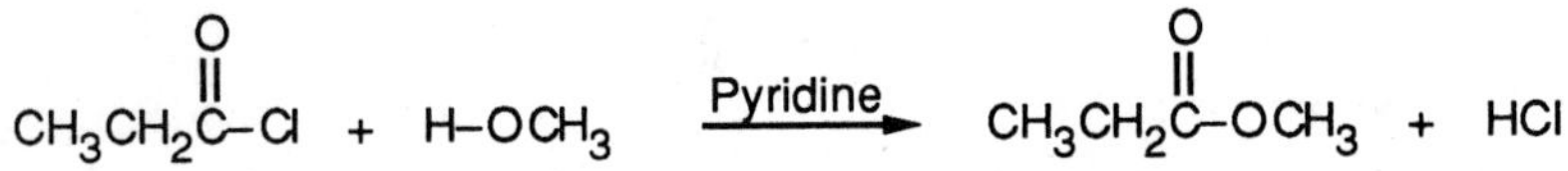

Methyl propanoate

b)

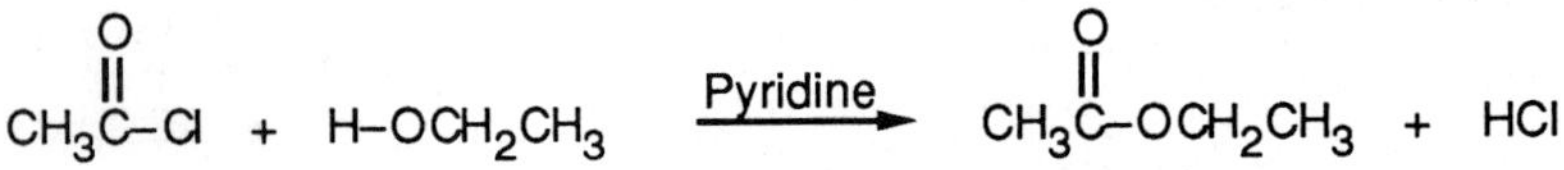

Ethyl acetate

c)

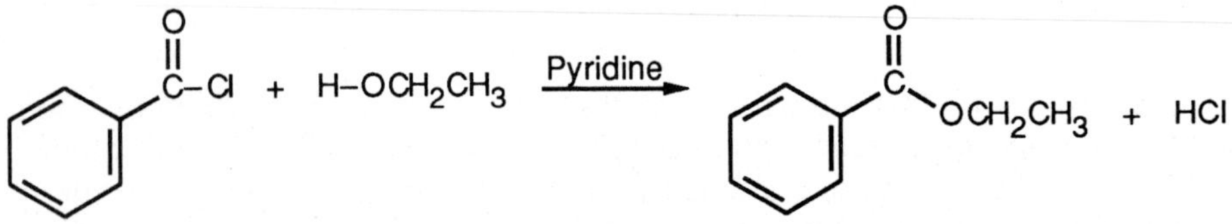

Ethyl benzoate

21.9   Cyclohexanol is a secondary alcohol, which for steric reasons is less reactive in the Fischer esterification reaction. Thus, reaction of cyclohexanol with benzoyl chloride is the preferred method for preparing cyclohexyl benzoate.

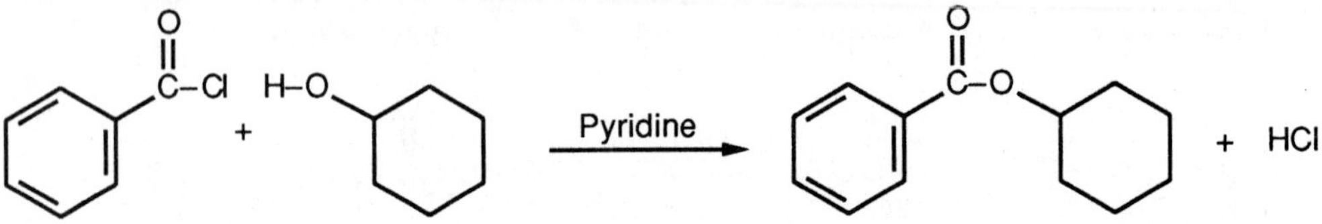

21.10

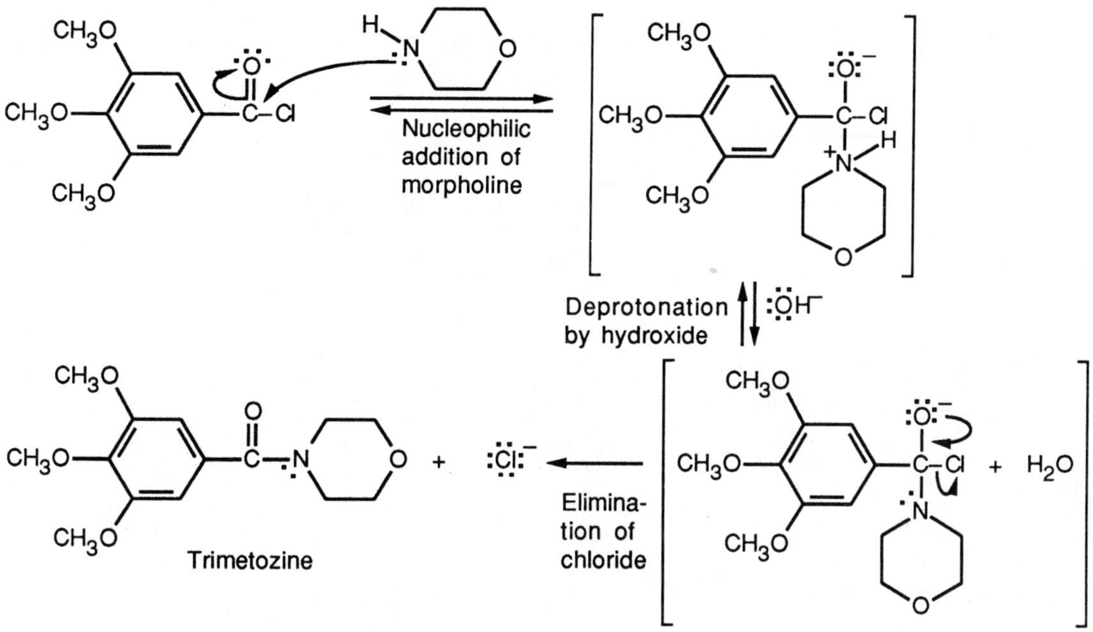

Trimetozine

21.11 For primary and secondary amines:

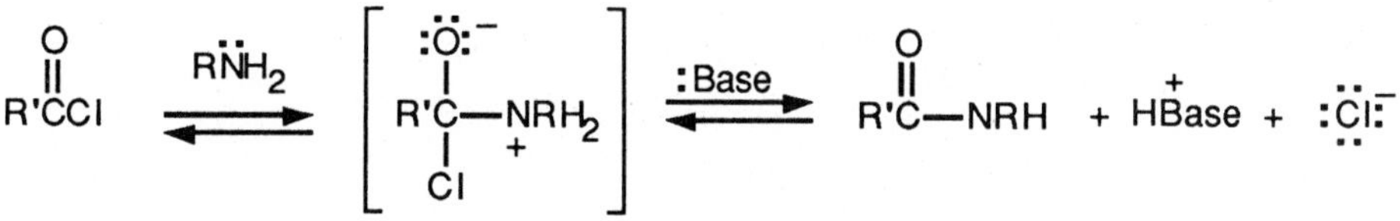

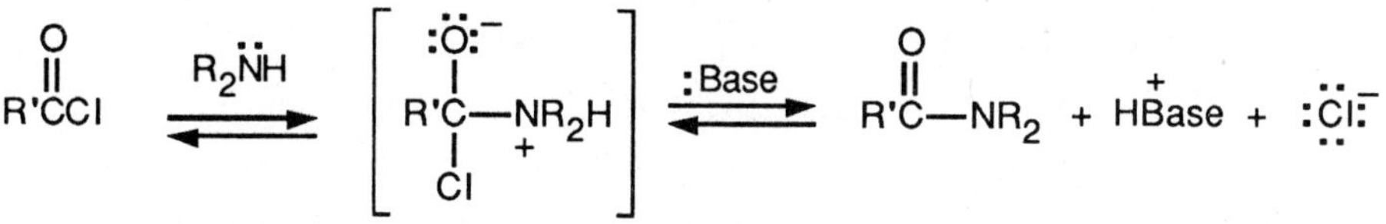

For triethylamine:

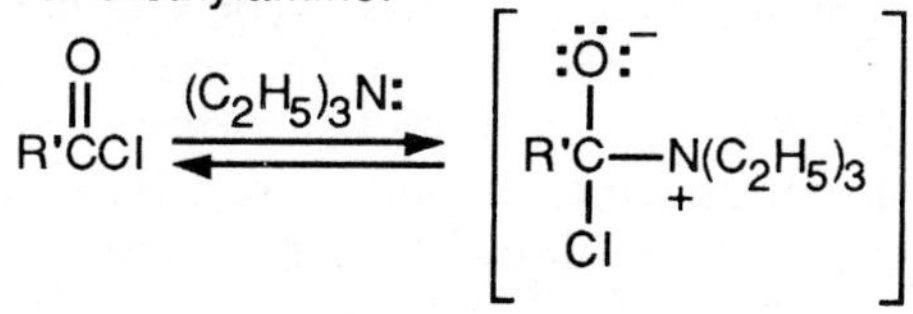

Triethylamine, like other amines, is a base whose lone pair of electrons can scavenge HCl. Unlike primary and secondary amines, triethylamine does not form an amide. The tetrahedral adduct formed by nucleophilic addition of triethylamine to an acid chloride reverts to starting material instead of forming an amide.

21.12

a)

$$CH_3CH_2\overset{O}{\underset{||}{C}}-Cl + H-NHCH_3 \xrightarrow{NaOH} CH_3CH_2\overset{O}{\underset{||}{C}}-NHCH_3 + H_2O + NaCl$$

N–Methylpropanamide

b)

$$\text{(benzene ring)}-\overset{O}{\underset{||}{C}}-Cl + H-N(CH_2CH_3)_2 \xrightarrow{NaOH} \text{(benzene ring)}-\overset{O}{\underset{||}{C}}-N(CH_2CH_3)_2 + H_2O + NaCl$$

N,N–Dimethylbenzamide

c)

$$CH_3CH_2\overset{O}{\underset{||}{C}}-Cl + 2\,H-NH_2 \longrightarrow CH_3CH_2\overset{O}{\underset{||}{C}}-NH_2 + NH_4{}^+Cl^-$$

Propanamide

21.13

a)

$$\text{(benzene ring)}-\overset{O}{\underset{||}{C}}-Cl + ((CH_3)_2CH)_2CuLi$$

or

$$[\text{(benzene ring)}]_2CuLi + Cl-\overset{O}{\underset{||}{C}}CH(CH_3)_2$$

$$\longrightarrow \text{(benzene ring)}-\overset{O}{\underset{||}{C}}-CH(CH_3)_2$$

b)

$$H_2C=CH\overset{O}{\underset{||}{C}}-Cl + (CH_3CH_2CH_2)_2CuLi$$

or

$$(CH_2=CH)_2CuLi + \overset{O}{\underset{||}{C}}HCCH_2CH_2CH_3$$

$$\longrightarrow H_2C=CH\overset{O}{\underset{||}{C}}CH_2CH_2CH_3$$

21.14 In the slow addition of Grignard reagent to a solution of an acid chloride, each drop of Grignard reagent is surrounded by a large amount of acid chloride. The Grignard reagent is consumed immediately in converting the reactive acid chloride to less reactive ketone. No Grignard reagent remains to react with the ketone to form a tertiary alcohol. If the acid chloride were slowly added to a solution of Grignard reagent, however, excess reagent would convert the initially formed ketone into a tertiary alcohol.

21.15

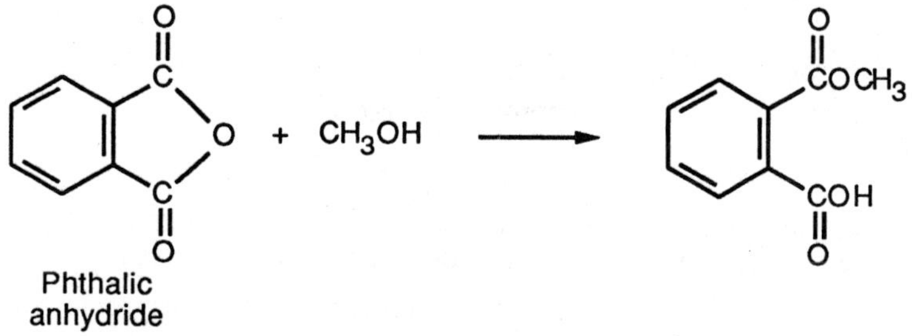

Phthalic
anhydride

The second half of a cyclic anhydride becomes a carboxylic acid.

21.16

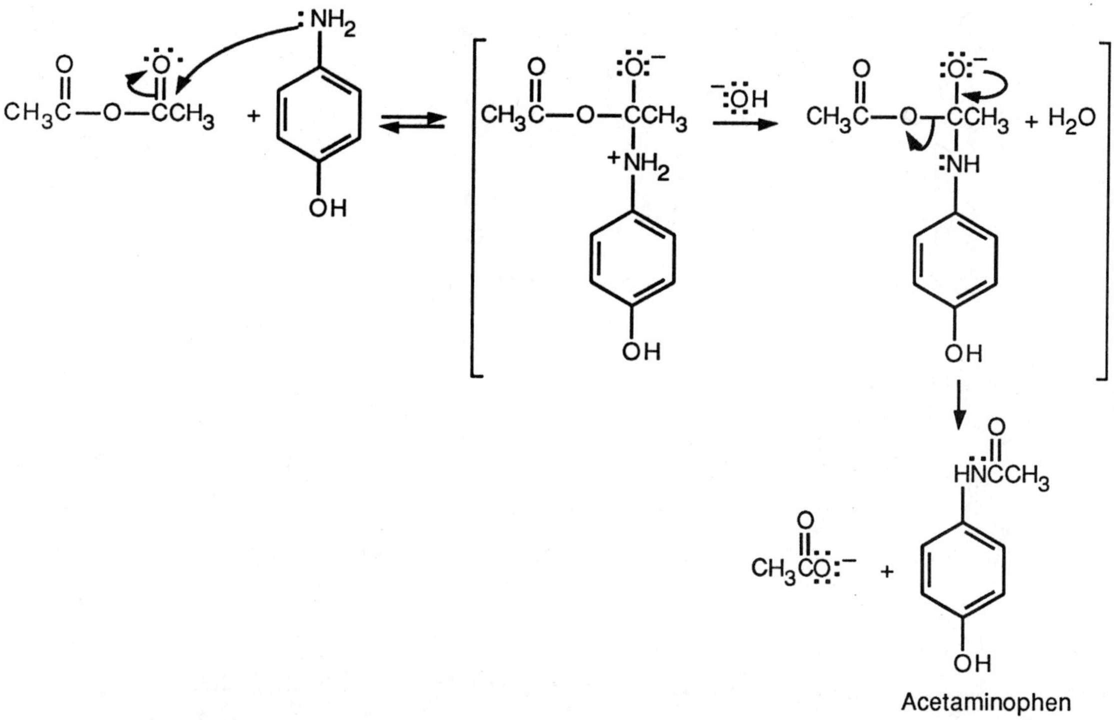

Acetaminophen

21.17 One equivalent of base must be added to the reaction mixture when an amine reacts with an anhydride. This base removes a proton from nitrogen after the formation of the initial tetrahedral intermediate. If base were not added, the amine starting material would serve as the base. Instead of going to completion, the reaction would stop when half of the starting amine has been converted to amide; the rest of the amine would be protonated and would no longer be nucleophilic.

21.18

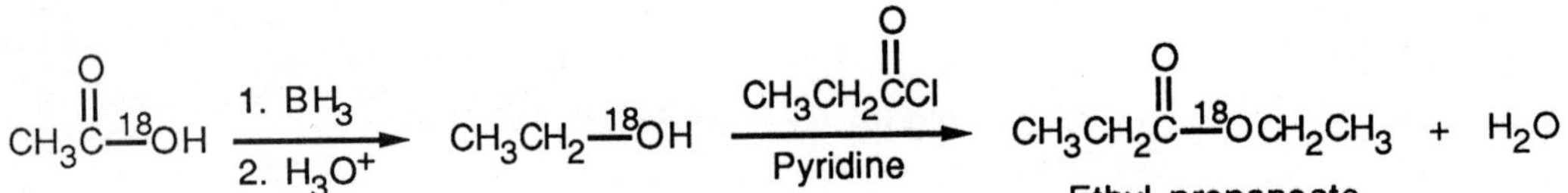

Ethyl propanoate

21.19 The products of acidic hydrolysis of an ester are a carboxylic acid and an alcohol. Under acidic conditions, these products can reform an ester by the Fischer esterification route.

The products of basic hydrolysis of an ester are an alcohol and a carboxylate anion, which is not attacked by nucleophiles because it already has a negative charge.

21.20

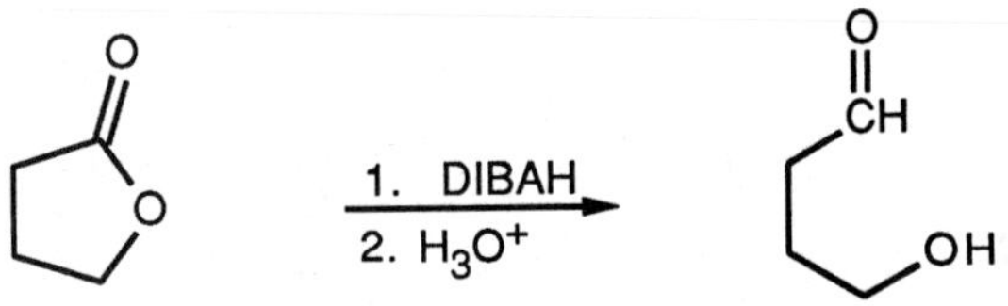

21.21

a)

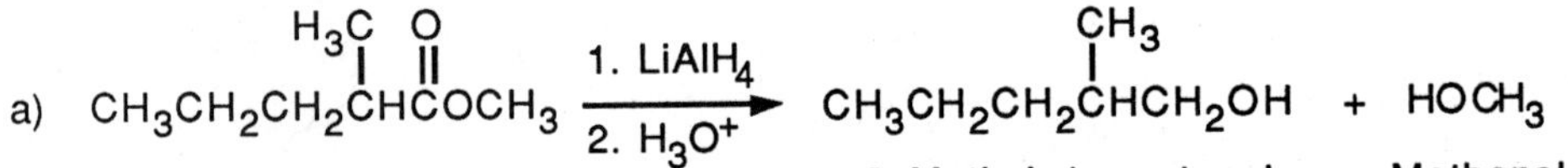

2–Methyl–1–pentanol Methanol

b)

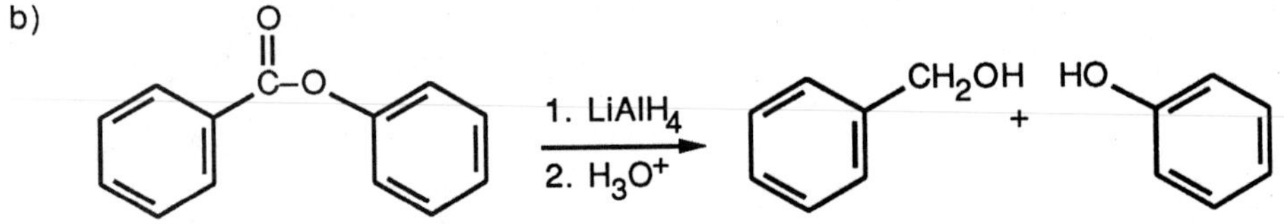

Benzyl alcohol Phenol

21.22

Ester — Grignard Reagent → Tertiary alcohol

a)

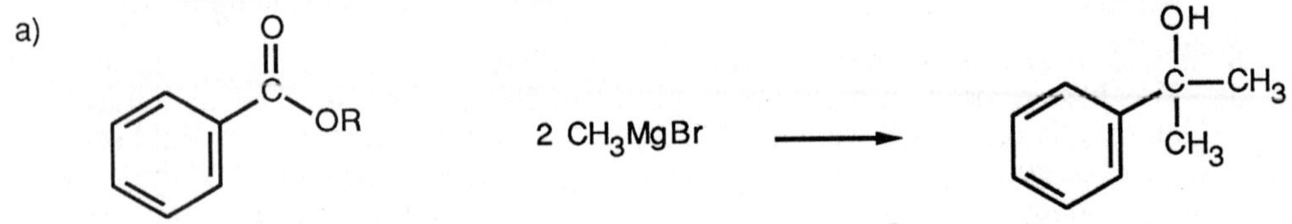

2–Phenyl–2–propanol

Ester   Grignard Reagent ⟶   Tertiary alcohol

b) 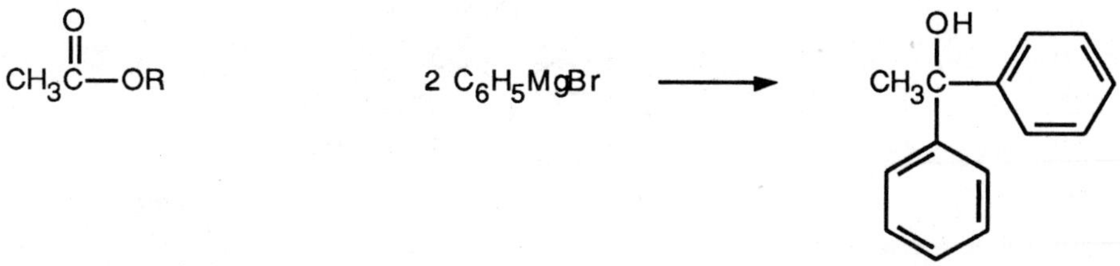 $CH_3\overset{O}{\overset{\|}{C}}-OR$ $2\ C_6H_5MgBr$ ⟶

1,1–Diphenylethanol

c) 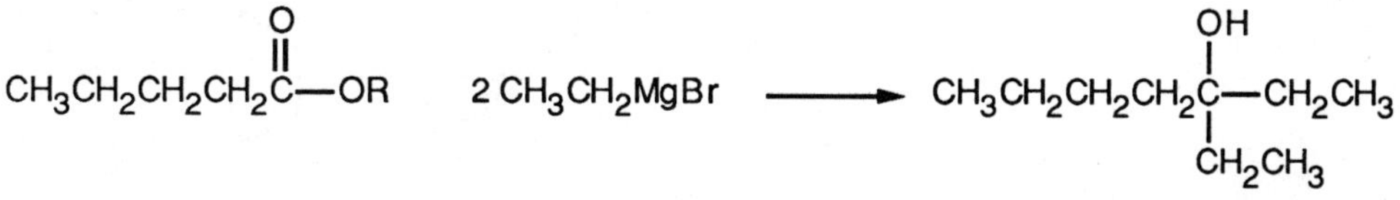 $CH_3CH_2CH_2CH_2\overset{O}{\overset{\|}{C}}-OR$ $2\ CH_3CH_2MgBr$ ⟶ $CH_3CH_2CH_2CH_2\overset{\overset{\displaystyle OH}{|}}{\underset{\underset{\displaystyle CH_2CH_3}{|}}{C}}-CH_2CH_3$

3–Ethyl–3–heptanol

21.23

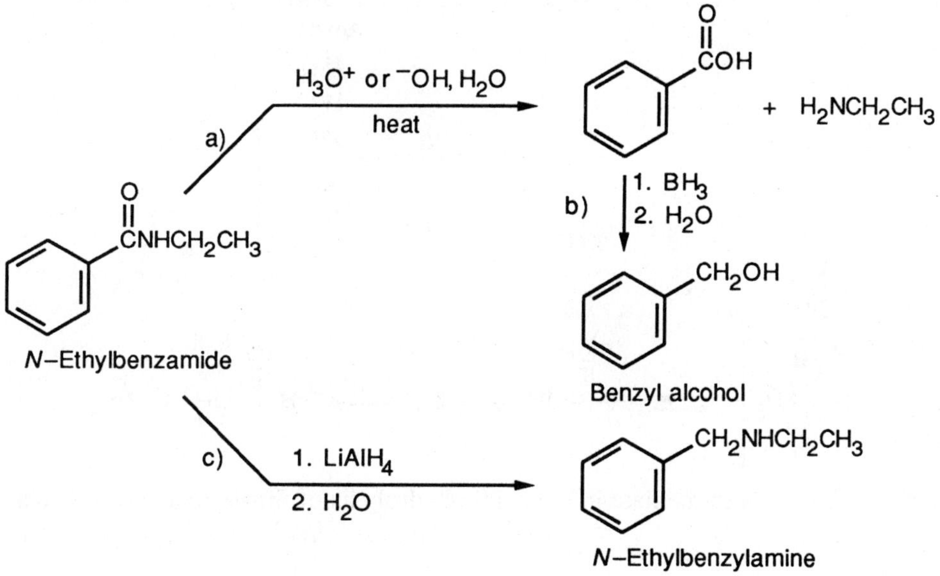

N–Ethylbenzamide

a) $\dfrac{H_3O^+ \text{ or } {}^-OH, H_2O}{heat}$ ⟶ $\underset{}{\overset{O}{\overset{\|}{C}}OH}$ $+\ H_2NCH_2CH_3$

b)   1. BH_3 2. H_2O

CH_2OH

Benzyl alcohol

c)   1. $LiAlH_4$ 2. H_2O ⟶

$CH_2NHCH_2CH_3$

N–Ethylbenzylamine

21.24

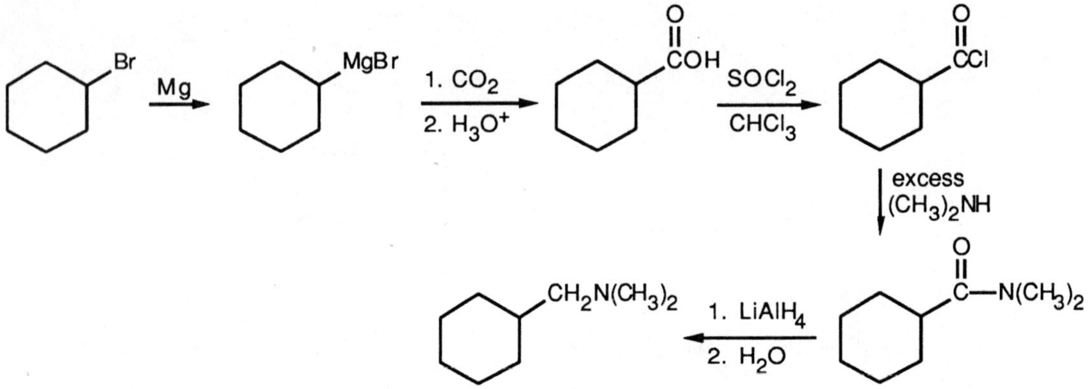

21.25

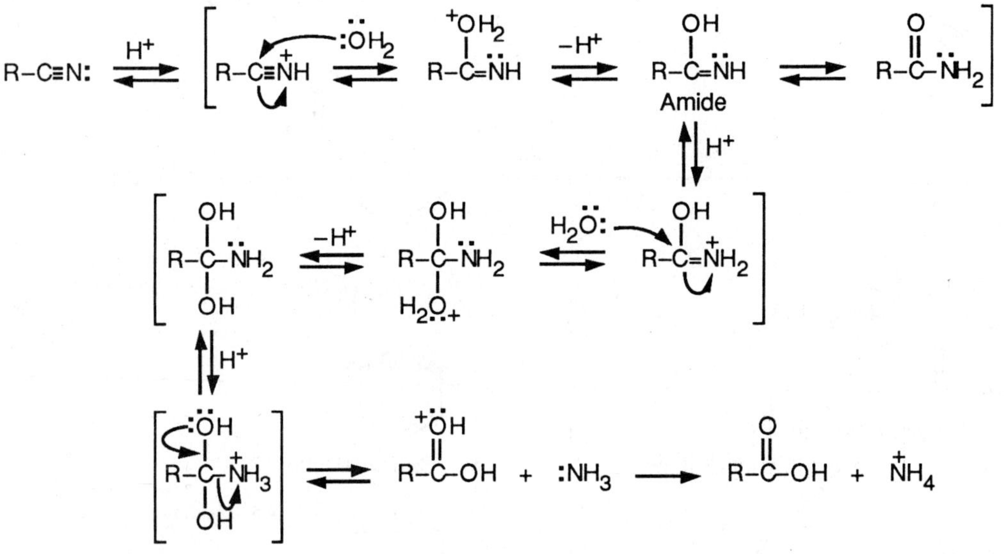

An amide is an intermediate in the acidic hydrolysis of nitrile to a carboxylic acid.

21.26

a) 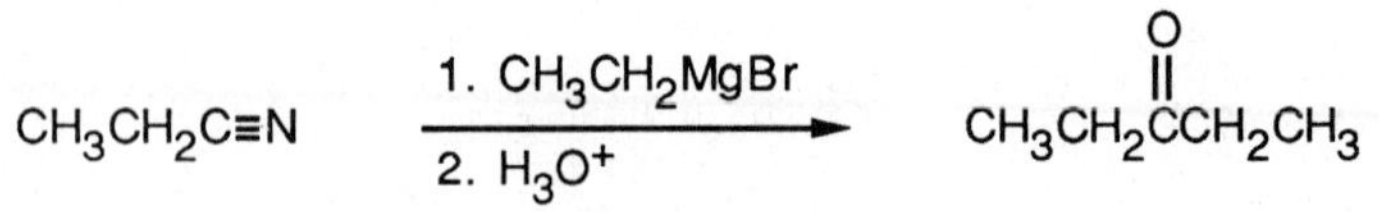 $CH_3CH_2C{\equiv}N$ $\xrightarrow[\text{2. }H_3O^+]{\text{1. }CH_3CH_2MgBr}$ $CH_3CH_2\overset{\displaystyle O}{\overset{\|}{C}}CH_2CH_3$

b) 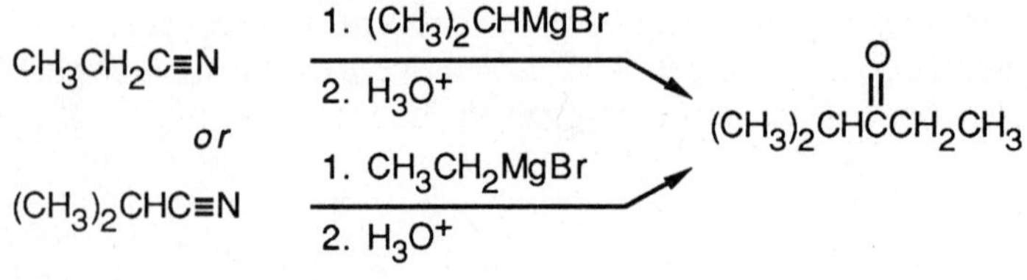 $CH_3CH_2C{\equiv}N$ $\xrightarrow[\text{2. }H_3O^+]{\text{1. }(CH_3)_2CHMgBr}$

   or

   $(CH_3)_2CHC{\equiv}N$ $\xrightarrow[\text{2. }H_3O^+]{\text{1. }CH_3CH_2MgBr}$ $(CH_3)_2CH\overset{\displaystyle O}{\overset{\|}{C}}CH_2CH_3$

c) 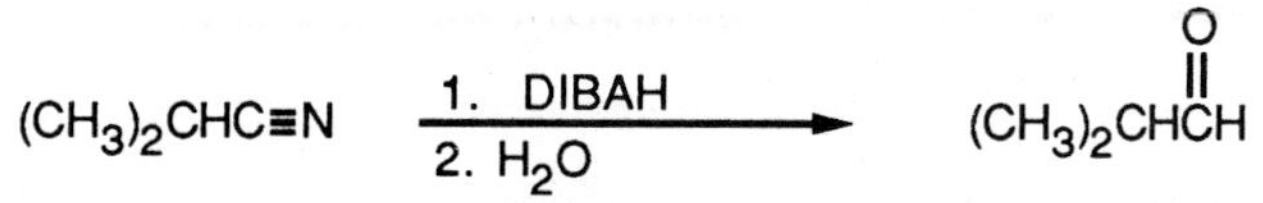 $(CH_3)_2CHC\equiv N$ $\xrightarrow{\begin{array}{l}1.\ \ DIBAH\\2.\ \ H_2O\end{array}}$ $(CH_3)_2CHCH$ (with C=O)

d)

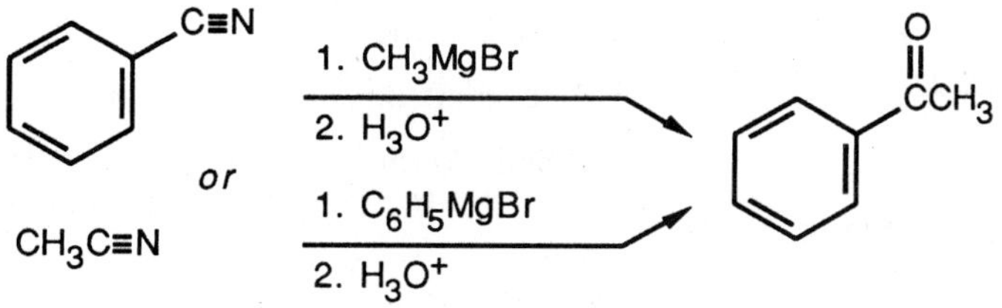

(benzene ring with $C\equiv N$)

$\xrightarrow{\begin{array}{l}1.\ \ CH_3MgBr\\2.\ \ H_3O^+\end{array}}$

or

$CH_3C\equiv N$

$\xrightarrow{\begin{array}{l}1.\ \ C_6H_5MgBr\\2.\ \ H_3O^+\end{array}}$

(benzene ring with CCH_3, C=O)

e)

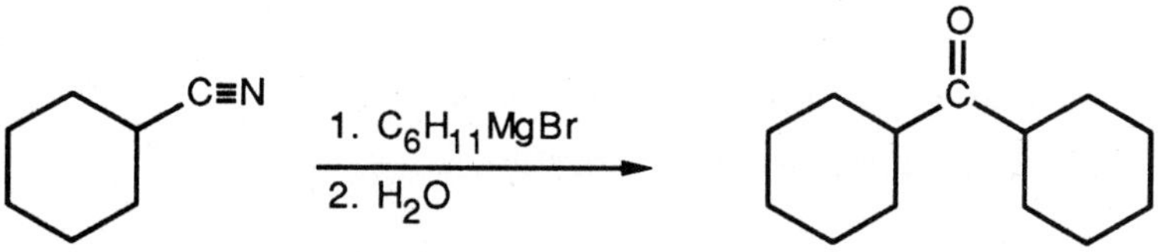

(cyclohexane ring with $C\equiv N$)

$\xrightarrow{\begin{array}{l}1.\ \ C_6H_{11}MgBr\\2.\ \ H_2O\end{array}}$

(two cyclohexane rings joined by C=O)

21.27

(benzene ring with CH_2Br) $\xrightarrow{NaCN}$ (benzene ring with CH_2CN) $\xrightarrow{\begin{array}{l}1.\ \ CH_3CH_2MgBr\\2.\ \ H_3O^+\end{array}}$ (benzene ring with $CH_2CCH_2CH_3$, C=O)

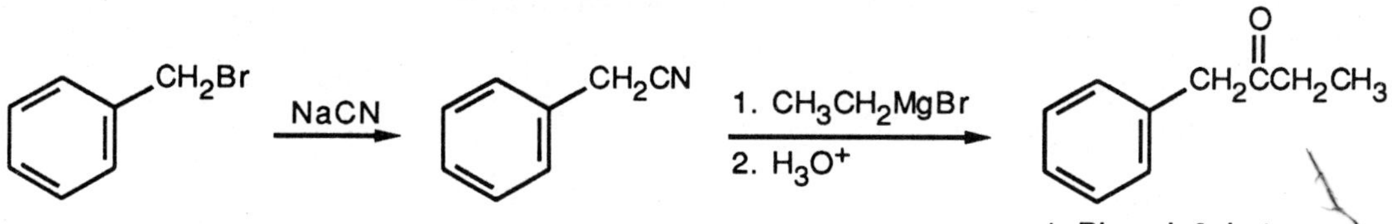

1–Phenyl–2–butanone

21.28 Write the mechanism for addition of each equivalent of Grignard reagent to the nitrile. The nitrogen atom of the imine anion bears the negative charge.

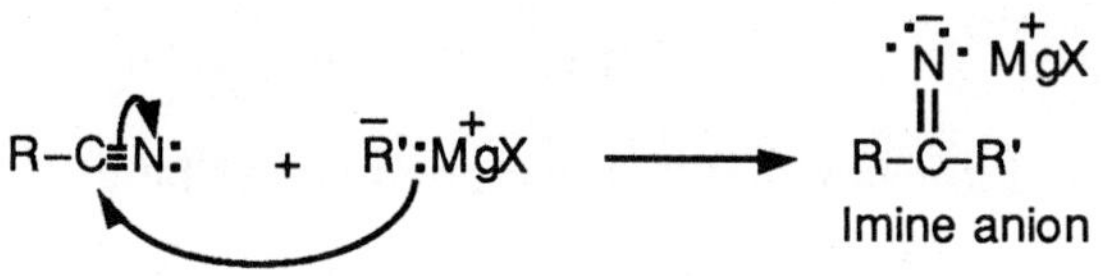

$R-C\equiv N:$ + $\bar{R}':\overset{+}{M}gX$ $\longrightarrow$ $R-\underset{\|}{C}-R'$ with $\overset{-}{\cdot}\overset{\cdot}{N}\cdot\ \overset{+}{M}gX$

Imine anion

A possible mechanism for addition of a second equivalent of Grignard reagent is:

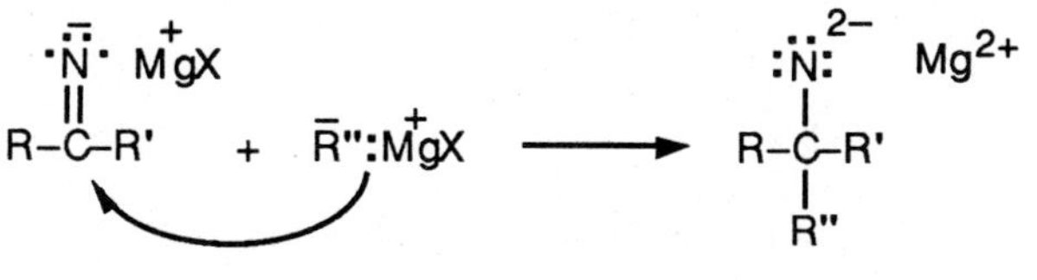

$R-\underset{\|}{C}-R'$ with $\overset{-}{\cdot}\overset{\cdot}{N}\cdot\ \overset{+}{M}gX$ + $\bar{R}'':\overset{+}{M}gX$ $\longrightarrow$ $R-\underset{|}{\overset{|}{C}}-R'$ with $:\overset{\cdot\cdot}{N}:^{2-}\ Mg^{2+}$ and R''

This second addition of Grignard reagent is unlikely to occur for two reasons:
(1) The imine carbon is relatively electron-rich and unreactive toward nucleophiles.
(2) Nitrogen would have to bear two negative charges in the intermediate.

21.29   <u>Absorption</u>   <u>Functional group present</u>

    a) $1735 \ cm^{-1}$ Aliphatic ester *or* 6-membered ring lactone

    b) $1810 \ cm^{-1}$ Aliphatic acid chloride

    c) $2500 - 3300 \ cm^{-1}$ and $1710 \ cm^{-1}$ Carboxylic acid

    d) $2250 \ cm^{-1}$ Aliphatic nitrile

    e) $1715 \ cm^{-1}$ Aliphatic ketone *or* 6-membered ring ketone

21.30 To solve this type of problem:

    1. Use the IR absorption to determine the functional group(s) present.

    2. Draw the functional group.

    3. Use the remaining atoms to complete the structure.

        a) 1. IR $2250 \ cm^{-1}$ corresponds to a nitrile.

           2. $-C{\equiv}N$

           3. CH_3CH_2CN is the structure of the compound.

        b) 1. IR $1735 \ cm^{-1}$ corresponds to an aliphatic ester.

           2.

$$-\overset{\displaystyle O}{\overset{\displaystyle \|}{C}}-O-$$

           3. The remaining five carbons and twelve hydrogens can be arranged in a number of ways to produce a satisfactory structure for this compound. For example:

$$CH_3CH_2CH_2\overset{\displaystyle O}{\overset{\displaystyle \|}{C}}OCH_2CH_3 \qquad or \qquad CH_3\overset{\displaystyle O}{\overset{\displaystyle \|}{C}}OCH_2CH_2CH_2CH_3$$

           The structural formula indicates that this compound can't be a lactone.

        c) $CH_3\overset{\displaystyle O}{\overset{\displaystyle \|}{C}}N(CH_3)_2$

        d) $CH_3CH{=}CH\overset{\displaystyle O}{\overset{\displaystyle \|}{C}}Cl$, $H_2C{=}C(CH_3)\overset{\displaystyle O}{\overset{\displaystyle \|}{C}}Cl$

21.31 a) *p*–Methylbenzamide   b) 4–Ethyl–2–hexenenitrile

    c) Dimethyl succinate *or*   d) Isopropyl 3–phenylpropanoate
       Dimethyl butanedioate

    e) Phenyl benzoate   f) *N*–Methyl–3–bromobutanamide

    g) 3,5–Dibromobenzoyl chloride   h) 1–Cyclopentenecarbonitrile

21.32

a)

p–Bromophenylacetamide

b)

m–Benzoylbenzonitrile

c)

CH$_3$CH$_2$CH$_2$CH$_2$C(CH$_3$)$_2$CNH$_2$

2,2–Dimethylhexanamide

d)

Cyclohexyl cyclohexane-
carboxylate

e)

2–Cyclobutenecarbonitrile

f)

COCl
|
CH$_3$CH$_2$CH$_2$CHCH$_2$COCl

1,2–Pentanedicarbonyl
dichloride

21.33 Many structures can be drawn for each part of this problem.

a)

Cyclopentanecarbonyl
chloride

(E)–2–Methyl–2–pentenoyl
chloride

H$_2$C=CHCHCCl
|
CH$_2$CH$_3$

2–Ethyl–3–butenoyl
chloride

b)

1–Cyclohexene-
carboxamide

CH$_3$CH$_2$CH$_2$C≡CCH$_2$CNH$_2$

3–Heptynamide

H$_2$C=CHCH=CHCN(CH$_3$)$_2$

N,N–Dimethyl–2,4–
pentadienamide

c)

Cyclobutanecarbonitrile

CH$_3$CH=CHCH$_2$C≡N

3–Pentenenitrile

CH$_3$
|
H$_2$C=CCH$_2$C≡N

3–Methyl–3–butenenitrile

21.34 The reactivity of esters in saponification reactions is influenced by steric factors. Branching in both the acyl and alkyl portions of an ester hinders attack of the hydroxide nucleophile. This effect is less pronounced in the alkyl portion of the ester than in the acyl portion because alkyl branching is one atom farther away from the site of attack. The reactivity order for saponification of alkyl acetates:

Most reactive ——————————————————→ Least reactive

$$CH_3\overset{O}{\overset{||}{C}}OCH_3 \;>\; CH_3\overset{O}{\overset{||}{C}}OCH_2CH_3 \;>\; CH_3\overset{O}{\overset{||}{C}}OCH(CH_3)_2 \;>\; CH_3\overset{O}{\overset{||}{C}}OC(CH_3)_3$$

21.35

2,4,6–Trimethylbenzoic acid

2,4,6–Trimethylbenzoic acid has two methyl groups *ortho* to the carboxylic acid functional group. These bulky methyl groups block the approach of the alcohol and prevent esterification from occurring under Fischer esterification conditions. Other possible routes to the methyl ester:

1. SOCl₂
2. CH₃OH, pyridine

CH₂N₂
Ether

1. NaOH
2. CH₃I

The last two routes succeed because reaction occurs farther away from the site of steric hindrance.

21.36

The tetrahedral intermediate **I** can eliminate any one of the three –OH groups to reform either the original carboxylic acid or labeled carboxylic acid. Further reaction of water with mono-labeled carboxylic acid leads to the doubly labeled product.

21.37

a)

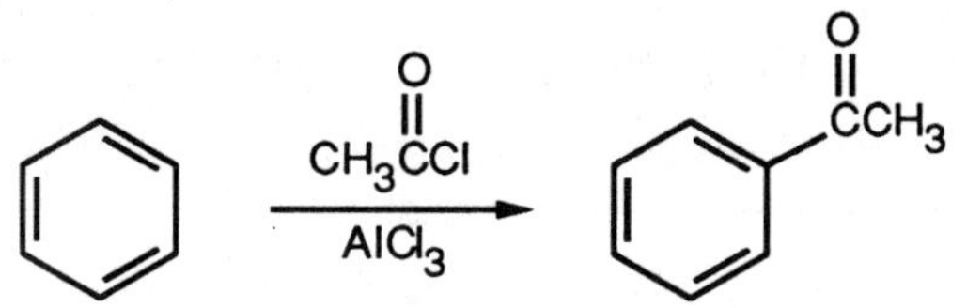

b)

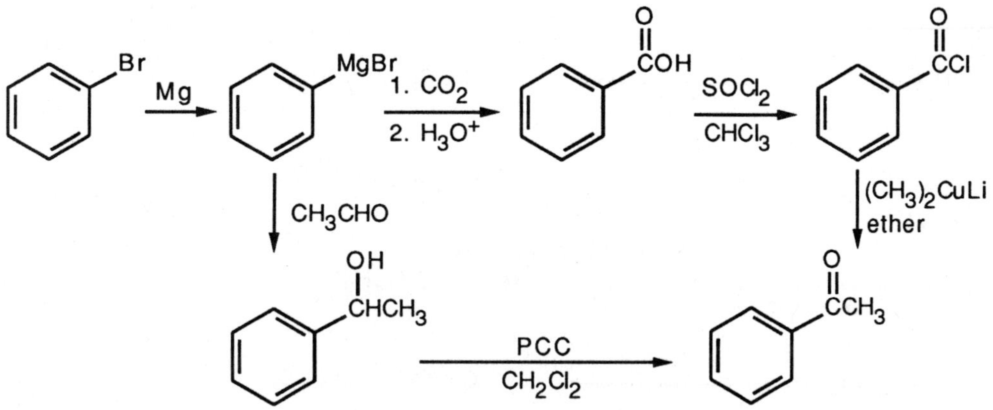

c)

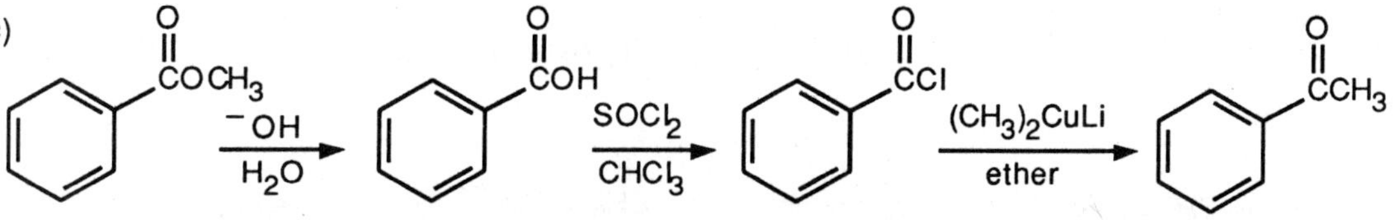

Reaction of an ester with Grignard reagent produces a tertiary alcohol, not a ketone.

d)

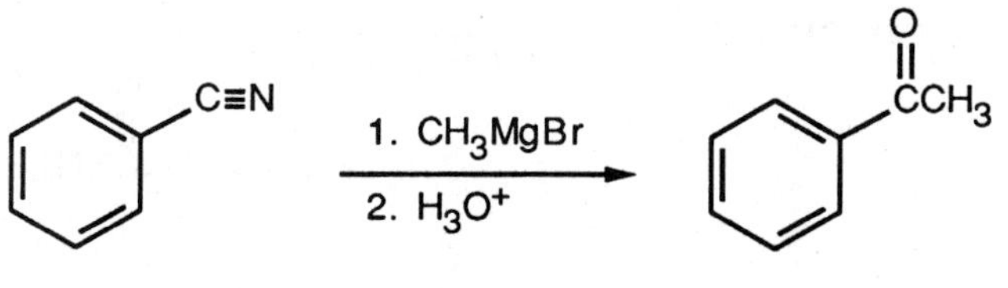

e)

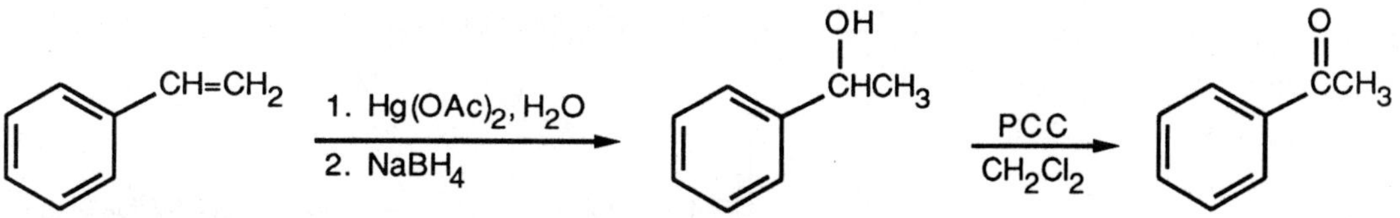

21.38 A negatively charged tetrahedral intermediate is initially formed when the nucleophile ⁻:OH attacks the carbonyl carbon of an ester. An electron-withdrawing substituent can stabilize the negatively charged tetrahedral intermediate and increase the rate of reaction. (Contrast this effect with substituent effects in electrophilic aromatic substitution, in which positive charge developed in the intermediate is stabilized by electron-*donating* substituents.) Substituents that are deactivating in electrophilic aromatic substitution are activating in ester hydrolysis, as the observed reactivity order shows. The substituents –CN and –CHO are electron-withdrawing; –NH$_2$ is strongly electron-donating.

Most reactive ⟶ Least reactive

X = –NO$_2$ > –CN > –CHO > –Br > –H > –CH$_3$ > –OCH$_3$ > –NH$_2$

21.39

a)   CH$_3$CH$_2$CH$_2$COOH $\xrightarrow[\text{2. H}_3\text{O}^+]{\text{1. BH}_3}$ CH$_3$CH$_2$CH$_2$CH$_2$OH

b)   CH$_3$CH$_2$CH$_2$CH$_2$OH $\xrightarrow{\text{PCC}}$ CH$_3$CH$_2$CH$_2$CHO

from (a)   *or*

CH$_3$CH$_2$CH$_2$COOH $\xrightarrow[\text{3. H}_3\text{O}^+]{\begin{array}{l}\text{1. SOCl}_2 \\ \text{2. LiAlH[OC(CH}_3)_3]_3\end{array}}$ CH$_3$CH$_2$CH$_2$CHO

c)   CH$_3$CH$_2$CH$_2$CH$_2$OH $\xrightarrow{\text{PBr}_3}$ CH$_3$CH$_2$CH$_2$CH$_2$Br

from (a)

d)   CH$_3$CH$_2$CH$_2$CH$_2$Br $\xrightarrow{\text{NaCN}}$ CH$_3$CH$_2$CH$_2$CH$_2$C≡N

from (c)

e)   CH$_3$CH$_2$CH$_2$CH$_2$Br $\xrightarrow[\text{Ethanol}]{\text{KOH}}$ CH$_3$CH$_2$CH=CH$_2$

from (c)

f)   CH$_3$CH$_2$CH$_2$CH$_2$C≡N $\xrightarrow{\text{H}_3\text{O}^+}$ CH$_3$CH$_2$CH$_2$CH$_2$COH ‖O $\xrightarrow[\text{2. CH}_3\text{NH}_2]{\text{1. SOCl}_2}$

from d)

CH$_3$CH$_2$CH$_2$CH$_2$CNHCH$_3$ ‖O

g)   CH$_3$CH$_2$CH$_2$CH$_2$C≡N $\xrightarrow[\text{2. H}_3\text{O}^+]{\text{1. CH}_3\text{MgBr}}$ CH$_3$CH$_2$CH$_2$CH$_2$CCH$_3$ ‖O

from (d)

h)   CH$_3$CH$_2$CH$_2$CCl ‖O $\xrightarrow[\text{AlCl}_3]{\text{C}_6\text{H}_6}$ $\xrightarrow[\text{Pd/C}]{\text{H}_2}$

from (f)

21.40 Dimethyl carbonate is a diester. Use your knowledge of the Grignard reaction to work your way through this problem.

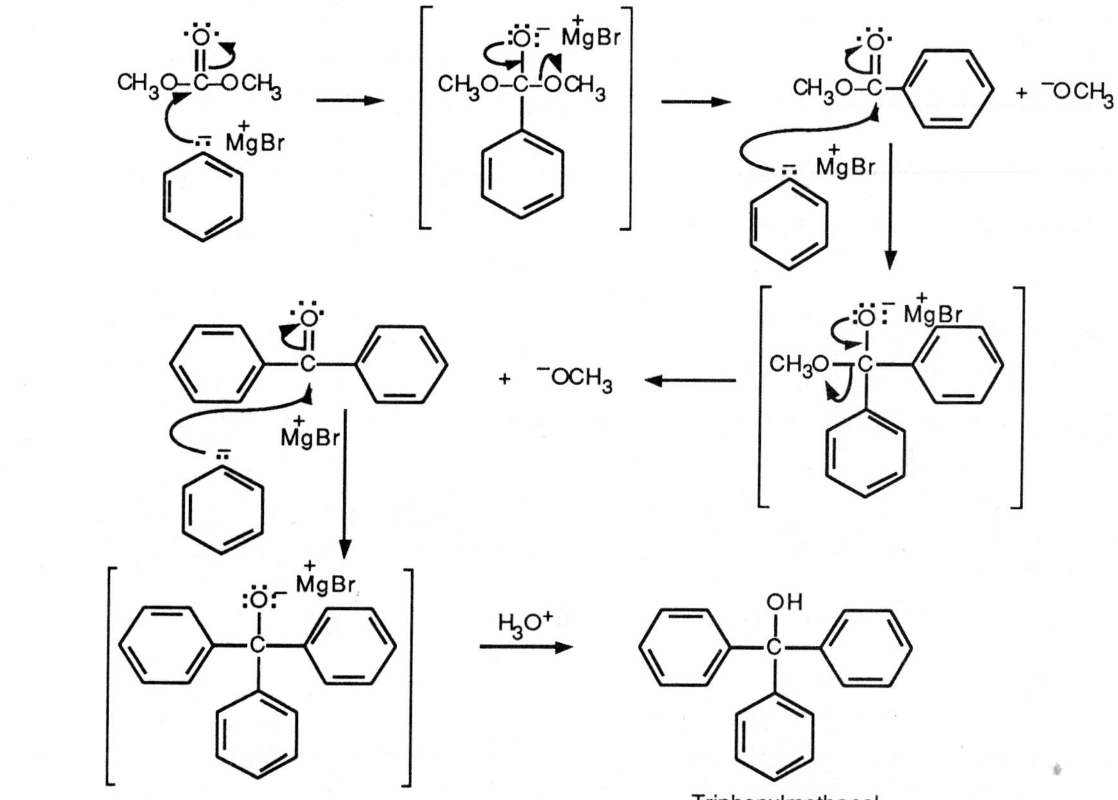

Triphenylmethanol

21.41

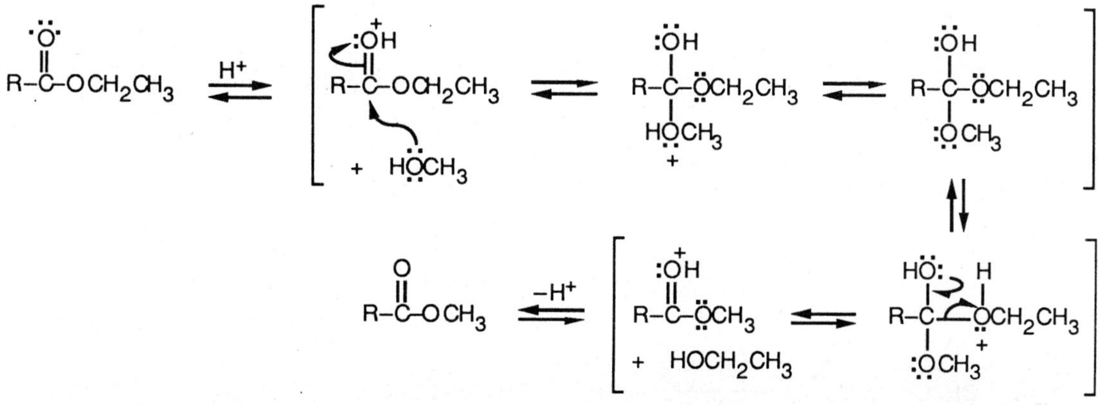

In acidic methanol, the ethyl ester can react to yield a methyl ester. Conversion of the ethyl ester to the methyl ester occurs because of the large excess of methanol, the solvent.

21.42

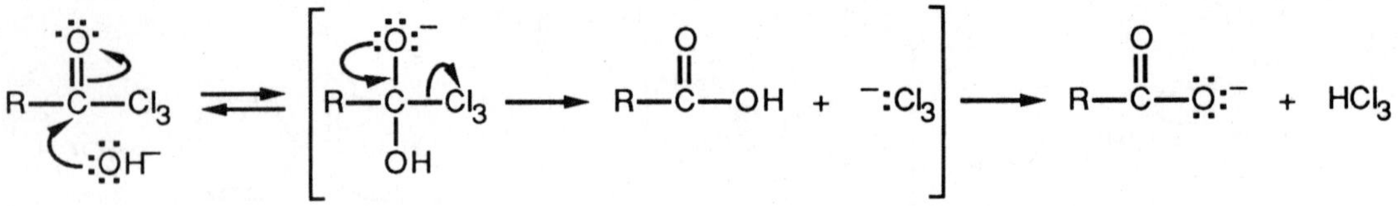

21.43 The nucleophile NH_3 replaces chlorine because chloride is a better leaving group than the methoxyl group.

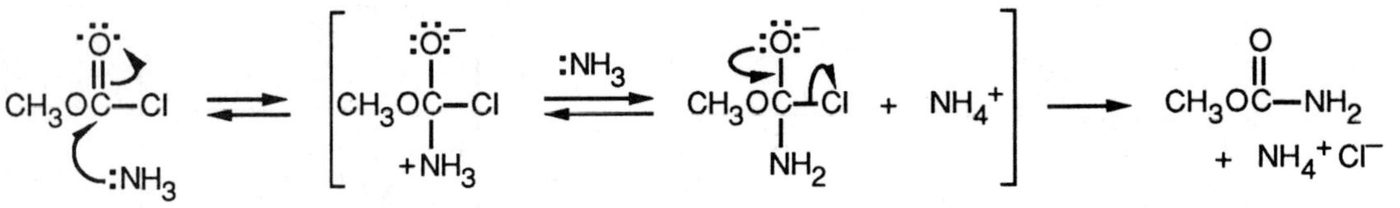

21.44

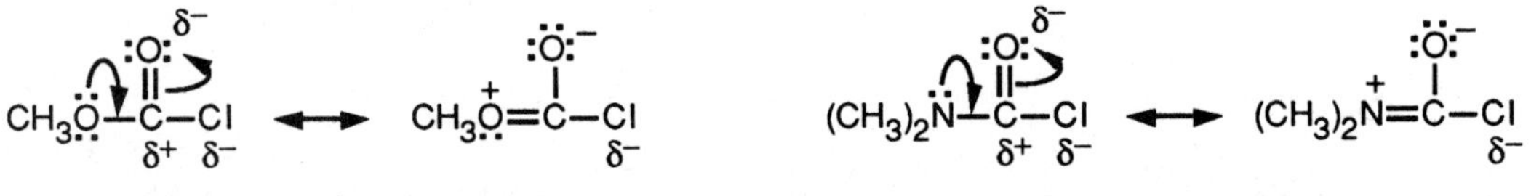

Both $-OCH_3$ and $-N(CH_3)_2$ are electron-donating substituents that make the carbonyl carbon less electrophilic. Since an $-OCH_3$ group is less strongly electron-donating, the carbonyl carbon of methoxycarbonyl chloride is less electron-rich and the carbonyl group is more polarized.

21.45 Methoxycarbonyl chloride, which has the more polarized carbonyl group, reacts faster with nucleophiles than does N,N–dimethylaminocarbonyl chloride.

21.46

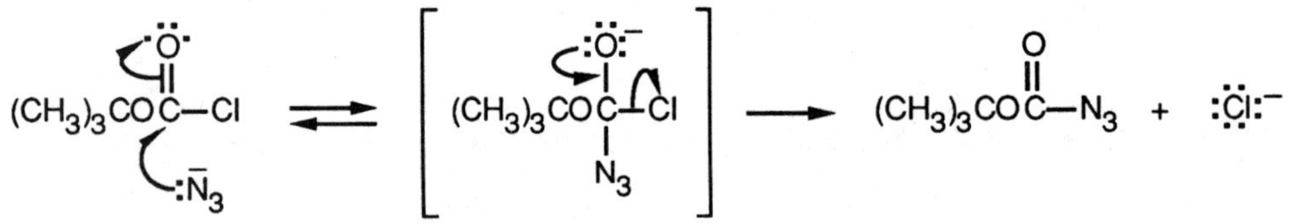

21.47

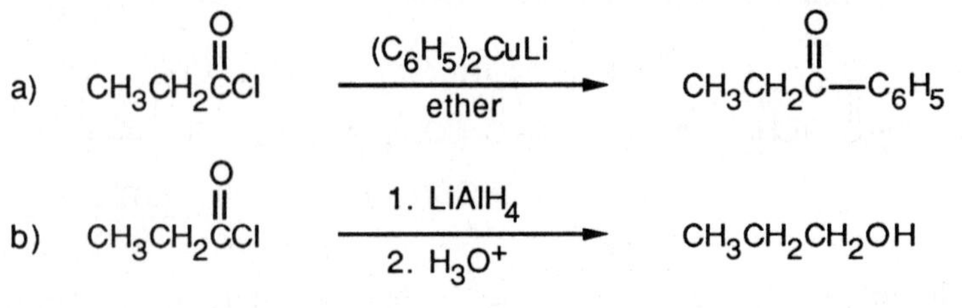

a) $CH_3CH_2\overset{\overset{O}{\|}}{C}Cl \xrightarrow[\text{ether}]{(C_6H_5)_2CuLi} CH_3CH_2\overset{\overset{O}{\|}}{C}-C_6H_5$

b) $CH_3CH_2\overset{\overset{O}{\|}}{C}Cl \xrightarrow[\text{2. }H_3O^+]{\text{1. LiAlH}_4} CH_3CH_2CH_2OH$

c) $CH_3CH_2\overset{\overset{\displaystyle O}{||}}{C}Cl$ $\xrightarrow[\text{2. } H_3O^+]{\text{1. } 2\,CH_3MgBr}$ $CH_3CH_2\overset{\overset{\displaystyle OH}{|}}{\underset{\underset{\displaystyle CH_3}{|}}{C}}CH_3$

d) $CH_3CH_2\overset{\overset{\displaystyle O}{||}}{C}Cl$ $\xrightarrow{\text{LiAlH}[OC(CH_3)_3]_3}$ $CH_3CH_2\overset{\overset{\displaystyle O}{||}}{C}H$

e) $CH_3CH_2\overset{\overset{\displaystyle O}{||}}{C}Cl$ $\xrightarrow{H_3O^+}$ $CH_3CH_2\overset{\overset{\displaystyle O}{||}}{C}OH$ +   HCl

f) $CH_3CH_2\overset{\overset{\displaystyle O}{||}}{C}Cl$ +   [cyclohexanol] $\xrightarrow{\text{Pyridine}}$ $CH_3CH_2\overset{\overset{\displaystyle O}{||}}{C}O-$[cyclohexyl]

g) $CH_3CH_2\overset{\overset{\displaystyle O}{||}}{C}Cl$ +   [aniline] $\xrightarrow{\text{NaOH}}$ $CH_3CH_2\overset{\overset{\displaystyle O}{||}}{C}-\underset{\underset{\displaystyle H}{|}}{N}$[phenyl]

h) $CH_3CH_2\overset{\overset{\displaystyle O}{||}}{C}Cl$ + $CH_3\overset{\overset{\displaystyle O}{||}}{C}\ddot{O}:^-\,Na^+$ $\longrightarrow$ $CH_3CH_2\overset{\overset{\displaystyle O}{||}}{C}-O\overset{\overset{\displaystyle O}{||}}{C}CH_3$

21.48 The reagents in parts a, d, f, and h do not react with methyl propanoate.

b) $CH_3CH_2\overset{\overset{\displaystyle O}{||}}{C}OCH_3$ $\xrightarrow[\text{2. } H_3O^+]{\text{1. LiAlH}_4}$ $CH_3CH_2CH_2OH$

c) $CH_3CH_2\overset{\overset{\displaystyle O}{||}}{C}OCH_3$ $\xrightarrow[\text{2. } H_3O^+]{\text{1. } 2\,CH_3MgBr}$ $CH_3CH_2\overset{\overset{\displaystyle OH}{|}}{\underset{\underset{\displaystyle CH_3}{|}}{C}}CH_3$

e) $CH_3CH_2\overset{\overset{\displaystyle O}{||}}{C}OCH_3$ $\xrightarrow{H_3O^+}$ $CH_3CH_2\overset{\overset{\displaystyle O}{||}}{C}OH$ + CH_3OH

g) $CH_3CH_2\overset{\overset{\displaystyle O}{||}}{C}OCH_3$ +   [aniline] $\xrightarrow{\text{NaOH}}$ $CH_3CH_2\overset{\overset{\displaystyle O}{||}}{C}-\underset{\underset{\displaystyle H}{|}}{N}$[phenyl]

21.49 The reagents in parts a, d, f, g, and h do not react with propanamide.

b) $CH_3CH_2\overset{\overset{\displaystyle O}{\|}}{C}NH_2$ $\xrightarrow[\text{2. H}_2\text{O}]{\text{1. LiAlH}_4}$ $CH_3CH_2CH_2NH_2$

c) $CH_3CH_2\overset{\overset{\displaystyle O}{\|}}{C}NH_2$ $\xrightarrow[\text{2. H}_3\text{O}^+]{\text{1. CH}_3\text{MgBr}}$ $CH_3CH_2\overset{\overset{\displaystyle O}{\|}}{C}\overset{\displaystyle \cdot\cdot}{N}HMgBr$ $+$ CH_4

e) $CH_3CH_2\overset{\overset{\displaystyle O}{\|}}{C}NH_2$ $\xrightarrow{\text{H}_3\text{O}^+}$ $CH_3CH_2\overset{\overset{\displaystyle O}{\|}}{C}OH$ $+$ $NH_4{}^+$

The reagent in parts a, d, f, g, and h do not react with propanenitrile.

b) $CH_3CH_2C\equiv N$ $\xrightarrow[\text{2. H}_2\text{O}]{\text{1. LiAlH}_4}$ $CH_3CH_2CH_2NH_2$

c) $CH_3CH_2C\equiv N$ $\xrightarrow[\text{2. H}_3\text{O}^+]{\text{1. CH}_3\text{MgBr}}$ $CH_3CH_2\overset{\overset{\displaystyle O}{\|}}{C}CH_3$

e) $CH_3CH_2C\equiv N$ $\xrightarrow{\text{H}_3\text{O}^+}$ $CH_3CH_2\overset{\overset{\displaystyle O}{\|}}{C}OH$ $+$ $^+NH_4$

21.50

a)

b) The trifluoroacetate portion of the mixed anhydride is an especially good leaving group. The electron-withdrawing fluorine atoms, which stabilize the negative charge of the trifluoroacetate anion, make the mixed anhydride more reactive than other anhydrides.

c) Because trifluoroacetate is a better leaving group than other carboxylate anions, the reaction proceeds as indicated.

21.51

a) $CH_3CH_2CH_2CH_2C\equiv N$ $\xrightarrow[\text{2. H}_2\text{O}]{\text{1. LiAlH}_4}$ $CH_3CH_2CH_2CH_2CH_2NH_2$

b) $CH_3CH_2CH_2CH_2C\equiv N$ $\xrightarrow{H_3O^+}$ $CH_3CH_2CH_2CH_2\overset{\overset{\displaystyle O}{\|}}{C}OH$

$\Big\downarrow$ SOCl$_2$ / CHCl$_3$

$CH_3CH_2CH_2CH_2\overset{\overset{\displaystyle O}{\|}}{C}N(CH_3)_2$ $\xleftarrow{2\ HN(CH_3)_2}$ $CH_3CH_2CH_2CH_2\overset{\overset{\displaystyle O}{\|}}{C}Cl$

$\Big\downarrow$ 1. LiAlH$_4$ 2. H$_2$O

$CH_3CH_2CH_2CH_2CH_2N(CH_3)_2$

c) $CH_3CH_2CH_2CH_2C\equiv N$ $\xrightarrow[\text{2. }H_3O^+]{\text{1. }CH_3MgBr}$ $CH_3CH_2CH_2CH_2\overset{\overset{\displaystyle O}{\|}}{C}CH_3$

$\Big\downarrow$ 1. CH$_3$MgBr 2. H$_3$O$^+$

$CH_3CH_2CH_2CH_2\overset{\overset{\displaystyle OH}{|}}{\underset{\underset{\displaystyle CH_3}{|}}{C}}CH_3$

d) $CH_3CH_2CH_2CH_2\overset{\overset{\displaystyle O}{\|}}{C}CH_3$ $\xrightarrow[\text{2. }H_3O^+]{\text{1. }NaBH_4}$ $CH_3CH_2CH_2CH_2\overset{\overset{\displaystyle OH}{|}}{\underset{\underset{\displaystyle H}{|}}{C}}CH_3$
(from part c)

e) $CH_3CH_2CH_2CH_2C\equiv N$ $\xrightarrow[\text{2. }H_2O]{\text{1. DIBAH}}$ $CH_3CH_2CH_2CH_2\overset{\overset{\displaystyle O}{\|}}{C}H$

21.52 First, convert cyclohexanol to bromocyclohexane by reaction with PBr$_3$. The first two transformations start with bromocyclohexane.

a)

b)

c)

d)

21.53

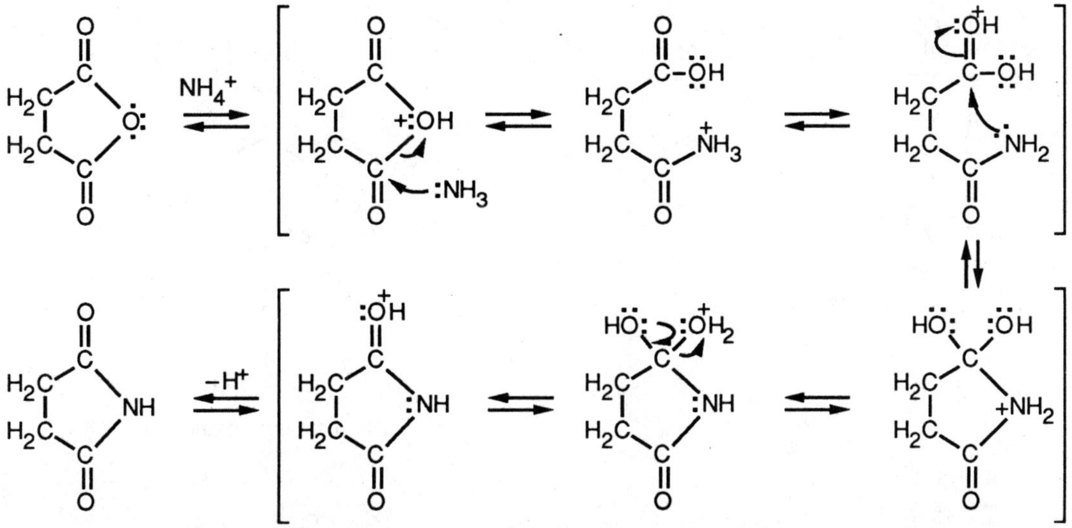

This reaction requires high temperatures because the intermediate amide is non-nucleophilic and the carboxylic acid carbonyl group is unreactive.

21.54 This synthesis requires a *nucleophilic aromatic substitution* reaction, which we studied in Section 16.9.

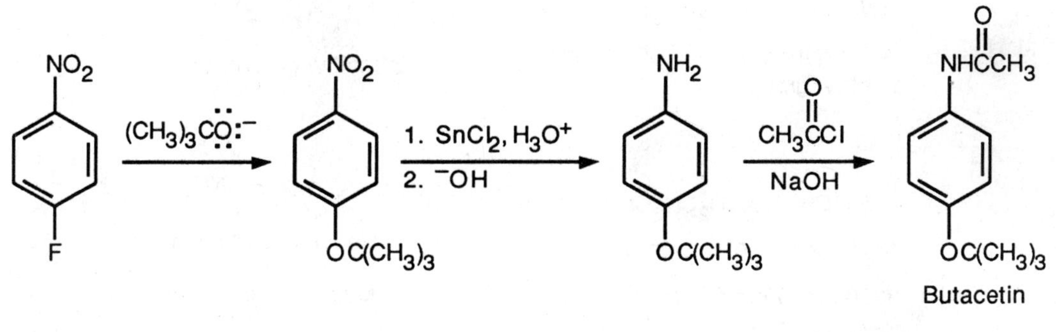

Butacetin

21.55

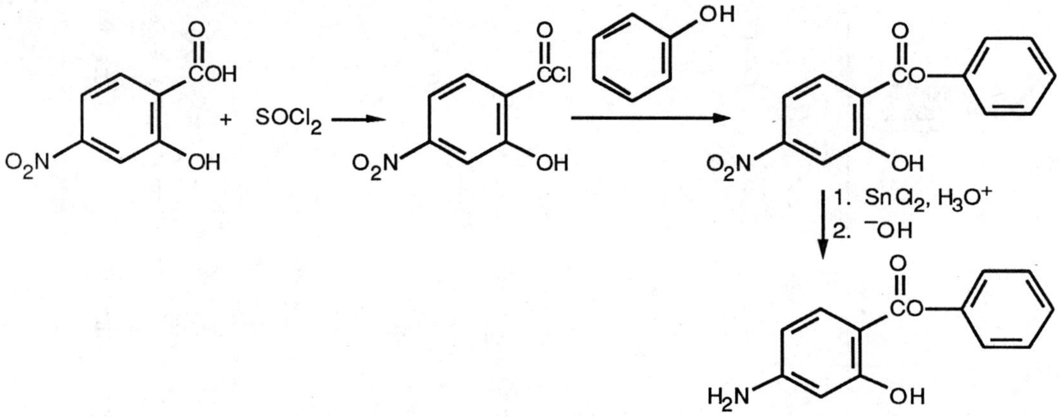

Phenyl 4–aminosalicylate

21.56

a)

$$CH_3CH_2\overset{\overset{\displaystyle O}{||}}{C}NHCH_3$$

N–Methylpropanamide

IR   1680 cm^{-1}
(*N*–substituted amide)

^{1}H NMR   1 methyl group
1 ethyl group

$$CH_3\overset{\overset{\displaystyle O}{||}}{C}N(CH_3)_2$$

N,N–Dimethylacetamide

1650 cm^{-1}
(*N,N*–disubstituted amide)

3 methyl groups

b)

$$HOCH_2CH_2CH_2CH_2C{\equiv}N$$

5–Hydroxypentanenitrile

IR   3300–3400 cm^{-1} (hydroxyl)
2250 cm^{-1} (nitrile)

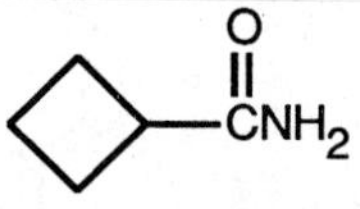

Cyclobutanecarboxamide

1690 cm^{-1} (amide)

c)

$$ClCH_2CH_2CH_2\overset{\overset{\displaystyle O}{||}}{C}OH$$

4–Chlorobutanoic acid

IR   1710 cm^{-1} (carboxylic acid)

$$CH_3OCH_2CH_2\overset{\overset{\displaystyle O}{||}}{C}Cl$$

3–Methoxypropanoyl chloride

1810 cm^{-1} (acid chloride)

d)

$$CH_3CH_2\overset{\overset{\displaystyle O}{\|}}{C}OCH_2CH_3$$

Ethyl propanoate

$$CH_3\overset{\overset{\displaystyle O}{\|}}{C}OCH_2CH_2CH_3$$

Propyl acetate

^{1}H NMR

2 triplets
2 quartets

1 singlet
1 triplet
1 quartet
1 multiplet

21.57

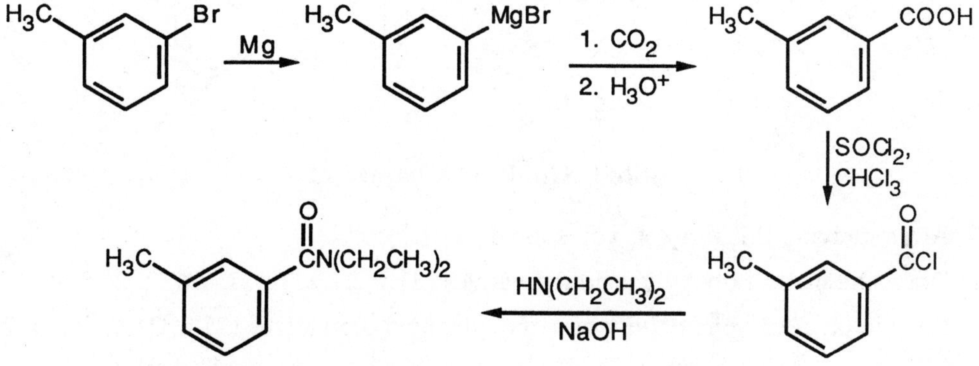

N,N-Dimethyl-m-toluamide

21.58

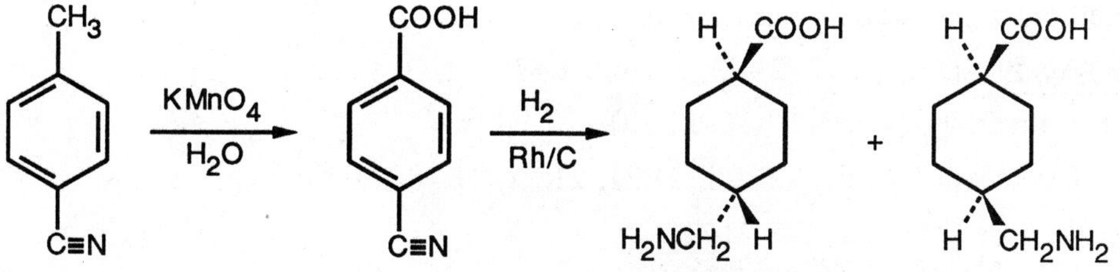

major product
Tranexamic acid

minor product

21.59

$$CH_3CHCl\overset{\overset{\displaystyle O}{\|}}{C}OCH_3$$

21.60

$$CH_3CH_2CH_2C\equiv N$$

21.61

$$CH_3CH_2\overset{\overset{\displaystyle O}{\|}}{C}OCH_2CH_3$$
$$a\quad b\qquad\quad c\quad d$$

a   1.15 δ
b   2.30 δ
c   4.15 δ
d   1.28 δ

Accidental overlap of the methyl group triplets a and d causes the absorptions to appear as a distorted quartet.

21.62

a)

$$CH_3CH_2CH_2\overset{\overset{\displaystyle O}{\|}}{C}Cl$$

b)

$$CH_3\overset{\displaystyle CH}{\underset{\underset{\displaystyle CN}{|}}{C}}\overset{\overset{\displaystyle O}{\|}}{C}OCH_2CH_3$$

c)

$$CH_3\overset{\overset{\displaystyle O}{\|}}{C}OCH(CH_3)_2$$

21.63

a)

$$ClCH_2CH_2\overset{\overset{\displaystyle O}{\|}}{C}OCH_2CH_3$$

b)

$$CH_3CH_2O\overset{\overset{\displaystyle O}{\|}}{C}CH_2\overset{\overset{\displaystyle O}{\|}}{C}OCH_2CH_3$$

c)

$$\text{(phenyl)}CH=CH\overset{\overset{\displaystyle O}{\|}}{C}OCH_2CH_3$$

Study Guide for Chapter 21

After studying this chapter, you should be able to:

(1)   Draw and name carboxylic acid derivatives (21.1, 21.2, 21.31, 21.32, 21.33).

(2)   Explain the relative reactivity of carboxylic acid derivatives (21.3, 21,4, 21.34, 21.44, 21.45).

(3)   Prepare carboxylic acid derivatives (21.5, 21.6, 21.8, 21.9, 21.18).

(4)   Predict the products of reactions of:

(a) Acid halides (21.12, 21.13, 21.14, 21.47).

(b) Anhydrides (21.15, 21.16, 21.17).

(c) Esters and lactones (21.20, 21.21, 21.22, 21.48, 21.53).

(d) Amides (21.23, 21.24, 21.49).

(e) Nitriles (21.25, 21.26, 21.27, 21.49, 21.51).

(5)   Formulate nucleophilic acyl substitution mechanisms for carboxylic acid derivatives (21.10, 21.11, 21.19, 21.28, 21.35, 21.36, 21.38, 21.40, 21.41, 21.42, 21.43, 21.46, 21.50).

(6)   Use nucleophilic acyl substitution reactions in synthesis (21.54, 21.55, 21.57, 21.58).

(7)   Identify carboxylic acid derivatives by spectroscopic techniques (21.29, 21.30, 21.56, 21.59, 21.60, 21.61, 21.62, 21.63).

22.1–22.2 Acidic hydrogens in the keto form of each of these compounds are underlined.

Keto Form	Enol Form	Number of Acidic Hydrogens
a)		4
b) H–C–CCl	H–C=C–Cl	3
c) H–C–COCH$_2$CH$_3$	H–C=COCH$_2$CH$_3$	3
d) CH$_3$–C–CH	CH$_3$–C=C–H	2
e) H–C–CO–H	H–C=C–OH	4
f)	–C=C–CH$_3$ or –C–C=CH$_2$	5
g)	–C=C–H	3

22.3

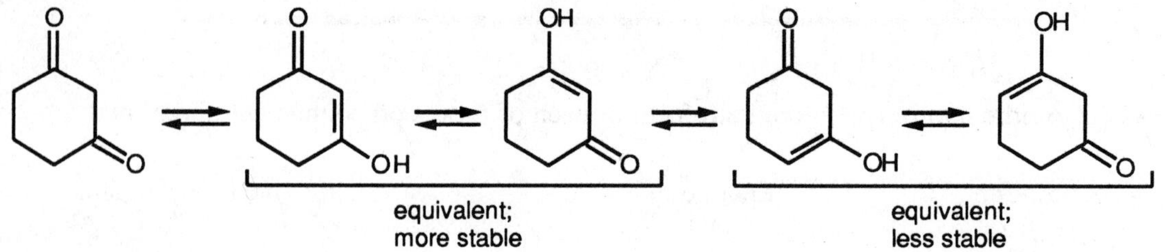

equivalent;
more stable

equivalent;
less stable

The first two mono-enols are more stable because the enol double bond is conjugated with the carbonyl group.

22.4

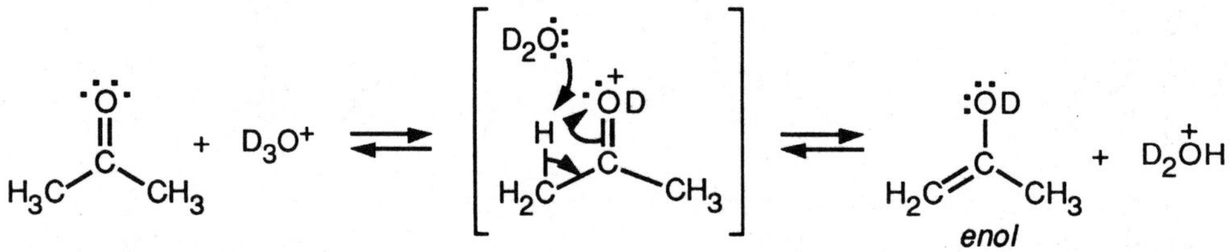

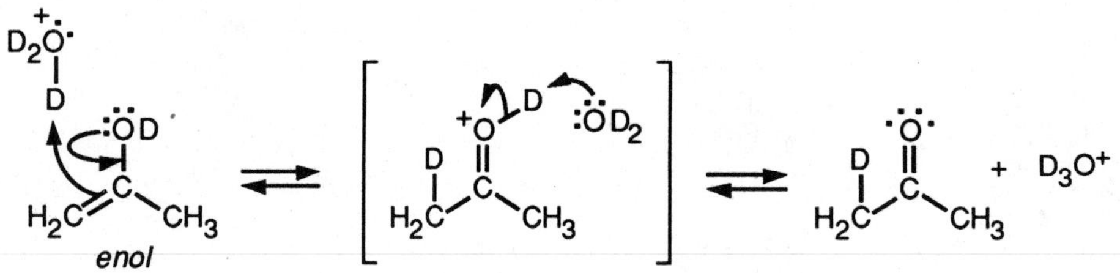

22.5

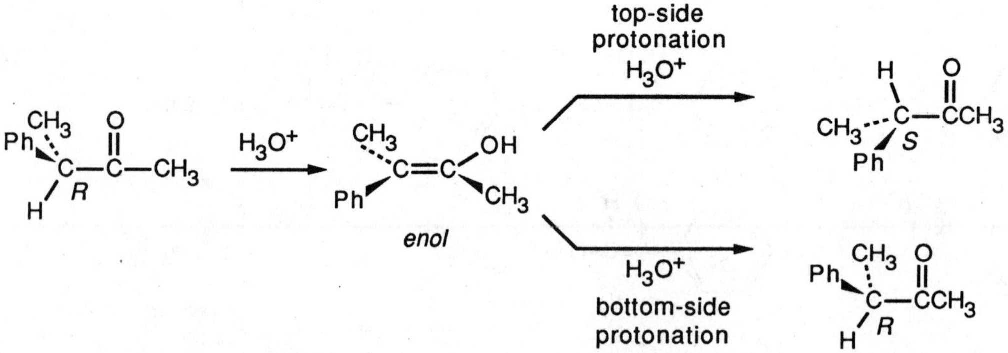

Loss of the proton at C3 during enolization produces a planar, sp^2–hybridized carbon at the original stereogenic center. Protonation occurs from either the top or the bottom side of the double bond to produce racemic 3–phenyl–2–butanone.

22.6

(R)–3–Methyl–3–phenyl–2–pentanone   *enol*

(R)–3–Methyl–3–phenyl–2–pentanone can enolize only toward the methyl group since there is no proton at carbon 3. Because the stereogenic center is not involved in the enolization process, the ketone is not racemized by acid treatment.

22.7

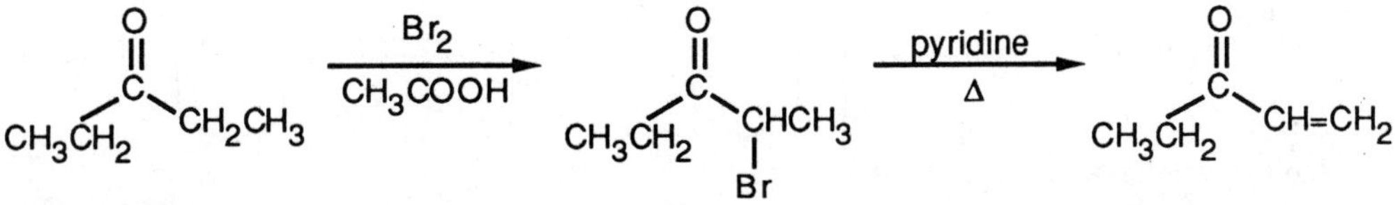

22.8   Hell-Volhard-Zelinskii bromination involves formation of an intermediate acid bromide enol. As in Problem 22.5, enolization results in loss of configuration at the stereogenic center. Bromination can occur from either side of the enol double bond, producing racemic 2–bromo–2–phenylpropanoic acid.

22.9

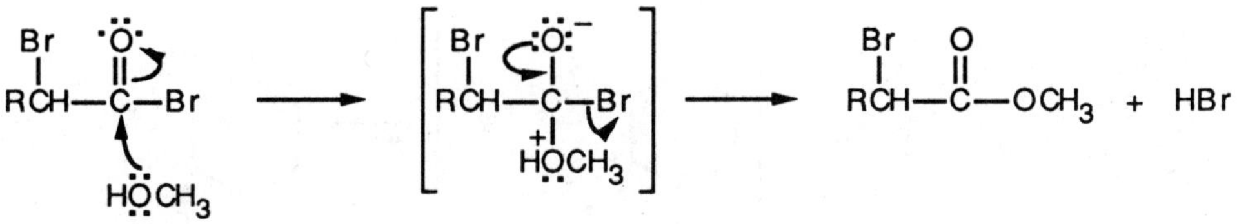

22.10

a) 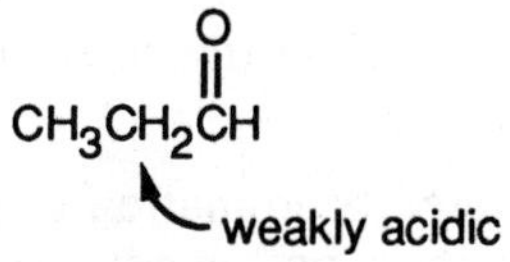 CH_3CH_2CH (with O double bond)
        weakly acidic

b) 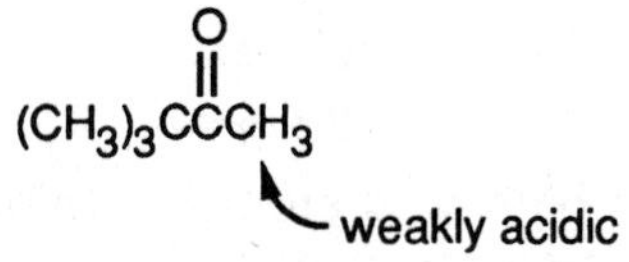 $(CH_3)_3CCCH_3$ (with O double bond)
        weakly acidic

c) 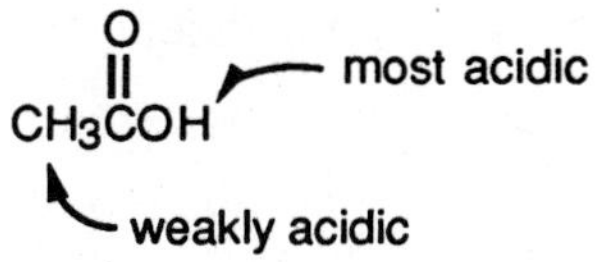 CH_3COH (with O double bond)
        most acidic
        weakly acidic

d) 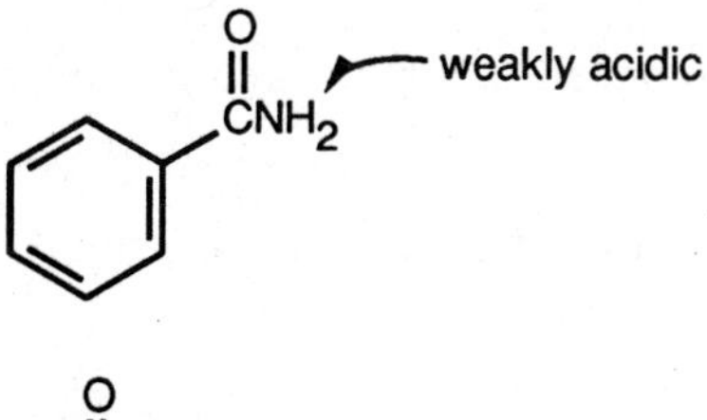   (benzamide structure) CNH_2 (with O double bond)
        weakly acidic

e) 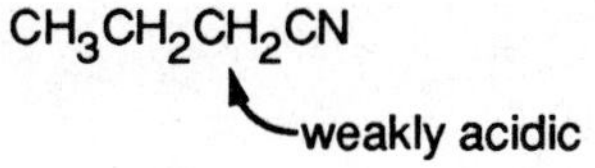 $CH_3CH_2CH_2CN$
        weakly acidic

f) 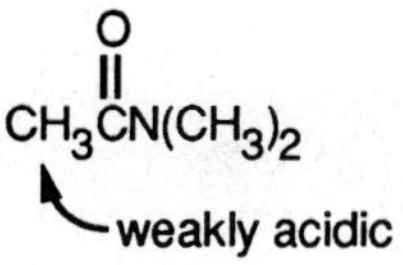 $CH_3CN(CH_3)_2$ (with O double bond)
        weakly acidic

g)

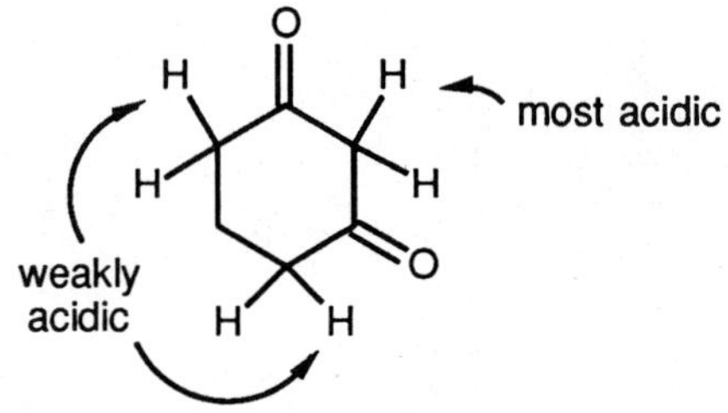

22.11

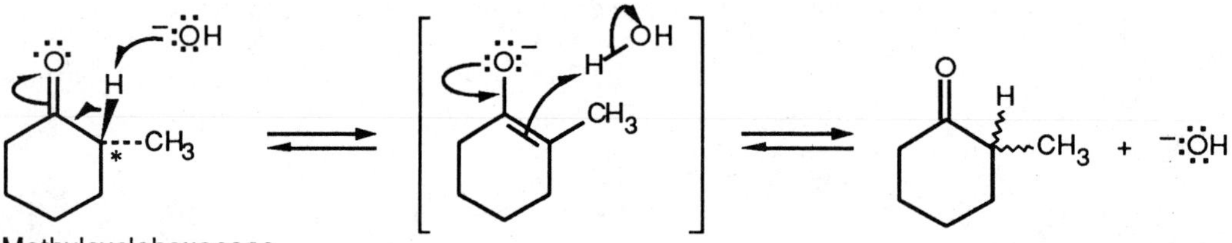

(R)–2–Methylcyclohexanone

Carbon 2 loses its chirality when the enolate ion double bond is formed. Protonation occurs with equal probability from either side of sp^2–hybridized carbon 2, resulting in racemic product.

22.12

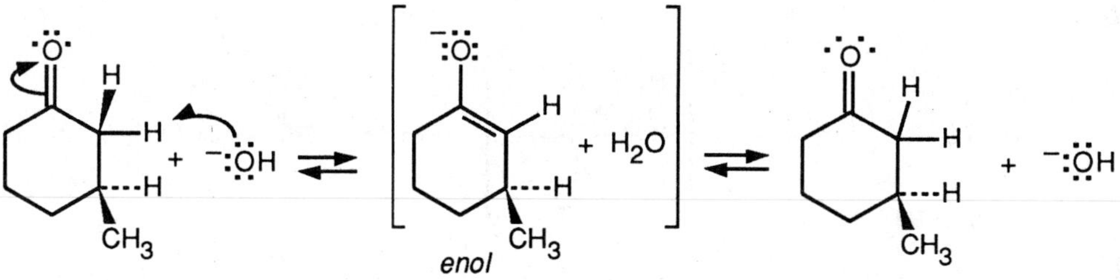

(S)–3–Methylcyclohexanone is not racemized by base because its stereogenic center is not involved in the enolization reaction.

22.13 Since both base-promoted chlorination and base-promoted bromination occur at the same rate, the step involving halogen must come after the slow, or rate-limiting, step. Formation of the enolate ion is the rate-limiting step and is dependent only on the concentrations of ketone and base.

$$\text{rate} = k \, [\text{base}] \, [\text{ketone}]$$

22.14 Halogenation in acid medium:

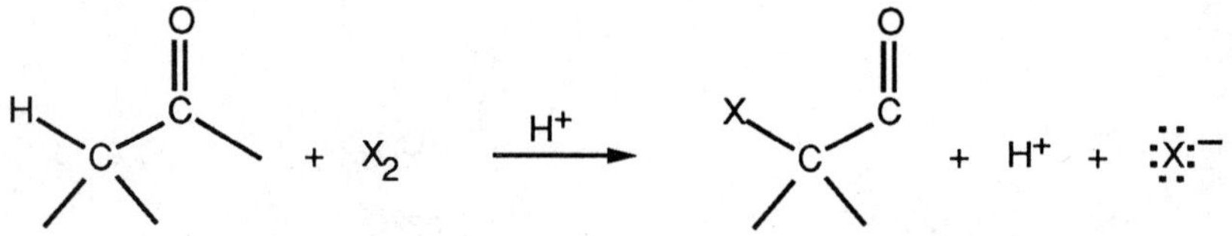

Halogenation in basic medium:

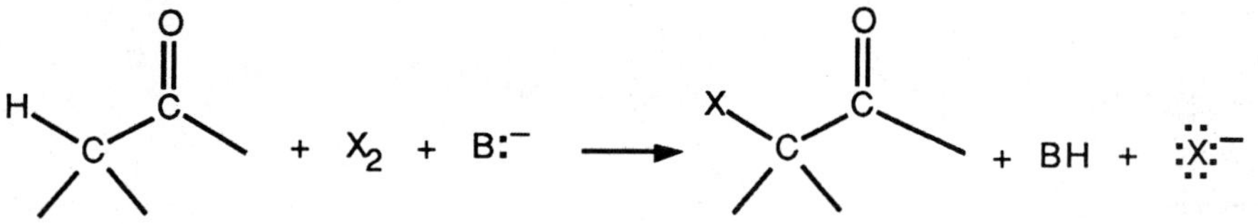

Ketone halogenation in acid medium requires a small amount of acid to form the enol initially. However, hydrogen ions are generated as a byproduct of halogenation, and they carry on the catalysis of enol formation.

In basic medium, the function of the base is to form the enolate anion by removing a proton from the ketone. Each molecule of ketone requires one molecule of base. Because base is consumed in stoichiometric amounts, the reaction in basic medium is not base-catalyzed.

22.15

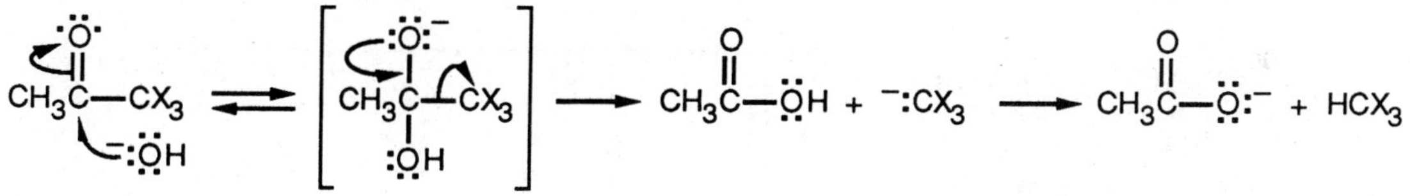

22.16

a)

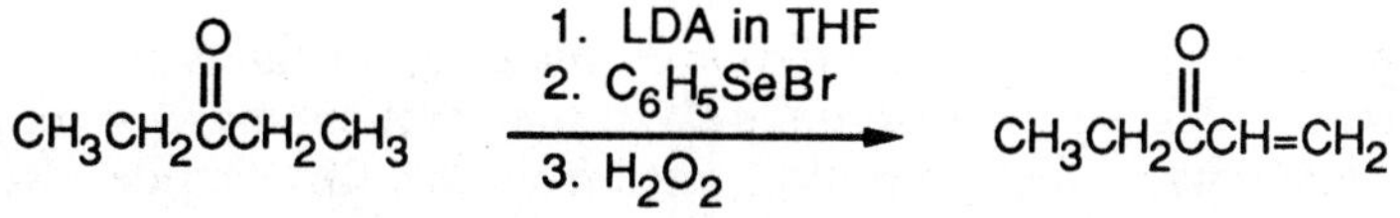

b)

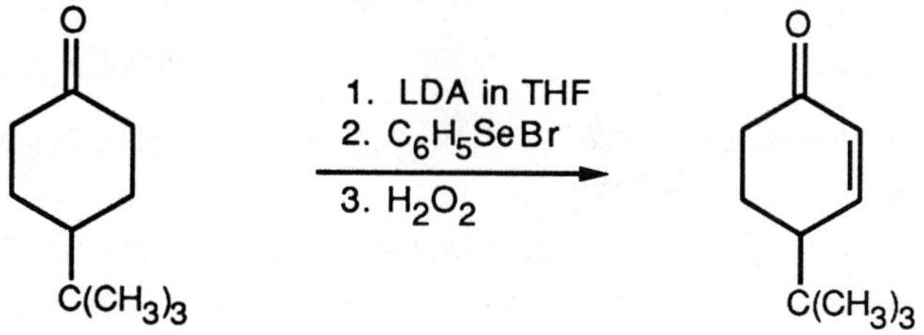

c)

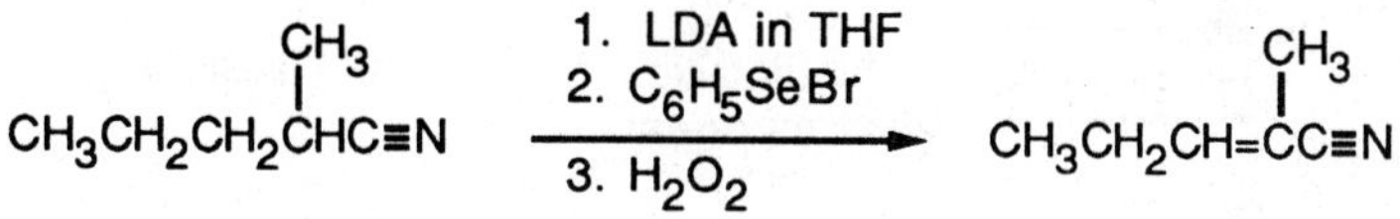

22.17

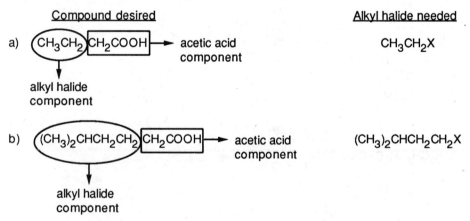

Acid catalyzes the formation of the enol.

The pi electrons from the enol double bond attack phenylselenenyl chloride.

Loss of a proton from oxygen yields the α –phenylseleno ketone. This reaction is similar to acid-catalyzed α –halogenation of ketones.

22.18 Keep in mind that the malonic ester synthesis produces substituted acetic acid compounds. Look for the acetic acid component of the compound you want to synthesize. The remainder of the molecule comes from the alkyl halide. Remember that the alkyl halide should be primary or methyl.

Compound desired		Alkyl halide needed
a) (CH₃CH₂)CH₂COOH →	acetic acid component	CH₃CH₂X
alkyl halide component		
b) ((CH₃)₂CHCH₂CH₂)CH₂COOH →	acetic acid component	(CH₃)₂CHCH₂CH₂X
alkyl halide component		

22.19

a) $CH_2(CO_2Et)_2$ $\xrightarrow[\text{2. PhCH}_2\text{Br}]{\text{1. Na}^+ \ ^-\text{OEt}}$ $PhCH_2CH(CO_2Et)_2$ + NaBr

$\downarrow H_3O^+, \Delta$

$PhCH_2CH_2CO_2H$ + CO_2 + $2\,C_2H_5OH$
3–Phenylpropanoic acid

b) $CH_2(CO_2Et)_2$ $\xrightarrow[\text{2. CH}_3\text{CH}_2\text{CH}_2\text{Br}]{\text{1. Na}^+ \ ^-\text{OEt}}$ $CH_3CH_2CH_2CH(CO_2Et)_2$ + NaBr

$\downarrow$ 1. Na$^+$ $^-$OEt
  2. CH$_3$Br

$CH_3CH_2CH_2CH(CH_3)CO_2H$ $\xleftarrow{H_3O^+, \Delta}$ $CH_3CH_2CH_2C(CH_3)(CO_2Et)_2$ + NaBr
2–Methylpentanoic acid
+ CO_2 + $2\,C_2H_5OH$

c) $CH_2(CO_2Et)_2$ $\xrightarrow[\text{2. (CH}_3)_2\text{CHCH}_2\text{Br}]{\text{1. Na}^+ \ ^-\text{OEt}}$ $(CH_3)_2CHCH_2CH(CO_2Et)_2$ + NaBr

$\downarrow H_3O^+, \Delta$

$(CH_3)_2CHCH_2CH_2CO_2H$ + CO_2 + $2\,C_2H_5OH$
4–Methylpentanoic acid

d) $CH_2(CO_2Et)_2$ $\xrightarrow[\text{2. BrCH}_2\text{CH}_2\text{CH}_2\text{Br}]{\text{1. 2 Na}^+ \ ^-\text{OEt}}$ [cyclobutane ring with CO₂Et, CO₂Et] + NaBr

$\downarrow H_3O^+, \Delta$

[cyclobutane]—CO₂CH₂CH₃ $\xleftarrow[\text{H}^+]{\text{CH}_3\text{CH}_2\text{OH}}$ [cyclobutane]—CO₂H + CO_2 + $2\,C_2H_5OH$

Ethyl cyclobutanecarboxylate

22.20 Since malonic ester has only two acidic hydrogen atoms, it can be alkylated only two times. Formation of trialkylated acetic acids is thus not possible.

22.21 As in the malonic acid synthesis, you should identify the structural fragments of the target compound. The "acetone component" comes from acetoacetic ester; the other component comes from a primary alkyl halide.

a)

$$(CH_3)_2CHCH_2 \overset{O}{\overset{\|}{CH_2CCH_3}} \longrightarrow \text{acetone component}$$

alkyl halide component

$$\underset{\underset{COOEt}{|}}{\overset{O}{\overset{\|}{CH_2CCH_3}}} \xrightarrow{Na^+ {}^-OEt} Na^+ {}^-:\underset{\underset{COOEt}{|}}{\overset{O}{\overset{\|}{CHCCH_3}}} \xrightarrow{(CH_3)_2CHCH_2Br} (CH_3)_2CHCH_2\underset{\underset{COOEt}{|}}{\overset{O}{\overset{\|}{CHCCH_3}}} + NaBr$$

$$\downarrow H_3O^+, \Delta$$

$$(CH_3)_2CHCH_2CH_2\overset{O}{\overset{\|}{CCH_3}} + CH_3CH_2OH$$

5–Methyl–2–hexanone   + CO_2

b)

$$C_6H_5CH_2CH_2 \overset{O}{\overset{\|}{CH_2CCH_3}} \longrightarrow \text{acetone component}$$

alkyl halide component

$$\underset{\underset{COOEt}{|}}{\overset{O}{\overset{\|}{CH_2CCH_3}}} \xrightarrow{Na^+ {}^-OEt} Na^+ {}^-:\underset{\underset{COOEt}{|}}{\overset{O}{\overset{\|}{CHCCH_3}}} \xrightarrow{C_6H_5CH_2CH_2Br} C_6H_5CH_2CH_2\underset{\underset{COOEt}{|}}{\overset{O}{\overset{\|}{CHCCH_3}}} + NaBr$$

$$\downarrow H_3O^+, \Delta$$

$$CH_3CH_2OH + C_6H_5CH_2CH_2CH_2\overset{O}{\overset{\|}{CCH_3}} + CO_2$$

5–Phenyl–2–pentanone

22.22

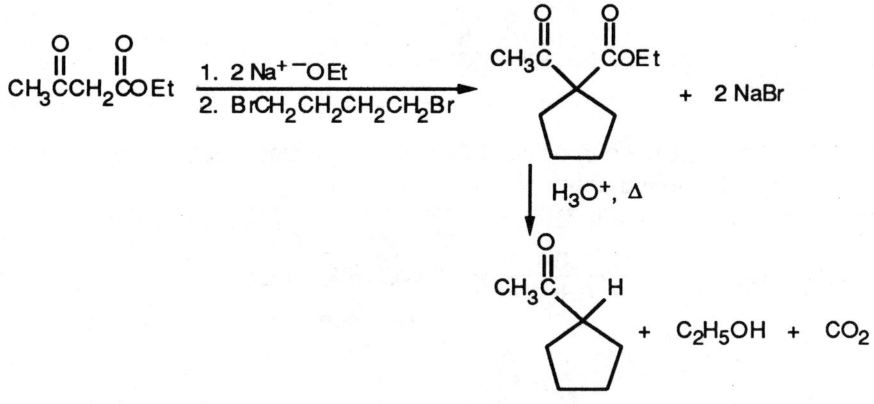

22.23 The acetoacetic ester synthesis can be used only if the desired products fulfill certain qualifications.

(1) Three carbons must originate from acetoacetic ester. In other words, compounds of the form $RCOCH_3$ can't be synthesized by the reaction of RX with acetoacetic ester.

(2) Alkyl halides must be primary or methyl.

(3) The acetoacetic ester synthesis can't be used to prepare compounds that are trisubstituted at the alpha position.

a)

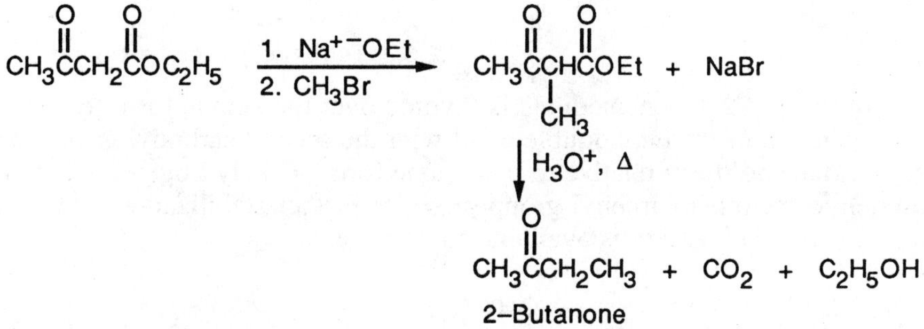

2–Butanone

b) Phenylacetone can't be produced by an acetoacetic ester synthesis because bromobenzene, the necessary halide, does not enter into S_N2 reactions. [See (2) above.]

c) Acetophenone can't be produced by an acetoacetic ester synthesis. [See (1) above.]

d) 3,3–Dimethyl–2–butanone can't be prepared because it is trisubstituted at the alpha position. [See (3) above.]

22.24

a)

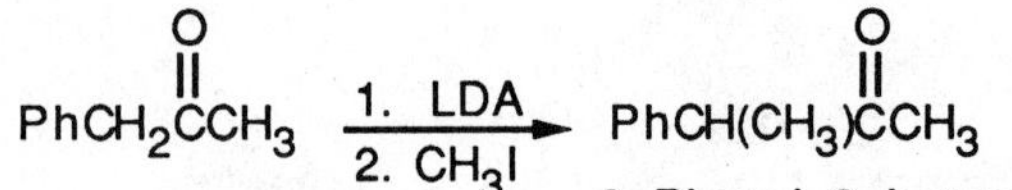

3–Phenyl–2–butanone

Alkylation occurs at the carbon next to the phenyl group because the phenyl group can stabilize the enolate anion intermediate.

b)

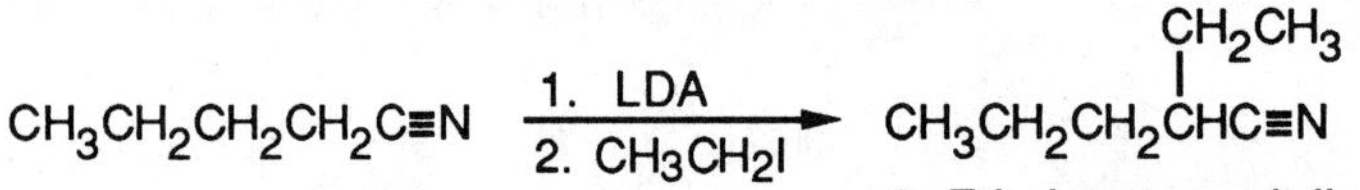

$$CH_3CH_2CH_2CH_2C\equiv N \xrightarrow[\text{2. } CH_3CH_2I]{\text{1. LDA}} CH_3CH_2CH_2\overset{\overset{\displaystyle CH_2CH_3}{|}}{C}HC\equiv N$$

2–Ethylpentanenitrile

c)

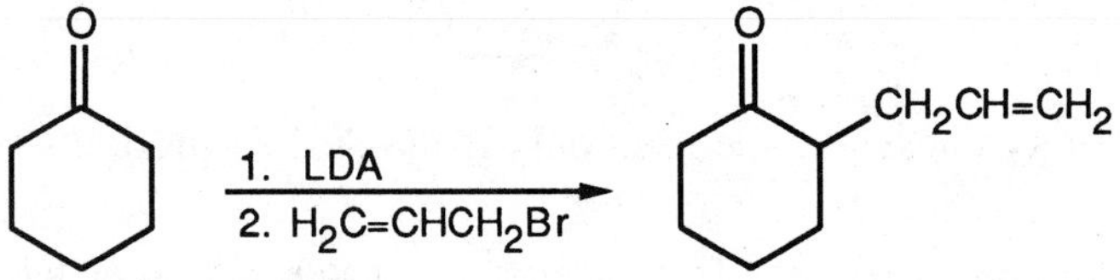

2–Allylcyclohexanone

d)

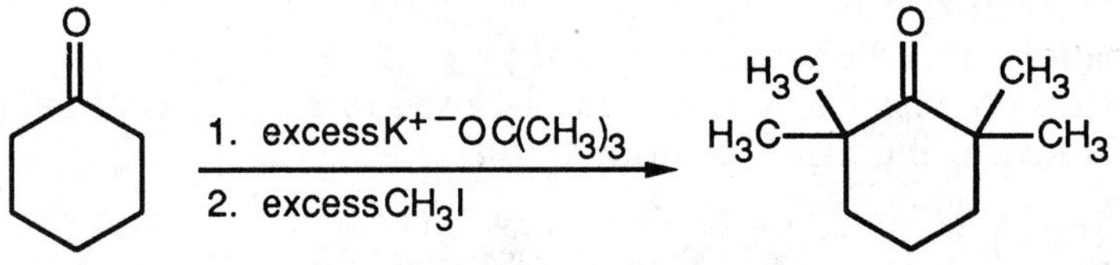

2,2,6,6–Tetramethylcyclohexanone

22.25 The enol form of 2,4–pentanedione is favored over the ketone form for two reasons. (1) Conjugation of the enol double bond with the second carbonyl group lowers the energy of the enol form relative to the ketone form. (2) Hydrogen bonding of the enol hydrogen to the other carbonyl group provides further stabilization. Neither of these types of enol stabilization is available to acetone.

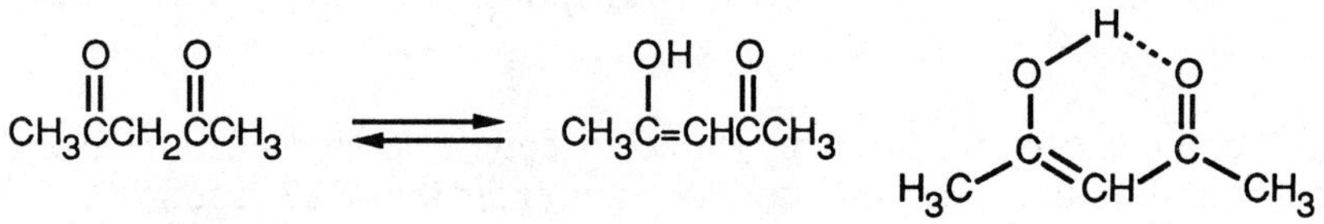

22.26

a)

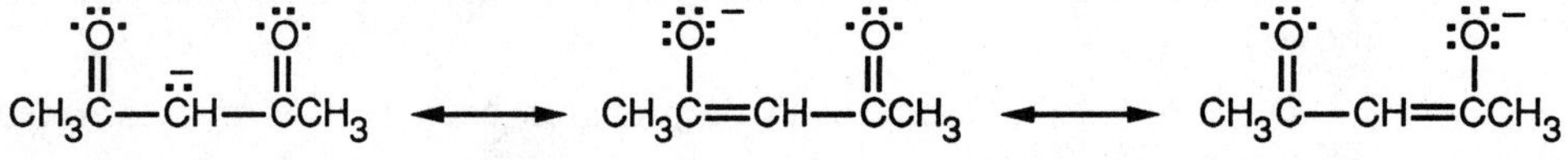

b)

c) 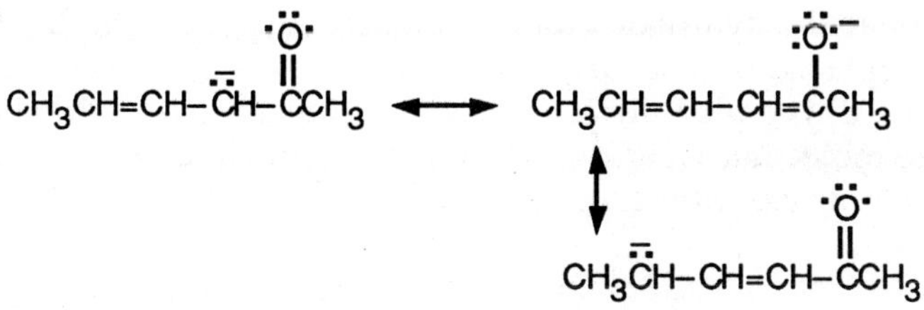 $CH_3CH=CH-\overset{..}{\underset{..}{C}}H-\overset{\overset{..}{\overset{..}{O}}}{\overset{\|}{C}}CH_3$ ⟷ $CH_3CH=CH-CH=\overset{:\overset{..}{O}:^-}{C}CH_3$

⟷

$CH_3\overset{..}{\underset{..}{C}}H-CH=CH-\overset{\overset{..}{\overset{..}{O}}}{\overset{\|}{C}}CH_3$

d) 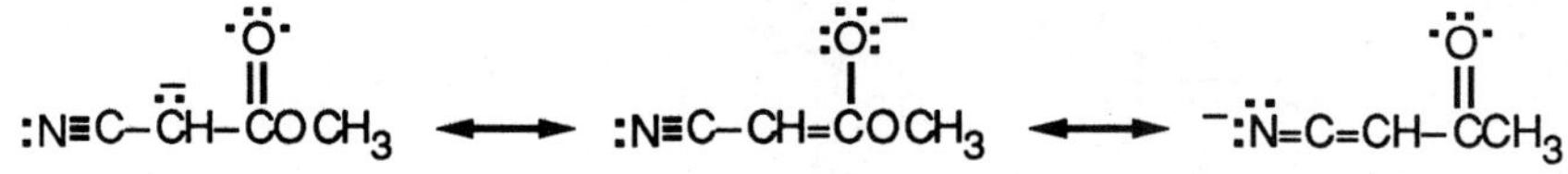 $:N≡C-\overset{..}{\underset{..}{C}}H-\overset{\overset{..}{\overset{..}{O}}}{\overset{\|}{C}}OCH_3$ ⟷ $:N≡C-CH=\overset{:\overset{..}{O}:^-}{C}OCH_3$ ⟷ $^-:\overset{..}{N}=C=CH-\overset{\overset{..}{\overset{..}{O}}}{\overset{\|}{C}}CH_3$

e)

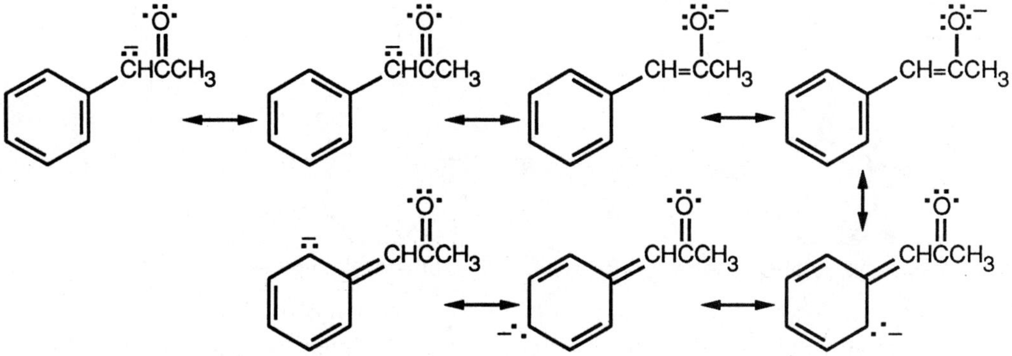

22.27 Acidic hydrogens are underlined.

a)

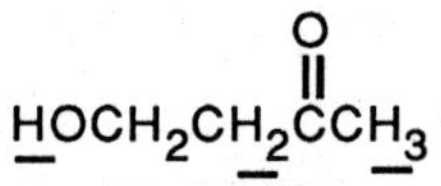

$\underline{H}OCH_2C\underline{H}_2\overset{\overset{O}{\|}}{C}C\underline{H}_3$

The alcohol hydrogen is more acidic than the alpha hydrogens

b)

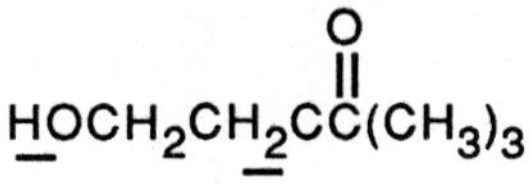

$\underline{H}OCH_2C\underline{H}_2\overset{\overset{O}{\|}}{C}C(CH_3)_3$

c)

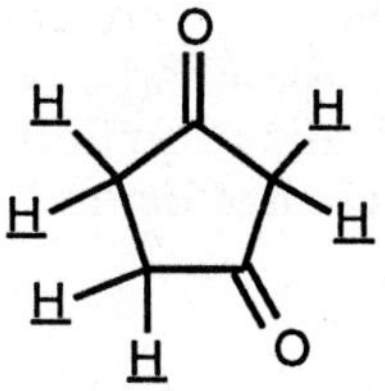

All hydrogens are somewhat acidic, although those between the carbonyls are the most acidic

d)

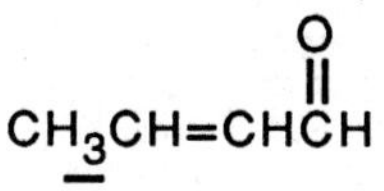

$CH_3C\underline{H}=CHC\overset{\overset{O}{\|}}{H}$

22.28 When a compound containing acidic hydrogen atoms is treated with NaOD in D_2O, all acidic protons are gradually replaced by deuterons. For each proton (atomic weight 1) lost, a deuteron (atomic weight 2) is added, for a molecular weight gain of 1 for each hydrogen replaced. Since the molecular weight of cyclohexanone increases from 98 to 102 after $NaOD/D_2O$ treatment, cyclohexanone contains four acidic hydrogen atoms.

22.29

2–Methylcycloheptanone   3–Methylcycloheptanone

2–Methylcycloheptanone contains three acidic hydrogen atoms; 3–methylcycloheptanone contains four. After treatment with NaOD in D_2O, the molecular weight of the 2–methyl isomer increases by 3 and that of the 3–methyl isomer increases by 4. A mass spectrum of the deuterated ketones can then distinguish between them.

22.30

Least acidic $\longrightarrow$ Most acidic

$$(CH_3CH_2)_2NH < CH_3\overset{O}{\overset{\|}{C}}CH_3 < CH_3CH_2OH < CH_3\overset{O}{\overset{\|}{C}}CH_2\overset{O}{\overset{\|}{C}}CH_3 < CH_3CH_2\overset{O}{\overset{\|}{C}}OH < Cl_3\overset{O}{\overset{\|}{C}}COH$$

22.31 A nitroso compound can be compared to a carbonyl compound. If there are protons *alpha* to the nitroso group, enolization similar to that observed for carbonyl compounds can occur, leading to formation of an oxime. Here, the enolization equilibrium favors the oxime. If no protons are adjacent to the nitroso group, enolization to the oxime cannot occur, and the nitroso compound is stable.

22.32

a) $CH_2(CO_2Et)_2$ $\xrightarrow[\text{2. } CH_3CH_2CH_2Br]{\text{1. } Na^{+-}OEt}$ $CH_3CH_2CH_2CH(CO_2Et)_2$ $+$ $NaBr$

$\downarrow H_3O^+, \Delta$

$CH_3CH_2CH_2CH_2CO_2C_2H_5$ $\xleftarrow[H^+]{C_2H_5OH}$ $CH_3CH_2CH_2CH_2CO_2H$ $+$ CO_2 $+$ $2\,C_2H_5OH$

Ethyl pentanoate

b)   This ester would be difficult to synthesize by the malonic ester route because the necessary alkyl halide, 2–bromopropane, is a secondary bromide that undergoes elimination as well as substitution.

c) $CH_2(CO_2Et)_2$ $\xrightarrow[\text{2. } CH_3CH_2Br]{\text{1. } Na^+ {}^-OEt}$ $CH_3CH_2CH(CO_2Et)_2$ + NaBr

$\xrightarrow[\text{2. } CH_3Br]{\text{1. NaOEt}}$

$CH_3CH_2\overset{\displaystyle CH_3}{\underset{\displaystyle |}{CH}}COOH$ + $2\,C_2H_5OH$ + CO_2 $\xleftarrow[\Delta]{H_3O^+}$ $CH_3CH_2\overset{\displaystyle CH_3}{\underset{\displaystyle |}{C}}(CO_2Et)_2$ + NaBr

$\downarrow \overset{\displaystyle C_2H_5OH}{\underset{\displaystyle H^+}{}}$

$CH_3CH_2\overset{\displaystyle CH_3}{\underset{\displaystyle |}{CH}}CO_2C_2H_5$

Ethyl 2–methyl–butanoate

d)   The malonic ester route can't be used to synthesize *alpha*–trisubstituted carboxylic acids.

22.33

a) $\underset{\displaystyle \underset{|}{COOEt}}{CH_2\overset{\displaystyle O}{\overset{\|}{C}}CH_3}$ $\xrightarrow[\text{2. } 2\,CH_3CH_2Br]{\text{1. } 2\,Na^+{}^-OEt}$ $(CH_3CH_2)_2\underset{\displaystyle \underset{|}{COOEt}}{\overset{\displaystyle O}{\overset{\|}{C}}CH_3}$

$\downarrow H_3O^+,\ \Delta$

$(CH_3CH_2)_2CH\overset{\displaystyle O}{\overset{\|}{C}}CH_3$ + CH_3CH_2OH + CO_2

b) $\underset{\displaystyle \underset{|}{COOEt}}{CH_2\overset{\displaystyle O}{\overset{\|}{C}}CH_3}$ $\xrightarrow[\text{2. } CH_3CH_2CH_2Br]{\text{1. } Na^+{}^-OEt}$ $CH_3CH_2CH_2\underset{\displaystyle \underset{|}{COOEt}}{CH}\overset{\displaystyle O}{\overset{\|}{C}}CH_3$

$\downarrow \begin{array}{l}\text{1. } Na^+{}^-OEt\\ \text{2. } CH_3Br\end{array}$

$CH_3CH_2CH_2\overset{\displaystyle H_3C\ \ O}{\overset{|\ \ \ \|}{CH}}CCH_3$ $\xleftarrow[\Delta]{H_3O^+,}$ $CH_3CH_2CH_2\underset{\displaystyle \underset{|}{COOEt}}{\overset{\displaystyle H_3C\ \ O}{\overset{|\ \ \ \|}{C}}}{-}CCH_3$

+ CH_3CH_2OH + CO_2

22.34 Use a *malonic ester synthesis* if the product you want is an α–substituted carboxylic acid or derivative. Use an *acetoacetic acid synthesis* if the product you want in an α–substituted methyl ketone.

a) $CH_2(CO_2Et)_2$ $\xrightarrow[\text{2. } 2\,CH_3Br]{\text{1. } 2\,Na^+{}^-OEt}$ $(CH_3)_2C(CO_2C_2H_5)_2$ + 2 NaBr

b) $CH_2\overset{O}{\overset{||}{C}}CH_3$ $\xrightarrow[\text{2. BrCH}_2(CH_2)_4CH_2Br]{\text{1. 2 Na}^+\ ^-OEt}$
     $|$
     $COOEt$

cycloheptane with $\overset{O}{\overset{||}{C}}CH_3$ and $COOEt$

$\downarrow$ H_3O^+
     Δ

cycloheptane with $-\overset{O}{\overset{||}{C}}CH_3$ + CH_3CH_2OH + CO_2

c) $CH_2(COOEt)_2$ $\xrightarrow[\text{2. BrCH}_2CH_2CH_2Br]{\text{1. 2 Na}^+\ ^-OEt}$

cyclobutane with $COOEt$ / $COOEt$

$\downarrow H_3O^+$
     Δ

cyclobutane $-COOH$ + CH_3CH_2OH + CO_2

d) $CH_3\overset{O}{\overset{||}{C}}CH_2\overset{O}{\overset{||}{C}}OEt$ $\xrightarrow[\text{2. H}_2C=CHCH_2Br]{\text{1. Na}^+\ ^-OEt}$ $H_2C=CHCH_2\overset{}{C}H\overset{O}{\overset{||}{C}}CH_3$ + NaBr
                                                                            $|$
                                                                         CO_2Et

$\downarrow H_3O^+, \Delta$

$H_2C=CHCH_2CH_2\overset{O}{\overset{||}{C}}CH_3$ + CO_2 + C_2H_5OH

22.35

a)

cyclohexane with $\overset{COOH}{\underset{COOH}{}}$ $\xrightarrow{\Delta}$ cyclohexane with $COOH$ + CO_2

b) $(CH_3)_2CHCO_2C_2H_5$ $\xrightarrow[\text{2. C}_6H_5SeBr]{\text{1. LDA}}$ $(CH_3)_2\overset{SeC_6H_5}{\underset{\mathbf{A}}{C}}CO_2C_2H_5$ $\xrightarrow{H_2O_2}$ $H_2C=\overset{CH_3}{\underset{\mathbf{B}}{C}}CO_2C_2H_5$
                                                                                                          + C_6H_5SeOH

c)

cyclopentane-1,3-dione $\xrightarrow[\text{2. CH}_3I]{\text{1. Na}^+\ ^-OEt}$ 2-methylcyclopentane-1,3-dione

d) $CH_3CH_2CH_2COOH$ $\xrightarrow[\text{PBr}_3]{\text{Br}_2}$ $CH_3CH_2CHBrCOBr$ $\xrightarrow{\text{H}_2\text{O}}$ $CH_3CH_2CHBrCOOH$

A   B

e)

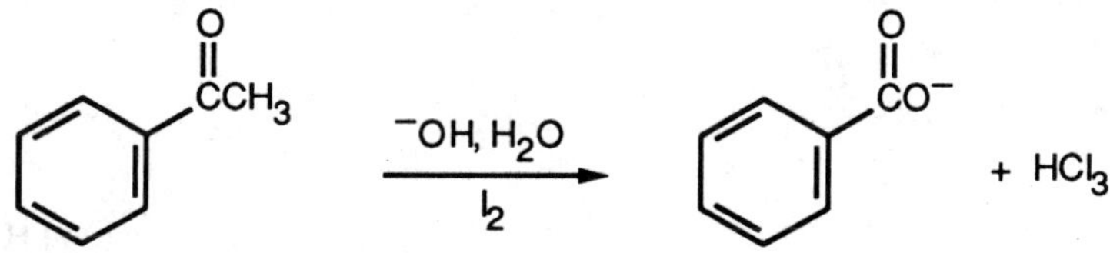

$\xrightarrow[\text{I}_2]{^-\text{OH, H}_2\text{O}}$

+ HCl_3

22.36

a)

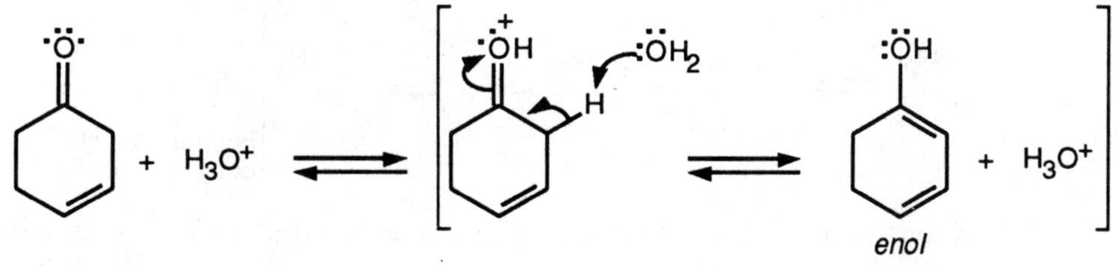

+ H_3O^+ ⇌ ... ⇌ ... + H_3O^+

enol

b)

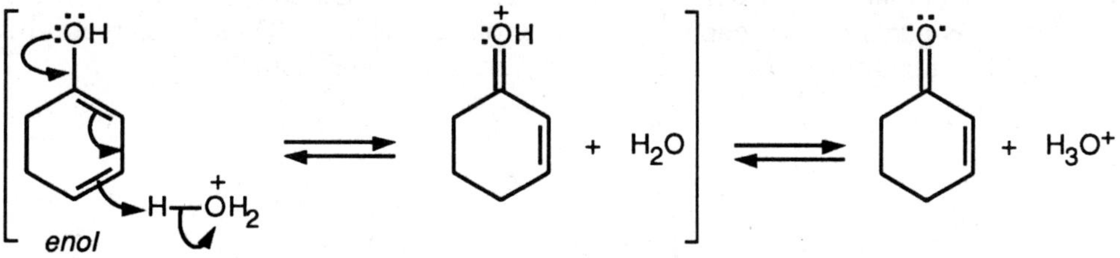

enol ⇌ ... + H_2O ⇌ ... + H_3O^+

22.37

+ BH

+ B:$^-$

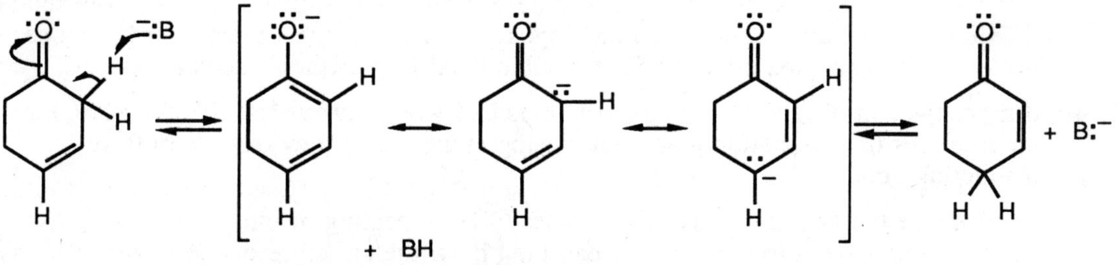

The enolate of 3–cyclohexenone can be reprotonated at three different positions. Reprotonation at the *gamma* position yields the α,ß–unsaturated ketone.

22.38

All protons in the five-membered ring can be exchanged by base treatment.

22.39 Protons *alpha* to a carbonyl group are acidic. We also know from Problem 22.37 that protons *gamma* to an enone carbonyl group are acidic. Thus for 2–methyl-2–cyclopentenone, protons at the starred positions are acidic.

Isomerization of a 2–substituted 2–cyclohexenone to a 6–substituted 2–cyclohexenone requires removal of a proton from the 5–position of the 2–substituted isomer. Since protons in this position are not acidic, double bond isomerization does not occur.

22.40 Start at the end of the sequence of reactions and work backwards. If necessary, cover up all pieces of information you are not using at the moment to keep them from distracting you.

a) Because the *keto acid* $C_9H_{13}NO_3$ loses CO_2 on heating, it must be a ß–keto acid. Neglecting stereoisomerism, we can then draw the structure of the ß–keto acid as:

b) When ecgonine ($C_9H_{15}NO_3$) is treated with CrO_3, the keto acid $C_9H_{13}NO_3$ is produced. Since CrO_3 is an oxidizing agent (used for oxidizing alcohols to carbonyl compounds), we can determine that ecgonine has the following structure. Again, the stereochemistry is unspecified.

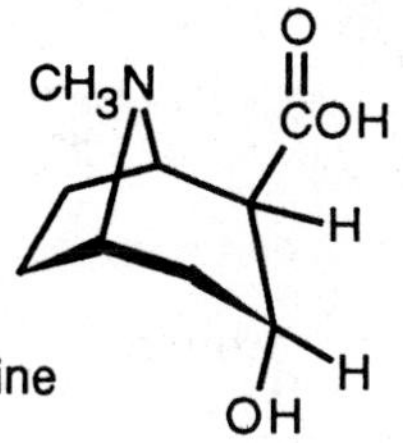

Ecgonine

c) Ecgonine contains carboxylic acid and alcohol functional groups. The other products of hydroxide treatment of cocaine are a carboxylic acid (benzoic acid) and an alcohol (methanol). Cocaine thus contains two ester functional groups, which are saponified on reaction with hydroxide.

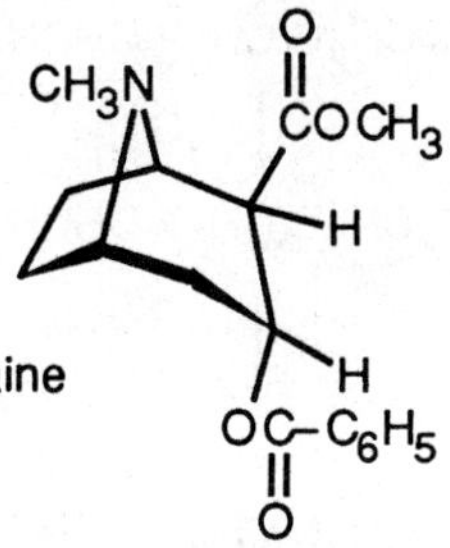

Cocaine

d) The complete reaction sequence:

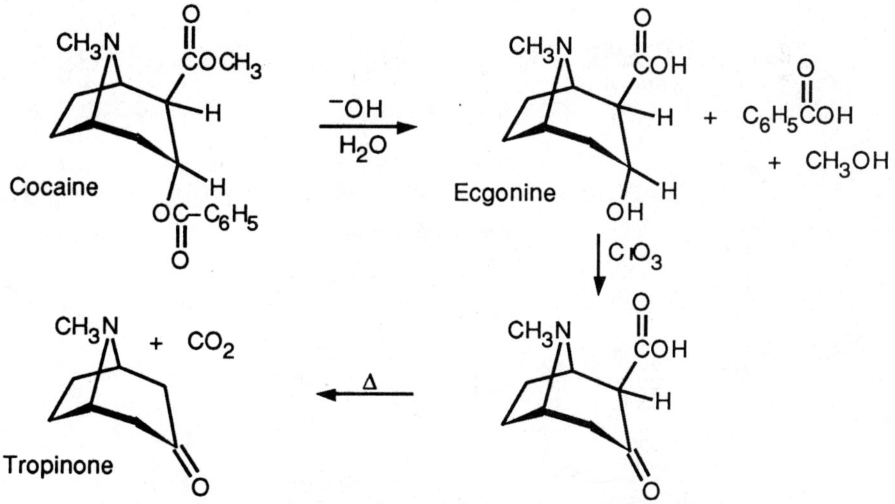

Cocaine

Ecgonine

Tropinone

22.41 The haloform reaction occurs only with methyl ketones.

Positive haloform reaction:

$$CH_3\overset{O}{\underset{||}{C}}CH_3, \ C_6H_5\overset{O}{\underset{||}{C}}CH_3$$

Negative haloform reaction:

$$CH_3CH_2CHO, \ CH_3COOH, \ CH_3C\equiv N$$

22.42 First treat geraniol with PBr$_3$ to form geranyl bromide,

$(CH_3)_2C=CHCH_2CH_2C(CH_3)=CHCH_2Br.$

a) CH_3CO_2Et $\xrightarrow[\begin{array}{c}2. \ \text{Geranyl}\\ \text{bromide}\end{array}]{1. \ \text{LDA}}$ $(CH_3)_2C=CHCH_2CH_2C(CH_3)=CHCH_2CH_2CO_2Et$

Ethyl geranylacetate

Alternatively:

$CH_2(CO_2Et)_2$ $\xrightarrow[\begin{array}{c}2. \ \text{Geranyl}\\ \text{bromide}\end{array}]{1. \ Na^+ \ ^-OEt}$ $(CH_3)_2C=CHCH_2CH_2C(CH_3)=CHCH_2CH(CO_2Et)_2$

$\downarrow H_3O^+, \ \Delta$

$(CH_3)_2C=CHCH_2CH_2C(CH_3)=CHCH_2CH_2COOH$

$\downarrow$ 1. SOCl$_2$
2. CH$_3$CH$_2$OH, pyridine

$(CH_3)_2C=CHCH_2CH_2C(CH_3)=CHCH_2CH_2CO_2Et$

Ethyl geranylacetate

b) $CH_3\overset{O}{\underset{||}{C}}CH_2CO_2Et$ $\xrightarrow[\begin{array}{c}2. \ \text{Geranyl}\\ \text{bromide}\end{array}]{1. \ Na^+ \ ^-OEt}$ $(CH_3)_2C=CHCH_2CH_2C(CH_3)=CHCH_2\underset{\underset{CO_2Et}{|}}{CH}\overset{O}{\underset{||}{C}}CH_3$

$\downarrow H_3O^+, \ \Delta$

$(CH_3)_2C=CHCH_2CH_2C(CH_3)=CHCH_2CH_2\overset{O}{\underset{||}{C}}CH_3$ + C_2H_5OH + CO_2

Geranylacetone

22.43

a)

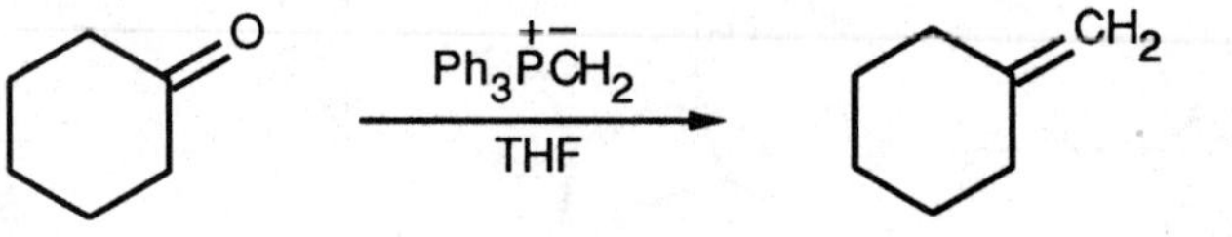

$\xrightarrow[\text{THF}]{Ph_3\overset{+-}{P}CH_2}$

b)   product of a) $\xrightarrow[\text{peroxides}]{HBr}$

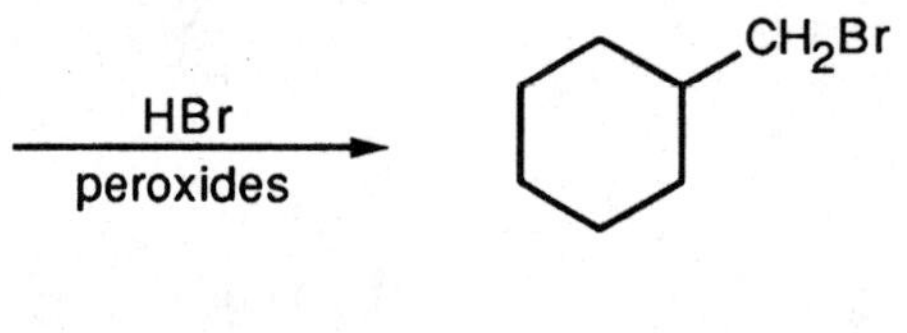

c)

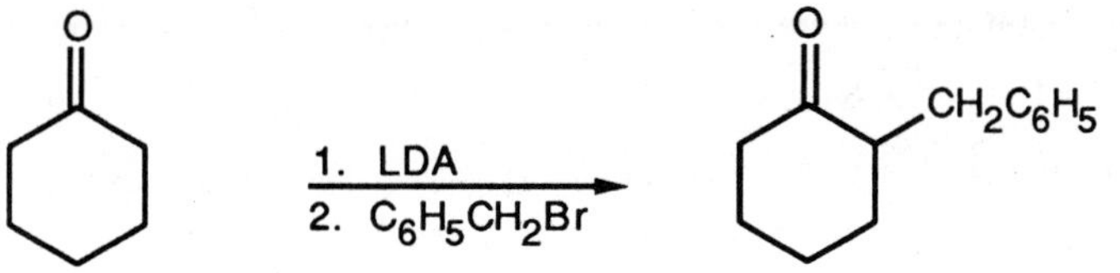

$$\text{1. LDA} \atop \text{2. } C_6H_5CH_2Br$$

d)

$$CH_2(CO_2Et)_2 \quad {\text{1. Na}^{+ -}\text{OEt} \atop \text{2. product} \atop \text{of b)}}$$

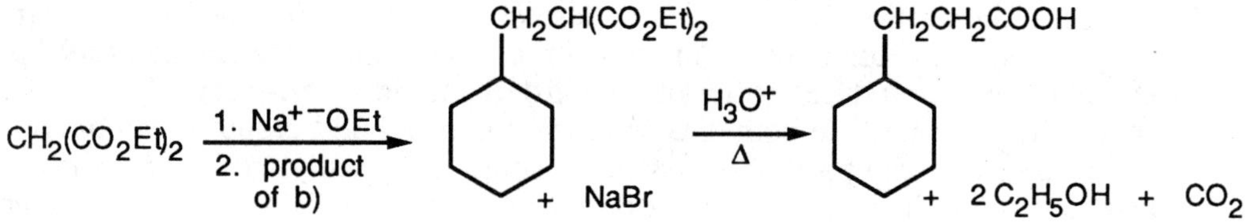

e)

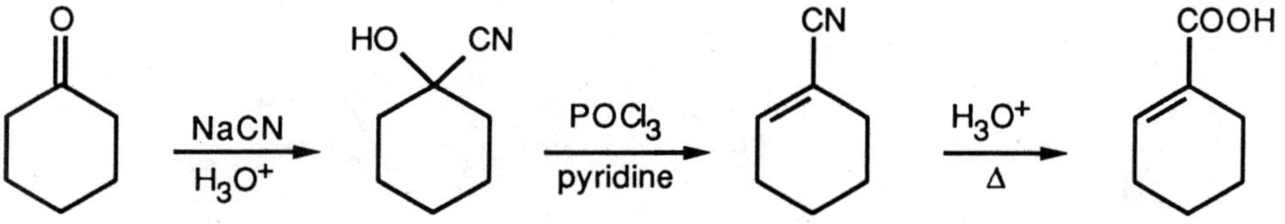

f)

Method 1:

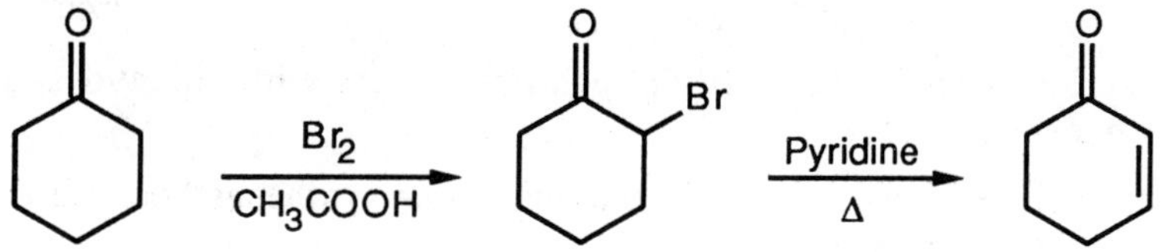

Method 2:

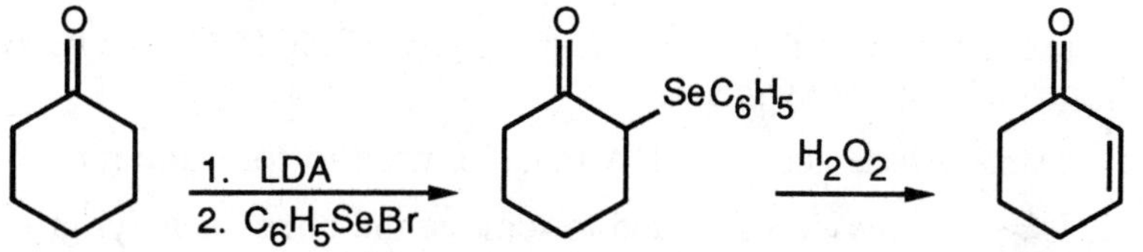

g)

product from d)

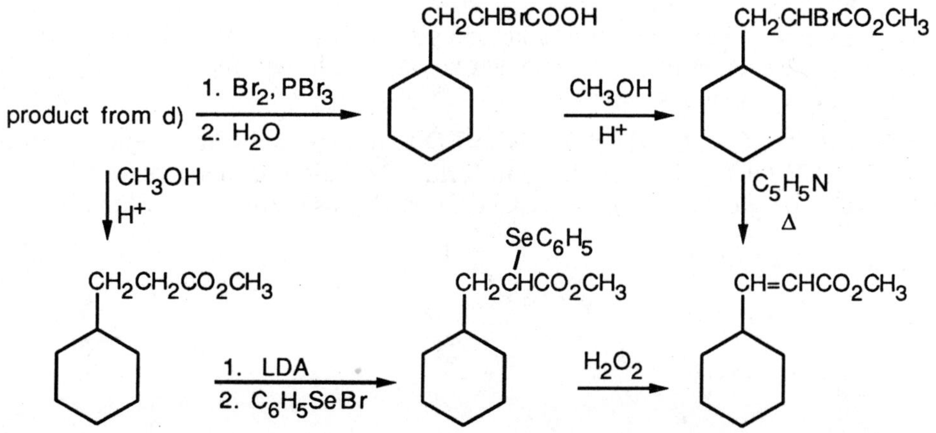

h)

$$\text{product from e)} \xrightarrow[\text{Pd/C}]{H_2}$$

22.44 Treatment of either the *cis* or *trans* with base causes enolization *alpha* to the carbonyl group and results in loss of configuration at the α–position. Reattachment of the proton at carbon 2 produces either of the diastereomeric 4–*tert*–butyl–2–methylcyclohexanones. In both diastereomers the *tert*–butyl group of carbon 4 occupies the equatorial position for steric reasons. The methyl group of the *cis* isomer is also equatorial; the methyl group of the *trans* isomer is axial. The *trans* isomer is less stable because of interactions of the axial methyl group with the ring protons.

22.45 a) Reaction with Br_2 at the *alpha* position occurs with aldehydes and ketones, not with esters.

b) Aryl halides can't be used in malonic ester syntheses because they don't undergo S_N2 displacement reactions.

c) The product of this reaction sequence, $H_2C=CHCH_2CH_2COCH_3$, is a methyl ketone, not a carboxylic acid.

d) (1) A strong base such as LDA must be used to form the nitrile anion;

(2) H_2O_2, not pyridine, is used to remove the phenylselenyl group and to form the double bond.

e) Base-promoted bromination of a ketone yields a mixture of mono– to tetra–bromo products. Reaction of the brominated ketones with pyridine and heat produces a mixture of compounds.

f) When $(CH_3)_2CHCOCH_3$ reacts with LDA, the less substituted enolate anion, $(CH_3)_2CHCOCH_2^-$ is more likely to form. Alkylation of this enolate with $C_6H_5CH_2Br$ yields $(CH_3)_2CHCOCH_2CH_2C_6H_5$ as the major product.

22.46

Enolization in the direction predicted for ß–diketones would produce an enol with a double bond at the "bridgehead" of the fused ring system. Since ring systems with bridgehead double bonds are highly strained, enolization occurs instead in the opposite direction. This diketone thus resembles a monoketone, rather than a diketone, in its pK_a and degree of enolization.

22.47 Both methylmagnesium bromide and *tert*–butylmagnesium bromide give the expected carbonyl addition products. The yield of the *tert*–butylmagnesium bromide addition product is very low, however, because of the difficulty of approach of the bulky *tert*-butyl Grignard reagent to the carbonyl carbon. More favorable is the acid-base reaction between the Grignard reagent and a carbonyl *alpha* proton.

When D_2O is added to the reaction mixture, the deuterated ketone is produced.

22.48 Laurene and its isomer differ only in the configuration at the carbon *alpha* to the methylene group. This chiral center, which is *alpha* to a carbonyl group in the starting material is epimerized by the basic Wittig reagent to yield an equilibrium mixture of two diastereomeric ketones. One of these ketones reacts preferentially with the Wittig reagent to give only one epimeric product.

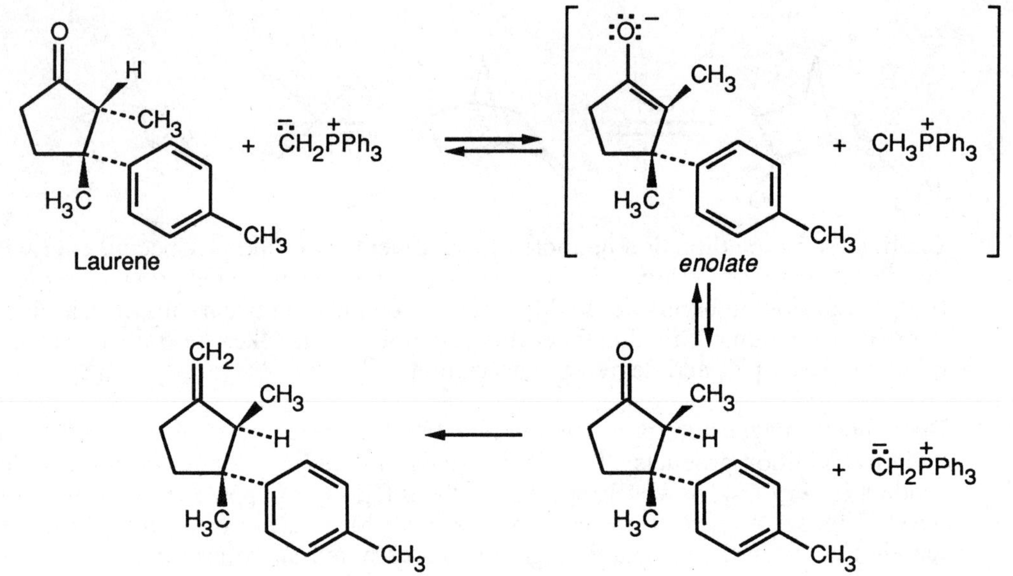

Laurene + $\bar{C}H_2\overset{+}{P}Ph_3$ ⇌ enolate + $CH_3\overset{+}{P}Ph_3$

22.49

$\overset{+}{K}\overset{-}{O}C(CH_3)_3$ in
$(CH_3)_3COH$
(an alpha-substitution reaction)

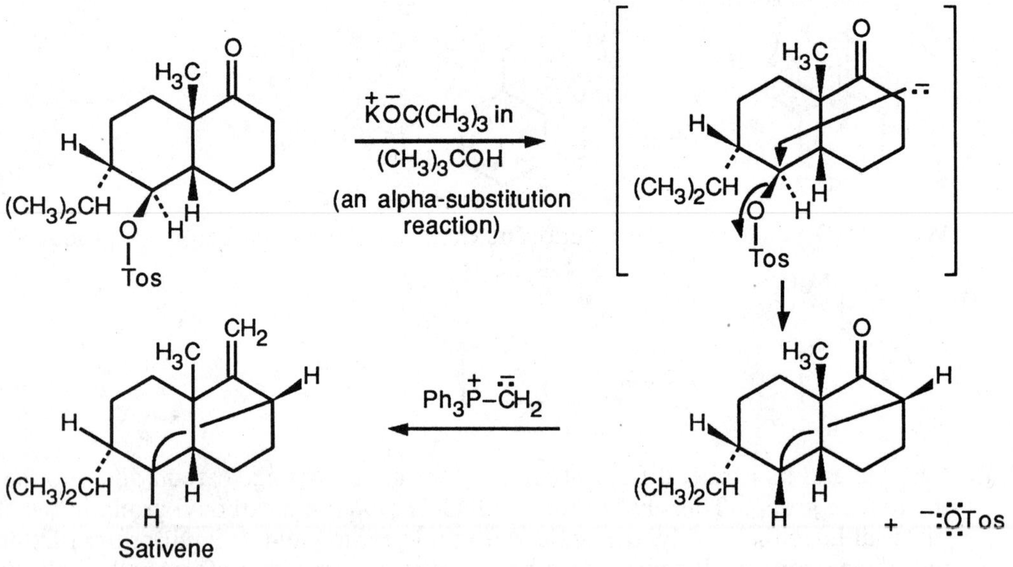

$Ph_3\overset{+}{P}-\overset{-}{\ddot{C}}H_2$

Sativene

$+ \quad ^-:\ddot{O}Tos$

22.50

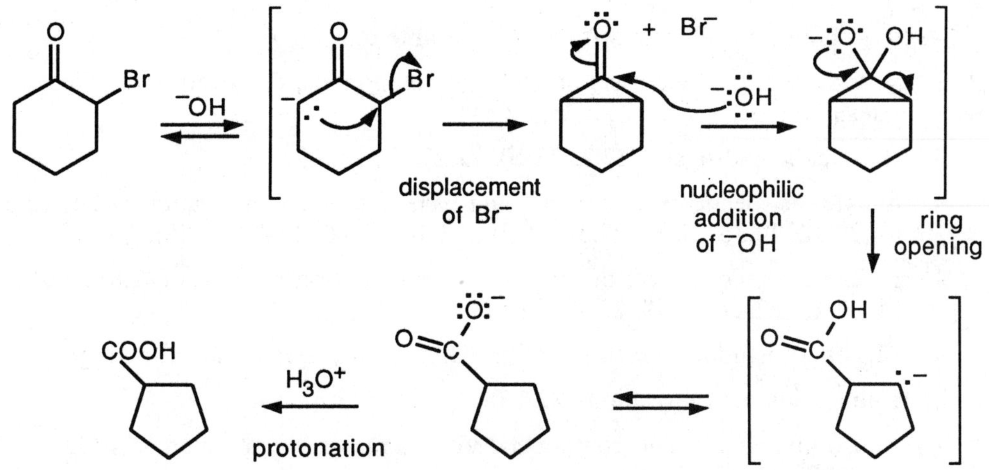

displacement
of Br⁻

nucleophilic
addition
of ⁻OH

ring
opening

COOH

H₃O⁺

protonation

22.51

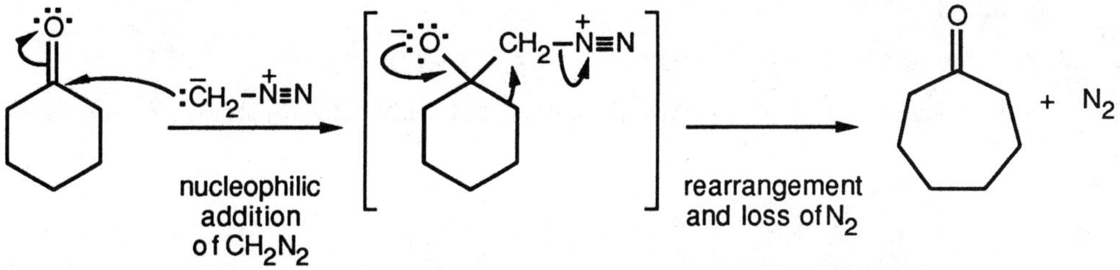

nucleophilic
addition
of CH₂N₂

rearrangement
and loss of N₂

+ N₂

Study Guide for Chapter 22

After studying this chapter, you should be able to:

(1)   Draw keto and enol tautomers of carbonyl compounds; draw resonance forms of enolates (22.1, 22.3, 22.25, 22.26, 22.31).

(2)   Identify acidic hydrogens (22.2, 22.10, 22.27, 22.28, 22.29).

(3)   Formulate the mechanisms of acid– and base–catalyzed enolization (22.4, 22.5, 22.6, 22.11, 22.12, 22.36, 22.37, 22.38, 22.39, 22.44, 22.46, 22.47, 22.48).

(4)   Formulate the mechanisms of other alpha–substitution reactions (22.8, 22.9, 22.13, 22.14, 22.15, 22.17, 22.49, 22.50, 22.51).

(5)   Use alpha–substitution reactions of enolates to synthesize the following:

(a) Enones via selenenylation (22.16).

(b) Substituted acetic acid compounds via the malonic ester synthesis (22.18, 22.19, 22.20, 22.32, 22.34).

(c) Substituted methyl ketones via the acetoacetic ester synthesis (22.21, 22.22, 22.23, 22.33, 22.34).

(d) Alkylated carbonyl compounds using LDA (22.24).

(6)   Predict the products of alpha–substitution reactions (22.35, 22.42, 22.43, 22.45).

23.1 When you are first learning the aldol condensation, write out all the steps.

(1) Form the enolate of one molecule of the carbonyl compound.

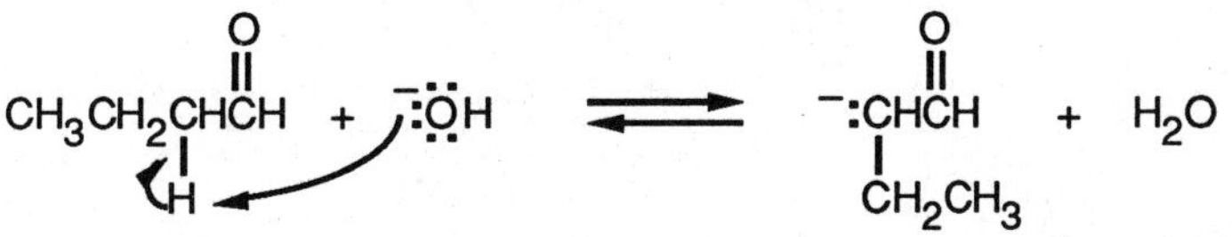

$$CH_3CH_2CHCH + \overset{..}{\underset{..}{:}}\overset{..}{O}H \rightleftharpoons \space ^-:CHCH + H_2O$$

(2) Have the enolate attack the electrophilic carbonyl of the other molecule.

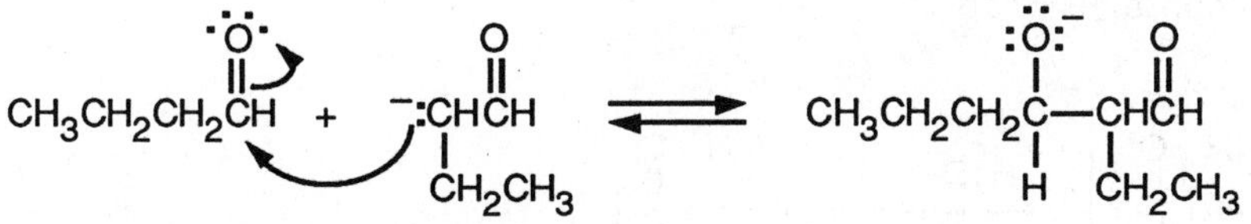

$$CH_3CH_2CH_2CH + \space ^-:CHCH \rightleftharpoons CH_3CH_2CH_2C-CHCH$$

(3) Protonate the anionic oxygen.

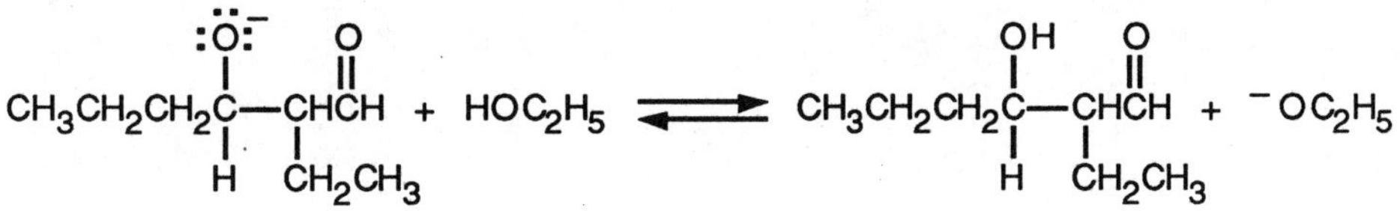

$$CH_3CH_2CH_2C-CHCH + HOC_2H_5 \rightleftharpoons CH_3CH_2CH_2C-CHCH + \space ^-OC_2H_5$$

Practice writing out these steps for the other aldol condensations.

b) 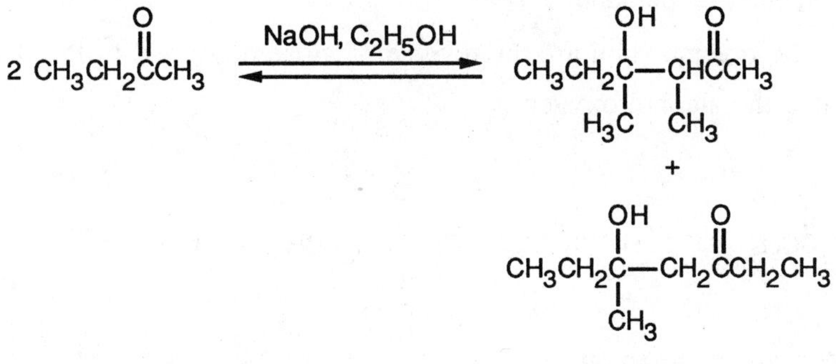 $2 \space CH_3CH_2CCH_3$ $\xrightarrow{\text{NaOH, } C_2H_5OH}$ $CH_3CH_2C-CHCCH_3$

$+$

$CH_3CH_2C-CH_2CCH_2CH_3$

c)

2 $\xrightarrow{\text{NaOH, } C_2H_5OH}$

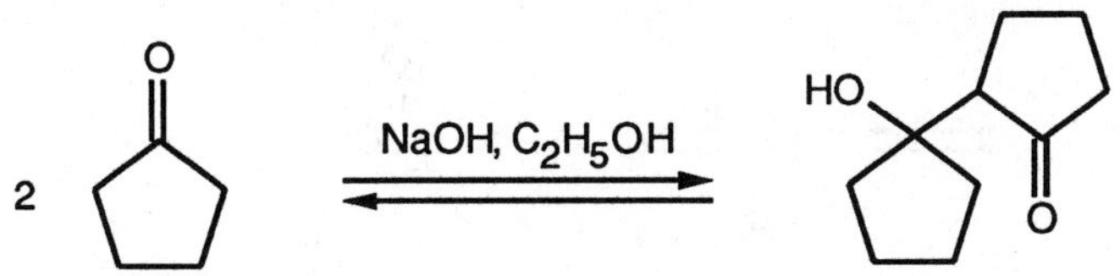

23.2   The reactive nucleophile in the acid-catalyzed aldol condensation is the *enol* of the carbonyl compound. The electrophile is the protonated carbonyl compound.

(1)

$$CH_3\overset{\overset{\displaystyle \cdot\cdot O\cdot}{\|}}{C}H + H_3O^+ \rightleftharpoons \left[\overset{\overset{\displaystyle +\cdot OH}{\|}}{\underset{\underset{\displaystyle H}{|}}{CH_2CH}} \overset{H_2O}{\rightleftharpoons} CH_2{=}\overset{\overset{\displaystyle OH}{|}}{CH} + H_3O^+ \right]$$

(2)

$$\left[\underset{\text{electrophile}}{\overset{\overset{\displaystyle +\cdot OH}{\|}}{CH_3CH}} + \underset{\text{nucleophile}}{\overset{\overset{\displaystyle \cdot OH}{|}}{CH_2{=}CH}} \rightleftharpoons CH_3\overset{\overset{\displaystyle \cdot\cdot OH}{|}}{C}CH_2\overset{\overset{\displaystyle +\cdot OH}{\|}}{\underset{\underset{\displaystyle H}{}}{C}}H \right]$$

(3)

$$\left[CH_3\overset{\overset{\displaystyle OH}{|}}{\underset{\underset{\displaystyle H}{|}}{C}}CH_2\overset{\overset{\displaystyle +OH}{\|}}{C}H \right] \overset{H_2O}{\rightleftharpoons} CH_3\overset{\overset{\displaystyle OH}{|}}{\underset{\underset{\displaystyle H}{|}}{C}}CH_2\overset{\overset{\displaystyle O}{\|}}{C}H + H_3O^+$$

23.3

$$CH_3\overset{\overset{\displaystyle OH}{|}}{\underset{\underset{\displaystyle CH_3}{|}}{C}}{-}CH_2\overset{\overset{\displaystyle O}{\|}}{C}CH_3$$

4–Methyl–4–hydroxy–2–pentanone

The steps for the reverse aldol are the reverse of those described in Problem 23.1.

(1) Deprotonate the alcohol oxygen.

$$CH_3\overset{\overset{\displaystyle :O{\diagup}H}{|}}{\underset{\underset{\displaystyle CH_3}{|}}{C}}CH_2\overset{\overset{\displaystyle O}{\|}}{C}CH_3 + {}^-{:}\overset{\cdot\cdot}{O}H \rightleftharpoons CH_3\overset{\overset{\displaystyle :O{:}^-}{|}}{\underset{\underset{\displaystyle CH_3}{|}}{C}}CH_2\overset{\overset{\displaystyle O}{\|}}{C}CH_3 + H_2O$$

(2) Eliminate the enolate anion.

$$CH_3\overset{\overset{\displaystyle :O{:}^-}{|}}{\underset{\underset{\displaystyle CH_3}{|}}{C}}{-}CH_2\overset{\overset{\displaystyle O}{\|}}{C}CH_3 \rightleftharpoons CH_3\overset{\overset{\displaystyle O}{\|}}{C}CH_3 + {}^-{:}CH_2\overset{\overset{\displaystyle O}{\|}}{C}CH_3$$

(3) Reprotonate the enolate anion.

$${}^-{:}CH_2\overset{\overset{\displaystyle O}{\|}}{C}CH_3 + H_2O \rightleftharpoons CH_3\overset{\overset{\displaystyle O}{\|}}{C}CH_3 + {}^-OH$$

23.4

a)

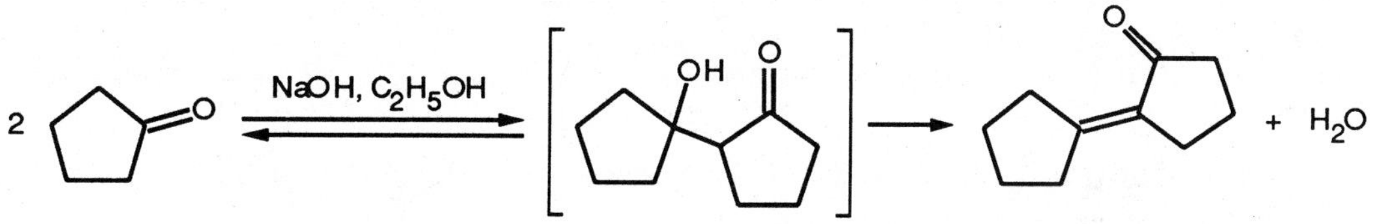

b)

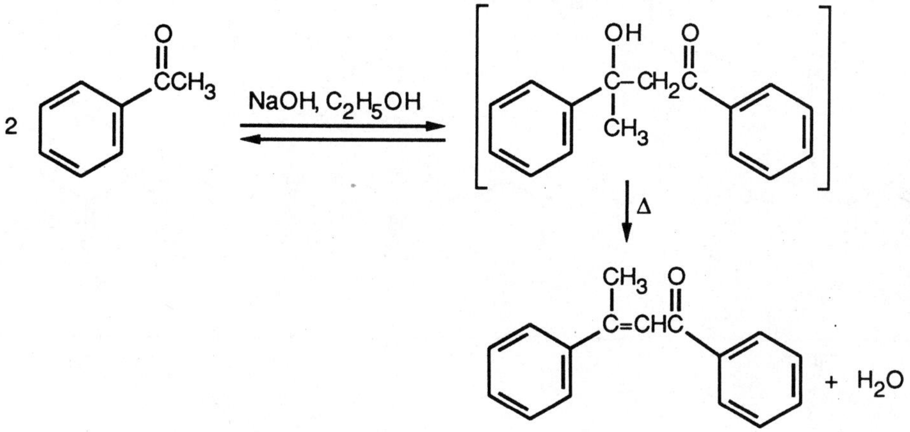

c)

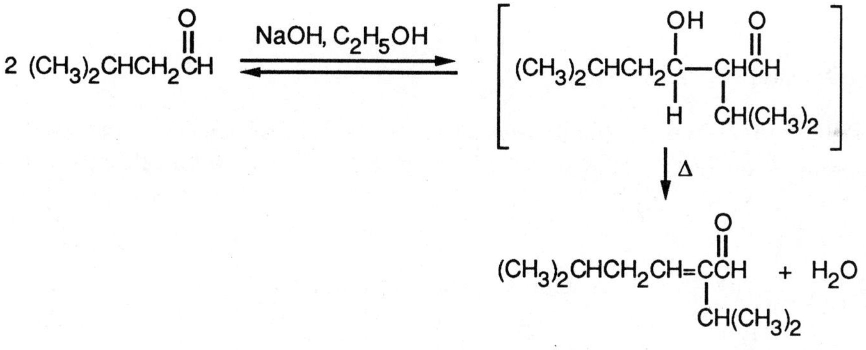

23.5

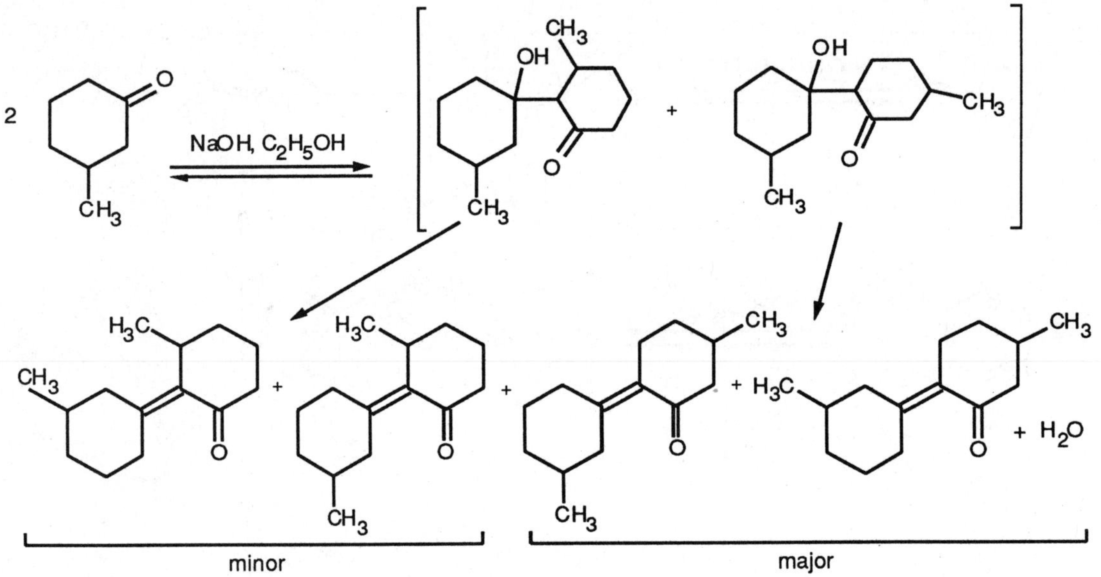

minor   major

23.6

a)

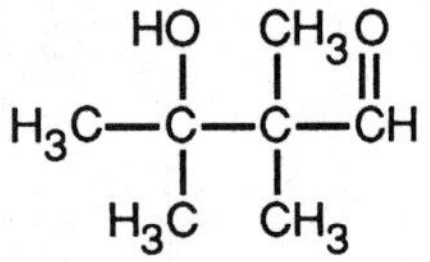

2,2,3–Trimethyl–3–hydroxybutanal is not an aldol self-condensation product.
(Note that no aldol *self*-condensation can yield a product with an odd number of
carbons.)

b)

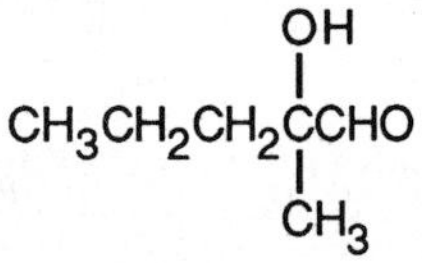

2–Methyl–2–hydroxypentanal

This is not an aldol product. The hydroxyl group in an aldol product must be ß, not
α, to the carbonyl group.

c)

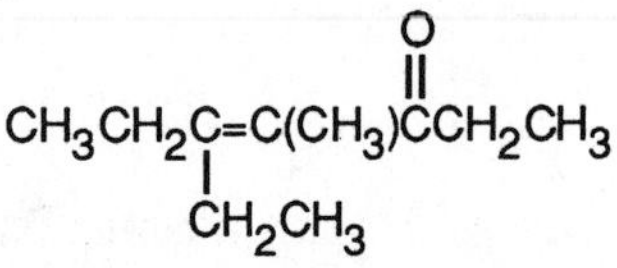

5–Ethyl–4–methyl–4–hepten–3–one

This product results from the aldol self-condensation of 3–pentanone, followed by
dehydration.

23.7

$$2\ CH_3CH \xrightarrow[\text{2.\ }\Delta]{\text{1.\ NaOH,\ EtOH}} CH_3CH{=}CHCH \xrightarrow[\text{2.\ }H_3O^+]{\text{1.\ LiAlH}_4} CH_3CH{=}CHCH_2OH \xrightarrow[\text{Pd/C}]{H_2} CH_3CH_2CH_2CH_2OH$$

1–Butanol

23.8

a)

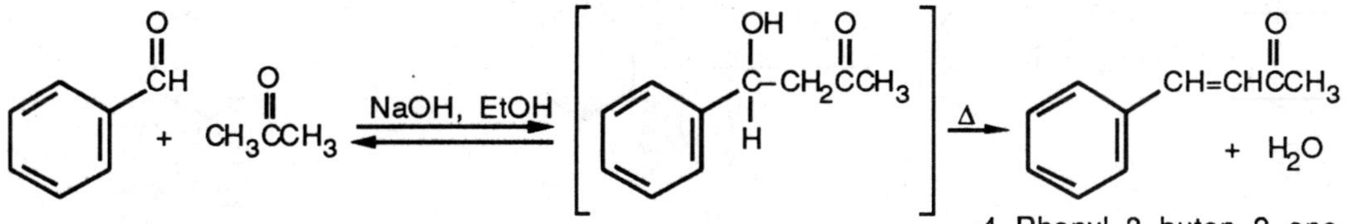

4–Phenyl–3–buten–2–one

This mixed aldol will succeed because one of the components, benzaldehyde, has no α–hydrogen atoms.

b)

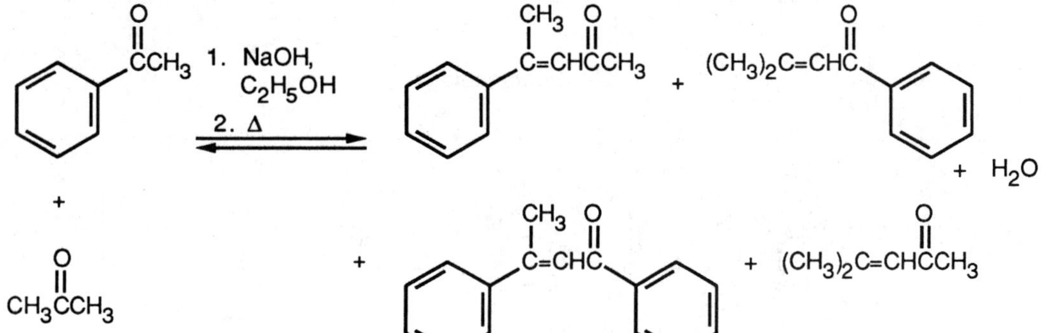

Four products result from the aldol condensation of acetone and acetophenone.

c)

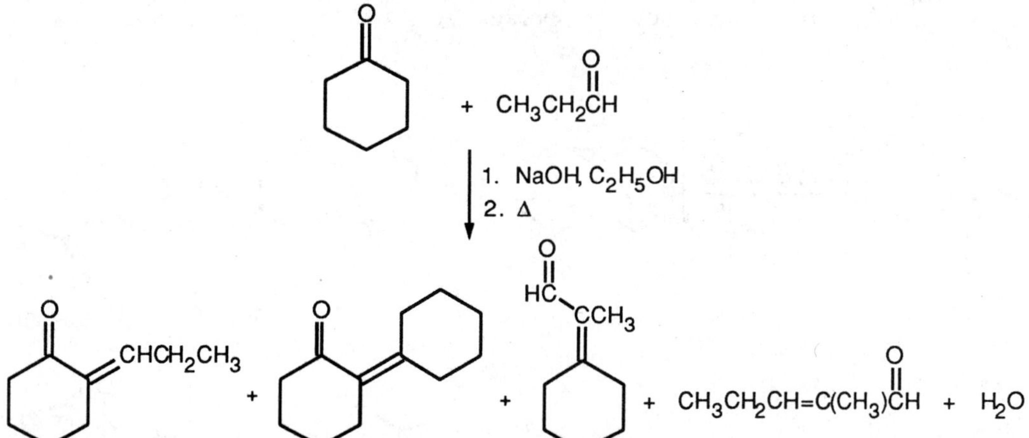

A mixture of products is formed because both carbonyl partners contain α–hydrogen atoms.

d)   This mixed aldol reaction succeeds because the anion of 2,4–pentanedione is formed more readily than the anion of cyclohexanone.

$$\text{cyclohexanone} + CH_2(\overset{O}{\overset{||}{C}}CH_3)_2 \underset{NaOH, C_2H_5OH}{\rightleftharpoons} \left[\text{HO}\text{—}\overset{CH(\overset{O}{\overset{||}{C}}CH_3)_2}{\underset{}{}}\text{cyclohexyl} \right] \overset{\Delta}{\longrightarrow} \overset{C(\overset{O}{\overset{||}{C}}CH_3)_2}{\text{cyclohexylidene}} + H_2O$$

23.9

$$H_2O + \left[CH_3\overset{O}{\overset{||}{C}}\overset{\cdot\cdot}{\underset{\cdot\cdot}{C}}H\overset{O}{\overset{||}{C}}CH_3 \right] \underset{}{\overset{^-:OH}{\rightleftharpoons}} CH_3\overset{O}{\overset{||}{C}}CH_2\overset{O}{\overset{||}{C}}CH_3 \overset{^-OH}{\rightleftharpoons} \left[\overset{\cdot\cdot}{\underset{\cdot\cdot}{C}}H_2\overset{O}{\overset{||}{C}}CH_2\overset{\overset{\cdot\cdot}{\overset{\cdot\cdot}{O}}\curvearrowright}{\overset{|}{C}}CH_3 \right] + H_2O$$

A   2,4–Pentanedione   B

2,4–Pentanedione is in equilibrium with two enolates after treatment with base. Enolate **B** can undergo internal aldol condensation to form cyclic product. Formation of enolate **A** is more likely, however, because **A** is a stabilized carbanion.

23.10

23.11

$$CH_2(CO_2C_2H_5)_2 + {}^-OC_2H_5 \rightleftharpoons {}^-{:}CH(CO_2C_2H_5)_2 + HOC_2H_5$$

[structure: benzaldehyde PhCH=O with arrow + $^-{:}CH(CO_2C_2H_5)_2$ $\rightleftharpoons$]

$$\left[Ph\text{-}\overset{\overset{:O:^-}{|}}{\underset{H}{C}}CH(CO_2C_2H_5)_2 \right]$$

$\updownarrow$ HOC_2H_5

$$\left[Ph\text{-}\overset{\overset{OH}{|}}{\underset{H}{C}}CH(CO_2C_2H_5)_2 \right]$$

$\updownarrow$ ${}^-OC_2H_5$

$$Ph\text{-}CH{=}C(CO_2C_2H_5)_2 + {}^-OH \rightleftharpoons \left[Ph\text{-}\overset{\overset{OH}{|}}{\underset{H}{C}}\text{-}\overset{\overset{:O:^-}{}}{\underset{CO_2C_2H_5}{C}}{-}COC_2H_5 \right]$$

23.12 Here is an outline of the Perkin reaction.

a) Acetate ion abstracts an α–proton from acetic anhydride to form the enolate anion.

$$CH_3\overset{\overset{O}{\|}}{C}O^- + \underset{H}{CH_2}\overset{\overset{O}{\|}}{C}O\overset{\overset{O}{\|}}{C}CH_3 \rightleftharpoons CH_3\overset{\overset{O}{\|}}{C}OH + {}^-{:}CH_2\overset{\overset{O}{\|}}{C}O\overset{\overset{O}{\|}}{C}CH_3$$

b) The enolate anion attacks the electrophilic carbon of benzaldehyde.

[structure: benzaldehyde PhCH=O + $^-{:}CH_2\overset{O}{\|}CO\overset{O}{\|}CCH_3$ $\rightleftharpoons$]

$$\left[Ph\text{-}\overset{\overset{:O:^-}{|}}{\underset{H}{C}}CH_2\overset{\overset{O}{\|}}{C}O\overset{\overset{O}{\|}}{C}CH_3 \right]$$

c) Protonation of the adduct, followed by heat, forms an unsaturated intermediate.

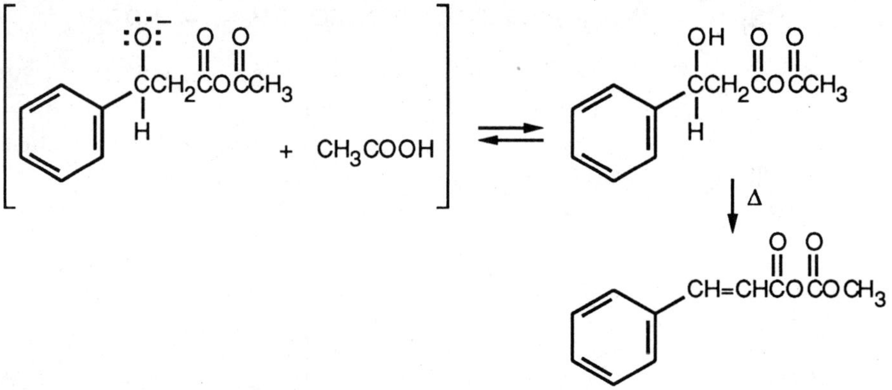

d) Treatment of the unsaturated intermediate with H_2O cleaves the anhydride and yields cinnamic acid.

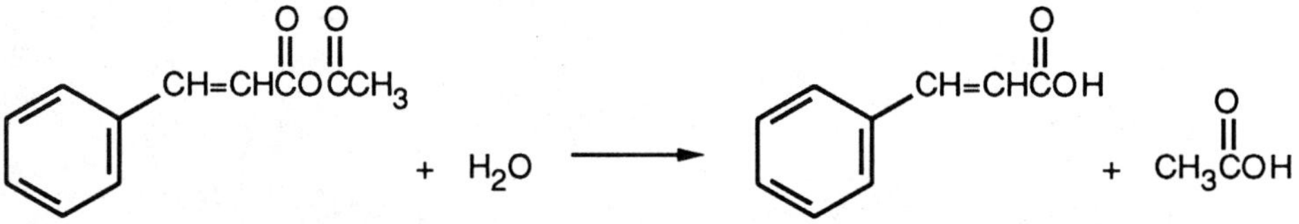

23.13 As in the aldol condensation, writing the two Claisen components in the correct orientation makes it easier to predict the product.

a)

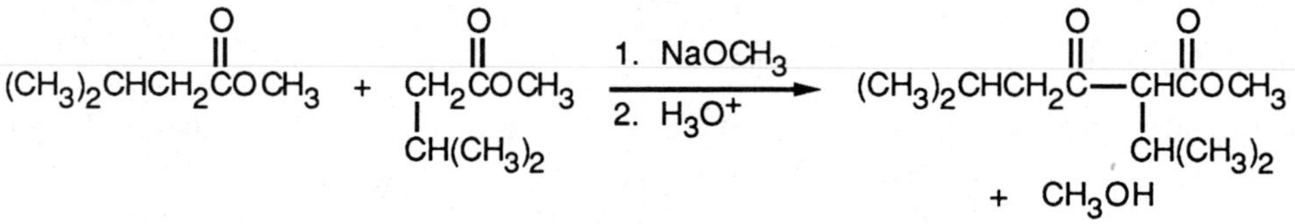

b)

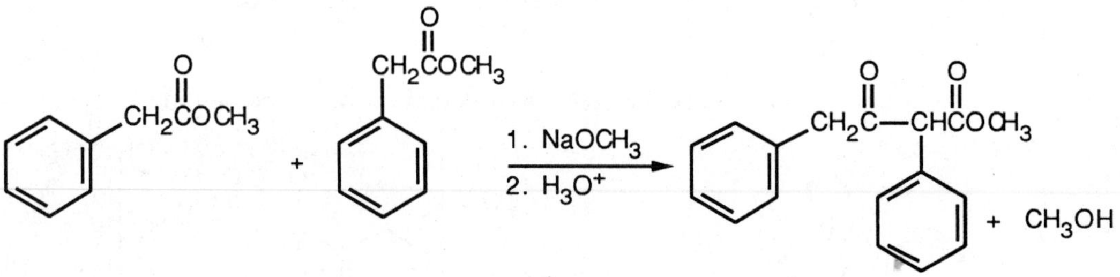

c)

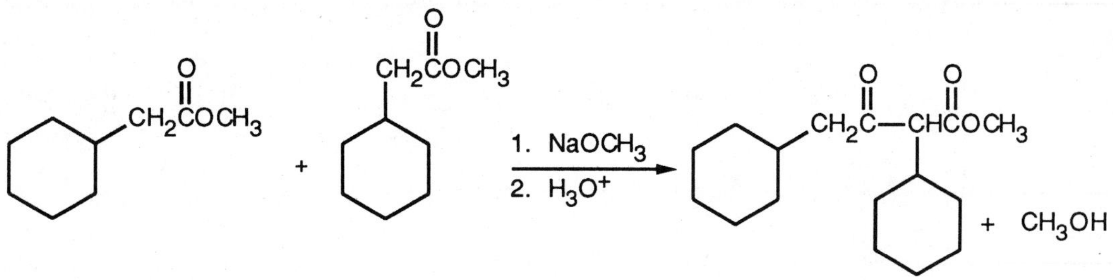

23.14

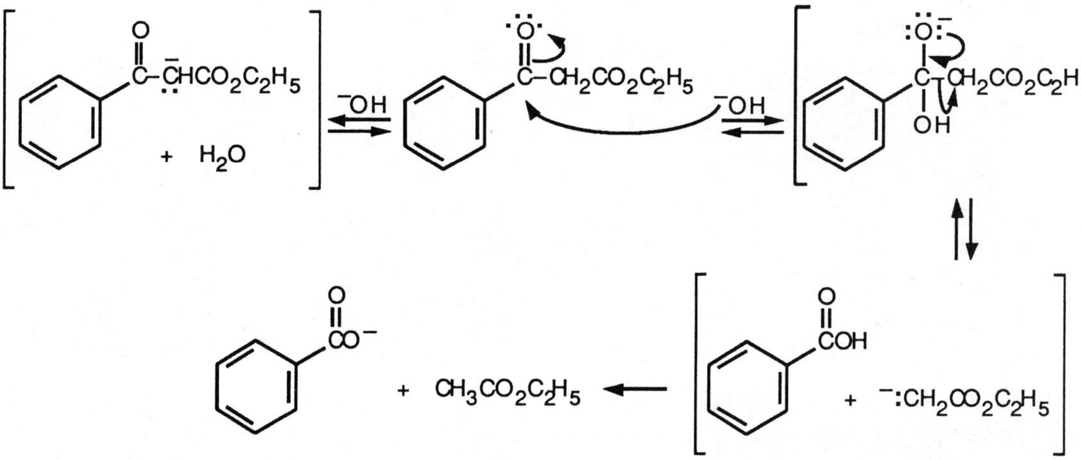

Hydroxide can react at two different sites of the ß–keto ester. Abstraction of the acidic α–proton is more favorable but is reversible and does not lead to product. Addition of hydroxide ion to the carbonyl group, followed by irreversible elimination of ethyl acetate, accounts for the observed product.

23.15 Diethyl oxalate gives good yields in mixed Claisen reactions because it has no *alpha* hydrogens.

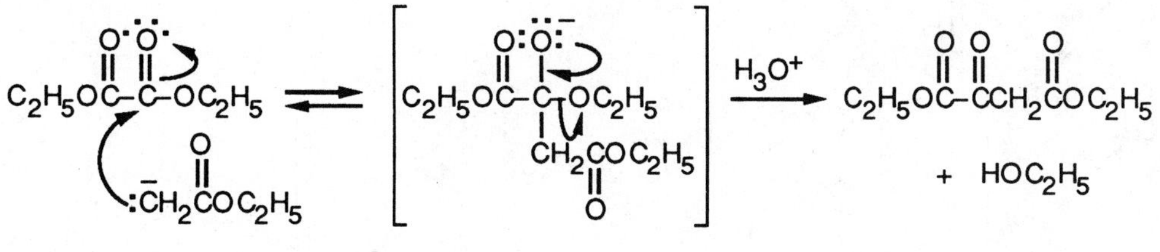

23.16

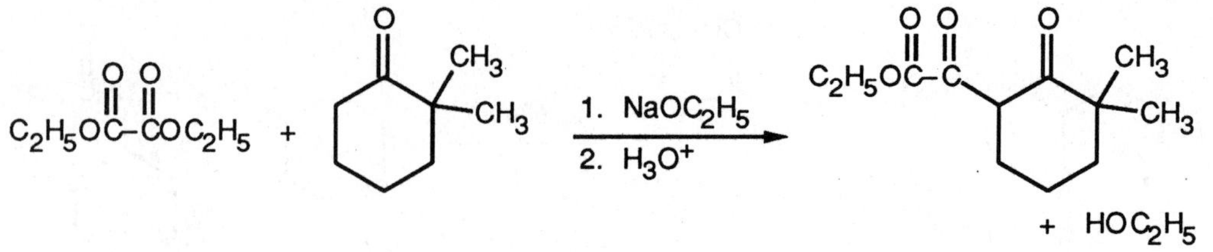

23.17

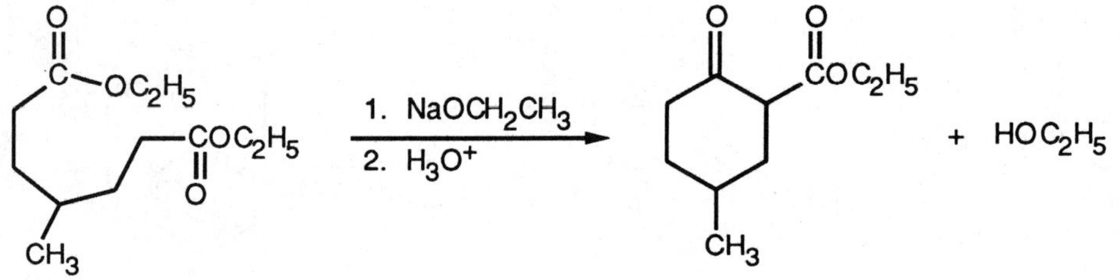

23.18

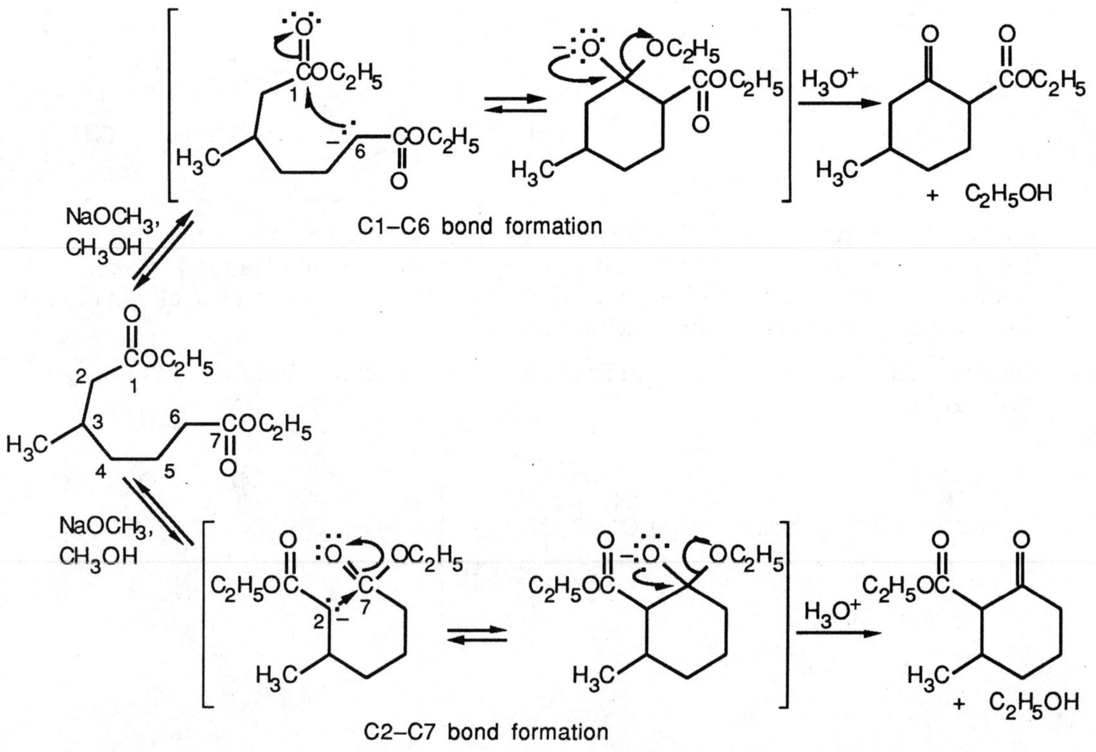

C1–C6 bond formation

C2–C7 bond formation

Unlike diethyl heptanedioate, diethyl 3–methylheptanedioate is unsymmetrical. Two different enolates can form, and each can cyclize to a different product.

23.19

	Michael Donor	Michael Acceptor	Product

a) $CH_3\overset{O}{\overset{\|}{C}}$ with CH_2 and $CH_3\overset{O}{\overset{\|}{C}}$ (pentane-2,4-dione) $+$ cyclohex-2-enone $\rightarrow$ product: 3-(1-acetyl-2-oxopropyl)cyclohexanone

b) $CH_3\overset{O}{\overset{\|}{C}}\!-\!CH_2\!-\!\overset{O}{\overset{\|}{C}}CH_3$ $+$ $H_2C=CHC\equiv N$ $\rightarrow$ $CH_3\overset{O}{\overset{\|}{C}}\!-\!CHCH_2CH_2C\equiv N\!-\!\overset{O}{\overset{\|}{C}}CH_3$

c) $CH_3\overset{O}{\overset{\|}{C}}\!-\!CH_2\!-\!\overset{O}{\overset{\|}{C}}CH_3$ $+$ $\underset{CH_3}{CH=CHCOCH_3}$ $\rightarrow$ $CH_3\overset{O}{\overset{\|}{C}}\!-\!CHCH\underset{CH_3}{CH_2}\overset{O}{\overset{\|}{C}}OCH_3\!-\!\overset{O}{\overset{\|}{C}}CH_3$

23.20

a) $C_2H_5O\overset{O}{\overset{\|}{C}}\!-\!CH_2\!-\!C_2H_5O\overset{O}{\overset{\|}{C}}$ $+$ $H_2C=CH\overset{O}{\overset{\|}{C}}CH_3$ $\rightarrow$ $C_2H_5O\overset{O}{\overset{\|}{C}}\!-\!CHCH_2CH_2\overset{O}{\overset{\|}{C}}CH_3\!-\!C_2H_5O\overset{O}{\overset{\|}{C}}$

b) 2-(methoxycarbonyl)cyclopentanone ($\overset{O}{\overset{\|}{C}}OCH_3$) $+$ $H_2C=CH\overset{O}{\overset{\|}{C}}CH_3$ $\rightarrow$ product with $CH_2CH_2\overset{O}{\overset{\|}{C}}CH_3$ and $\overset{O}{\overset{\|}{C}}OCH_3$

c) O_2NCH_3 $+$ $H_2C=CH\overset{O}{\overset{\|}{C}}CH_3$ $\rightarrow$ $O_2NCH_2CH_2CH_2\overset{O}{\overset{\|}{C}}CH_3$

23.21 Michael reactions occur between stabilized enolate anions and α,ß–unsaturated carbonyl compounds. Learn to locate these components in possible Michael products. Usually, it is easier to recognize the enolate nucleophile; in a) the nucleophile is the ethyl acetoacetate anion.

$$CH_3\overset{O}{\overset{||}{C}}CH-CH_2CH_2\overset{O}{\overset{||}{C}}C_6H_5 \quad \text{comes from} \quad CH_3\overset{O}{\overset{||}{C}}CH_2$$
$$\underset{CO_2CH_3}{|} \qquad\qquad\qquad\qquad \underset{CO_2CH_3}{|}$$

The rest of the compound is the Michael acceptor. Draw a double bond in conjugation with an electron-withdrawing group in this part of the molecule.

$$CH_3\overset{O}{\overset{||}{C}}CH-CH_2CH_2\overset{O}{\overset{||}{C}}C_6H_5 \quad \text{comes from} \quad CH_2=CH\overset{O}{\overset{||}{C}}C_6H_5$$
$$\underset{CO_2CH_3}{|}$$

$$CH_3\overset{O}{\overset{||}{C}}CH_2 \;+\; CH_2=CH\overset{O}{\overset{||}{C}}C_6H_5 \quad \xrightarrow[\text{2. } H_3O^+]{\text{1. } NaOC_2H_5} \quad CH_3\overset{O}{\overset{||}{C}}CHCH_2CH_2\overset{O}{\overset{||}{C}}C_6H_5$$
$$\underset{CO_2CH_3}{|} \qquad\qquad\qquad\qquad\qquad\qquad\qquad\qquad \underset{CO_2CH_3}{|}$$

Michael donor   Michael acceptor

b) Since Michael products are often decarboxylated after the addition reaction, it is more difficult to recognize the original enolate anion.

$$CH_3\overset{O}{\overset{||}{C}}CH_2 \;+\; CH_2=CH\overset{O}{\overset{||}{C}}CH_3 \quad \xrightarrow[\text{2. } H_3O^+]{\text{1. } NaOC_2H_5} \quad CH_3\overset{O}{\overset{||}{C}}CHCH_2CH_2\overset{O}{\overset{||}{C}}CH_3$$
$$\underset{CO_2CH_3}{|} \qquad\qquad\qquad\qquad\qquad\qquad\qquad\qquad \underset{CO_2CH_3}{|}$$

Michael donor   Michael acceptor

$$CH_3\overset{O}{\overset{||}{C}}CHCH_2CH_2\overset{O}{\overset{||}{C}}CH_3 \quad \xrightarrow[\Delta]{H_3O^+} \quad CH_3\overset{O}{\overset{||}{C}}CH_2CH_2CH_2\overset{O}{\overset{||}{C}}CH_3$$
$$\underset{CO_2CH_3}{|}$$

c) $(CH_3O_2C)_2CH_2 \;+\; CH_2=CHCN \quad \xrightarrow[\text{2. } H_3O^+]{\text{1. } NaOC_2H_5} \quad (CH_3O_2C)_2CHCH_2CH_2CN$

Michael donor   Michael acceptor

d) $CH_3CH_2NO_2 \;+\; CH_2=CH\overset{O}{\overset{||}{C}}OCH_3 \quad \xrightarrow[\text{2. } H_3O^+]{\text{1. } NaOC_2H_5} \quad CH_3\overset{NO_2}{\underset{|}{C}}HCH_2CH_2\overset{O}{\overset{||}{C}}OCH_3$

Michael donor   Michael acceptor

e) $(CH_3O_2C)_2CH_2 \;+\; CH_2=CHNO_2 \quad \xrightarrow[\text{2. } H_3O^+]{\text{1. } NaOC_2H_5} \quad (CH_3O_2C)_2CHCH_2CH_2NO_2$

Michael donor   Michael acceptor

23.22

a)

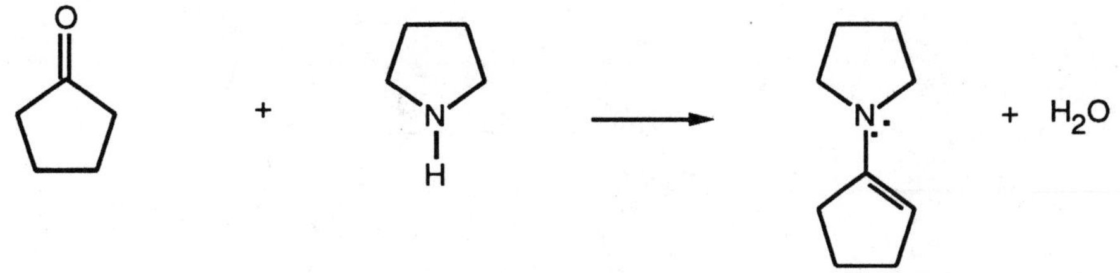

+   H$_2$O

b)

+   H$_2$O

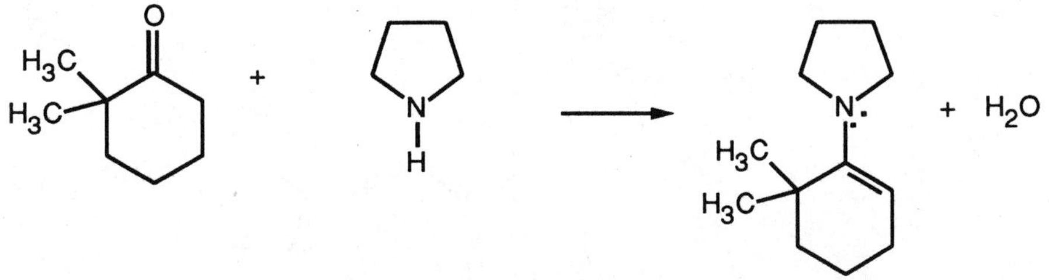

23.23

	Enamine	Michael Acceptor	Product (after hydrolysis)

a)

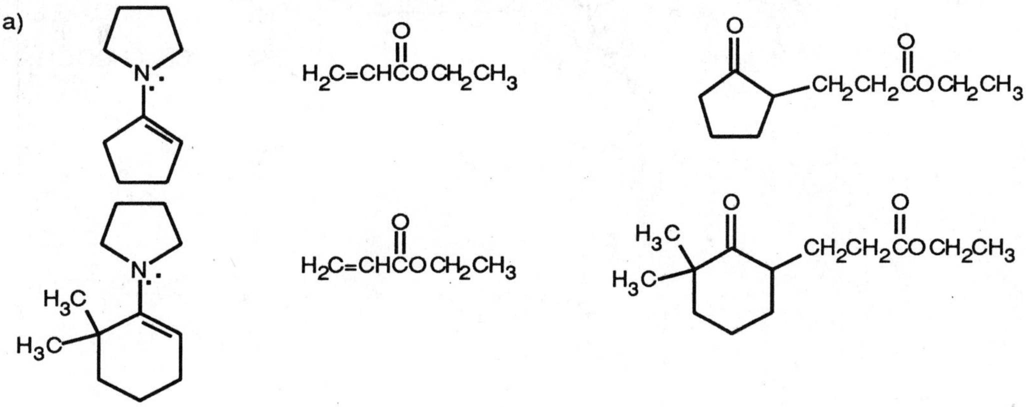

b)

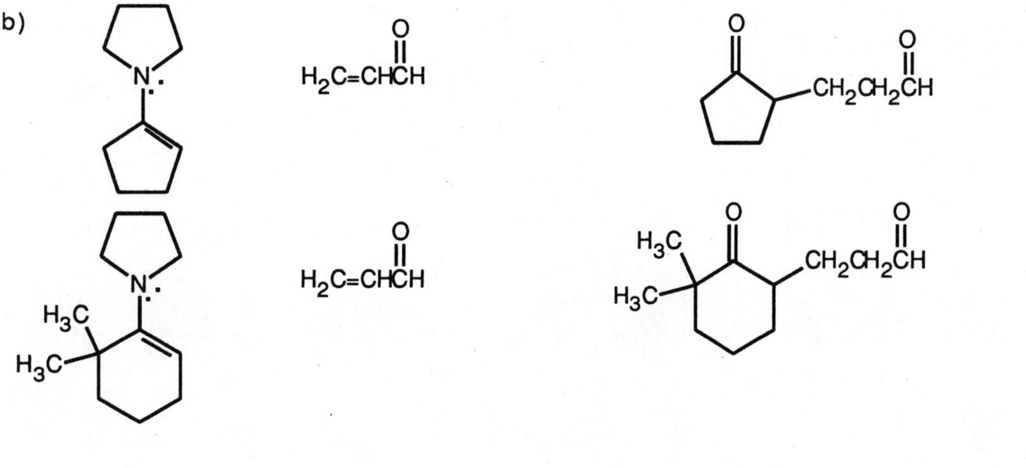

23.24

a)

b)

23.25

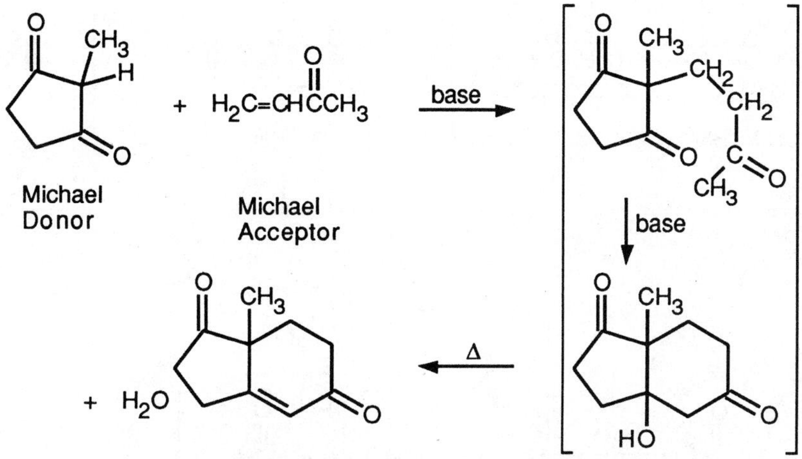

23.26

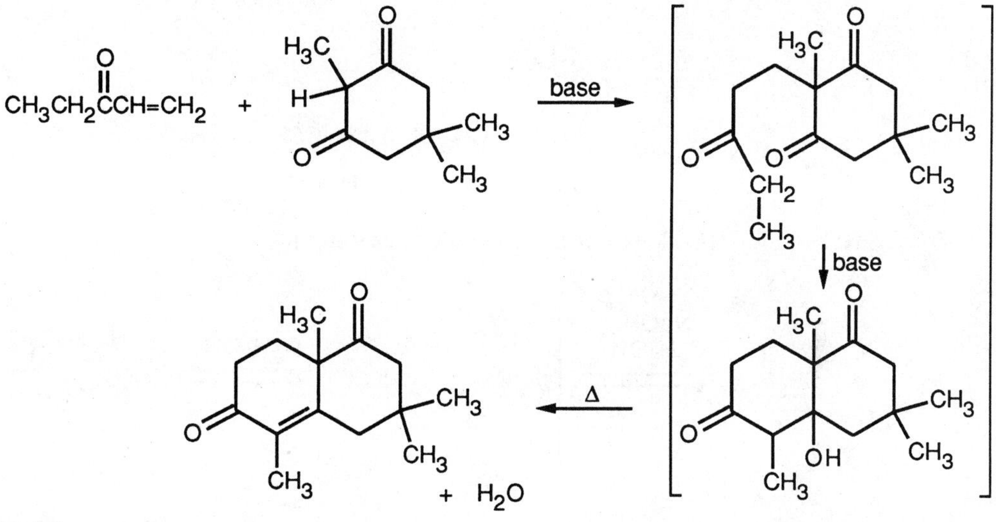

23.27–23.28

a) (CH₃)₃CCHO has no *alpha* protons and does not undergo aldol self-condensation.

b)

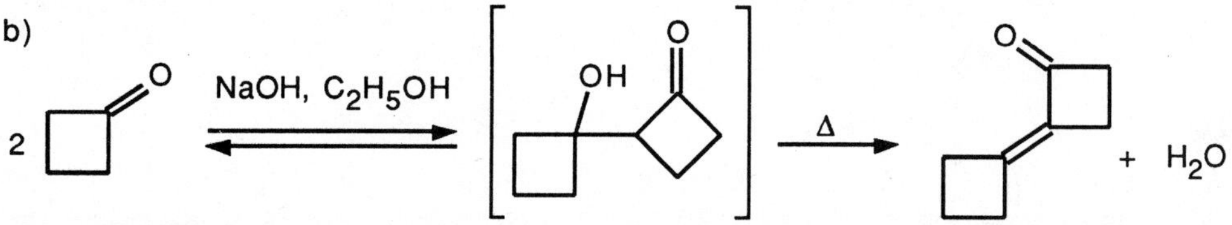

c) Benzophenone does not undergo aldol self-condensation.

d)

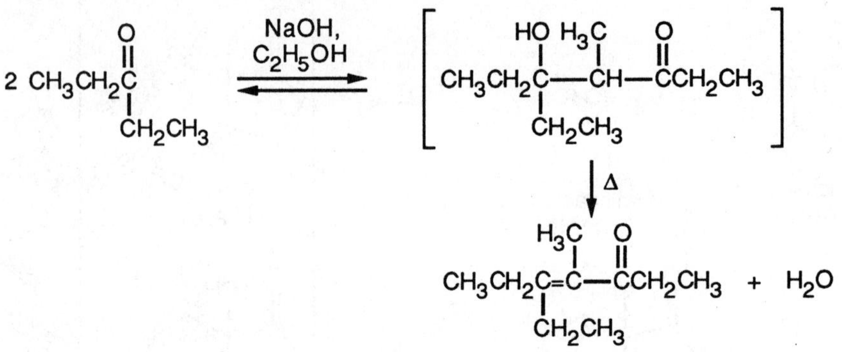

$$2\ CH_3CH_2\overset{O}{\underset{CH_2CH_3}{\overset{\|}{C}}} \quad \xrightarrow[C_2H_5OH]{NaOH,} \quad \left[CH_3CH_2\overset{HO}{\underset{CH_2CH_3}{C}}-\overset{H_3C}{\underset{}{CH}}-\overset{O}{\overset{\|}{C}}CH_2CH_3 \right]$$

$$\downarrow \Delta$$

$$CH_3CH_2\overset{H_3C}{\underset{CH_2CH_3}{C}}=\overset{}{C}-\overset{O}{\overset{\|}{C}}CH_2CH_3 \quad + \quad H_2O$$

e)

$$2\ CH_3(CH_2)_8\overset{O}{\overset{\|}{C}}H \quad \xrightarrow[C_2H_5OH]{NaOH,} \quad \left[CH_3(CH_2)_8\overset{OH}{\underset{H\ \ (CH_2)_7CH_3}{C}}-\overset{O}{\overset{\|}{C}}HCH \right]$$

$$\downarrow \Delta$$

$$CH_3(CH_2)_7CH=\overset{O}{\underset{CH_3(CH_2)_7}{\overset{\|}{C}}}CH \quad + \quad H_2O$$

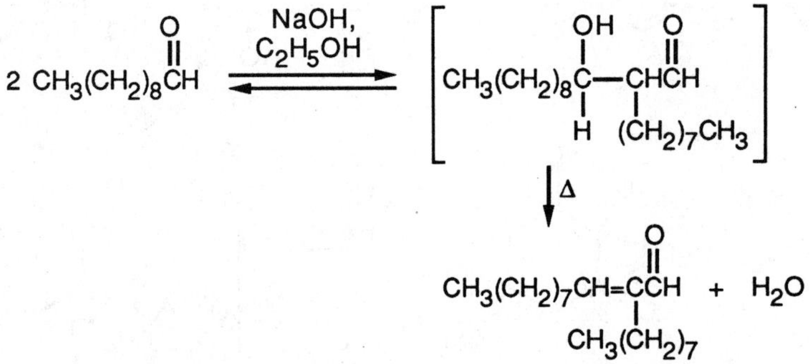

f) $C_6H_5CH=CHCHO$ does not undergo aldol reactions.

23.29

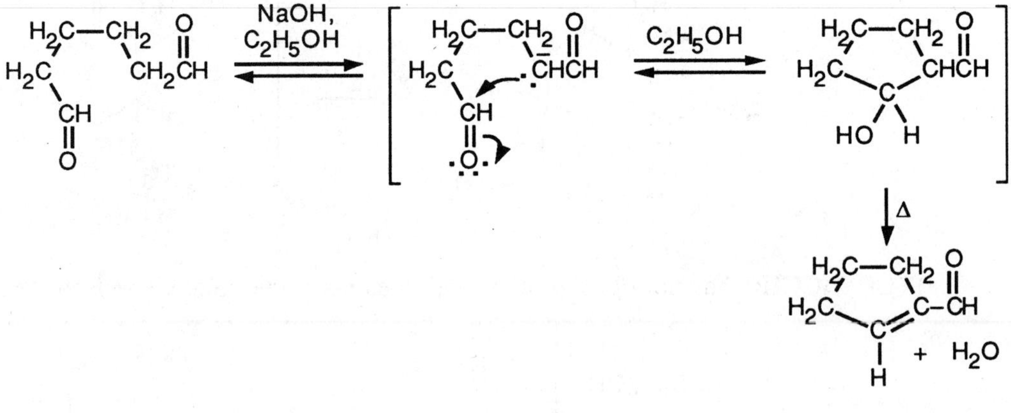

23.30

a)

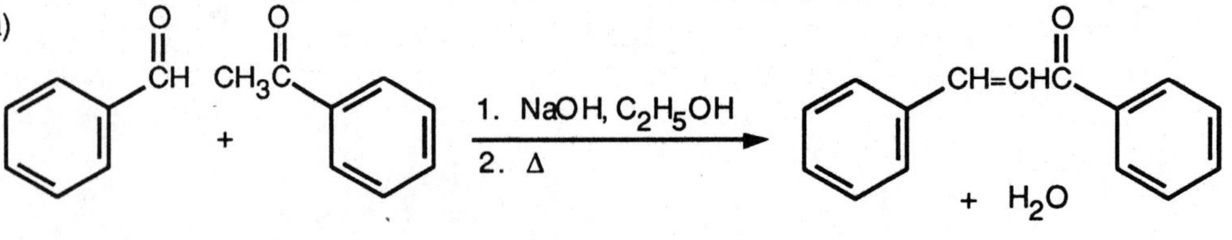

1. NaOH, C_2H_5OH

2. Δ

+ H_2O

b)

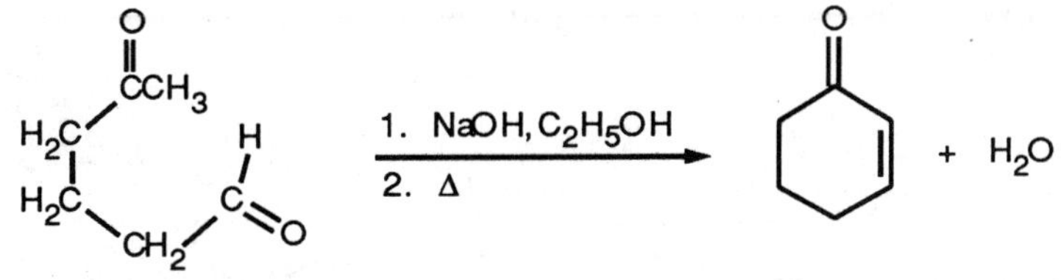

1. NaOH, C$_2$H$_5$OH
2. Δ

+ H$_2$O

c)

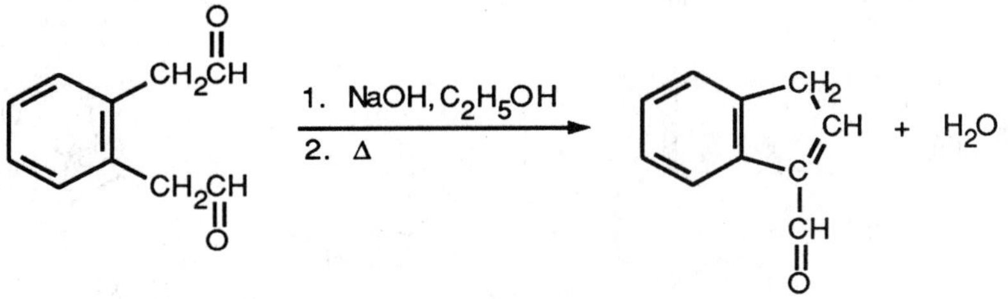

1. NaOH, C$_2$H$_5$OH
2. Δ

+ H$_2$O

d)

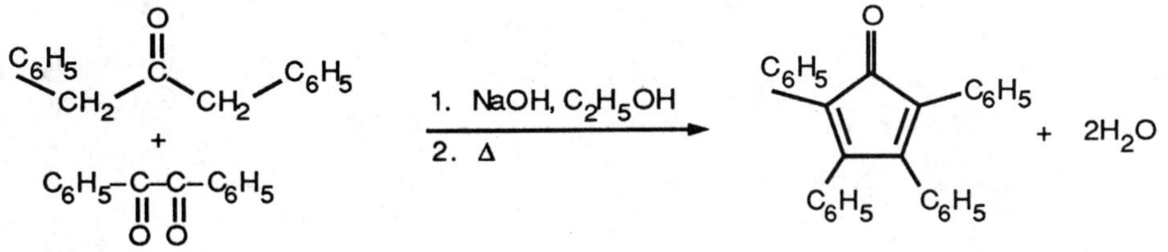

+

C$_6$H$_5$-C-C-C$_6$H$_5$

1. NaOH, C$_2$H$_5$OH
2. Δ

+ 2H$_2$O

23.31

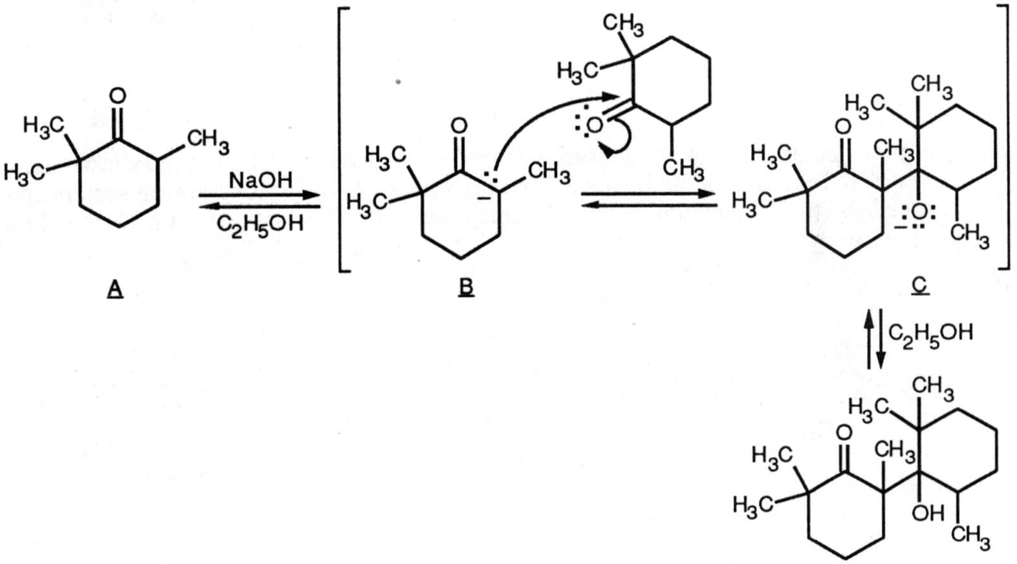

A

NaOH

C$_2$H$_5$OH

B

C

C$_2$H$_5$OH

D

An aldol condensation involves a series of reversible equilibrium steps. In general, formation of product is favored by the dehydration of the ß–hydroxy ketone to form a conjugated enone. Here dehydration to form *conjugated* product can't occur with **D**. In addition, the **B** ⇌ **C** equilibrium favors **B** because of steric hindrance in the nucleophilic addition step.

23.32

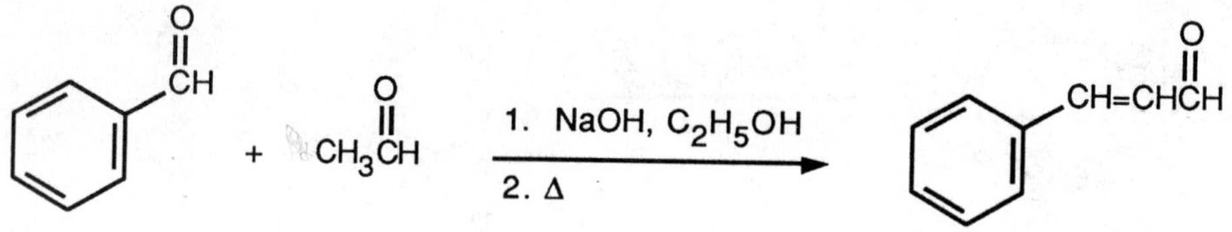

23.33

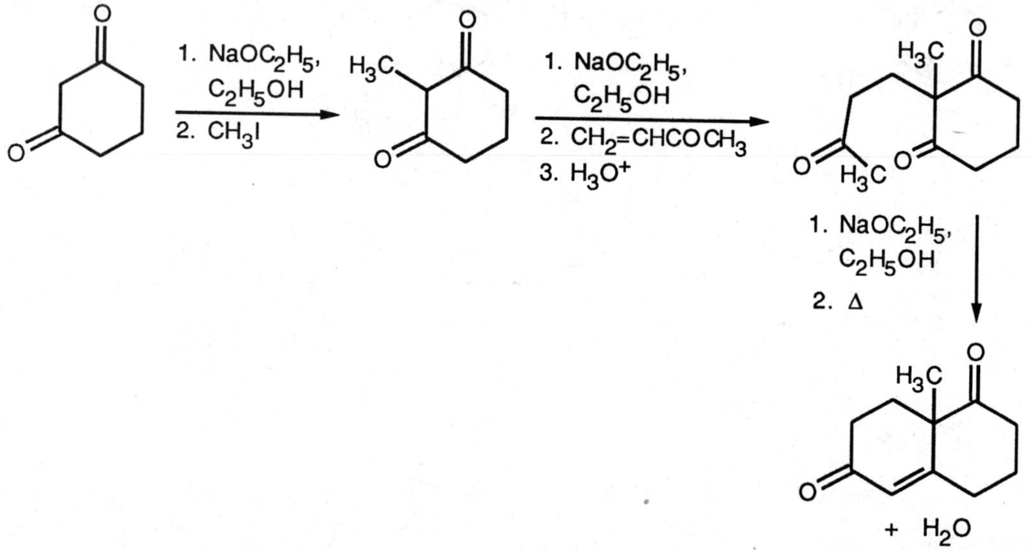

23.34 The first step of an aldol condensation is enolate formation. The ketone shown here does not enolize because double bonds at the bridgehead of small bicyclic ring systems are too strained to form. Since the bicyclic ketone does not enolize, it doesn't undergo aldol condensation.

23.35

a)

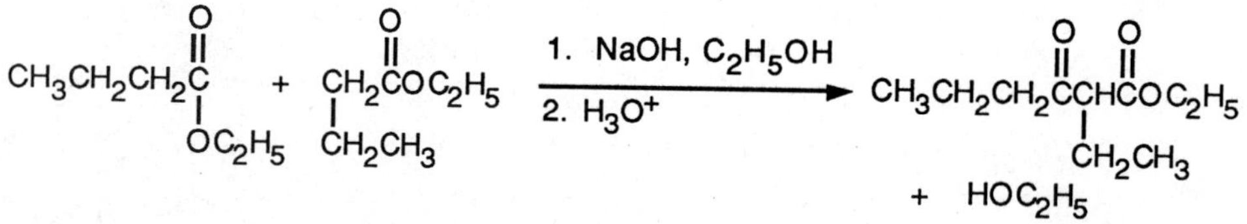

b)

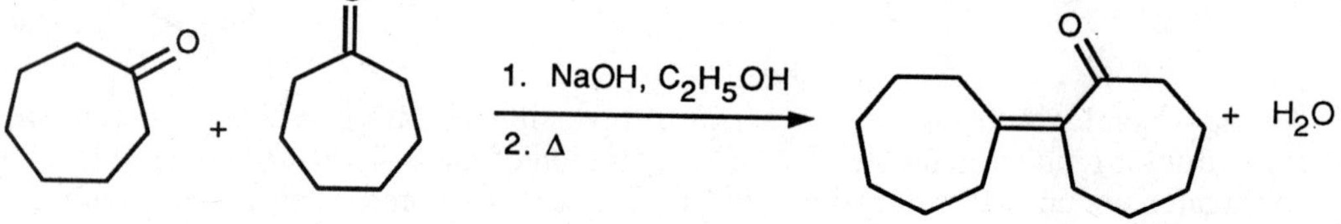

c)

1. NaOH, C_2H_5OH
2. Δ

major

d)

$C_6H_5CH_2CH_2\overset{\overset{\displaystyle O}{\|}}{C}H$ + $\underset{\underset{\displaystyle CH_2C_6H_5}{|}}{CH_2}\overset{\overset{\displaystyle O}{\|}}{C}H$

1. NaOH, C_2H_5OH
2. Δ

$C_6H_5CH_2CH_2CH=\underset{\underset{\displaystyle CH_2C_6H_5}{|}}{C}\overset{\overset{\displaystyle O}{\|}}{C}H$

23.36

a)
$$CH_3CO_2CH_3 \;+\; CH_3CH_2CO_2CH_3$$

$\downarrow$ $NaOC_2H_5$, C_2H_5OH

$CH_3\overset{\overset{\displaystyle O}{\|}}{C}CH_2CO_2CH_3$

$CH_3\underset{\underset{\displaystyle CH_3}{|}}{\overset{\overset{\displaystyle O}{\|}}{C}CH}CO_2CH_3$

+   +   +

$CH_3CH_2\underset{\underset{\displaystyle CH_3}{|}}{\overset{\overset{\displaystyle O}{\|}}{C}CH}CO_2CH_3$

$CH_3CH_2\overset{\overset{\displaystyle O}{\|}}{C}CH_2CO_2CH_3$

self-condensation products   mixed condensation products

Approximately equal amounts of each product will form.

b)
$$C_6H_5CO_2CH_3 \;+\; C_6H_5CH_2CO_2CH_3$$

$\downarrow$ $NaOC_2H_5$, C_2H_5OH

$C_6H_5CH_2\underset{\underset{\displaystyle C_6H_5}{|}}{\overset{\overset{\displaystyle O}{\|}}{C}CH}CO_2CH_3$ + $C_6H_5\underset{\underset{\displaystyle C_6H_5}{|}}{\overset{\overset{\displaystyle O}{\|}}{C}CH}CO_2CH_3$

self-condensation product   mixed condensation product

The mixed condensation product predominates.

c)

$CH_3O\overset{\overset{\displaystyle O}{\|}}{C}OCH_3$ +   cyclohexanone

$\xrightarrow[\displaystyle C_2H_5OH]{\displaystyle NaOC_2H_5}$

This is the only *Claisen* monocondensation product.

d)

$$C_6H_5\overset{O}{\overset{\|}{C}}H + CH_3CO_2CH_3 \xrightarrow[C_2H_5OH]{NaOC_2H_5} CH_3\overset{O}{\overset{\|}{C}}CH_2CO_2CH_3 + C_6H_5CH=CHCO_2CH_3$$

self-condensation   mixed condensation
product   product

The mixed Claisen product is the major product.

23.37 a) Several other products are formed in addition to the one pictured. Self-condensation of acetaldehyde and acetone (less likely) can occur, and an additional mixed product is formed.

b) *Conjugate* addition occurs between a carbon nucleophile and an α,ß–unsaturated carbonyl electrophile (the Michael reaction). Addition occurs at the double bond, not at the carbonyl group.

$$CH_3\overset{O}{\overset{\|}{C}}CH=CH_2 + CH_2(CO_2C_2H_5)_2 \xrightarrow{base} CH_3\overset{O}{\overset{\|}{C}}CH_2CH_2CH(CO_2C_2H_5)_2$$

c) There are two problems with this reaction. (1) Michael reactions occur in low yield with mono-ketones. Formation of the enamine, followed by the Michael reaction, gives a higher yield of product. (2) Addition can occur on either side of the ketone to give a mixture of products.

d) Internal aldol condensation of 2,6–heptanedione can product a four-membered ring or a six-membered ring. The six-membered ring is more likely to form because it is less strained.

e) Michael addition occurs at the most acidic position of the Michael donor.

23.38 If cyclopentanone and base are mixed first, aldol self-condensation of cyclopentanone can occur before ethyl formate is added. If both carbonyl components are mixed together before adding base, the more favorable mixed Claisen condensation occurs with little competition from the aldol self-condensation reaction.

23.39

$$CH_3\overset{\overset{\displaystyle ..}{\underset{\displaystyle ..}{O}}}{\underset{}{\overset{\|}{C}}}C(CH_3)_2\overset{O}{\overset{\|}{C}}OC_2H_5 \;\rightleftharpoons\; \left[CH_3\overset{\overset{\displaystyle :\ddot{O}:^-}{|}}{\underset{\underset{OC_2H_5}{|}}{C}}{-}C(CH_3)_2\overset{O}{\overset{\|}{C}}OC_2H_5 \right] \;\rightleftharpoons\; CH_3\overset{O}{\overset{\|}{C}}OC_2H_5$$

$$+$$

$$(CH_3)_2\overset{O}{\overset{\|}{\underset{..}{C}}}COC_2H_5$$

$$(CH_3)_2\overset{O}{\overset{\|}{\underset{..}{C}}}COC_2H_5 \;+\; C_2H_5OH \;\rightleftharpoons\; (CH_3)_2CH\overset{O}{\overset{\|}{C}}OC_2H_5 \;+\; C_2H_5O^-$$

23.40 Two different reactions are possible when ethyl acetoacetate reacts with ethoxide anion. One reaction involves attack of ethoxide ion on the carbonyl carbon, followed by elimination of the anion of ethyl acetate. This scheme is similar to the one illustrated in 23.39.

$$CH_3\overset{\overset{\displaystyle ..}{\underset{\displaystyle ..}{O}}}{\overset{\|}{C}}CH_2\overset{O}{\overset{\|}{C}}OC_2H_5 \;\rightleftharpoons\; \left[CH_3\overset{\overset{\displaystyle :\ddot{O}:^-}{|}}{\underset{\underset{OC_2H_5}{|}}{C}}{-}CH_2\overset{O}{\overset{\|}{C}}OC_2H_5 \right] \;\rightleftharpoons\; CH_3\overset{O}{\overset{\|}{C}}OC_2H_5 \;+\; {}^-{:}CH_2\overset{O}{\overset{\|}{C}}OC_2H_5$$

$${}^-{:}CH_2\overset{O}{\overset{\|}{C}}OC_2H_5 \;+\; C_2H_5OH \;\rightleftharpoons\; CH_3\overset{O}{\overset{\|}{C}}OC_2H_5 \;+\; C_2H_5\ddot{O}{:}^-$$

This is a reverse Claisen reaction. More likely, however, is the acid-base reaction of ethoxide ion and an activated α–proton of ethyl acetoacetate.

$$CH_3\overset{O}{\overset{\|}{C}}CH_2\overset{O}{\overset{\|}{C}}OC_2H_5 \;+\; {}^-{:}\ddot{O}C_2H_5 \;\rightleftharpoons\; CH_3\overset{O}{\overset{\|}{C}}\overset{O}{\underset{}{\overset{\|}{C}}}HCOC_2H_5 \;+\; HOC_2H_5$$

The resonance-stabilized acetoacetate anion is no longer reactive toward nucleophiles, and no further reaction occurs at room temperature. Elevated temperatures are required to make the cleavage reaction proceed.

23.41

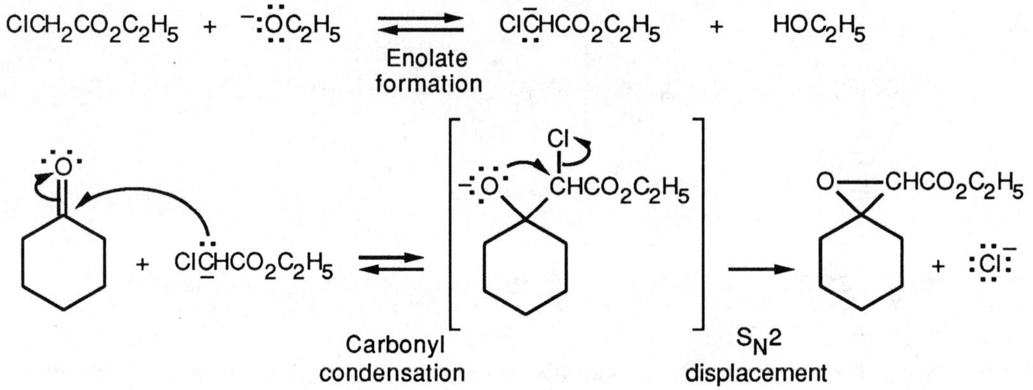

ClCH2CO2C2H5 + −:ÖC2H5
Enolate formation
ClC̄HCO2C2H5 + HOC2H5
ClC̄HCO2C2H5
Cl
CHCO2C2H5
Carbonyl condensation
SN2 displacement
CHCO2C2H5
:Cl:

23.42

a)

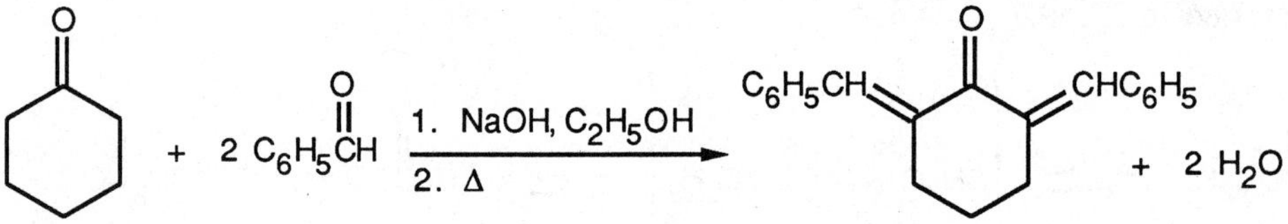

2 C6H5CH
1. NaOH, C2H5OH
2. Δ
C6H5CH
CHC6H5
2 H2O

b)

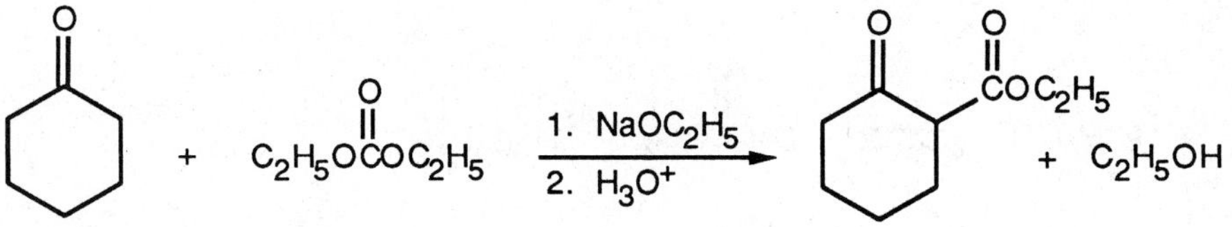

C2H5OCOC2H5
1. NaOC2H5
2. H3O+
COC2H5
C2H5OH

c)

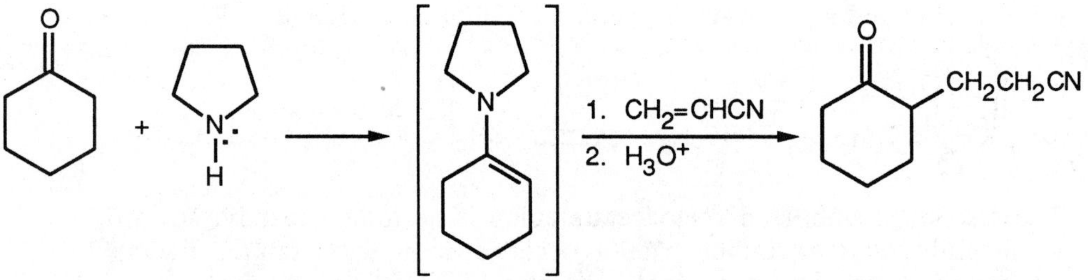

N
H
1. CH2=CHCN
2. H3O+
CH2CH2CN

d)

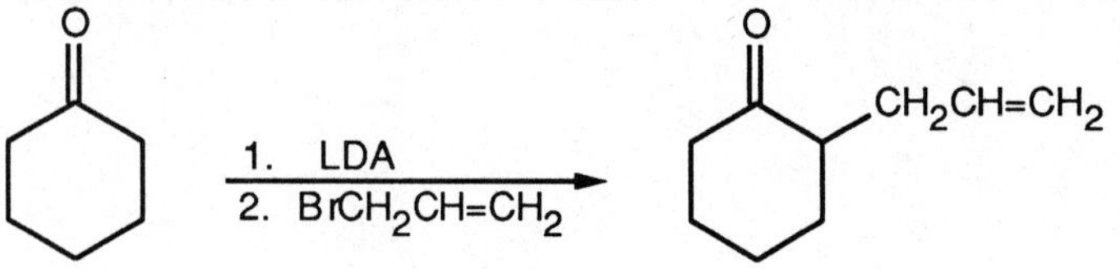

1. LDA
2. BrCH2CH=CH2
CH2CH=CH2

23.43

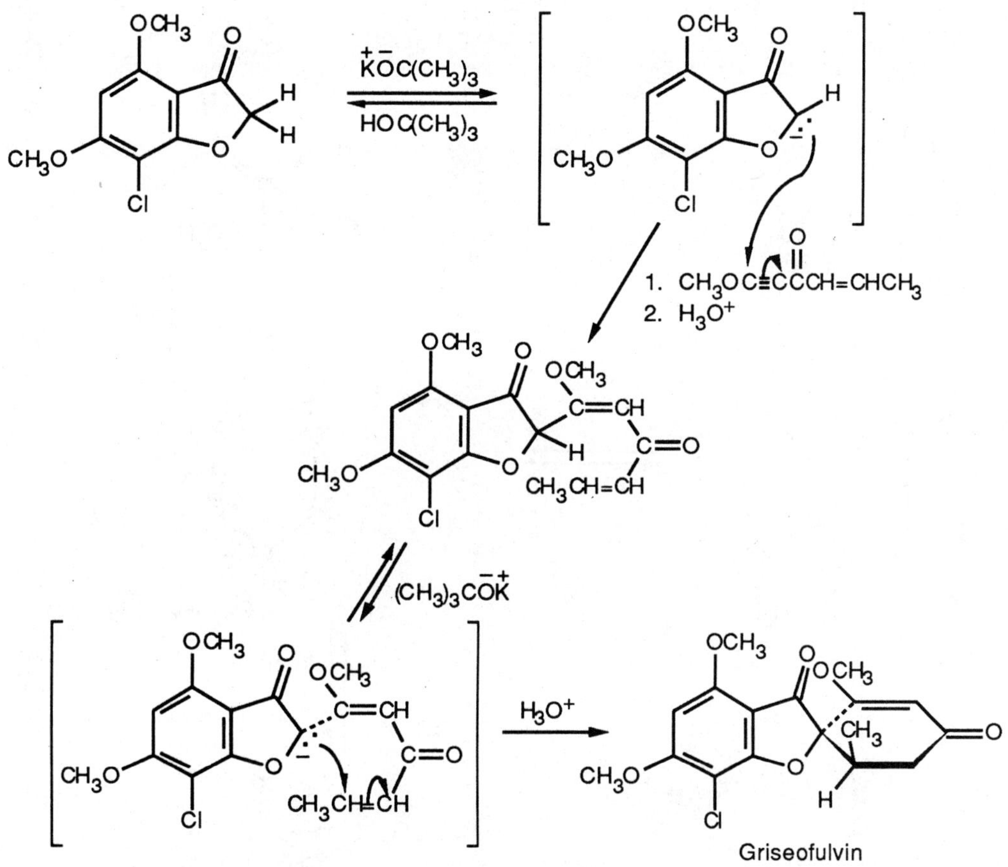

Two Michael reactions are involved in the key step that forms the *spiro* ring.

23.44

a)

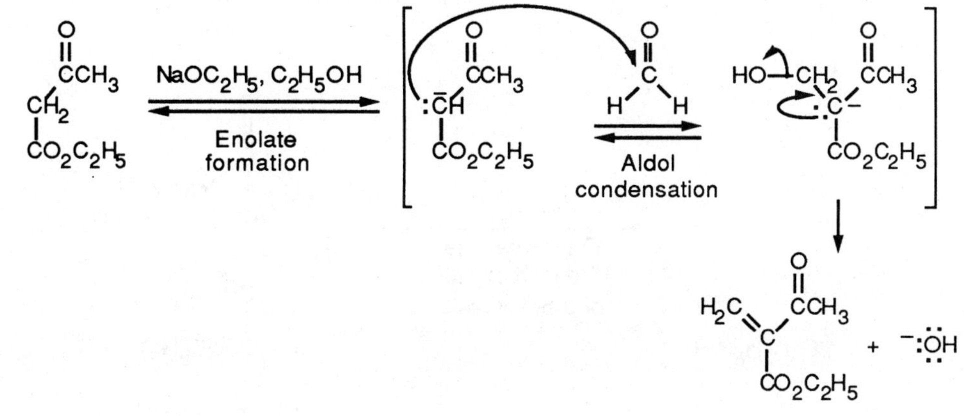

b)

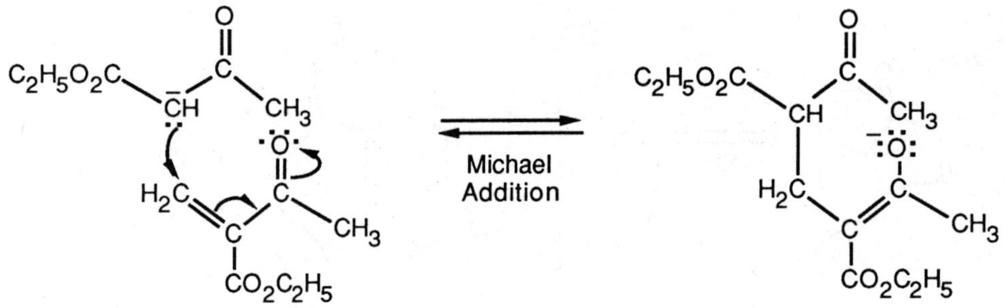

Michael
Addition

23.45

c)

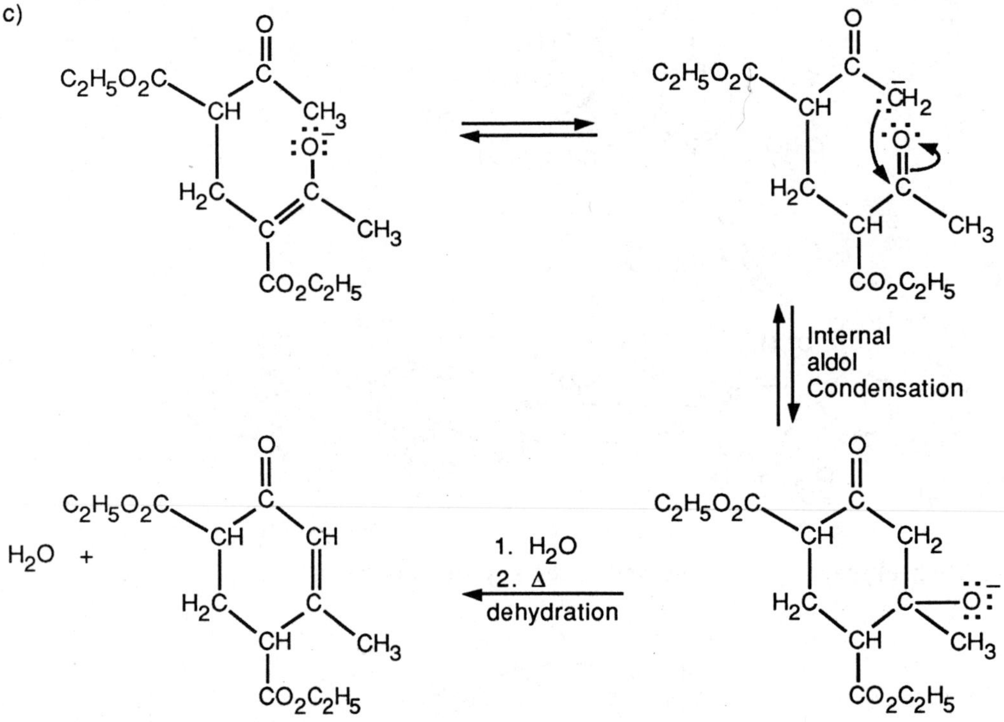

Internal
aldol
Condensation

1. H_2O
2. Δ
dehydration

H_2O +

d)

H_3O^+, Δ

Ester cleavage
and decarboxylation
of a ß–ketoacid

+ CH_3CH_2OH
+ CO_2

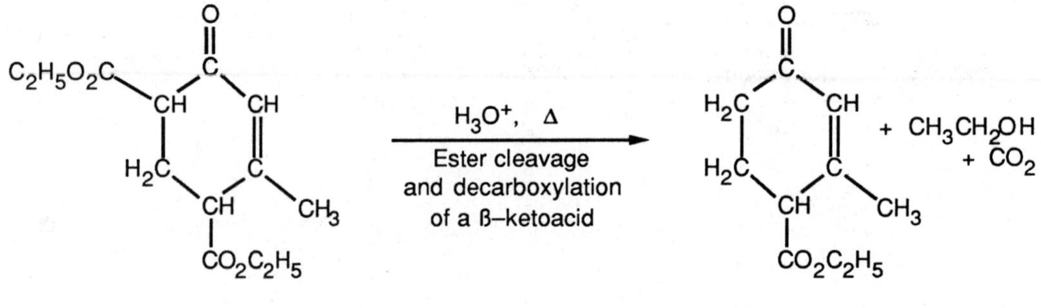

Hagemann's ester

23.46

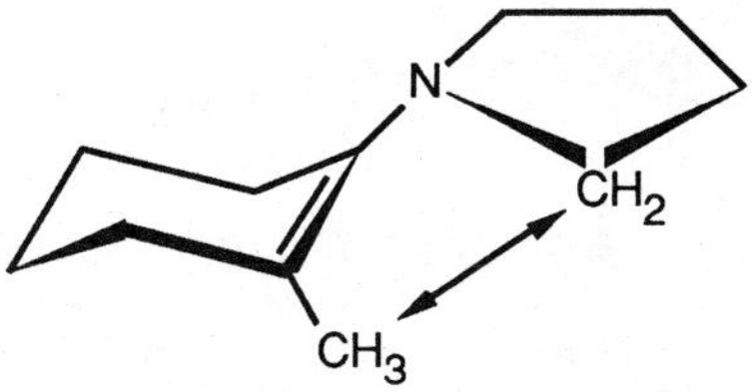

Crowding between the methyl group and the pyrrolidine ring disfavors this enamine.

A

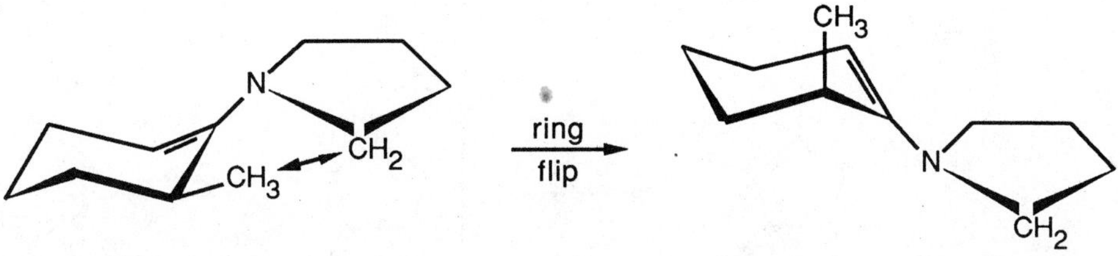

The crowding in this enamine can be relieved by a ring-flip, which puts the methyl group in an axial position. This enamine is the only one formed.

23.47

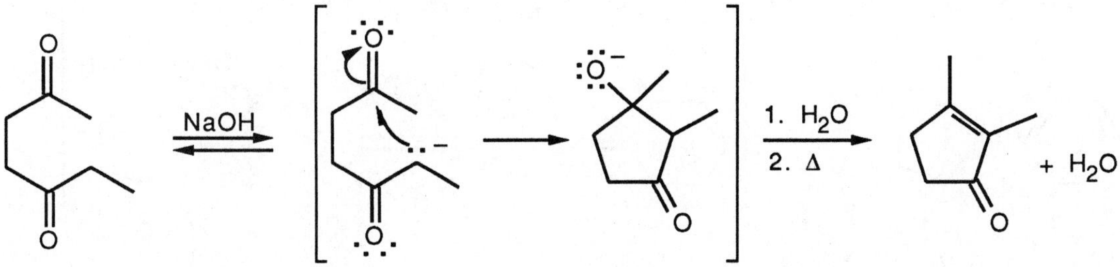

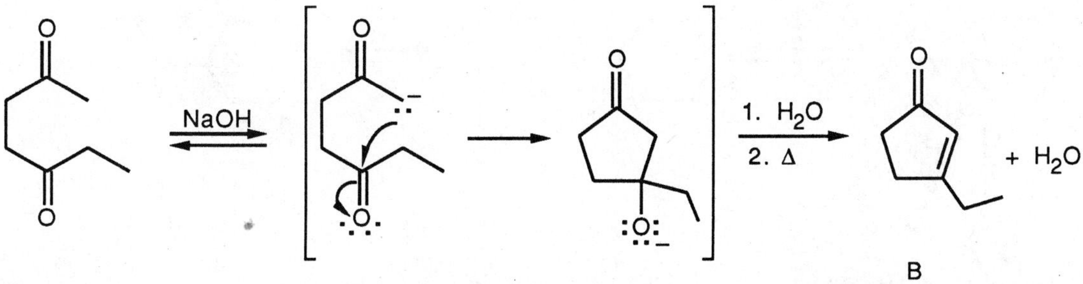

B

23.48 Product A, which has two singlet methyl groups and no vinylic protons in its ^{1}H NMR, is the major product of the intramolecular cyclization of 2,5–heptanedione.

23.49

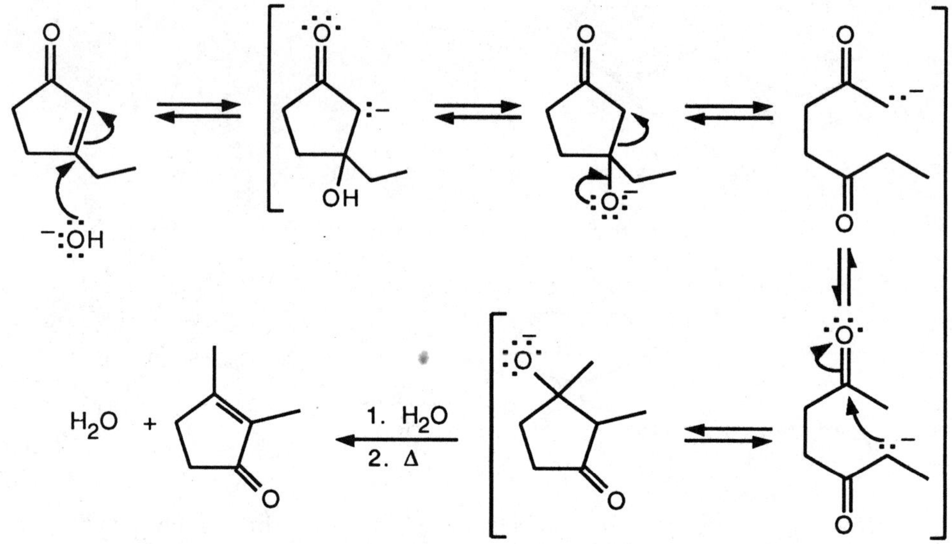
O
⁻:ÖH
OH
H₂O +
1. H₂O
2. Δ

23.50

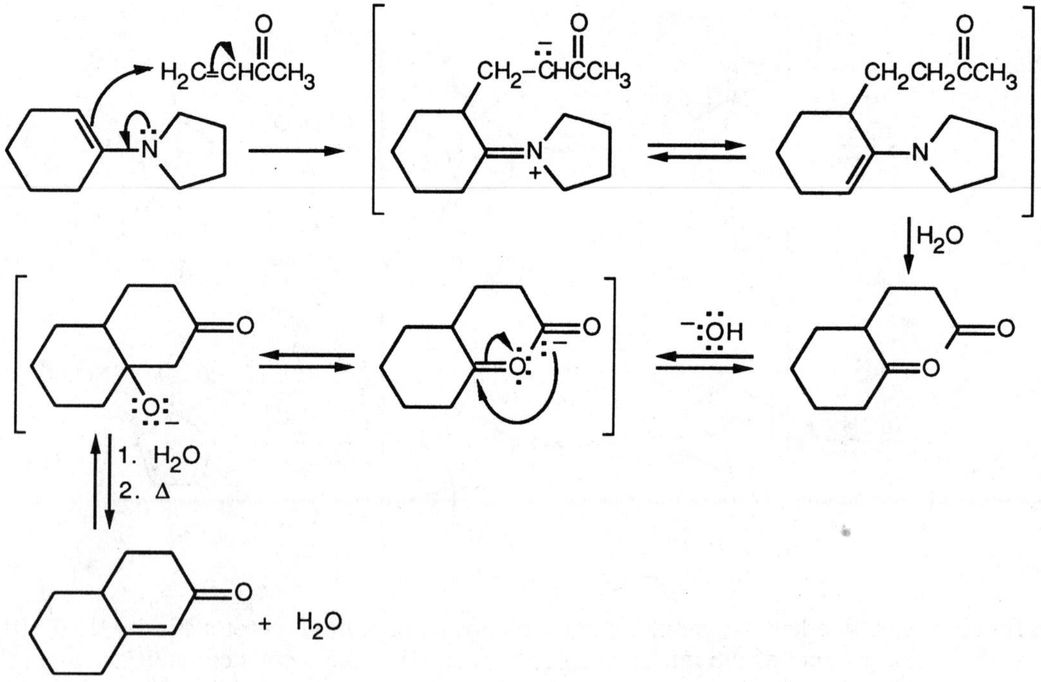
H₂C=CHCCH₃
CH₂–CHCCH₃
CH₂CH₂CCH₃
N
H₂O
⁻:ÖH
1. H₂O
2. Δ
+ H₂O

23.51

a)

b)

23.52 This problem becomes easier if you draw the starting material so that it resembles the product.

23.53

23.54

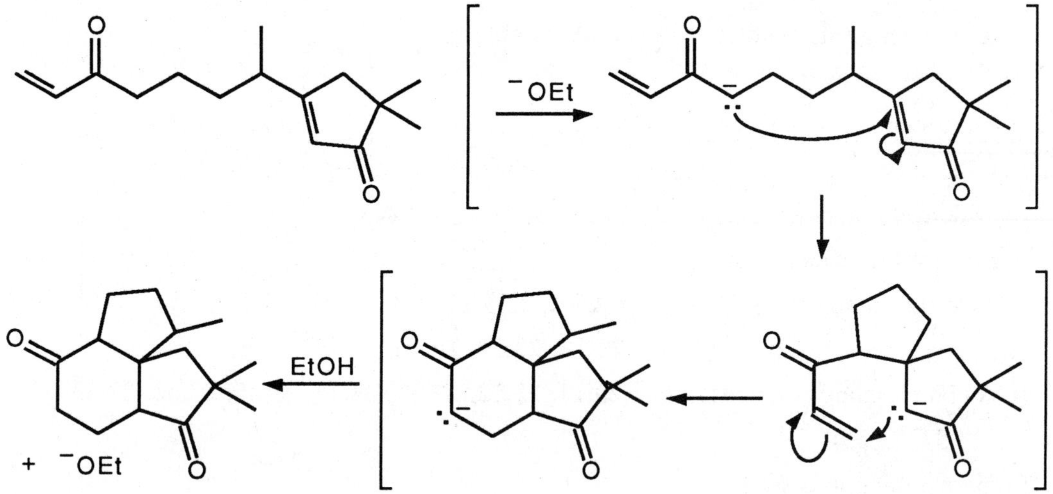

23.55

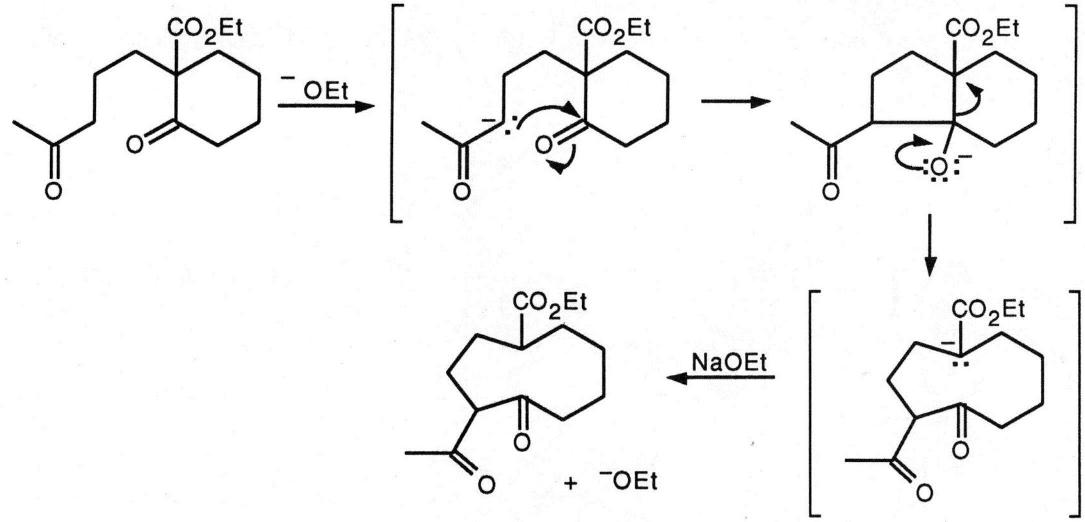

Study Guide for Chapter 23

After studying this chapter, you should be able to:

(1)   Predict the products of:

    (a) Aldol condensations (23.1, 23.4, 23.5, 23.7, 23.10, 23.27, 23.28, 23.29);

    (b) Claisen condensations (23.13, 23.16, 23.17, 23.18, 23.36, 23.38);

    (c) Michael reactions (23.19, 23.20, 23.23, 23.25, 23.26).

(2)   Formulate the mechanism of:

    (a) Aldol condensations (23.2, 23.3, 23.9, 23.31, 23.34, 23.47, 23.49);

    (b) Claisen condensations (23.14, 23.15, 23.39, 23.40);

    (c) Other carbonyl condensations (23.11, 23.12, 23.41, 23.44, 23.45, 23.52, 23.53, 23.54, 23.55).

(3)   Prepare compounds by:

    (a) Aldol condensations (23.6, 23.8, 23.30, 23.32);

    (b) Michael reactions (23.21, 23.24).

(4)   Use carbonyl condensation reactions in synthesis (23.33, 23.35, 23.42, 23.43, 23.50, 23.51).

24.1

a) 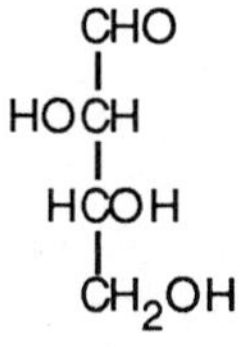
```
      CHO
   HOCH
    HCOH
    CH₂OH
```
Threose
an *aldotetrose*

b) 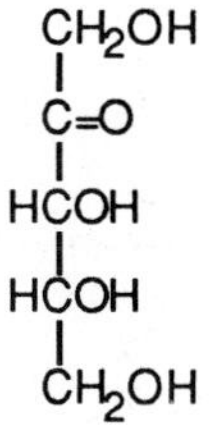
```
    CH₂OH
     C=O
    HCOH
    HCOH
    CH₂OH
```
Ribulose
a *ketopentose*

c) 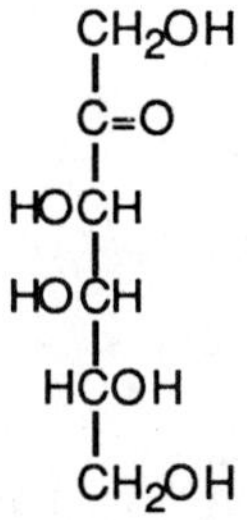
```
    CH₂OH
     C=O
   HOCH
   HOCH
    HCOH
    CH₂OH
```
Tagatose
a *ketohexose*

d) 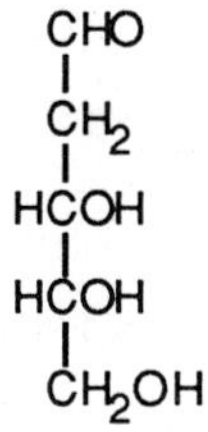
```
     CHO
     CH₂
    HCOH
    HCOH
    CH₂OH
```
2–Deoxyribose
an *aldopentose*

24.2 Projections A, B and C represent the *S* enantiomer of glyceraldehyde; projection D represents the *R* enantiomer.

24.3

a)

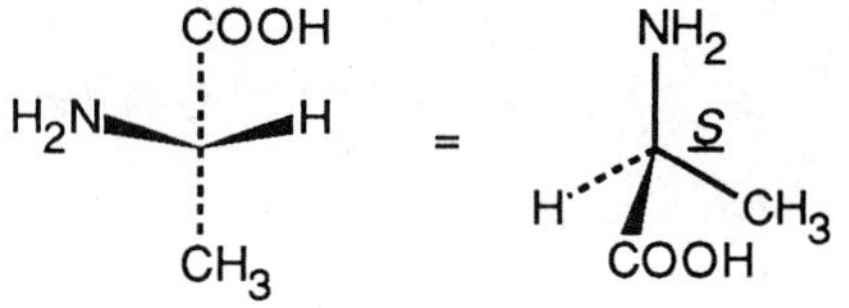

= (with COOH, H₂N, H, CH₃ / *S* configuration, H, CH₃, COOH)

b)

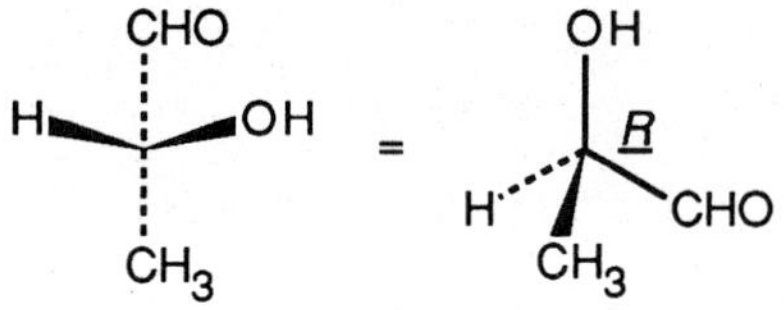

= (with CHO, H, OH, CH₃ / *R* configuration, OH, H, CH₃, CHO)

c)

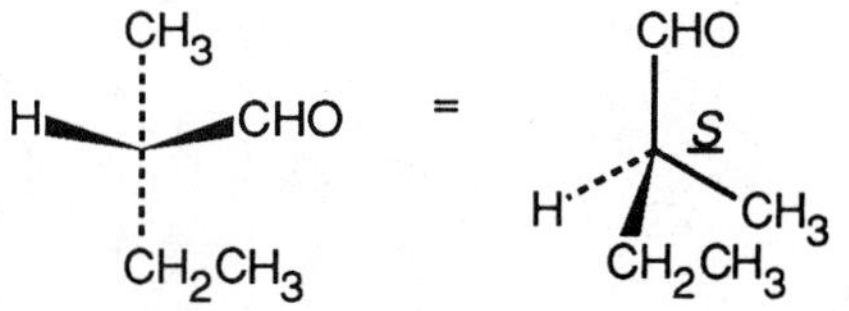

= (with CH₃, H, CHO, CH₂CH₃ / *S* configuration, H, CH₃, CH₂CH₃)

24.4

a) 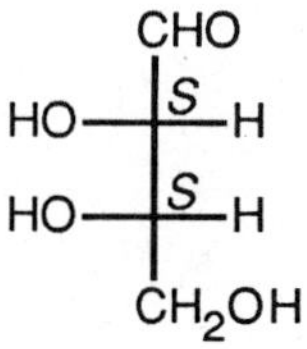
```
        CHO
   HO——S——H
   HO——S——H
        CH₂OH
```
L–Erythrose

b) 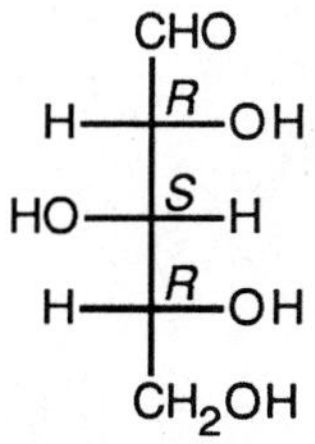
```
        CHO
    H——R——OH
   HO——S——H
    H——R——OH
        CH₂OH
```
D–Xylose

c) 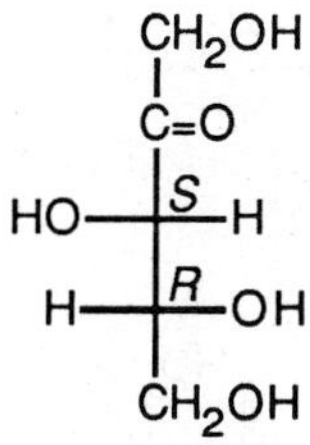
```
        CH₂OH
         C=O
   HO——S——H
    H——R——OH
        CH₂OH
```
D–Xylulose

24.5

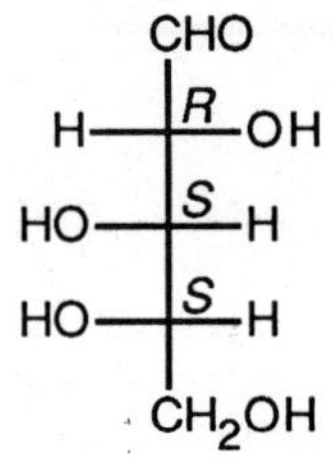

```
        CHO
        |
H——————|R|——OH
        |
HO—————|S|——H
        |
HO—————|S|——H
        |
       CH₂OH
```

L–(+)–Arabinose

24.6

a)

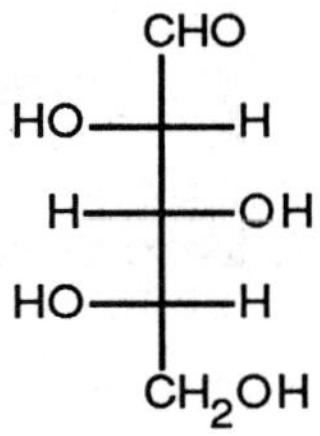

```
        CHO
        |
HO——————H
        |
 H——————OH
        |
HO——————H
        |
       CH₂OH
```

L–Xylose

b)

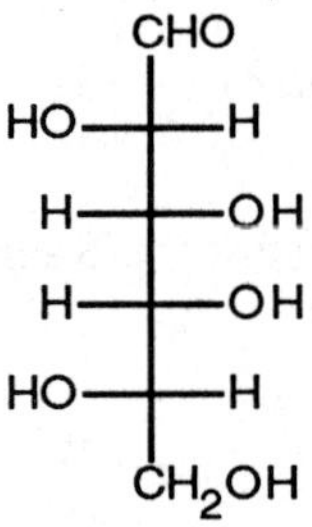

```
        CHO
        |
HO——————H
        |
 H——————OH
        |
 H——————OH
        |
HO——————H
        |
       CH₂OH
```

L–Galactose

c)

```
        CHO
        |
HO——————H
        |
 H——————OH
        |
HO——————H
        |
HO——————H
        |
       CH₂OH
```

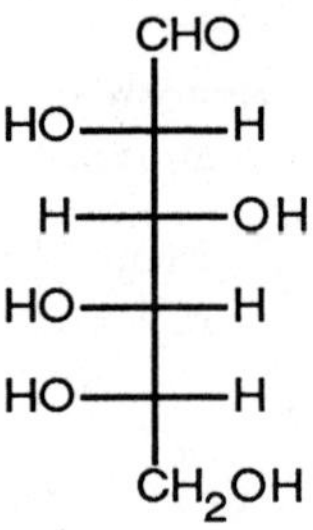

L–Glucose

24.7   There are 32 aldoheptoses -- sixteen D and sixteen L.

24.8

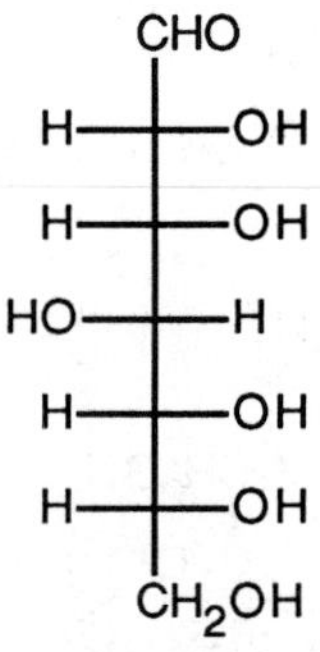

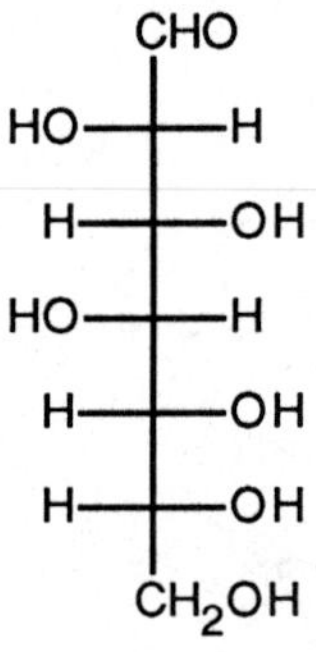

```
        CHO
        |
 H——————OH
        |
 H——————OH
        |
HO——————H
        |
 H——————OH
        |
 H——————OH
        |
       CH₂OH
```

```
        CHO
        |
HO——————H
        |
 H——————OH
        |
HO——————H
        |
 H——————OH
        |
 H——————OH
        |
       CH₂OH
```

24.9

```
        CHO
        |
 H——————OH
        |
HO——————H
        |
HO——————H
        |
 H——————OH
        |
       CH₂OH
```

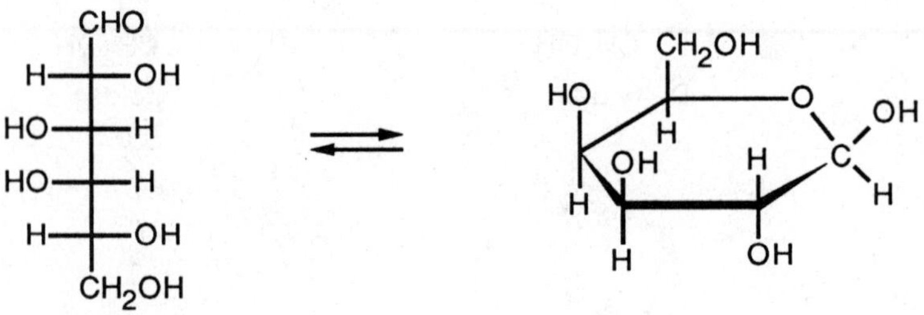

D–Galactose

⇌

24.10

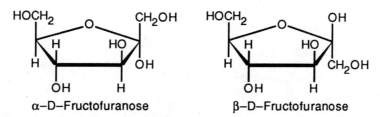

L–Glucose   D–Ribose

24.11

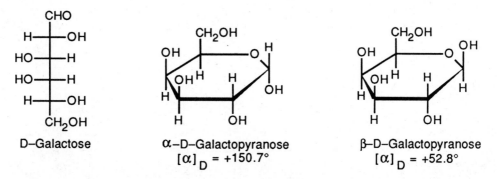

α–D–Fructofuranose   β–D–Fructofuranose

24.12 Let x be the fraction of D–glucose present as the α anomer, and let y be the fraction of D–glucose present as the ß anomer.

Then: $112.2° x + 18.7° y = 52.6°$ $x + y = 1; \quad y = 1 - x$

$112.2°x + 18.7°(1-x) = 52.6°$

$93.5°x = 33.9°$

$x = 0.362$

$y = 0.638$

Thus, 36.2% of glucose is present as the α anomer and 63.8% is present as the ß anomer.

24.13

D–Galactose   α–D–Galactopyranose   β–D–Galactopyranose
                              $[\alpha]_D = +150.7°$ $[\alpha]_D = +52.8°$

Let x be the fraction of D–galactose present as the α anomer and y be the fraction of D–Galactose present as the ß anomer.

$$150.7° \, x + 52.8°y = 80.2° \qquad x + y = 1; \quad y = 1 - x$$

$$150.7°x + 52.8°(1-x) = 80.2°$$

$$97.9°x = 27.4°$$

$$x = 0.280$$

$$y = 0.720$$

28.0% of D–galactose is present as the α anomer, and 72.0% is present as the ß anomer.

24.14

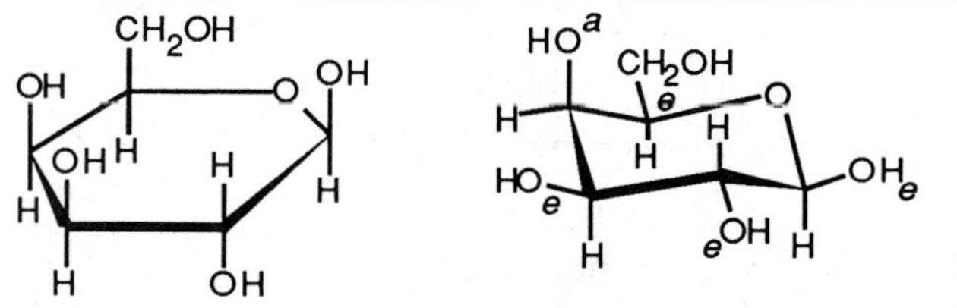

ß–D–Galactopyranose

ß–D–Mannopyranose

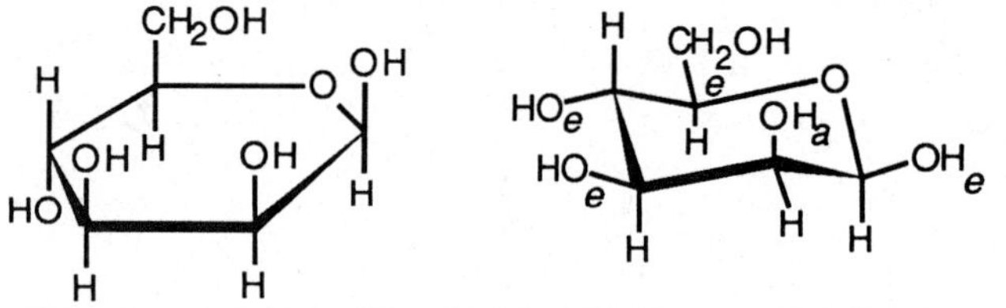

ß–D–Galactopyranose and ß–D–mannopyranose each have one hydroxyl group in the axial position. Galactose and mannose are of similar stability.

24.15

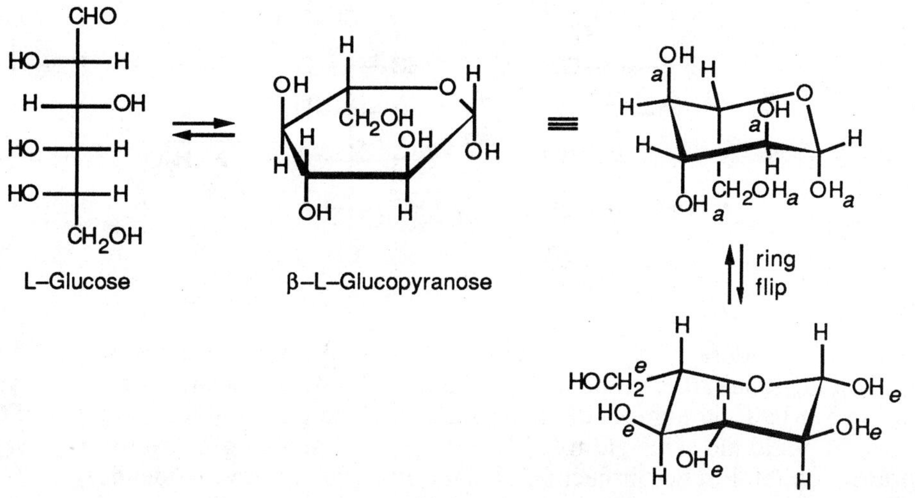

L-Glucose ⇌ β–L–Glucopyranose ≡

ring flip

All substituents are equatorial in the more stable conformation of ß–L–glucopyranose.

24.16

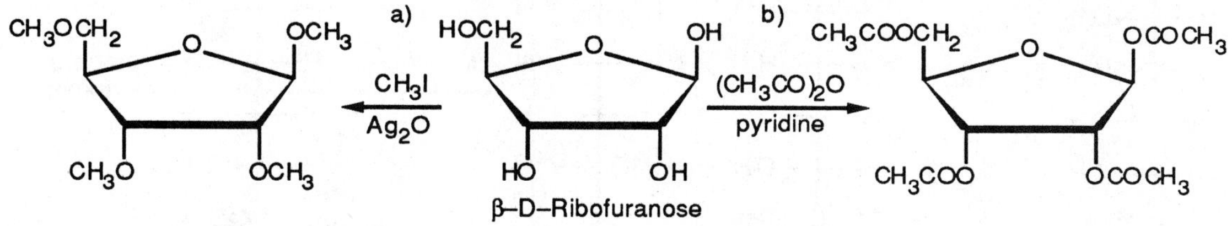

a) CH_3I / Ag_2O

β–D–Ribofuranose

b) $(CH_3CO)_2O$ / pyridine

24.17

D–Galactose

1. $NaBH_4$
2. H_2O

Galactitol

- - - - - - - - - Plane of symmetry

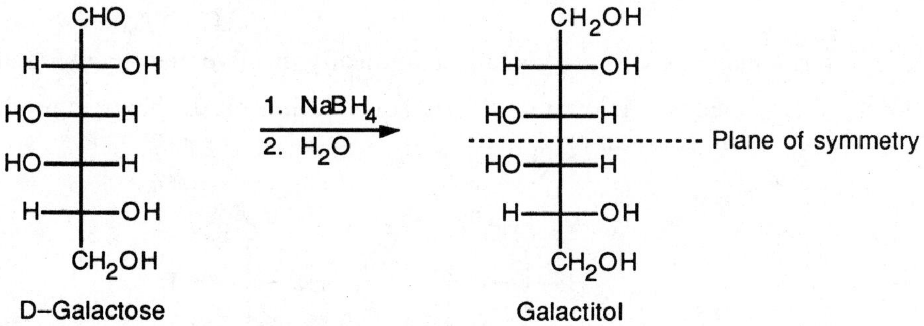

Reaction of D–galactose with $NaBH_4$ yields an alditol that has a plane of symmetry. Galactitol is a *meso* compound.

24.18

CHO — H—OH, HO—H, H—OH, H—OH, CH₂OH (D–Glucose)

1. NaBH₄
2. H₂O

→ CH₂OH — H—OH, HO—H, H—OH, H—OH, CH₂OH (D–Glucitol)

≡ CH₂OH — HO—H, HO—H, H—OH, HO—H, CH₂OH (L–Gulitol)

← 1. NaBH₄ 2. H₂O

CHO — HO—H, HO—H, H—OH, HO—H, CH₂OH (L–Gulose)

Reaction of an aldose with $NaBH_4$ produces a polyol (alditol). Because an alditol has the same functional group at both ends, the number of stereoisomers of an n–carbon alditol is one-half the number of stereoisomers of the parent aldose, and two different aldoses can yield the same alditol. Here, L–glucose and D–glucose form the same alditol (rotate the Fischer projection of L–gulitol 180° to see the identity).

24.19

CHO — H—OH, HO—H, H—OH, H—OH, CH₂OH (D–Glucose)

dil HNO_3, Δ

→ COOH — H—OH, HO—H, H—OH, H—OH, COOH (D–Glucaric acid)

CHO — H—OH, H—OH, H—OH, H—OH, CH₂OH (D–Allose)

dil HNO_3, Δ

→ COOH — H—OH, H—OH, H—OH, H—OH, COOH (Allaric acid)

······· plane of symmetry

Allaric acid has a plane of symmetry and is an optically inactive *meso* compound.

24.20 D–Allose and D–galactose yield *meso* aldaric acids. All other D–hexoses produce optically active aldaric acids on oxidation.

24.21

CHO — H—OH, H—OH, H—OH, CH₂OH (D–Ribose)

1. HCN
2. Pd, BaSO₄
3. H₃O⁺

→ CHO — H—OH, H—OH, H—OH, H—OH, CH₂OH (D–Allose)

+

CHO — HO—H, H—OH, H—OH, H—OH, CH₂OH (D–Altrose)

24.22

```
      CHO                                    CHO                      CHO
HO ──┼── H          1.  HCN            H ──┼── OH          HO ──┼── H
 H ──┼── OH         2.  Pd, BaSO4     HO ──┼── H           HO ──┼── H
HO ──┼── H          ─────────────►    H ──┼── OH      +     H ──┼── OH
      │             3.  H3O+          HO ──┼── H           HO ──┼── H
    CH2OH                                 CH2OH                 CH2OH

   L–Xylose                              L–Idose              L–Gulose
```

24.23

```
      CHO                CHO           1.  H2NOH                    CHO
 H ──┼── OH        HO ──┼── H         2.  (CH3CO)2O,          HO ──┼── H
HO ──┼── H   or    HO ──┼── H             NaO2CCH3             H ──┼── OH
 H ──┼── OH         H ──┼── OH         ──────────────►             CH2OH
      │                 │             3.  NaOCH3
    CH2OH             CH2OH

  D–Xylose           D–Lyxose                                  D–Threose
```

24.24

```
      CHO                CHO
HO ──┼── H         HO ──┼── H
 H ──┼── OH        HO ──┼── H
 H ──┼── OH         H ──┼── OH
      │                 │
    CH2OH             CH2OH

 D–Arabinose        D–Lyxose
```

D–Glucose has the same configuration at C3, C4, and C5 as D–arabinose. D–Lyxose is the only other aldopentose that yields an optically active aldaric acid on nitric acid oxidation.

24.25

```
a)    CHO          b)    CHO          c)    CHO          d)    CHO
 H ──┼── OH         H ──┼── OH         H ──┼── OH        HO ──┼── H
 H ──┼── OH        HO ──┼── H          H ──┼── OH         H ──┼── OH
 H ──┼── OH         H ──┼── OH              CH2OH              CH2OH
      │                 │
    CH2OH             CH2OH

  D–Ribose           D–Xylose         D–Erythrose        D–Threose
```

24.26

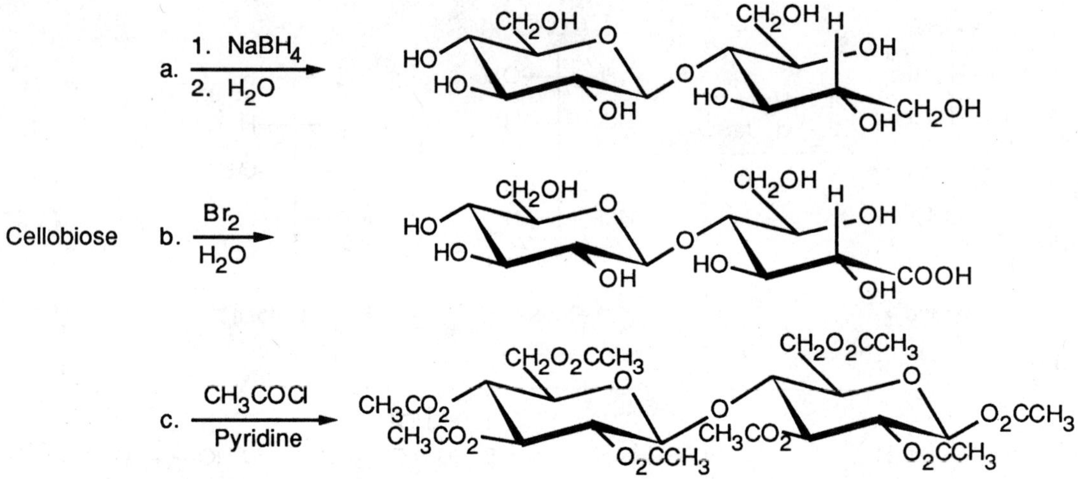

a. $\xrightarrow[\text{2. H}_2\text{O}]{\text{1. NaBH}_4}$

Cellobiose   b. $\xrightarrow[\text{H}_2\text{O}]{\text{Br}_2}$

c. $\xrightarrow[\text{Pyridine}]{\text{CH}_3\text{COCl}}$

24.27

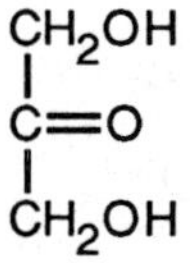

CH$_2$OH
|
C=O
|
CH$_2$OH

a ketotriose

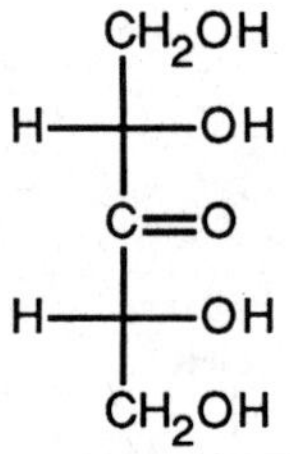

CH$_2$OH
H——OH
C=O
H——OH
CH$_2$OH

a ketopentose

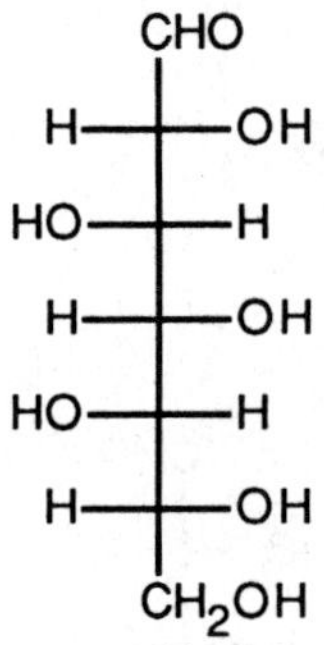

CHO
H——OH
HO——H
H——OH
HO——H
H——OH
CH$_2$OH

an aldoheptose

24.28

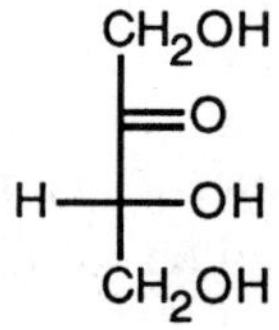

CH$_2$OH
=O
H——OH
CH$_2$OH

a ketotetrose

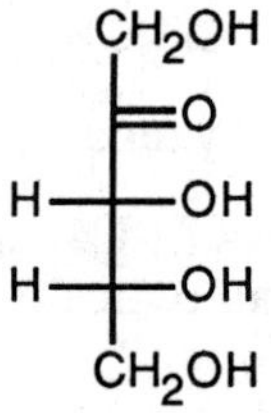

CH$_2$OH
=O
H——OH
H——OH
CH$_2$OH

a ketopentose

24.29

O=C—H
HO——H
H——OH
H——H
H——OH
CH$_2$OH

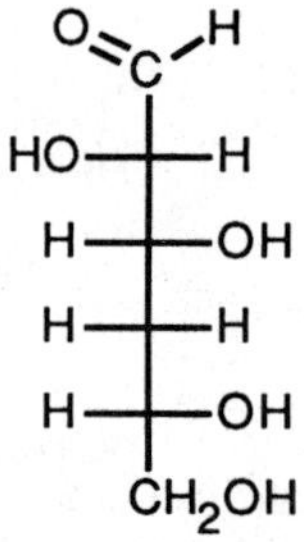

a deoxyaldohexose

24.30

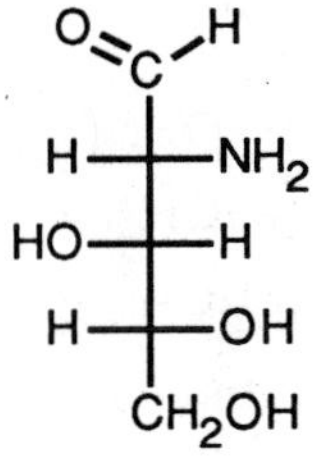

a five-carbon amino sugar

24.31–24.32

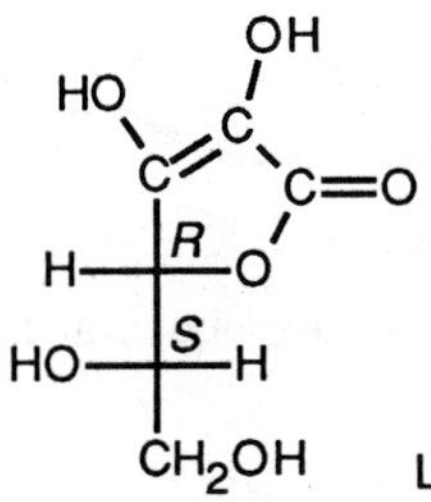

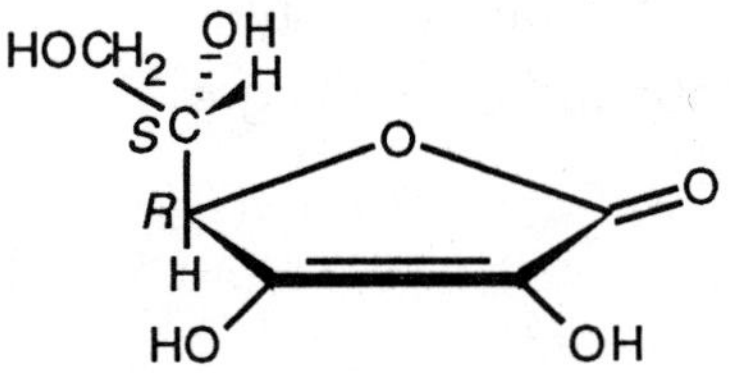

L–Ascorbic Acid

24.33

Definition		Example

a) A *monosaccharide* is a carbohydrate that cannot be hydrolyzed into smaller units.

D–Glucose

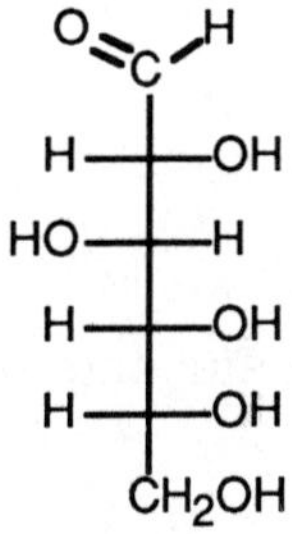

b) An *anomeric center* is a chiral center formed when an open chain nonosaccharide cyclizes to a furanose or pyranose ring.

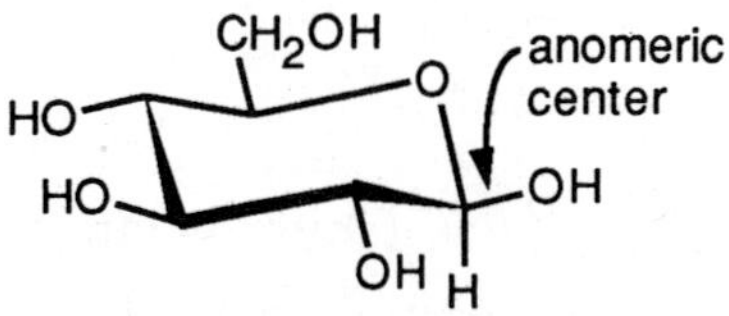

anomeric center

β–D–Glucopyranose

c) A *Haworth projection* is a drawing of a pyranose or furanose in which the ring is drawn as flat. This projection allows the relationship of the ring substituents to be viewed more easily.

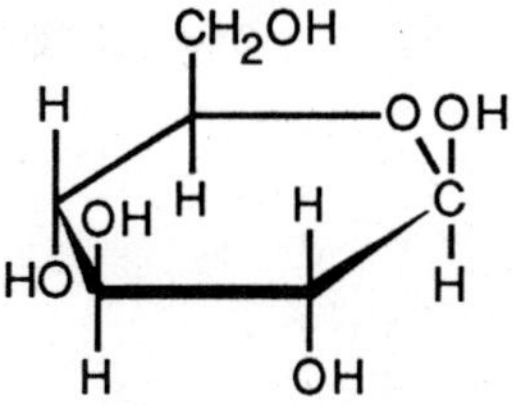

β–D–Glucopyranose

d)   A *Fischer projection* is a drawing of a
      carbohydrate in which each chiral center
      is represented as a pair of perpendicular
      lines. Vertical lines represent bonds
      going into the page, and horizontal lines
      represent bonds coming out of the page.

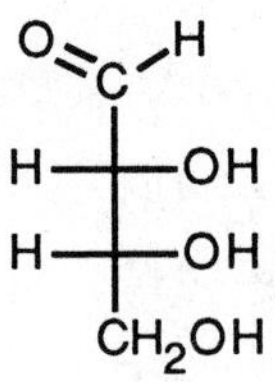

D-Erythrose

e)   A *glycoside* is a acetal of a carbohydrate,
      formed when an anomeric hydroxyl group
      reacts with another compound containing a
      hydroxyl group.

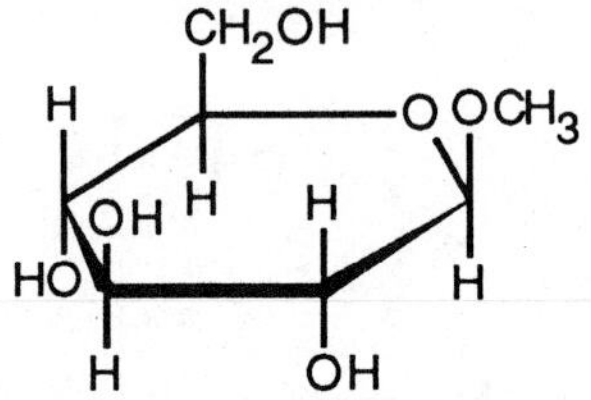

Methyl β–D–glucopyranoside

f)   A *reducing sugar* is a sugar that reacts with
      any of several reagents to yield an oxidized
      sugar plus reduced reagent.

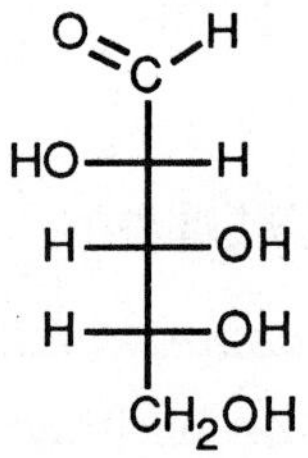

D-Arabinose

g)   A *pyranose* is a six-membered cyclic
      hemiacetal ring form of a monosaccharide.

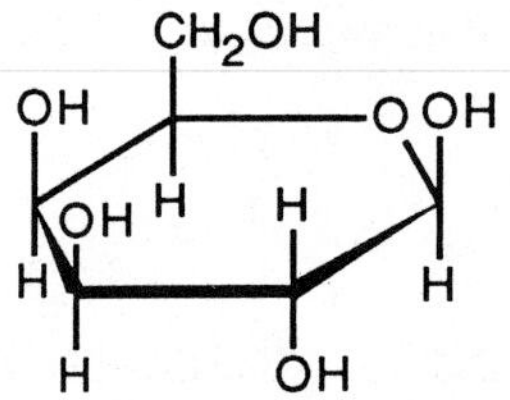

β–D–Galactopyranose

h)   A *1,4' link* occurs when the anomeric hydroxyl
      group (at carbon 1) of a pyranose or furanose
      forms a glycosidic bond with the hydroxyl group
      at carbon 4 of a second nonosaccharide.

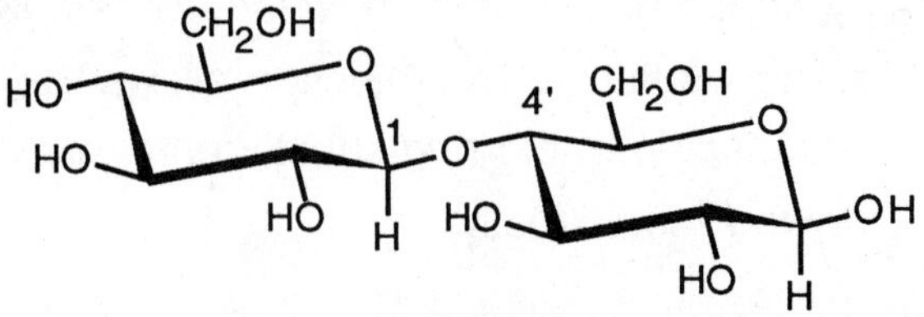

Cellobiose

i)   A *D–sugar* is a sugar in which the hydroxyl
     group farthest from the carbonyl group
     points to the right in a Fischer projection.

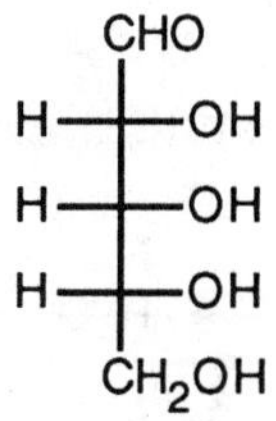

D–Ribose

24.34

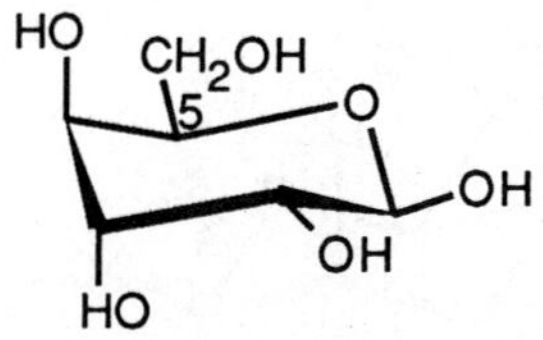

β–D–Gulopyranose

This structure is a pyranose (6-membered
ring) and is a β anomer (the C–1 hydroxyl
group and the –CH₂OH groups are *cis*).
It is a D–sugar because the –O– at C5
is on the right in the uncoiled form.

24.35

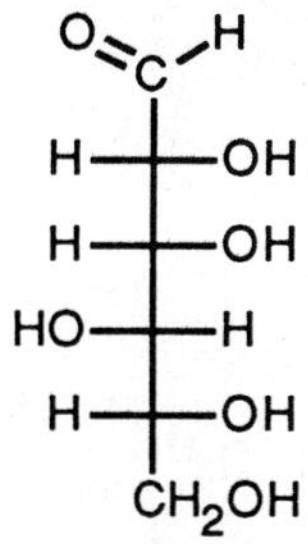

D–Gulose

24.36

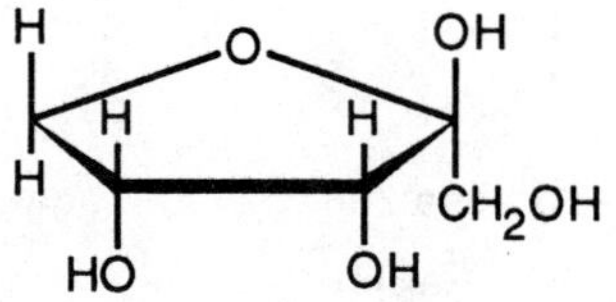

D–Ribulofuranose (β anomer)

24.37–24.38

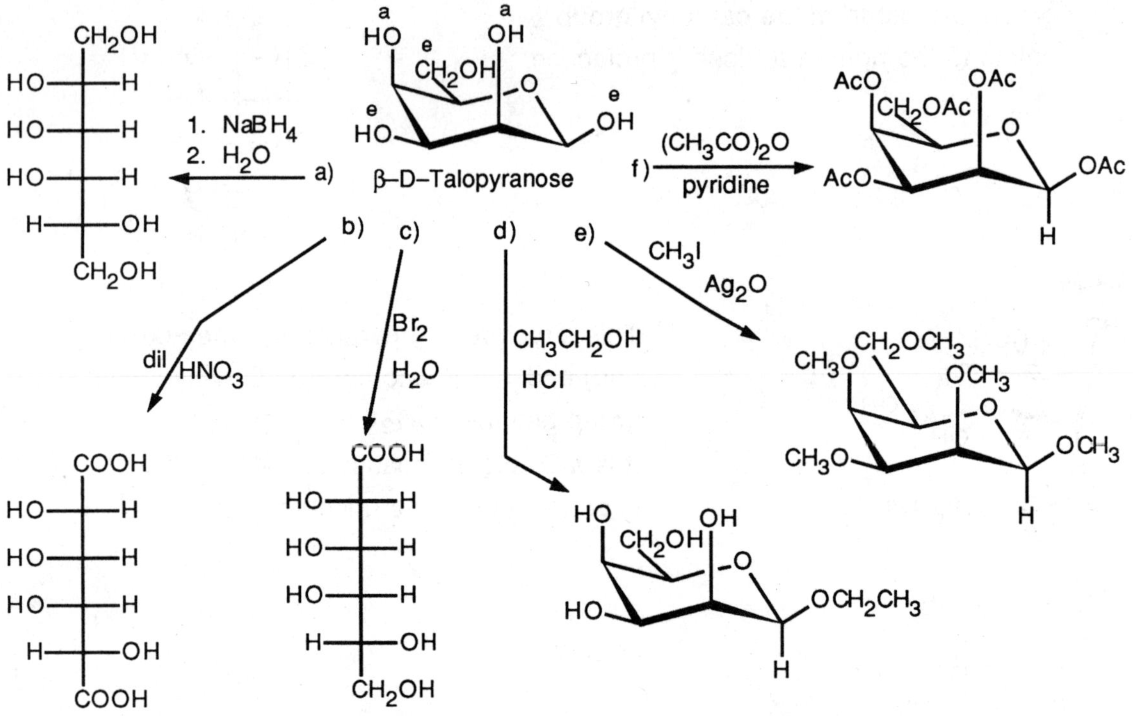

a) CH$_2$OH / HO—H / HO—H / HO—H / H—OH / CH$_2$OH

β–D–Talopyranose

1. NaBH$_4$
2. H$_2$O

f) (CH$_3$CO)$_2$O / pyridine

b) c) d) e)

b) dil HNO$_3$

COOH / HO—H / HO—H / HO—H / H—OH / COOH

c) Br$_2$ / H$_2$O

COOH / HO—H / HO—H / HO—H / H—OH / CH$_2$OH

d) CH$_3$CH$_2$OH / HCl

e) CH$_3$I / Ag$_2$O

24.39–24.41 Four D–2–ketohexoses are possible.

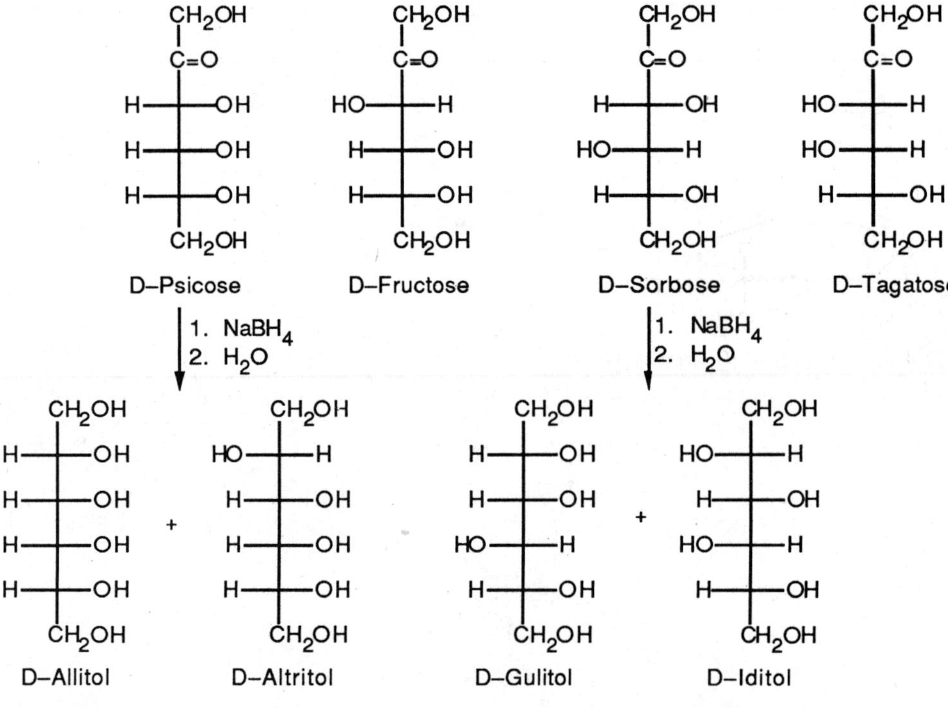

CH$_2$OH / C=O / H—OH / H—OH / H—OH / CH$_2$OH

D–Psicose

CH$_2$OH / C=O / HO—H / H—OH / H—OH / CH$_2$OH

D–Fructose

CH$_2$OH / C=O / H—OH / HO—H / H—OH / CH$_2$OH

D–Sorbose

CH$_2$OH / C=O / HO—H / HO—H / H—OH / CH$_2$OH

D–Tagatose

1. NaBH$_4$
2. H$_2$O

1. NaBH$_4$
2. H$_2$O

CH$_2$OH / H—OH / H—OH / H—OH / H—OH / CH$_2$OH

D–Allitol

+

CH$_2$OH / HO—H / H—OH / H—OH / H—OH / CH$_2$OH

D–Altritol

CH$_2$OH / H—OH / H—OH / HO—H / H—OH / CH$_2$OH

D–Gulitol

+

CH$_2$OH / HO—H / H—OH / HO—H / H—OH / CH$_2$OH

D–Iditol

24.42

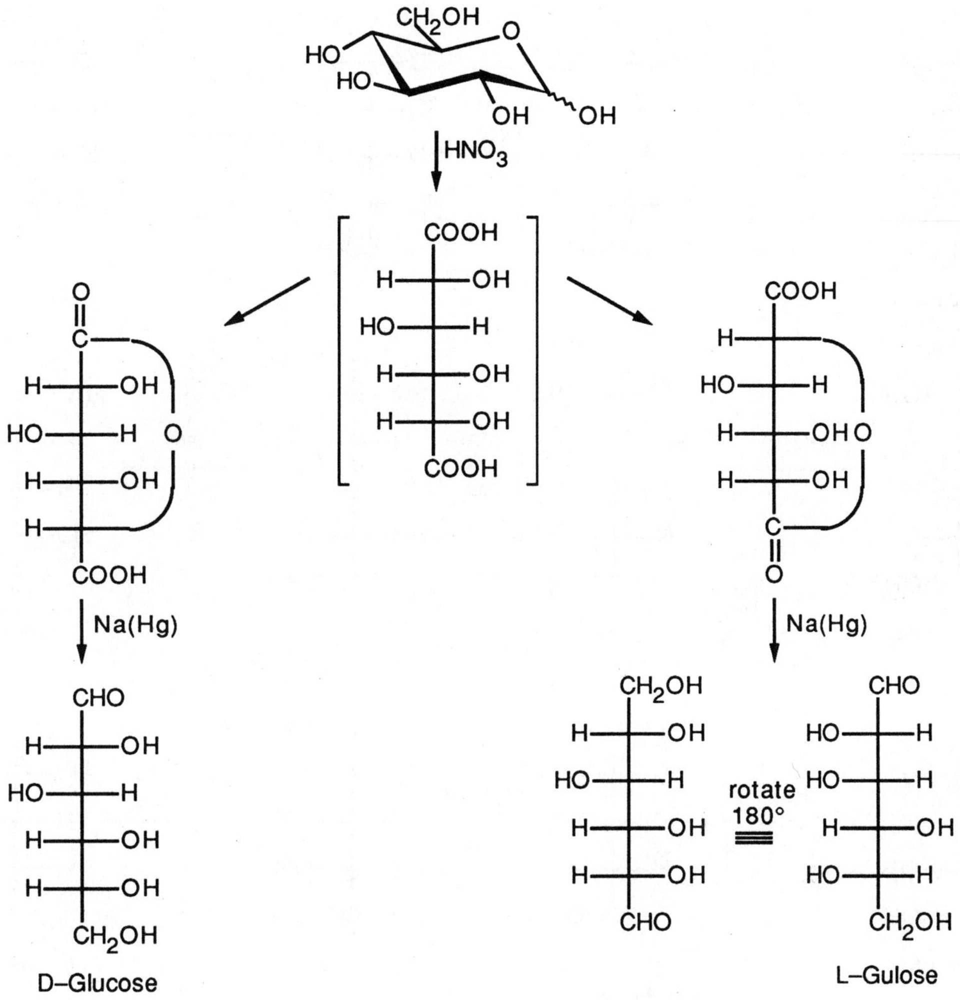
HNO3
Na(Hg)
Na(Hg)
rotate 180°
D–Glucose
L–Gulose

24.43

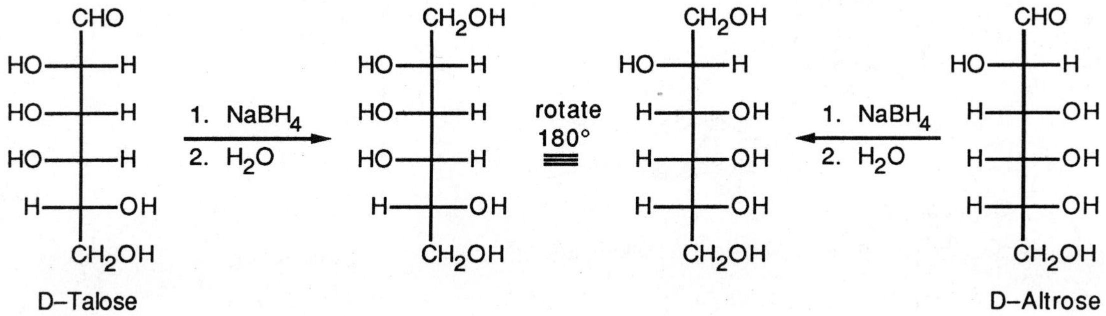
1. NaBH4
2. H2O
rotate 180°
1. NaBH4
2. H2O
D–Talose
D–Altrose

24.44

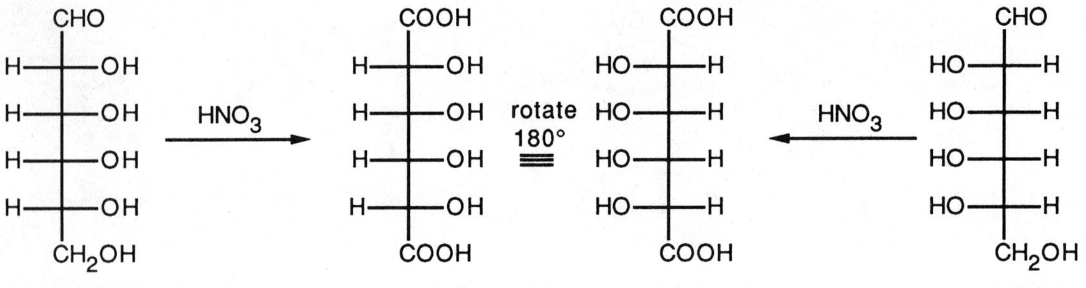

CHO / H—OH / H—OH / H—OH / H—OH / CH₂OH

D–Allose

HNO₃ →

COOH / H—OH / H—OH / H—OH / H—OH / COOH

rotate 180° ≡

COOH / HO—H / HO—H / HO—H / HO—H / COOH

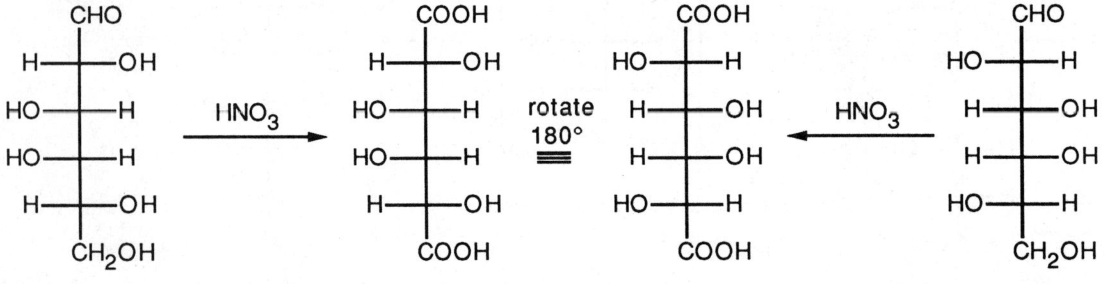

← HNO₃

CHO / HO—H / HO—H / HO—H / HO—H / CH₂OH

L–Allose

CHO / H—OH / HO—H / HO—H / H—OH / CH₂OH

D–Galactose

HNO₃ →

COOH / H—OH / HO—H / HO—H / H—OH / COOH

rotate 180° ≡

COOH / HO—H / H—OH / H—OH / HO—H / COOH

← HNO₃

CHO / HO—H / H—OH / H—OH / HO—H / CH₂OH

L–Galactose

24.45

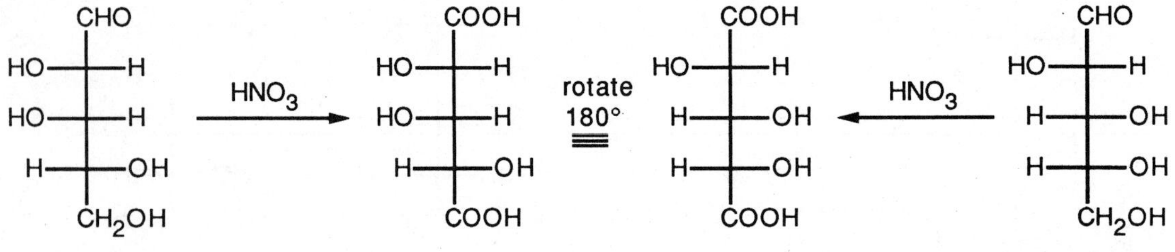

CHO / HO—H / HO—H / H—OH / CH₂OH

D–Lyxose

HNO₃ →

COOH / HO—H / HO—H / H—OH / COOH

rotate 180° ≡

COOH / HO—H / H—OH / H—OH / COOH

← HNO₃

CHO / HO—H / H—OH / H—OH / CH₂OH

D–Arabinose

24.46

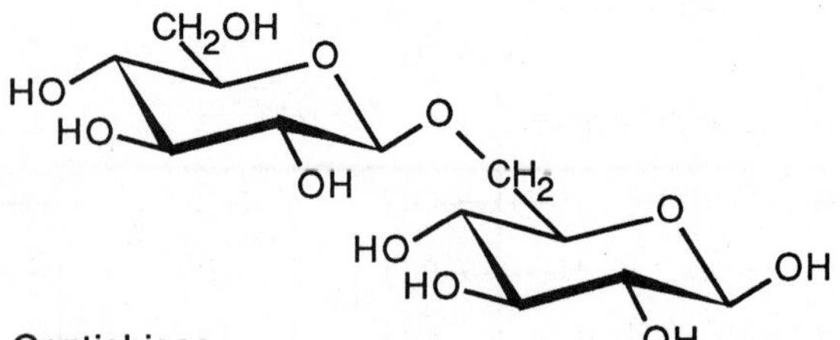

Gentiobiose
6–*O*–(β–D–Glucopyranosyl)–β–D–glucopyranose

24.47

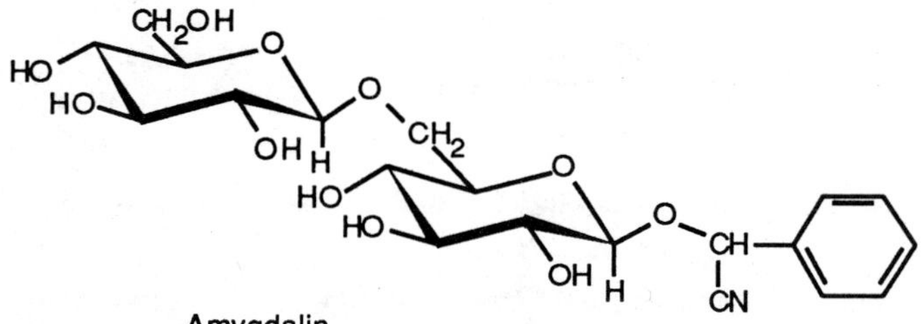

Amygdalin

24.48 There are three possible structures for trehalose. The two glucopyranose rings can be connected (α,α), (ß,ß), or (α,ß).

24.49 Since trehalose is not cleaved by ß–glycosidases, it must have an α,α glycosidic linkage.

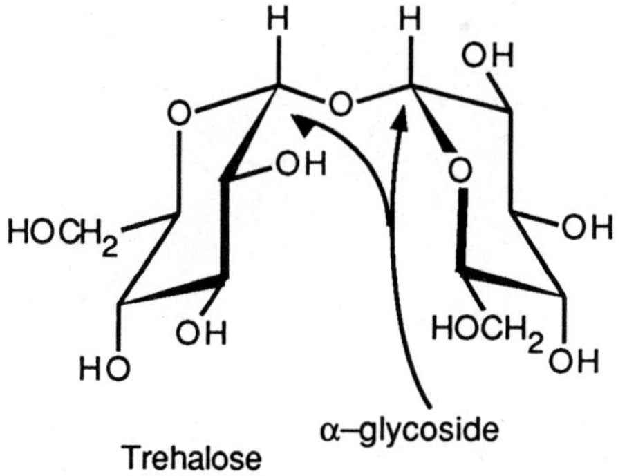

α–glycoside

Trehalose
1–*O*–(α–D–Glucopyranosyl)–α–D–glucopyranose

24.50

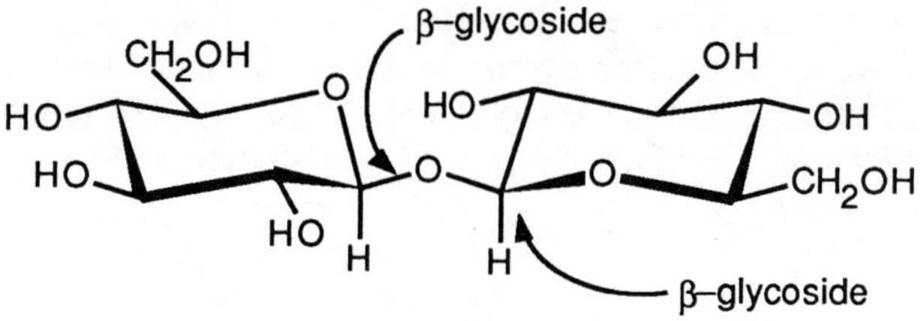

β–glycoside
β–glycoside

Neotrehalose
1–*O*–(β–D–Glucopyranosyl)–β–D–glucopyranose

α–glycoside
β–glycoside

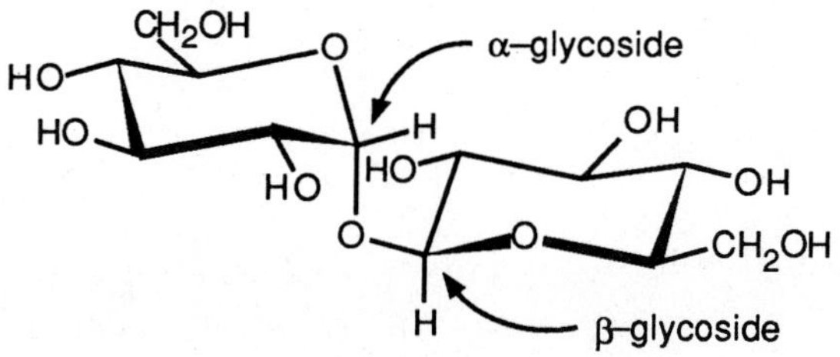

Isotrehalose
1–*O*–(α–D–Glucopyranosyl)–β–D–glucopyranose

24.51

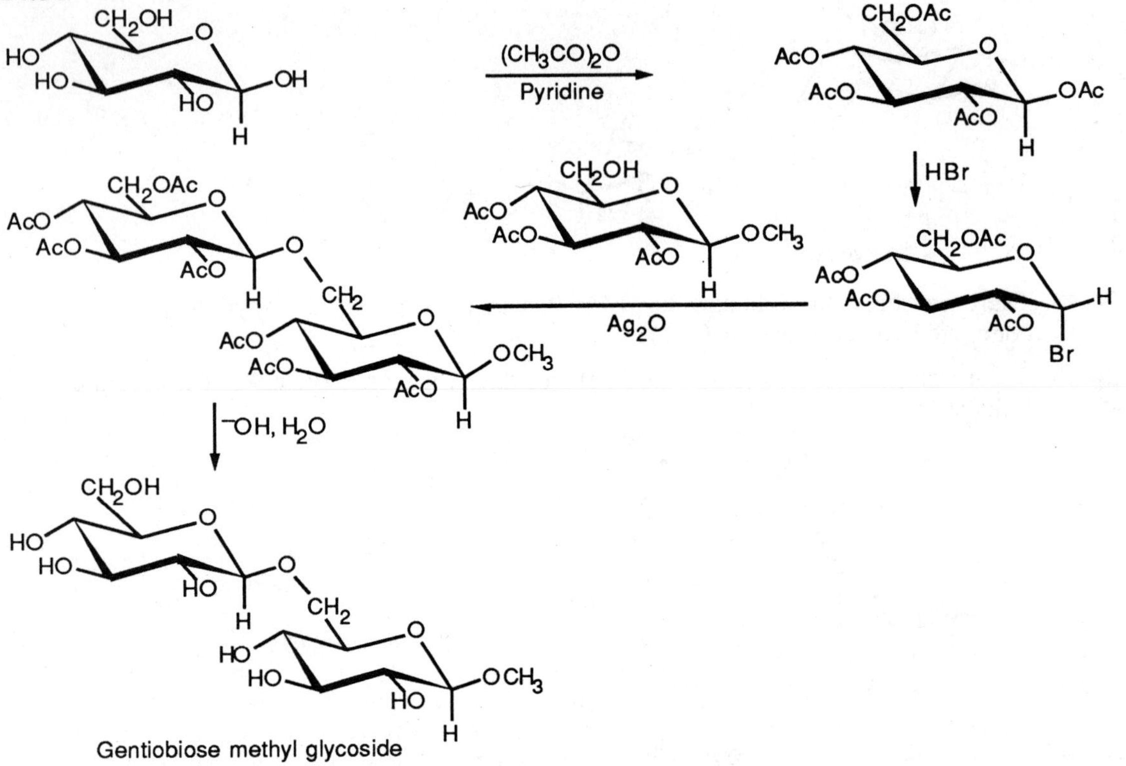

Gentiobiose methyl glycoside

The key step in this synthesis is the Koenigs-Knorr reaction (Sec. 24.8). Treatment of penta–*O*–acetyl–ß–D–glucopyranose with HBr converts the anomeric ß–acetate into an α–glycosyl bromide, which reacts with methyl–ß–D–2,3,4–tri-*O*–acetylgluco-pyranoside in the presence of Ag$_2$O. Reaction occurs at the 6–hydroxyl position of the methyl pyranoside because all other positions are blocked. Cleavage of the acetyl groups does not affect the glycosidic bonds because they are stable to base.

24.52

Glucopyranose is in equilibrium with glucofuranose.

Reaction with two equivalents of acetone occurs *via* the mechanism we learned for acetal formation (Sec. 19.14).

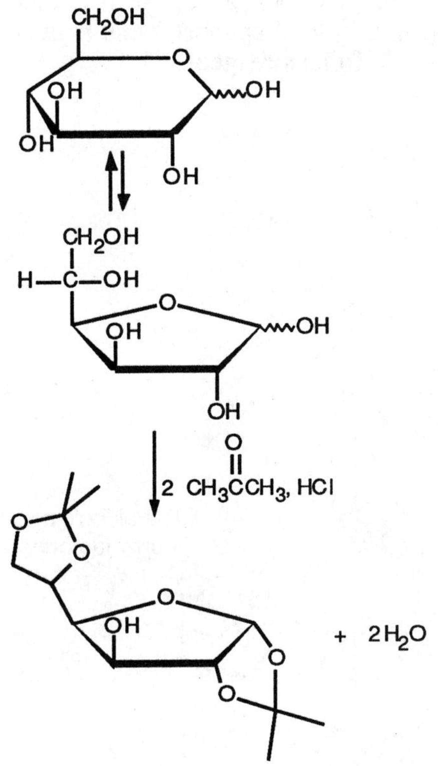

A five-membered acetal ring forms much more readily when the hydroxyl groups that are to be part of the acetal ring are *cis* to one another. In glucofuranose the C3 hydroxyl is *trans* to the C2 hydroxyl, and acetal formation occurs between acetone and the C1 and C2 hydroxyls. Since the C1 hydroxyl group is part of the acetone acetal, the furanose is no longer in equilibrium with the free aldehyde, and the diacetone derivative is not a reducing sugar.

24.53

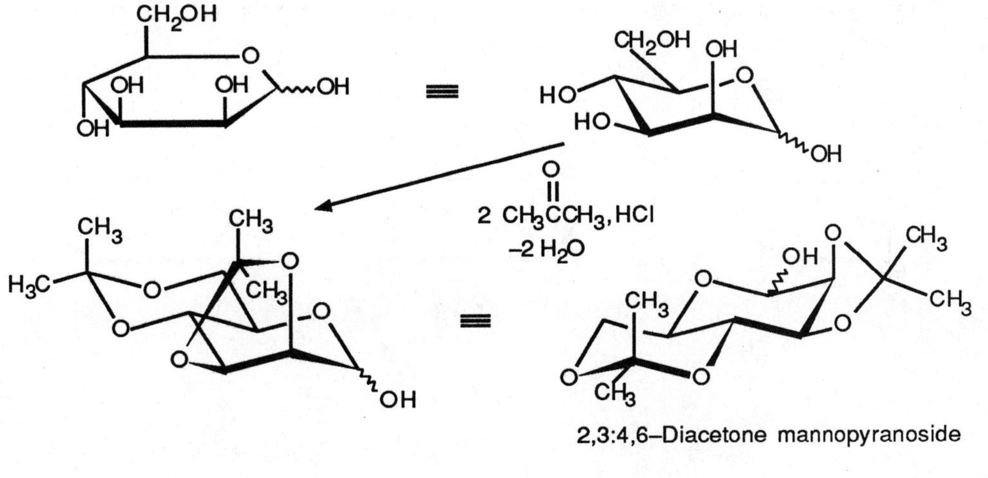

2,3:4,6–Diacetone mannopyranoside

Acetone forms an acetal with the hydroxyl groups at C2 and C3 of D–mannopyranoside because the hydroxyl groups at these positions are *cis* to one another. The
pyranoside ring is still a hemiacetal that is in equilibrium with free aldehyde, which is
reducing toward Tollens reagent.

24.54

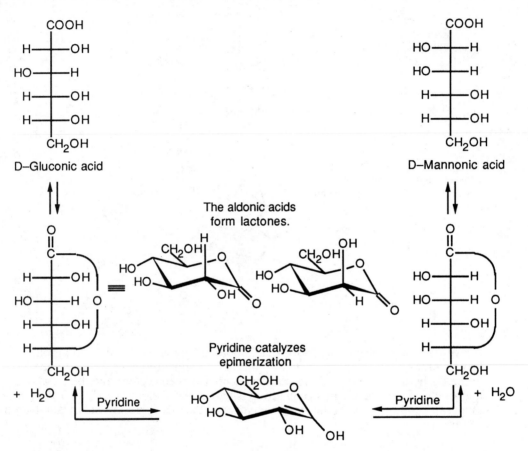

Epimerization at C2 occurs because the enediol can be reprotonated on either side of
the double bond.

24.55

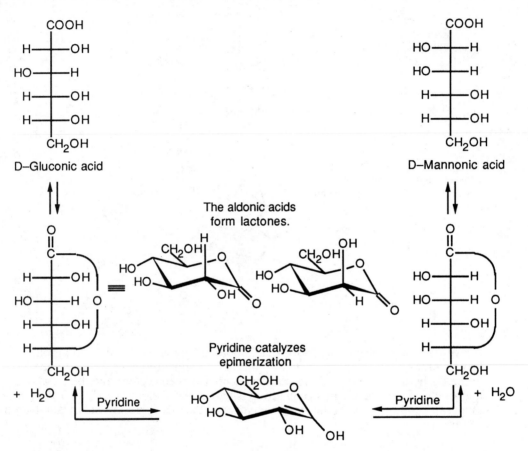

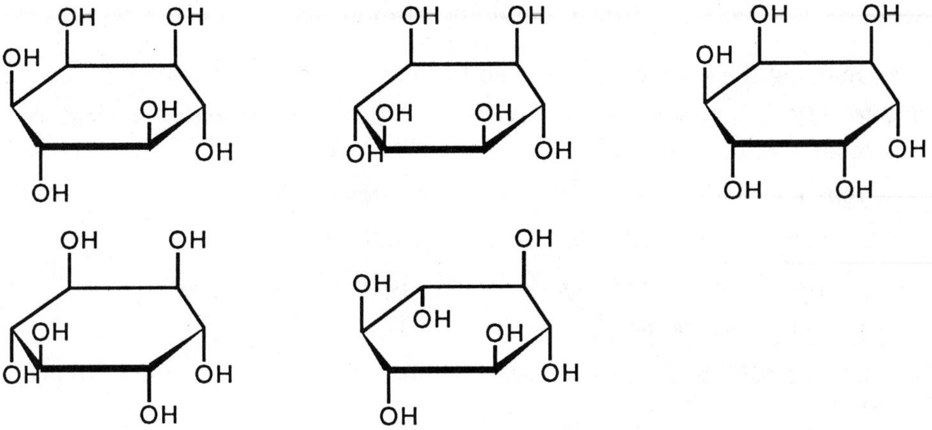

24.56

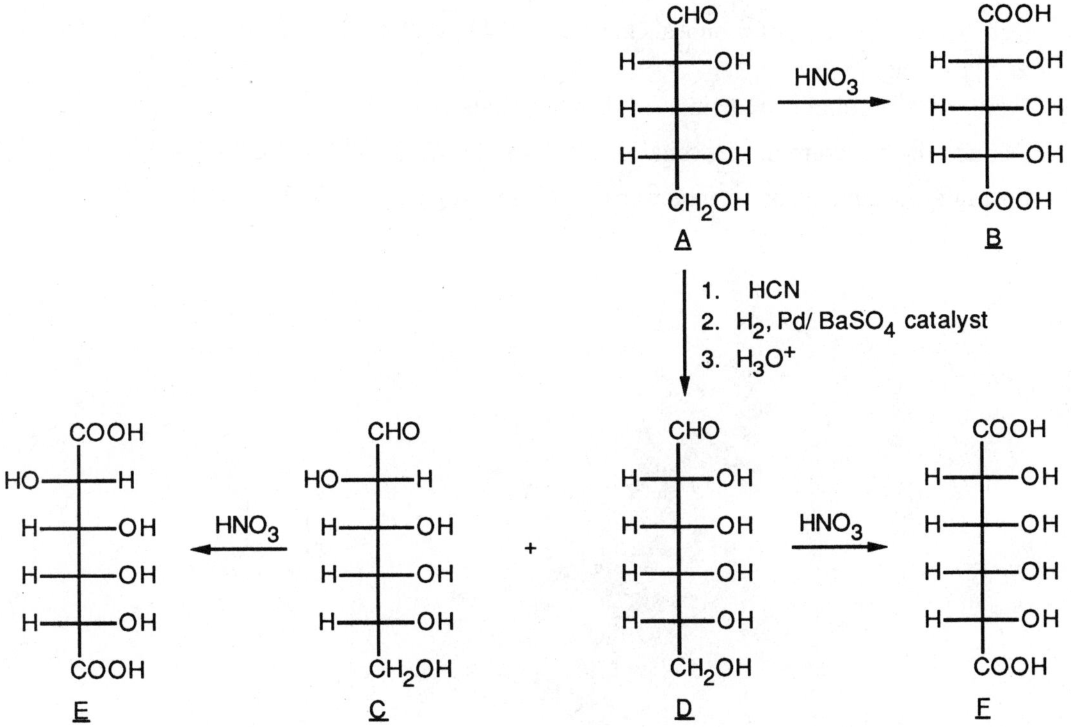

Top row of cyclohexane (Haworth/chair) ring structures with OH substituents.

Reaction scheme:

Compound A (CHO, H—OH, H—OH, H—OH, CH₂OH) → HNO₃ → B (COOH, H—OH, H—OH, H—OH, COOH)

A:
$$\begin{array}{c} CHO \\ H\!-\!OH \\ H\!-\!OH \\ H\!-\!OH \\ CH_2OH \end{array}$$

→ HNO₃ →

B:
$$\begin{array}{c} COOH \\ H\!-\!OH \\ H\!-\!OH \\ H\!-\!OH \\ COOH \end{array}$$

A → 1. HCN 2. H₂, Pd/BaSO₄ catalyst 3. H₃O⁺

C:
$$\begin{array}{c} CHO \\ HO\!-\!H \\ H\!-\!OH \\ H\!-\!OH \\ H\!-\!OH \\ CH_2OH \end{array}$$

+

D:
$$\begin{array}{c} CHO \\ H\!-\!OH \\ H\!-\!OH \\ H\!-\!OH \\ H\!-\!OH \\ CH_2OH \end{array}$$

C → HNO₃ → E:
$$\begin{array}{c} COOH \\ HO\!-\!H \\ H\!-\!OH \\ H\!-\!OH \\ H\!-\!OH \\ COOH \end{array}$$

D → HNO₃ → E:
$$\begin{array}{c} COOH \\ H\!-\!OH \\ H\!-\!OH \\ H\!-\!OH \\ H\!-\!OH \\ COOH \end{array}$$

Study Guide for Chapter 24

After studying this chapter, you should be able to:

(1)   Classify carbohydrates as aldoses, ketoses, D or L sugars, monosaccharides or polysaccharides (24.1, 24.2, 24.4, 24.27, 24.28, 24.29, 24.30, 24.34).

(2)   Draw monosaccharides in the following projections:

(a) Fischer projection (24.5, 24.6, 24.8, 24.35).

(b) Haworth projection (24.9, 24.10, 24.11, 24.32, 24.36, 24.55).

(c) Chair conformation (24.14, 24.15, 24.37).

(3)   Calculate the equilibrium percentage of anomers from their specific rotations (24.12, 24.13).

(4)   Predict the products of reactions of monosaccharides (24.16, 24.17, 24.18, 24.19, 24.20, 24.21, 24.22, 24.23, 24.38, 24.51).

(5)   Deduce the structure of monosaccharides (24.24, 24.25, 24.40, 24.41, 24.42, 24.43, 24.44, 24.45, 24.56).

(6)   Predict the product of reactions of disaccharides (24.26).

(7)   Deduce the structure of disaccharides (24.46, 24.47, 24.48, 24.49, 24.50).

(8)   Formulate mechanisms for reactions involving sugars (24.52, 24.53, 24.54).

Chapter 25 – Aliphatic Amines

25.1

a)

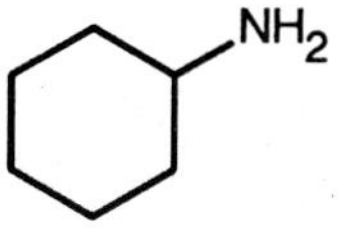
NH2

Primary amine

b)

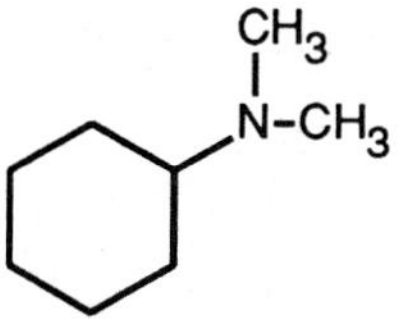
CH3
N-CH3

Tertiary amine

c)

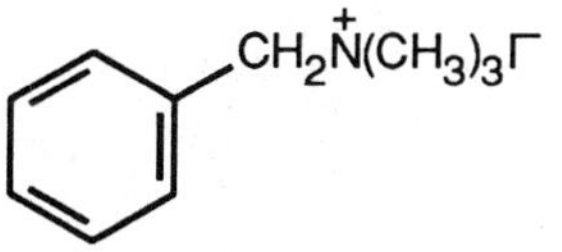
CH2N(CH3)3I

Quaternary ammonium salt

d) $[(CH_3)_2CH]_2NH$

Secondary amine

e)

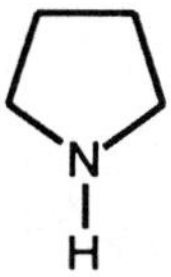
N
H

Secondary amine

25.2

a)

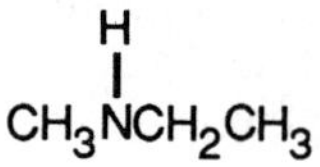
H
CH3NCH2CH3

N–Methylethylamine

b)

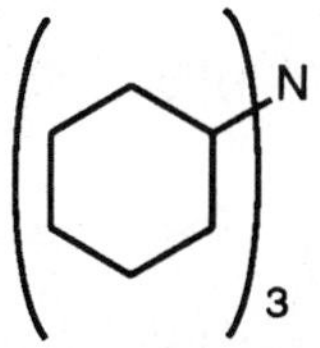
N
3

Tricyclohexylamine

c)

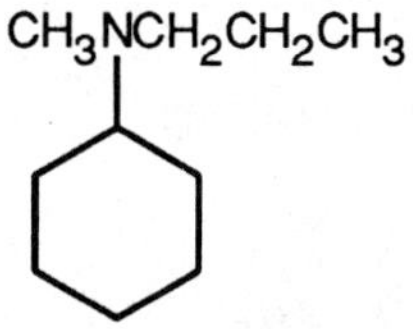
CH3NCH2CH2CH3

N–Methyl–*N* –propylcyclohexylamine

d)

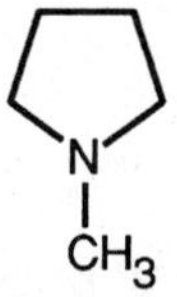
N
CH3

N–Methylpyrrolidine

e) $[(CH_3)_2CH]_2NH$

Diisopropylamine

f)

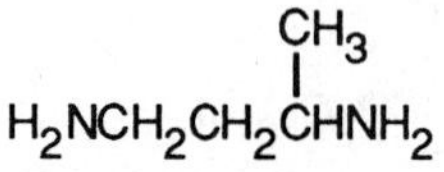
CH3
H2NCH2CH2CHNH2

1,3–Butanediamine

25.3

a) $(CH_3CH_2)_3N$

Triethylamine

b) $(H_2C{=}CHCH_2)_3N$

Triallylamine

c)

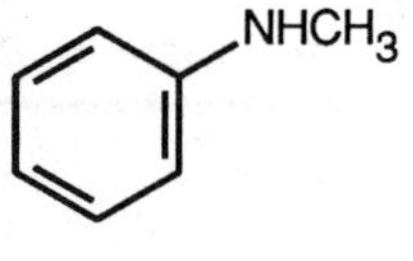

N–Methylaniline

d)

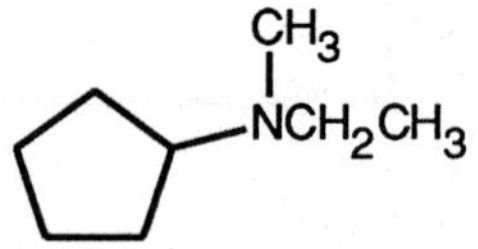

N-Ethyl–N–methylcyclopentylamine

e)

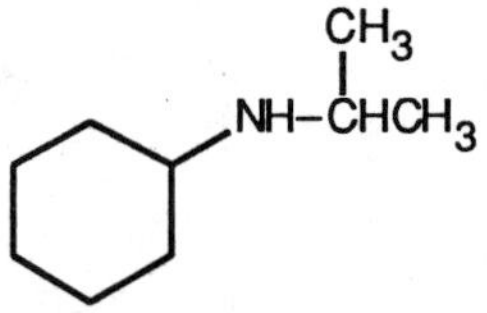

N–Isopropylcyclohexylamine

f)

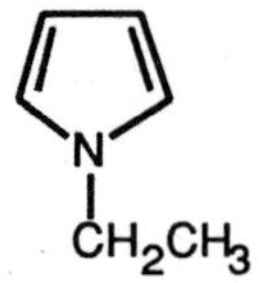

N–Ethylpyrrole

25.4

a)

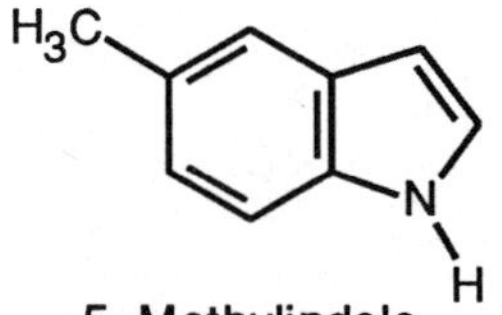

5–Methylindole

b)

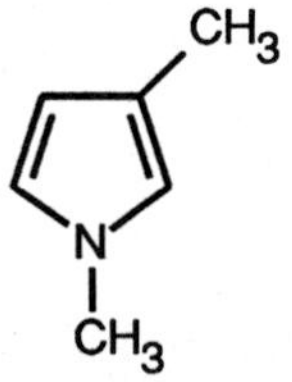

1,3–Dimethylpyrrole

c)

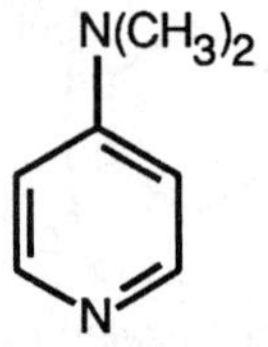

4–(N,N–Dimethylamino)pyridine

d)

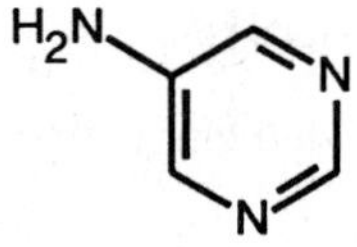

5–Aminopyrimidine

25.5

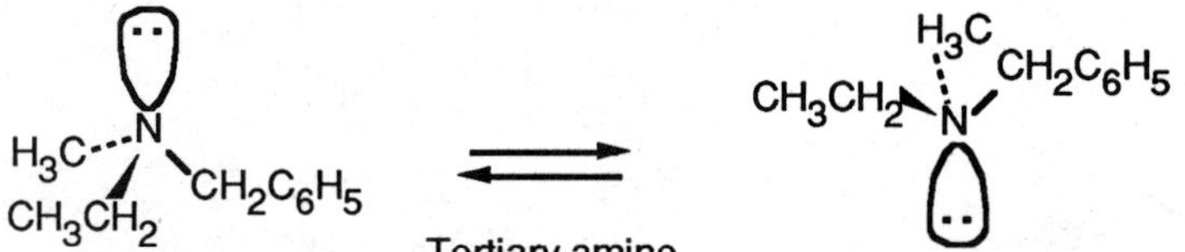

Tertiary amine

Quaternary ammonium salt

Tertiary and quaternary amines such as those pictured above are chiral. Most tertiary amines undergo pyramidal inversion, a relatively low-energy process that interconverts the enantiomers. For quaternary amines, interconversion, which would require breaking and reforming bonds, does not occur. Thus, each enantiomer is stable and resolvable.

25.6

More Basic	Less Basic
a) $CH_3CH_2NH_2$	$CH_3CH_2CONH_2$
amine	amide
b) NaOH	CH_3NH_2
hydroxide	amine
c) CH_3NHCH_3	CH_3OCH_3
amine	ether

25.7

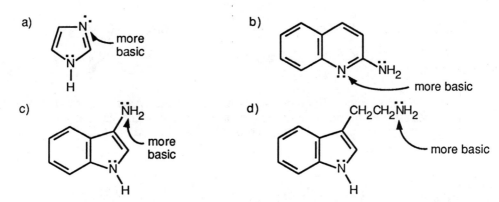

a) more basic

b) more basic

c) more basic

d) more basic

In parts a), c), and d), the less basic nitrogen has a lone electron pair that is part of an aromatic ring pi system. In part b), the less basic nitrogen is conjugated to an aromatic ring.

25.8

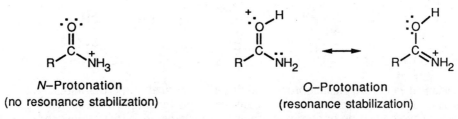

N–Protonation
(no resonance stabilization)

O–Protonation
(resonance stabilization)

Protonation occurs at oxygen because an *O*–protonated amide can be stabilized by resonance.

25.9

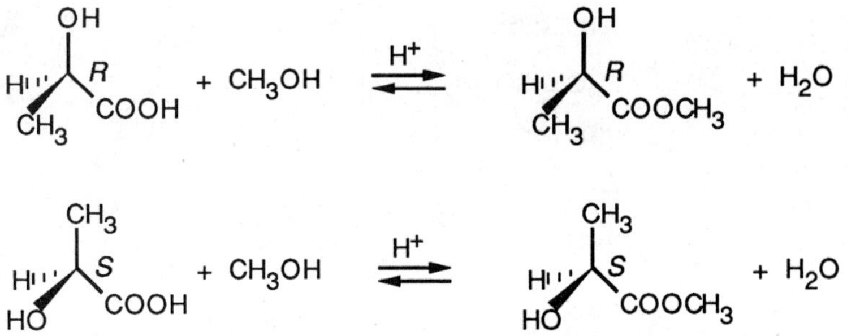

The product is a racemic mixture of R and S methyl lactates.

25.10

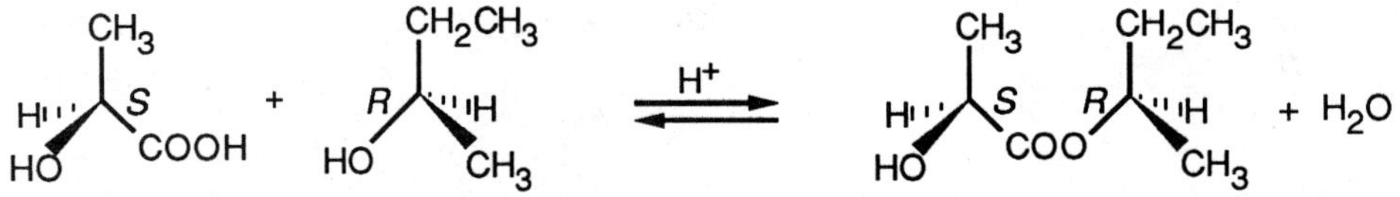

The product is enantiomerically pure R–sec–butyl S–lactate. As in the previous problem, the stereochemistry at the stereogenic centers of the product is the same as the stereochemistry of the reactants because no bonds were formed or broken at the stereogenic center.

25.11

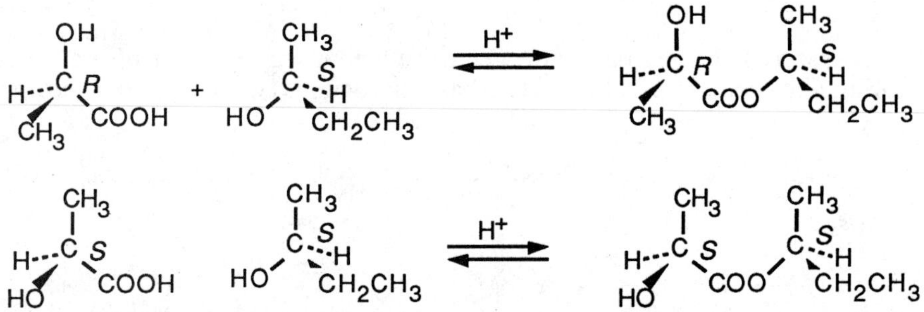

The product esters are diastereomers and differ in physical properties and chemical behavior. It should be possible to separate them by a technique such as distillation, fractional crystallization or chromatography. After separation, the esters can be treated with NaOH, which regenerates (R) or (S)–lactic acid and (S)–2–butanol.

25.12

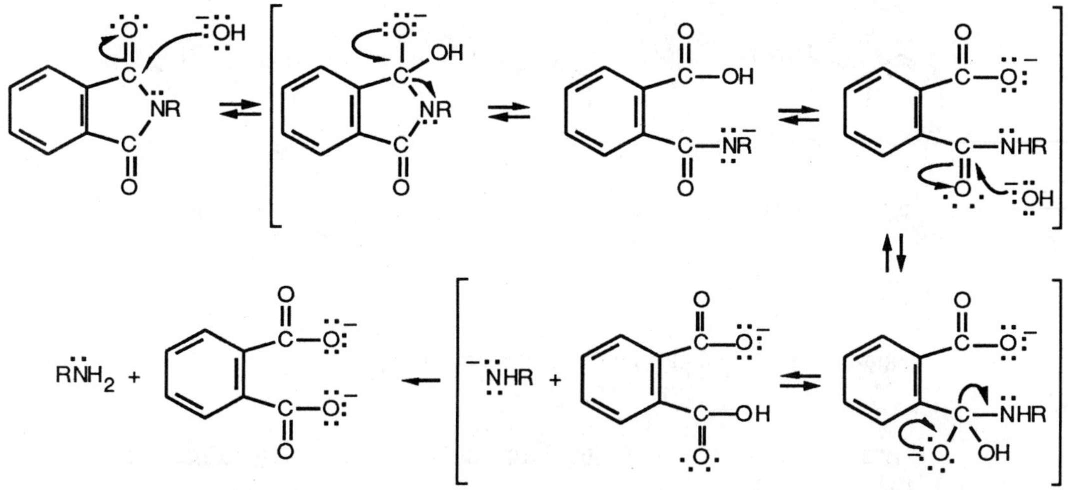

25.13

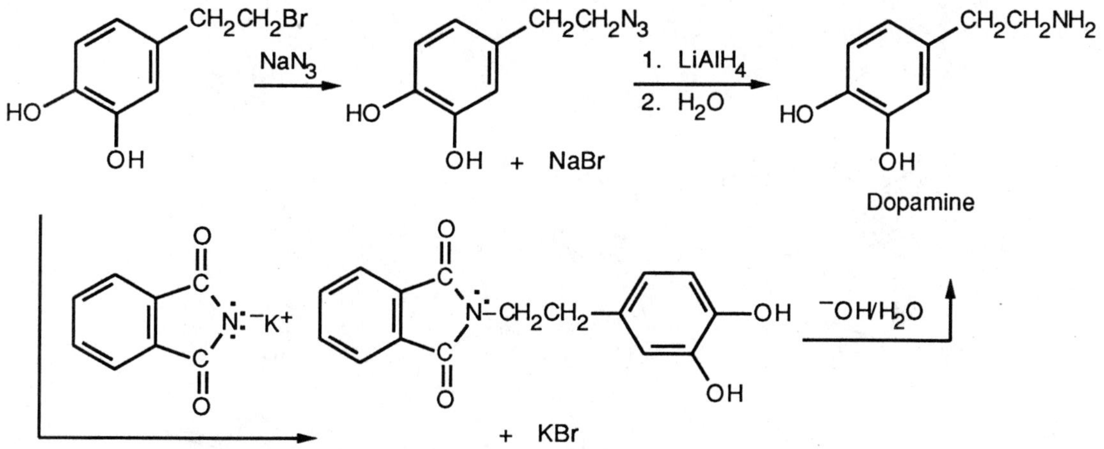

Dopamine

25.14

	Amine	Nitrile Precursor	Amide Precursor
a)	$CH_3CH_2CH_2NH_2$	CH_3CH_2CN	$CH_3CH_2\overset{\displaystyle O}{\overset{\|}{C}}NH_2$
b)	$(CH_3CH_2CH_2)_2NH$		$CH_3CH_2\overset{\displaystyle O}{\overset{\|}{C}}NHCH_2CH_2CH_3$

Amine	Nitrile Precursor	Amide Precursor

c)   [structure: benzyl amine, CH₂NH₂ on benzene]   [structure: benzonitrile, CN on benzene]   [structure: benzamide, C(=O)NH₂ on benzene]

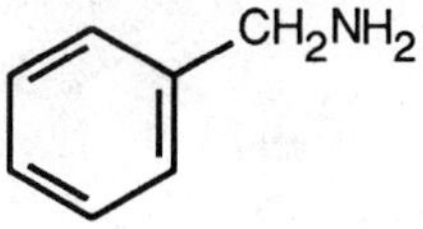

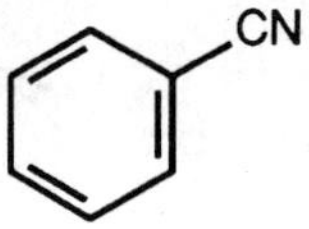

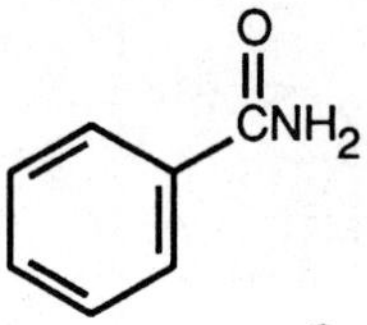

d)   [structure: N-ethylaniline, NHCH₂CH₃ on benzene]   [structure: acetanilide, NHC(=O)CH₃ on benzene]

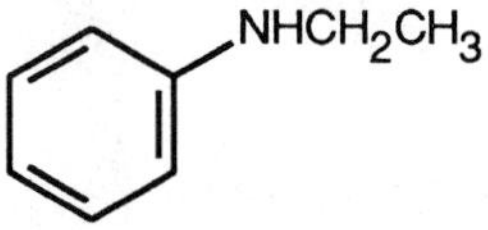

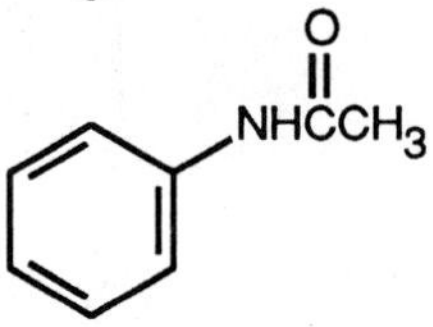

The compounds in parts b and d can't be prepared in one step by reduction of a nitrile.

25.15

Amine	Amine Precursor	Carbonyl Precursor

a) $CH_3CH_2NHCH(CH_3)_2$ $CH_3CH_2NH_2$ + $CH_3\overset{\displaystyle O}{\overset{\|}{C}}CH_3$

or

$H_2NCH(CH_3)_2$ + $CH_3\overset{\displaystyle O}{\overset{\|}{C}}H$

b)   [structure: N-ethylaniline, NHCH₂CH₃ on benzene]   [structure: aniline, NH₂ on benzene]   + $H\overset{\displaystyle O}{\overset{\|}{C}}CH_3$

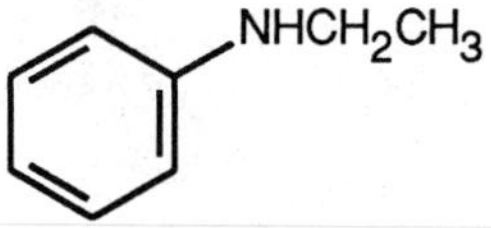

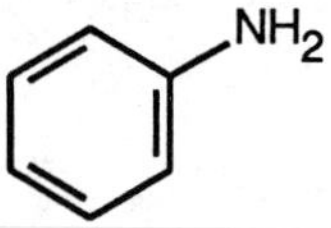

c)   [structure: N-methylcyclopentylamine, NHCH₃ on cyclopentane]   [structure: cyclopentylamine, NH₂ on cyclopentane]   + $H\overset{\displaystyle O}{\overset{\|}{C}}H$

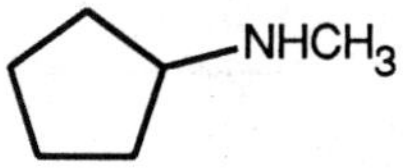

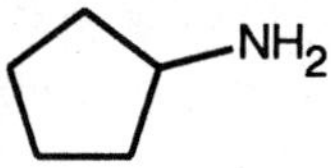

or

H_2NCH_3 +   [structure: cyclopentanone]

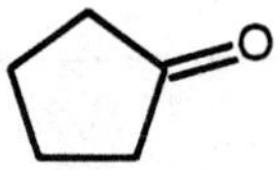

25.16

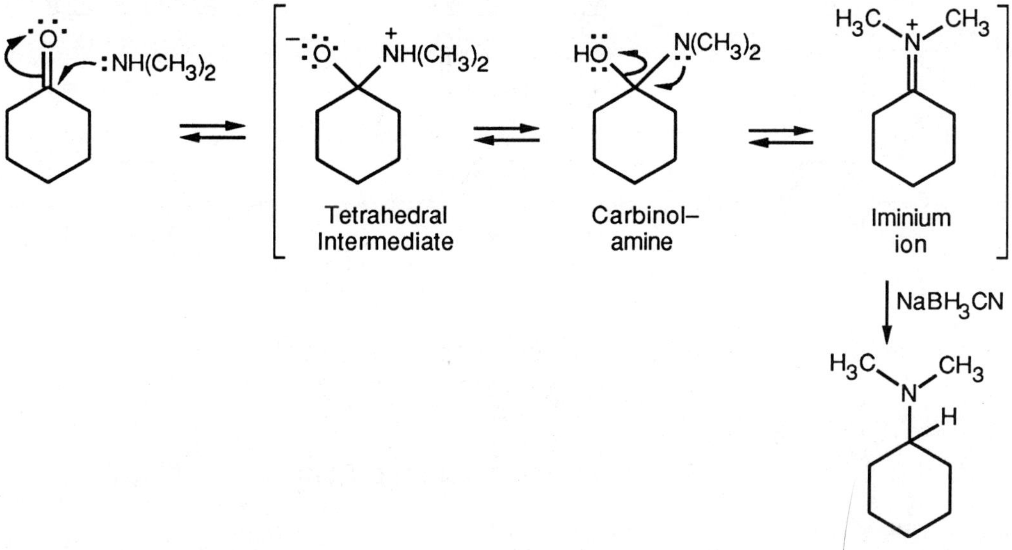

Tetrahedral
Intermediate

Carbinol–
amine

Iminium
ion

NaBH₃CN

25.17

Ephedrine

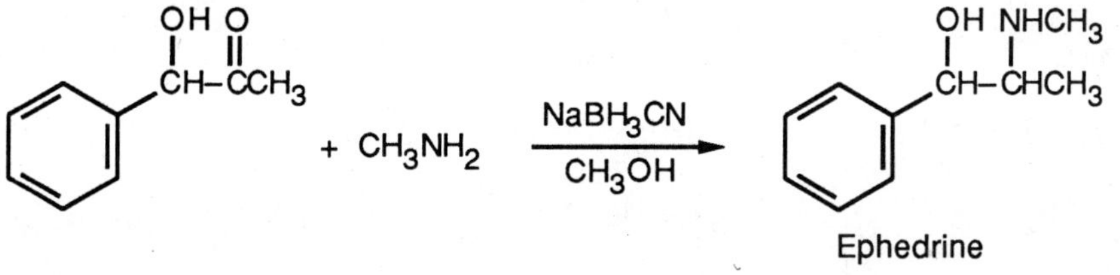

25.18

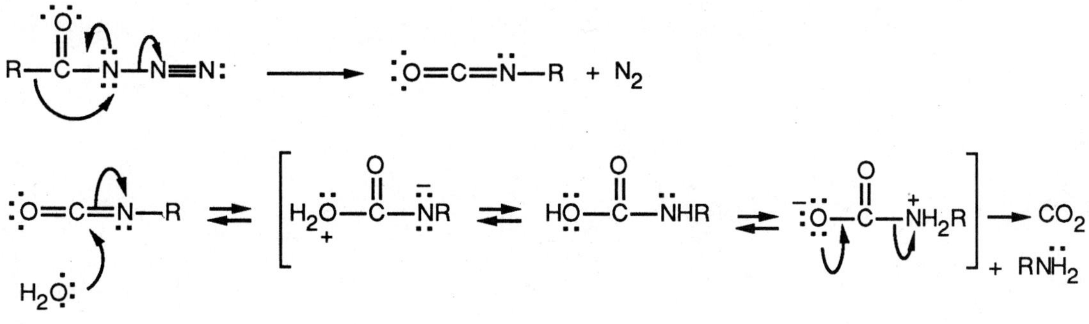

25.19

	Amine	Amide Precursor	Acyl Azide Precursor
a)	$(CH_3)_3CCH_2CH_2NH_2$	$(CH_3)_3CCH_2CH_2\overset{O}{\overset{\|}{C}}NH_2$	$(CH_3)_3CCH_2CH_2\overset{O}{\overset{\|}{C}}N_3$

b)

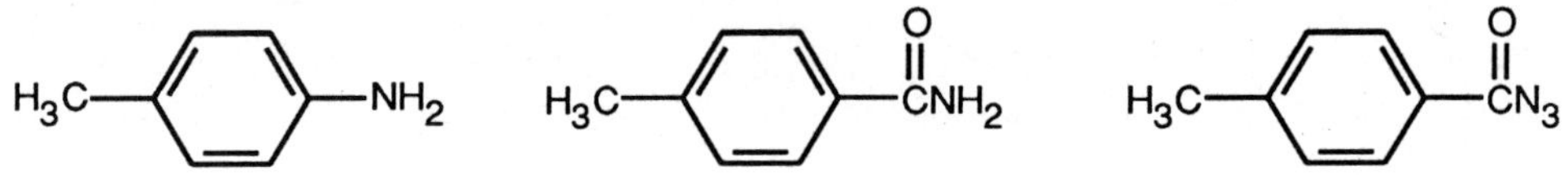

25.20

	Amine	Alkene Products	Amine Products
a)	$CH_3CH_2CH_2\overset{NH_2}{\underset{\|}{C}H}CH_2CH_2CH_2CH_3$	$CH_3CH_2CH=CHCH_2CH_2CH_2CH_3$ *or* $CH_3CH_2CH_2CH=CHCH_2CH_2CH_3$	$+ (CH_3)_3N:$ $+ (CH_3)_3N:$

b)

$+ (CH_3)_3N:$

	Amine		
c)	$CH_3CH_2CH_2\overset{NH_2}{\underset{\|}{C}H}CH_2CH_2CH_3$	$CH_3CH_2CH=CHCH_2CH_2CH_3$	$+ (CH_3)_3N:$

d)

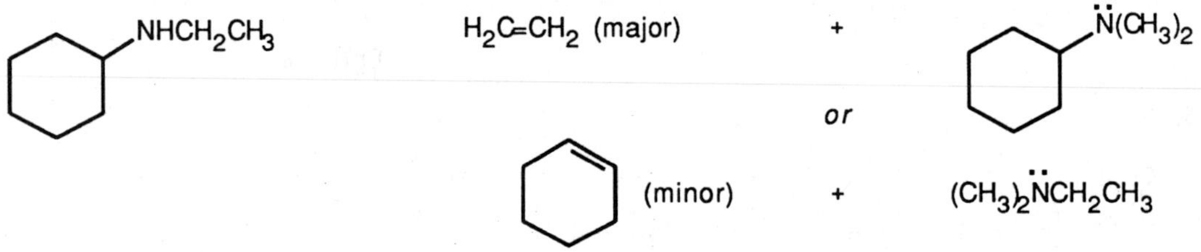

$H_2C=CH_2$ (major)   +

or

(minor)   +

$(CH_3)_2\overset{..}{N}CH_2CH_3$

25.21

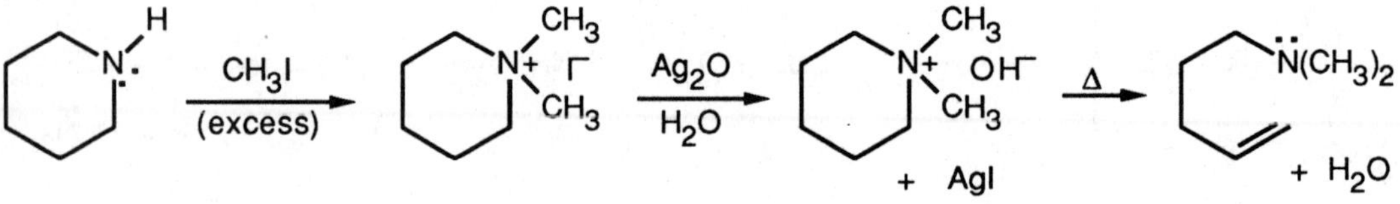

25.22

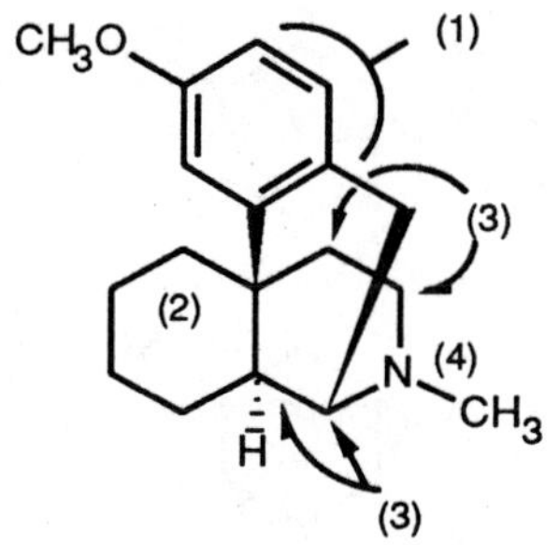

Dextromethorphan

(1) aromatic ring
(2) quaternary carbon
(3) two carbons
(4) tertiary amine

25.23

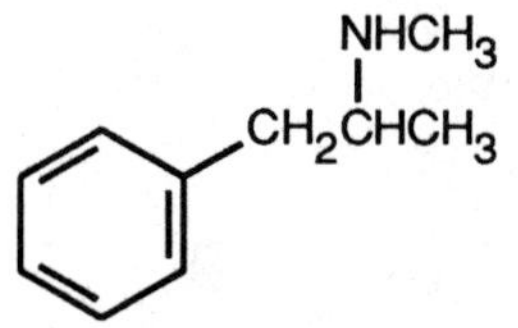

Methamphetamine

Mass spectrum: The odd-numbered molecular ion indicates an odd number of nitrogen atoms.

Infrared: The weak band at 3350 cm^{-1} shows that methamphetamine is a secondary amine.

^1H *NMR:*

Absorption	Due to
7.15 δ (5H)	monosubstituted benzene ring
2.65 δ (3H)	two benzylic protons one proton adjacent to nitrogen
2.35 δ (3H)	N–methyl group
1.20 δ (1H)	N–H proton (disappears when D$_2$O is added)
1.00 δ (3H)	methyl group split by one adjacent proton

25.24

a)

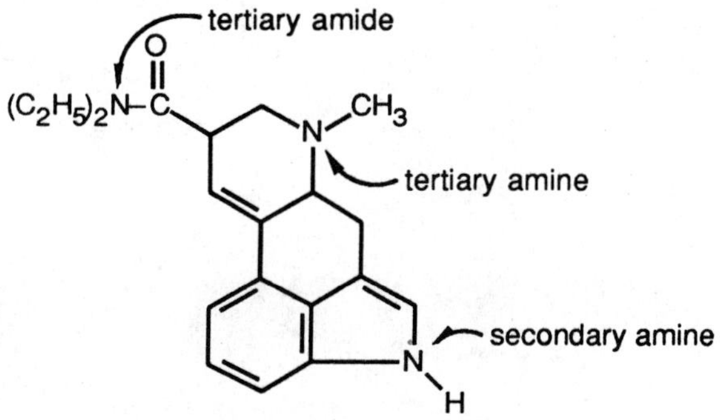

Lysergic acid diethylamide

b)

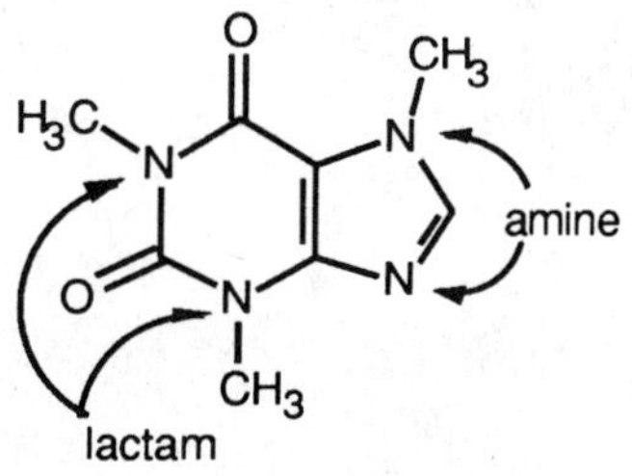

all nitrogens are tertiary

Caffeine

25.25

a) 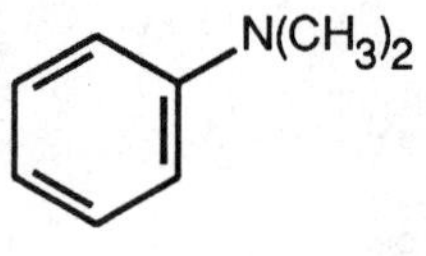 N(CH$_3$)$_2$

b) 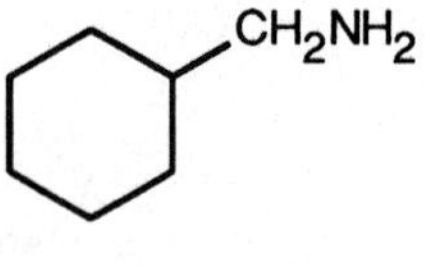 CH$_2$NH$_2$

c) 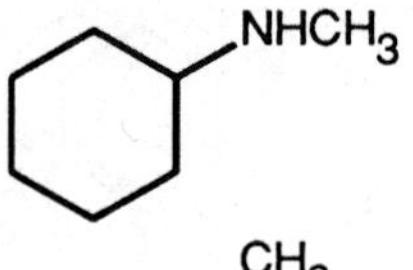 NHCH$_3$

d) 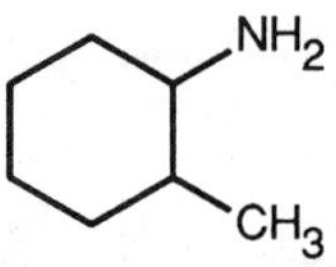 NH$_2$ CH$_3$

e) (CH$_3$)$_2$NCH$_2$CH$_2$COOH

f) 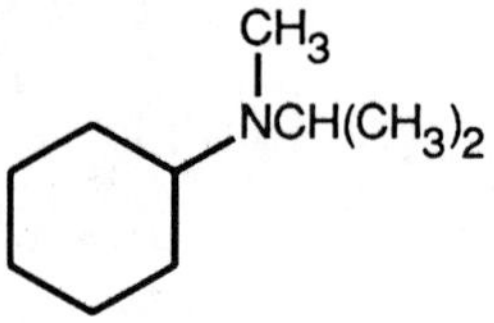 CH$_3$ NCH(CH$_3$)$_2$

25.26 a) 2,4–Dibromoaniline   b) (2–Cyclopentyl)ethylamine

c) *N*–Ethylcyclopentylamine   d) *N,N*–Dimethylcyclopentylamine

e) *N*–Propylpyrrolidine   f) 4–Aminobutanenitrile

25.27

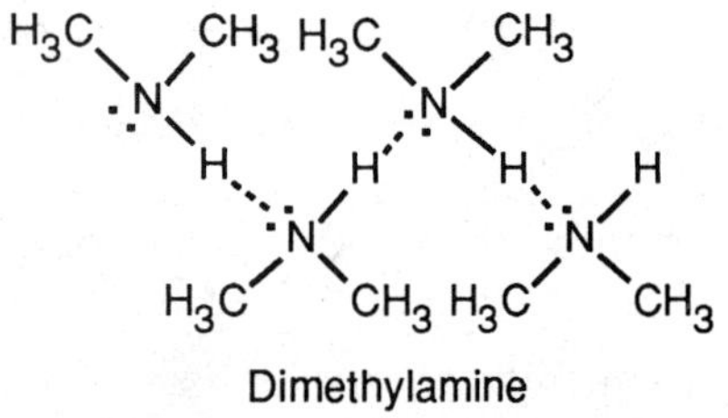

Dimethylamine   (CH$_3$)$_3$N:

                                                   Trimethylamine

Even though dimethylamine has a lower molecular weight than trimethylamine, it boils at a higher temperature. Liquid dimethylamine forms hydrogen bonds that must be broken in the boiling process. Since extra energy must be added to break these hydrogen bonds, dimethylamine has a higher boiling point than trimethylamine, which does not form hydrogen bonds.

25.28

a) CH$_3$CH$_2$CH$_2$CH$_2$OH $\xrightarrow{\text{PBr}_3}$ CH$_3$CH$_2$CH$_2$CH$_2$Br $\xrightarrow{\text{NaN}_3}$ CH$_3$CH$_2$CH$_2$CH$_2$N$_3$

$\downarrow$ 1. LiAlH$_4$
    2. H$_2$O

CH$_3$CH$_2$CH$_2$CH$_2$NH$_2$
Butylamine

b) $CH_3CH_2CH_2CH_2OH$ $\xrightarrow{\text{Jones'}}$ $CH_3CH_2CH_2\overset{\displaystyle O}{\overset{\|}{C}}OH$ $\xrightarrow[\text{CHCl}_3]{\text{SOCl}_2}$ $CH_3CH_2CH_2\overset{\displaystyle O}{\overset{\|}{C}}-Cl$

$\downarrow$ $\begin{array}{c} CH_3CH_2CH_2CH_2NH_2 \\ \text{[from a)]} \\ \text{NaOH} \end{array}$

$(CH_3CH_2CH_2CH_2)_2NH$ $\xleftarrow[\text{2. H}_2O]{\text{1. LiAlH}_4}$ $CH_3CH_2CH_2\overset{\displaystyle O}{\overset{\|}{C}}-NHCH_2CH_2CH_2CH_3$

Dibutylamine

c) $CH_3CH_2CH_2CH_2OH$ $\xrightarrow{\text{Jones'}}$ $CH_3CH_2CH_2\overset{\displaystyle O}{\overset{\|}{C}}OH$ $\xrightarrow{\text{SOCl}_2}$ $CH_3CH_2CH_2\overset{\displaystyle O}{\overset{\|}{C}}-Cl$

$\downarrow$ $2\ NH_3$

$CH_3CH_2CH_2NH_2$ $\xleftarrow[\text{H}_2O,\ \Delta]{\text{Br}_2,\ ^-\text{OH}}$ $CH_3CH_2CH_2\overset{\displaystyle O}{\overset{\|}{C}}-NH_2$

Propylamine

or $CH_3CH_2CH_2\overset{\displaystyle O}{\overset{\|}{C}}-Cl$ $\xrightarrow{\text{NaN}_3}$ $CH_3CH_2CH_2\overset{\displaystyle O}{\overset{\|}{C}}N_3$ $\xrightarrow[\text{2. H}_2O]{\text{1. }\Delta}$ $CH_3CH_2CH_2NH_2$

d) $CH_3CH_2CH_2CH_2OH$ $\xrightarrow{\text{PBr}_3}$ $CH_3CH_2CH_2CH_2Br$ $\xrightarrow{\text{NaCN}}$ $CH_3CH_2CH_2CH_2CN$

$\downarrow$ $\begin{array}{l} \text{1. LiAlH}_4 \\ \text{2. H}_2O \end{array}$

$CH_3CH_2CH_2CH_2CH_2NH_2$

Pentylamine

e) $CH_3CH_2CH_2CH_2OH$ $\xrightarrow{\text{PCC}}$ $CH_3CH_2CH_2\overset{\displaystyle O}{\overset{\|}{C}}H$ $\xrightarrow[\text{NaBH}_3\text{CN}]{(CH_3)_2NH}$ $CH_3CH_2CH_2CH_2N(CH_3)_2$

N,N–Dimethylbutylamine

f) $CH_3CH_2CH_2NH_2$
[from c)] $\xrightarrow[\text{CH}_3\text{I}]{\text{excess}}$ $CH_3CH_2CH_2\overset{+}{N}(CH_3)_3I^-$ $\xrightarrow[\text{H}_2O]{\text{Ag}_2O}$ $CH_3CH_2CH_2\overset{+}{N}(CH_3)_3\ ^-OH$

$\downarrow \Delta$

$CH_3CH=CH_2\ +\ (CH_3)_3N$

Propene

25.29

a)

$$CH_3CH_2CH_2CH_2\overset{\overset{O}{\|}}{C}OH \xrightarrow{SOCl_2} CH_3CH_2CH_2CH_2\overset{\overset{O}{\|}}{C}-Cl$$

$$\downarrow 2\ NH_3$$

$$CH_3CH_2CH_2CH_2\overset{\overset{O}{\|}}{C}NH_2$$
Pentanamide

b)

$$CH_3CH_2CH_2CH_2\overset{\overset{O}{\|}}{C}NH_2 \xrightarrow[H_2O,\ \Delta]{Br_2,\ {}^-OH} CH_3CH_2CH_2CH_2NH_2$$
[from a)]   Butylamine

c)

$$CH_3CH_2CH_2CH_2\overset{\overset{O}{\|}}{C}NH_2 \xrightarrow[2.\ H_2O]{1.\ LiAlH_4} CH_3CH_2CH_2CH_2CH_2NH_2$$
[from a)]   Pentylamine

d)

$$CH_3CH_2CH_2CH_2\overset{\overset{O}{\|}}{C}OH \xrightarrow[2.\ H_2O]{1.\ Br_2,\ PBr_3} CH_3CH_2CH_2CH_2\overset{\overset{Br}{|}}{C}H\overset{\overset{O}{\|}}{C}OH$$
2–Bromopentanoic acid

e)

$$CH_3CH_2CH_2CH_2\overset{\overset{O}{\|}}{C}OH \xrightarrow[2.\ H_3O^+]{1.\ BH_3} CH_3CH_2CH_2CH_2CH_2OH$$

$$\downarrow PBr_3$$

$$CH_3CH_2CH_2CH_2CH_2CN \xleftarrow{NaCN} CH_3CH_2CH_2CH_2CH_2Br$$
Hexanenitrile

f)

$$CH_3CH_2CH_2CH_2CH_2CN \xrightarrow[2.\ H_2O]{1.\ LiAlH_4} CH_3CH_2CH_2CH_2CH_2CH_2NH_2$$
[from e)]   Hexylamine

25.30

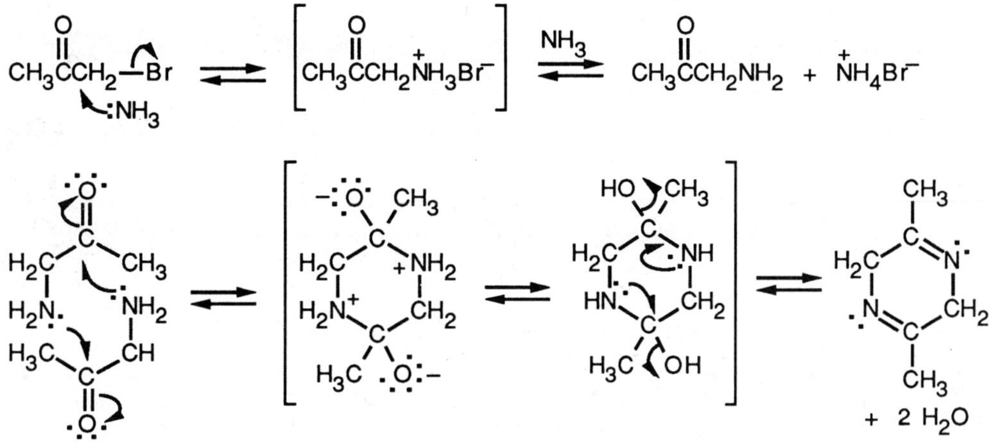

25.31

a)

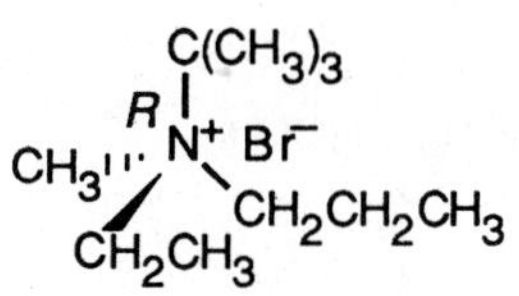

R-tert-Butylethylmethyl–
propylammonium bromide

b)

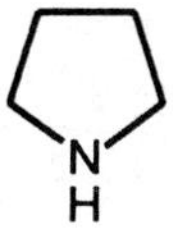

Pyrrolidine

c)

$$CH_2=CHCH_2\overset{\overset{\displaystyle H}{|}}{N}CH_2CH=CH_2$$

Diallylamine

25.32

a) $CH_3CH_2CH_2CH_2\overset{\overset{\displaystyle O}{\|}}{C}NH_2 \xrightarrow[\text{2. } H_2O]{\text{1. LiAlH}_4} CH_3CH_2CH_2CH_2CH_2NH_2$

b) $CH_3CH_2CH_2CH_2CN \xrightarrow[\text{2. } H_2O]{\text{1. LiAlH}_4} CH_3CH_2CH_2CH_2CH_2NH_2$

c) $CH_3CH_2CH=CH_2 \xrightarrow[\text{peroxides}]{\text{HBr}} CH_3CH_2CH_2CH_2Br \xrightarrow{\text{NaCN}} CH_3CH_2CH_2CH_2CN$

$\downarrow$ 1. LiAlH$_4$
   2. H$_2$O

$CH_3CH_2CH_2CH_2CH_2NH_2$

d) $CH_3CH_2CH_2CH_2CH_2\overset{\overset{\displaystyle O}{\|}}{C}NH_2 \xrightarrow[\text{H}_2\text{O, } \Delta]{\text{Br}_2, \text{ }^-\text{OH}} CH_3CH_2CH_2CH_2CH_2NH_2$

e) $CH_3CH_2CH_2CH_2OH$ $\xrightarrow{PBr_3}$ $CH_3CH_2CH_2CH_2Br$ $\xrightarrow{NaCN}$ $CH_3CH_2CH_2CH_2CN$

$$\downarrow \begin{array}{l} 1.\ LiAlH_4 \\ 2.\ H_2O \end{array}$$

$CH_3CH_2CH_2CH_2CH_2NH_2$

f) $CH_3CH_2CH_2CH_2CH=CHCH_2CH_2CH_2CH_3$ $\xrightarrow[\text{2. Zn, H}_3O^+]{\text{1. O}_3}$ $2\ CH_3CH_2CH_2CH_2\overset{\overset{\displaystyle O}{\|}}{C}H$

$CH_3CH_2CH_2CH_2\overset{\overset{\displaystyle O}{\|}}{C}H$ $\xrightarrow[\text{NaBH}_3CN]{\text{NH}_3}$ $CH_3CH_2CH_2CH_2CH_2NH_2$

g) $CH_3CH_2CH_2CH_2\overset{\overset{\displaystyle O}{\|}}{C}OH$ $\xrightarrow{SOCl_2}$ $CH_3CH_2CH_2CH_2\overset{\overset{\displaystyle O}{\|}}{C}Cl$

$$\downarrow 2\ NH_3$$

$CH_3CH_2CH_2CH_2CH_2NH_2$ $\xleftarrow[\text{2. H}_2O]{\text{1. LiAlH}_4}$ $CH_3CH_2CH_2CH_2\overset{\overset{\displaystyle O}{\|}}{C}NH_2$

25.33

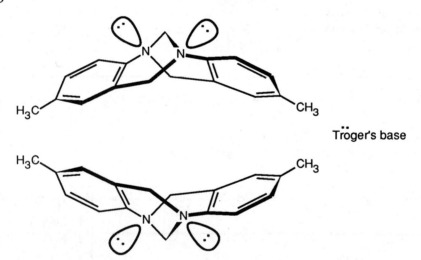

Tröger's base

The enantiomers of Tröger's base do not interconvert. Because of the rigid ring system, the substituents bonded to nitrogen can't be forced into the planar sp^2 geometry necessary for inversion at nitrogen to occur. Since inversion is not possible, the enantiomers are resolvable.

25.34

a)

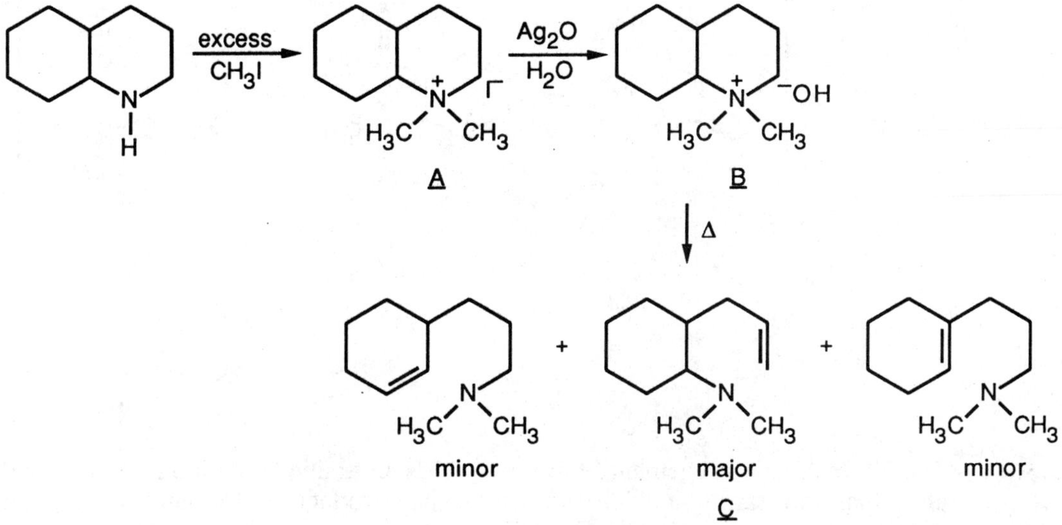
excess
CH_3I
Ag_2O
H_2O
A
B
Δ
minor
major
minor
C

b)

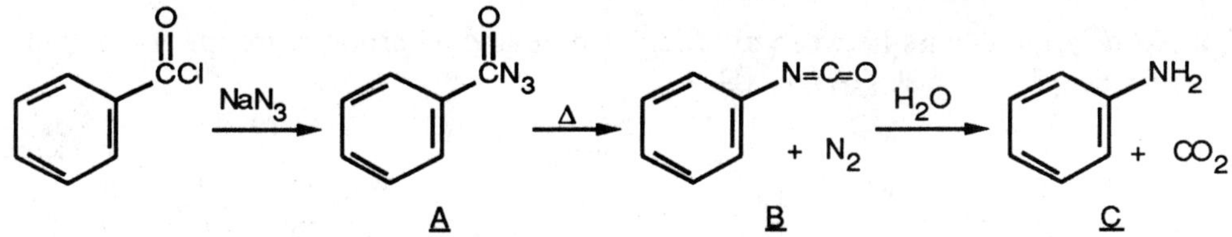
NaN_3
Δ
H_2O
+ N_2
+ CO_2
A
B
C

c)

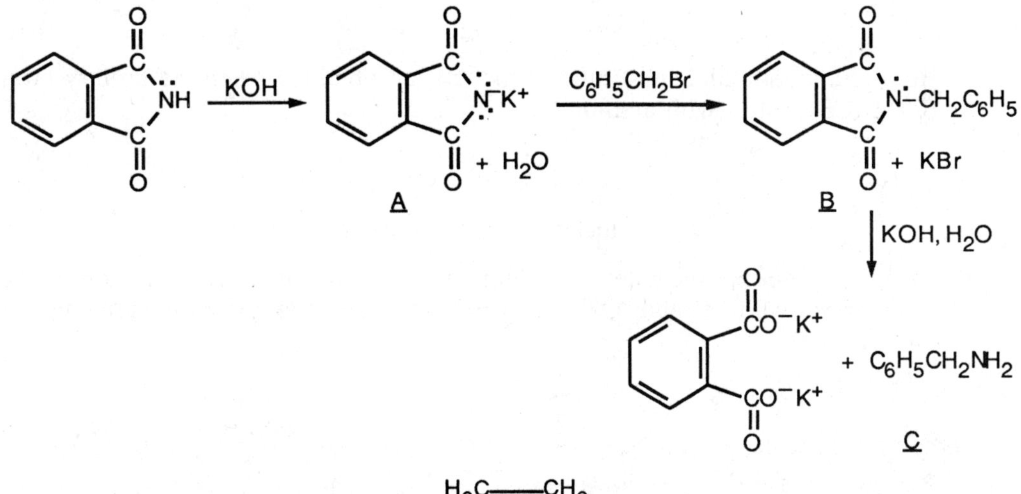
KOH
$C_6H_5CH_2Br$
+ H_2O
+ KBr
A
B
KOH, H_2O
+ $C_6H_5CH_2NH_2$
C

d)

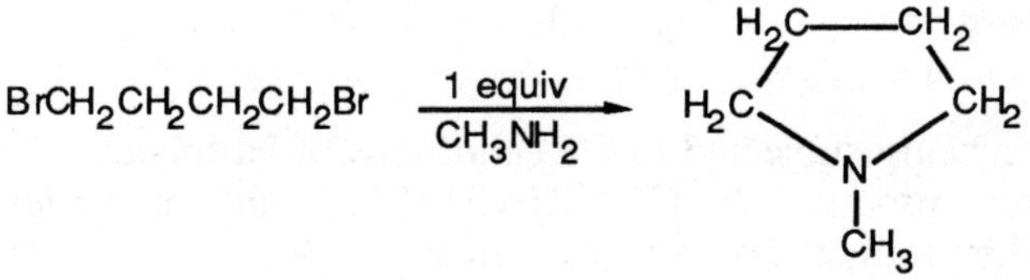
$BrCH_2CH_2CH_2CH_2Br$
1 equiv
CH_3NH_2

25.35

25.36 a) The Hofmann rearrangement of amides yields an amine containing one less carbon atom than the starting amide. In this case, the product of Hofmann rearrangement is $CH_3CH_2NH_2$, not $CH_3CH_2CH_2NH_2$.

b) No reaction occurs between a tertiary amine and a carbonyl group. To obtain the product shown, use $(CH_3)_2NH$.

c) Both elimination and substitution products are obtained when the tertiary bromide $(CH_3)_3CBr$ reacts with ammonia.

d) An isocyanate intermediate in the Hofmann rearrangement results from treatment of an amide with Br_2 and ^-OH. Heat is used to decarboxylate the carbamic acid intermediate.

e) The amine in this problem is being subjected to the conditions of Hofmann elimination. The major alkene product, $CH_3CH_2CH_2CH=CH_2$, contains the *less* substituted double bond. The product shown is the minor product.

25.37 The base-insoluble Hinsberg product indicates that coniine is a secondary amine. The structural formula indicates that coniine has one double bond or ring; the Hofmann elimination product shows that the nitrogen atom is part of a ring.

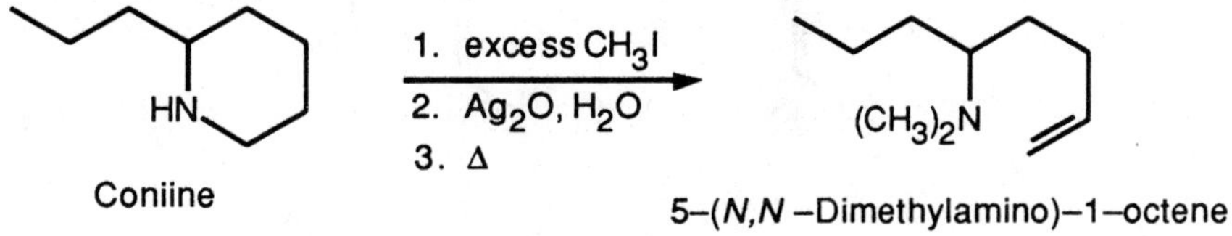
HN
Coniine
1. excess CH_3I
2. Ag_2O, H_2O
3. Δ
$(CH_3)_2N$
5-(*N,N*-Dimethylamino)-1-octene

25.38

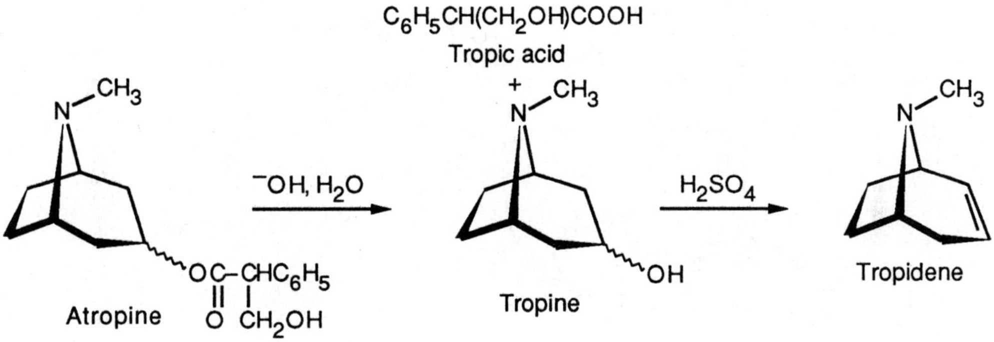
$C_6H_5CH(CH_2OH)COOH$
Tropic acid
+
N
CH_3
Atropine
$OC-CHC_6H_5$
O CH_2OH
^-OH, H_2O
Tropine
OH
H_2SO_4
Tropidene

25.39

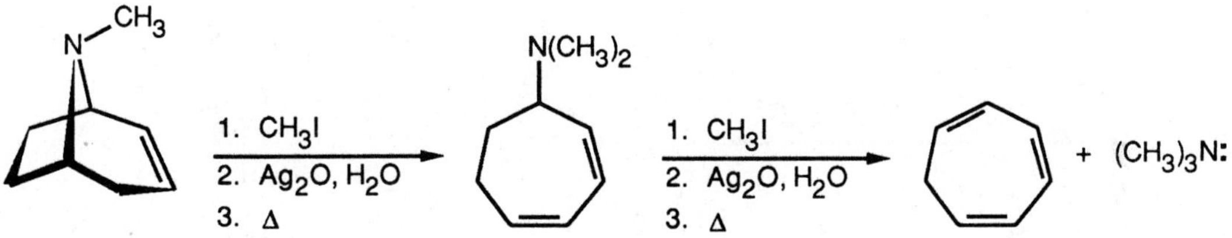
N
CH_3
1. CH_3I
2. Ag_2O, H_2O
3. Δ
$N(CH_3)_2$
1. CH_3I
2. Ag_2O, H_2O
3. Δ
+ $(CH_3)_3N$:

25.40

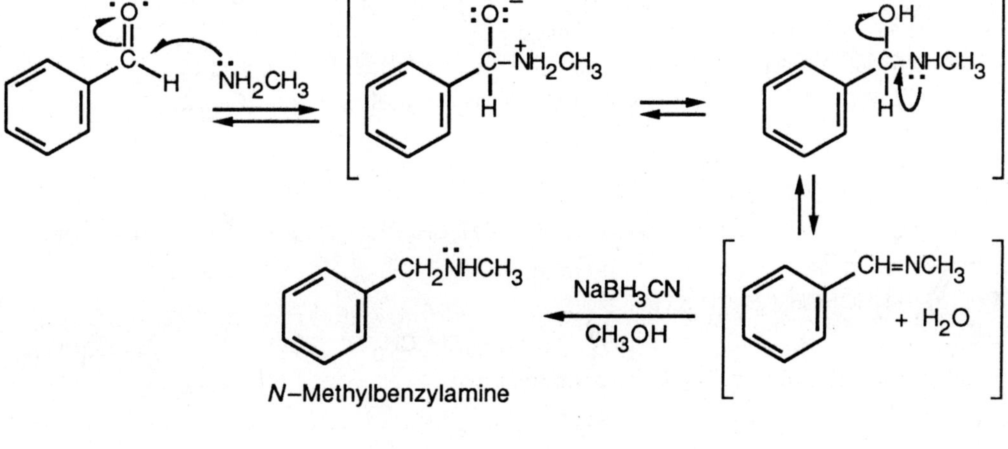
O
C
H
NH_2CH_3
O
$C-NH_2CH_3$
H
OH
$C-NHCH_3$
H
CH_2NHCH_3
N-Methylbenzylamine
$NaBH_3CN$
CH_3OH
$CH=NCH_3$
+ H_2O

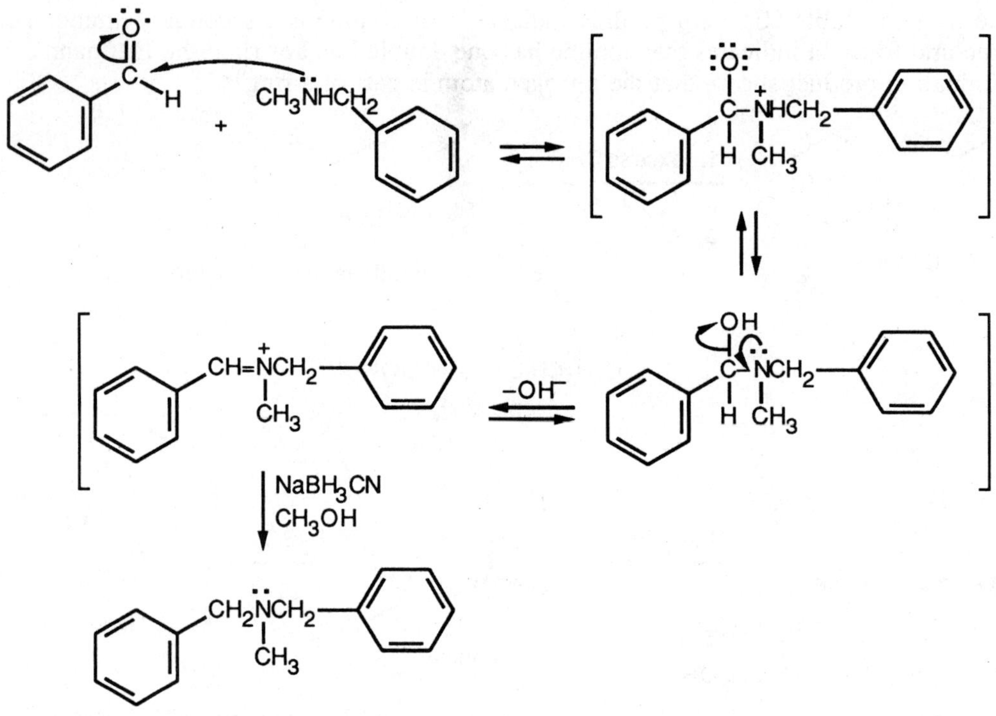

+
NaBH$_3$CN
CH$_3$OH
–OH$^-$

N–Methyldibenzylamine

25.41

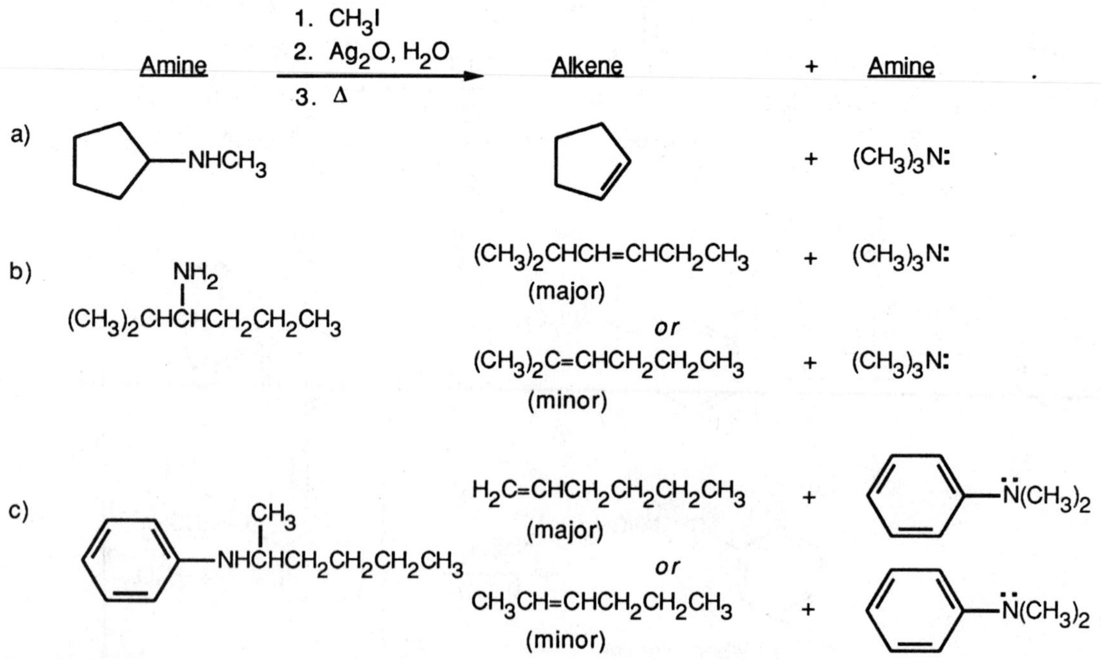

1. CH$_3$I
2. Ag$_2$O, H$_2$O
3. Δ
Amine
Alkene
Amine
a)
b)
c)
(major)
(minor)
or

25.42

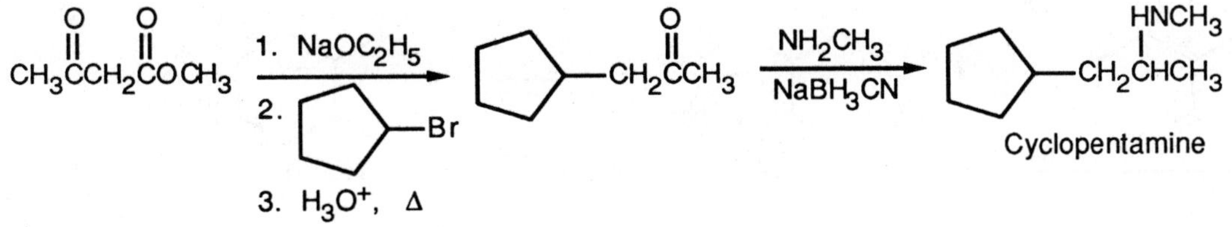

25.43

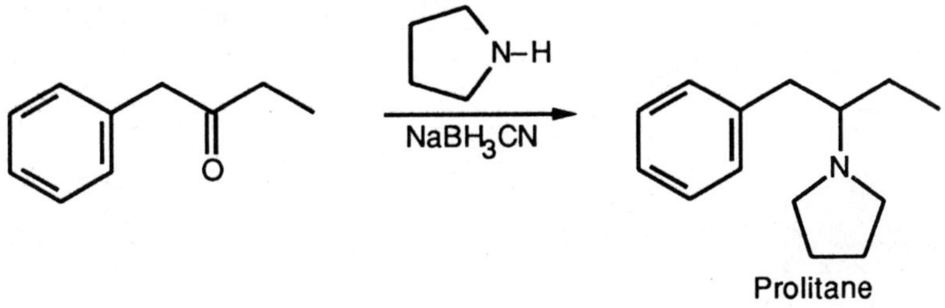

Prolitane

25.44

a)

Tetracaine

b)

c)

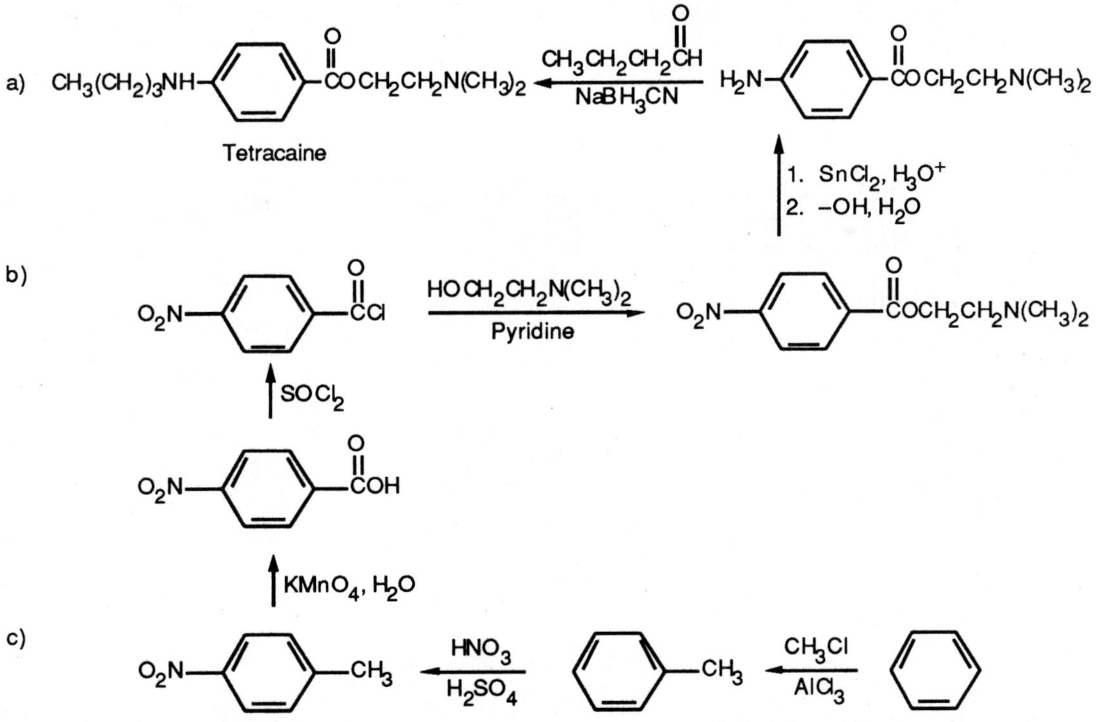

25.45

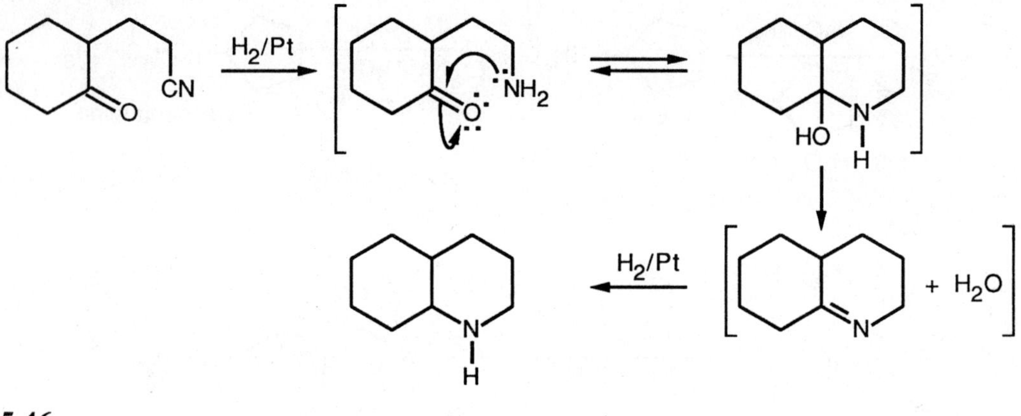

25.46

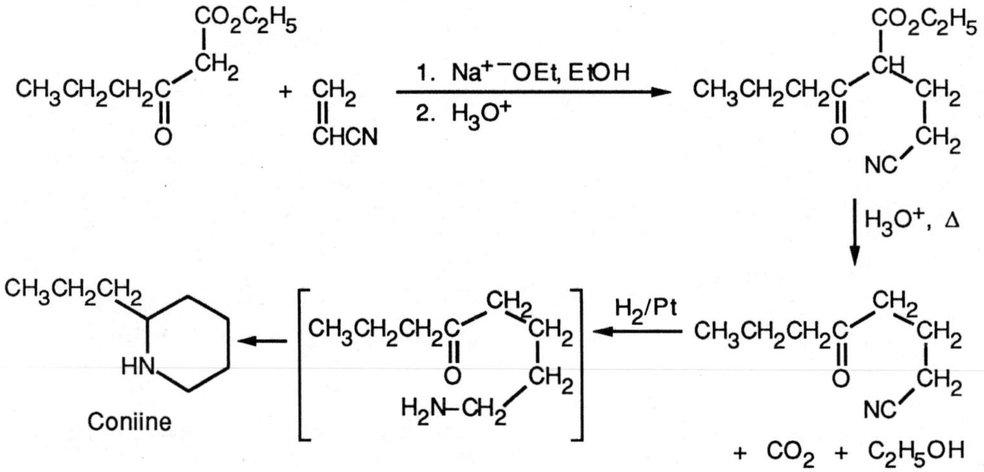

25.47

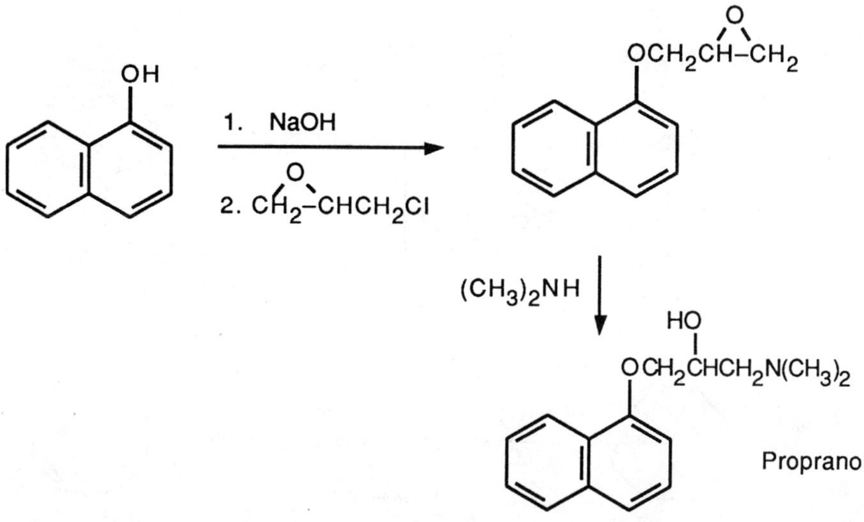

25.48 During pyramidal inversion of nitrogen, the hybridization at nitrogen changes momentarily from sp^3 to sp^2. For an aziridine, this sp^2–hybridized, planar intermediate is of high energy, due to severe ring strain in the three-membered aziridine ring. The energy barrier for inversion is relatively high, and the rate of inversion of (+)–1–chloro–2,2–diphenylaziridine is slow.

25.49

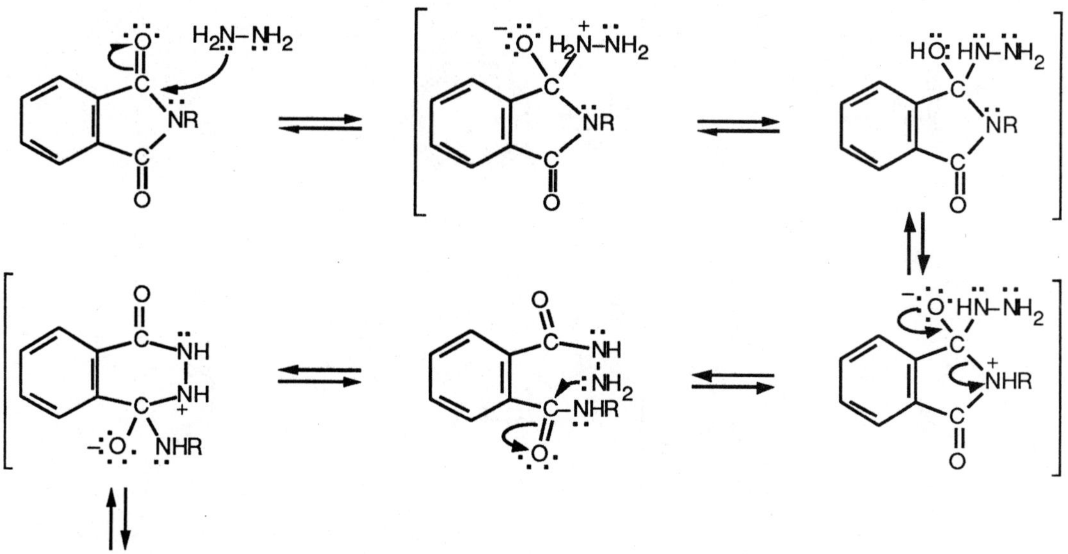

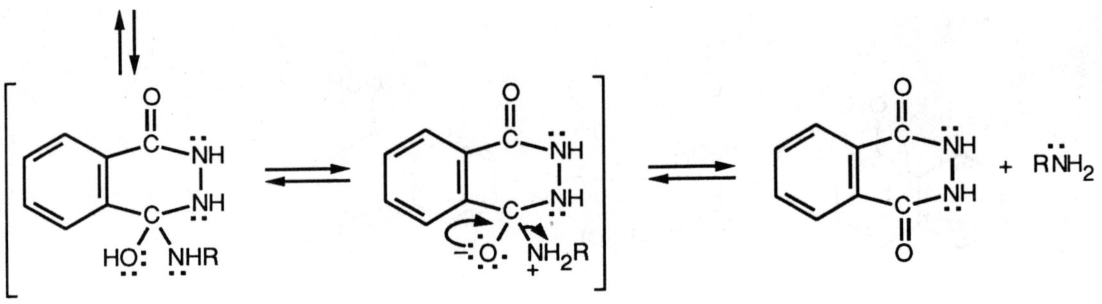
HO: NHR
NH₂R
+ RNH₂

25.50

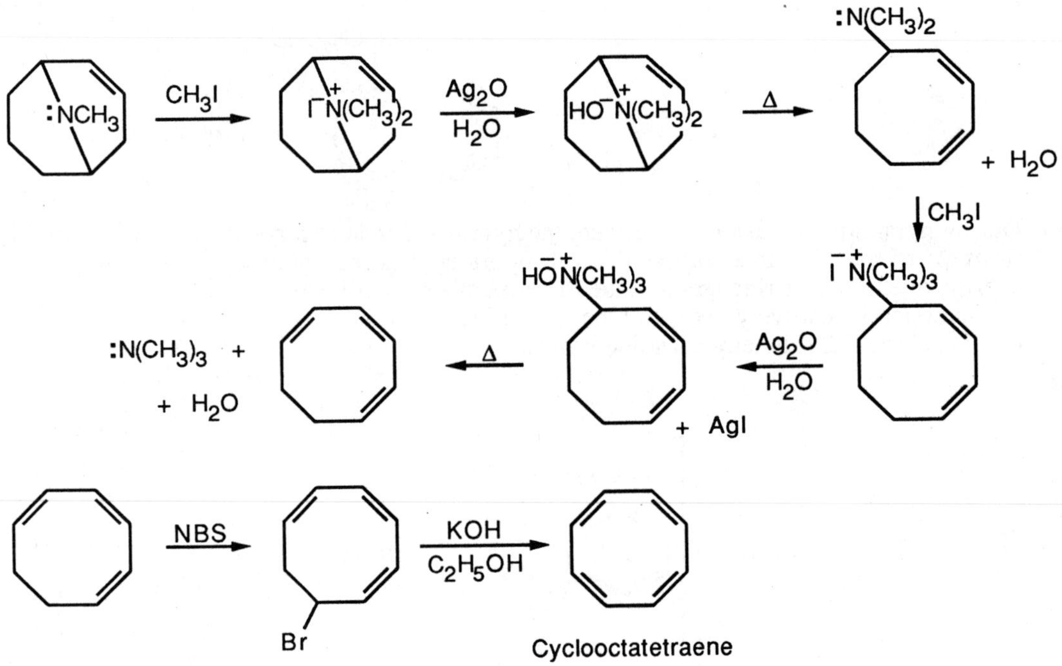
CH₃I
Ag₂O
H₂O
Δ
+ H₂O
+ AgI
NBS
KOH
C₂H₅OH
Br
Cyclooctatetraene

25.51

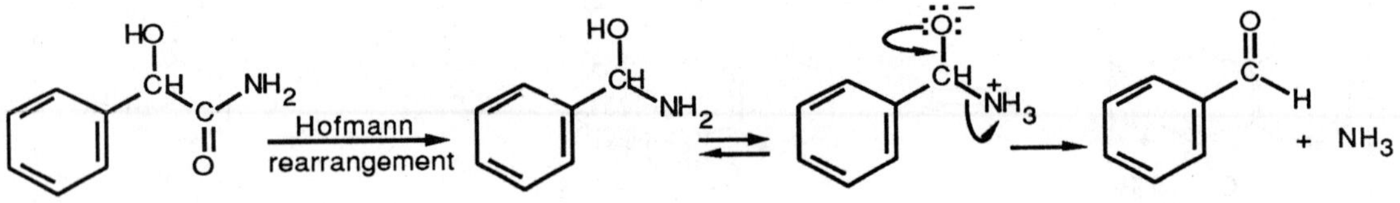
Hofmann
rearrangement
+ NH₃

25.52

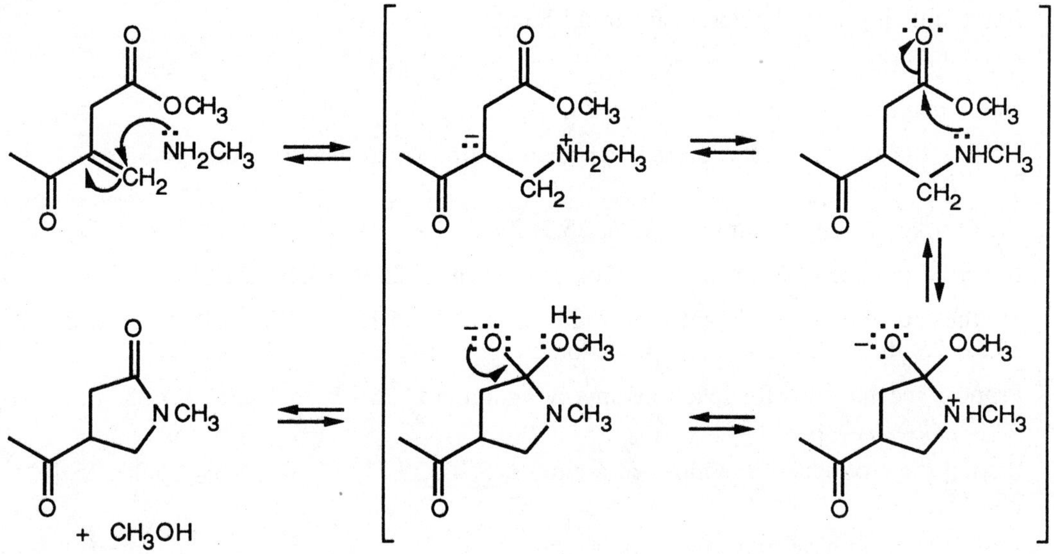

25.53

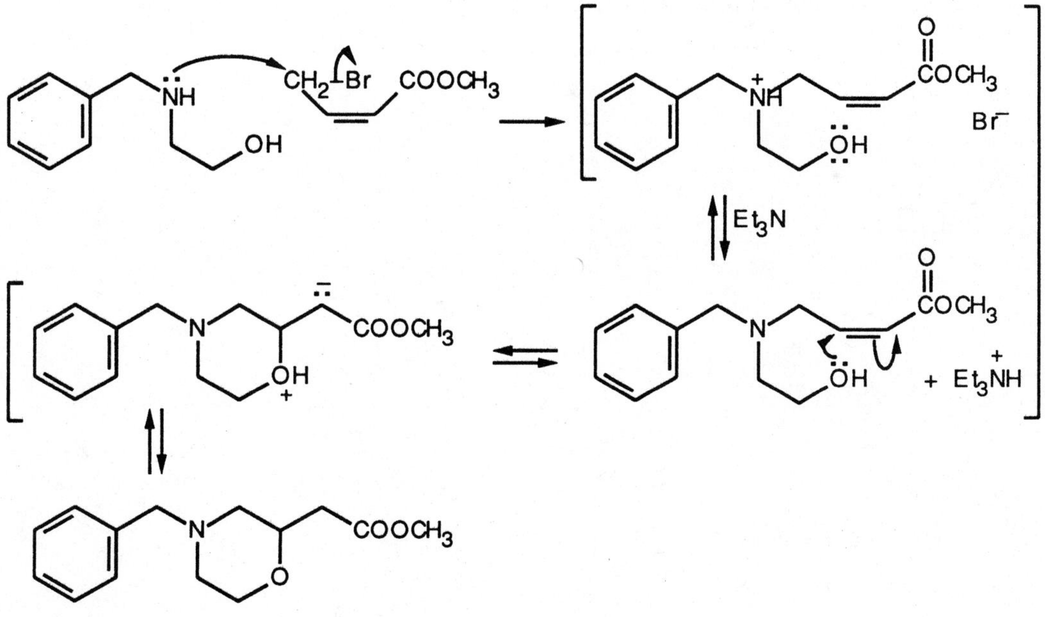

25.54

a) $(CH_3)_2CHCH_2NH_2$ b) $HOCH_2CH_2CH_2NH_2$ c) $(CH_3O)_2CHCH_2NH_2$

Note the accidental overlap of $-NH_2$ and $-OH$ protons in spectrum b.

Study Guide for Chapter 25

After studying this chapter, you should be able to:

(1)   Classify amines as primary, secondary tertiary and quaternary (25.1, 25.24).

(2)   Name and draw amines (25.2, 25.3, 25.4, 25.25, 25.26, 25.31).

(3)   Understand the geometry, stereochemistry and physical properties of amines (25.5, 25.27, 25.33, 25.48).

(4)   Predict the basicity of amines (25.6, 25.7, 25.8).

(5)   Describe the use of amines to resolve enantiomers (25.9, 25.10, 25.11).

(6)   Synthesize amines by several routes (25.13, 25.14, 25.15, 25.17, 25.19, 25.28, 25.29, 25.32, 25.36, 25.42, 25.43, 25.44, 25.45, 25.46, 25.47).

(7)   Propose mechanisms for reactions involving amines (25.12, 25.16, 25.18, 25.33, 25.41, 25.49, 25.51, 25.52, 25.53).

(8)   Predict the products of reactions of amines (25.20, 25.21, 25.30, 25.34, 25.39, 25.40, 25.50).

(9)   Apply the morphine rule to alkaloids (25.22).

(10)   Identify amines by using:

   (a) The Hofmann elimination (25.37).

   (b) Spectroscopic techniques (25.23, 25.54).

Chapter 26 – Arylamines and Phenols

26.1

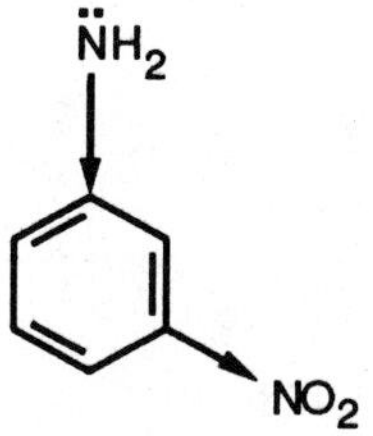

The inductive effect of the electron-withdrawing nitro group makes the amine nitrogen of *m*–nitroaniline less electron-rich and less basic than aniline.

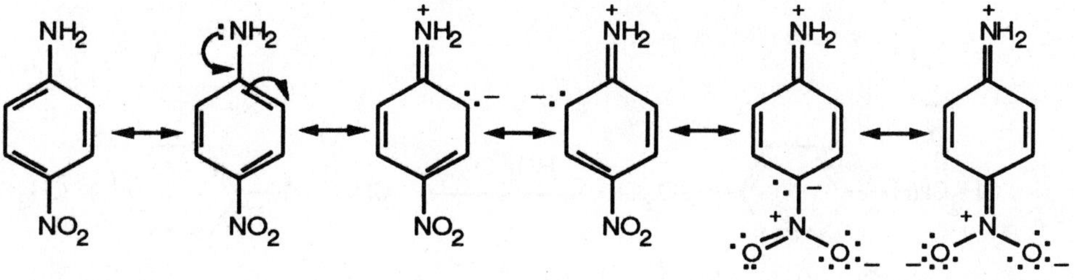

When the nitro group is *para* to the amino group, conjugation of the amino group with the nitro group can also occur. *p*–Nitroaniline is thus even less basic than *m*–nitroaniline.

26.2

Least basic ⟶ Most basic

a)

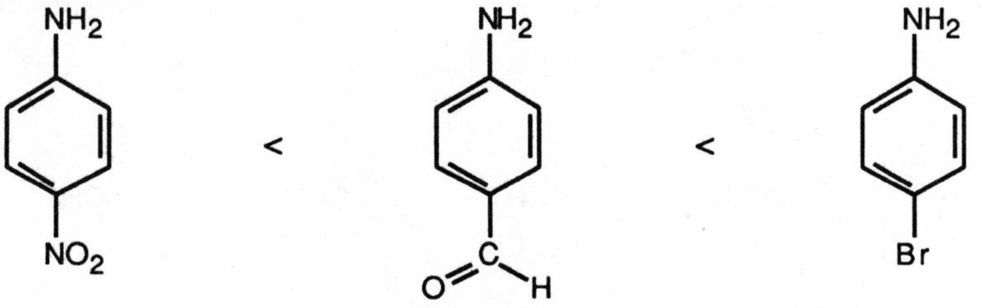

b)

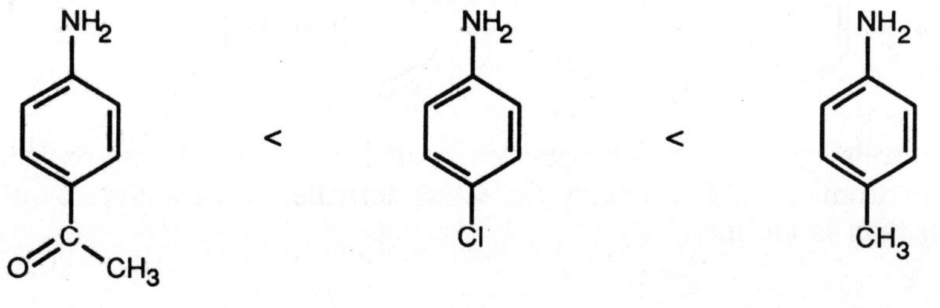

c)

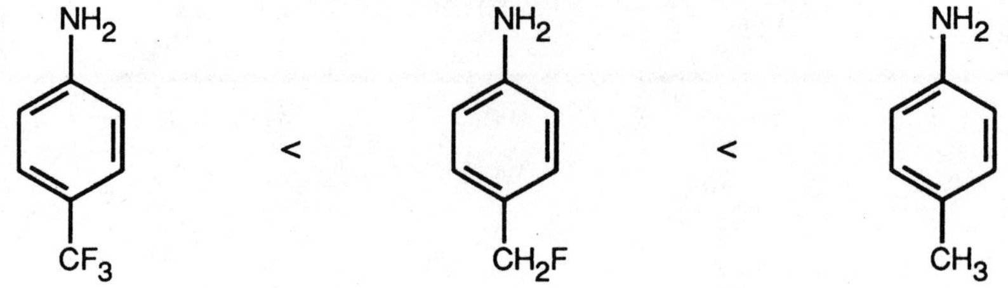

26.3

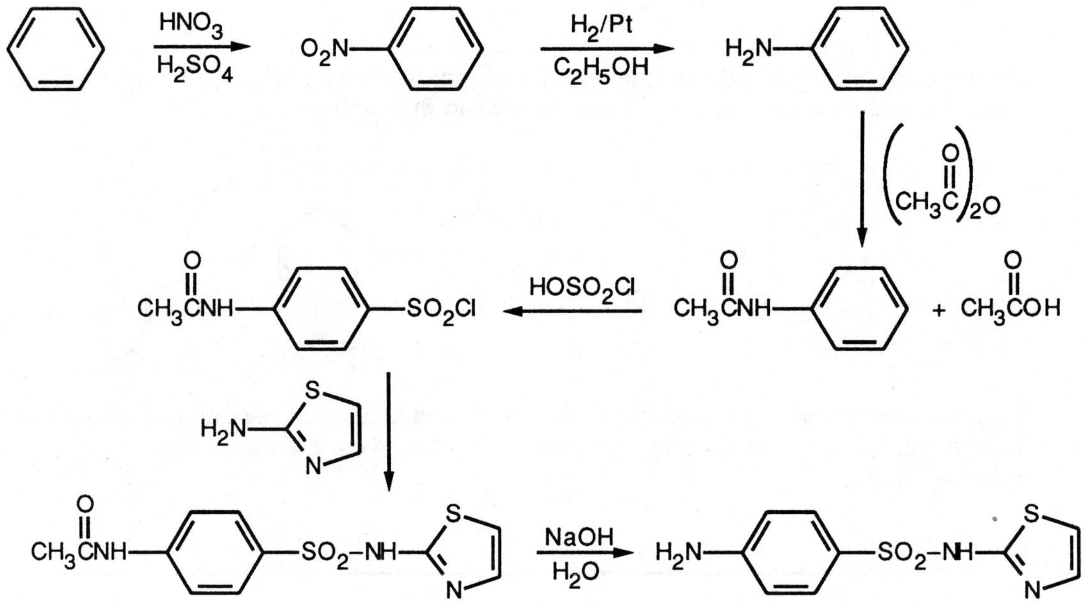

Sulfathiazole

26.4

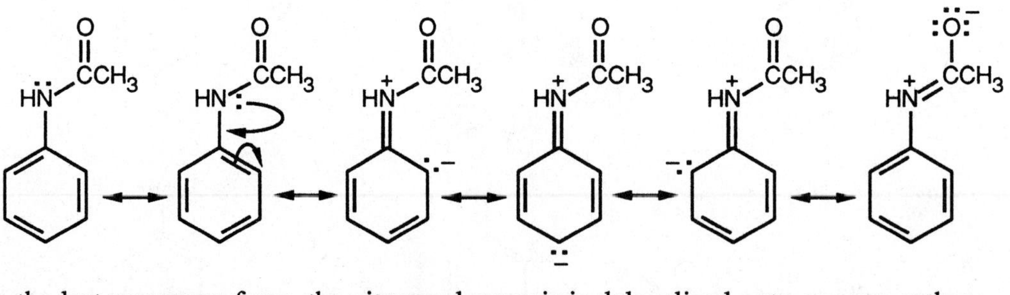

In the last resonance form, the nitrogen lone pair is delocalized onto oxygen, rather than into the aromatic ring. Acetanilide is less activated toward electrophilic aromatic substitution than is aniline.

26.5

a)

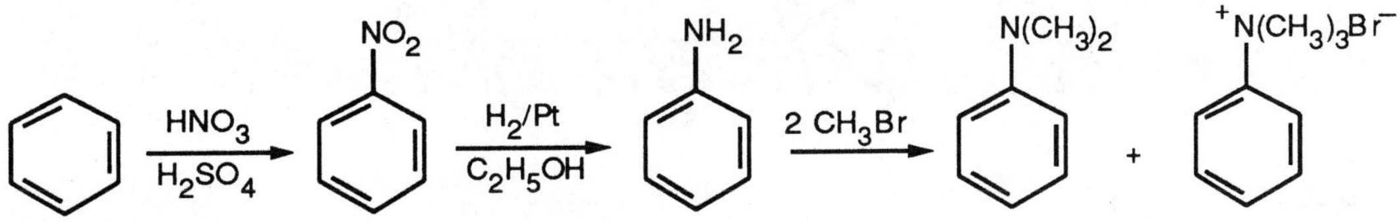

b)

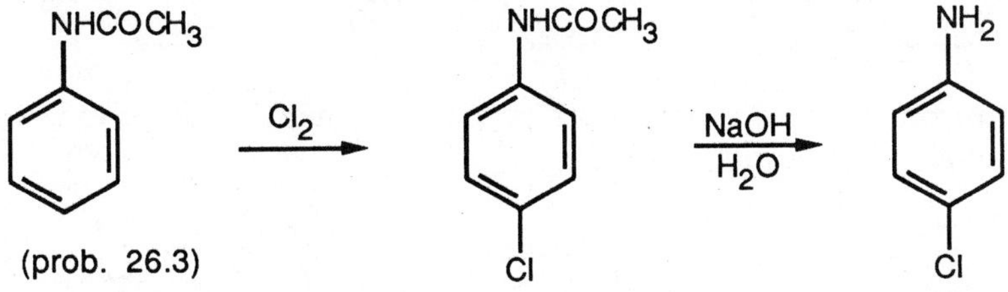

(prob. 26.3)   p–Chloroaniline

c)

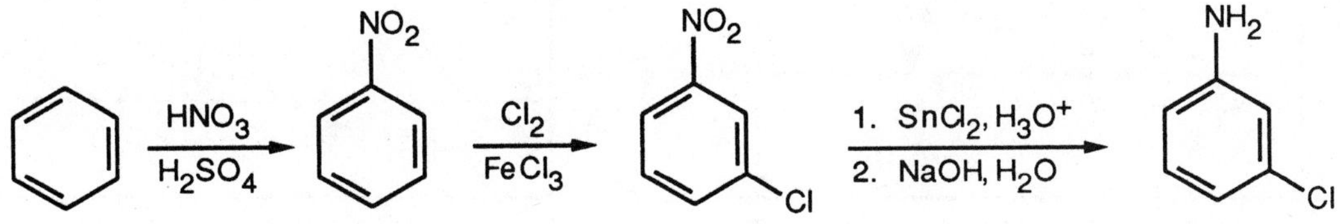

m–Chloroaniline

d)

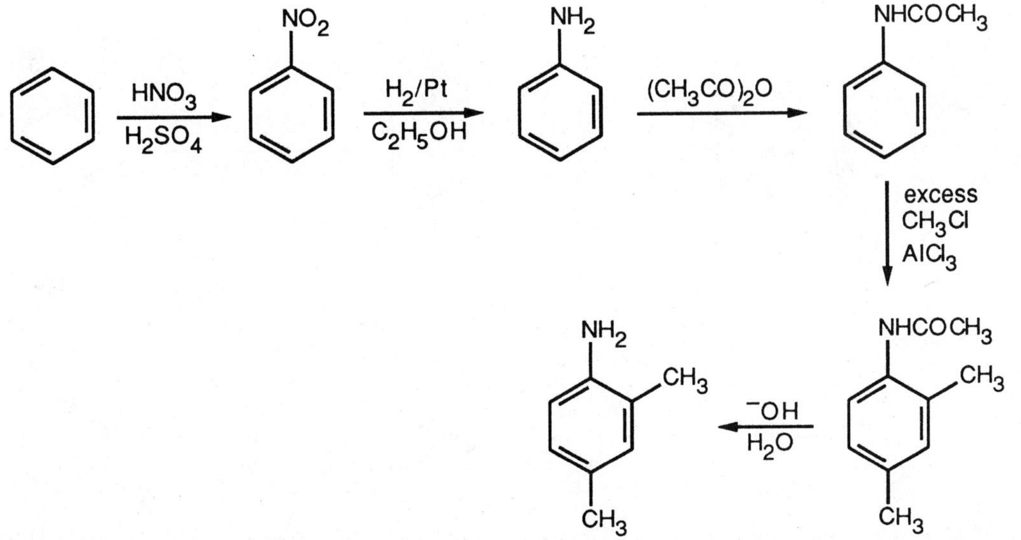

26.6 Aryldiazonium salts are more stable than alkyldiazonium salts for two reasons: (1) Overlap of the nitrogen pi electrons with the aromatic ring stabilizes an aryldiazonium salt relative to an alkyldiazonium salt.

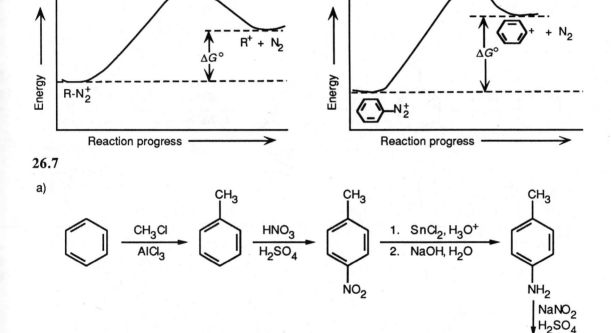

(2) Loss of nitrogen from an aryldiazonium salt produces a phenyl carbocation, which is less stable than an alkyl carbocation.

26.7

a)

Other routes to *p*–bromobenzoic acid are also possible.

b)

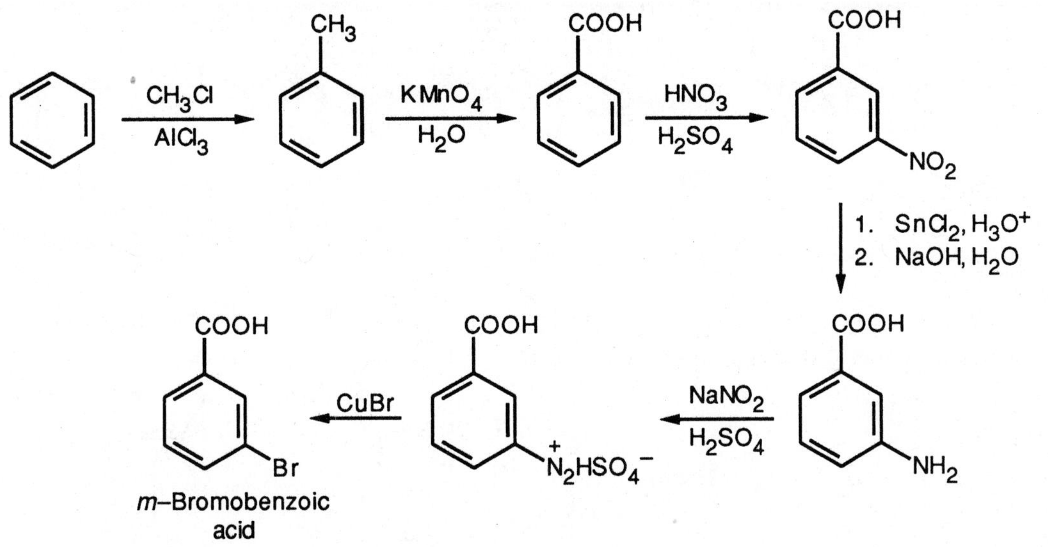
CH3
COOH
COOH
CH3Cl
AlCl3
KMnO4
H2O
HNO3
H2SO4
NO2
1. SnCl2, H3O+
2. NaOH, H2O
COOH
COOH
COOH
CuBr
NaNO2
H2SO4
Br
+N2HSO4−
NH2
m–Bromobenzoic acid

c)

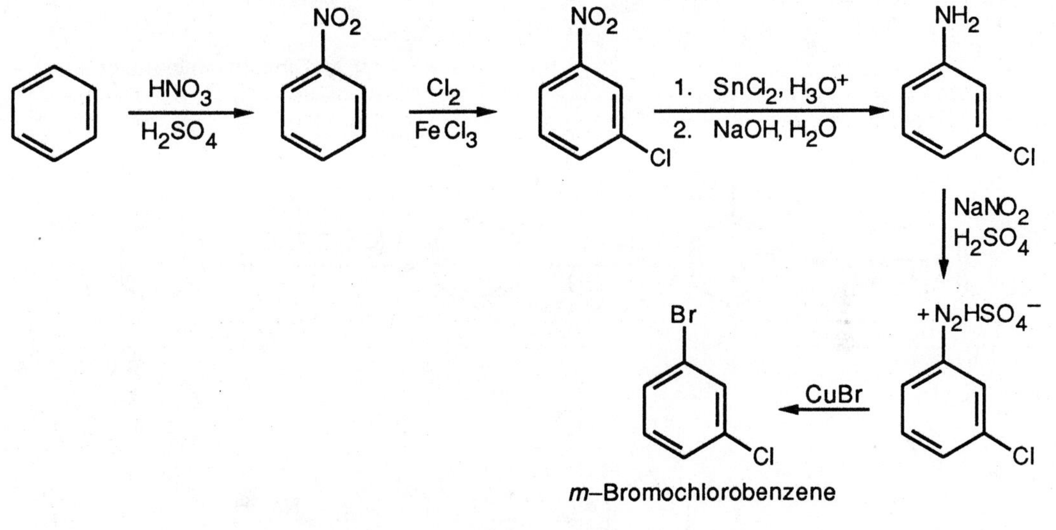
NO2
NO2
NH2
HNO3
H2SO4
Cl2
FeCl3
1. SnCl2, H3O+
2. NaOH, H2O
Cl
Cl
NaNO2
H2SO4
Br
+N2HSO4−
CuBr
Cl
Cl
m–Bromochlorobenzene

d)

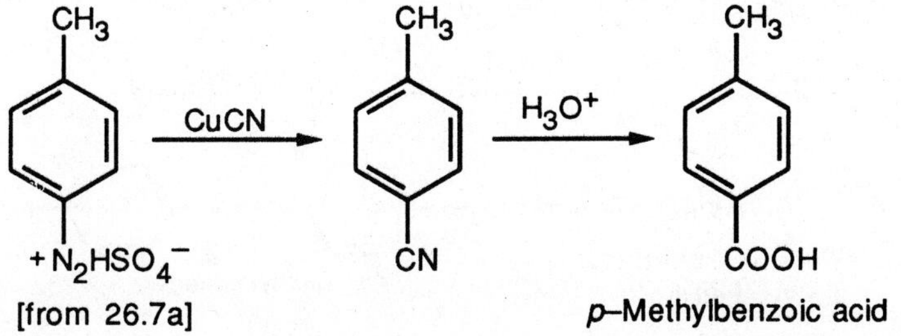
CH3
CH3
CH3
CuCN
H3O+
+N2HSO4−
CN
COOH
[from 26.7a]
p–Methylbenzoic acid

e)

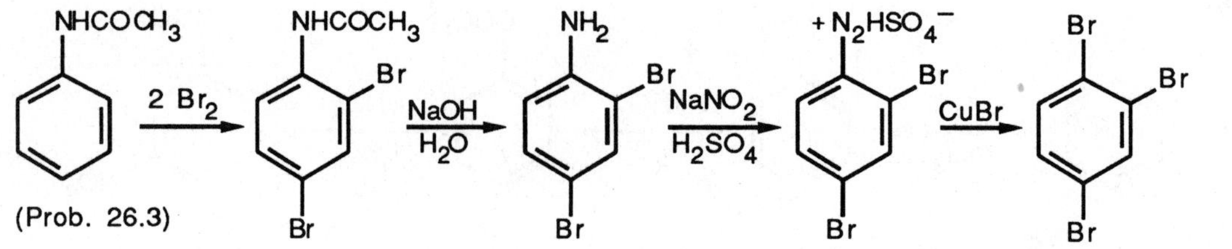

(Prob. 26.3)

1,2,4–Tribromobenzene

26.8

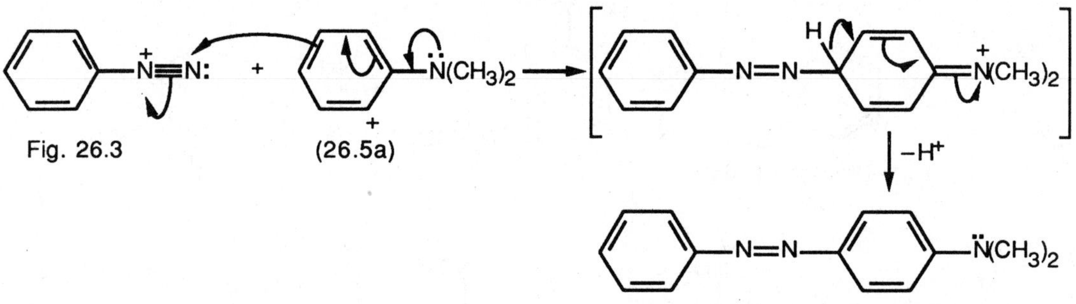

Fig. 26.3   (26.5a)

p–(N,N–Dimethylamino)azobenzene

26.9

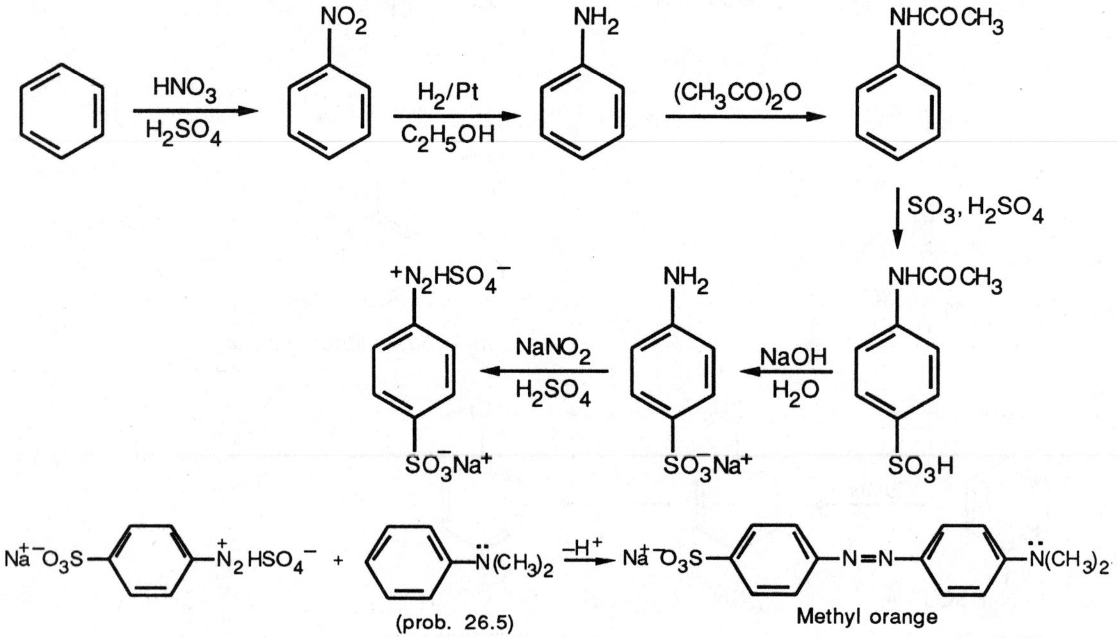

(prob. 26.5)

Methyl orange

26.10

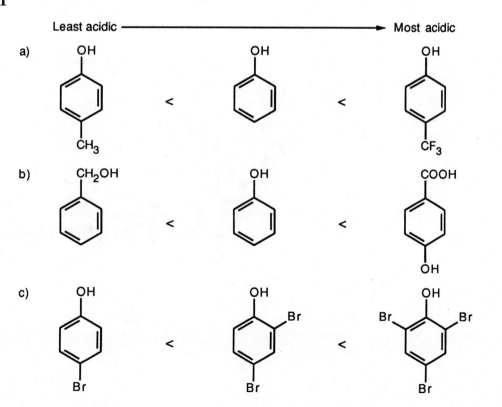

2,4–Dichlorophenoxyacetic acid

26.11

26.12

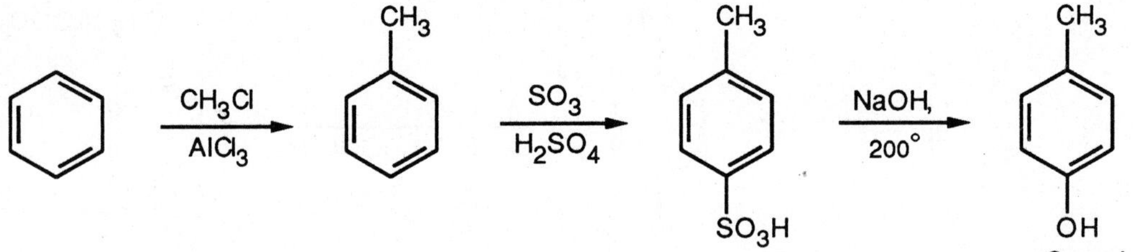

26.13

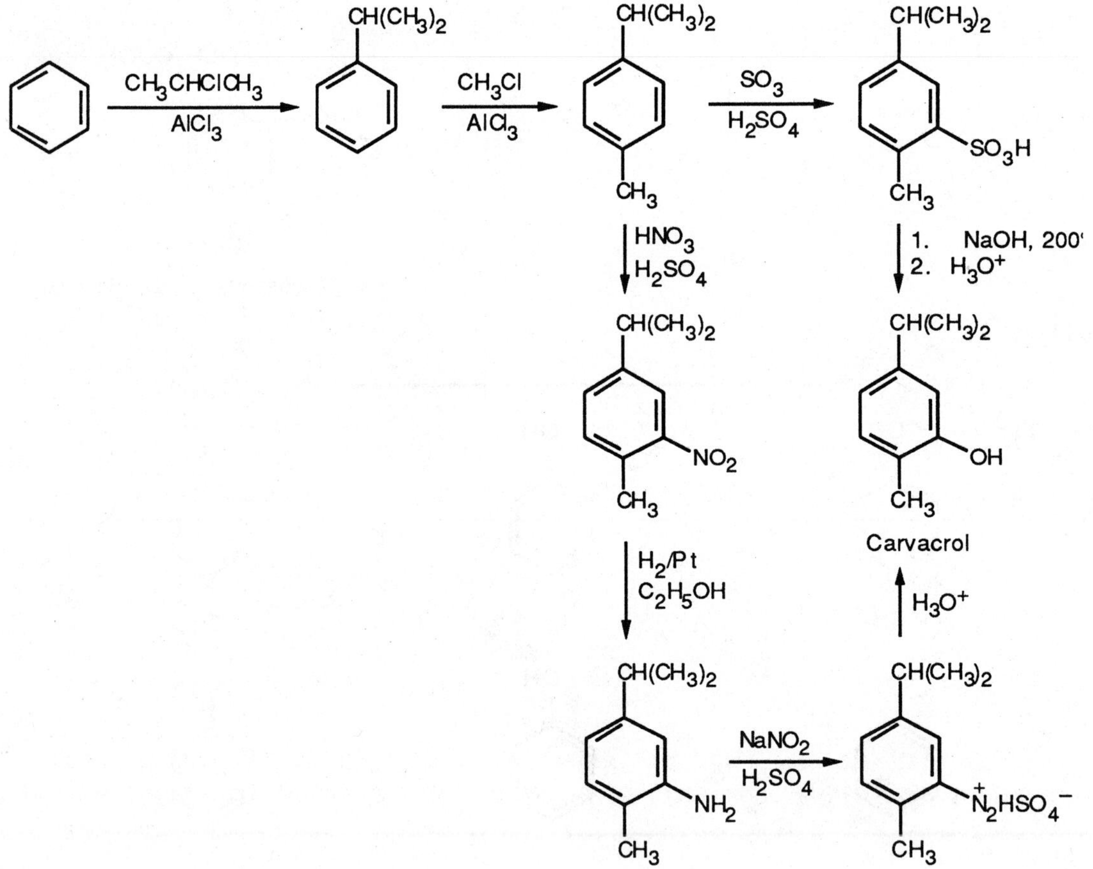

26.14

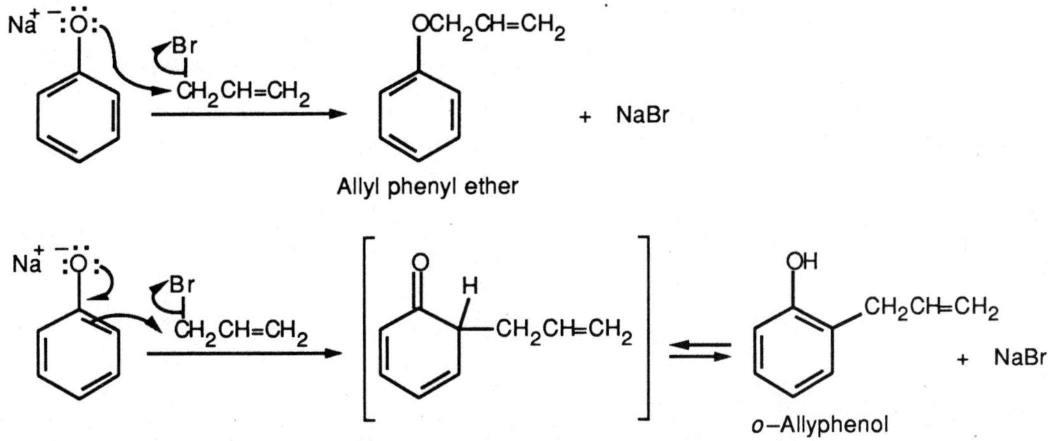

Allyl phenyl ether

o–Allyphenol

26.15

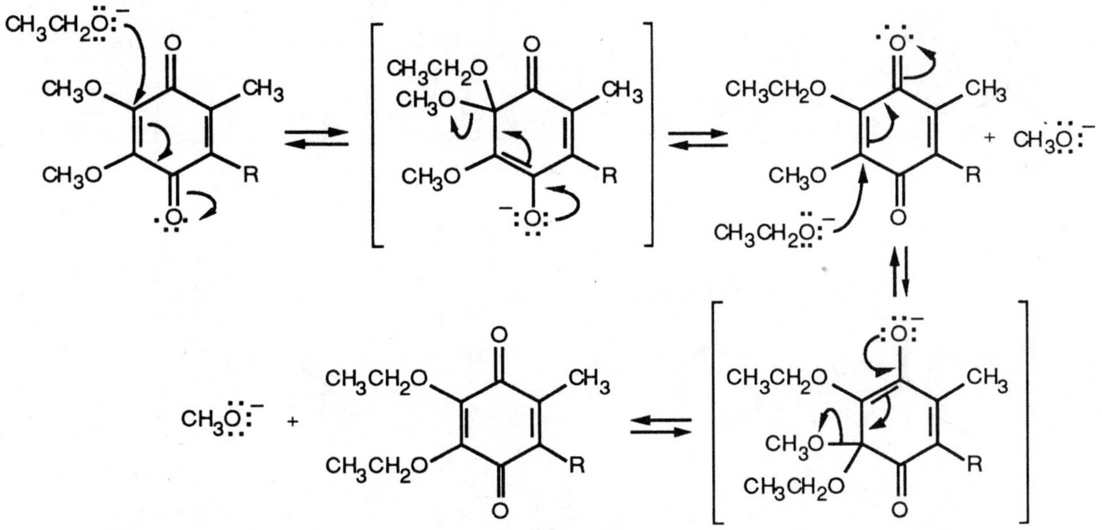

Conjugate addition of ethoxide to the double bond produces an intermediate that eliminates methoxide anion. The diethoxyl product predominates because of the large excess of ethanol.

26.16

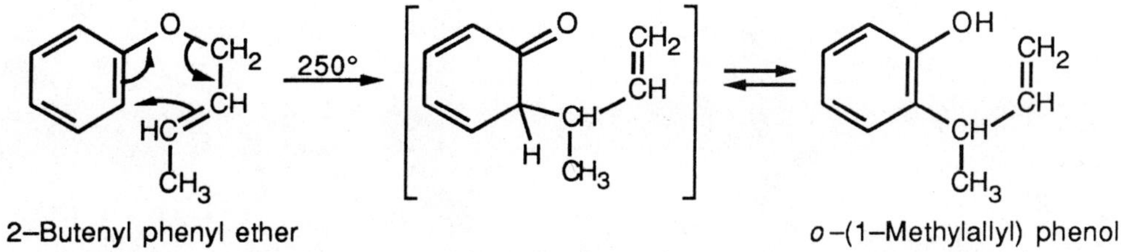

2–Butenyl phenyl ether   o–(1–Methylallyl) phenol

26.17 a) *p*–Bromophenol

b) 2,3–Dichloro–*N*–methylaniline

c) 2–Methyl–1,4–benzenediamine

d) 3–Methoxy–5–methylphenol

e) 1,3,5–Benzenetriol

f) *N*,3–Diethyl–*N*,5–dimethylaniline

26.18

a)

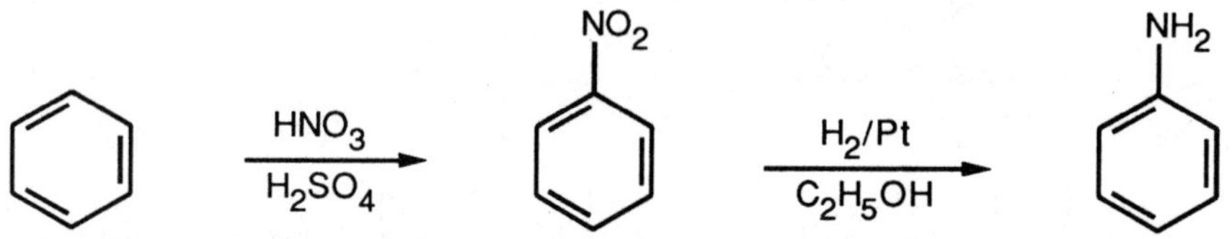

b)

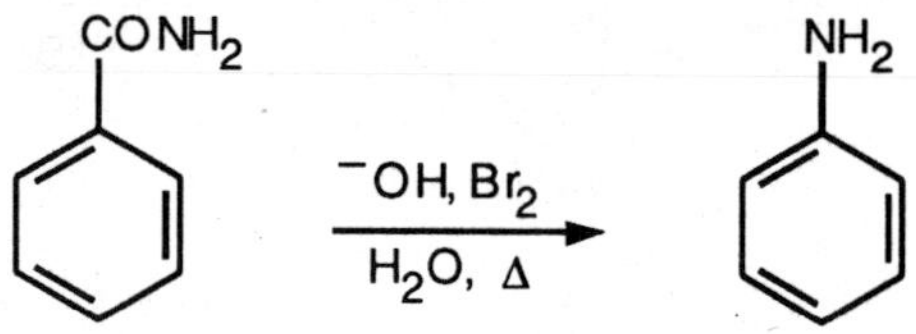

c)

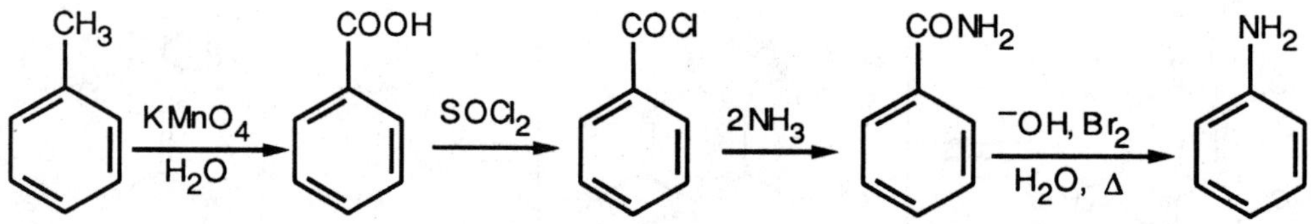

26.19

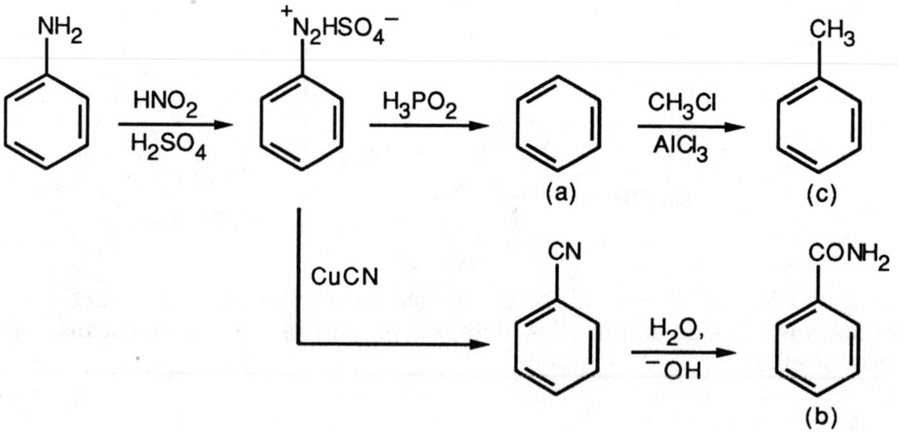

26.20

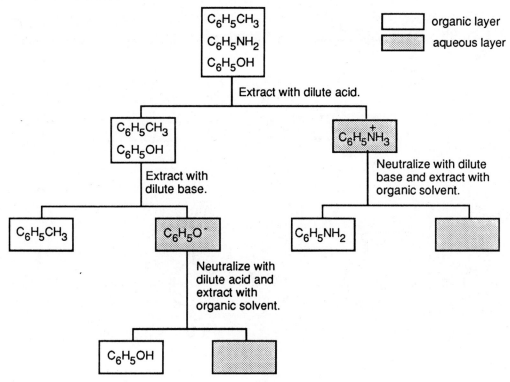

26.21

a)

m-Toluidine $\xrightarrow[\text{1 mole}]{Br_2}$ (two products shown)

b)

$\xrightarrow[H_2O]{(KSO_3)_2NO}$

c)

$\xrightarrow{\text{excess } CH_3I}$

d)

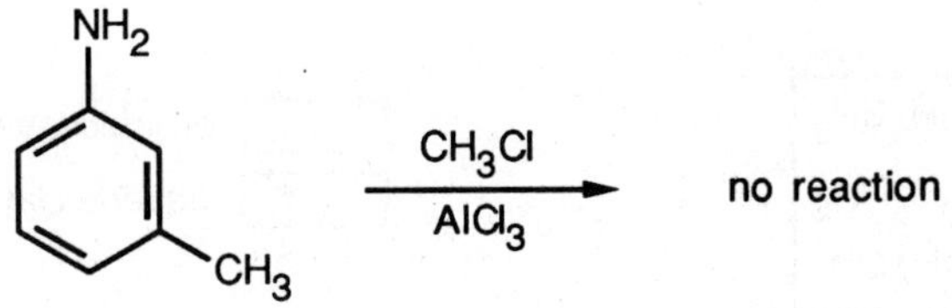
NH2
CH3Cl
AlCl3
no reaction
CH3

e)

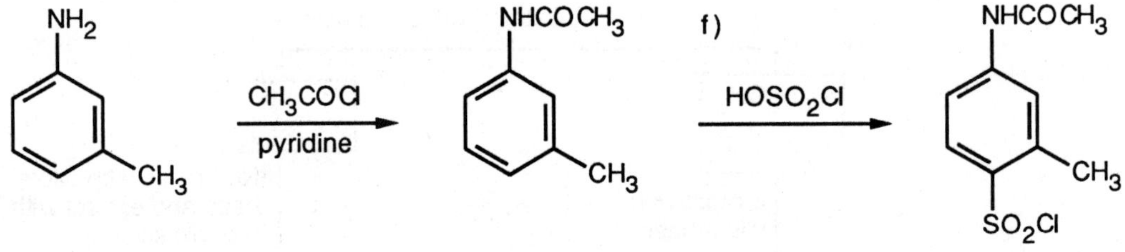
NH2
CH3COCl
pyridine
CH3
NHCOCH3
f)
HOSO2Cl
NHCOCH3
CH3
SO2Cl

26.22

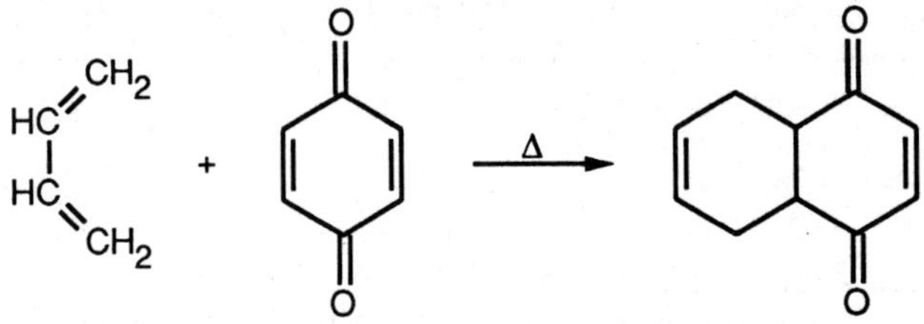

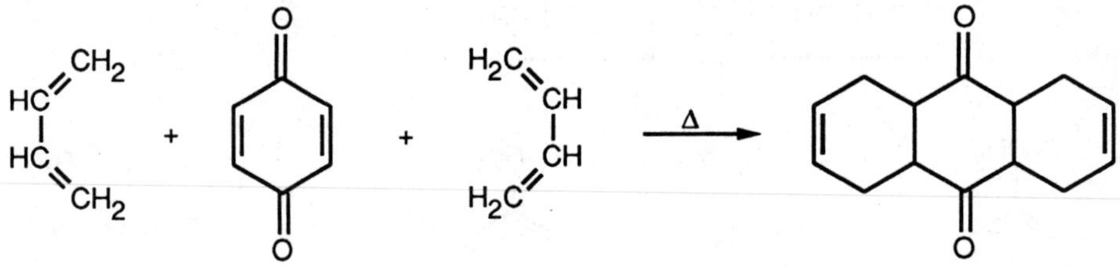

26.23

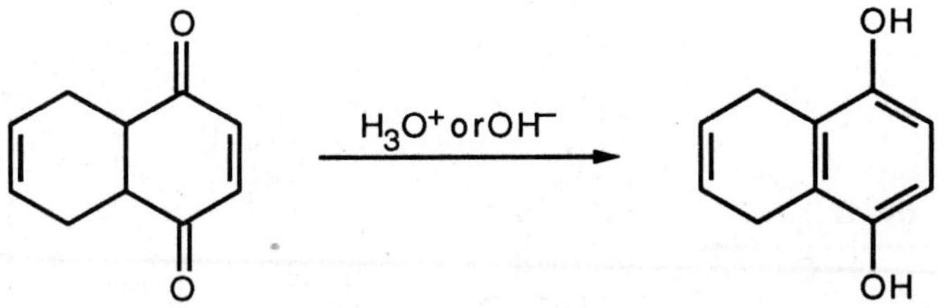
H3O+ or OH−
OH
OH

A B

The two-proton singlet at 6.7 δ is due to the aromatic ring protons.

26.24

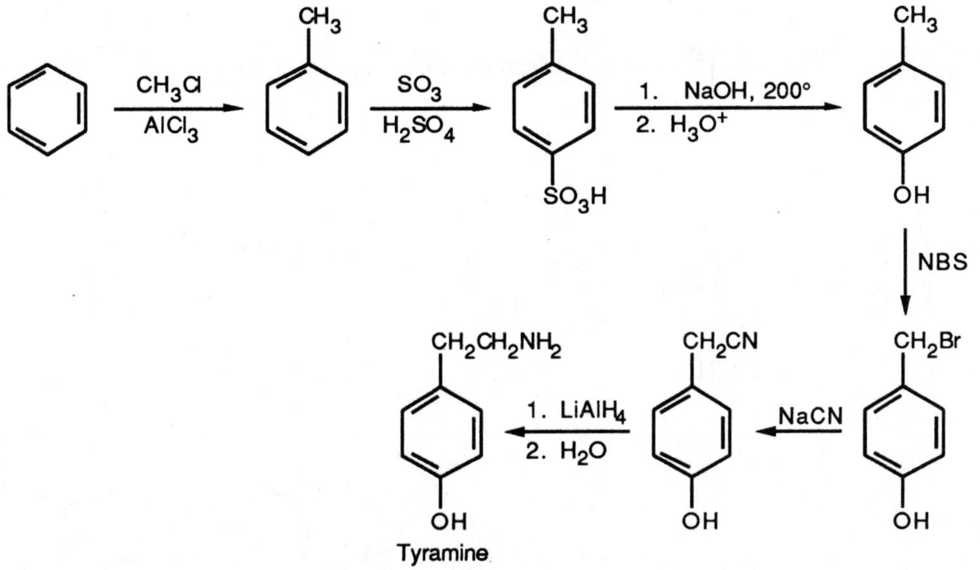

Tyramine

26.25 The nitrogen lone pair of electrons of diphenylamine can overlap the pi electron system of either ring. Electron delocalization occurs to an even greater extent for diphenylamine than for aniline. Because the energy difference between non-protonated and protonated amine is much greater for diphenylamine than for aniline, diphenylamine is non-basic.

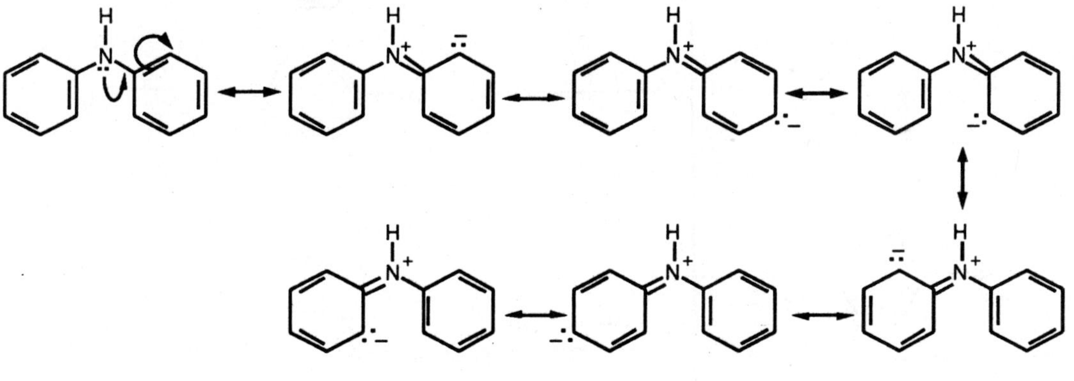

26.26

a)

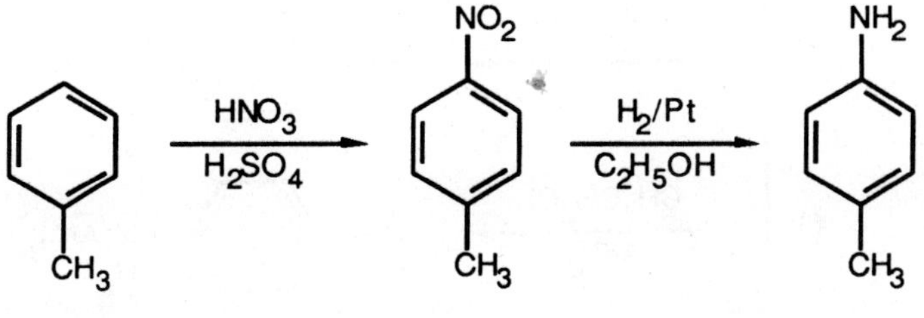

b)

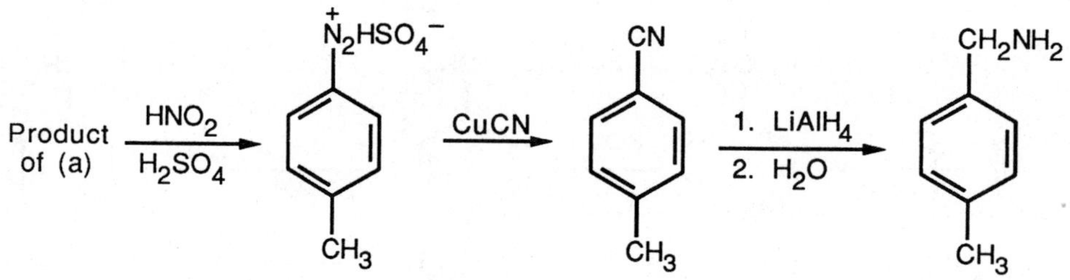

c)

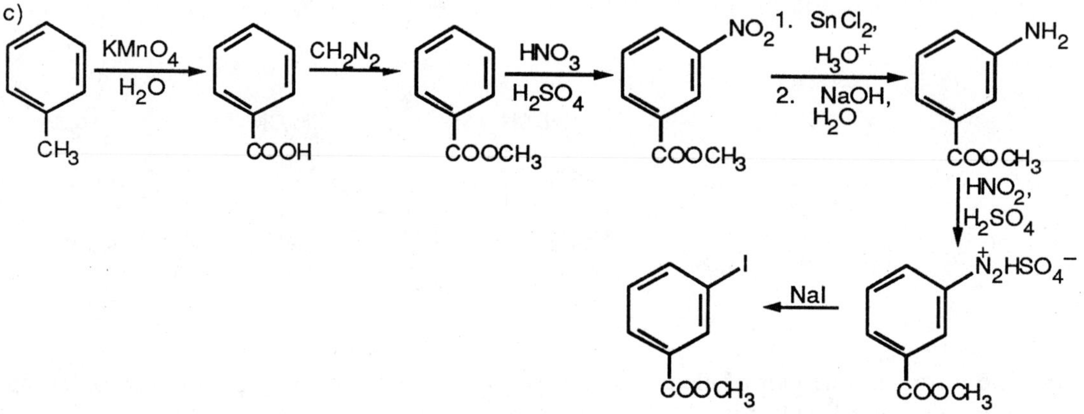

26.27

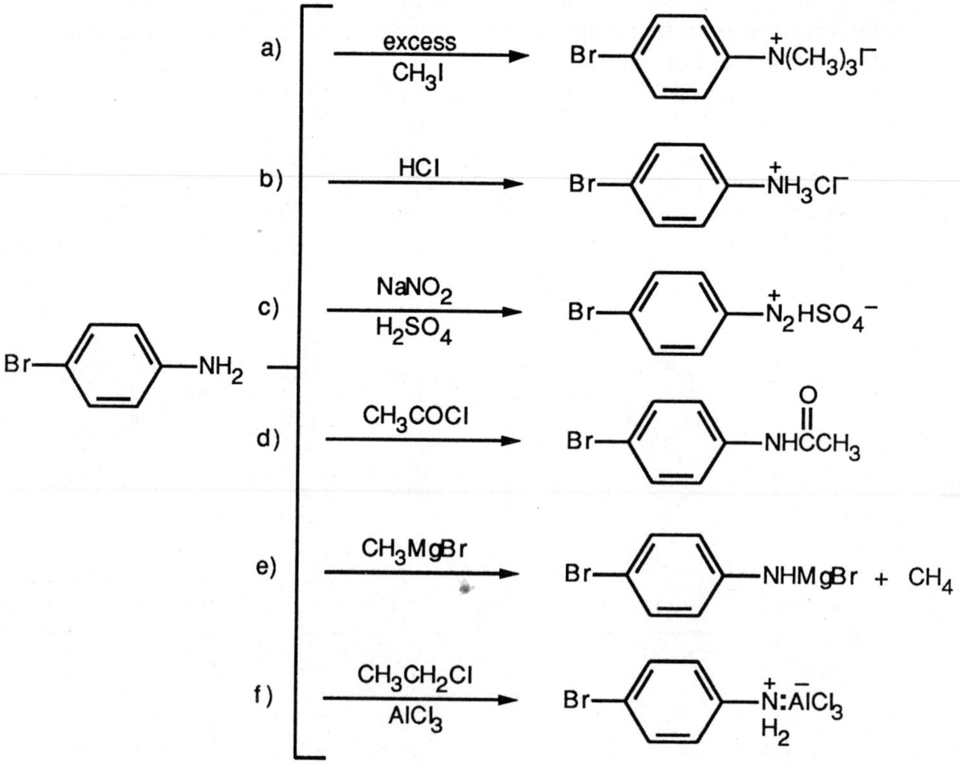

a) $\xrightarrow[CH_3I]{\text{excess}}$ Br—⟨benzene⟩—$\overset{+}{N}(CH_3)_3 I^-$

b) $\xrightarrow{HCl}$ Br—⟨benzene⟩—$\overset{+}{N}H_3 Cl^-$

c) $\xrightarrow[H_2SO_4]{NaNO_2}$ Br—⟨benzene⟩—$\overset{+}{N}_2 HSO_4^-$

d) $\xrightarrow{CH_3COCl}$ Br—⟨benzene⟩—$NH\overset{O}{\overset{\|}{C}}CH_3$

e) $\xrightarrow{CH_3MgBr}$ Br—⟨benzene⟩—$NHMgBr + CH_4$

f) $\xrightarrow[AlCl_3]{CH_3CH_2Cl}$ Br—⟨benzene⟩—$\overset{+}{N}:\bar{A}lCl_3$ H_2

Br—⟨benzene⟩—NH_2

g) 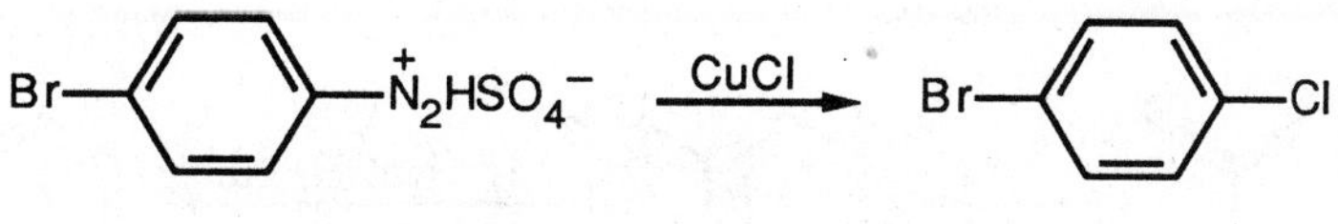
Br
$\overset{+}{N}_2HSO_4^-$
CuCl
Br
Cl

h) 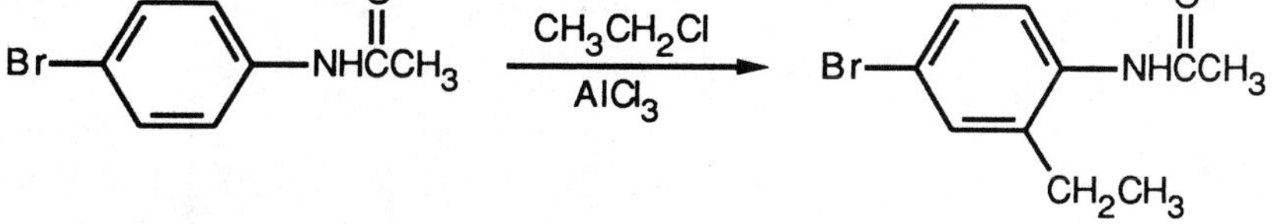
Br
NHCCH$_3$
CH_3CH_2Cl
$AlCl_3$
Br
NHCCH$_3$
CH_2CH_3

26.28

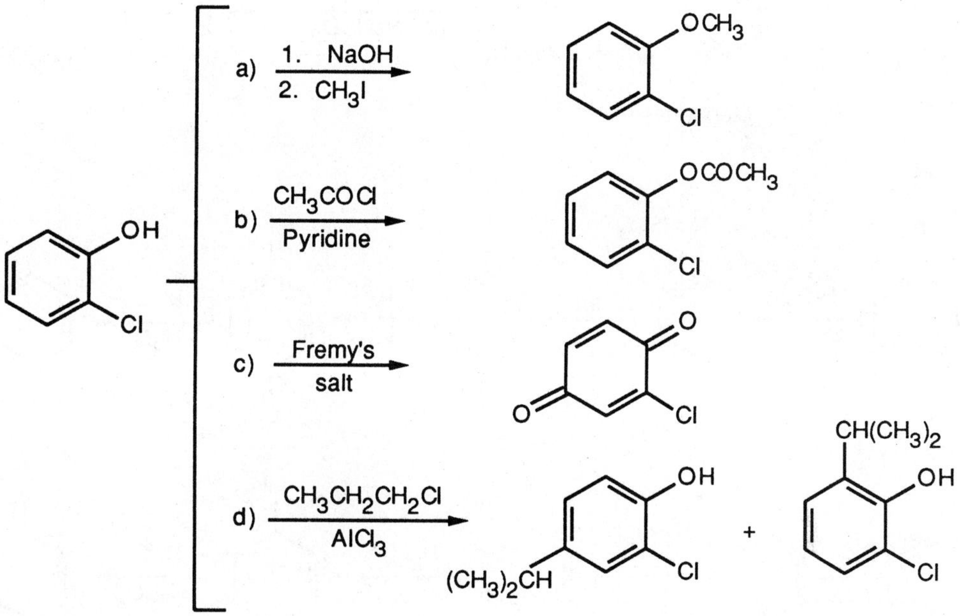
OH
Cl
a)
1. NaOH
2. CH_3I
OCH_3
Cl
b)
CH_3COCl
Pyridine
$OCOCH_3$
Cl
c)
Fremy's salt
O
O
Cl
d)
$CH_3CH_2CH_2Cl$
$AlCl_3$
OH
Cl
$(CH_3)_2CH$
+
$CH(CH_3)_2$
OH
Cl

26.29

a) 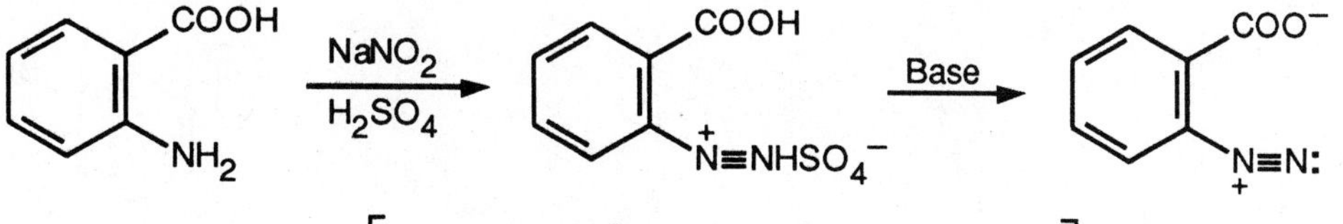
COOH
NH_2
$NaNO_2$
H_2SO_4
COOH
$\overset{+}{N}\equiv NHSO_4^-$
Base
COO^-
$N\equiv N:$

b) 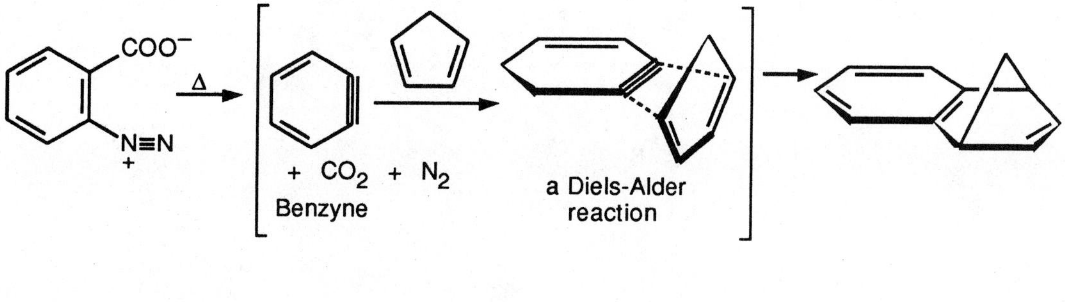
COO^-
$N\equiv N$
Δ
+ CO_2 + N_2
Benzyne
a Diels-Alder reaction

26.30

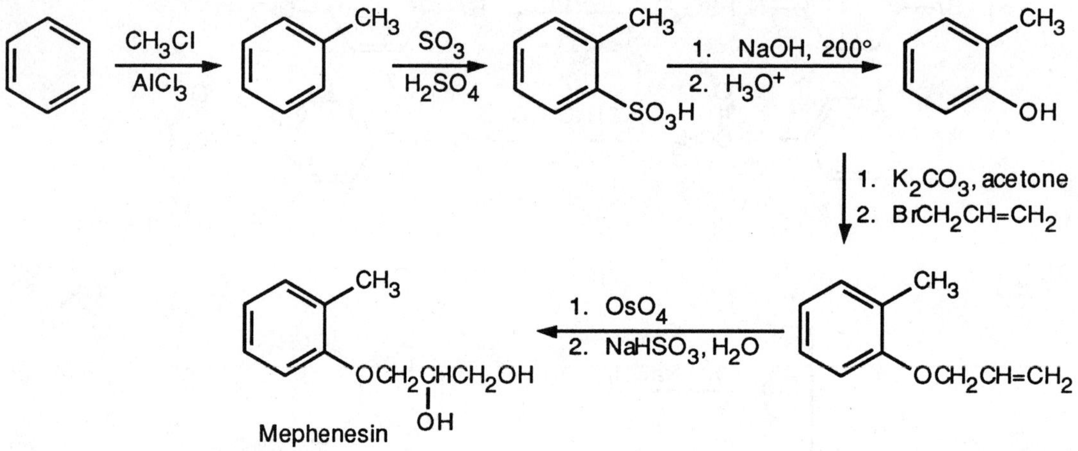

26.31

a)

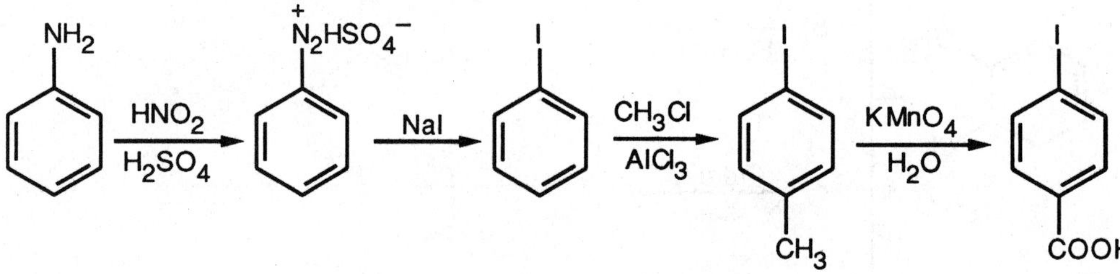

b)

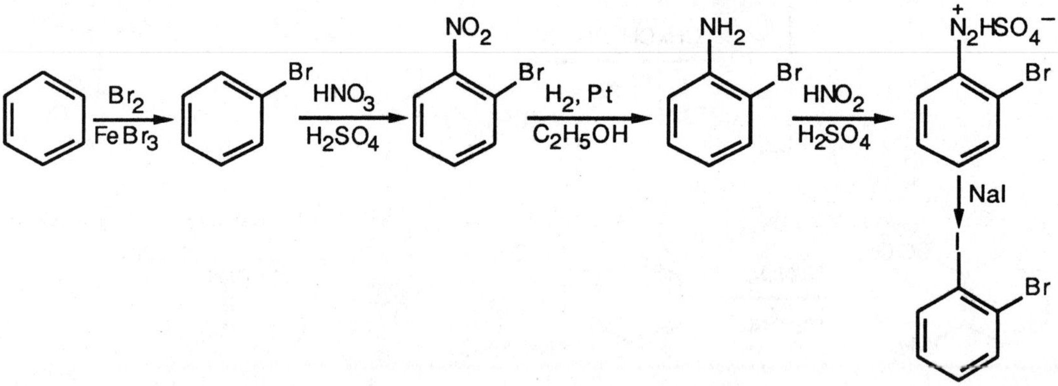

26.32

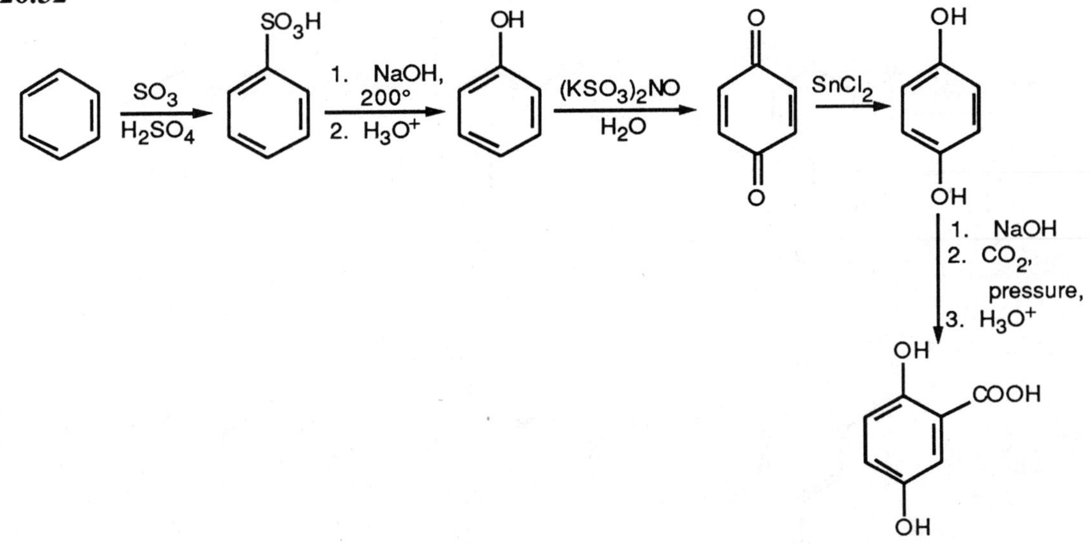

Gentisic acid

26.33

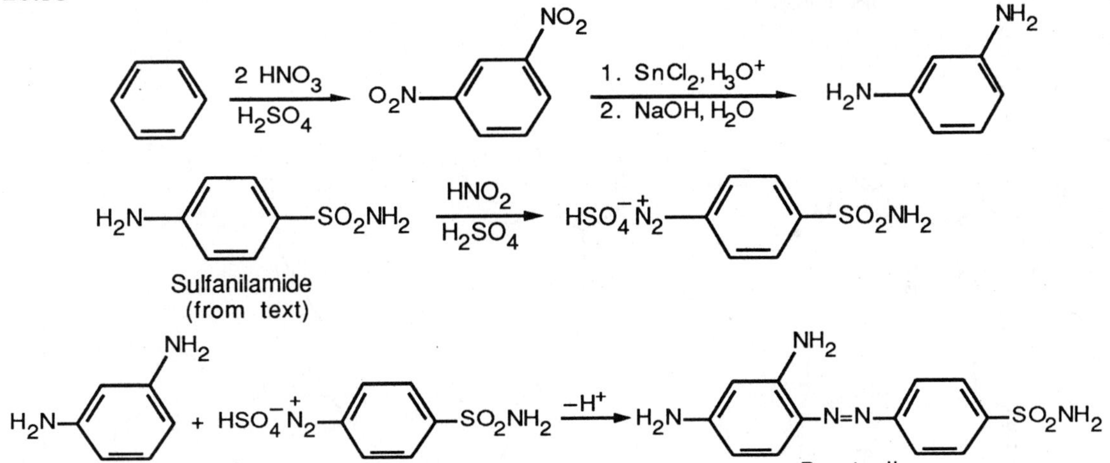

Sulfanilamide
(from text)

Prontosil

26.34

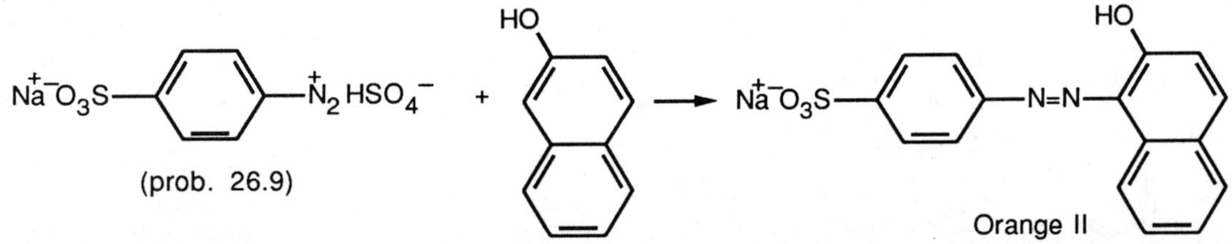

(prob. 26.9)

Orange II

Electrophilic substitution on ß–naphthol occurs between the hydroxyl group and the fused ring.

26.35

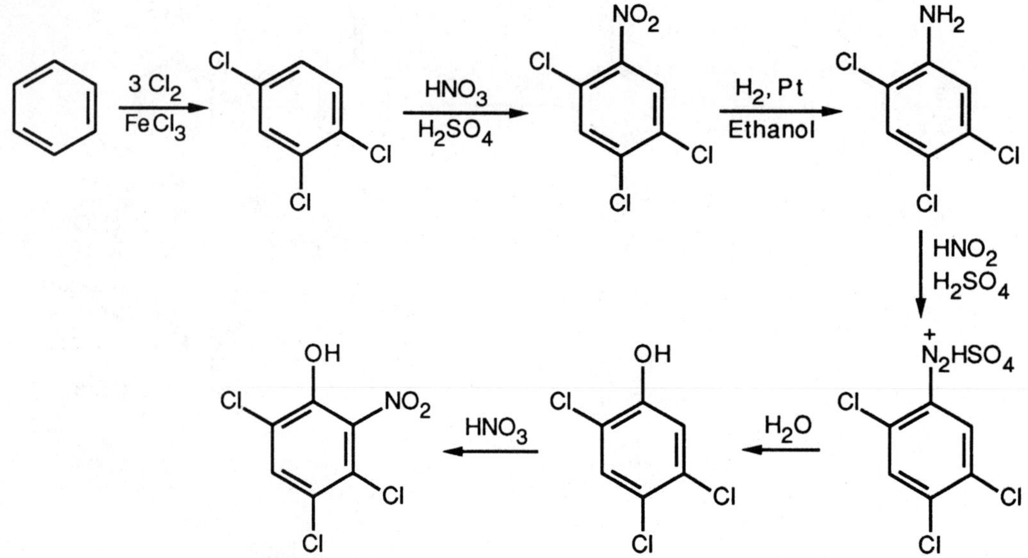

2–Nitro–3,4,6–
trichlorophenol

26.36

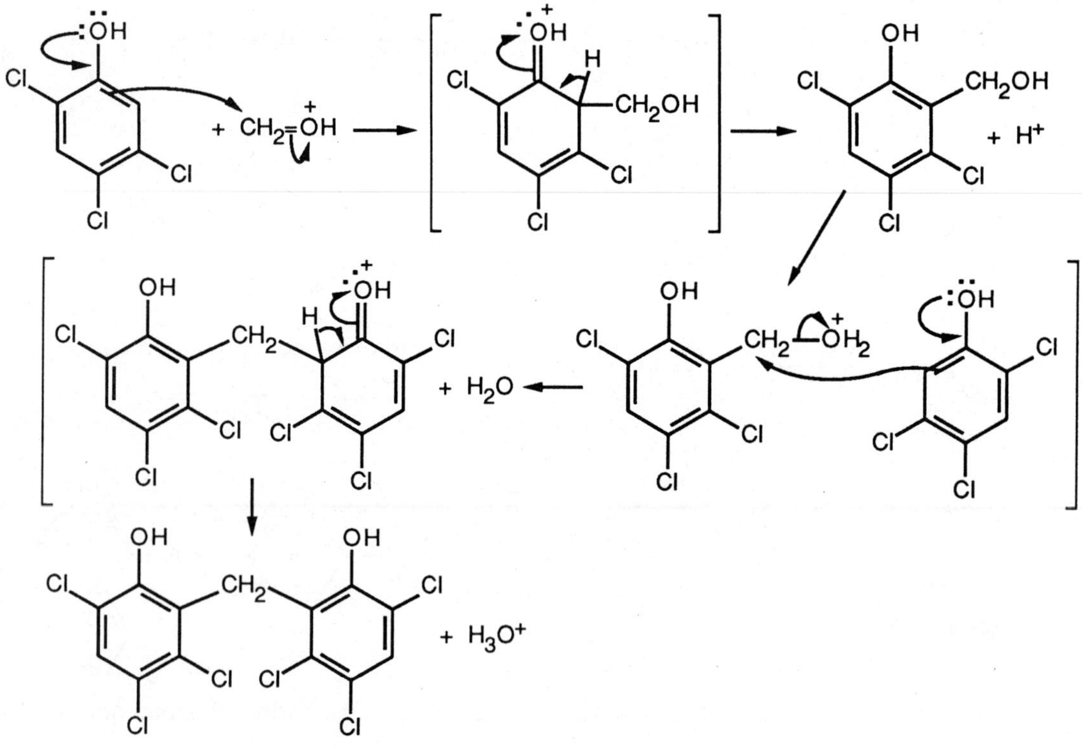

Hexachlorophene

26.37

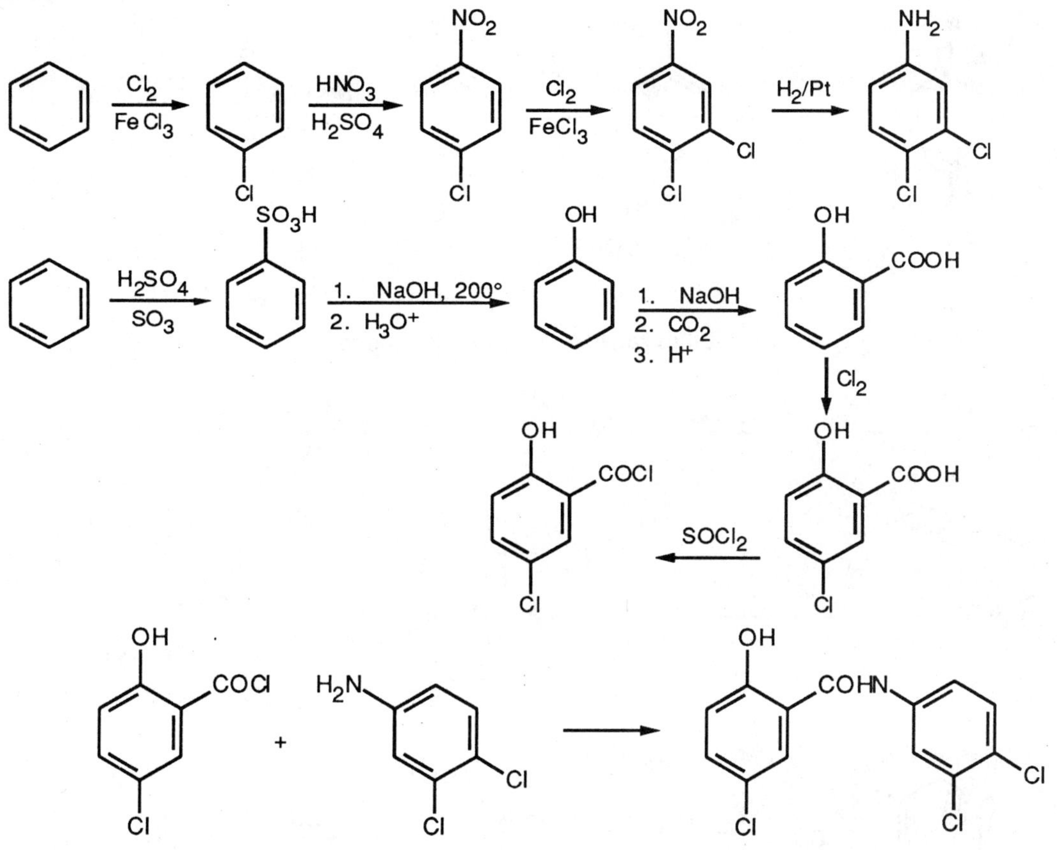

Trichlorosalicylanilide

26.38 Structural formula: $C_8H_{10}O$ contains 4 multiple bonds and/or rings.

Infrared: The broad band at 3500 cm^{-1} indicates a hydroxyl group. The absorbances at 1500 cm^{-1} and 1600 cm^{-1}, as well as at 830 cm^{-1}, are due to an aromatic ring. Compound A is probably a phenol.

^{1}H NMR: The triplet at 1.16 δ (3H) is coupled with the quartet at 2.54 δ (2H). These two absorptions are due to an ethyl group.

The peaks at 6.80 δ (4H) are due to an aromatic ring. The symmetrical splitting pattern of these peaks indicate that the aromatic ring is p–disubstituted.

The absorption at 5.50 δ (1H) is due to an –OH proton.

Compound A

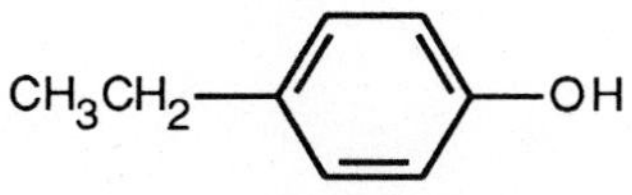

CH_3CH_2——⟨ ⟩——OH

p-Ethylphenol

26.39

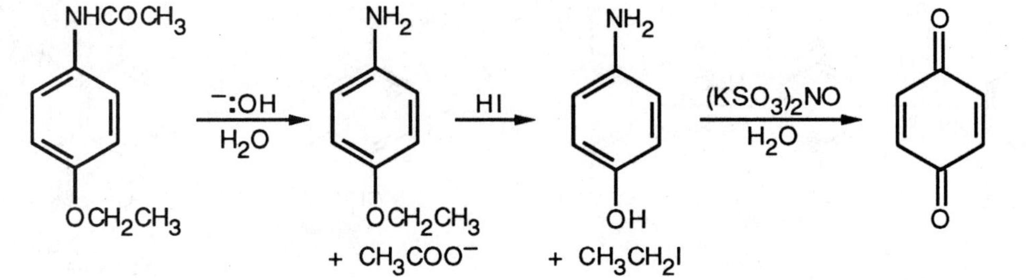

$$C_{10}H_{13}NO_2 \qquad\qquad C_8H_{11}NO \qquad\qquad C_6H_7NO$$

Phenacetin   *p*–Ethoxyaniline   *p*–Aminophenol   Benzoquinone

26.40

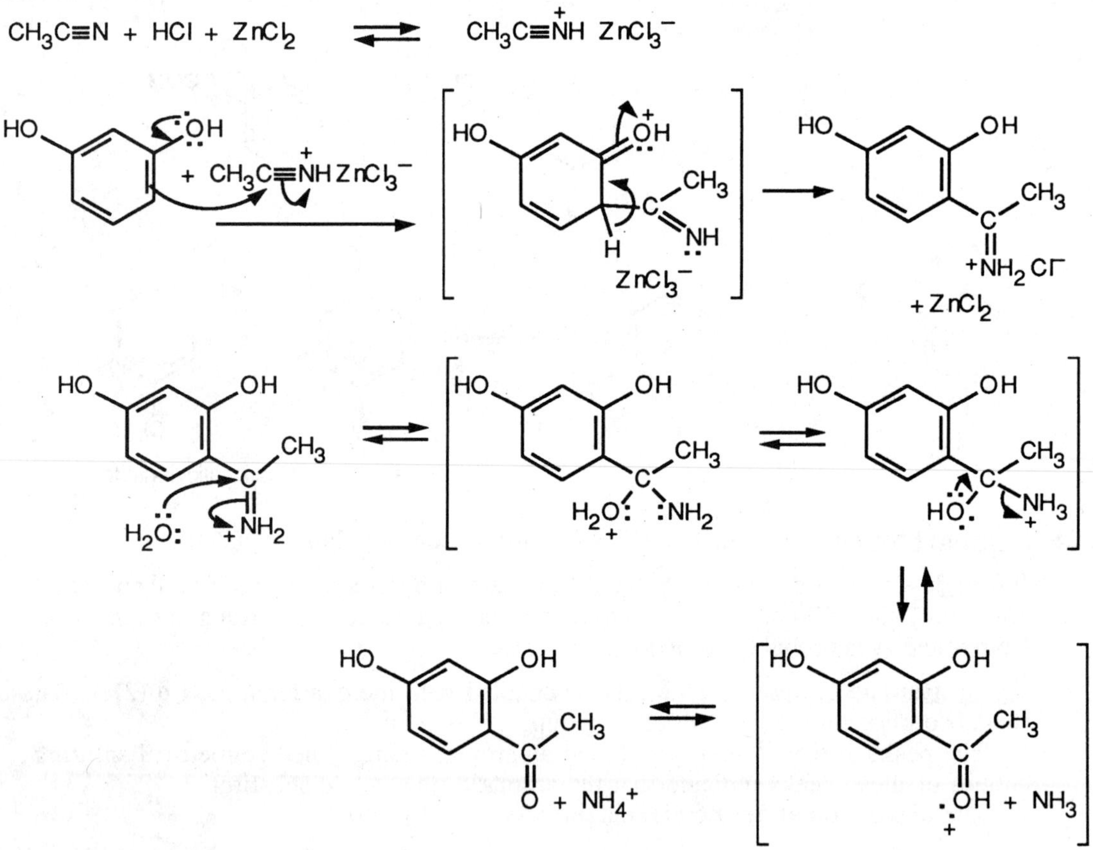

$$CH_3C{\equiv}N + HCl + ZnCl_2 \;\rightleftharpoons\; CH_3C{\equiv}\overset{+}{N}H\ ZnCl_3^-$$

The Hoesch reaction is mechanistically similar to Friedel–Crafts acylation.

26.41

a)

OH
(CH$_3$)$_3$C C(CH$_3$)$_3$

CH$_3$

(BHT)

b)

N(CH$_3$)$_2$

CH$_3$

26.42

a)

CH$_3$CH$_2$—N
   |
   CH$_2$

b)

HO—⬡—C(=O)—CH$_2$CH$_2$CH$_2$Cl

Study Guide for Chapter 26

After studying this chapter, you should be able to:

(1)   Draw and name aromatic amines and phenols (26.17).

(2)   Predict the effects of substituents on the acidity and basicity of

   a) Aromatic amines (26.1, 26.2, 26.25).

   b) Phenols (26.11).

(3)   Synthesize, by several different methods

   a) Aromatic amines (26.3, 26.5, 26.18).

   b) Phenols (26.10, 26.12, 26.13, 26.30, 26.32, 26.35, 26.36, 26.37).

(4)   Formulate mechanisms for reactions involving aromatic amines and phenols (26.40).

(5)   Use diazonium salts in the synthesis of substituted aromatic compounds (26.7, 26.26, 26.29, 26.31).

(6)   Synthesize simple dyes using diazo coupling reactions (26.8, 26.9, 26.33, 26.34).

(7)   Predict the products of reactions involving:

   a) Aromatic amines (26.19, 26.21, 26.24, 26.27).

   b) Phenols (26.14, 26.15, 26.16, 26.22, 26.28).

(8)   Identify aromatic amines and phenols spectroscopically (26.23, 26.38, 26.39, 26.41, 26.42).

27.1 Amino Acids with aromatic rings: Phe, Tyr, Trp, His.
Amino acids containing sulfur: Cys, Met.
Amino acids that are alcohols: Ser, Thr. (Tyr is a phenol.)
Amino acids having hydrocarbon side chains: Ala, Ile, Leu, Val.

27.2

$$H_2N \overset{\overset{\displaystyle H}{|}}{\underset{\underset{\displaystyle COOH}{|}}{C}} R$$

A projection of the alpha carbon of an L–amino acid is pictured above.

For most L–amino acids:

Group	Priority
$-NH_2$	1
$-COOH$	2
$-R$	3
$-H$	4

For cysteine:

Group	Priority
$-NH_2$	1
$-CH_2SH$	2
$-COOH$	3
$-H$	4

27.3–27.4

$$\begin{array}{c} COOH \\ H_2N \overset{S}{-\!\!\!-} H \\ H \overset{R}{-\!\!\!-} OH \\ CH_3 \end{array}$$

L–Threonine

$$\begin{array}{c} COOH \\ H_2N \overset{S}{-\!\!\!-} H \\ HO \overset{S}{-\!\!\!-} H \\ CH_3 \end{array} \qquad \begin{array}{c} COOH \\ H \overset{R}{-\!\!\!-} NH_2 \\ H \overset{R}{-\!\!\!-} OH \\ CH_3 \end{array}$$

Diastereomers of L–Threonine

27.5

Phenylalanine

Serine

Proline

27.6

a)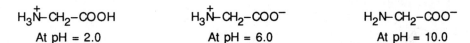

b)

c)

27.7

$$H_3\overset{+}{N}-CH_2-COOH$$
At pH = 2.0

$$H_3\overset{+}{N}-CH_2-COO^-$$
At pH = 6.0

$$H_2N-CH_2-COO^-$$
At pH = 10.0

27.8   a) *Amino acid   Isoelectric point*

Amino acid	Isoelectric point
Val	6.0
Glu	3.2
His	7.6

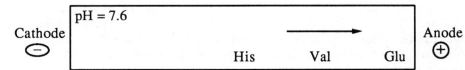

b) *Amino Acid   Isoelectric point*

Amino Acid	Isoelectric point
Gly	6.0
Phe	5.5
Ser	5.7

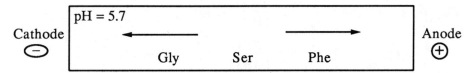

c)

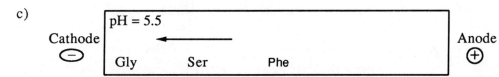

d)

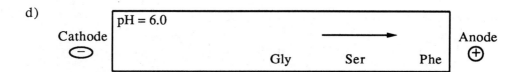

27.9

The indole nitrogen lone pair of electrons is part of the pi electron system of the two rings. Protonation of the indole nitrogen is unlikely to occur because it would disrupt the aromaticity of the indole ring system. Tryptophan is not basic, and its isoelectric point is near neutral pH.

Nitrogen A is not basic because its lone pair is part of the imidazole ring pi system. Nitrogen B is basic because its lone pair electrons lie in the plane of the imidazole ring and can be protonated. Nitrogen B is less basic than an aliphatic amine nitrogen, however. Thus, histidine is weakly basic, and its isoelectric point is slightly higher than neutral.

27.10 The Henderson-Hasselbalch equation states:

$$\log \frac{[A^-]}{[HA]} = pH - pKa$$

For pK_1 of Threonine : $\log \dfrac{[A^-]}{[HA]} = pH - 2.1$

At pH = 1.5 : $\log \dfrac{[A^-]}{[HA]} = 1.5 - 2.1 = -0.6$

or $\log \dfrac{[HA]}{[A^-]} = 0.6$; $\dfrac{[HA]}{[A^-]} = 4.0$

At pH = 1.5 :

For pK_2 of Threonine : $\quad \log \dfrac{[A^-]}{[HA]} = pH - 9.1$

At pH $= 10.0$: $\quad \log \dfrac{[A^-]}{[HA]} = 10.0 - 9.1 = 0.9; \quad \dfrac{[A^-]}{[HA]} = 7.9$

At pH $= 10.0$:

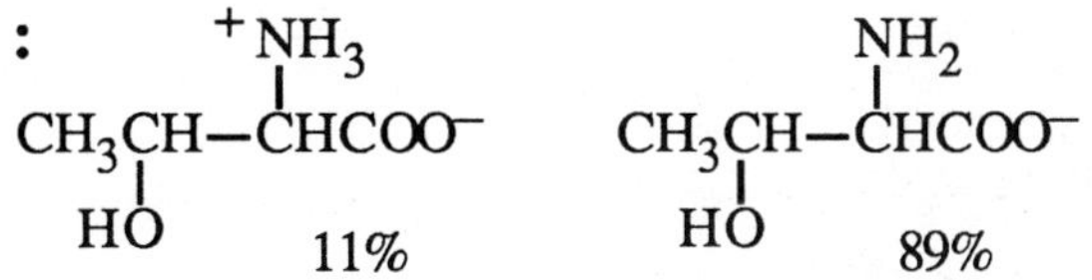

$$\underset{HO}{\overset{\overset{+}{N}H_3}{CH_3CH-CHCOO^-}} \quad 11\% \qquad \underset{HO}{\overset{NH_2}{CH_3CH-CHCOO^-}} \quad 89\%$$

27.11

a) $C_6H_5CH_2CH_2COOH$ $\xrightarrow[\text{2. } H_2O]{\text{1. } Br_2, PBr_3}$ $\underset{Br}{C_6H_5CH_2CHCOOH}$ $\xrightarrow[\text{excess}]{NH_3}$ $\underset{NH_2}{C_6H_5CH_2CHCOOH}$

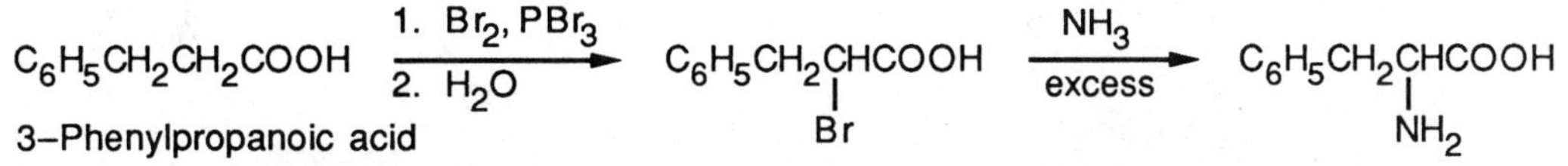

3–Phenylpropanoic acid

b) $(CH_3)_2CHCH_2COOH$ $\xrightarrow[\text{2. } H_2O]{\text{1. } Br_2, PBr_3}$ $\underset{Br}{(CH_3)_2CHCHCOOH}$ $\xrightarrow[\text{excess}]{NH_3}$ $\underset{NH_2}{(CH_3)_2CHCHCOOH}$

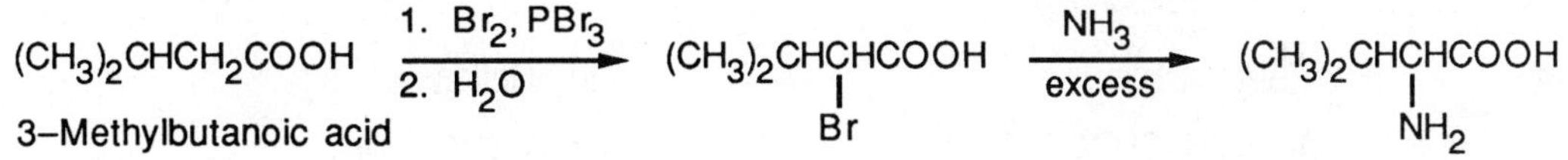

3–Methylbutanoic acid

27.12

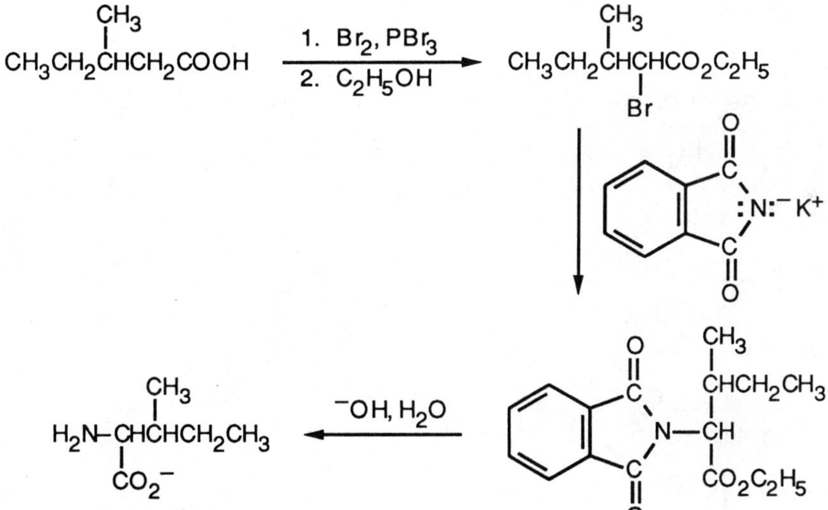

$$\underset{}{\overset{CH_3}{CH_3CH_2CHCH_2COOH}} \xrightarrow[\text{2. } C_2H_5OH]{\text{1. } Br_2, PBr_3} \underset{Br}{\overset{CH_3}{CH_3CH_2CHCHCO_2C_2H_5}}$$

$$\underset{\text{Isoleucine}}{\overset{CH_3}{H_2N-CHCHCH_2CH_3}} \xleftarrow[]{^-OH,\ H_2O} $$

27.13

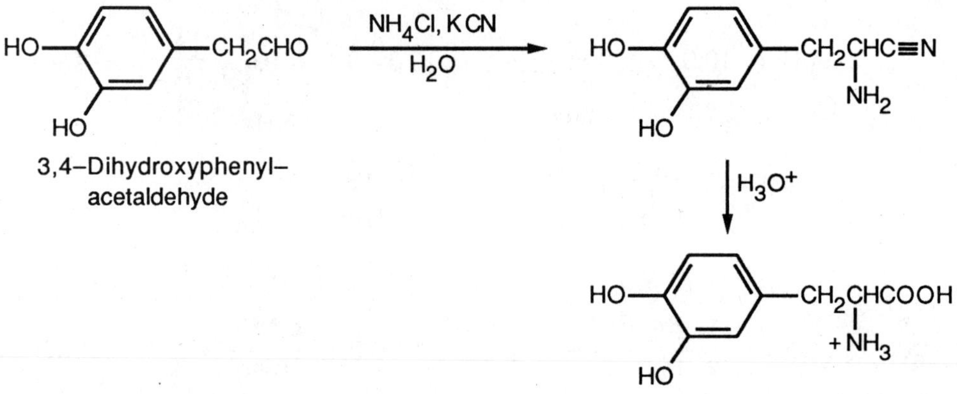

3,4–Dihydroxyphenyl–
acetaldehyde

NH_4Cl, KCN / H_2O

H_3O^+

L–Dopa

27.14

Amino Acid	Halide
a) $(CH_3)_2CHCH_2CHCOOH$ with NH_2 below — Leucine	$(CH_3)_2CHCH_2Br$
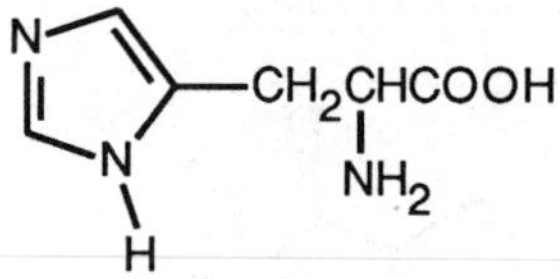 b) Histidine: imidazole ring–$CH_2CHCOOH$ with NH_2	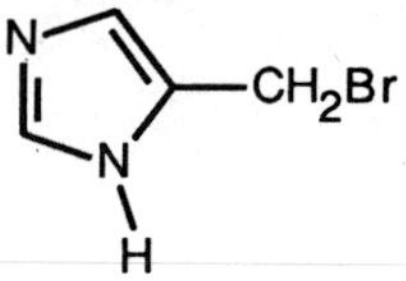 imidazole ring–CH_2Br
c) Tryptophan: indole ring–$CH_2CHCOOH$ with NH_2	indole ring–CH_2Br
d) $CH_3SCH_2CH_2CHCOOH$ with NH_2 below — Methionine	$CH_3SCH_2CH_2Br$

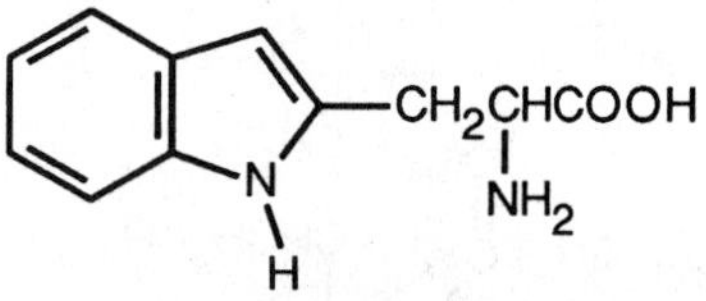

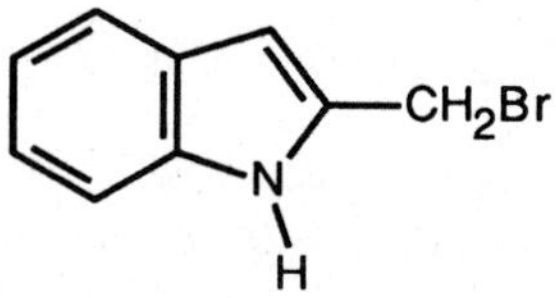

27.15

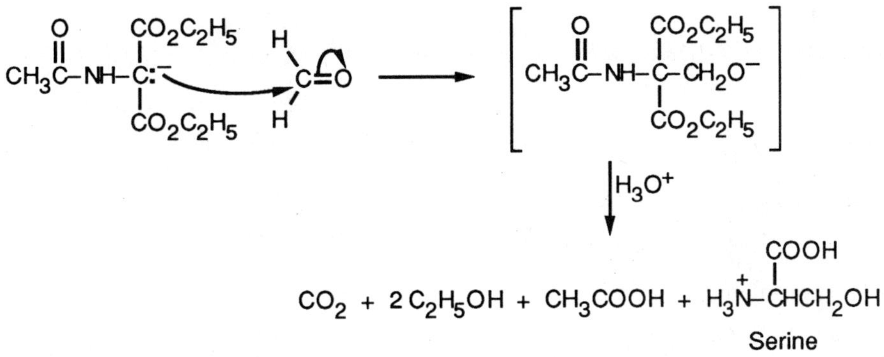

$$CO_2 + 2\,C_2H_5OH + CH_3COOH + \overset{+}{H_3N}-\overset{\displaystyle COOH}{\underset{}{CHCH_2OH}}$$

Serine

The synthesis of serine is a variation of the Knoevenagel reaction (Sec. 23.8).

27.16 H–Val–Tyr–Gly–OH   H–Tyr–Gly–Val–OH   H–Gly–Val–Tyr–OH

H–Val–Gly–Tyr–OH   H–Tyr–Val–Gly–OH   H–Gly–Tyr–Val–OH

27.17

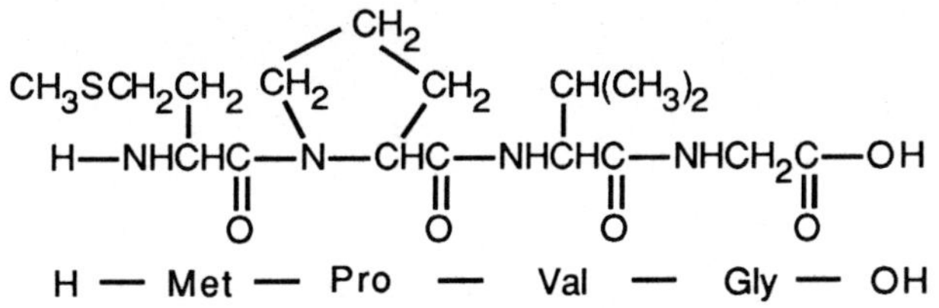

H — Met — Pro — Val — Gly — OH

27.18

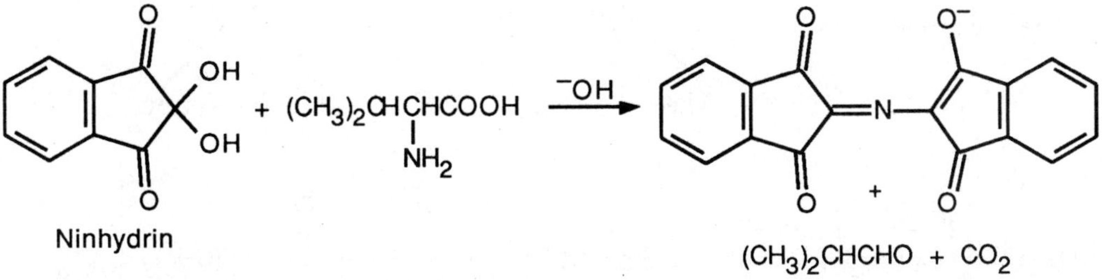

Ninhydrin

$(CH_3)_2CHCHO + CO_2$

27.19 Only primary amines can form the extensively conjugated purple ninhydrin product. A secondary amine such as proline yields a product containing a shorter system of conjugated bonds, which absorbs at a shorter wavelength (440 nm vs. 570 nm).

27.20 Trypsin cleaves peptide bonds at the carboxyl side of *lysine* and *arginine*.

Chymotrypsin cleaves peptide bonds at the carboxyl side of *phenylalanine, tyrosine* and *tryptophan*.

$$\text{H–Asp–Arg–Val–Tyr–Ile–His–Pro–Phe–OH} \xrightarrow{\text{Trypsin}} \text{H–Asp–Arg–OH} \quad + \quad \text{H–Val–Tyr–Ile–His–Pro–Phe–OH}$$

$$\xrightarrow{\text{Chymotrypsin}} \text{H–Asp–Arg–Val–Tyr–OH} \quad + \quad \text{H–Ile–His–Pro–Phe–OH}$$

27.21 a) H–Arg–Pro–OH

H–Pro–Leu–Gly–OH

H–Gly–Ile–Val–OH

The complete sequence:

H–Arg–Pro–Leu–Gly–Ile–Val–OH

b) H–Val–Met–Trp–OH

H–Trp–Asp–Val–OH

H–Val–Leu–OH

The complete sequence:

H–Val–Met–Trp–Asp–Val–Leu–OH

27.22 H–Pro–Leu–Gly–OH

H–Gly–Pro–Arg–OH

H–Arg–Pro–OH

The complete sequence:

H–Pro–Leu–Gly–Pro–Arg–Pro–OH

27.23   The tripeptide is cyclic.

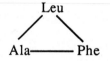

27.24

$$(CH_3)_3CO-\overset{:O:}{\underset{}{\overset{\|}{C}}}-O-\overset{O}{\underset{}{\overset{\|}{C}}}OC(CH_3)_3 \;+\; NH_2CHCOOH \;\rightleftharpoons\;$$

$$\left[(CH_3)_3CO-\overset{:O:^-}{\underset{}{\overset{}{C}}}-O-\overset{O}{\underset{}{\overset{\|}{C}}}-OC(CH_3)_3 \;+\; NH_2CHRCOOH \right.$$

$$\updownarrow (CH_3CH_2)_3N:$$

$$\left. (CH_3)_3CO-\overset{:O:^-}{\underset{}{\overset{}{C}}}-O-\overset{O}{\underset{}{\overset{\|}{C}}}-OC(CH_3)_3 \;\;\; NHCHRCOO^- \;+\; (CH_3CH_2)_3\overset{+}{N}H \right]$$

$$\longleftarrow \; (CH_3)_3CO-\overset{O}{\underset{}{\overset{\|}{C}}}-NHCHCOO^- \;+\; CO_2 \;+\; HOC(CH_3)_3 \quad (R)$$

27.25

$$Leu = \begin{matrix} CH(CH_3)_2 \\ | \\ CH_2 \\ | \\ H_2N-CH-COOH \end{matrix} \qquad R = \begin{matrix} CH(CH_3)_2 \\ | \\ CH_2 \\ | \\ \end{matrix}$$

1. Protect the amino group of leucine.

$$(CH_3)_3COCOCOC(CH_3)_3 + H_2NCHCOOH \xrightarrow{(C_2H_5)_3N} (CH_3)_3COC-NH-CHCOOH$$
$$\text{Leu} \qquad\qquad + CO_2 + (CH_3)_3COH$$

2. Protect the carboxylic acid group of alanine.

$$\underset{\text{Ala}}{H_2NCHCOOH} + CH_3OH \xrightarrow[\text{catalysis}]{\text{acid}} H_2NCHCOOCH_3 + H_2O$$

with CH_3 groups on the α-carbons.

3. Couple the protected acids with DCC.

$$(CH_3)_3COCNHCHCOOH + H_2N-CHCOOCH_3 + \text{[N=C=N with two cyclohexyl groups]}$$

$$\downarrow$$

$$(CH_3)_3COCNHCHC-NHCHCOOCH_3 + \text{[NH-C-NH with two cyclohexyl groups, C=O]}$$

4. Remove the leucine protecting group.

$$(CH_3)_3COCNHCHC-NHCHCOOCH_3 \xrightarrow{CF_3COOH} H_3NCHC-NHCHCOOCH_3$$
$$+ (CH_3)_2C=CH_2 + CO_2$$

5. Remove the alanine protecting group.

$$H_3NCHC-NHCHCOOCH_3 \xrightarrow[\text{2. } H_3O^+]{\text{1. } ^-OH, H_2O} H_3NCHC-NHCHCOOH + CH_3OH$$

$$H-Leu-Ala-OH$$

27.26

a)

1. H–Leu–Ala–OH + $(CH_3)_3CO\overset{\overset{O}{\|}}{C}O\overset{\overset{O}{\|}}{C}OC(CH_3)_3$ $\xrightarrow{(CH_3CH_2)_3N}$ BOC–Leu–Ala–OH
    (Prob. 27.25)

2. H–Gly–OH + CH_3OH $\xrightarrow{H^+}$ H–Gly–OCH$_3$

3. BOC–Leu–Ala–OH + H–Gly–OCH$_3$ $\xrightarrow{DCC}$ BOC–Leu–Ala–Gly–OCH$_3$

4. BOC–Leu–Ala–Gly–OCH$_3$ $\xrightarrow{CF_3COOH}$ H–Leu–Ala–Gly–OCH$_3$

5. H–Leu–Ala–Gly–OCH$_3$ $\xrightarrow[\text{2. } H_3O^+]{\text{1. } ^-OH, H_2O}$ H–Leu–Ala–Gly–OH

b)

1. H–Gly–OH + $(CH_3)_3CO\overset{\overset{O}{\|}}{C}O\overset{\overset{O}{\|}}{C}OC(CH_3)_3$ $\xrightarrow{(CH_3CH_2)_3N}$ BOC–Gly–OH

2. BOC–Gly–OH + H–Leu–Ala–OCH$_3$ $\xrightarrow{DCC}$ BOC–Gly–Leu–Ala–OCH$_3$
    (Prob. 27.25, part 4)

3. BOC–Gly–Leu–Ala–OCH$_3$ $\xrightarrow{CF_3COOH}$ H–Gly–Leu–Ala–OCH$_3$

4. H–Gly–Leu–Ala–OCH$_3$ $\xrightarrow[\text{2. } H_3O^+]{\text{1. } ^-OH, H_2O}$ H–Gly–Leu–Ala–OH

27.27 A proline residue in a polypeptide chain interrupts α–helix formation. The amide nitrogen of proline has no hydrogen that can contribute to the hydrogen-bonded structure of an α–helix. In addition, the pyrrolidine ring of proline restricts rotation about the C–N bond and reduces flexibility in the polypeptide chain.

27.28 a) Pyruvate decarboxylase is a lyase.

b) Chymotrypsin is a hydrolase.

c) Alcohol dehydrogenase is an oxidoreductase.

27.29

```
     COOH                    COOH
      |                       |
 H----+----NH2          H----+----NH2
      |                       |
    CH2OH                    CH3
  (R)–Serine              (R)–Alanine
```

27.30

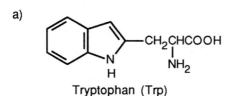

(S)– Proline

27.31

a)

-CH₂CHCOOH

Tryptophan (Trp)

b)

CH₃CH₂CHCHCOOH

Isoleucine (Ile)

c) HS CH₂CHCOOH
         NH₂

Cysteine (Cys)

d)

-CH₂CHCOOH

Histidine (His)

27.32

a) HO— ⟨ ⟩ —CH₂CHCO⁻
                    ⁺NH₃

Tyrosine

b) CH₃CHCHCO⁻
      OH O
          ⁺NH₃

Threonine

27.33 Water is a polar solvent; chloroform is a non-polar solvent. Since charged species are less stable in non-polar solvents than in polar solvents, amino acids exist as the non-ionic amino carboxylic acid form in chloroform.

27.34 *Amino Acid   Isoelectric point*

Amino Acid	Isoelectric point
Histidine	7.6
Serine	5.7
Glutamic acid	3.2

The optimum pH for the electrophoresis of three amino acid occurs at the isoelectric point of the amino acid intermediate in acidity. At this pH one amino acid migrates toward the cathode (the least acidic), one migrates toward the anode (the most acidic), and the amino acid intermediate in acidity does not migrate. In this example, electrophoresis at pH = 5.7 allows the maximum separation of the three amino acids.

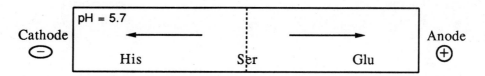

27.35 a) *Amphoteric* compounds can react either as acids or as bases, depending on the circumstances.

b) The *isoelectric point* is the pH at which a solution of an amino acid or protein is electronically neutral.

c) A *zwitterion* is a compound that contains both positively charged and negatively charged portions.

27.36 a) H–Val–Leu–Ser–OH   H–Ser–Val–Leu–OH

H–Val–Ser–Leu–OH   H–Leu–Val–Ser–OH

H–Ser–Leu–Val–OH   H–Leu–Ser–Val–OH

b) H–Ser–Leu–Leu–Pro–OH   H–Leu–Leu–Ser–Pro–OH

H–Ser–Leu–Pro–Leu–OH   H–Leu–Leu–Pro–Ser–OH

H–Ser–Pro–Leu–Leu–OH   H–Leu–Ser–Leu–Pro–OH

H–Pro–Leu–Leu–Ser–OH   H–Leu–Ser–Pro–Leu–OH

H–Pro–Leu–Ser–Leu–OH   H–Leu–Pro–Leu–Ser–OH

H–Pro–Ser–Leu–Leu–OH   H–Leu–Pro–Ser–Leu–OH

27.37   100 g of Cytochrome C contains 0.43 g iron.

100 g of Cytochrome C contains:

$$\frac{0.43 \text{ g Fe}}{55.8 \text{ g/mol Fe}} = 0.0077 \text{ moles Fe}$$

$$\frac{100 \text{ g Cytochrome C}}{0.0077 \text{ moles Fe}} = \frac{X \text{ g Cytochrome C}}{1 \text{ mole Fe}}$$

13,000 g/mol Fe = X

Cytochrome C has a minimum molecular weight of 13,000.

27.38

a) (CH$_3$)$_2$CHCH–COO$^-$ $\xrightarrow{\text{CH}_3\text{CH}_2\text{OH, H}^+}$ (CH$_3$)$_2$CHCHCOOCH$_2$CH$_3$
    | |
    $^+$NH$_3$ $^+$NH$_3$
    L–Valine

b) (CH$_3$)$_2$CHCH–COO$^-$ $\xrightarrow[\text{(CH}_3\text{CH}_2)_3\text{N:}]{\overset{\overset{\text{O}}{\|}\;\overset{\text{O}}{\|}}{\text{(CH}_3)_3\text{OCOCOC(CH}_3)_3}}$ (CH$_3$)$_2$CHCHCOO$^-$
    | |
    $^+$NH$_3$ NHCOC(CH$_3$)$_3$
                                                                                  $\|$
                                                                                  O

c) $(CH_3)_2CHCH-COO^-$ $\xrightarrow{\text{KOH, H}_2\text{O}}$ $(CH_3)_2CHCHCOO^-$
            | |
          $+NH_3$ NH_2

d) $(CH_3)_2CHCH-COO^-$ $\xrightarrow[\text{2. H}_2\text{O}]{\text{1. CH}_3\text{COCl, Pyr}}$ $(CH_3)_2CHCH-COO^-$
            | |
          $+NH_3$ $NHCOCH_3$

27.39

a)

H — Val — Phe — Cys — Ala — OH

b)

H — Glu — Pro — Ile — Leu — OH

27.40

1. $(CH_3)_3COCOCOC(CH_3)_3$ + H–Val–OH $\xrightarrow{(CH_3CH_2)_3N\text{:}}$ $(CH_3)_3COC$–Val–OH
                                                                    (BOC–Val–OH)

2.   BOC–Val–OH + Cl–CH$_2$–(Polymer) $\longrightarrow$ BOC–Val–OCH$_2$–(Polymer)

3.   BOC–Val–OCH$_2$–(Polymer) $\xrightarrow[\text{2. CF}_3\text{COOH}]{\text{1. Wash}}$ H–Val–OCH$_2$–(Polymer)

4.   BOC–Ala–OH + H–Val–OCH$_2$–(Polymer) $\xrightarrow[\text{2. Wash}]{\text{1. DCC}}$ BOC–Ala–Val–OCH$_2$–(Polymer)

5.   BOC–Ala–Val–OCH$_2$–(Polymer) $\xrightarrow{\text{CF}_3\text{COOH}}$ H–Ala–Val–OCH$_2$–(Polymer)

6.   BOC–Phe–OH + H–Ala–Val–OCH$_2$–(Polymer)

        | 1. DCC
        | 2. Wash
        ↓

   BOC–Phe–Ala–Val–OCH$_2$–(Polymer)

7. BOC–Phe–Ala–Val–OCH$_2$–(Polymer)

$\downarrow$ HF

H–Phe–Ala–Val–OH   +   HOCH$_2$–(Polymer)

27.41

Peptide	$\xrightarrow{\text{PITC}}$	Phenylthiohydantoin	Shortened Peptide
a)	H–Ile–Leu–Pro–Phe–OH	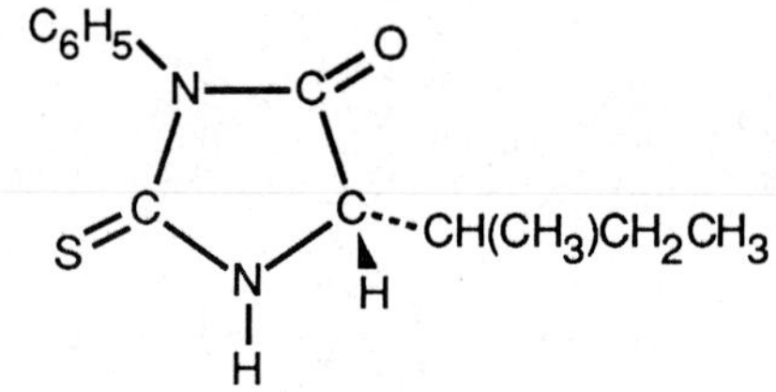	H–Leu–Pro–Phe–OH
b)	H–Asp–Thr–Ser–Gly–Ala–OH	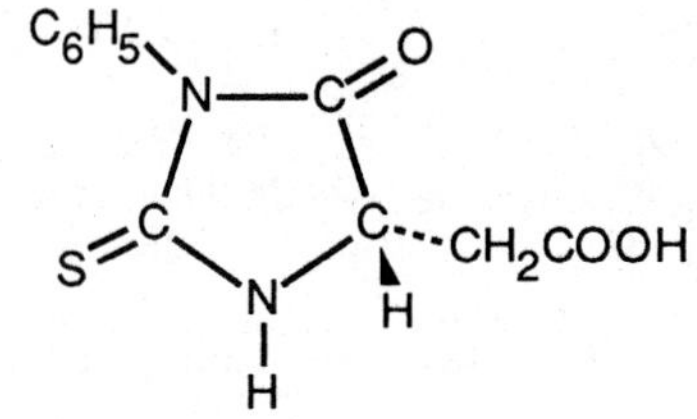	H–Thr–Ser–Gly–Ala–OH

27.42

CH$_3$OCH$_2$Cl + SnCl$_4$ $\rightleftharpoons$ CH$_3\overset{+}{O}$=CH$_2$SnCl$_5^-$

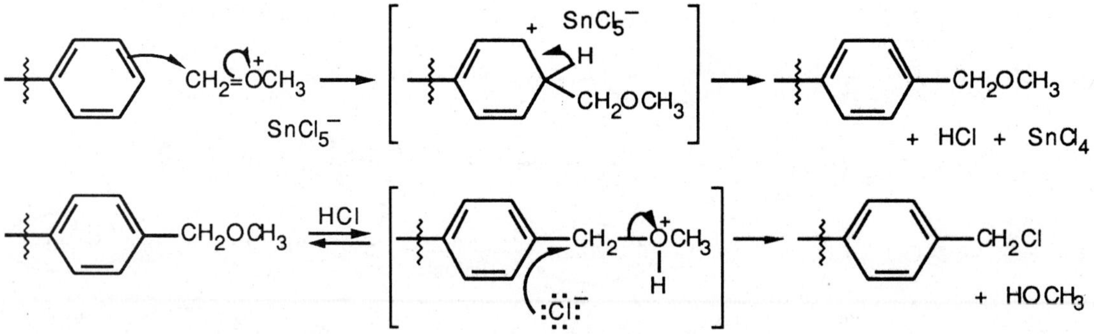

27.43

H—Phe⌇Leu—Met—Lys⦙Tyr⌇Asp—Gly—Gly—Arg⦙Val—Ile—Pro—Tyr—OH

Cleaved by Trypsin = ···········

Cleaved by Chymotrypsin = ⋁⋁⋁⋁⋁

27.44

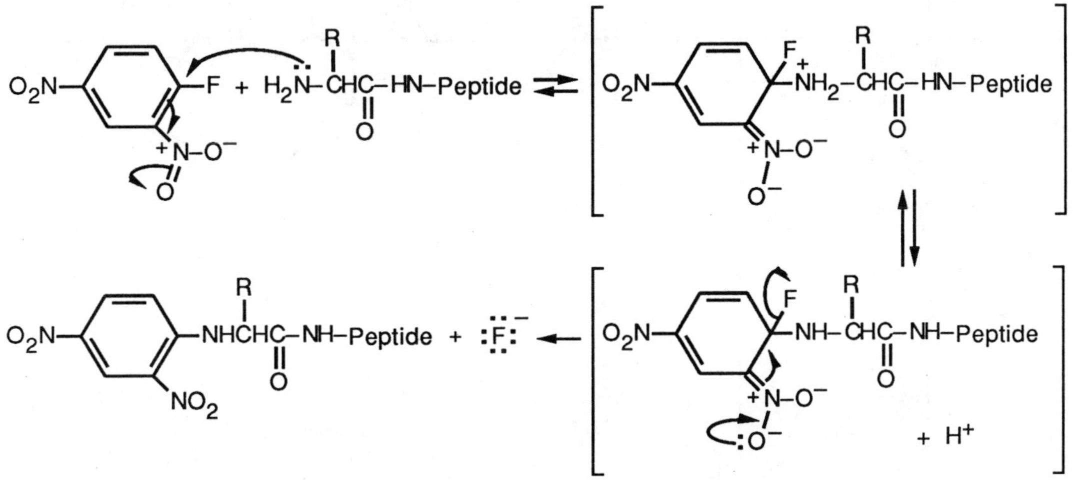

This reaction proceeds via a nucleophilic aromatic substitution mechanism.

27.45 2,4–Dinitrofluorobenzene (Sanger's reagent) reacts not only with the *N*–terminal amino group of a protein, but also with the terminal –NH_2 group of amino acids such as lysine or arginine. In order to use Sanger's reagent for end-group analysis in a protein containing lysine or arginine, these amino acids must first have their basic groups protected.

27.46

a)

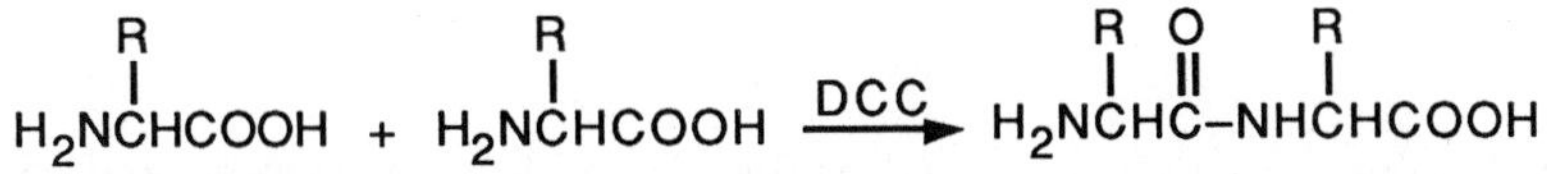

In this sequence of steps, a dipeptide is formed from two equivalents of

$$H_2NCHCOOH$$
(R)

The mechanism is pictured in Fig. 27.8.

b)

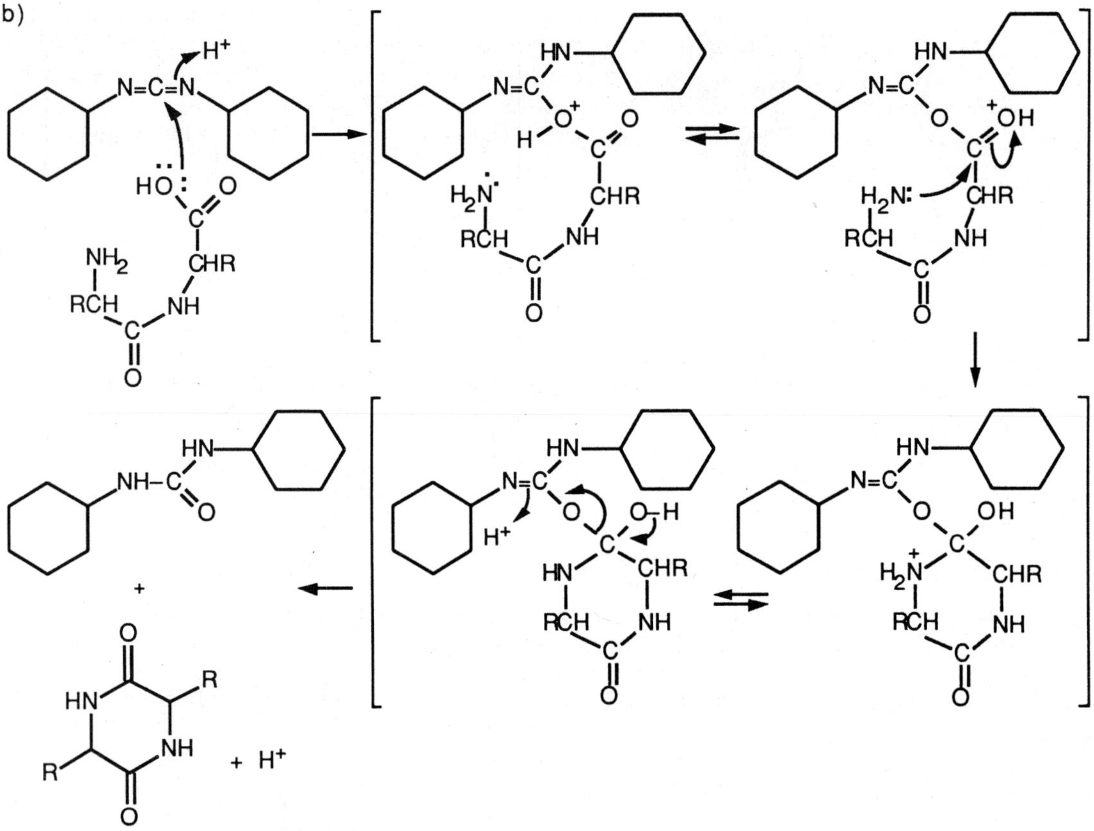

DCC couples the carboxylic acid end of the dipeptide to the amino end to yield the 2,4-diketopiperazine.

27.47

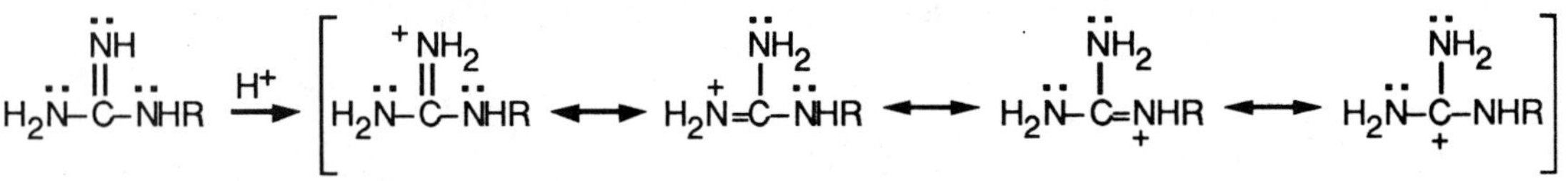

The protonated guanidino group can be stabilized by resonance.

27.48 ^{1}H NMR shows that the two methyl groups of N,N-dimethylformamide are non-equivalent at room temperature. If rotation around the CO–N bond were unrestricted, the methyl groups would be interconvertible, and their ^{1}H NMR absorptions would coalesce into a single signal.

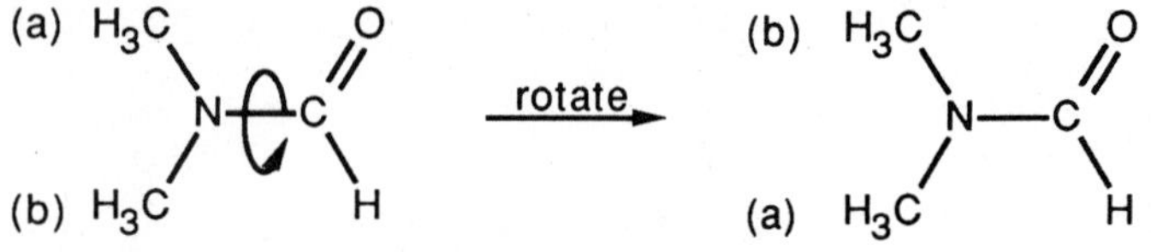

The presence of two absorptions shows that there is a barrier to rotation around the CO–N bond. This barrier is due to the partial double-bond character of the CO–N bond, as indicated by the two resonance forms. Heating to 180° supplies enough energy to allow rapid rotation and to cause the two NMR absorptions to merge.

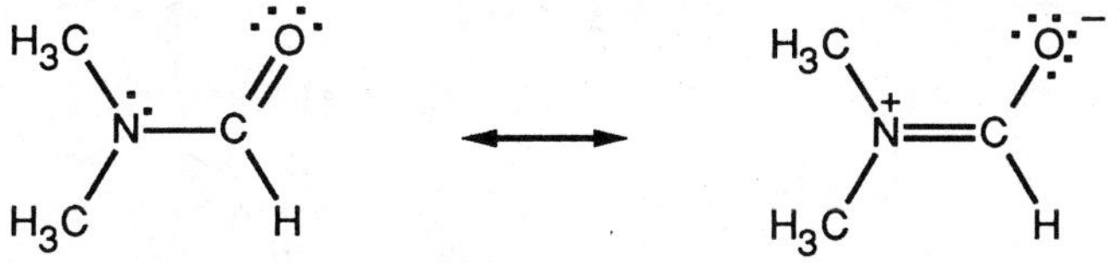

27.49 H–Gly–Gly–Asp–Phe–Pro–Val–Pro–Leu–OH

27.50

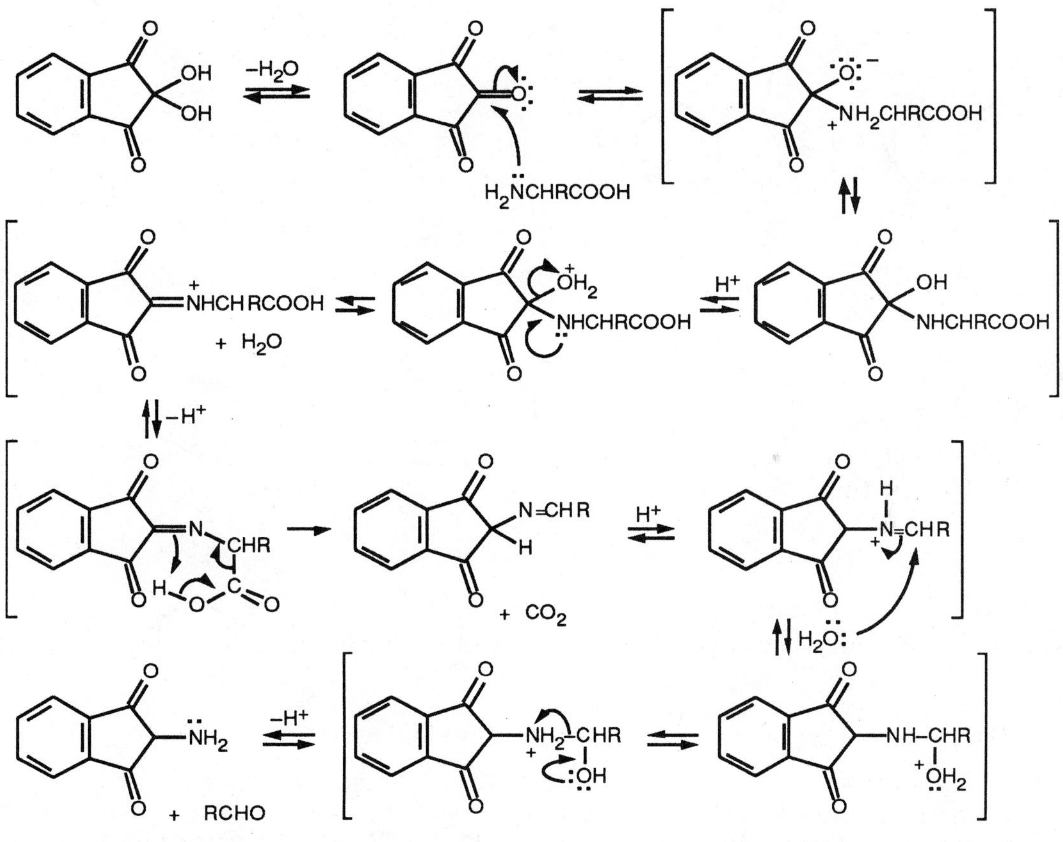

(continued on next page)

27.51

It is also possible to draw many other resonance forms that involve the pi electrons of the aromatic 6-membered rings.

27.52

H
|
Gly
|
Ile
|
Val
|
Glu
|
Gln–CyS–CyS–Thr–Ser–Ile–CyS–Ser–Leu–Tyr⁓Gln–Leu–Glu–Asn–Tyr⁓CyS–Asn–OH

His–Leu–CyS–Gly–Ser–His–Leu–Val–Glu–Ala–Leu–Tyr⁓Leu–Val–CyS
| |
Glu   Gly
| |
Asn   Glu
| |
Val   Arg
| |
Phe   HO–Thr⁝Lys–Pro–Thr⁓Tyr⁓Phe⁓Phe–Gly
|
H

Cleaved by Trypsin = ------------
Cleaved by Chymotrypsin = ⟿⟿⟿

27.53 H–Ser–Ile–Arg–Val–Val–Pro–Tyr–Leu–Arg–OH

27.54 H–Cys–Tyr–Ile–Gln–Asn–Cys–Pro–Leu–Gly–OH
           Reduced Oxytocin   (NH$_2$)

```
        ╱ Asn–Cys–Pro–Leu–Gly–OH
       ╱         |         (NH$_2$)
     Gln         S
      |          |
     Ile         S
       ╲         |
        ╲ Tyr–Cys–H
         Oxidized Oxytocin
```

The C–terminal end of oxytocin is an amide, but this can't be determined from the information given.

27.55

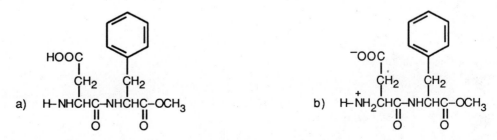

a)

b)

c)

27.56

Study Guide for Chapter 27

After studying this chapter, you should be able to:

(1) Identify the common amino acids and draw them with correct stereochemistry and in dipolar form (27.1, 27.2, 27.3, 27.4, 27.5, 27.6, 27.29, 27.30, 27.31, 27.47).

(2) Understand that acid-base behavior of amino acids (27.7, 27.8, 27.9, 27.10 27.29, 27.30, 27.32, 27.33, 27.52, 27.55).

(3) Synthesize amino acids (27.11, 27.12, 27.13, 27.14, 27.15).

(4) Draw the structure of simple peptides (27.16, 27.17, 27.36, 27.39).

(5) Determine the structure of peptides and proteins (27.19, 27.20, 27.21, 27.22, 27.23, 27.43, 27.49, 27.52, 27.53, 27.54).

(6) Outline the synthesis of peptides (27.25, 27.26, 27.40).

(7) Understand the classification of proteins and the levels of structure of peptides (27.27, 27.28, 27.48).

(8) Draw structures of reaction products of amino acids and peptides (27.18, 27.38, 27.41, 27.44, 27.45, 27.46, 27.51).

28.1

$$CH_3(CH_2)_{18}\overset{\displaystyle O}{\overset{\|}{C}}-O(CH_2)_{31}CH_3$$

28.2

a)

$$CH_2O\overset{\displaystyle O}{\overset{\|}{C}}(CH_2)_{14}CH_3$$
$$|$$
$$CHO\overset{\displaystyle O}{\overset{\|}{C}}(CH_2)_{14}CH_3$$
$$|$$
$$CH_2O\overset{\displaystyle O}{\overset{\|}{C}}(CH_2)_{14}CH_3$$

Glyceryl tripalmitate

b)

$$CH_2O\overset{\displaystyle O}{\overset{\|}{C}}(CH_2)_7CH=CH(CH_2)_7CH_3 \quad (cis)$$
$$|$$
$$CHO\overset{\displaystyle O}{\overset{\|}{C}}(CH_2)_7CH=CH(CH_2)_7CH_3 \quad (cis)$$
$$|$$
$$CH_2O\overset{\displaystyle O}{\overset{\|}{C}}(CH_2)_7CH=CH(CH_2)_7CH_3 \quad (cis)$$

Glyceryl trioleate

Glyceryl tripalmitate is higher melting because it is saturated.

28.3

$$CH_3(CH_2)_7C\equiv C(CH_2)_7COOH \xrightarrow[\text{2. Zn, } CH_3COOH]{\text{1. } O_3} CH_3(CH_2)_7COOH + HOOC(CH_2)_7COOH$$

Stearolic acid   Nonanoic acid   Nonanedioic acid

28.4

$$CH_3(CH_2)_7CH=CH(CH_2)_7\overset{\displaystyle O}{\overset{\|}{C}}O^-Mg^{++}{}^-O\overset{\displaystyle O}{\overset{\|}{C}}(CH_2)_7CH=CH(CH_2)_7CH_3$$

Magnesium oleate

The double bonds are *cis*.

28.5

This is an example of a nucleophilic acyl substitution reaction.

28.6

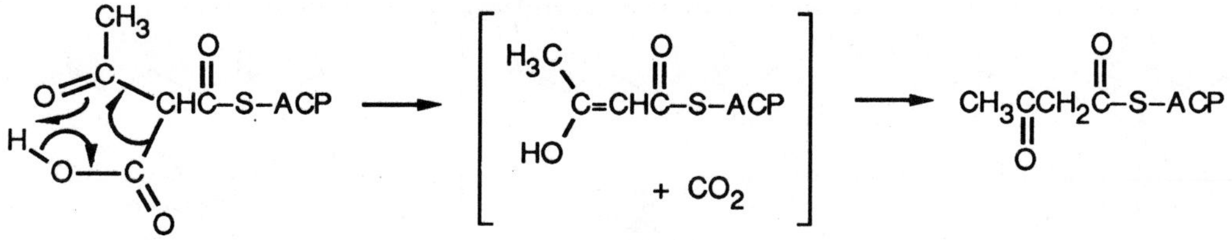

28.7 A fatty acid synthesized from $^{14}CH_3COOH$ has an alternating labeled and unlabeled carbon chain. The carboxylic acid carbon is unlabeled.

$$\overset{*}{C}H_3CH_2\overset{*}{C}H_2CH_2\overset{*}{C}H_2CH_2\overset{*}{C}H_2CH_2\overset{*}{C}H_2CH_2\overset{*}{C}H_2CH_2\overset{*}{C}H_2CH_2COOH$$

28.8

a)

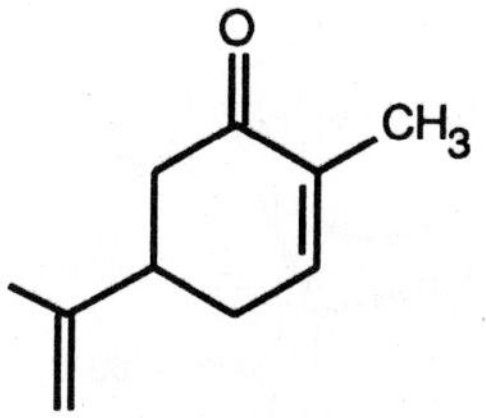

Carvone

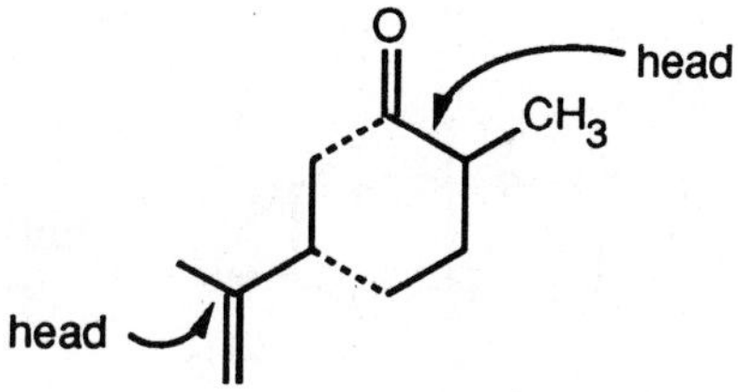

b)

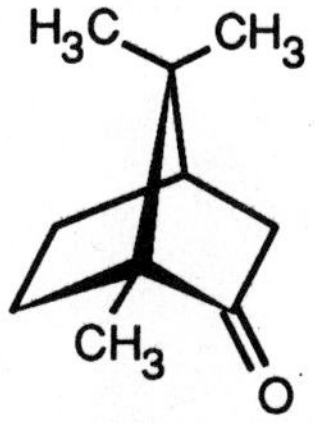

Camphor

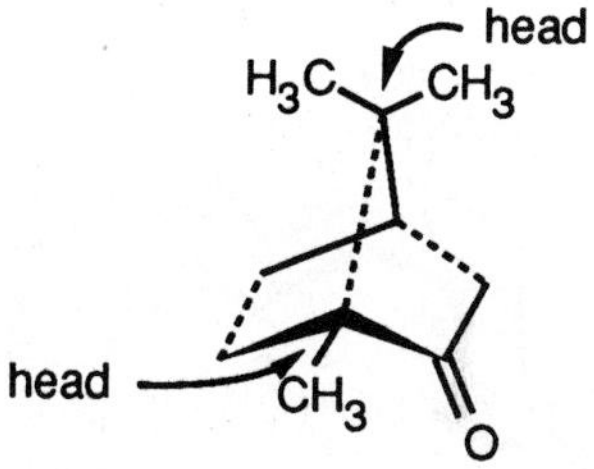

c)

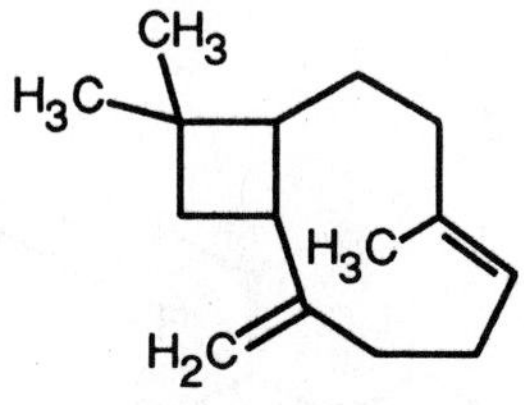

Caryophyllene
(a sesquiterpene)

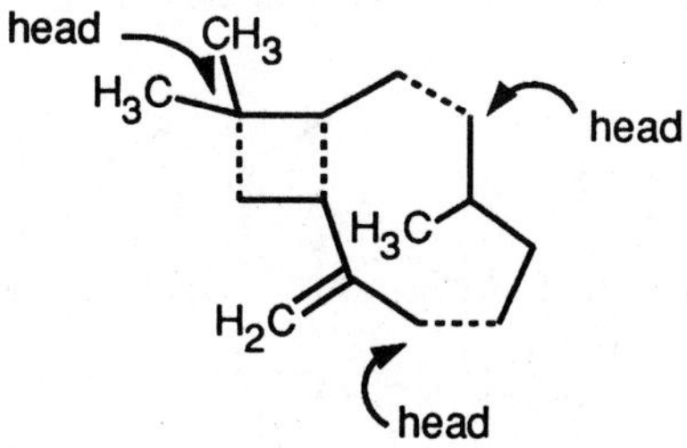

28.9

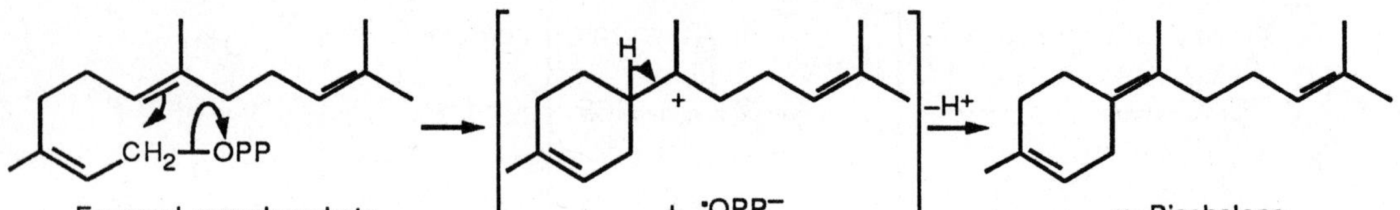

Farnesyl pyrophosphate   + :OPP⁻   γ–Bisabolene

28.10

a)

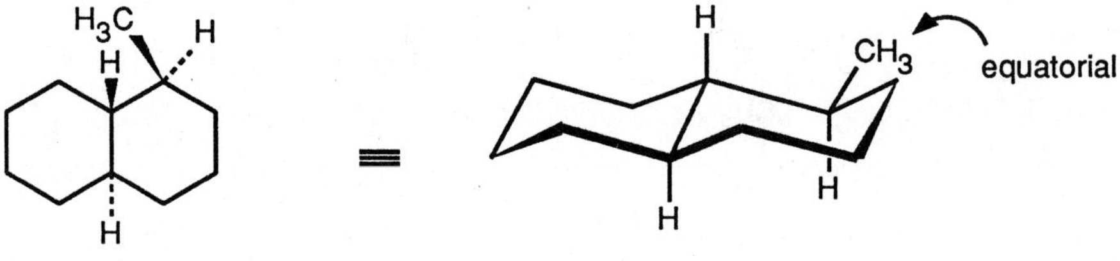

≡   equatorial

b)

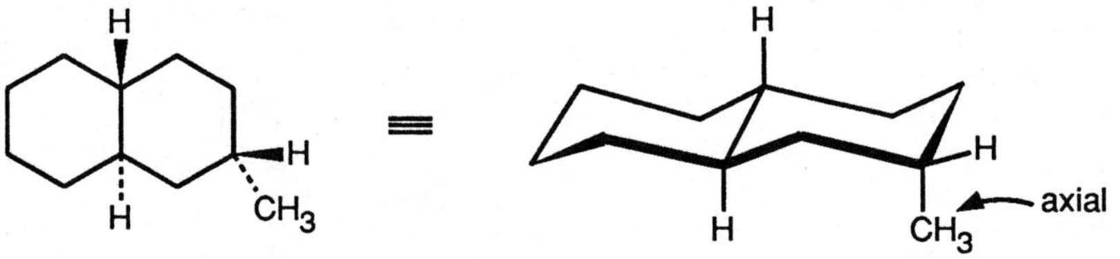

≡   axial

28.11

equatorial

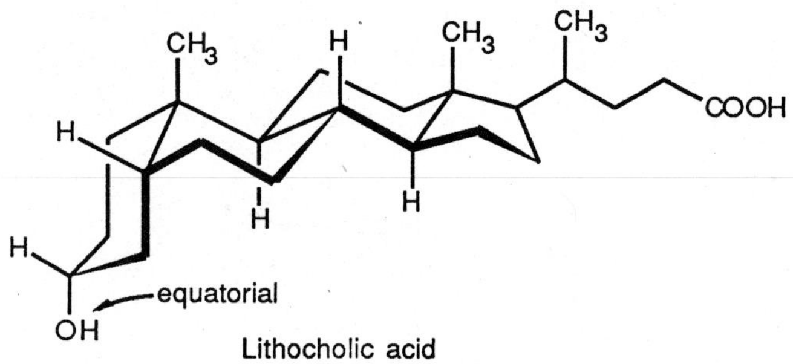

Lithocholic acid

28.12

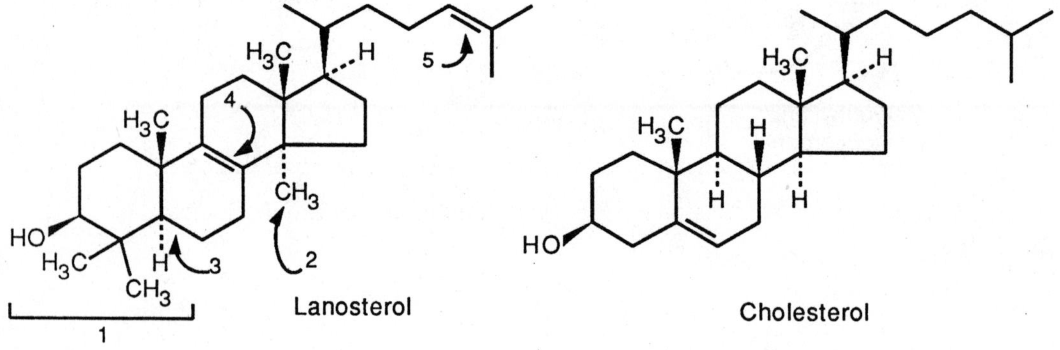

Lanosterol   Cholesterol

Differences between lanosterol and cholesterol:

Lanosterol	*Cholesterol*
1. Two methyl groups at C4.	1. Two hydrogens at C4.
2. One methyl group at C14.	2. One hydrogen at C14.
3. C5–C6 single bond.	3. C5–C6 double bond
4. C8–C9 double bond.	4. C8–C9 single bond.
5. Double bond in side chain	5. Saturated side chain.

28.13

$$CH_2O\overset{O}{\overset{\|}{C}}(CH_2)_{16}CH_3$$
$$CHO\overset{O}{\overset{\|}{C}}(CH_2)_7CH=CH(CH_2)_7CH_3 \quad or$$
$$CH_2O\overset{O}{\overset{\|}{C}}(CH_2)_{16}CH_3$$

optically inactive

$$CH_2O\overset{O}{\overset{\|}{C}}(CH_2)_{16}CH_3$$
$$*CHO\overset{O}{\overset{\|}{C}}(CH_2)_{16}CH_3$$
$$CH_2O\overset{O}{\overset{\|}{C}}(CH_2)_7CH=CH(CH_2)_7CH_3$$

optically active

1. $^-OH, H_2O$
2. H_3O^+

$$CH_2OH$$
$$CHOH + 2\ CH_3(CH_2)_{16}COOH + CH_3(CH_2)_7CH=CH(CH_2)_7COOH$$
$$CH_2OH \qquad \text{Stearic acid} \qquad\qquad \text{Oleic acid}$$

Four different groups are bonded to the central glycerol carbon atom in the optically active fat.

28.14

$$CH_3(CH_2)_{14}\overset{O}{\overset{\|}{C}}-OCH_2(CH_2)_{14}CH_3 \qquad \text{Cetyl Palmitate}$$

28.15 Fats, lecithins, cephalins and plasmalogens are all esters of a glycerol molecule that has carboxylic acid ester groups at C1 and C2. The third group bonded to glycerol, however, differs with the type of lipid.

Lipid	*Functional group at C3 of glycerol*
fat	carboxylic acid ester
cephalin	phosphate ester (also bonded to an amino alcohol)
lecithin	phosphate ester (also bonded to an amino alcohol)
plasmalogen	vinyl ether

28.16

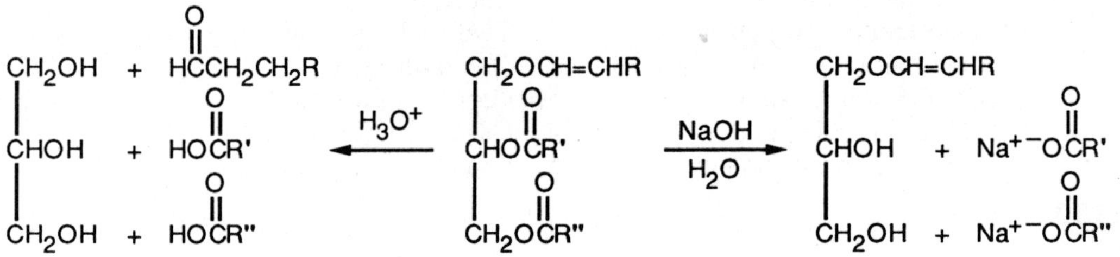

28.17

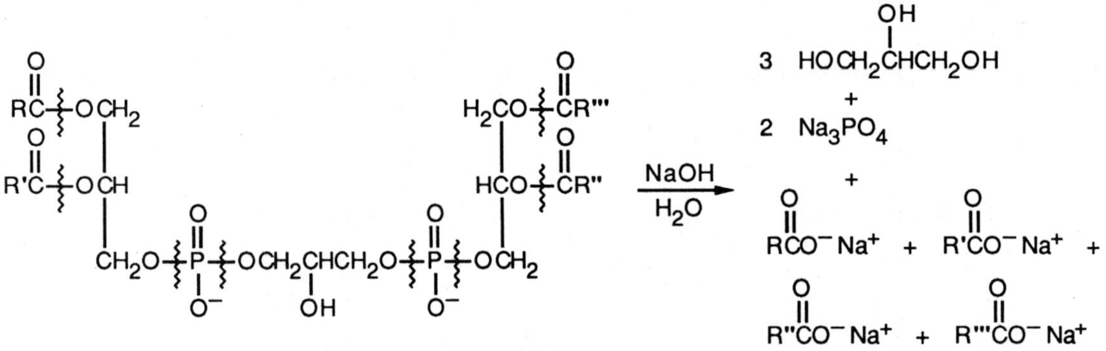

28.18

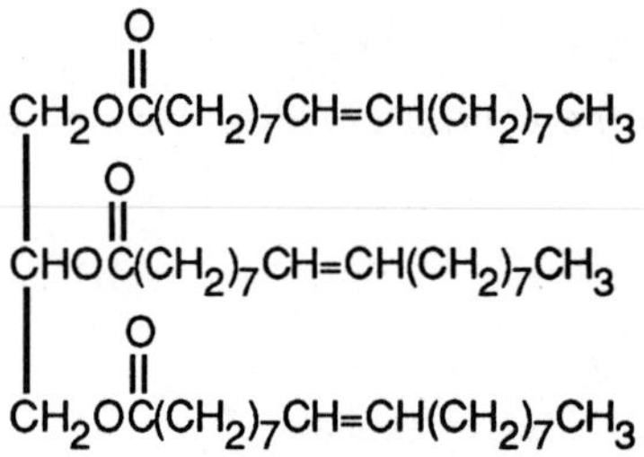

Glyceryl trioleate

a)

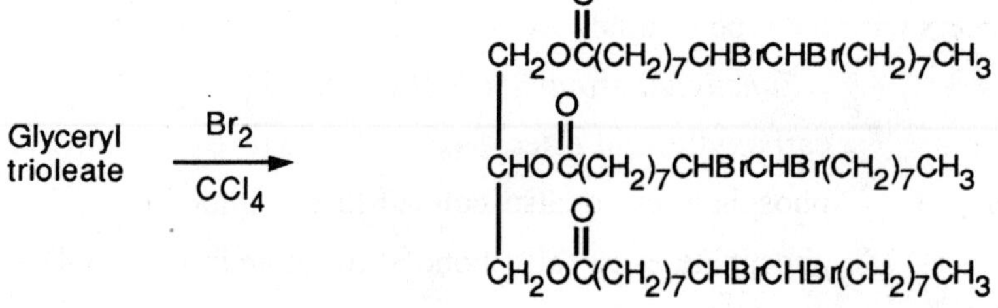

b)

Glyceryl trioleate $\xrightarrow{\text{H}_2\ /\text{Pd}}$

$$
\begin{array}{l}
\text{CH}_2\text{O}\overset{\displaystyle O}{\overset{\|}{\text{C}}}(\text{CH}_2)_{16}\text{CH}_3 \\[1em]
\text{CHO}\overset{\displaystyle O}{\overset{\|}{\text{C}}}(\text{CH}_2)_{16}\text{CH}_3 \\[1em]
\text{CH}_2\text{O}\overset{\displaystyle O}{\overset{\|}{\text{C}}}(\text{CH}_2)_{16}\text{CH}_3
\end{array}
$$

c)

Glyceryl trioleate $\xrightarrow{\text{NaOH, H}_2\text{O}}$

$$
\begin{array}{l}
\text{CH}_2\text{OH} \\
\text{CHOH} \quad + \quad 3\ \text{CH}_3(\text{CH}_2)_7\text{CH=CH}(\text{CH}_2)_7\text{COO}^-\,\text{Na}^+ \\
\text{CH}_2\text{OH}
\end{array}
$$

d)

Glyceryl trioleate $\xrightarrow[\text{2. Zn, CH}_3\text{COOH}]{\text{1. O}_3}$

$$
\begin{array}{l}
\text{CH}_2\text{O}\overset{\displaystyle O}{\overset{\|}{\text{C}}}(\text{CH}_2)_7\overset{\displaystyle O}{\overset{\|}{\text{C}}}\text{H} \\[1em]
\text{CHO}\overset{\displaystyle O}{\overset{\|}{\text{C}}}(\text{CH}_2)_7\overset{\displaystyle O}{\overset{\|}{\text{C}}}\text{H} \quad + \quad 3\ \text{CH}_3(\text{CH}_2)_7\overset{\displaystyle O}{\overset{\|}{\text{C}}}\text{H} \\[1em]
\text{CH}_2\text{O}\overset{\displaystyle O}{\overset{\|}{\text{C}}}(\text{CH}_2)_7\overset{\displaystyle O}{\overset{\|}{\text{C}}}\text{H}
\end{array}
$$

e)

Glyceryl trioleate $\xrightarrow[\text{2. H}_3\text{O}^+]{\text{1. LiAlH}_4}$

$$
\begin{array}{l}
\text{CH}_2\text{OH} \\
\text{CHOH} \quad + \quad 3\ \text{CH}_3(\text{CH}_2)_7\text{CH=CH}(\text{CH}_2)_7\text{CH}_2\text{OH} \\
\text{CH}_2\text{OH}
\end{array}
$$

f)

Glyceryl trioleate $\xrightarrow[\text{2. H}_3\text{O}^+]{\text{1. CH}_3\text{MgBr}}$

$$
\begin{array}{l}
\text{CH}_2\text{OH} \qquad\qquad\qquad\qquad\qquad\qquad \overset{\displaystyle \text{OH}}{} \\
\text{CHOH} \quad + \quad 3\ \text{CH}_3(\text{CH}_2)_7\text{CH=CH}(\text{CH}_2)_7\overset{\displaystyle \text{OH}}{\underset{}{\text{C}}}(\text{CH}_3)_2 \\
\text{CH}_2\text{OH}
\end{array}
$$

g)

Glyceryl trioleate $\xrightarrow[\text{2. CH}_2\text{N}_2]{\text{1. NaOH, H}_2\text{O}}$

$$
\begin{array}{l}
\text{CH}_2\text{OH} \qquad\qquad\qquad\qquad\qquad\qquad\quad \overset{\displaystyle O}{} \\
\text{CHOH} \quad + \quad 3\ \text{CH}_3(\text{CH}_2)_7\text{CH=CH}(\text{CH}_2)_7\overset{\displaystyle O}{\overset{\|}{\text{C}}}\text{OCH}_3 \\
\text{CH}_2\text{OH}
\end{array}
$$

28.19

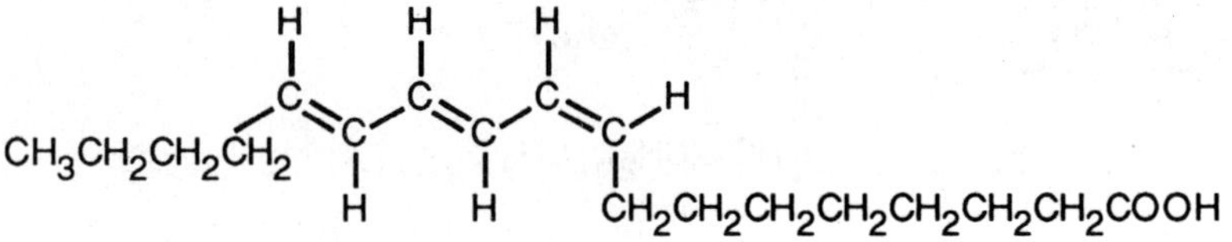

(9Z,11E,13E,)–Octadecatrienoic acid

(Eleostearic acid)

1. O_3
2. Zn, CH_3COOH

$$CH_3CH_2CH_2CH_2\overset{O}{\overset{||}{C}}H + H\overset{O}{\overset{||}{C}}-\overset{O}{\overset{||}{C}}H + H\overset{O}{\overset{||}{C}}-\overset{O}{\overset{||}{C}}H + H\overset{O}{\overset{||}{C}}CH_2CH_2CH_2CH_2CH_2CH_2CH_2COOH$$

The stereochemistry of the double bonds can't be determined from the information given.

28.20

a)

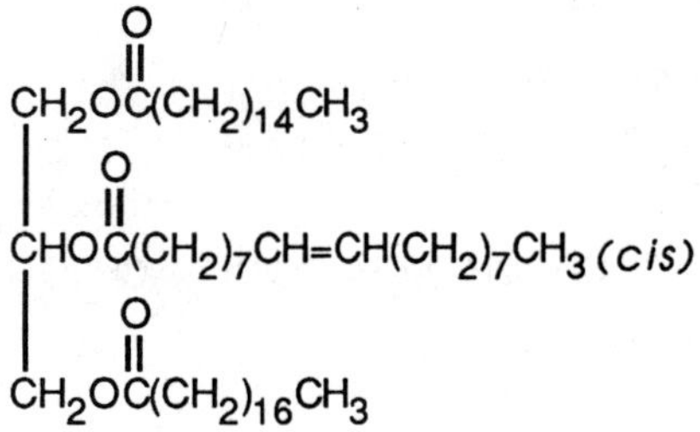

$$CH_2O\overset{O}{\overset{||}{C}}(CH_2)_{14}CH_3$$

$$CHO\overset{O}{\overset{||}{C}}(CH_2)_7CH=CH(CH_2)_7CH_3 \ (cis)$$

$$CH_2O\overset{O}{\overset{||}{C}}(CH_2)_{16}CH_3$$

a fat

b)

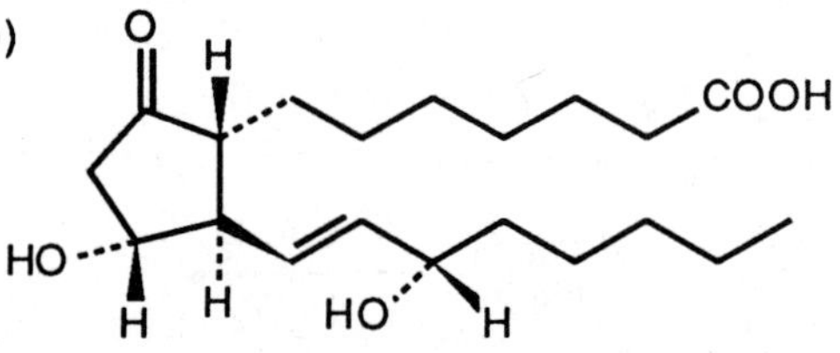

a prostaglandin (prostaglandin E_1)

c)

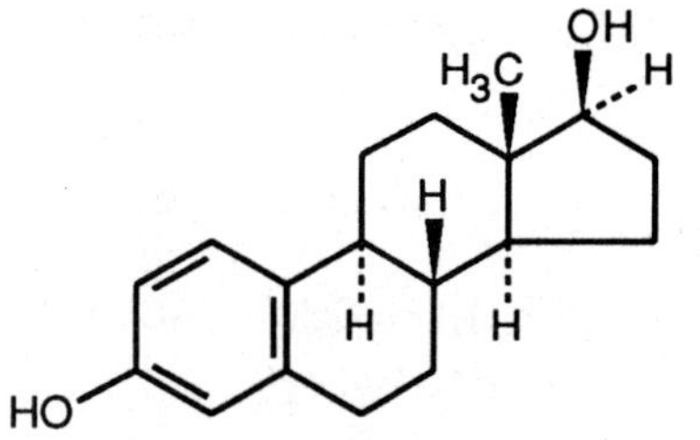

a steroid (Estradiol)

28.21

cis–$CH_3(CH_2)_7CH=CH(CH_2)_7COOH$

Oleic acid

a)   Oleic acid $\xrightarrow{CH_2N_2 \ or \ CH_3OH, \ HCl}$ $CH_3(CH_2)_7CH=CH(CH_2)_7COOCH_3$

Methyl oleate

b) Methyl oleate $\xrightarrow[\text{Pd/C}]{\text{H}_2}$ $CH_3(CH_2)_{16}COOCH_3$
   (part a)   Methyl stearate

c) Oleic acid $\xrightarrow[\text{2. Zn, CH}_3\text{COOH}]{\text{1. O}_3}$ $CH_3(CH_2)_7CHO$ + $OHC(CH_2)_7COOH$

   Nonanal   9–Oxononanoic
                         acid

d) 9–Oxononanoic $\xrightarrow[\text{NH}_4\text{OH}]{\text{Ag}_2\text{O}}$ $HOOC(CH_2)_7COOH$
   acid   Nonanedioic acid
   (part c)

e) Oleic acid $\xrightarrow[\text{CCl}_4]{\text{Br}_2}$ $CH_3(CH_2)_7CHBrCHBr(CH_2)_7COOH$

   $\downarrow$ 1. 3 NaNH$_2$/NH$_3$
   2. H$_3$O$^+$

   $CH_3(CH_2)_7C{\equiv}C(CH_2)_7COOH$
   Stearolic acid

f) Oleic acid $\xrightarrow[\text{Pd/C}]{\text{H}_2}$ $CH_3(CH_2)_{15}CH_2COOH$ $\xrightarrow[\text{2. H}_2\text{O}]{\text{1. Br}_2\text{, PBr}_3}$ $CH_3(CH_2)_{15}CHBrCOOH$

   Stearic acid   2–Bromostearic acid

g) Methyl stearate $\xrightarrow[\text{2. H}_3\text{O}^+]{\text{1. DIBAH}}$ $CH_3(CH_2)_{16}CHO$
   (part b)

   Stearic acid $\xrightarrow[\text{Br}_2]{\text{HgO}}$ $CH_3(CH_2)_{15}CH_2Br$ $\xrightarrow{\text{Mg}}$ $CH_3(CH_2)_{15}CH_2MgBr$

   $CH_3(CH_2)_{16}CHO$ $\xrightarrow[\text{2. H}_3\text{O}^+]{\text{1. CH}_3(CH_2)_{15}CH_2MgBr}$ $CH_3(CH_2)_{16}\overset{\displaystyle OH}{\underset{|}{CH}}(CH_2)_{16}CH_3$

   $\downarrow$ Jones'
   (CrO$_3$, H$_2$O, H$_2$SO$_4$)

   $CH_3(CH_2)_{16}\overset{\displaystyle O}{\overset{||}{C}}(CH_2)_{16}CH_3$
   18–Pentatriacontanone

an alternate method:

$$2 \ CH_3(CH_2)_{16}COOCH_3 \xrightarrow[\text{2. } H_3O^+]{\text{1. } NaOCH_3} CH_3(CH_2)_{16}\overset{\overset{O}{\|}}{C}\overset{|}{\underset{COOCH_3}{C}}H(CH_2)_{15}CH_3 \ + \ HOCH_3$$

Methyl stearate
(part b)

$$\downarrow H_3O^+, \ \Delta$$

$$CH_3(CH_2)_{16}\overset{\overset{O}{\|}}{C}CH_2(CH_2)_{15}CH_3 \ + \ CO_2 \ + \ CH_3OH$$

This synthesis uses a Claisen condensation, followed by a ß–keto ester decarboxylation.

28.22

$$CH_3(CH_2)_7C{\equiv}CH \xrightarrow{NaNH_2} CH_3(CH_2)_7C{\equiv}C{:}^- \ Na^+ \ + \ NH_3$$

$$\downarrow I{-}CH_2(CH_2)_5CH_2Cl$$

$$CH_3(CH_2)_7C{\equiv}C(CH_2)_6CH_2CN \xleftarrow{NaCN} CH_3(CH_2)_7C{\equiv}C(CH_2)_6CH_2Cl \ + \ NaI$$

$$\downarrow H_3O^+$$

$$CH_3(CH_2)_7C{\equiv}C(CH_2)_7COOH$$

Stearolic acid

S_N2 displacement by acetylide occurs at iodine, rather than at chlorine, because I^- is a better leaving group than Cl^-.

28.23

$$\xrightarrow[\substack{\text{2. Zn,}\\ CH_3COOH}]{\text{1. } O_3} CH_3(CH_2)_5\overset{\overset{O}{\|}}{C}H \ + \ H\overset{\overset{O}{\|}}{C}(CH_2)_9COOH$$

Heptanal   11–Oxoundecanoic
acid

Vaccenic acid

$$\xrightarrow[\text{Zn/Cu}]{CH_2I_2}$$

Lactobacillic acid

28.24–28.26 Only a fraction of the possible stereoisomers of these compounds are found in nature or can be synthesized. Some stereoisomers have highly strained ring fusions; others contain high energy 1,3–diaxial interactions.

a)

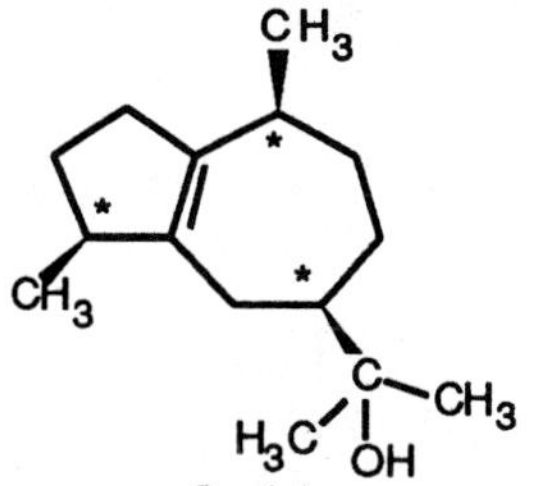

Guaiol
(8 possible stereoisomers)

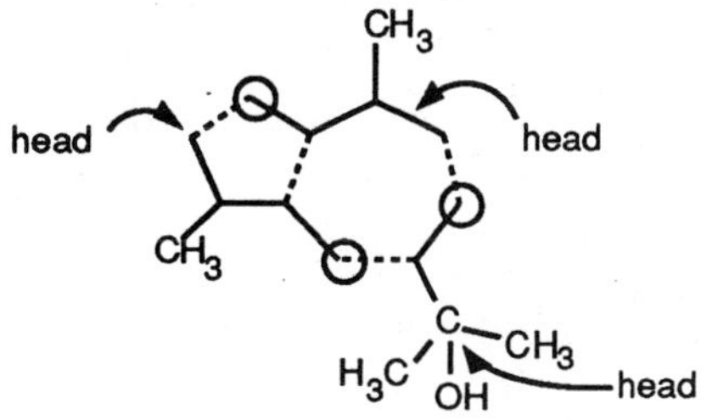

b)

Sabinene
(4 possible stereoisomers)

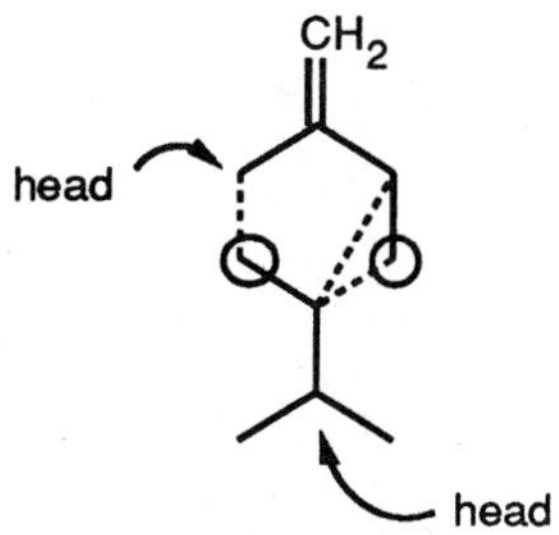

c)

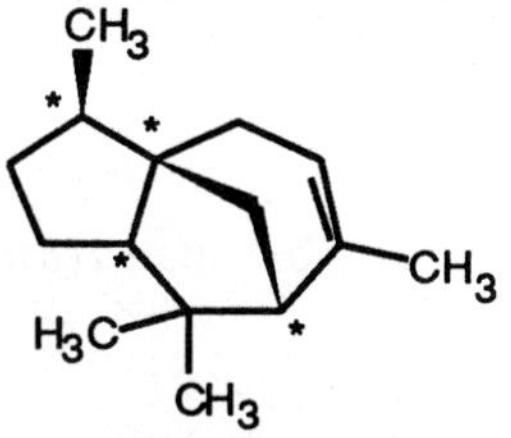

Cedrene
(16 possible stereoisomers)

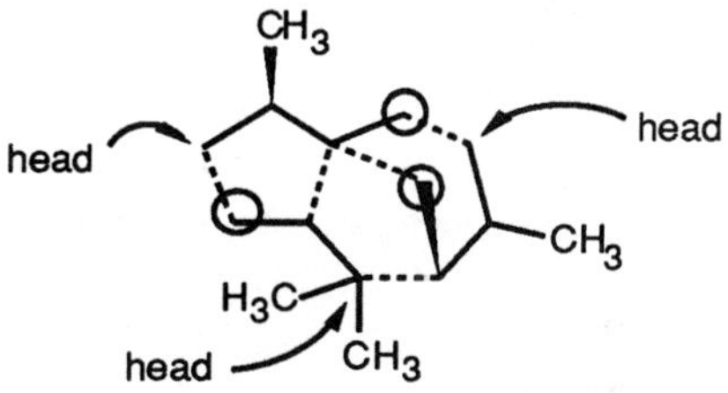

If carbon 1 of each pyrophosphate were isotopically labeled, the labels would appear at the circled positions of the terpenes.

28.27

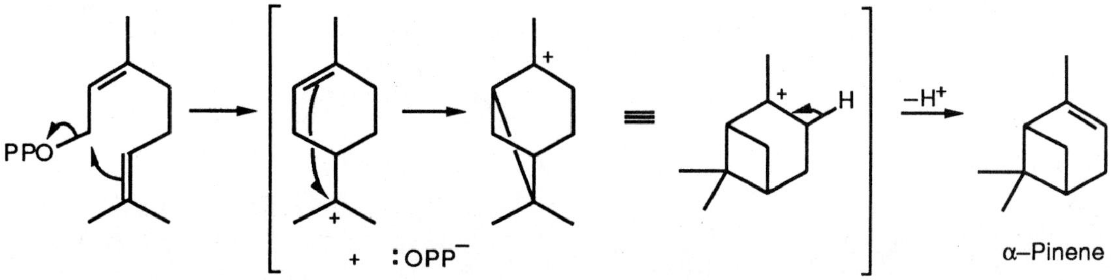

α–Pinene

28.28

ψ–Ionone

β–Ionone

28.29

trans–Decalin

cis–Decalin

Three 1,3–diaxial interactions cause *cis*–decalin to be less stable than *trans*–decalin.

28.30

Dihydrocarvone

28.31

Menthol

28.32–28.33

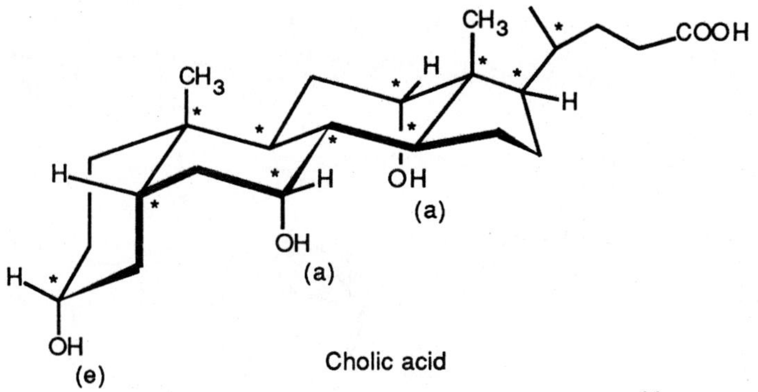

Cholic acid

Cholic acid has eleven stereogenic centers and $2^{11} = 2048$ possible stereoisomers (most of these are very strained).

28.34

a)

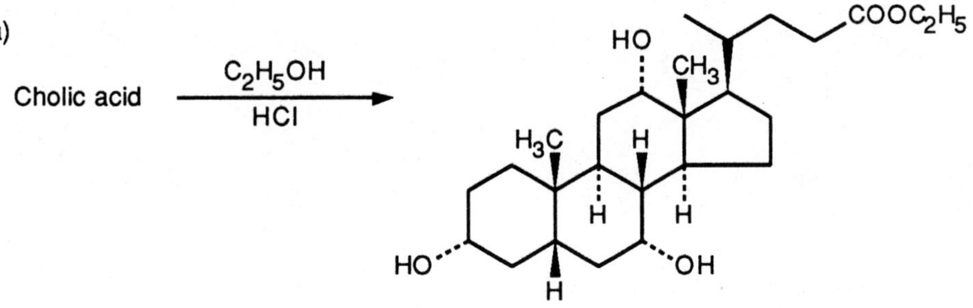

Cholic acid $\xrightarrow[\text{HCl}]{\text{C}_2\text{H}_5\text{OH}}$

b)

Cholic acid $\xrightarrow[\text{CH}_2\text{Cl}_2]{\text{excess PCC}}$

c)

Cholic acid $\xrightarrow[\text{2. H}_3\text{O}^+]{\text{1. BH}_3}$

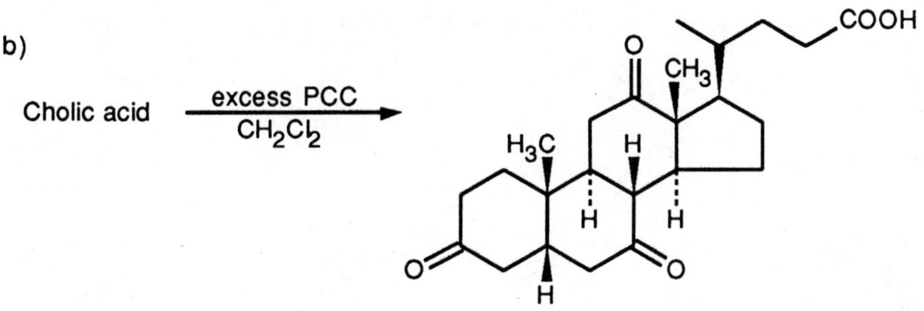

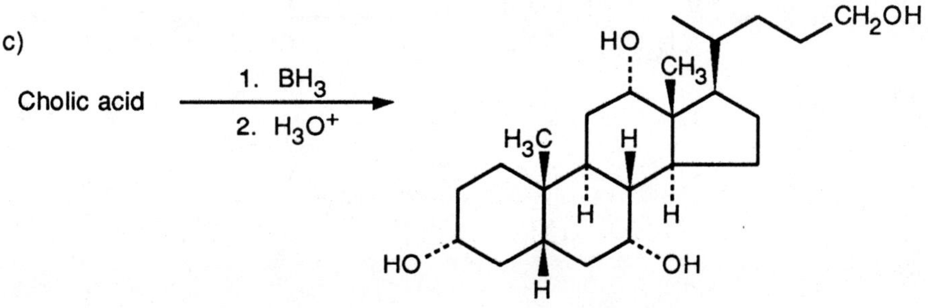

28.35

a)

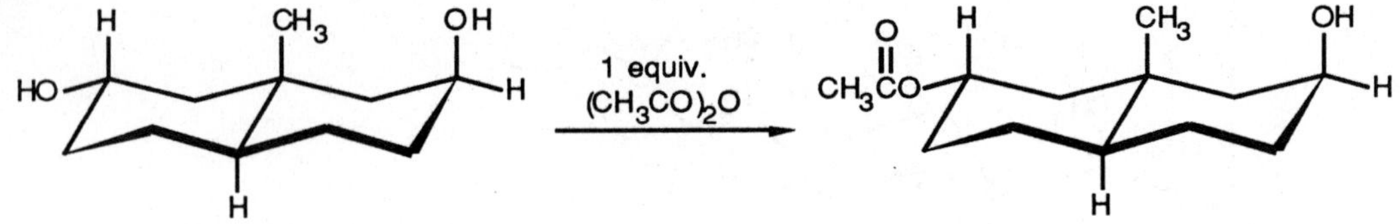

b)

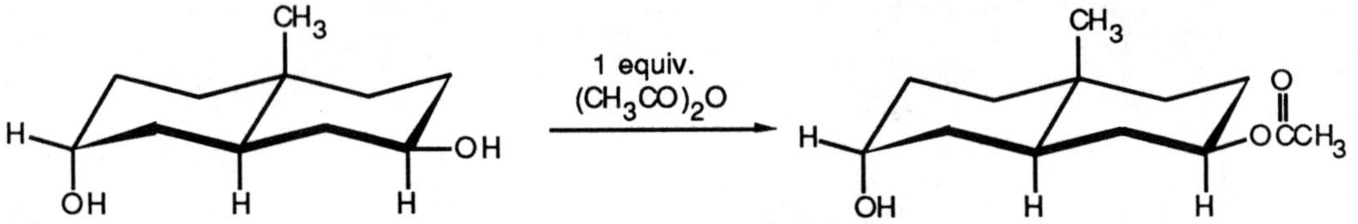

28.36

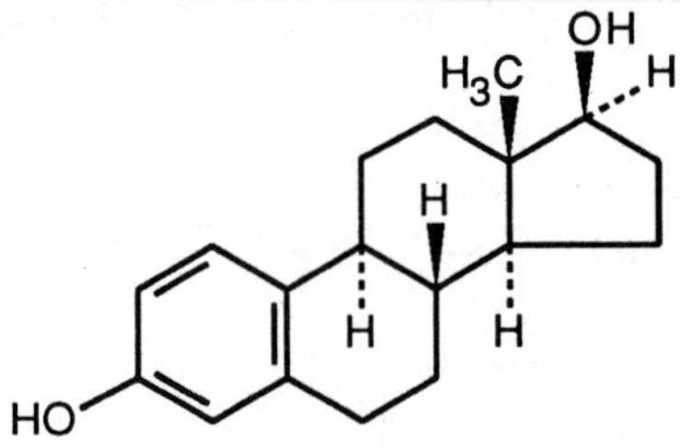

Estradiol

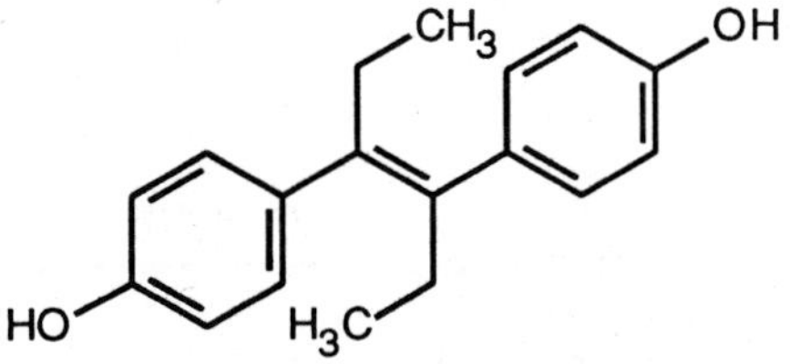

Diethylstilbestrol

28.37

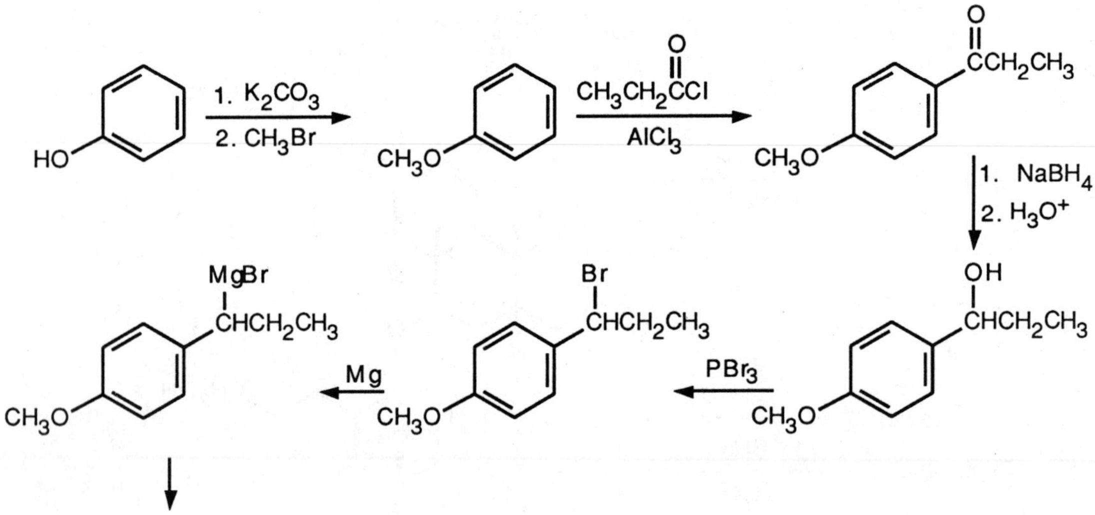

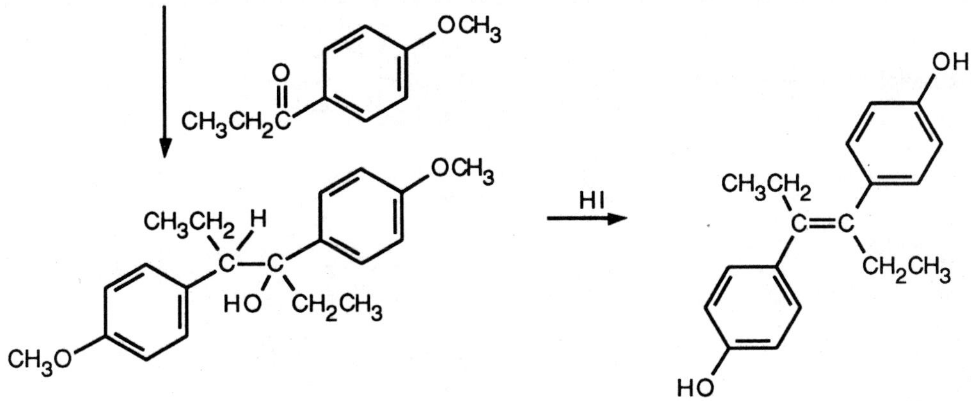

Diethylstilbestrol

28.38

a) 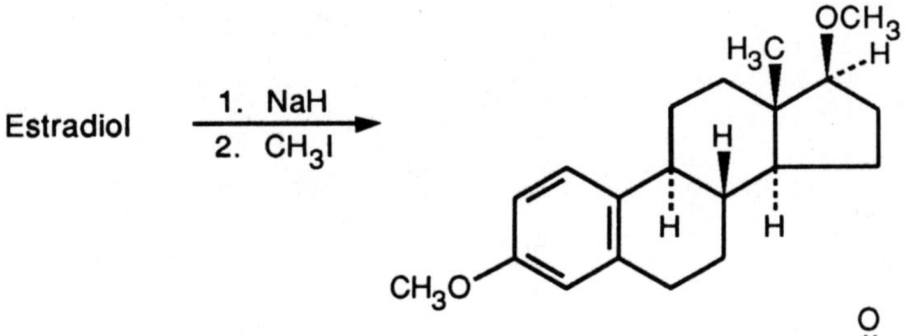   Estradiol $\xrightarrow{\text{1. NaH} \atop \text{2. CH}_3\text{I}}$

b) 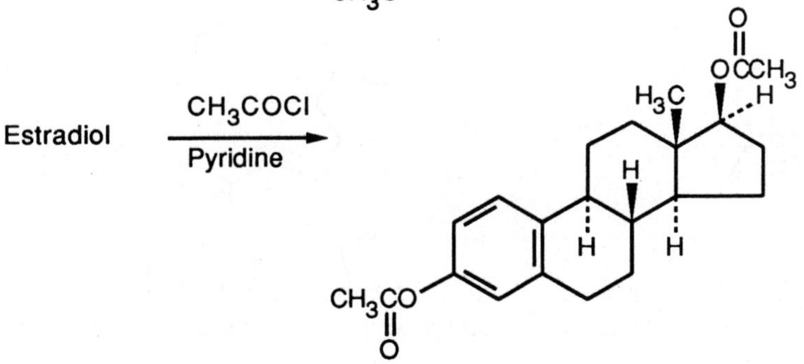   Estradiol $\xrightarrow{\text{CH}_3\text{COCl} \atop \text{Pyridine}}$

c) 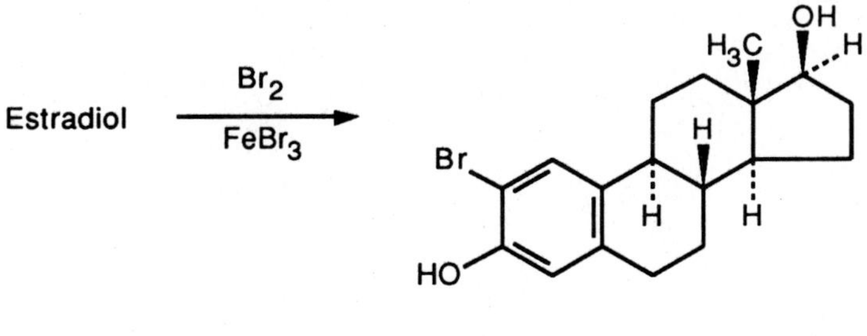   Estradiol $\xrightarrow{\text{Br}_2 \atop \text{FeBr}_3}$

d)   Estradiol $\xrightarrow[\text{CH}_2\text{Cl}_2]{\text{PCC}}$

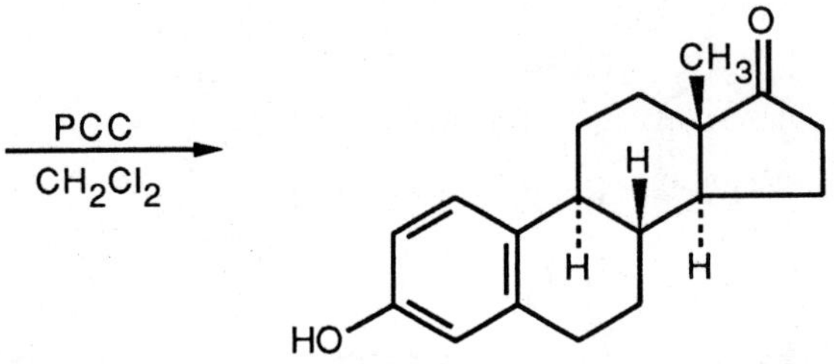

28.39

Cembrene

1 equiv H_2

Dihydrocembrene

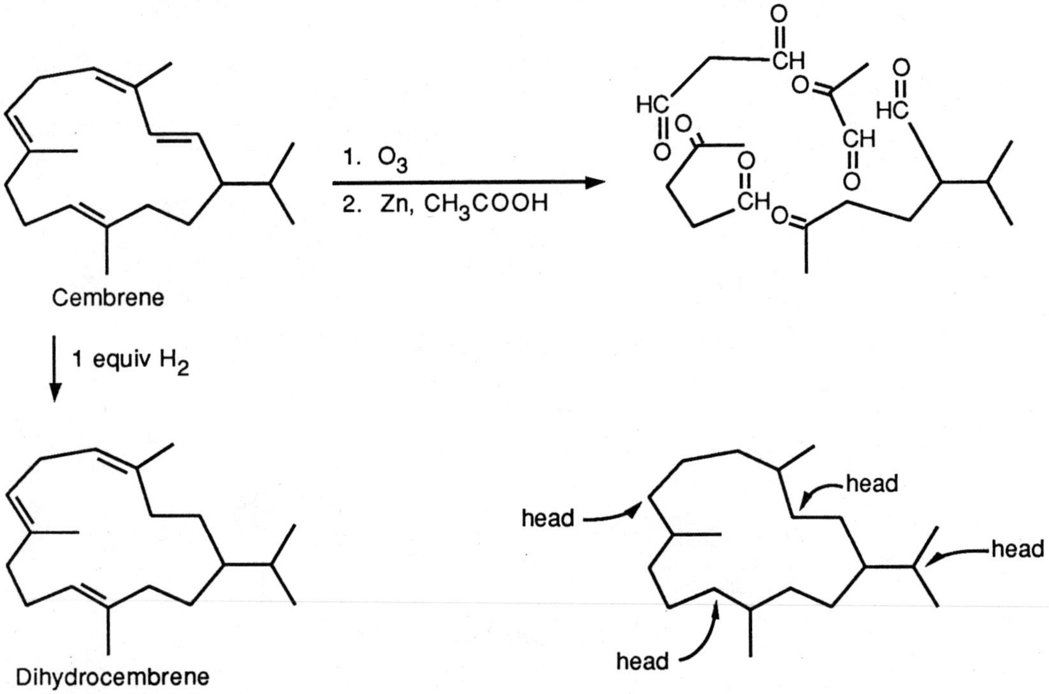

One equivalent of H_2 hydrogenates the least substituted double bond.
Dihydrocembrene exhibits no ultraviolet absorption because it is not conjugated.

Study Guide for Chapter 28

After studying this chapter, you should be able to:

(1)   Draw the structures of fats, oils, steroids, and other lipids (28.1, 28.2, 28.4, 28.14, 28.20).

(2)   Determine the structure of a fat (28.3, 28.13, 28.19, 28.23, 28.39).

(3)   Understand the mechanism of fatty acid biosynthesis (28.5, 28.6, 28.7).

(4)   Predict the products of reactions of fats and steroids (28.18, 28.21, 28.22, 28.34, 28.35, 28.38).

(5)   Locate the isoprene units in a terpene (28.8, 28.24, 28.39).

(6)   Understand the mechanism of terpene and steroid biosynthesis (28.9, 28.12, 28.26, 28.27, 28.28).

(7)   Draw the structures and conformations of steroids and other fused-ring systems (28.10, 28.11, 28.29, 28.30, 28.31, 28.32, 28.36).

29.1

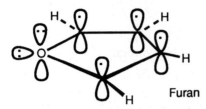

Furan

One oxygen lone pair is in a p orbital that is part of the pi electron system of furan. The other oxygen lone pair is in an sp^2 orbital that lies in the plane of the furan ring.

29.2

1.8 D

The dipole moment of pyrrole points in the direction indicated because the ring carbons are more electron-rich than nitrogen.

29.3

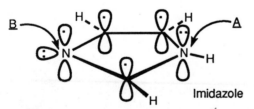

2–Deuteriopyrrole

29.4

Imidazole

Nitrogen atom <u>B</u> is more basic because its lone pair of electrons lies in an sp^2 orbital and is more available for donation to a Lewis acid than the lone pair of electrons of nitrogen <u>A</u>, which is part of the ring pi system.

29.5

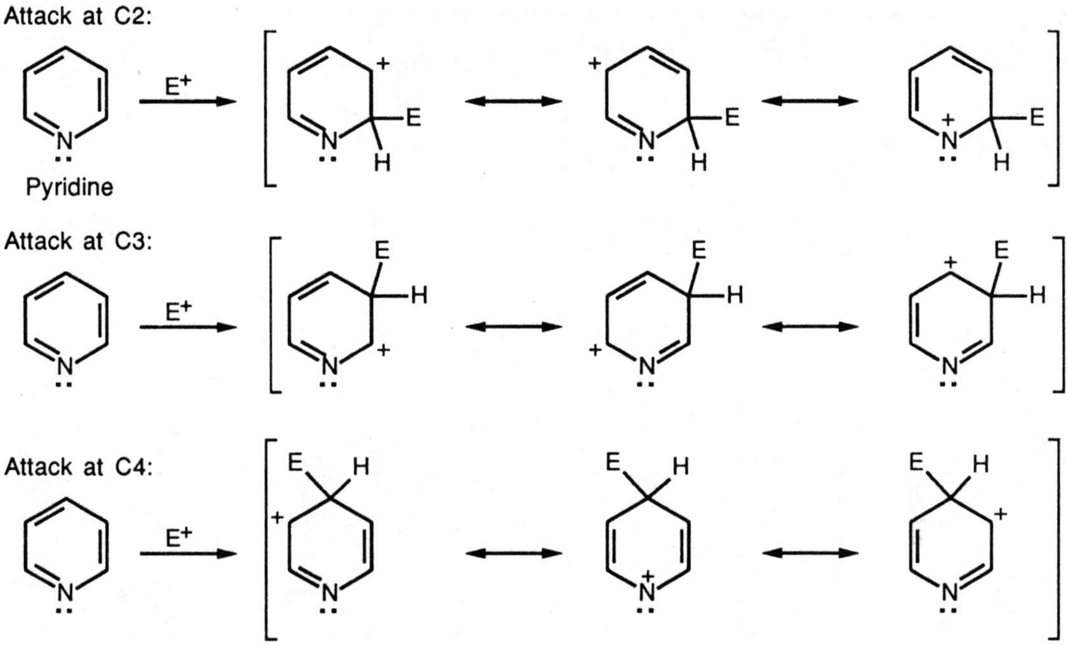

Attack at C2:

Pyridine

Attack at C3:

Attack at C4:

C3 attack is favored over C2 or C4 attack. The positive charge of the cationic intermediate of C3 attack is delocalized onto three carbon atoms, rather than onto two carbons and the electronegative pyridine nitrogen. Electrophilic attack at nitrogen is also possible.

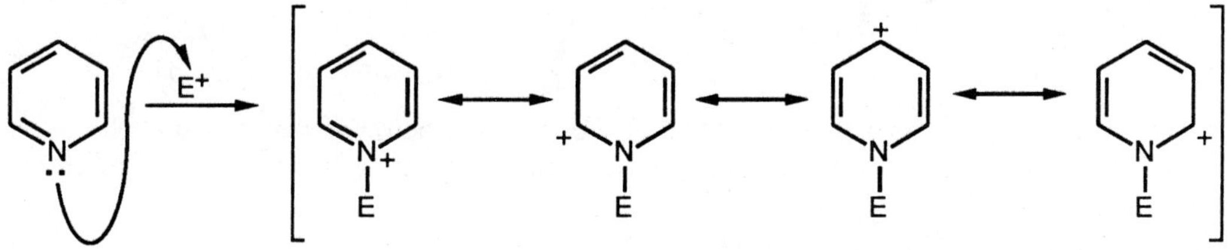

29.6

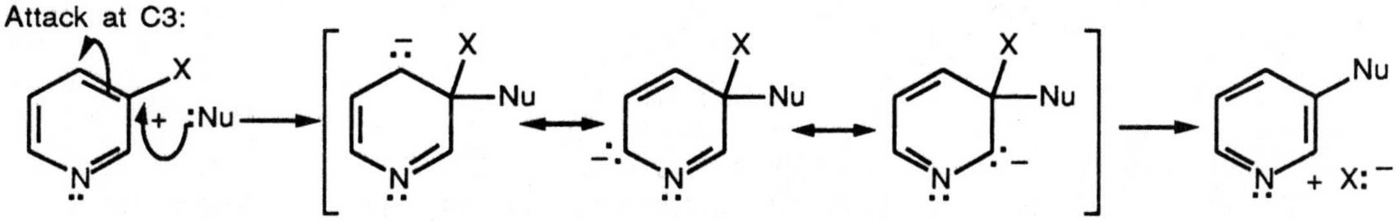

Attack at C3:

Attack at C4:

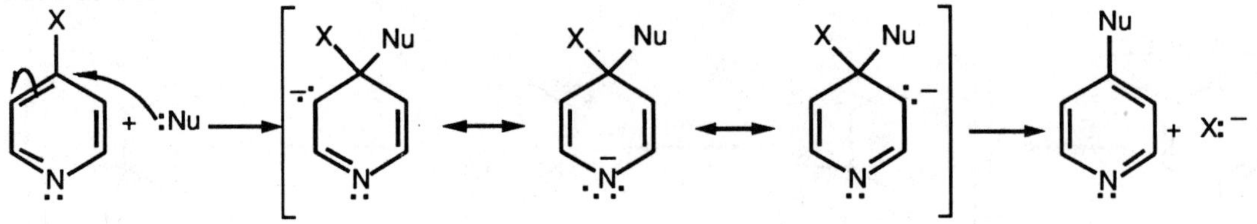

The negative charge resulting from C4 attack can be stabilized by nitrogen. Since no such stabilization is possible for C3 attack, reaction at C3 does not occur.

29.7

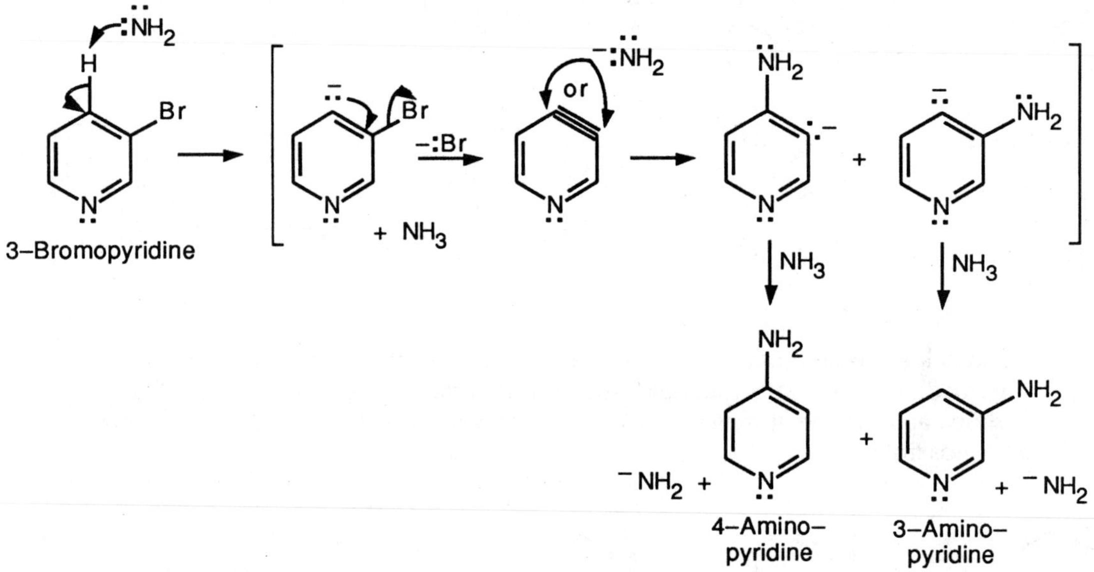

3–Bromopyridine

4–Amino–
pyridine

3–Amino–
pyridine

Reaction of 3–bromopyridine with $NaNH_2$ occurs by a benzyne mechanism. Since $^-{:}NH_2$ can add to either end of the triple bond of the benzyne intermediate, a mixture of products is formed.

29.8

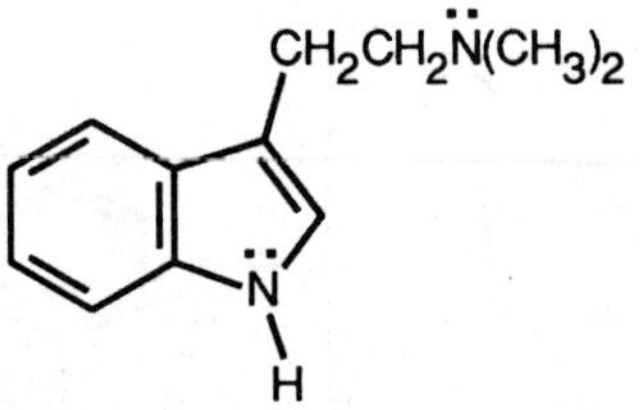

The aliphatic nitrogen atom of *N,N*–dimethyltryptamine is more basic because its lone electron pair is more available for donation to a Lewis acid. The aromatic nitrogen electron lone pair is part of the ring pi electron system.

29.9

C2 attack:

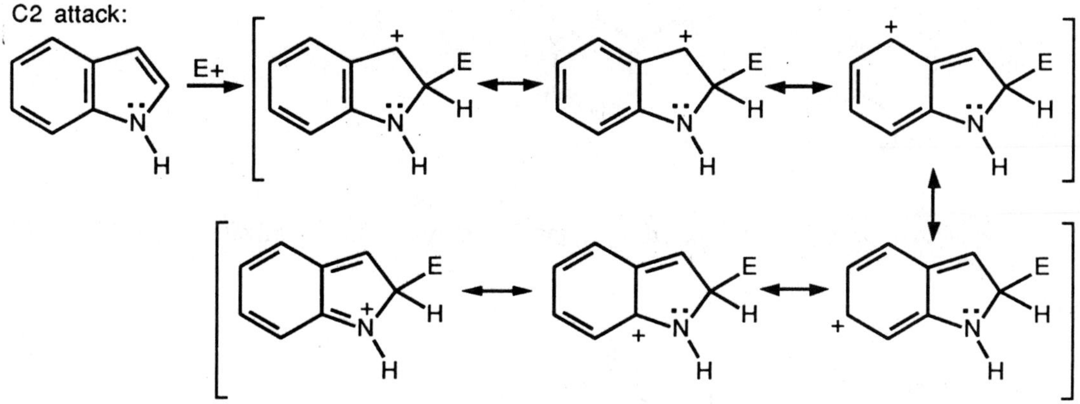

C3 attack:

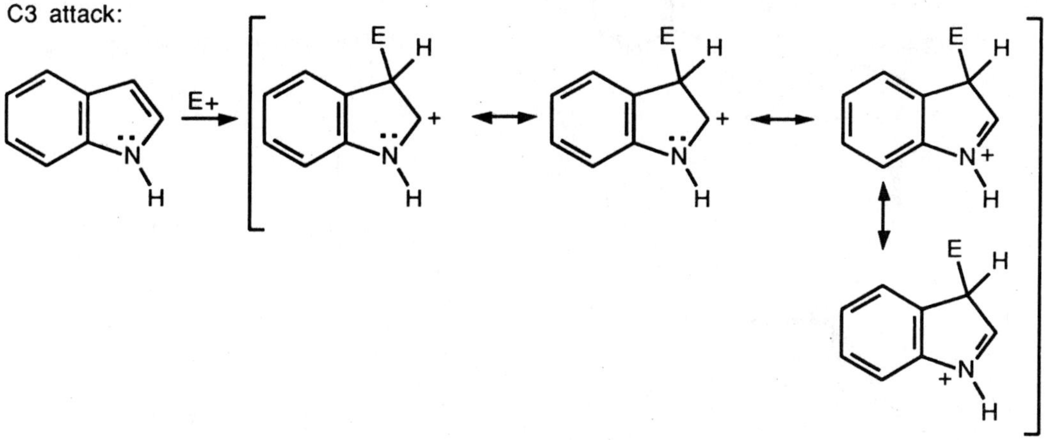

Positive charge can be stabilized by the nitrogen lone pair electrons in both C2 and C3 attack. In C2 attack, however, stabilization by nitrogen also destroys the aromaticity of the fused benzene ring. Reaction at C3 is favored, even though the cationic intermediate has fewer resonance forms, because the aromaticity of the six-membered-ring is preserved in the most favored resonance form.

29.10

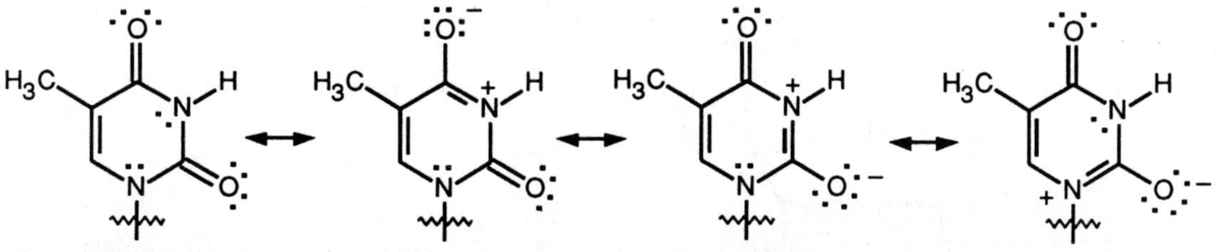

Lactam

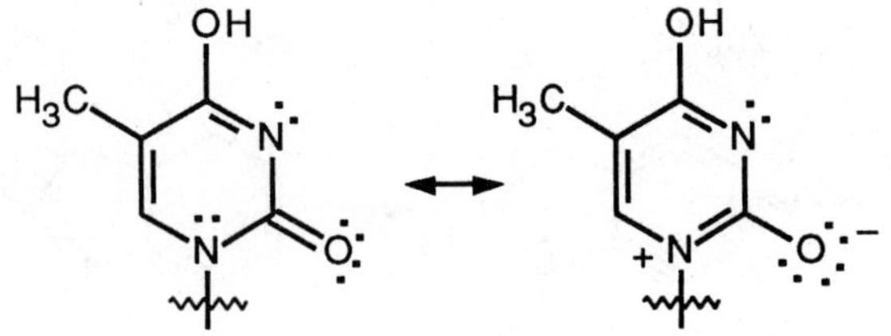

Lactim

The lactam form of 2'–deoxythymidine has greater resonance stabilization than the lactim form.

29.11

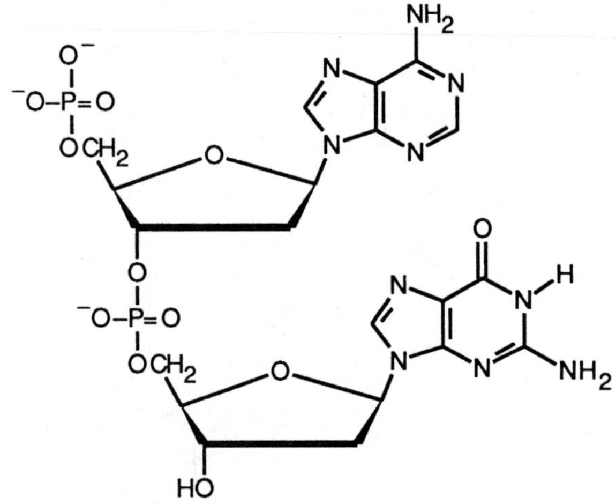

2'–Deoxyadenosine (A)

2'–Deoxyguanosine (G)

29.12

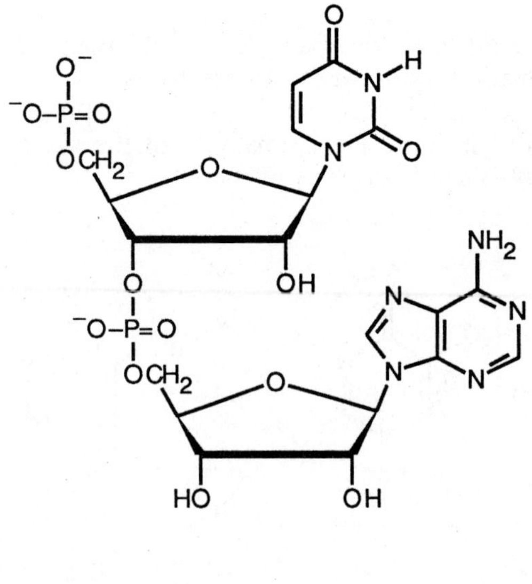

Uridine (U)

Adenosine (A)

29.13 DNA G–G–C–T–A–A–T–C–C–G–T is complementary to

DNA C–C–G–A–T–T–A–G–G–C–A

29.14

Uracil   Adenine

29.15 DNA G–A–T–T–A–C–C–G–T–A is complementary to

RNA C–U–A–A–U–G–G–C–A–U

29.16–29.17 Several different codons can code for the same amino acid. The corresponding anticodon follows each codon.

Amino Acid:	Ala	Phe	Leu	Tyr
Codon sequence/	GCU/CGA	UUU/AAA	UUA/AAU	UAU/AUA
tRNA anticodon:	GCC/CGG	UUC/AAG	UUG/AAC	UAC/AUG
	GCA/CGU		CUU/GAA	
	GCG/CGC		CUC/GAG	
			CUA/GAU	
			CUG/GAC	

29.18–29.20

The mRNA base sequence:	CUU–AUG–GCU–UGG–CCC–UAA
The amino acid sequence:	Leu—Met—Ala—Trp—Pro—OH (stop)
The tRNA sequence:	GAA UAC CGA ACC GGG AUU
The DNA sequence:	GAA–TAC–CGA–ACC–GGG–ATT

29.21 Remember:

1. Only a few of the many possible splittings occur in each reaction.

2. Cleavage occurs at both sides of the reacting nucleotide.

^{32}P–A–A–C–A–T–G–G–C–G–C–T–T–A–T–G–A–C–G–A

Reaction	Fragments
a) A	^{32}P
	^{32}P–A
	^{32}P–A–A–C
	^{32}P–A–A–C–A–T–G–G–C–G–C–T–T
	^{32}P–A–A–C–A–T–G–G–C–G–C–T–T–A–T–G

³²P–A–A–C–A–T–G–G–C–G–C–T–T–A–T–G–A–C–G

³²P–A–A–C–A–T–G–G–C–G–C–T–T–A–T–G–A–C–G–A

b) G   ³²P–A–A–C–A–T

³²P–A–A–C–A–T–G

³²P–A–A–C–A–T–G–G–C

³²P–A–A–C–A–T–G–G–C–G–C–T–T–A–T

³²P–A–A–C–A–T–G–G–C–G–C–T–T–A–T–G–A–C

³²P–A–A–C–A–T–G–G–C–G–C–T–T–A–T–G–A–C–G–A

c) C   ³²P–A–A

³²P–A–A–C–A–T–G–G

³²P–A–A–C–A–T–G–G–C–G

³²P–A–A–C–A–T–G–G–C–G–C–T–T–A–T–G–A

³²P–A–A–C–A–T–G–G–C–G–C–T–T–A–T–G–A–C–G–A

d) C + T   ³²P–A–A

³²P–A –A –C –A

³²P–A–A–C–A–T–G–G

³²P–A–A–C–A–T–G–G–C–G

³²P–A–A–C–A–T–G–G–C–G–C

³²P–A–A–C–A–T–G–G–C–G–C–T

³²P–A–A–C–A–T–G–G–C–G–C–T–T–A

³²P–A–A–C–A–T–G–G–C–G–C–T–T–A–T–G–A

³²P–A–A–C–A–T–G–G–C–G–C–T–T–A–T–G–A–C–G–A

29.22

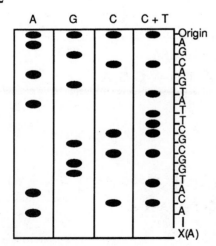

29.23 X–A–T–C–A–G–C–G–A–T–T–C–G–G–T–A–C

29.24

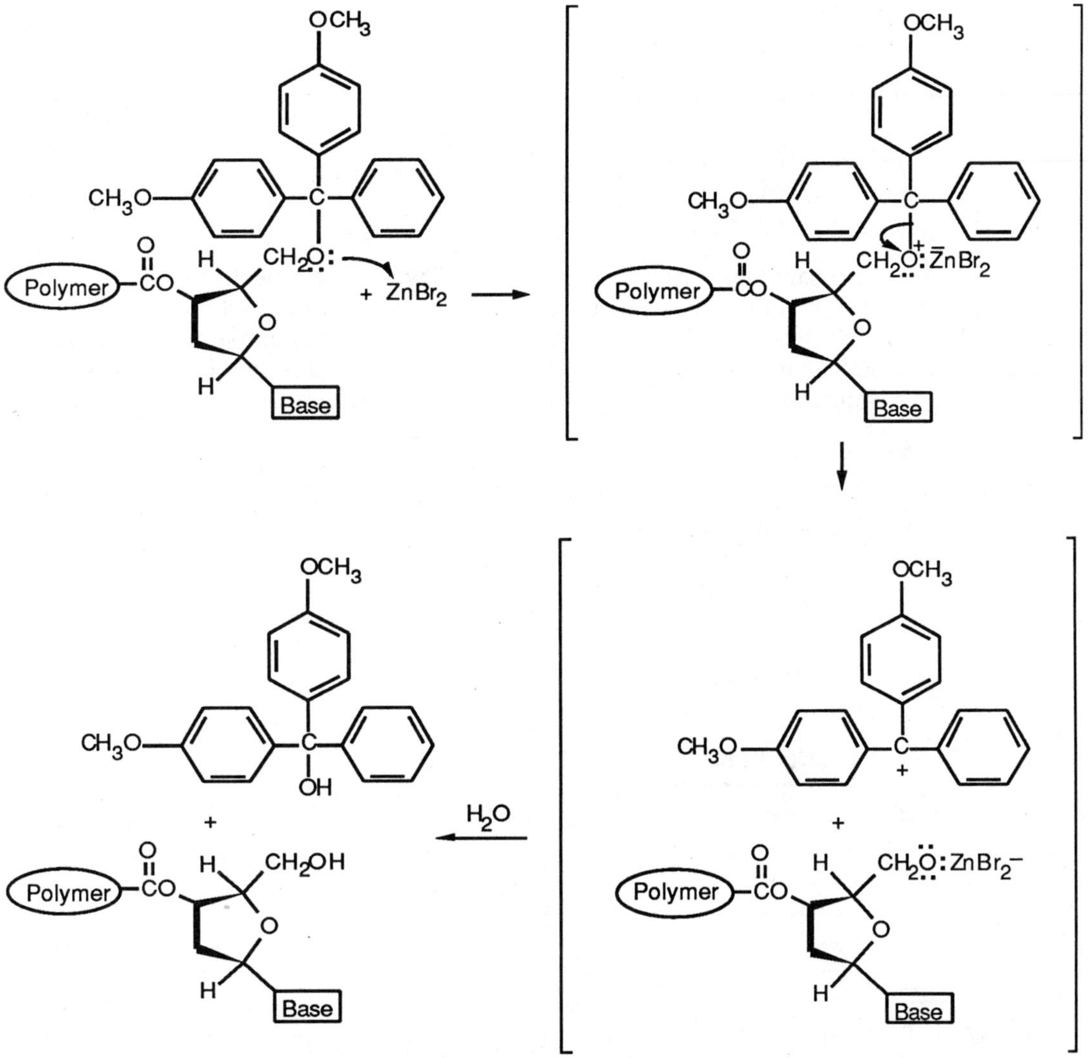

Cleavage of DMT ethers proceeds by an S_N1 mechanism and is rapid because the DMT cation formed is unusually stable.

29.25

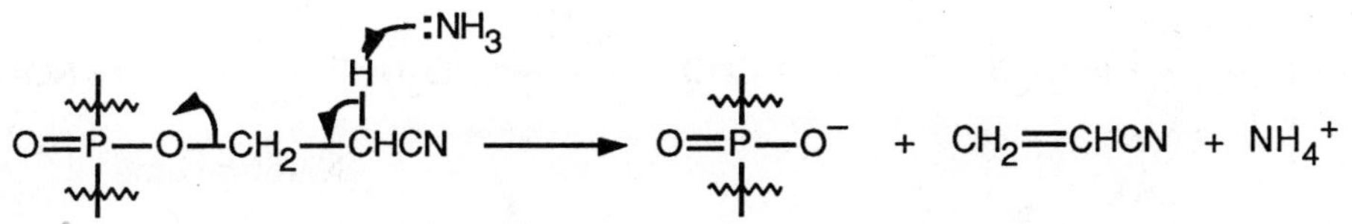

This is an E2 elimination reaction.

29.26 The pyrrole anion, $C_4H_4N:^-$, is a 6 pi electron species that is isoelectronic with the cyclopentadienyl anion. Both of these anions possess the stability of 6 pi electron systems.

29.27

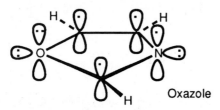

Oxazole

Oxazole is an aromatic 6 pi electron heterocycle. Two oxygen electrons and one nitrogen electron are in p orbitals that are part of the pi electron system of the ring, along with one electron from each carbon. An oxygen lone pair and a nitrogen lone pair are in sp^2 orbitals that lie in the plane of the ring. Since the nitrogen lone pair is available for donation to acids, oxazole is more basic than pyrrole.

29.28

a)

$$\xrightarrow[\text{dioxane}]{Br_2}$$

b)

$$\xrightarrow[\text{(CH}_3\text{CO}_2)\text{O}]{HNO_3}$$

c)

$$\xrightarrow[\text{SnCl}_4]{CH_3COCl}$$

d)

$$\xrightarrow[\text{Pd}]{H_2}$$

e)

$$\xrightarrow[\text{pyridine}]{SO_3}$$

29.29

Furfural $\xrightarrow[\substack{\text{acetic}\\\text{anhydride}}]{HNO_3}$ O_2N—CHO $\xrightarrow[H^+]{H_2NOH}$ O_2N—CH=NOH

Nitrofuroxime

29.30

29.31

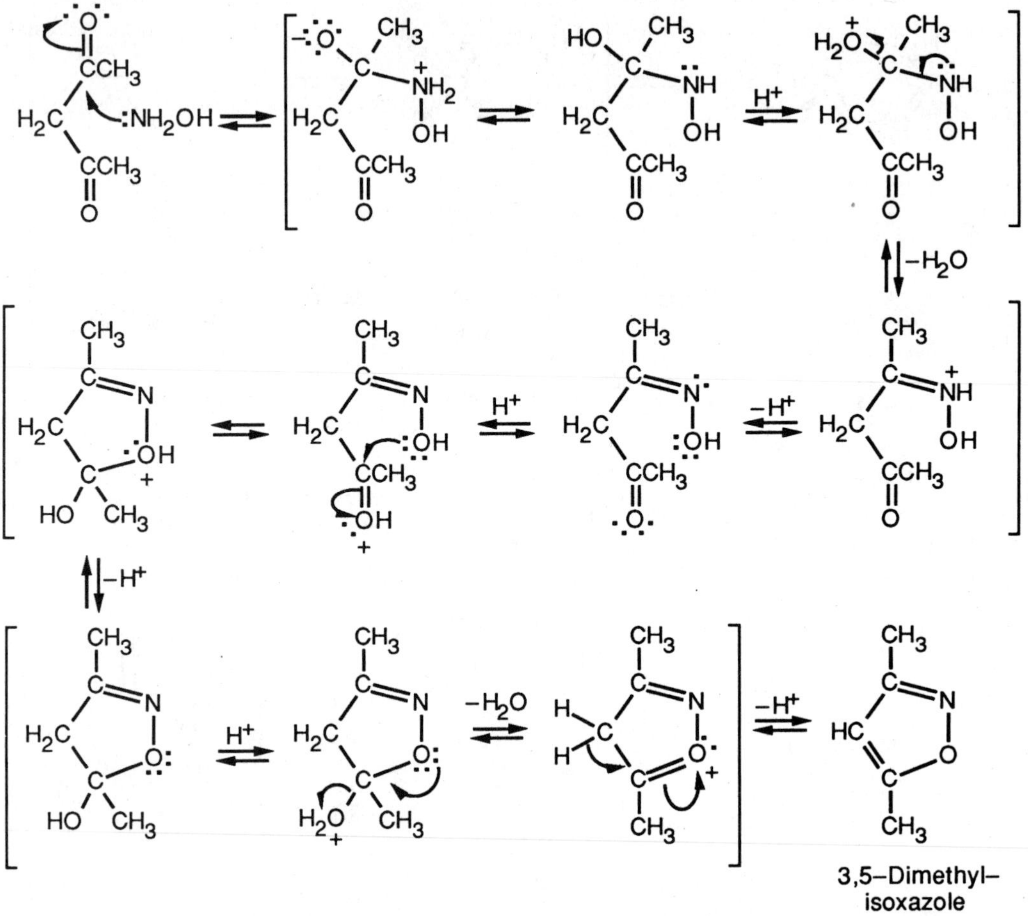

3,5–Dimethyl–
isoxazole

29.32 a) A *heterocycle* is a cyclic compound whose ring framework is composed of one or more different elements in addition to carbon.

b) *DNA* is a biological polymer whose monomeric units are nucleotides. A nucleotide is composed of a heterocyclic amine base, deoxyribose, and a phosphate group. DNA is the transmitter of the genetic code of all complex living organisms.

c) A *base pair* is a specific pair of heterocyclic amine bases that hydrogen-bond to each other in a DNA double helix and during protein synthesis.

d) *Transcription* is the process by which the genetic message contained in DNA is read by RNA and carried from the nucleus to the ribosomes.

e) *Translation* is the process by which RNA decodes the genetic message and uses the information to synthesize proteins.

f) *Replication* is the enzyme-catalyzed process by which a new molecule of DNA is produced. A DNA double helix separates into two strands, and nucleotide monomers line up and base-pair with each strand. Two new DNA double helices are produced, each one identical to the original DNA helix.

g) A *codon* is a sequence of three mRNA bases that specifies a particular amino acid to be used in protein synthesis.

h) An *anticodon* is a sequence of three tRNA bases that is complementary to the codon sequence. Each tRNA covalently bonds to a specific amino acid. The anticodon brings the amino acid into correct position for protein synthesis by hydrogen-bonding to its particular codon.

29.33–29.35

		a) AAU	b) GAG	c) UCC	d) CAU
mRNA codon:					
Amino acid:		Asn	Glu	Ser	His
DNA sequence:		TTA	CTC	AGG	GTA
tRNA anticodon:		UUA	CUC	AGG	GUA

29.36–29.37 UAC is the codon for tyrosine. It was transcribed from ATG of the DNA chain.

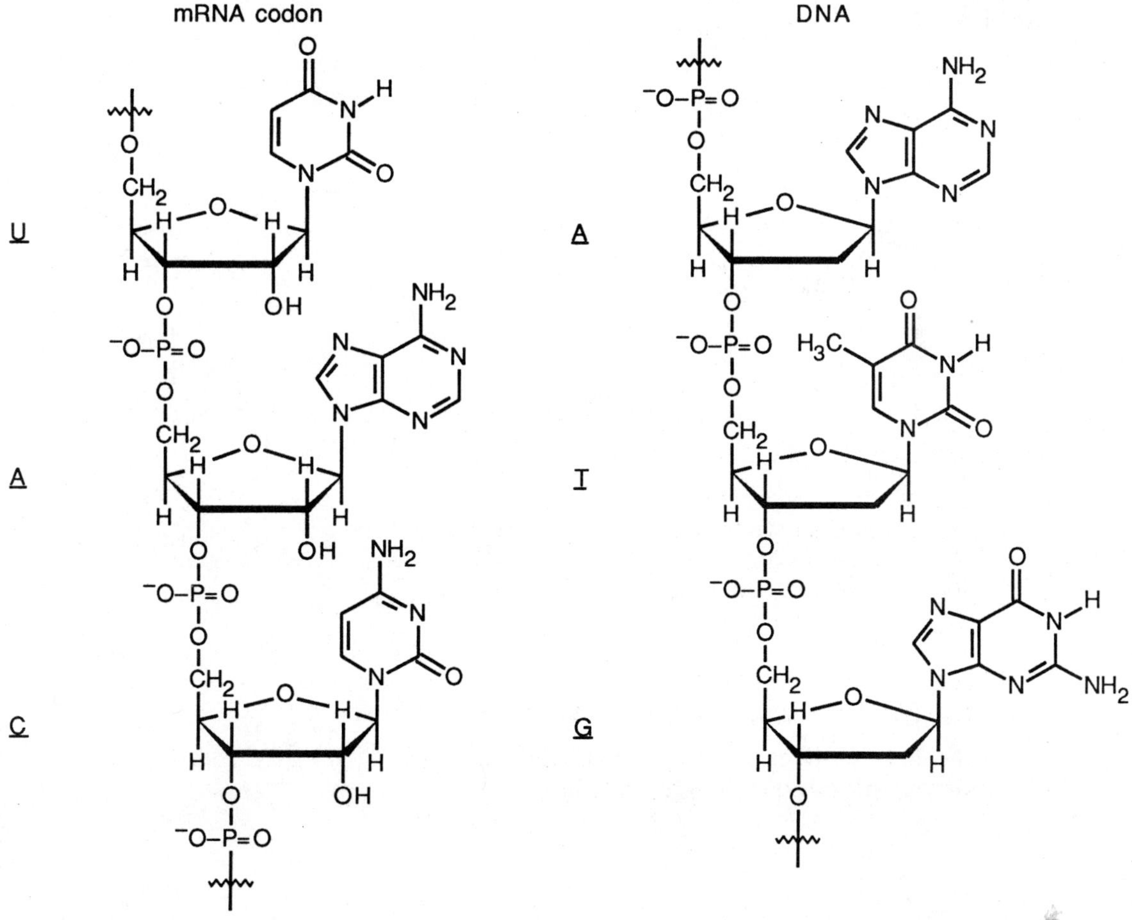

mRNA codon   DNA

U

A

C

A

I

G

29.38 H–Tyr———Gly———Gly———Phe———Met–OH (stop)   is coded by

(UAC)	(GGU)	(GGU)	(UUU)	(AUG)	(UAA)
(UAU)	(GGC)	(GGC)	(UUC)		(UAG)
	(GGA)	(GGA)			(UGA)
	(GGG)	(GGG)			

29.39 Angiotensin II: H—Asp—Arg——Val—Tyr—Ile—His—Pro——Phe—OH

mRNA sequence: GAU CGU GUU UAU AUU CAU CCU UUU UAA

GAC CGC GUC UAC AUC CAC CCC UUC UAG

CGA GUA   AUA   CCA   UGA

CGG GUG   CCG

AGA

AGG

29.40 mRNA sequence: CUA—GAC—CGU—UCC—AAG—UGA

Amino Acid: Leu——Asp—Arg——Ser—Lys—OH

29.41

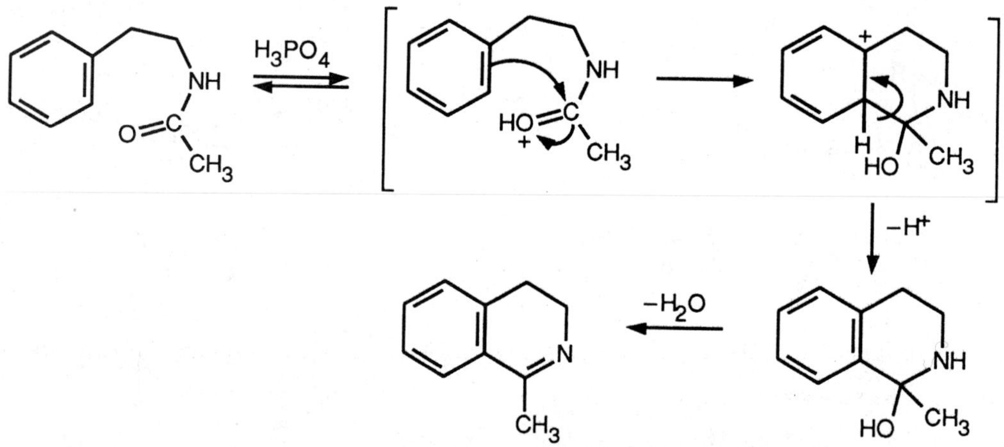

The cyclization is an electrophilic aromatic substitution.

29.42

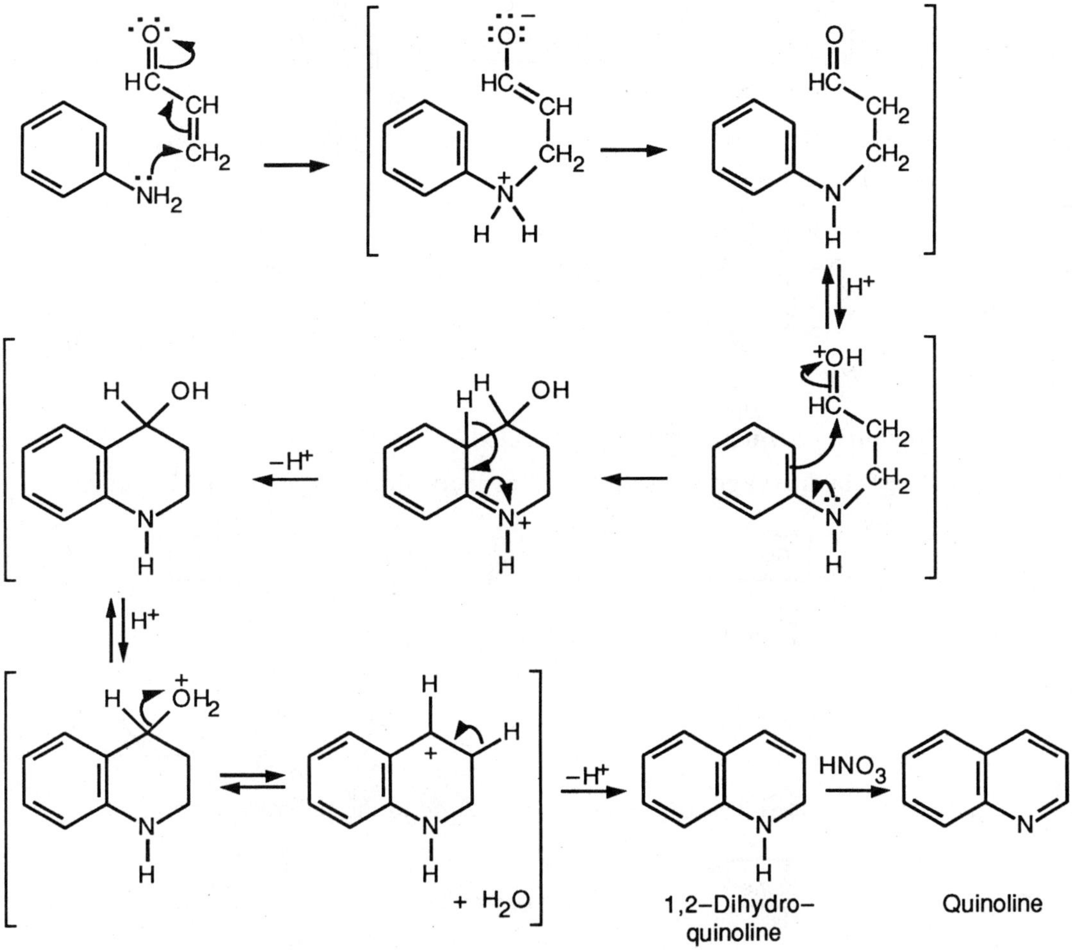

Conjugate addition of aniline is followed by electrophilic aromatic substitution.

29.43 1. First, protect the nucleotides.

a) Bases are protected by amide formation.

Adenine

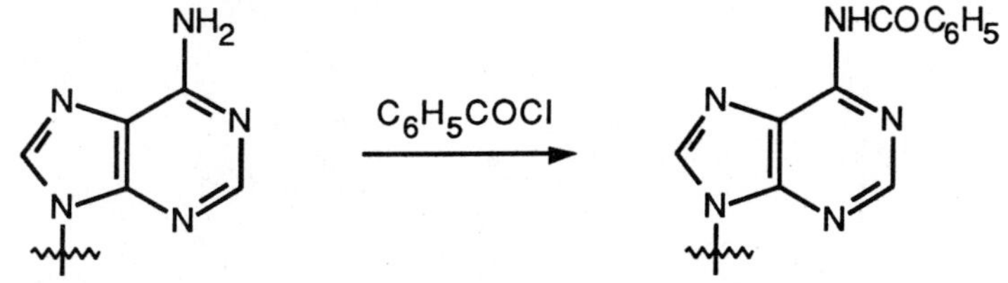

Cytosine

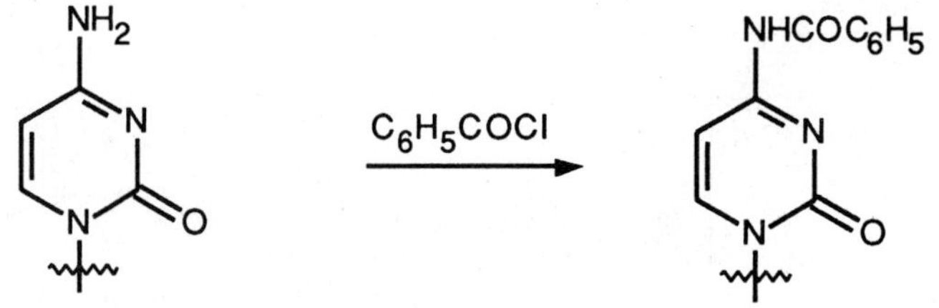
NH2
N
N
O
C6H5COCl
NHCOC6H5
N
N
O

Guanine

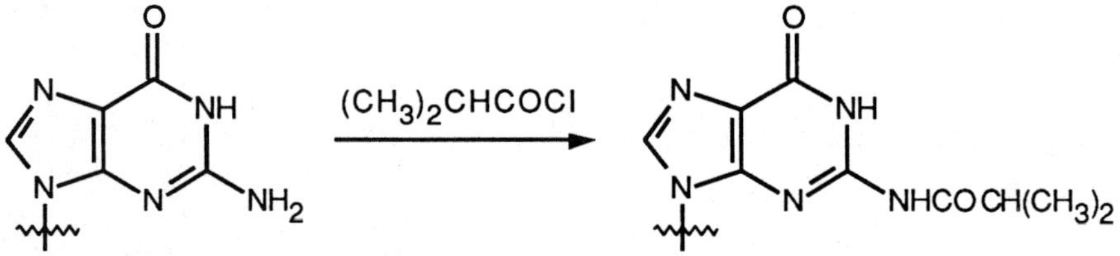
O
N
NH
N
N
NH2
(CH3)2CHCOCl
O
N
NH
N
N
NHCOCH(CH3)2

Thymine does not need to be protected.

b) The 5' hydroxyl group is protected as its *p*–dimethoxytrityl (DMT) ether.

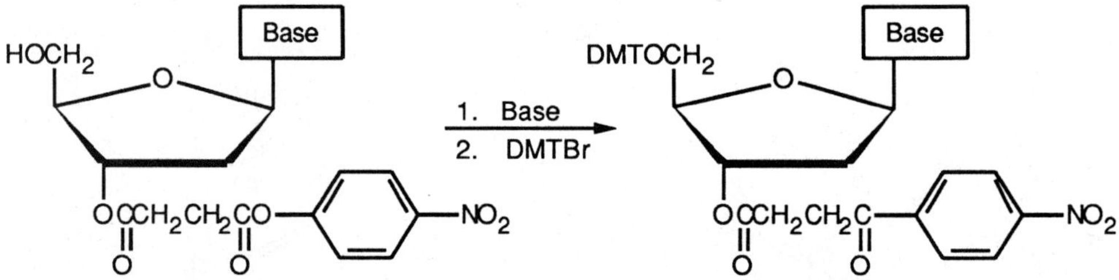
HOCH2
Base
O
OCCH2CH2CO
O
O
NO2
1. Base
2. DMTBr
DMTOCH2
Base
O
OCCH2CH2C
O
O
NO2

2. Attach a protected 2–deoxycytidine nucleo*side* to the polymer support.

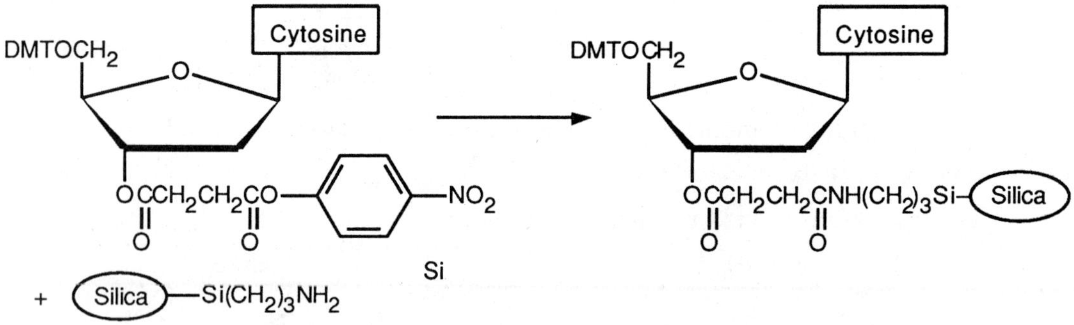
DMTOCH2
Cytosine
O
OCCH2CH2CO
O
O
NO2
Si
+
Silica
Si(CH2)3NH2
DMTOCH2
Cytosine
O
OCCH2CH2CNH(CH2)3Si
O
O
Silica

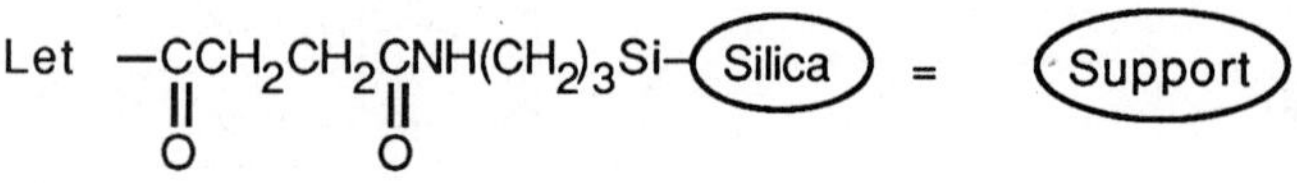
Let
CCH2CH2CNH(CH2)3Si
O
O
Silica
=
Support

3. Cleave the DMT ether.

4. Couple protected 2'–deoxythymidine to the polymer–2'–deoxycytidine. (The nucleosides have a phosphoramidite group at the 3' position.

5. Oxidize the phosphite product to a phosphate triester, using iodine.

6. Repeat steps 3–5 with protected 2'–deoxyadenosine and protected 2'–deoxyguanosine.

7. Cleave all protecting groups with aqueous ammonia, to yield the desired sequence.

29.44

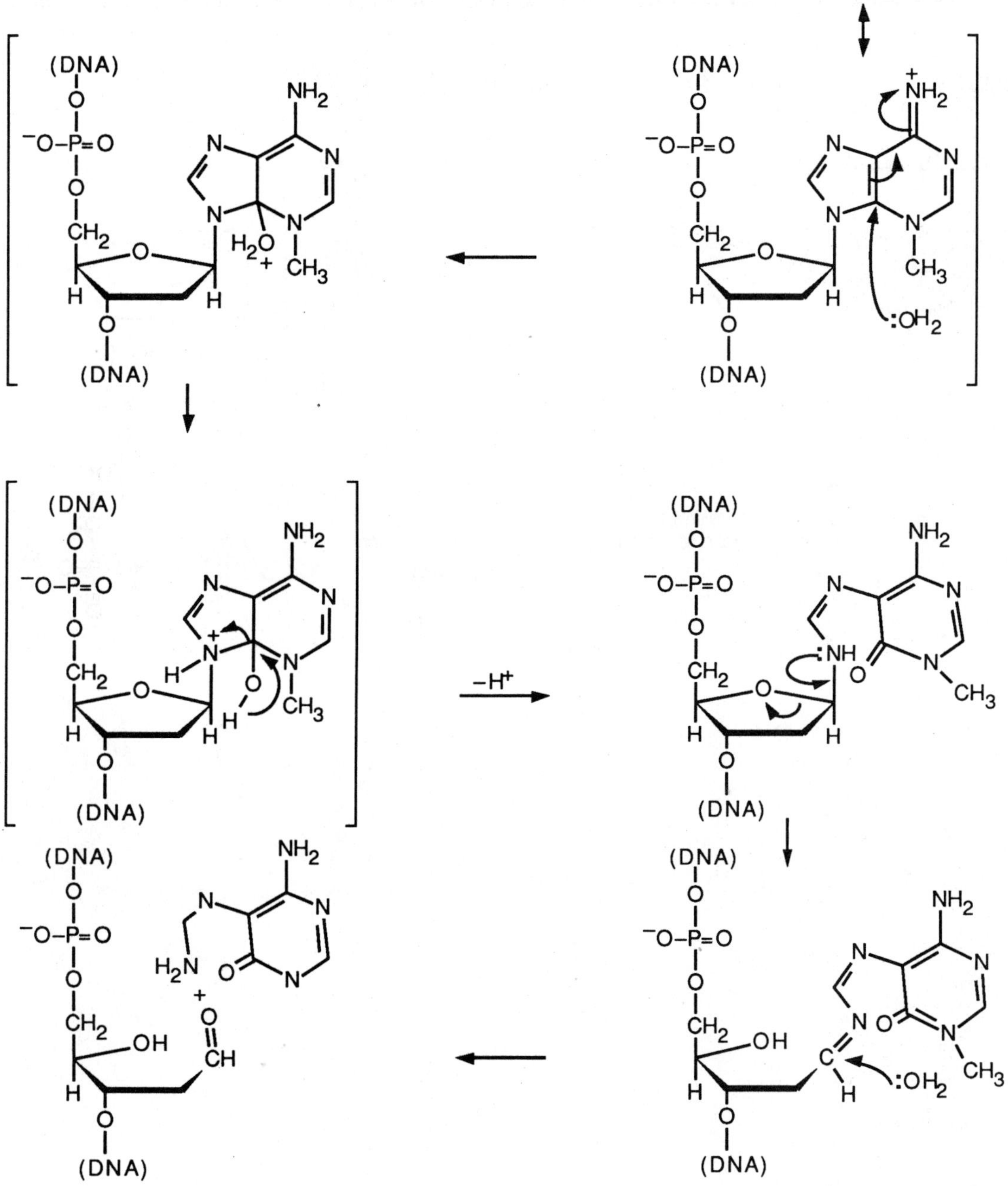
(DNA)
NH_2
$^-O-P=O$
CH_2
H_2O^+
CH_3
$:OH_2$
$-H^+$
OH
CH
H_2N

This intermediate under-
goes further reaction.

Study Guide for Chapter 29

After studying this chapter, you should be able to:

(1)   Draw orbital pictures of heterocycles and explain their acid-base properties (29.1, 29.2, 29.4, 29.8, 29.26, 29.27).

(2)   Explain orientation and reactivity in aromatic heterocycle reactions (29.5, 29.6, 29.7, 29.9).

(3)   Predict the products of reactions of heterocycles (29.28, 29.29).

(4)   Formulate mechanisms of reactions of heterocycles (29.3, 29.31, 29.41, 29.42).

(5)   Draw purines, pyrimidines, nucleosides, nucleotides, and representative segments of DNA (29.10, 29.11, 29.12, 29.36, 29.37).

(6)   Given a DNA or RNA strand, draw its complementary strand (29.13, 29.15, 29.20, 29.34).

(7)   List the codon sequence for a given amino acid or peptide (and *vice versa*) (29.16, 29.33, 29.38)

(8)   Deduce an amino acid sequence from a given mRNA base sequence (and *vice versa*) (29.18, 29.39, 29.40).

(9)   Draw the anticodon sequence of tRNA, given the mRNA sequence (29.17, 29.19, 29.35).

(10)   Outline the process of DNA sequencing, and deduce a DNA sequence from an electrophoresis pattern (29.21, 29.22, 29.23).

(11)   Outline the method of DNA synthesis, and formulate mechanisms of synthetic steps (29.24, 29.25, 29.43, 29.44).

(12)   Define the important terms in this chapter (29.32).

30.1

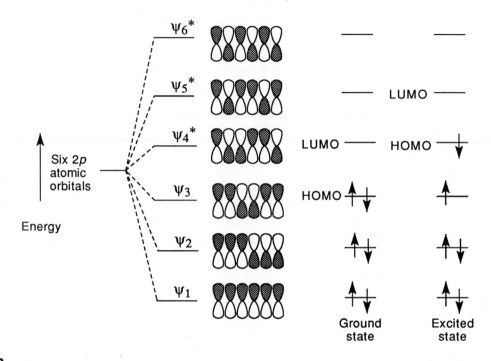

30.2

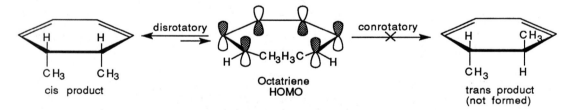

The symmetry of the octatriene HOMO predicts that ring closure will occur by a disrotatory path and that only *cis* product will be formed.

30.3   Note: *Trans*–3,4–dimethylcyclobutene is chiral; the *S,S* enantiomer will be used for this argument.

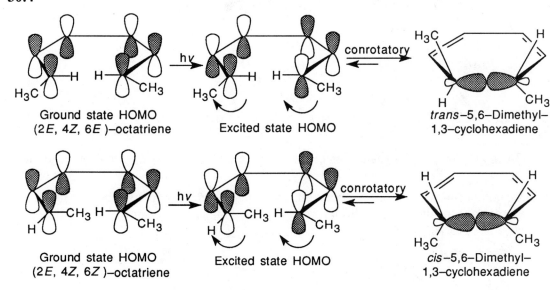

Conrotatory ring opening of *trans*–3,4–dimethylcyclobutene can occur in either a clockwise or a counterclockwise manner. Clockwise opening (path <u>A</u>) yields the *E,E* isomer; counterclockwise opening (path <u>B</u>) yields the *Z,Z* isomer. Production of (2 *Z*,4 *Z*)–hexadiene is disfavored because of unfavorable steric interactions between the methyl groups in the transition state leading to ring-opened product.

30.4

Ground state HOMO
(2*E*, 4*Z*, 6*E*)–octatriene

Excited state HOMO

trans–5,6–Dimethyl–
1,3–cyclohexadiene

Ground state HOMO
(2*E*, 4*Z*, 6*Z*)–octatriene

Excited state HOMO

cis–5,6–Dimethyl–
1,3–cyclohexadiene

Photochemical electrocyclic reactions of 6 pi electron systems always occur in a conrotatory manner.

30.5

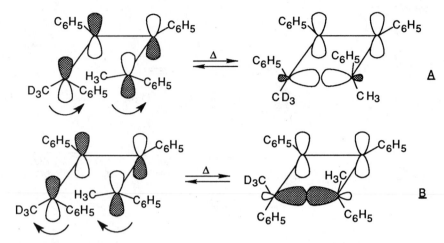

A

B

The diene can cyclize by either of two conrotatory paths to form cyclobutenes $\underline{A}$ and $\underline{B}$. Using $\underline{B}$ as an example:

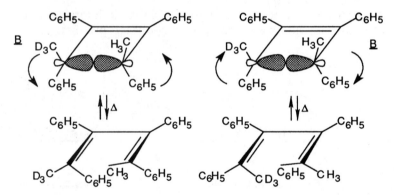

Opening of each cyclobutene ring can occur by either of two conrotatory routes to yield the isomeric dienes.

30.6 A photochemical electrocyclic reaction involving two electron pairs proceeds in a *disrotatory* manner (Table 30.1).

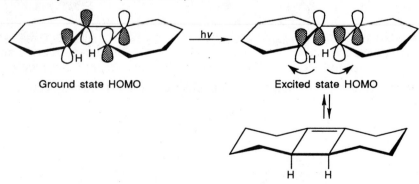

Ground state HOMO   Excited state HOMO

The two hydrogen atoms in the four-membered ring are *cis* to each other in the cyclobutene product.

30.7

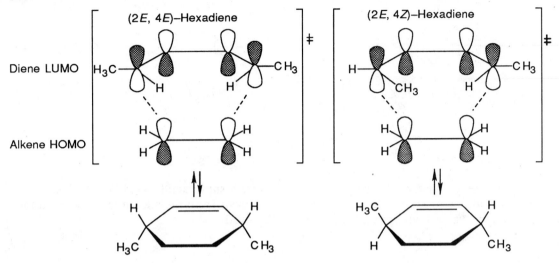

The Diels-Alder reaction is a thermal [4+2] cycloaddition, which occurs with suprafacial geometry. The stereochemistry of the diene is maintained in the product.

30.8

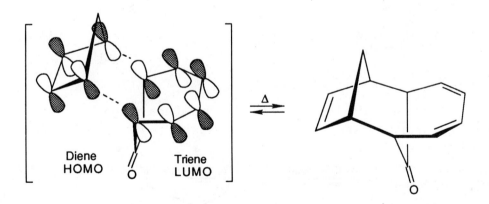

The reaction of cyclopentadiene and cycloheptatrienone is a [6+4] cycloaddition. This thermal cycloaddition proceeds with suprafacial geometry since five electron pairs are involved in the concerted process. The pi electrons of the carbonyl group do not take part in the reaction.

30.9

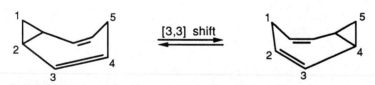

Formation of the bicyclic ring system occurs by a suprafacial [4+2] Diels-Alder cycloaddition process. Only one pair of pi electrons from the alkyne is involved in the reaction; the carbonyl pi electrons are not involved.

Loss of CO_2 is a reverse Diels-Alder [4+2] cycloaddition reaction.

30.10

The ^{13}C NMR spectrum of homotropilidene would show five peaks if rearrangement were slow. In fact, rearrangement occurs at a rate that is too fast for NMR to detect. The ^{13}C NMR spectrum taken at room temperature is an average of the two equilibrating forms, in which positions 1 and 5 are equivalent, as are positions 2 and 4. Thus, only three distinct types of carbons are visible in the ^{13}C NMR spectrum of homotropilidene.

30.11 Scrambling of the deuterium label of 1–deuterioindene occurs by a series of [1,5] sigmatropic rearrangements. This thermal reaction involves three electron pairs—one pair of pi electrons from the six-membered ring, the pi electrons from the five-membered ring, and two electrons from a carbon-deuterium (or hydrogen) single bond—and proceeds with suprafacial geometry.

30.12 This [1,7] sigmatropic reaction proceeds with antarafacial geometry since four electron pairs are involved in the arrangement.

30.13

The Claisen arrangement of an unsubstituted allyl phenyl ether is a [3,3] sigmatropic rearrangement in which the allyl group usually ends up in the position *ortho* to oxygen. In this problem both *ortho* positions are occupied by methyl groups. The Claisen intermediate undergoes a second [3,3] rearrangement, and the final product is *p*–allyl phenol.

30.14

Type of reaction	Number of electron pairs	Stereochemistry
a) Thermal electrocyclic	four	conrotatory
b) Photochemical electrocyclic	four	disrotatory
c) Photochemical cycloaddition	four	suprafacial
d) Thermal cycloaddition	four	antarafacial
e) Photochemical sigmatropic rearrangement	four	suprafacial

30.15 a) An *electrocyclic reaction* is a reversible, pericyclic process in which a ring is formed by the reorganization of the pi electrons of a conjugated polyene.

b) *Conrotatory* motion in a pericyclic reaction occurs when the lobes of two orbitals both rotate in either a clockwise or a counterclockwise fashion.

c) A *suprafacial* pericyclic reaction occurs between orbital lobes on the same face of one component and orbital lobes on the same face of the other component.

d) An *antarafacial* pericyclic reaction occurs between orbital lobes on the same face of one component and orbital lobes on opposite faces of the other component.

e) *Disrotatory motion* occurs when the lobes of one orbital in an electrocyclic reaction rotate clockwise and the lobes of the other orbital rotate counterclockwise.

f) A *sigmatropic rearrangement* is a pericyclic process in which a sigma-bonded group migrates across a pi electron system.

30.16

a)

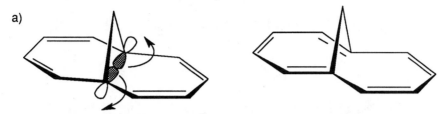

Rotation of the orbitals in the 6 electron system occurs in a disrotatory fashion. According to the rules in Table 30.1, the reaction should be carried out under thermal conditions.

b)

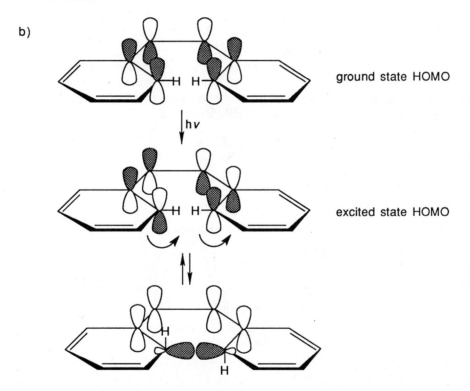

ground state HOMO

excited state HOMO

In order for the hydrogens to be *trans* in the product, rotation must occur in a conrotatory manner. This can happen only if the HOMO has the symmetry pictured. For a 6 pi electron system, this HOMO must arise from photochemical excitation of a pi electron. To obtain a product having the correct stereochemistry, the reaction must be carried out under photochemical conditions.

30.17

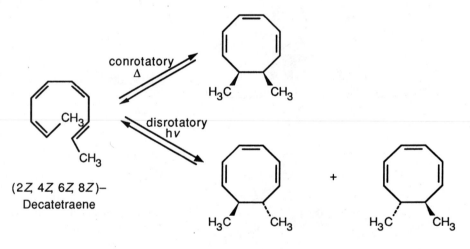

Four electron pairs undergo reorganization in this electrocyclic reaction. The thermal reaction occurs with conrotatory motion to yield a pair of enantiomeric *trans*–7,8-dimethyl–1,3,5–cyclooctatrienes. The photochemical cyclization occurs with disrotatory motion to yield the *cis*–7,8–dimethyl isomer.

30.18

30.19

Type of reaction	Number of electron pairs	Stereochemistry
a) Photochemical [1,5] sigmatropic rearrangement	3	antarafacial
b) Thermal [4+6] cycloaddition	5	suprafacial
c) Thermal [1,7] sigmatropic rearrangement	4	antarafacial
d) Photochemical [2+6] cycloaddition	4	suprafacial

30.20

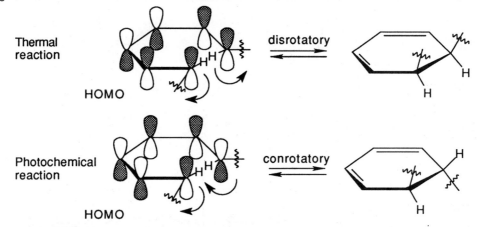

Thermal reaction

HOMO

disrotatory

Photochemical reaction

HOMO

conrotatory

Two electrocyclic reactions, involving three electron pairs each, occur in this isomerization. The thermal reaction is a disrotatory process that yields two *cis*-fused six-membered rings. The photochemical reaction yields the *trans*-fused isomer. The two pairs of pi electrons in the eight-membered ring do not take part in the electrocyclic reaction.

30.21

This reaction is a reverse [4+2] cycloaddition. The reacting orbitals have the correct symmetry for the reaction to take place by a favorable suprafacial process.

This [2+2] reverse cycloaddition is not likely to occur as a concerted process because the required antarafacial geometry for the thermal reaction is not possible for a four-electron system.

30.22

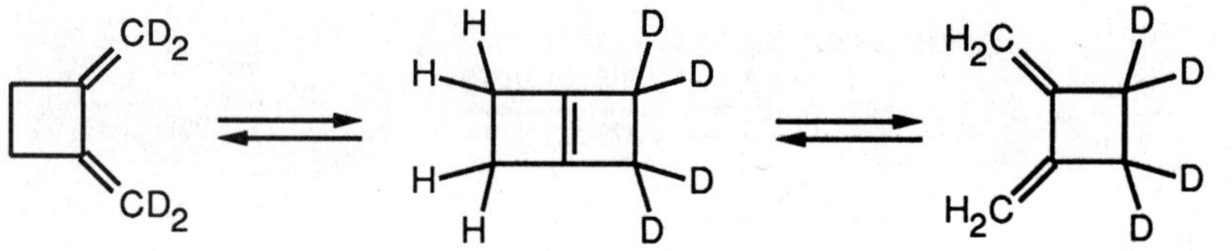

Each electrocyclic reaction involves two pairs of electrons and proceeds in a conrotatory manner.

30.23

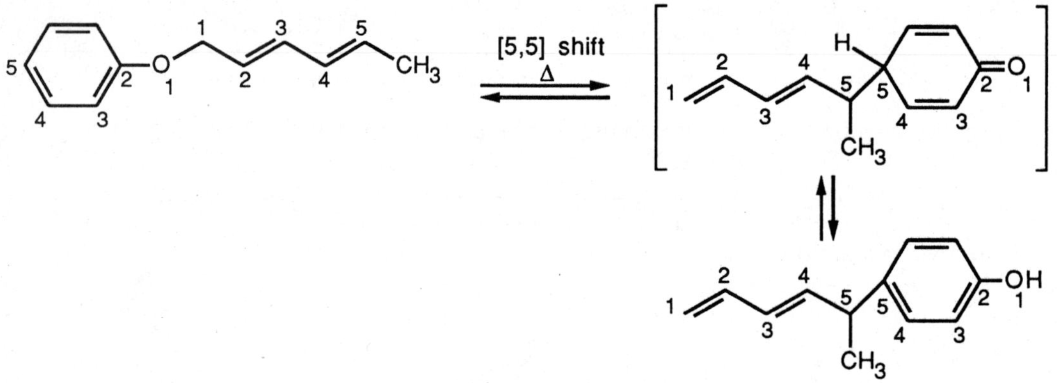

[5,5] shift
Δ

This thermal sigmatropic rearrangement is a suprafacial process since five electron pairs are involved in the reaction.

30.24

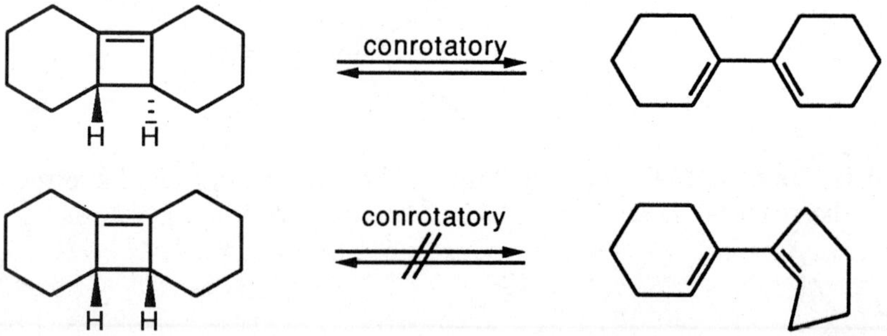

conrotatory

conrotatory

The observed product can be formed by a four-electron *pericyclic* process only if the four-membered ring geometry is *trans*. Ring-opening of the *cis* isomer by a concerted process would form a severely strained six-membered ring containing a *trans* double bond. Reaction of the *cis* isomer to yield the observed product occurs instead by a higher energy, non-concerted path.

30.25

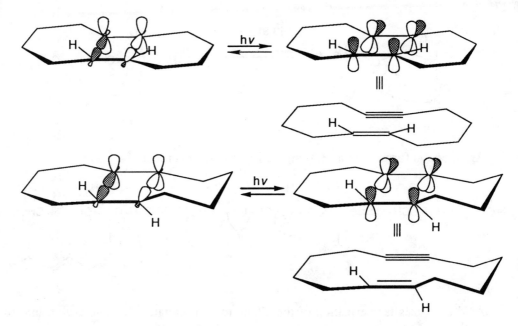

Both of these reactions are reverse [2+2] photochemical cycloadditions, which occur with suprafacial geometry.

30.26

The first reaction is a Diels-Alder [4+2] cycloaddition, which proceeds with suprafacial geometry.

The second reaction is a reverse Diels-Alder [4+2] cycloaddition

30.27

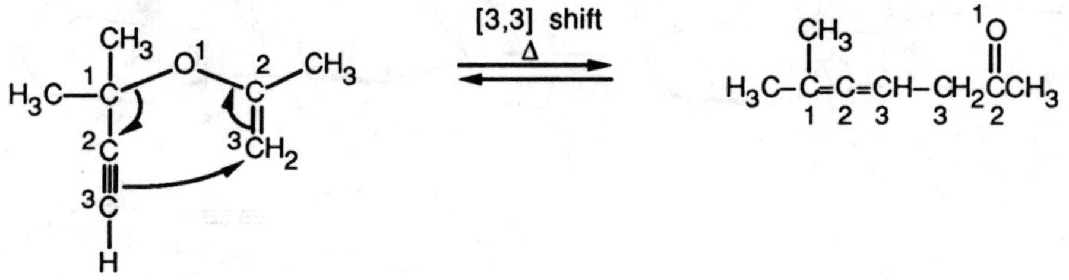

An allene is formed by a [3,3] sigmatropic rearrangement.

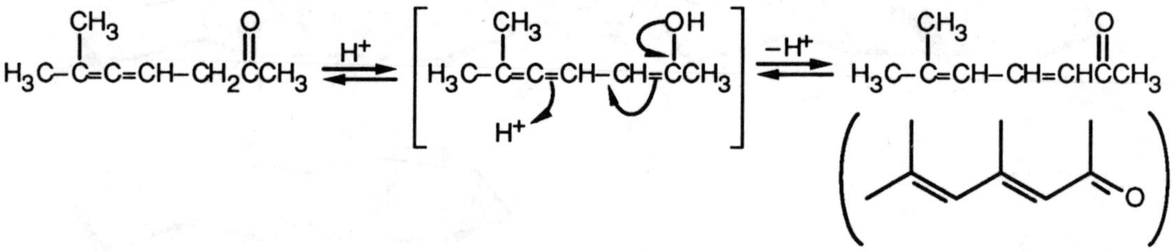

Acid catalyzes isomerization of the allene to a conjugated dienone *via* an intermediate enol.

30.28

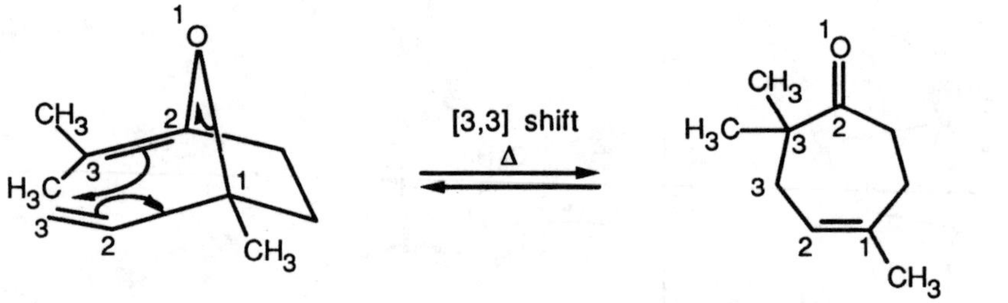

Karahanaenone is formed by a [3,3] sigmatropic rearrangement (Claisen rearrangement).

30.29

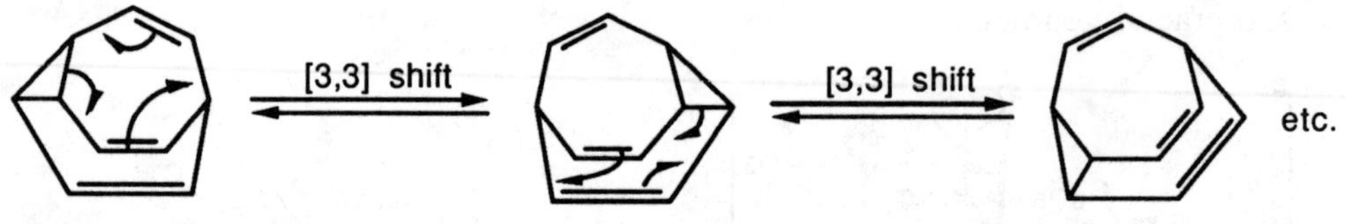

Bullvalene can undergo [3,3] sigmatropic rearrangements in all directions. At 100°, the rate of rearrangement is fast enough to make all hydrogen atoms equivalent, and only one signal is seen in the ^{1}H NMR spectrum.

30.30 Suprafacial shift:

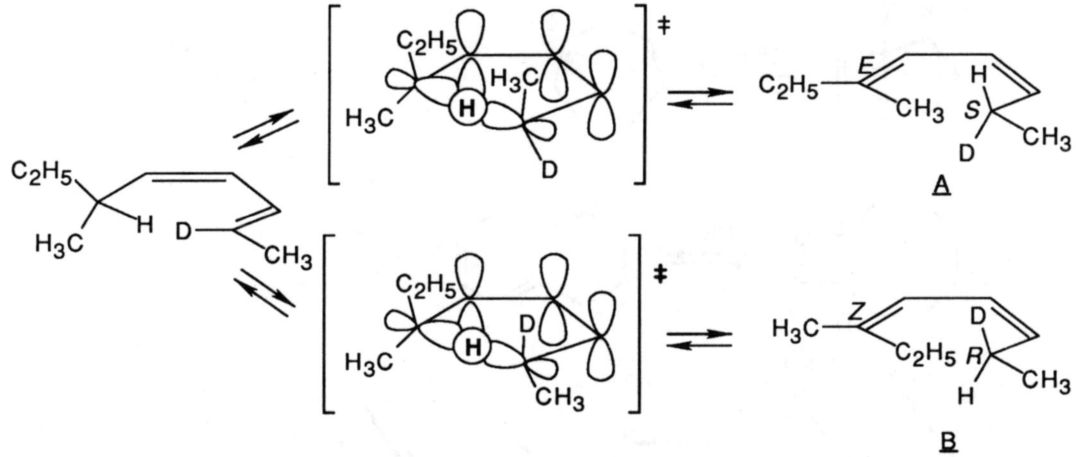

Antarafacial shift:

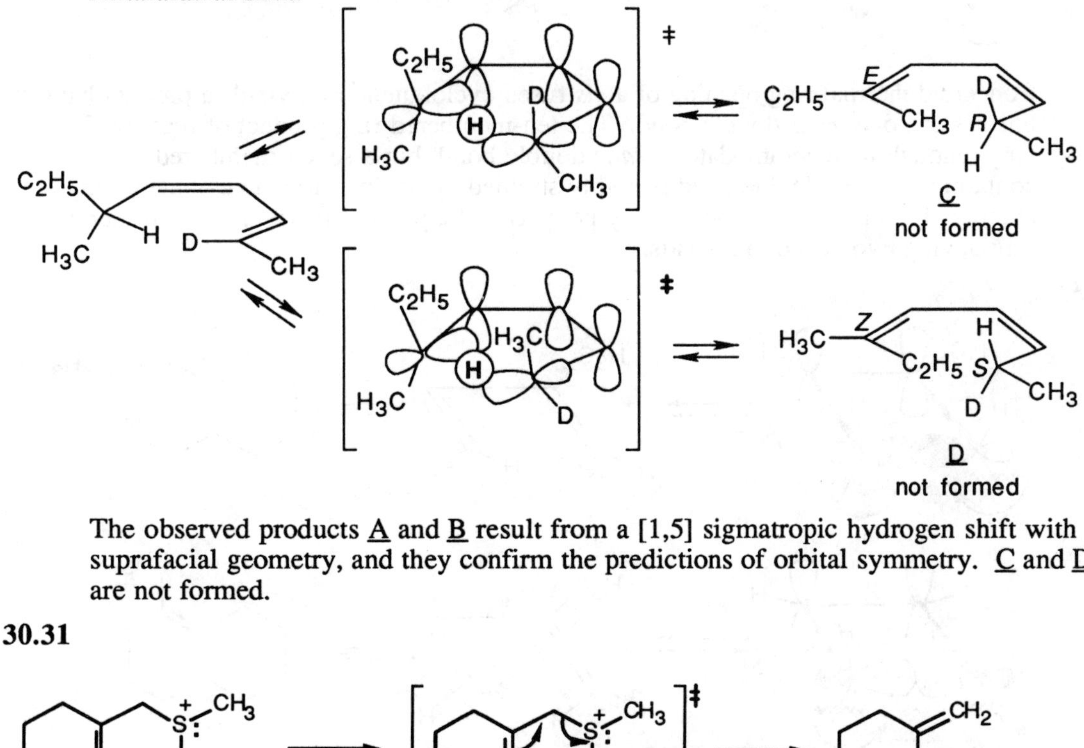

The observed products <u>A</u> and <u>B</u> result from a [1,5] sigmatropic hydrogen shift with suprafacial geometry, and they confirm the predictions of orbital symmetry. <u>C</u> and <u>D</u> are not formed.

30.31

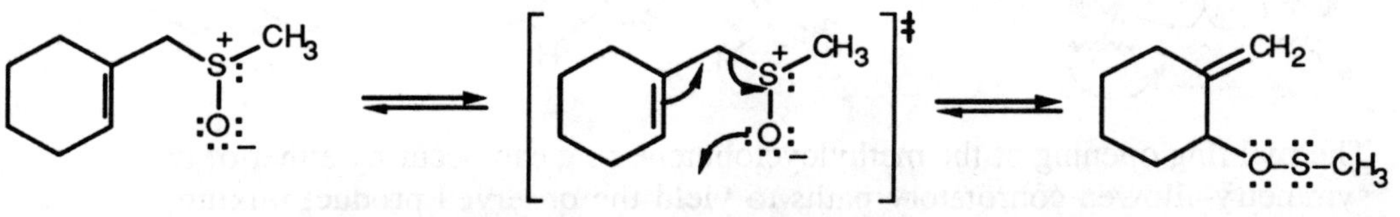

This [2,3] sigmatropic rearrangement involves three electron pairs and should occur with suprafacial geometry.

30.32

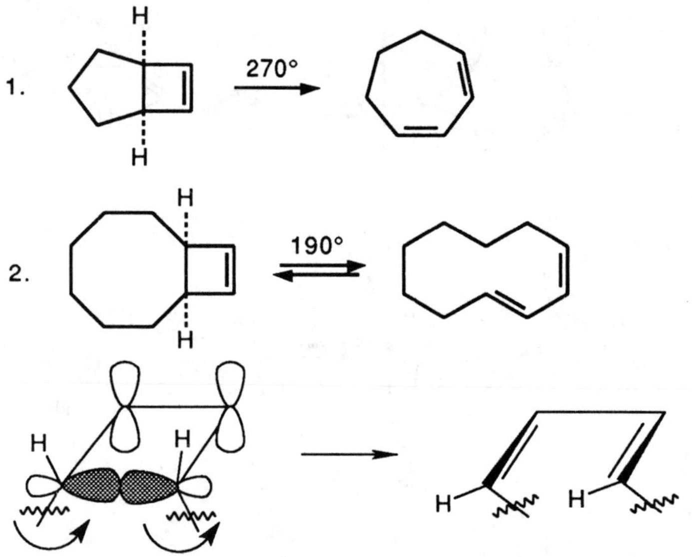

Concerted thermal ring opening of a *cis* fused cyclobutene ring yields a product having one *cis* and one *trans* double bond. The ten-membered ring product of reaction 2 is large enough to accommodate a *trans* double bond, but a seven-membered ring containing a *trans* double bond is highly strained. Opening of the cyclobutene ring in reaction 1 occurs by a higher energy nonpericyclic process to yield a seven-membered ring having two *cis* double bonds.

30.33

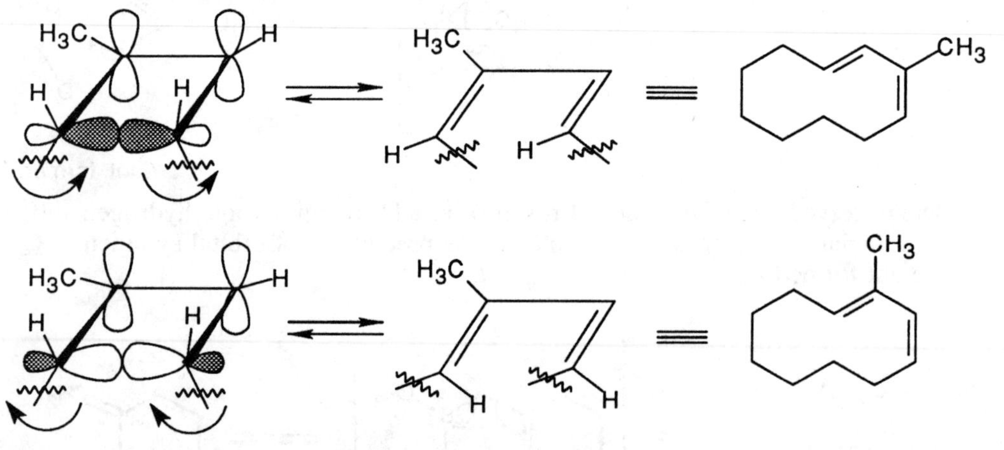

Thermal ring opening of the methylcyclobutene ring can occur by either of two symmetry-allowed conrotatory paths to yield the observed product mixture.

30.34

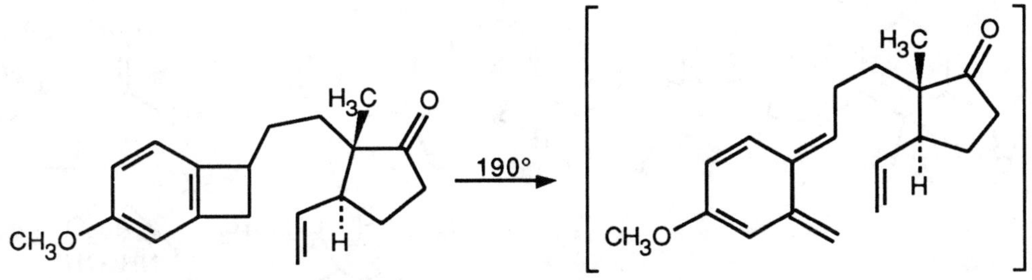

The first reaction is an electrocyclic opening of a cyclobutene ring.

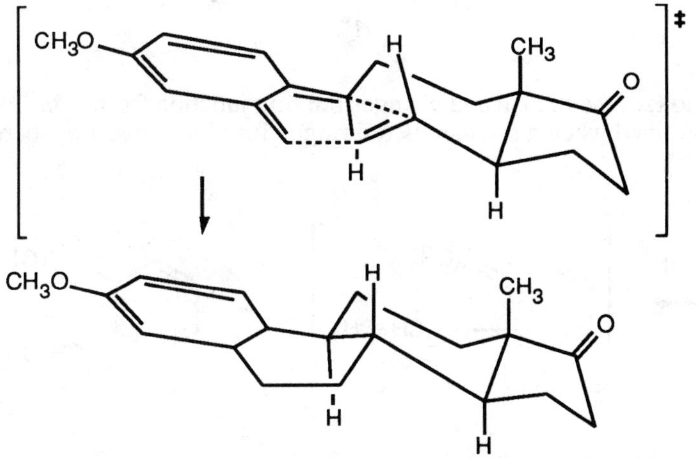

Formation of estrone methyl ether occurs by a Diels-Alder [4+2] cycloaddition.

30.35

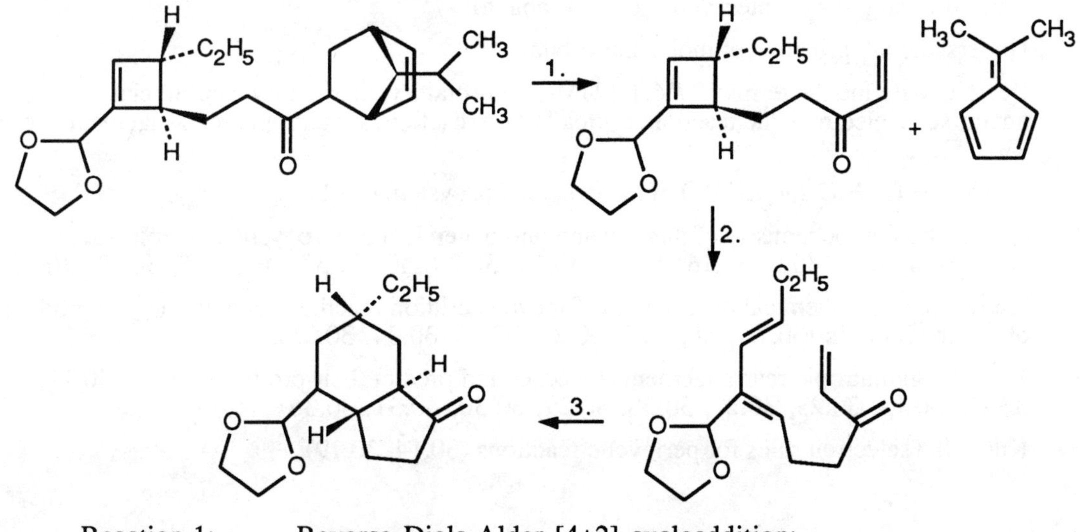

Reaction 1:   Reverse Diels-Alder [4+2] cycloaddition;
Reaction 2:   Conrotatory electrocyclic opening of a cyclobutene ring;
Reaction 3:   Diels-Alder [4+2] cycloaddition.

Treatment with base enolizes the ketone and changes the ring junction from *trans* to *cis*. A *cis* ring fusion is less strained when a six-membered ring is fused to a five-membered ring.

30.36

Study Guide for Chapter 30

After studying this chapter, you should be able to:

(1)   Understand the principle of molecular orbitals.

(2)   Be able to define the terms HOMO, LUMO, conrotatory, disrotatory, suprafacial, antarafacial, electrocyclic reaction, cycloaddition reaction, sigmatropic rearrangement (30.15).

(3)   Locate the HOMO and LUMO of a conjugated pi system (30.1).

(4)   Predict the stereochemistry of thermal and photochemical electrocyclic reactions (30.2, 30.3, 30.4, 30.5, 30.6, 30.16, 30.17, 30.18, 30.20, 30.22, 30.24, 30.25, 30.32, 30.33).

(5)   Know the stereochemical requirements for cycloaddition reactions, and predict the products of cycloadditions (30.7, 30.8, 30.9, 30.21, 30.26, 30.34, 30.35).

(6)   Classify sigmatropic rearrangements by order and predict their products (30.10, 30.11, 30.12, 30.13, 30.23, 30.27, 30.28, 30.29, 30.30, 30.31, 30.36).

(7)   Know the selection rules for pericyclic reactions (30.14, 30.19).

31.1

Monomer	Polymer

a) $CH_2=CHOCH_3$
Methyl vinyl ether

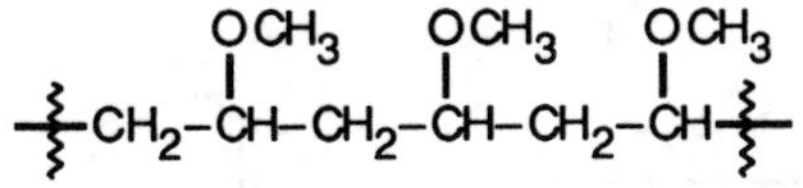

$$\left\{ CH_2-\underset{\underset{OCH_3}{|}}{CH}-CH_2-\underset{\underset{OCH_3}{|}}{CH}-CH_2-\underset{\underset{OCH_3}{|}}{CH} \right\}$$

b) CH_2 ⟨ring⟩ $=CH_2$

$$\left\{ CH_2-\bigcirc-CH_2-CH_2-\bigcirc-CH_2 \right\}$$

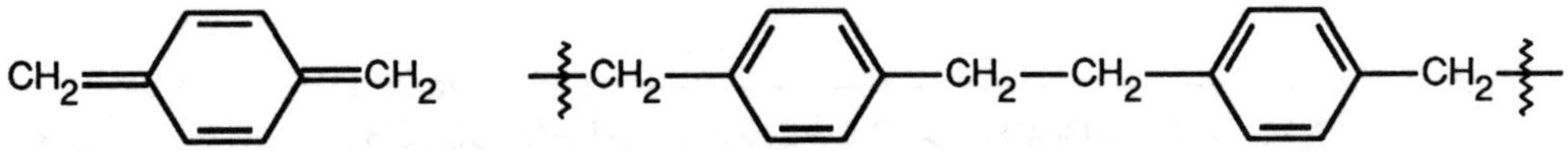

c)   CHCl=CHCl
1,2–Dichloroethylene

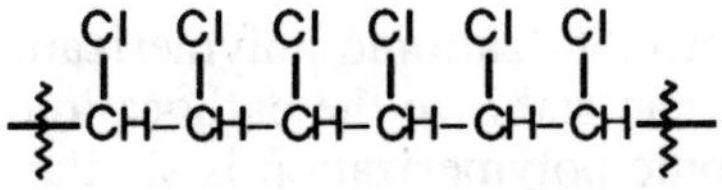

$$\left\{ \underset{\underset{Cl}{|}}{CH}-\underset{\underset{Cl}{|}}{CH}-\underset{\underset{Cl}{|}}{CH}-\underset{\underset{Cl}{|}}{CH}-\underset{\underset{Cl}{|}}{CH}-\underset{\underset{Cl}{|}}{CH} \right\}$$

31.2 Addition of an initiator to a styrene double bond produces a radical that can be stabilized by the phenyl ring.

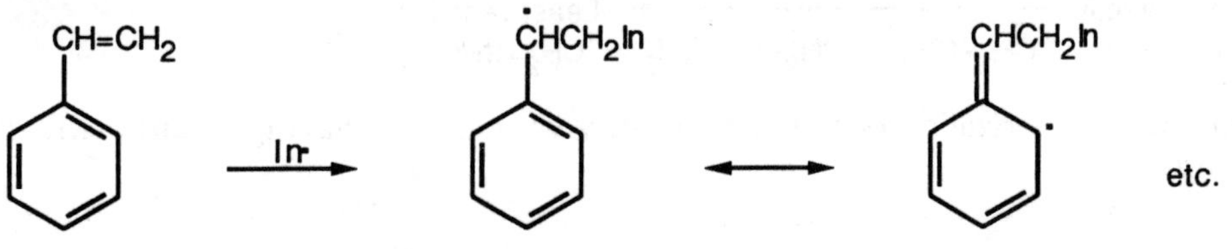

If this radical adds to the $CH_2=$ end of another styrene double bond, the new radical can also be stabilized by a phenyl ring.

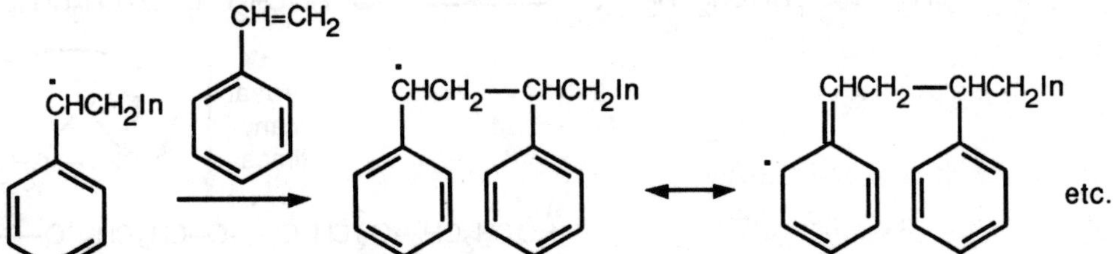

If addition occurs at the PhCH= end of the double bond, the product radical can't be stabilized. This product, with phenyl groups on neighboring carbons, is not formed.

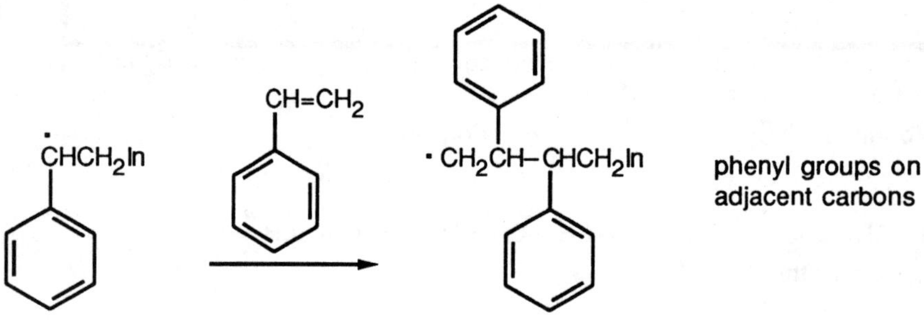

phenyl groups on
adjacent carbons

31.3

Most reactive ⟶ Least reactive
$CH_2=CHC_6H_5 > CH_2=CHCH_3 > CH_2=CHCl > CH_2=CHCO_2CH_3$

The alkenes most reactive to cationic polymerization contain electron-donating functional groups that can stabilize the carbocation intermediate. The reactivity order of substituents in cationic polymerization is similar to the reactivity order of substituted benzenes in electrophilic aromatic substitution.

31.4

Most reactive ⟶ Least reactive
$CH_2=CHCN > CH_2=CF_2 > CH_2=CHC_6H_5 > CH_2=CHCH_3$

Anionic polymerization occurs most readily with alkenes having electron-withdrawing groups.

31.5

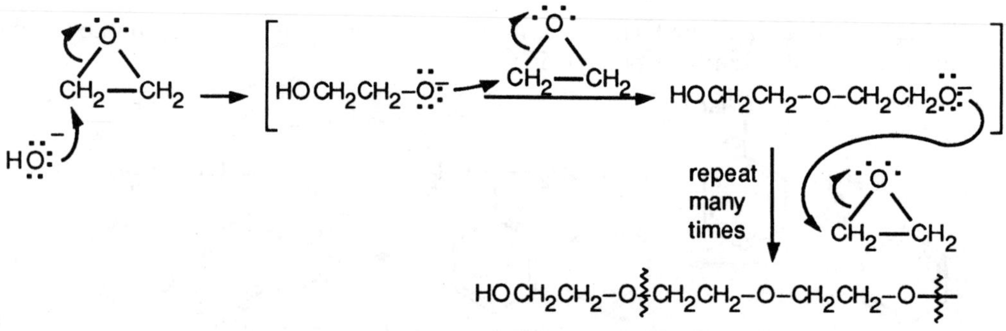

31.6

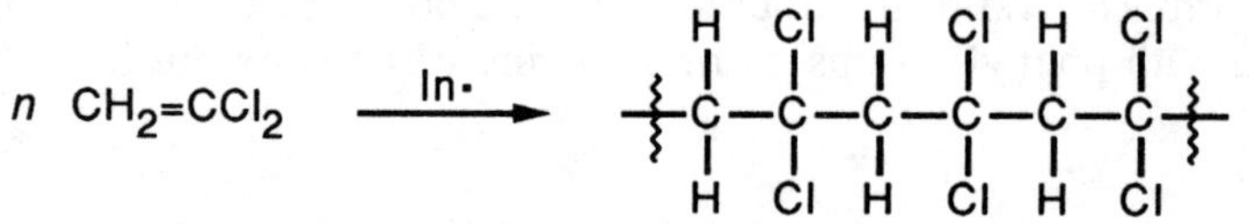

Vinylidene chloride does not polymerize in isotactic, syndiotactic, or atactic forms because no asymmetric centers are formed during polymerization.

31.7 None of the polypropylenes rotate plane-polarized light. If an optically inactive reagent and an achiral compound react, the product must be optically inactive. For every stereogenic center generated, an enantiomeric stereogenic center is also generated, and the resulting polymer mixture is racemic.

31.8 A vinyl branch in a diene polymer is the result of an occasional 1,2–double bond addition to the polymer chain.

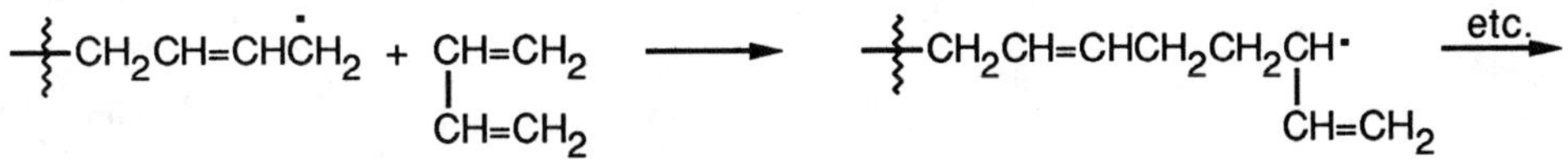

31.9

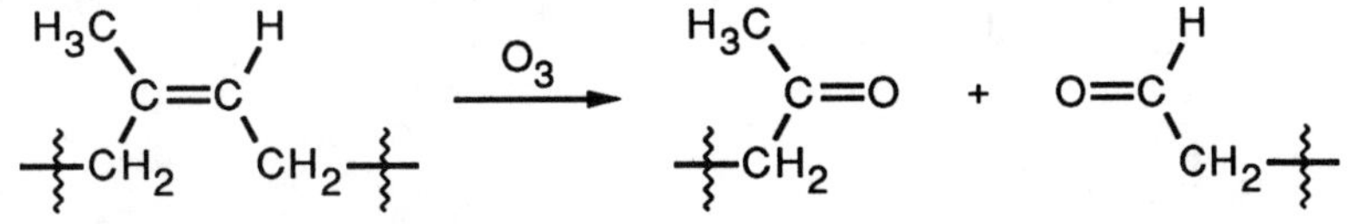

Ozone causes oxidative cleavage of the double bonds in rubber and breaks the polymer chain.

31.10

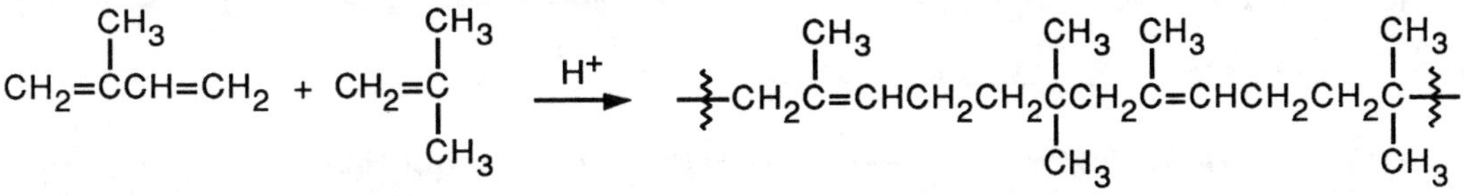

2–Methyl–1,3–butadiene   2–methyl–propene

31.11

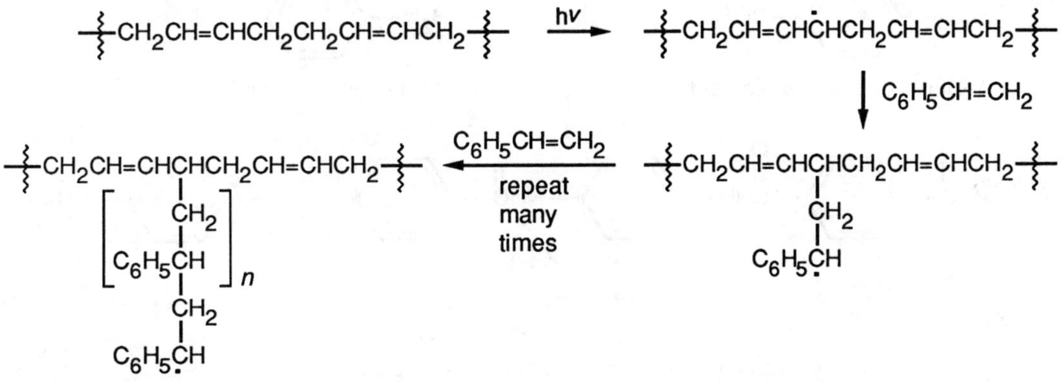

Irradiation homolytically cleaves an allylic C–H bond because it has the lowest bond energy. The resulting radical adds to styrene to produce a polystyrene graft.

31.12 Either aqueous acid or aqueous base hydrolyzes the amide bonds of nylon. The bonds of orlon and polyethylene are inert to acid and to base.

acid:

$$\text{\{-(CH}_2\text{)}_5\text{NHC(CH}_2\text{)}_5\text{-\}} \xrightarrow{H_3O^+} \left[\text{\{-(CH}_2\text{)}_5\text{NHC(CH}_2\text{)}_5\text{-\}}\right] \longrightarrow$$

$$\text{\{-(CH}_2\text{)}_5\overset{+}{\text{N}}\text{H}_3}$$
$$+$$
$$\text{HOOC(CH}_2\text{)}_5\text{-\}}$$

base:

$$\text{\{-(CH}_2\text{)}_5\text{NHC(CH}_2\text{)}_5\text{-\}} \xrightarrow{-OH} \left[\text{\{-(CH}_2\text{)}_5\text{NHC(CH}_2\text{)}_5\text{-\}}\right] \xrightarrow{H_2O}$$

$$\text{\{-(CH}_2\text{)}_5\text{NH}_2}$$
$$+$$
$$^-\text{OOC(CH}_2\text{)}_5\text{-\}}$$

31.13

a)　$BrCH_2CH_2CH_2Br + HOCH_2CH_2CH_2OH \xrightarrow{Base} \text{\{-CH}_2CH_2CH_2OCH_2CH_2CH_2O\text{-\}}$

b)　$HOCH_2CH_2OH + HOC(CH_2)_6COH \xrightarrow{H_2SO_4} \text{\{-OCH}_2CH_2OC(CH_2)_6COCH_2CH_2OC(CH_2)_6C\text{-\}}$

c)　$H_2N(CH_2)_6NH_2 + ClC(CH_2)_4CCl \longrightarrow \text{\{-NH(CH}_2)_6NHC(CH_2)_4CNH(CH_2)_6NHC(CH_2)_4C\text{-\}}$

31.14

HOC—⟨benzene⟩—COH　　+　　H₂N—⟨benzene⟩—NH₂

1,4–Benzenedicarboxylic acid　　　　　1,4–Benzenediamine

⟨-C—⟨benzene⟩—CNH—⟨benzene⟩—NHC—⟨benzene⟩—CNH—⟨benzene⟩—NH-⟩

Kevlar

31.15

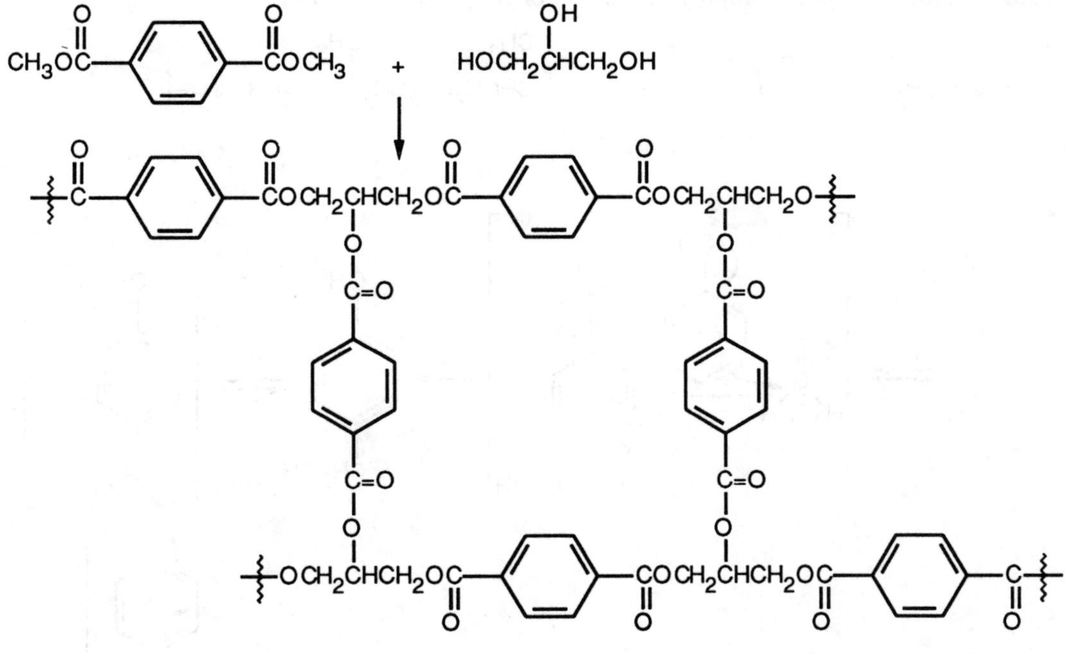

The product of the reaction of dimethyl terephthalate with glycerol has a high degree of cross-linking. This polymer is more rigid than Dacron.

31.16

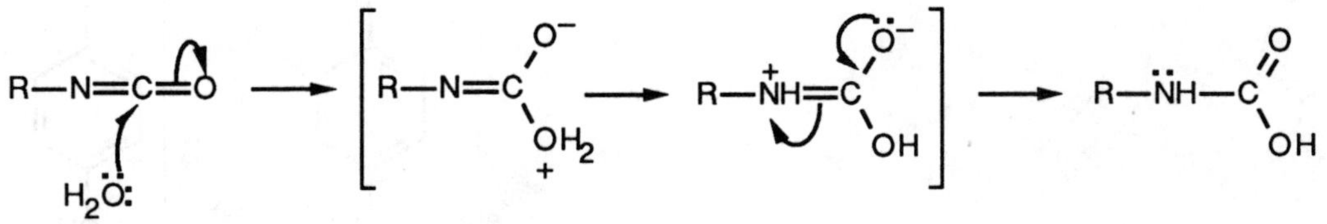

31.17

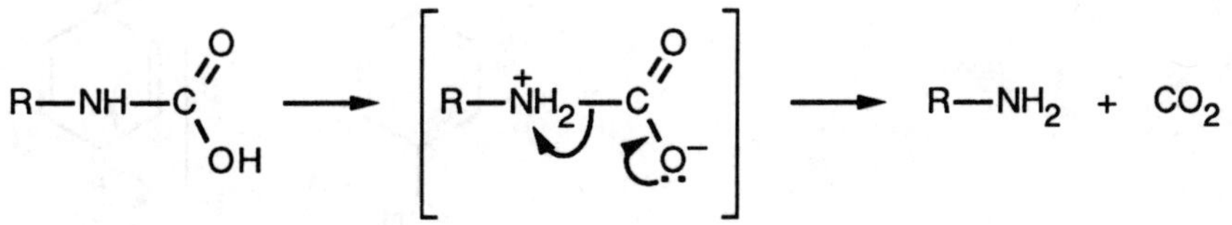

This is an elimination reaction.

31.18

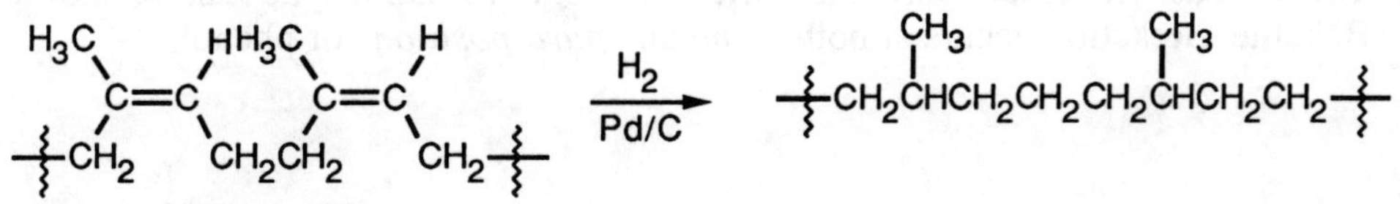

Natural rubber

The product of catalytic hydrogenation of natural rubber is atactic. This product also results from the radical copolymerization of propene with ethylene.

$$CH_3CH=CH_2 + CH_2=CH_2 \xrightarrow{\text{Irr}} \substack{CH_3 \\ |} \quad \substack{CH_3 \\ |}$$

31.19

This product can react many times with additional formaldehyde and phenol to yield Bakelite. Reaction occurs at both *ortho* and *para positions* of phenol.

31.20–31.21

Polymer	Monomer unit	Type of polymer
a) ─(CH₂CH–CH₂CH)─ with NO₂, NO₂	CH₂=CH with NO₂	chain-growth
b) ─(CFCl–CF₂–CFCl–CF₂)─	CFCl=CF₂	chain-growth
c) ─(CH₂–O–CH₂–O–CH₂–O)─	CH₂=O	chain-growth
d)	HO─⬡─CO₂H	step-growth
e) =CH(CH₂)₄CH=N–(CH₂)₆–N=	H₂N(CH₂)₆NH₂ and OHC(CH₂)₄CHO	step-growth

31.22

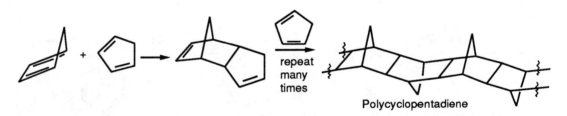

Polycyclopentadiene

Polycyclopentadiene is the product of successive Diels-Alder additions of cyclopentadiene to a growing polymer chain. Strong heat causes depolymerization of the chain and reversion to cyclopentadiene monomer units.

31.23

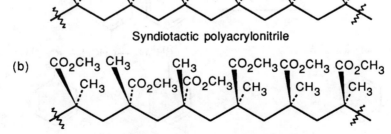

(a) Syndiotactic polyacrylonitrile

(b) Atactic poly(methyl methacrylate)

(c)

Isotactic poly(vinyl chloride)

31.24 1. *p*–Divinylbenzene is incorporated into the growing polystyrene chain.

2. Another growing polystyrene chain reacts with the second double bond of *p*–divinylbenzene.

3. The final product contains polystyrene chains cross-linked by *p*–divinylbenzene units.

31.25

$$CH_2=CHCH=CH_2 \xrightarrow[\text{1,4–addition}]{Br_2, CCl_4} BrCH_2CH=CHCH_2Br$$

1,3–Butadiene

$$\downarrow \begin{array}{c} 2\ NaCN, \\ H_3O^+ \end{array}$$

$$H_2NCH_2CH_2CH_2CH_2CH_2CH_2NH_2 \xleftarrow[\text{Pd}]{H_2} NCCH_2CH=CHCH_2CN + 2\ NaBr$$

1,6–Hexanediamine

31.26 The white coating on the distillation flask is caused by the thermal polymerization of nitroethylene.

$$n\ CH_2=\overset{\overset{\displaystyle NO_2}{|}}{CH} \xrightarrow{\text{heat}} \left(CH_2\overset{\overset{\displaystyle NO_2}{|}}{CH}\right)$$

31.27

$$n\ CH_2=\overset{\overset{\displaystyle O\overset{\displaystyle \|}{C}CH_3}{|}}{CH} \xrightarrow{\text{In}\cdot} \left(CH_2\overset{\overset{\displaystyle O\overset{\displaystyle \|}{C}CH_3}{|}}{CH}-CH_2\overset{\overset{\displaystyle O\overset{\displaystyle \|}{C}CH_3}{|}}{CH}\right) \xrightarrow[H_2O]{OH^-} \left(CH_2\overset{\overset{\displaystyle OH}{|}}{CH}-CH_2\overset{\overset{\displaystyle OH}{|}}{CH}\right)$$

Poly(vinyl alcohol) is formed by radical polymerization of vinyl acetate, followed by hydrolysis of the acetate groups.

$$\left(CH_2\overset{\overset{\displaystyle}{|}}{CH}-CH_2\overset{\overset{\displaystyle}{|}}{CH}\right)_{OH\ \ OH} + \overset{\overset{\displaystyle O}{\|}}{\underset{H\ \ n-C_3H_7}{C}} \xrightarrow{H^+} \left(CH_2CH-CH_2CH\right)$$

Reaction of poly(vinyl alcohol) with *n*–butanal produces poly(vinyl butyral).

31.28 The polyimide pictured is a step-growth polymer of a benzene tetracarboxylic acid and an aromatic diamine.

1,2,4,5–Benzene–   1,4–Benzene–   a polyimide
tetracarboxylic acid   diamine

31.29

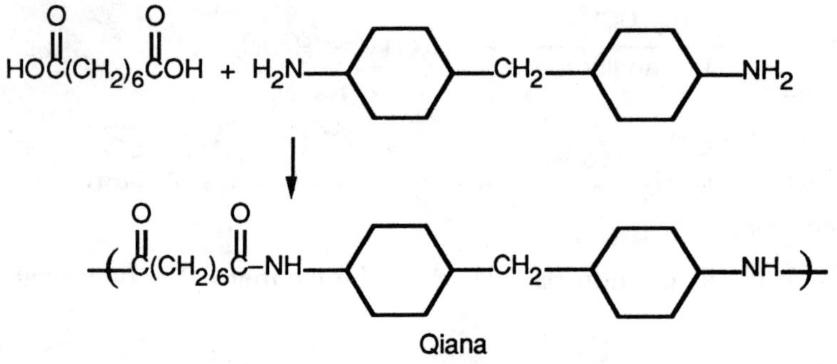

31.30

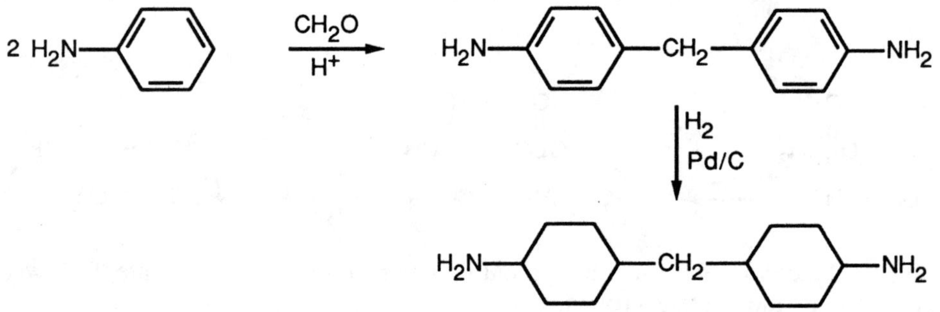

31.31

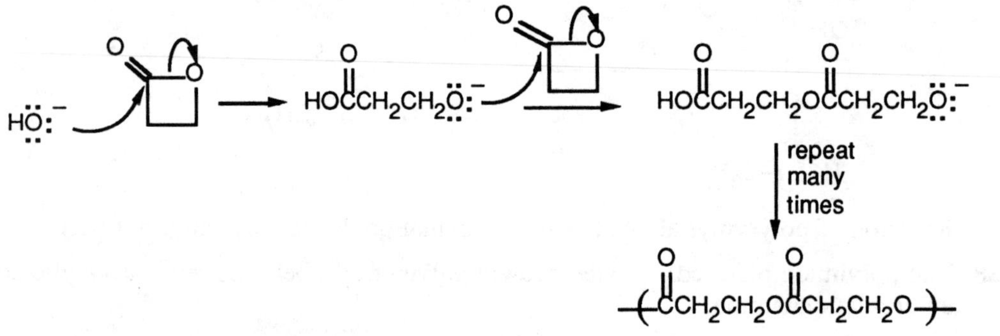

This polymer is a polyester.

31.32

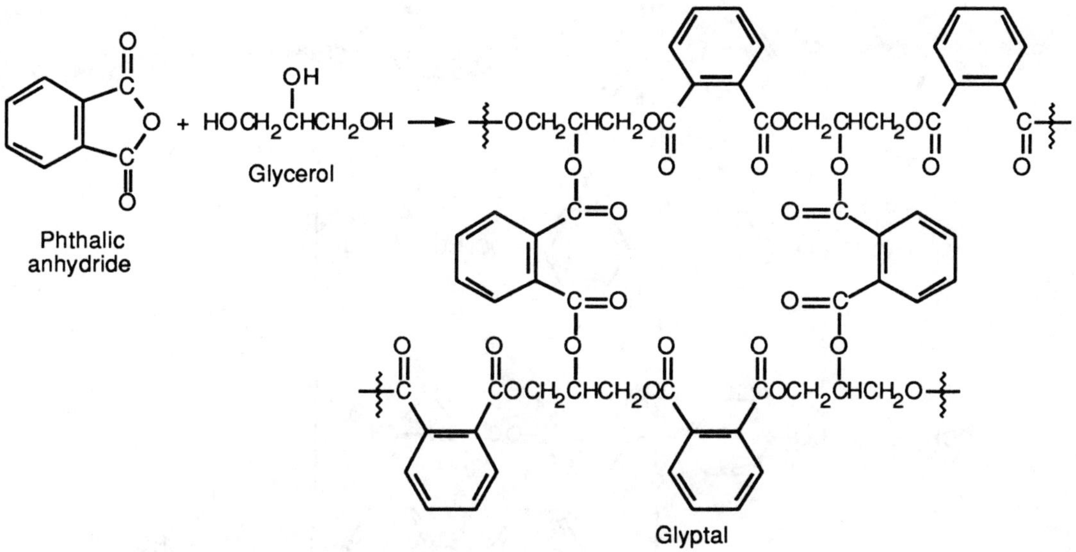

Phthalic anhydride + Glycerol → Glyptal

This polymer and the one in problem 31.15 are closely related.

31.33

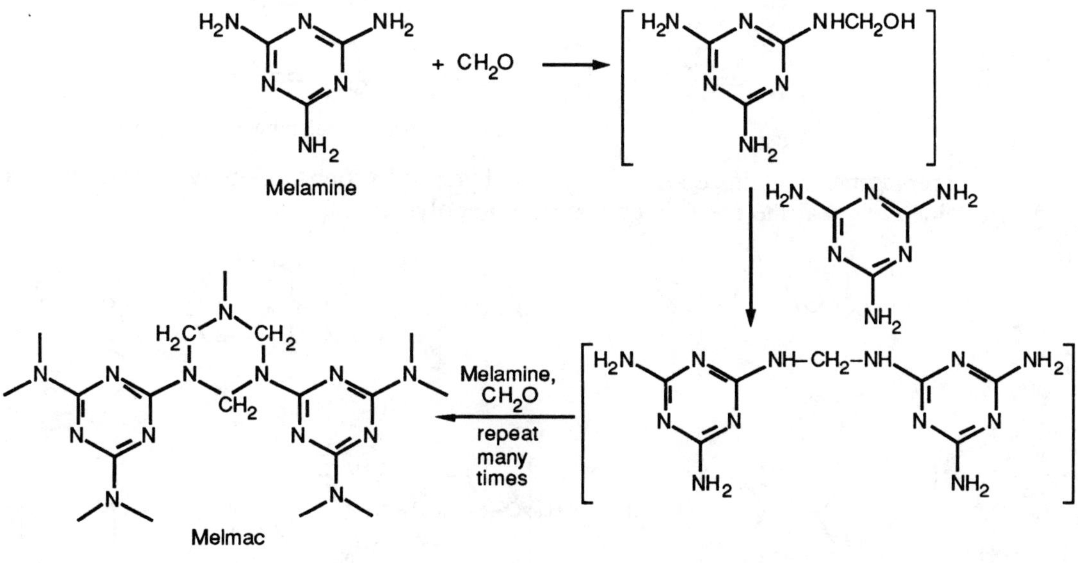

Melamine + CH_2O → ... Melamine, CH_2O, repeat many times → Melmac

31.34

a)

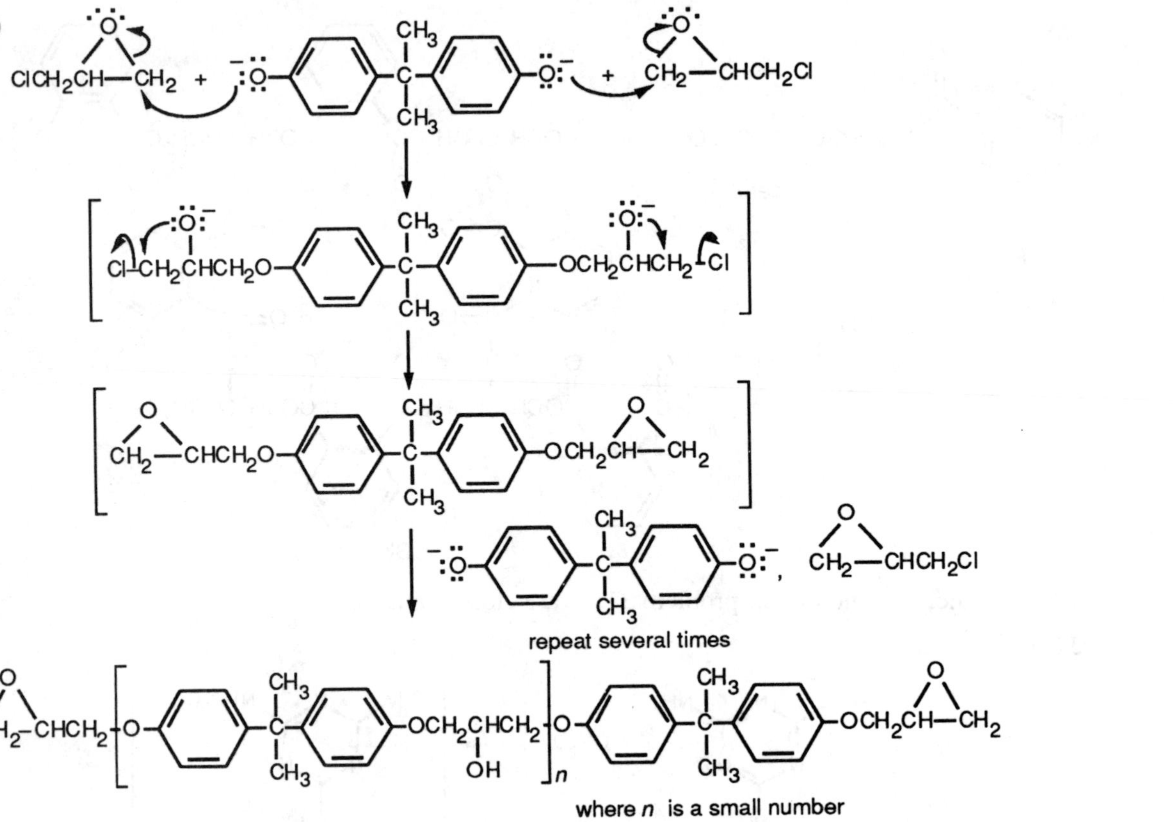
repeat several times
where *n* is a small number

The prepolymer contains epoxide rings and hydroxyl groups. Copolymerization with a triamine occurs at the epoxide ends of the prepolymer.

b)

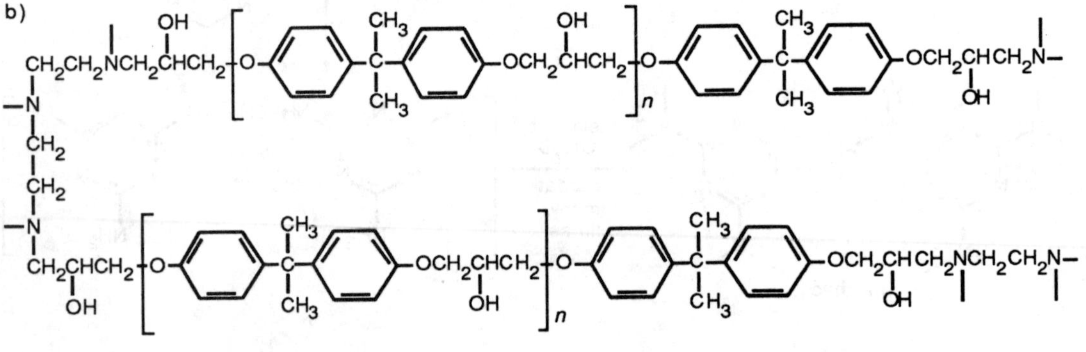

31.35 1. Formation of the diamine.

2.   Reaction of the diamine with two equivalents of phosgene.

MDI   + 2 HCl

31.36

31.37

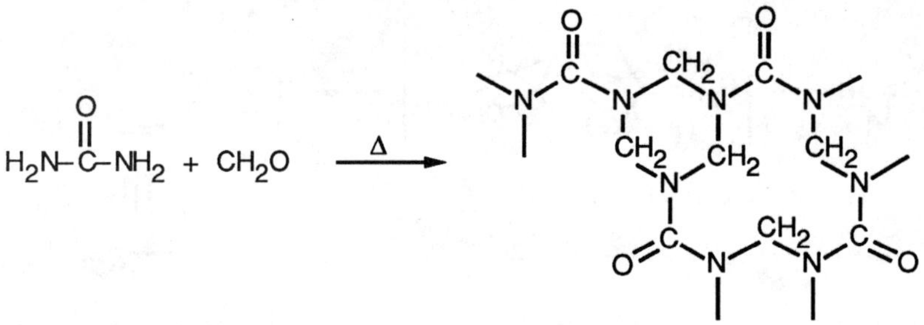

31.38

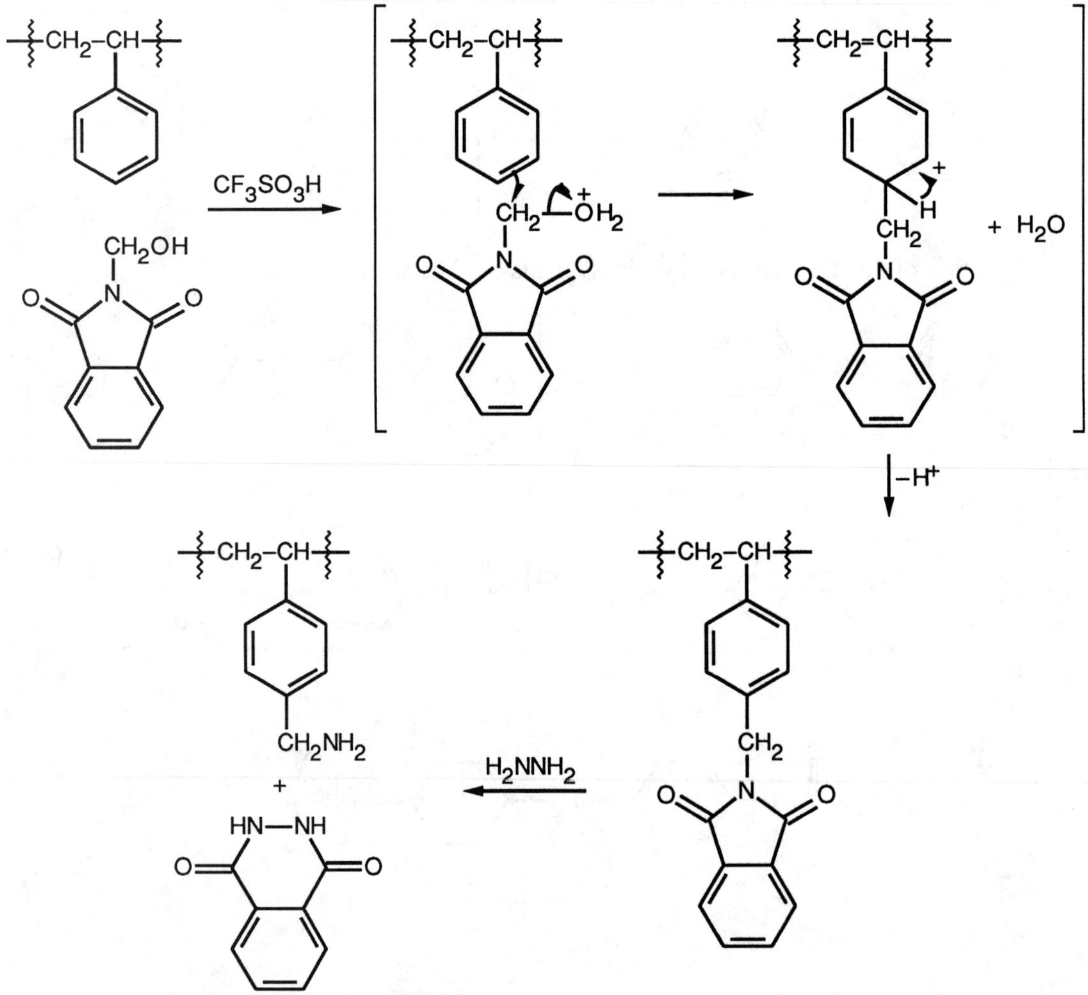

The mechanism for hydrolysis of the phthalimide is illustrated in Problem 25.49.

Study Guide for Chapter 31

After studying this chapter, you should be able to:

(1)   Locate the monomer units of a polymer, and deduce the structure of a polymer, given its monomer units (31.1, 31.10, 31.13, 31.14, 31.15, 31.20, 31.21, 31.22, 31.26, 31.28, 319, 31.30, 31.31, 31.32, 31.33, 31.36, 31.37).

(2)   Understand and formulate the mechanisms of

(a) Radical, cationic and anionic polymerizations (31.2, 31.3, 31.4, 31.5, 31.8, 31.9, 31.11, 31.24, 31.27).

(b) Step-growth polymerizations (31.12, 31.19, 31.34, 31.35, 31.38).

(3)   Understand the stereochemistry of polymerization and the structures of isotactic, syndiotactic and atactic polymers (31.6, 31.7, 31.18, 31.23).

(4)   Understand copolymerization, graft polymerization, and block polymerization.

Functional-Group Synthesis

The following table summarizes the synthetic methods by which important functional groups can be prepared. The functional groups are listed alphabetically, followed by reference to the appropriate text section and a brief description of each synthetic method.

Acetals, $R_2C(OR')_2$
(Sec. 19.14)   from ketones and aldehydes by acid-catalyzed reaction with alcohols

Acid anhydrides, RCOOCOR'
(Sec. 21.4)   from dicarboxylic acids by heating
(Sec. 21.6)   from acid chlorides by reaction with carboxylate salts

Acid bromides, RCOBr
(Sec. 21.5)   from carboxylic acids by reaction with PBr_3

Acid chlorides, RCOCl
(Sec. 21.4)   from carboxylic acids by reaction with either $SOCl_2$, PCl_3, or oxalyl chloride

Alcohols, ROH
(Sec. 7.4)   from alkenes by oxymercuration/demercuration
(Sec. 7.5)   from alkenes by hydroboration/oxidation
(Sec. 7.8)   from alkenes by hydroxylation with either OsO_4 or $KMnO_4$
(Sec. 11.4, 11.5)   from alkyl halides and tosylates by S_N2 reaction with hydroxide ion
(Sec. 18.6)   from ethers by acid-induced cleavage
(Sec. 18.8)   from epoxides by acid-catalyzed ring opening with either H_2O or HX
(Sec. 18.8)   from epoxides by base-induced ring opening
(Sec. 17.6, 19.11)   from ketones and aldehydes by reduction with $NaBH_4$ or $LiAlH_4$
(Sec. 17.7, 19.10)   from ketones and aldehydes by addition of Grignard reagents
(Sec. 20.8)   from carboxylic acids by reduction with either $LiAlH_4$ or BH_3
(Sec. 21.5)   from acid chlorides by reduction with $LiAlH_4$
(Sec. 21.5)   from acid chlorides by reaction with Grignard reagents
(Sec. 21.6)   from acid anhydrides by reduction with $LiAlH_4$
(Sec. 17.6, 21.7)   from esters by reduction with $LiAlH_4$
(Sec. 17.7, 21.7)   from esters by reaction with Grignard reagents

Aldehydes, RCHO
(Sec. 7.9)   from disubstituted alkenes by ozonolysis
(Sec. 7.10)   from 1,2-diols by cleavage with sodium periodate
(Sec. 8.6)   from terminal alkynes by hydroboration with disiamylborane followed by oxidation
(Sec. 17.9, 19.3)   from primary alcohols by oxidation
(Sec. 21.5)   from acid chlorides by partial reduction with $LiAl(O\text{-}t\text{-}Bu)_3H$
(Sec. 19.3, 21.7)   from esters by reduction with DIBAH [$H\text{-}Al(i\text{-}Bu)_2$]
(Sec. 21.9)   from nitriles by partial reduction with DIBAH

Alkanes, RH
(Sec. 7.7)	from alkenes by catalytic hydrogenation
(Sec. 10.9)	from alkyl halides by protonolysis of Grignard reagents
(Sec. 10.10)	from alkyl halides by coupling with Gilman reagents
(Sec. 19.13)	from ketones and aldehydes by Wolff-Kishner reaction
(Sec. 19.13)	from ketones and aldehydes by Clemmensen reduction
(Sec. 19.15)	from ketones and aldehydes by reduction of dithioacetals with Raney nickel

Alkenes, $R_2C=CR_2$
(Sec. 7.1, 11.11)	from alkyl halides by treatment with strong base (E2 reaction)
(Sec. 7.1, 17.8)	from alcohols by dehydration
(Sec. 8.6)	from alkynes by hydroboration followed by protonolysis
(Sec. 8.7)	from alkynes by catalytic hydrogenation using the Lindlar catalyst
(Sec. 8.7)	from alkynes by reduction with lithium in liquid ammonia
(Sec. 19.16)	from ketones and aldehydes by treatment with alkylidenetriphenyl-phosphoranes (Wittig reaction)
(Sec. 22.3)	from α-bromo ketones by heating with pyridine
(Sec. 22.8)	from saturated ketones by phenylselenenylation, followed by oxidative elimination
(Sec. 25.8)	from amines by methylation and Hofmann elimination

Alkynes, RC≡CR
(Sec. 8.3)	from dihalides by base-induced double dehydrohalogenation
(Sec. 8.9)	from terminal alkynes by alkylation of acetylide anions

Amides, $RCONH_2$
(Sec. 21.4)	from carboxylic acids by heating with ammonia
(Sec. 21.5, 25.8)	from acid chlorides by treatment with an amine or ammonia
(Sec. 21.6)	from acid anhydrides by treatment with an amine or ammonia
(Sec. 21.7)	from esters by treatment with an amine or ammonia
(Sec. 21.9)	from nitriles by partial hydrolysis with either acid or base
(Sec. 27.11)	from a carboxylic acid and an amine by treatment with dicyclohexyl-carbodiimide (DCC)

Amines, RNH_2
(Sec. 19.18)	from conjugated enones by addition of primary or secondary amines
(Sec. 21.8, 25.7)	from amides by reduction with $LiAlH_4$
(Sec. 21.9, 25.7)	from nitriles by reduction with $LiAlH_4$
(Sec. 25.7)	from primary alkyl halides by treatment with ammonia
(Sec. 25.7)	from primary alkyl halides by Gabriel synthesis
(Sec. 25.7)	from acid chlorides by Curtius rearrangement of acyl azides
(Sec. 25.7)	from primary amides by Hofmann rearrangement
(Sec. 25.7)	from primary alkyl azides by reduction with $LiAlH_4$
(Sec. 25.7)	from ketones and aldehydes by reductive amination with an amine and $NaBH_3CN$

Amino Acids, RCH(NH₂)COOH

(Sec. 27.4)	from α-bromo acids by S_N2 reaction with ammonia
(Sec. 27.4)	from aldehydes by reaction with KCN and ammonia (Strecker synthesis)
(Sec. 27.4)	from α-keto acids by reductive amination
(Sec. 27.4)	from primary alkyl halides by alkylation with diethyl acetamidomalonate

Arenes, Ar-R

(Sec. 16.3)	from arenes by Friedel-Crafts alkylation with a primary alkyl halide
(Sec. 16.12)	from aryl alkyl ketones by catalytic reduction of the keto group
(Sec. 26.3)	from arenediazonium salts by treatment with phosphorous acid

Arylamines, Ar-NH₂

(Sec. 16.2, 26.2)	from nitroarenes by reduction with either Fe, Sn, or H_2/Pd.

Arenediazonium salts, Ar-N₂⁺ X⁻

(Sec. 26.3)	from arylamines by reaction with nitrous acid

Arenesulfonic acids Ar-SO₃H

(Sec. 16.2)	from arenes by electrophilic aromatic substitution with SO_3/H_2SO_4

Azides, R-N₃

(Sec. 11.4, 25.7)	from primary alkyl halides by S_N2 reaction with azide ion

Carboxylic acids, RCOOH

(Sec. 7.9)	from mono- and 1,2-disubstituted alkenes by ozonolysis
(Sec. 16.11)	from arenes by side-chain oxidation with $Na_2Cr_2O_7$ or $KMnO_4$
(Sec. 19.5)	from aldehydes by oxidation
(Sec. 20.6)	from alkyl halides by conversion into Grignard reagents followed by reaction with CO_2
(Sec. 20.6, 21.9)	from nitriles by vigorous acid or base hydrolysis
(Sec. 21.5)	from acid chlorides by reaction with aqueous base
(Sec. 21.6)	from acid anhydrides by reaction with aqueous base
(Sec. 21.7)	from esters by hydrolysis with aqueous base
(Sec. 21.8)	from amides by hydrolysis with aqueous base
(Sec. 22.7)	from methyl ketones by reaction with halogen and base (haloform reaction)
(Sec. 26.8)	from phenols by treatment with CO_2 and base (Kolbe carboxylation)

Cyanohydrins, RCH(OH)CN

(Sec. 19.9)	from aldehydes and ketones by reaction with HCN

Cycloalkanes

(Sec. 7.11)	from alkenes by addition of dichlorocarbene
(Sec. 7.10)	from alkenes by reaction with CH_2I_2 and Zn/Cu (Simmons-Smith reaction)
(Sec. 16.12)	from arenes by rhodium-catalyzed hydrogenation

Disulfides, R-SS-R'
(Sec. 17.12)   from thiols by oxidation with bromine

Enamines, RCH=CRNR$_2$
(Sec. 19.12)   from ketones or aldehydes by reaction with secondary amines

Epoxides, $R_2C\overset{\displaystyle O}{\overset{/\ \backslash}{-}}CR_2$
(Sec. 18.7)   from alkenes by treatment with a peroxyacid
(Sec. 18.7)   from halohydrins by treatment with base

Esters, RCOOR'
(Sec. 21.4)   from carboxylic acid salts by S$_N$2 reaction with primary alkyl halides
(Sec. 21.4)   from carboxylic acids by acid-catalyzed reaction with an alcohol (Fischer esterification)
(Sec. 21.4)   from carboxylic acids by reaction with diazomethane
(Sec. 21.5)   from acid chlorides by base-induced reaction with an alcohol
(Sec. 21.6)   from acid anhydrides by base-induced reaction with an alcohol
(Sec. 22.9)   from alkyl halides by alkylation with diethyl malonate
(Sec. 22.9)   from esters by treatment of their enolate ions with alkyl halides
(Sec. 26.8)   from phenols by base-induced reaction with an acid chloride

Ethers, R-O-R'
(Sec. 16.9)   from activated haloarenes by reaction with alkoxide ions
(Sec. 16.10)   from unactivated haloarenes by reaction with alkoxide ions via benzyne intermediates
(Sec. 18.4)   from primary alkyl halides by S$_N$2 reaction with alkoxide ions (Williamson ether synthesis)
(Sec. 18.5)   from alkenes by alkoxymercuration/demercuration
(Sec. 18.7)   from alkenes by epoxidation with peroxyacids
(Sec. 26.8)   from phenols by reaction of phenoxide ions with primary alkyl halides

Halides, alkyl, R$_3$C-X
(Sec. 6.9)   from alkenes by electrophilic addition of HX
(Sec. 7.2)   from alkenes by addition of halogen
(Sec. 7.3)   from alkenes by electrophilic addition of hypohalous acid (HOX) to yield halohydrins
(Sec. 7.6)   from alkenes by radical-catalyzed addition of HBr
(Sec. 8.4)   from alkynes by addition of halogen
(Sec. 8.4)   from alkynes by addition of HX
(Sec. 10.5)   from alkenes by allylic bromination with N-bromosuccinimide (NBS)
(Sec. 10.8)   from alcohols by reaction with HX
(Sec. 10.8)   from alcohols by reaction with SOCl$_2$
(Sec. 10.8)   from alcohols by reaction with PBr$_3$
(Sec. 11.4, 11.5) from alkyl tosylates by S$_N$2 reaction with halide ions
(Sec. 16.11)   from arenes by benzylic bromination with N-bromosuccinimide (NBS)
(Sec. 18.6)   from ethers by cleavage with either HX

(Sec. 20.9)   from carboxylic acids by treatment with Ag_2O and bromine (Huns-diecker reaction)

(Sec. 22.3)   from ketones by alpha-halogenation with bromine

(Sec. 22.4)   from carboxylic acids by alpha-halogenation with phosphorus and bromine (Hell-Volhard-Zelinskii reaction)

Halides, aryl, Ar-X

(Sec. 16.1, 16.2) from arenes by electrophilic aromatic substitution with halogen

(Sec. 26.3)   from arenediazonium salts by reaction with cuprous halides (Sandmeyer reaction)

(Sec. 29.3)   from aromatic heterocycles by electrophilic aromatic substitution with halogen

Halohydrins, $R_2CXC(OH)R_2$

(Sec. 7.3)   from alkenes by electrophilic addition of hypohalous acid (HOX)

(Sec. 18.8)   from epoxides by acid-induced ring opening with HX

Imines. $R_2C=NR'$

(Sec. 19.12)   from ketones or aldehydes by reaction with primary amines

Ketones, $R_2C=O$

(Sec. 7.9)   from alkenes by ozonolysis

(Sec. 7.10)   from 1,2-diols by cleavage reaction with sodium periodate

(Sec. 8.5)   from alkynes by mercuric-ion-catalyzed hydration

(Sec. 8.6)   from alkynes by hydroboration/oxidation

(Sec. 16.4)   from arenes by Lewis-acid-catalyzed reaction with an acid chloride (Friedel-Crafts acylation)

(Sec. 17.9, 19.4) from secondary alcohols by oxidation

(Sec. 19.4, 21.5) from acid chlorides by reaction with lithium diorganocopper (Gilman) reagents

(Sec. 19.18)   from conjugated enones by addition of lithium diorganocopper reagents

(Sec. 21.9)   from nitriles by reaction with Grignard reagents

(Sec. 22.9)   from primary alkyl halides by alkylation with ethyl acetoacetate

(Sec. 22.9)   from ketones by alkylation of their enolate ions with primary alkyl halides

Nitriles, R-C≡N

(Sec. 11.5, 20.6) from primary alkyl halides by S_N2 reaction with cyanide ion

(Sec. 19.18)   from conjugated enones by addition of HCN

(Sec. 21.9)   from primary amides by dehydration with $SOCl_2$

(Sec. 22.9)   from nitriles by alkylation of their alpha-anions with primary alkyl halides

(Sec. 26.3)   from arenediazonium ions by treatment with CuCN

Nitroarenes, Ar-NO$_2$

(Sec. 16.2)   from arenes by electrophilic aromatic substitution with nitric/sulfuric acids

Organometallics, R-M

(Sec. 10.9)	formation of Grignard reagents from organohalides by treatment with magnesium
(Sec. 10.10)	formation of organolithium reagents from organohalides by treatment with lithium
(Sec. 10.10)	formation of lithium diorganocopper reagents (Gilman reagents) from organolithium reagents by treatment with cuprous halides

Phenols, Ar-OH

(Sec. 16.2, 26.7)	from arenesulfonic acids by fusion with KOH
(Sec. 26.3)	from arenediazonium salts by reaction with aqueous acid
(Sec. 16.9)	from aryl halides by nucleophilic aromatic substitution with hydroxide ion

Quinones, $O=\langle\!\!\!\!\bigcirc\!\!\!\!\rangle=O$

(Sec. 26.8)	from phenols by oxidation with Fremy's salt [$(KSO_3)_2NO$]
(Sec. 26.8)	from arylamines by oxidation with Fremy's salt

Sulfides, R-S-R'

(Sec. 18.11)	from thiols by S_N2 reaction of thiolate ions with primary alkyl halides

Sulfones, R-SO$_2$-R'

(Sec. 18.11)	from sulfides or sulfoxides by oxidation with peroxyacids

Sulfoxides, R-SO-R'

(Sec. 18.11)	from sulfides by oxidation with H_2O_2

Thioacetals, R$_2$C(SR')$_2$

(Sec. 19.15)	from ketones and aldehydes by acid-catalyzed reaction with thiols

Thiols, R-SH

(Sec. 10.5)	from primary alkyl halides by S_N2 reaction with hydrosulfide anion
(Sec. 17.12)	from primary alkyl halides by S_N2 reaction with thiourea followed by hydrolysis

The following table summarizes the reactions that important functional groups undergo. The functional groups are listed alphabetically, followed by a reference to the appropriate text section.

Acetals
1. Hydrolysis to yield a ketone or aldehyde plus alcohol (Sec. 19.14)

Acid anhydrides
1. Hydrolysis to yield a carboxylic acid (Sec. 21.6)
2. Alcoholysis to yield an ester (Sec. 21.6)
3. Aminolysis to yield an amide (Sec. 21.6)
4. Reduction to yield a primary alcohol (Sec. 21.6)

Acid chlorides
1. Hydrolysis to yield a carboxylic acid (Sec. 21.5)
2. Alcoholysis to yield an ester (Sec. 21.5)
3. Aminolysis to yield an amide (Sec. 21.5)
4. Reduction to yield a primary alcohol (Sec. 21.5)
5. Partial reduction to yield an aldehyde (Sec. 21.5)
6. Grignard reaction to yield a tertiary alcohol (Sec. 21.5)
7. Reaction with lithium diorganocuprates to yield ketones (Sec. 21.5)

Alcohols
1. Acidity (Sec. 17.4)
2. Oxidation (Sec. 17.9)
   a. Primary alcohol
   b. Secondary alcohol
3. Reaction with carboxylic acids to yield esters (Sec. 21.4)
4. Reaction with acid chlorides to yield esters (Sec. 21.5)
5. Dehydration to yield alkenes (Sec. 17.9)
6. Reaction with primary alkyl halides to yield ethers (Sec. 18.4)
7. Conversion into alkyl halides (Sec. 17.8)
   a. Reaction of primary and secondary alcohols with $SOCl_2$
   b. Reaction of primary and secondary alcohols with PBr_3

Aldehydes
1. Oxidation to yield carboxylic acids (Sec. 19.5)
2. Nucleophilic addition reactions
   a. Reduction to yield primary alcohols (Sec. 17.6)
   b. Reaction with Grignard reagents to yield secondary alcohols (Sec. 17.7)
   c. Grignard reaction of formaldehyde to yield primary alcohols (Sec. 17.7)
   d. Reaction with HCN to yield a cyanohydrin (Sec. 19.9)
   e. Wolff-Kishner reaction with hydrazine to yield alkanes (Sec. 19.13)
   f. Clemmensen reduction with zinc amalgam to yield alkanes (Sec. 19.13)

g. Reaction with alcohol to yield an acetal (Sec. 19.14)
h. Reaction with a thiol to yield a thioacetal (Sec. 19.15)
i. Wittig reaction to yield alkenes (Sec. 19.16)
j. Reaction with an amine to yield an imine (Sec. 19.12)
3. Aldol reaction to yield a β-hydroxy aldehyde (Sec. 23.2)
4. Alpha bromination of aldehydes (Sec. 22.3)

Alkanes
1. Radical halogenation of methane (Sec. 10.4)

Alkenes
1. Electrophilic addition of HX to yield an alkyl halide (Secs. 6.9 - 6.13)
   Markovnikov regiochemistry is observed. H adds to the less highly substituted carbon and X adds to the more highly substituted one.
2. Electrophilic addition of halogen to yield a 1,2-dihalide (Sec. 7.2)
   Anti stereochemistry is observed
3. Oxymercuration to yield an alcohol (Sec. 7.4)
   Markovnikov regiochemistry is observed, yielding the more highly substituted alcohol.
4. Hydroboration/oxidation to yield an alcohol (Section 7.5)
5. Hydrogenation of alkenes to yield alkanes (Sec. 7.7)
6. Hydroxylation of alkenes to yield 1,2-diols (Sec. 7.8)
7. Radical addition of HBr to yield an alkyl bromide (Sec. 7.6)
8. Oxidative cleavage of alkenes to yield carbonyl compounds (Sec. 7.10)
9. Reaction with peroxyacids to yield epoxides (Sec. 18.7)
10. Simmons-Smith reaction with CH_2I_2 to yield a cyclopropane (Sec. 7.11)

Alkynes
1. Electrophilic addition of HX to yield a vinylic halide (Sec. 8.4)
2. Electrophilic addition of halogen to yield a dihalide (Sec. 8.4)
3. Mercuric-sulfate-catalyzed hydration to yield a methyl ketone (Sec. 8.5)
4. Hydroboration with disiamylborane to yield an aldehyde (Sec. 8.6)
5. Alkylation of alkyne anions (Sec. 8.9)
6. Reduction of alkynes (Sec. 8.7)
   a. Hydrogenation to yield a cis alkene
   b. Reduction to yield a trans alkene

Amides
1. Hydrolysis to yield carboxylic acids (Sec. 21.8)
2. Reduction with $LiAlH_4$ to yield amines (Sec. 21.8)
3. Dehydration to yield nitriles (Section 21.9)

Amines
1. S_N2 alkylation of alkyl halides to yield amines (Sec. 25.8)
2. Nucleophilic acyl substitution reactions
   a. Reaction with acid chlorides to yield amides (Sec. 21.5)
   b. Reaction with acid anhydrides to yield amides (Sec. 21.6)
   c. Reaction with esters to yield amides (Sec. 21.7)
3. Hofmann elimination to yield alkenes (Sec. 21.8)
4. Formation of arenediazonium salts (Sec. 26.3)

Arenes
1. Oxidation of the alkylbenzene side chain (Sec. 16.11)
2. Catalytic reduction to yield a cyclohexane (Sec. 16.12)
3. Birch reduction to yield a cyclohexadiene (Sec. 16.12)
3. Reduction of aryl alkyl ketones (Sec. 16.12)
4. Electrophilic aromatic substitution (Secs. 16.1 - 16.4)
    a. Bromination (Sec. 16.1)
    b. Chlorination (Sec. 16.2)
    c. Iodination (Sec. 16.2)
    d. Nitration (Sec. 16.2)
    e. Sulfonation (Sec. 16.2)
    f. Friedel-Crafts alkylation (Sec. 16.3)
        Aromatic ring must be at least as reactive as a halobenzene
        R must be methyl, ethyl, secondary, or tertiary
    g. Friedel-Crafts acylation (Sec. 16.4)

Arenediazonium salts
1. Conversion into aryl chlorides (Sec. 26.3)
2. Conversion into aryl bromides (Sec. 26.3)
3. Conversion into aryl iodides (Sec. 26.3)
4. Conversion into aryl cyanides (Sec. 26.3)
5. Conversion into phenols (Sec. 26.3)
6. Conversion into arenes (Sec. 26.3)

Arenesulfonic acids
1. Conversion into phenols (Sec. 16.2)

Carboxylic acids
1. Acidity (Secs. 20.3 - 20.5)
2. Reduction to yield a primary alcohol (Sec. 17.6)
    a. Reduction with $LiAlH_4$
    b. Reduction with BH_3
3. Nucleophilic acyl substitution reactions (Sec. 21.4)
    a. Conversion into an acid chloride
    b. Conversion into an acid anhydride
    c. Conversion into an ester
        (1) Fischer esterification
        (2) S_N2 reaction with an alkyl halide
        (3) Reaction with diazomethane

Epoxides
1. Acid-catalyzed ring opening with HX to yield a halohydrin (Sec. 18.8)
2. Ring opening with aqueous acid to yield a 1,2-diol (Sec. 18.8)

Esters

1. Hydrolysis to yield a carboxylic acid (Sec. 21.7)
2. Aminolysis to yield an amide (Sec. 21.7)
3. Reduction to yield a primary alcohol (Sec. 17.6)
4. Partial reduction with DIBAH to yield an aldehyde (Sec. 21.7)
5. Grignard reaction to yield a tertiary alcohol (Sec. 17.7)
6. Claisen condensation to yield a β-keto ester (Sec. 23.9)

Ethers

1. Acid-induced cleavage (Sec. 18.6)

Halides, alkyl

1. Reaction with magnesium to form Grignard reagents (Sec. 10.9)
2. Reduction to yield alkanes (Sec. 10.9)
3. Nucleophilic substitution (S_N1 or S_N2) (Secs. 11.1 - 11.9)
4. Dehydrohalogenation (E1 or E2) (Secs. 11.10-11.14)

Halohydrins

1. Conversion into epoxides (Sec. 18.7)

Ketones

1. Nucleophilic addition reactions
   a. Reduction to yield a secondary alcohol (Secs. 17.6)
   b. Reaction with Grignard reagents to yield a tertiary alcohol (Sec. 17.7)
   c. Wolff-Kishner reaction with hydrazine to yield an alkane (Sec. 19.13)
   d. Reaction with HCN to yield a cyanohydrin (Sec. 19.9)
   e. Reaction with alcohol to yield an acetal (Sec. 19.14)
   f. Reaction with a thiol to yield a thioacetal (Sec. 19.15)
   g. Wittig reaction to yield an alkene (Sec. 19.16)
   h. Reaction with an amine to yield an imine (Sec. 19.12)
2. Aldol reaction to yield a β-hydroxy ketone (Sec. 23.2)
3. Alpha bromination of ketones (Sec. 22.3)
4. Clemmensen reduction with zinc amalgam to yield alkanes (Sec. 19.13)

Nitriles

1. Hydrolysis to yield a carboxylic acid (Sec. 21.9)
2. Reduction to yield a primary amine (Sec. 21.9)
3. Partial reduction with DIBAH to yield an aldehyde (Sec. 21.9)
4. Reaction with Grignard reagents to yield ketones (Sec. 21.9)

Nitroarenes

1. Reduction to yield arylamines (Sec. 26.2)

Organometallics

1. Reduction by treatment with acid (Sec. 10.9)
2. Nucleophilic addition to carbonyl compounds (Secs. 17.7)
3. Conjugate addition of lithium diorganocuprates to α,β-unsaturated ketones (Sec. 19.18)
4. Coupling reaction of lithium diorganocuprates with alkyl halides (Sec. 10.10)
5. Reaction with carbon dioxide to yield a carboxylic acid (Sec. 20.6)

Phenols
1. Acidity (Sec. 26.6)
2. Reaction with acid chlorides to yield esters (Sec. 26.8)
3. Reaction with alkyl halides to yield ethers (Sec. 26.8)
4. Oxidation to yield quinones (Sec. 26.8)

Quinones

1. Reduction to yield hydroquinones (Sec. 26.8)

Sulfides
1. Reaction with alkyl halides to yield sulfonium salts (Sec. 18.11)
2. Oxidation to yield sulfoxides (Sec. 18.11)
3. Oxidation to yield sulfones, (Sec. 18.11)

Thiols
1. Reaction with alkyl halides to yields sulfides (Sec. 17.12)
2. Oxidation to yield disulfides (Sec. 17.12)

Reagents in Organic Chemistry

The following table summarizes the uses of some important reagents in organic chemistry. The reagents are listed alphabetically, followed by a brief description of the uses of each and references to the appropriate text sections.

Acetic acid, CH$_3$COOH: Reacts with vinylic organoboranes to yield alkenes. The net effect of alkyne hydroboration followed by protonolysis with acetic acid is reduction of the alkyne to a cis alkene (Section 8.6).

Acetic anhydride, (CH$_3$CO)$_2$O: Reacts with alcohols to yield acetate esters (Sections 21.6, 24.8).

Aluminum chloride, AlCl$_3$: Acts as a Lewis acid catalyst in Friedel-Crafts alkylation and acylation reactions of aromatic-ring compounds (Sections 16.3, 16.4).

Ammonia, NH$_3$: A solvent for the reduction of alkynes by lithium metal to yield trans alkenes (Section 8.7).
- Reacts with acid chlorides to yield amides (Section 21.5).

Borane, BH$_3$: Adds to alkenes, giving alkylboranes that can be oxidized with alkaline hydrogen peroxide to yield alcohols (Section 7.5).
- Adds to alkynes, giving vinylic organoboranes that can either be treated with acetic acid to yield cis alkenes (Section 8.6) or can be oxidized with alkaline hydrogen peroxide to yield aldehydes (Section 8.6).
- Reduces carboxylic acids to yield primary alcohols (Section 20.8).

Bromine, Br$_2$: Adds to alkenes yielding 1,2-dibromides (Sections 7.2, 14.5).
- Adds to alkynes yielding either 1,2-dibromoalkenes or 1,1,2,2-tetrabromoalkanes (Section 8.4).
- Reacts with arenes in the presence of ferric bromide catalyst to yield bromoarenes (Section 16.1).
- Reacts with ketones in acetic acid solvent to yield α-bromo ketones (Section 22.3).
- Reacts with carboxylic acids in the presence of phosphorus tribromide to yield α-bromo carboxylic acids (Hell-Volhard-Zelinskii reaction; Section 22.4).
- Reacts with methyl ketones in the presence of sodium hydroxide to yield carboxylic acids and bromoform (Haloform reaction; Section 22.7).
- Oxidizes aldoses to yield aldonic acids (Section 24.8).

N-Bromosuccinimide (NBS), (CH$_2$CO)$_2$NBr: Reacts with alkenes in the presence of aqueous dimethylsulfoxide to yield bromohydrins (Section 7.3).
- Reacts with alkenes in the presence of light to yield allylic bromides (Wohl-Ziegler reaction; Section 10.5).
- Reacts with alkylbenzenes in the presence of light to yield benzylic bromides; (Section 16.11).

Di-*tert*-butoxy dicarbonate, (t-BuOCO)$_2$O: Reacts with amino acids to give t-BOC protected amino acids suitable for use in peptide synthesis (Section 27.11).

Butyllithium, CH$_3$CH$_2$CH$_2$CH$_2$Li: A strong base that reacts with alkynes to yield acetylide anions that can be alkylated (Section 8.8).
- Reacts with dialkylamines to yield lithium dialkylamide bases such as LDA [lithium diisopropylamide] (Section 22.5).

Carbon dioxide, CO$_2$: Reacts with Grignard reagents to yield carboxylic acids (Section 20.6).
- Reacts with phenoxide anions to yield o-hydroxybenzoic acids (Kolbe-Schmitt carboxylation reaction; Section 26.8).

Chlorine, Cl$_2$: Adds to alkenes to yield 1,2-dichlorides (Section 7.2, 14.5).
- Reacts with alkanes in the presence of light to yield chloroalkanes by a radical chain reaction pathway (Section 10.4).
- Reacts with arenes in the presence of ferric chloride catalyst to yield chloroarenes (Section 16.2).

m-Chloroperoxybenzoic acid, m-ClC$_6$H$_4$CO$_3$H: Reacts with alkenes to yield epoxides (Section 18.7).

Chlorotrimethylsilane, (CH$_3$)$_3$SiCl: Reacts with alcohols to add the trimethylsilyl protecting group (Section 17.10).

Chromium trioxide, CrO$_3$: Oxidizes alcohols in aqueous sulfuric acid (Jones reagent) to yield carbonyl-containing products. Primary alcohols yield carboxylic acids and secondary alcohols yield ketones (Sections 17.9, 19.5, 20.6).

Cuprous bromide, CuBr: Reacts with arenediazonium salts to yield bromoarenes (Sandmeyer reaction; Section 26.3).

Cuprous chloride, CuCl: Reacts with arenediazonium salts to yield chloroarenes (Sandmeyer reaction; Section 26.3).

Cuprous cyanide, CuCN: Reacts with arenediazonium salts to yield substituted benzonitriles (Sandmeyer reaction; Section 26.3).

Cuprous iodide, CuI: Reacts with organolithiums to yield lithium diorganocopper reagents (Gilman reagents; Section 10.10).

Diazomethane, CH$_2$N$_2$: Reacts with carboxylic acids to yield methyl esters (Section 21.4).

Dichloroacetic acid, Cl$_2$CHCOOH: Acts as a reagent for cleaving DMT protecting groups in DNA synthesis (Section 29.16).

Dicyclohexylcarbodiimide (DCC), C$_6$H$_{11}$-N=C=N-C$_6$H$_{11}$: Couples an amine with a carboxylic acid to yield an amide. DCC is often used in peptide synthesis (Section 27.11).

Diethyl acetamidomalonate, $CH_3CONHCH(CO_2Et)_2$: Reacts with alkyl halides in a common method of α-amino acid synthesis (Section 27.4).

Diethylaluminum cyanide, $(Et)_2AlCN$: Reacts with α,β-unsaturated ketones to yield β-keto nitriles (Section 19.18).

Diiodomethane, CH_2I_2: Reacts with alkenes in the presence of zinc-copper couple to yield cyclopropanes (Simmons-Smith reaction; Section 7.11).

Diisobutylaluminum hydride (DIBAH), $(i\text{-}Bu)_2AlH$: Reduces esters to yield aldehydes (Sections 19.3, 21.7).
- Reduces nitriles to yield aldehydes (Section 21.9).

2,4-Dinitrophenylhydrazine, $2,4\text{-}(NO_2)_2C_6H_3NHNH_2$: Reacts with ketones and aldehydes to yield 2,4-DNPs that serve as useful crystalline derivatives (Section 19.12).

Disiamylborane, $[(CH_3)_2CHCH(CH_3)]_2BH$: A hindered dialkylborane that adds to terminal alkynes, giving trialkylboranes that can be oxidized with alkaline hydrogen peroxide to yield aldehydes (Section 8.6).

1,2-Ethanedithiol, $HSCH_2CH_2SH$: Reacts with ketones or aldehydes in the presence of an acid catalyst to yield dithioacetals that can be reduced with Raney nickel to yield alkanes (Section 19.15).

Ethylene glycol, $HOCH_2CH_2OH$: Reacts with ketones or aldehydes in the presence of a acid catalyst to yield acetals that serve as useful carbonyl protecting groups (Section 19.14).

Ferric bromide, $FeBr_3$: Acts as a catalyst for the reaction of arenes with bromine to yield bromoarenes (Section 16.1).

Ferric chloride, $FeCl_3$: Acts as a catalyst for the reaction of arenes with chlorine to yield chloroarenes (Section 16.2).

Grignard reagent, $RMgX$: Adds to carbonyl-containing compounds (ketones, aldehydes, esters) to yield alcohols (Section 19.10).
- Adds to nitriles to yield ketones (Section 21.9).

Hydrazine, H_2NNH_2: Reacts with ketones or aldehydes in the presence of potassium hydroxide to yield the corresponding alkanes (Wolff-Kishner reaction; Section 19.13).

Hydrogen bromide, HBr: Adds to alkenes to yield alkyl bromides. Markovnikov regiochemistry is observed (Sections 6.9, 14.5).
- Adds to alkenes in the presence of a peroxide catalyst to yield alkyl bromides. Non-Markovnikov regiochemistry is observed (Section 7.6).
- Adds to alkynes to yield either bromoalkenes or 1,1-dibromoalkanes (Section 8.4).
- Reacts with alcohols to yield alkyl bromides (Sections 10.8, 17.8).
- Cleaves ethers to yield alcohols and alkyl bromides (Section 18.6).

Hydrogen chloride, HCl: Adds to alkenes to yield alkyl chlorides. Markovnikov regio-chemistry is observed (Sections 6.9, 14.5).
- Adds to alkynes to yield either chloroalkenes or 1,1-dichloroalkanes (Section 8.4).
- Reacts with alcohols to yield alkyl chlorides (Sections 10.8, 17.8).

Hydrogen cyanide, HCN: Adds to ketones and aldehydes to yield cyanohydrins (Section 19.9).

Hydrogen iodide, HI: Reacts with alcohols to yield alkyl iodides (Section 17.8).
- Cleaves ethers to yield alcohols and alkyl iodides (Section 18.6).

Hydrogen peroxide, H_2O_2: Oxidizes organoboranes to yield alcohols. Used in conjunction with addition of borane to alkenes, the overall transformation effects syn Markovnikov addition of water to an alkene (Section 7.5).
- Oxidizes vinylic boranes to yield aldehydes. Since the vinylic borane starting materials are prepared by hydroboration of a terminal alkyne with disiamyl borane, the overall transformation is the hydration of a terminal alkyne to yield an aldehyde (Section 8.6).
- Oxidizes sulfides to yield sulfoxides (Section 18.11).
- Reacts with α-phenylselenenyl ketones to yield α,β-unsaturated ketones (Section 22.8).

Hydroxylamine, NH_2OH: Reacts with ketones and aldehydes to yield oximes (Section 19.12).
- Reacts with aldoses to yield oximes as the first step in the Wohl degradation of aldoses (Section 24.8).

Hypophosphorous acid, H_3PO_2: Reacts with arenediazonium salts to yield arenes (Section 26.3).

Iodine, I_2: Reacts with arenes in the presence of cupric chloride or hydrogen peroxide to yield iodoarenes (Section 16.2).
- Reacts with carboxylic acids in the presence of lead tetraacetate to yield alkyl iodides (Hunsdiecker reaction; Section 20.9).
- Reacts with methyl ketones in the presence of aqueous sodium hydroxide to yield carboxylic acids and iodoform (Section 22.7).

Iodomethane, CH_3I: Reacts with alkoxide anions to yield methyl ethers (Section 18.4).
- Reacts with carboxylate anions to yield methyl esters (Section 21.7).
- Reacts with enolate ions to yield α-methylated carbonyl compounds (Section 22.9).
- Reacts with amines to yield methylated amines (Section 25.7).

Iron, Fe: Reacts with nitroarenes in the presence of mineral acid to yield anilines (Section 26.2).

Lead tetraacetate, $Pb(OCOCH_3)_4$: Reacts with carboxylic acids in the presence of iodine to yield alkyl iodides and carbon dioxide (Hunsdiecker reaction; Section 20.9).

Lindlar catalyst: Acts as a catalyst for the hydrogenation of alkynes to yield cis alkenes (Section 8.7).

Lithium, Li: Reduces alkynes in liquid ammonia solvent to yield trans alkenes (Section 8.7).
- Reacts with organohalides to yield organolithium compounds (Section 10.10).
- Reacts with methoxysubstituted benzenes to yield cyclohexadienes (Birch reduction; Section 16.12.

Lithium aluminum hydride, LiAl$_4$: Reduces ketones, aldehydes, esters, and carboxylic acids to yield alcohols (Section 17.6).
- Reduces amides to yield amines (Section 21.8).
- Reduces alkyl azides to yield amines (Section 25.7).
- Reduces nitriles to yield amines (Sections 21.9 and 25.7).

Lithium diisopropylamide (LDA), LiN(i-Pr)$_2$: Reacts with carbonyl compounds (aldehydes, ketones, esters) to yield enolate ions (Sections 22.5, 22.9).

Lithium diorganocopper reagent (Gilman reagent), LiR$_2$Cu: Couples with alkyl halides to yield alkanes (Section 10.10).
- Adds to α,β-unsaturated ketones to give 1,4-addition products (Section 19.18).

Lithium tri-*tert*-butoxyaluminum hydride, LiAl(O-t-Bu)$_3$H: Reduces acid chlorides to yield aldehydes (Section 21.5).

Magnesium, Mg: Reacts with organohalides to yield Grignard reagents (Section 10.9).

Magnesium monoperoxyphthalate (MMPH): Reacts with alkenes to yield epoxides (Section 18.7).

Mercuric acetate, Hg(OCOCH$_3$)$_2$: Adds to alkenes in the presence of water, giving α-hydroxy organomercury compounds that can be reduced with sodium borohydride to yield alcohols. The overall reaction effects the Markovnikov hydration of an alkene (Section 7.4).

Mercuric oxide, HgO: Reacts with carboxylic acids in the presence of bromine to yield alkyl bromides and carbon dioxide (Hunsdiecker reaction; Section 20.9).

Mercuric sulfate, HgSO$_4$: Acts as a catalyst for the addition of water to alkynes in the presence of aqueous sulfuric acid, yielding ketones (Section 8.5).

Mercuric trifluoroacetate, Hg(OCOCF$_3$)$_2$: Adds to alkenes in the presence of alcohol, giving α-alkoxy organomercury compounds that can be reduced with sodium borohydride to yield ethers. The overall reaction effects a net addition of an alcohol to an alkene (Section 18.5).

Methyl sulfate, (CH$_3$O)$_2$SO$_2$: A reagent used to methylate heterocyclic amine bases during Maxam-Gilbert DNA sequencing (Section 29.16).

Nitric acid, HNO$_3$: Reacts with arenes in the presence of sulfuric acid to yield nitroarenes (Section 16.2).
- Oxidizes aldoses to yield aldaric acids (Section 24.8).

Nitronium tetrafluoroborate, NO$_2$BF$_4$: Reacts with arenes to yield nitroarenes (Section 16.2).

Nitrous acid, HNO$_2$: Reacts with amines to yield diazonium salts (Section 26.3).

Osmium tetraoxide, OsO$_4$: Adds to alkenes to yield 1,2-diols (Section 7.8).
- Reacts with alkenes in the presence of periodic acid to cleave the carbon-carbon double bond, yielding ketone or aldehyde fragments (Section 7.10).

Oxalyl chloride ClCOCOCl: Reacts with carboxylic acids yielding acid chlorides (Section 21.4).

Ozone, O$_3$: Adds to alkenes to cleave the carbon-carbon double bond and give ozonides. The ozonides can then be reduced with zinc in acetic acid to yield carbonyl compounds (Section 7.9).

Palladium on barium sulfate, Pd/BaSO$_4$: Acts as a hydrogenation catalyst for nitriles in the Kiliani-Fischer chain-lengthening reaction of carbohydrates (Section 24.8).

Palladium on carbon, Pd/C: Acts as a hydrogenation catalyst for reducing carbon-carbon multiple bonds. Alkenes and alkynes are reduced to yield alkanes (Sections 7.7, 8.7).
- Acts as a hydrogenation catalyst in the reduction of aryl ketones to yield alkylbenzenes (Section 16.12).
- Acts as a hydrogenation catalyst in the reduction of nitroarenes to yield anilines (Section 26.2).

Periodic acid, HIO$_4$: Reacts with 1,2-diols to yield carbonyl-containing cleavage products (Section 7.10).

Peroxyacetic acid, CH$_3$CO$_3$H: Oxidizes sulfoxides to yield sulfones (Section 18.11)

Phenyl isothiocyanate, C$_6$H$_5$.N=C=S: A reagent used in the Edman degradation of peptides to identify N-terminal amino acids (Section 27.9).

Phenylselenenyl bromide, C$_6$H$_5$SeBr: Reacts with enolate ions to yield α-phenylselenenyl ketones. On oxidation of the product with hydrogen peroxide, an α,β-unsaturated ketone is produced (Section 22.8).

Phosphorus oxychloride, POCl$_3$: Reacts with secondary and tertiary and alcohols to yield alkene dehydration products (Section 17.8).

Phosphorus tribromide, PBr$_3$: Reacts with alcohols to yield alkyl bromides (Section 10.8).
- Reacts with carboxylic acids to yield acid bromides (Section 21.5).
- Reacts with carboxylic acids in the presence of bromine to yield α-bromo carboxylic acids (Hell-Volhard-Zelinskii reaction; Section 22.4).

Phosphorus trichloride, PCl$_3$: Reacts with carboxylic acids to yield acid chlorides (Section 21.4).

Platinum oxide (Adam's catalyst), PtO$_2$: Acts as a hydrogenation catalyst in the reduction of alkenes and alkynes to yield alkanes (Section 7.7).

Potassium hydroxide, KOH: Reacts with alkyl halides to yield alkenes by an elimination reaction (Sections 7.1, 11.10-11.14).
- Reacts with 1,1- or 1,2-dihaloalkanes to yield alkynes by a twofold elimination reaction (Section 8.1).

Potassium nitrosodisulfonate (Fremy's salt), K(SO$_3$)$_2$NO: Oxidizes phenols and anilines to yield quinones (Section 26.8).

Potassium permanganate, KMnO$_4$: oxidizes alkenes under alkaline conditions to yield 1,2-diols (Section 7.8).
- Oxidizes alkenes under neutral or acidic conditions to give carboxylic acid double-bond cleavage products (Sections 7.9).
- Oxidizes alkynes to give carboxylic acid triple-bond cleavage products (Section 8.10).
- Oxidizes arenes to yield benzoic acids (Section 16.11).

Potassium phthalimide, C$_6$H$_4$(CO)$_2$NK: Reacts with alkyl halides to yield an *N*-alkyl-phthalimide that is hydrolyzed by aqueous sodium hydroxide to yield an amine (Gabriel amine synthesis; Section 25.7).

Potassium *tert*-butoxide, KO-*t*-Bu: Reacts with alkyl halides to yield alkenes (Sections 11.10-11.14).
- Reacts with allylic halides to yield conjugated dienes (Section 14.1).
- Reacts with chloroform in the presence of an alkene to yield a dichlorocyclopropane (Section 7.11).

Pyridine, C$_5$H$_5$N: Acts as a catalyst for the reaction of alcohols with acid chlorides to yield esters (Section 21.5).
- Acts as a catalyst for the reaction of alcohols with acetic anhydride to yield acetate esters (Section 21.6).
- Reacts with α-bromo ketones to yield α,β-unsaturated ketones (Section 22.3).

Pyridinium chlorochromate (PCC), C$_5$H$_6$NCrO$_3$Cl: Oxidizes primary alcohols to yield aldehydes and secondary alcohols to yield ketones (Sections 17.9, 19.3, 19.4).

Pyrrolidine, C$_4$H$_8$N: Reacts with ketones to yield enamines for use in the Stork enamine reaction (Sections 19.12, 23.13).

Raney nickel, Ni: Reduces dithioacetals to yield alkanes by a desulfurization reaction (Section 19.15).

Rhodium on carbon, Rh/C: Acts as a hydrogenation catalyst in the reduction of benzene rings to yield cyclohexanes (Section 16.12).

Silver oxide, Ag$_2$O: Oxidizes primary alcohols in aqueous ammonia solution to yield aldehydes (Tollens oxidation; Sections 19.5, 20.6).
- Catalyzes the reaction of alcohols with alkyl halides to yield ethers (Section 24.8).

Sodium amide, NaNH$_2$:   Reacts with terminal alkynes to yield acetylide anions (Section 8.8).
- Reacts with 1,1-or 1,2-dihalides to yield alkynes by a twofold elimination reaction (Section 8.1).
- Reacts with aryl halides to yield anilines by a benzyne aromatic substitution mechanism (Section 16.10).

Sodium azide, NaN$_3$:   Reacts with alkyl halides to yield alkyl azides (Section 25.7).
- Reacts with acid chlorides to yield acyl azides. On heating in the presence of water, acyl azides yield amines and carbon dioxide (Section 25.7).

Sodium bisulfite, NaHSO$_3$:   Reduces osmate esters, prepared by treatment of an alkene with osmium tetraoxide, to yield 1,2-diols (Section 7.8).

Sodium borohydride, NaBH$_4$:   Reduces organomercury compounds, prepared by oxymercuration of alkenes, to convert the C-Hg bond to C-H (Section 7.4).
- Reduces ketones and aldehydes to yield alcohols (Sections 17.6, 19.11).
- Reduces quinones to yield hydroquinones (Section 26.8).

Sodium cyanide, NaCN: Reacts with alkyl halides to yield alkanenitriles (Sections 20.6, 21.9).

Sodium   cyanoborohydride,   NaBH$_3$CN: Reacts with ketones and aldehydes in the presence of ammonia to yield an amine by a reductive amination process (Section 25.7).

Sodium dichromate, Na$_2$Cr$_2$O$_7$:   Oxidizes primary alcohols to yield carboxylic acids and secondary alcohols to yield ketones (Sections 17.9, 19.5).
- Oxidizes alkylbenzenes to yield benzoic acids (Section 16.11).

Sodium hydride, NaH: Reacts with alcohols to yield alkoxide anions (Section 17.4).

Sodium hydroxide, NaOH: Reacts with arenesulfonic acids at high temperature to yield phenols (Section 16.2).
- Reacts with aryl halides to yield phenols by a benzyne aromatic substitution mechanism (Section 16.10).
- Catalyzes the reaction of acid chlorides with alcohols to yield esters - the Schotten-Baumann reaction (Section 21.5).
- Catalyzes the reaction of acid chlorides with amines to yields amides (Section 21.5).
- Reacts with methyl ketones in the presence of iodine to yield carboxylic acids and iodoform (Section 22.7).

Sodium iodide, NaI :   Reacts with arenediazonium salts to yield aryl iodides (Section 26.3).

Stannous chloride, SnCl$_2$:   Reduces nitroarenes to yield anilines (Sections 16.2, 26.2).
- Reduces quinones to yield hydroquinones (Section 26.8).

Sulfur trioxide, SO$_3$:   Reacts with arenes in sulfuric acid solution to yield arenesulfonic acids (Section 16.2).

Sulfuric acid, H_2SO_4: Reacts with alcohols to yield alkenes (Sections 7.1, 17.8).
- Reacts with alkynes in the presence of water and mercuric acetate to yield ketones (Section 8.5).
- Catalyzes the reaction of nitric acid with aromatic rings to yield nitroarenes (Section 16.2).
- Catalyzes the reaction of SO_3 with aromatic rings to yield arenesulfonic acids (Section 16.2).

Tetrazole: Acts as a coupling reagent for use in DNA synthesis (Section 29.16).

Thionyl chloride, $SOCl_2$: Reacts with primary and secondary alcohols to yield alkyl chlorides (Section 10.8).
- Reacts with carboxylic acids to yield acid chlorides (Section 21.4).

Thiourea, H_2NCSNH_2: Reacts with primary alkyl halides to yield thiols (Section 17.12).

***p*-Toluenesulfonyl chloride, *p*-$CH_3C_6H_4SO_2Cl$:** Reacts with alcohols to yield tosylates (Sections 11.2, 17.8).

Trifluoroacetic acid, CF_3COOH: Acts as a catalyst for cleaving *tert*-butyl ethers, yielding alcohols and 2-methylpropene (Section 18.6).
- Acts as a catalyst for cleaving the *t*-BOC protecting group from amino acids in peptide synthesis (Section 27.11).

Triphenylphosphine, $(C_6H_5)_3P$: Reacts with primary alkyl halides to yield the alkyltriphenylphosphonium salts used in Wittig reactions (Section 19.16).

Zinc, Zn: Reduces ozonides, produced by addition of ozone to alkenes, to yield ketones and aldehydes (Section 7.9).
- Reduces disulfides to yield thiols (Section 17.12).
- Reduces ketones and aldehydes in the presence of aqueous HCl to yield alkanes (Clemmensen reduction; Section 19.13).

Zinc-copper couple, Zn-Cu: Reacts with diiodomethane in the presence of alkenes to yield cyclopropanes (Simmons-Smith reaction; Section 7.11).

Name Reactions

Acetoacetic ester synthesis (Section 22.9): a multi-step reaction sequence for converting a primary alkyl halide into a methyl ketone having three more carbon atoms in the chain.

$$RCH_2X + CH_3-\overset{O}{\overset{\|}{C}}-\overset{..}{\overset{..}{C}}H-\overset{O}{\overset{\|}{C}}-OCH_3 \xrightarrow[\text{2. } H_3O^+, \text{ heat}]{\text{1. heat}} RCH_2-CH_2\overset{O}{\overset{\|}{C}}CH_3 + CO_2 + CH_3OH$$

Adams' catalyst (Section 7.7): PtO_2, a catalyst used for the hydrogenation of carbon-carbon double bonds.

Aldol condensation reaction (Section 23.2): the nucleophilic addition of an enol or enolate ion to a ketone or aldehyde, yielding a β-hydroxy ketone.

$$2\ R-\overset{O}{\overset{\|}{C}}-\overset{|}{\underset{|}{C}}-H \xrightarrow{NaOH} R-\overset{O}{\overset{\|}{C}}-\overset{|}{\underset{\underset{R}{|}}{C}}-\overset{OH}{\overset{|}{\underset{|}{C}}}-\overset{|}{\underset{|}{C}}-H$$

Amidomalonate amino acid synthesis (Section 27.4): a multi-step reaction sequence, similar to the malonic ester synthesis, for converting a primary alkyl halide into an amino acid.

$$RCH_2X + {}^-:CH(NHAc)(CO_2Et)_2 \xrightarrow[\text{2. } H_3O^+]{\text{1. mix}} RCH_2-\underset{\underset{NH_2}{|}}{CH}\overset{O}{\overset{\|}{C}}OH + CO_2 + 2\ EtOH$$

Birch reduction (Section 16.12): the reduction of a methoxy-substituted benzene to yield a cyclohexadiene on reaction with lithium or sodium in liquid ammonia. Subsequent hydrolysis of the product yields a cyclohexenone.

Cannizzaro reaction (Section 19.17): the disproportionation reaction that occurs when a nonenolizable aldehyde is treated with base.

$$2\ R_3C\overset{O}{\overset{\|}{C}}H \xrightarrow[\text{2. } H_3O^+]{\text{1. } HO^-} R_3C\overset{O}{\overset{\|}{C}}OH + R_3CCH_2OH$$

Claisen condensation reaction (Section 23.8): a nucleophilic acyl substitution reaction that occurs when an ester enolate ion attacks the carbonyl group of a second ester molecule. The product is a β-keto ester.

$$2\ R\text{--}CH_2\text{--}\overset{\overset{\displaystyle O}{\|}}{C}\text{--}OCH_3 \quad \xrightarrow[\text{2. }H_3O^+]{\text{1. }Na^+\ {}^-OCH_3} \quad R\text{--}CH_2\text{--}\overset{\overset{\displaystyle O}{\|}}{C}\text{--}\underset{\underset{\displaystyle R}{|}}{C}H\text{--}\overset{\overset{\displaystyle O}{\|}}{C}\text{--}OCH_3 \quad + \quad CH_3OH$$

Claisen rearrangement (Sections 26.9 and 30.12): the thermal [3.3] sigmatropic rearrangement of an allyl vinyl ether or an allyl phenyl ether.

Clemmenson reduction (Section 19.13): a method for reducing a ketone or aldehyde to an alkane by treatment with amalgamated zinc and aqueous HCl.

Cope rearrangement (Section 30.12): the thermal [3.3] sigmatropic rearrangement of a 1,5-diene to a new 1,5-diene.

Curtius rearrangement (Section 25.7): the thermal rearrangement of an acyl azide to an isocyanate, followed by hydrolysis to yield an amine.

$$R\text{--}\overset{\overset{\displaystyle O}{\|}}{C}\text{--}N{=}\overset{+}{N}{=}\overset{-}{N} \quad \xrightarrow[\text{2. }H_2O]{\text{1. heat}} \quad RNH_2\ +\ CO_2\ +\ N_2$$

Diazonium coupling reaction (Section 26.4): the coupling reaction between an aromatic diazonium salt and a phenol or aniline.

Dieckmann reaction (Section 23.11): the intramolecular Claisen condensation reaction of a 1,6- or 1,7-diester, yielding a cyclic β-keto ester.

Diels-Alder cycloaddition reaction (Sections 14.7 and 30.8): the thermal reaction between a diene and a dienophile to yield a cyclohexene ring.

Edman degradation (Section 27.9): a method for cleaving the N-terminal amino acid from a peptide by treatment of the peptide with N-phenylisothiocyanate.

Fehling's test (Section 24.8): a chemical test for aldehydes, involving treatment with cupric ion.

Fischer esterification reaction (Section 21.4): the acid-catalyzed reaction between a carboxylic acid and an alcohol, yielding the ester.

Friedel-Crafts reaction (Section 16.3-4): the alkylation or acylation of an aromatic ring by treatment with an alkyl- or acyl chloride in the presence of a Lewis-acid catalyst.

Gabriel amine synthesis (Section 25.7): a multi-step sequence for converting a primary alkyl halide into a primary amine by alkylation with potassium phthalimide, followed by hydrolysis.

Gilman reagent (Section 10.10): a lithium dialkylcopper reagent, R_2CuLi, prepared by treatment of a cuprous salt with an alkyllithium. Gilman reagents undergo a coupling reaction with alkyl halides, and undergo a 1,4-addition reaction with α,β-unsaturated ketones.

Grignard reaction (Section 19.10): the nucleophilic addition reaction of an alkylmagnesium halide to a ketone, aldehyde, or ester carbonyl group.

Grignard reagent (Section 10.9): an organomagnesium halide, RMgX, prepared by reaction between an organohalide and magnesium metal.

Haloform reaction (Section 22.7): the conversion of a methyl ketone to a carboxylic acid and haloform by treatment with halogen and base.

Hell-Volhard-Zelinskii reaction (Section 22.4): the α-bromination of carboxylic acids by treatment with bromine and phosphorus.

Hinsberg test (Section 25.8): a chemical means of distinguishing among primary, secondary, and tertiary amines by observing their reactions with benzenesulfonyl chloride.

Hofmann elimination (Section 25.8): a method for effecting the elimination reaction of an amine to yield an alkene. The amine is first treated with excess iodomethane, and the resultant quaternary ammonium salt is heated with silver oxide.

Hofmann rearrangement (Section 25.7): the rearrangement of an *N*-bromoamide to a primary amine by treatment with aqueous base.

Hunsdiecker reaction (Section 20.9): a reaction for converting a carboxylic acid into an alkyl halide by treatment with mercuric oxide and halogen.

Jones reagent (Section 17.9): a solution of CrO_3 in acetone/aqueous sulfuric acid. This reagent oxidizes primary and secondary alcohols to carbonyl compounds under mild conditions.

Kiliani-Fischer synthesis (Section 24.8): a multi-step sequence for chain-lengthening an aldose into the next higher homolog.

Koenigs-Knorr reaction (Section 24.8): a method for synthesizing glycosides by reaction between an alcohol and a bromo-substituted carbohydrate.

Kolbe-Schmitt carboxylation reaction (Section 26.8): a method for introducing a carboxyl group in the ortho position of a phenol by treatment of the phenoxide anion with CO_2.

Malonic ester synthesis (Section 22.9): a multi-step sequence for converting an alkyl halide into a carboxylic acid with the addition of two carbon atoms to the chain.

Maxam-Gilbert DNA sequencing (Section 29.16): a rapid and efficient method for sequencing long chains of DNA by employing selective cleavage reactions.

McLafferty rearrangement (Section 19.20): a general mass spectral fragmentation pathway for carbonyl compounds having a hydrogen three carbon atoms away from the carbonyl carbon.

Meisenheimer complex (Section 16.9): an intermediate formed in the nucleophilic aryl substitution reaction of a base with a nitro-substituted aromatic ring.

Merrifield solid-phase peptide synthesis (Section 27.12): a rapid and efficient means of peptide synthesis in which the growing peptide chain is attached to an insoluble polymer support.

Michael reaction (Section 23.12): the 1,4-addition reaction of a stabilized enolate anion such as that from a 1,3-diketone to an α,β-unsaturated carbonyl compound.

Nagata hydrocyanation reaction (Section 19.18): a reaction for effecting the conjugate (1,4)-addition of HCN to an α,β-unsaturated ketone by treatment with diethylaluminum cyanide.

Raney nickel (Section 19.15): a specially prepared form of nickel that is used for desulfurization of thioacetals.

Robinson annulation reaction (Section 23.14): a multi-step sequence for building a new cyclohexenone ring onto a ketone. The sequence involves an initial Michael reaction of the ketone followed by an internal aldol cyclization.

Sandmeyer reaction (Section 26.4): a method for converting aryldiazonium salts into aryl halides by treatment with cuprous halide.

Schotten-Baumann reaction (Section 21.5): a method for preparing esters by treatment of an acid chloride with an alcohol in the presence of aqueous base.

$$R-\overset{\overset{\displaystyle O}{\|}}{C}-Cl + R'OH \quad\xrightarrow{\text{NaOH, }H_2O}\quad R-\overset{\overset{\displaystyle O}{\|}}{C}-OR'$$

Simmons-Smith reaction (Section 7.11): a method for preparing cyclopropanes by treating an alkene

Stork enamine reaction (Section 23.13): a multi-step sequence whereby ketones are converted into enamines by treatment with a secondary amine, and the enamines are then used in Michael reactions.

Strecker amino acid synthesis (Section 27.4): a multi-step sequence for converting an aldehyde into an amino acid by initial treatment with ammonium cyanide, followed by hydrolysis.

$$R-\overset{\overset{\displaystyle O}{\|}}{C}-H \quad\xrightarrow{NH_3,\ KCN}\quad R-\overset{\overset{\displaystyle NH_2}{|}}{C}H-CN \quad\xrightarrow{H_3O^+}\quad R-\overset{\overset{\displaystyle NH_2}{|}}{C}H-COOH$$

Tollen's test (Section 19.5): a chemical test for detecting aldehydes by treatment with ammoniacal silver nitrate. A positive test is signaled by formation of a silver mirror on the walls of the reaction vessel.

Walden inversion (Section 11.1): the inversion of stereochemistry at a stereogenic center during S_N2 reactions.

Williamson ether synthesis (Section 18.4): a method for preparing ethers by treatment of a primary alkyl halide with an alkoxide ion.

$$R-O^- Na^+ + R'CH_2Br \quad\longrightarrow\quad R-O-CH_2R' + NaBr$$

Wittig reaction (Section 19.16): a general method of alkene synthesis by treatment of a ketone or aldehyde with an alkylidenetriphenylphosphorane.

Wohl degradation (Section 24.8): a multi-step reaction sequence for degrading an aldose into the next lower homolog.

Wohl-Ziegler reaction (Section 10.5): a reaction for effecting allylic bromination by treatment of an alkene with *N*-bromosuccinimide.

Wolff-Kishner reaction (Section 19.13): a method for converting a ketone or aldehyde into the corresponding hydrocarbon by treatment with hydrazine and strong base.

Woodward-Hoffmann orbital symmetry rules (Section 30.13): a series of rules for predicting the stereochemistry of pericyclic reactions. Even-electron species react thermally through either antarafacial or conrotatory pathways, whereas odd-electron species react thermally through either suprafacial or disrotatory pathways (even-antara-con; odd-supra-dis).

Ziegler-Natta polymerization (Section 31.6): a method for carrying out the stereoregular polymerization of alkenes by using titanium-aluminum catalysts.

Abbreviations

Å   symbol for Angstrom unit (10^{-8} cm)

Ac-   acetyl group, $CH_3\overset{\overset{\displaystyle O}{\|}}{C}-$

Ar-   aryl group

at. no.   atomic number

at. wt.   atomic weight

$[\alpha]_D$ specific rotation

BOC   *tert*-butoxycarbonyl group, $(CH_3)_3CO\overset{\overset{\displaystyle O}{\|}}{C}-$

bp   boiling point

n-Bu   *n*-butyl group, $CH_3CH_2CH_2CH_2-$

sec-Bu   *sec*-butyl group, $CH_3CH_2CH(CH_3)-$

t-Bu   *tert*-butyl group, $(CH_3)_3C-$

cm   centimeter

cm^{-1} wavenumber or reciprocal centimeter

D   stereochemical designation of carbohydrates and amino acids

DCC   dicyclohexylcarbodiimide, $C_6H_{11}-N=C=N-C_6H_{11}$

δ   chemical shift in ppm downfield from TMS

Δ   symbol for heat; also symbol for change

ΔH   heat of reaction

dm   decimeter (0.1 m)

DMF   dimethylformamide, $(CH_3)_2NCHO$

DMSO   dimethyl sulfoxide, $(CH_3)_2SO$

DNA   deoxyribonucleic acid

DNP   dinitrophenyl group, as in 2,4-DNP (2,4-dinitrophenylhydrazone)

(*E*)   entgegen, stereochemical designation of double bond geometry

E_{act} activation energy

E1   unimolecular elimination reaction

E2   bimolecular elimination reaction

Et   ethyl group, CH_3CH_2-

g   gram

$h\nu$	symbol for light
Hz	Hertz, or cycles per second
i-	iso
IR	infrared
J	Joule
J	symbol for coupling constant
K	Kelvin temperature
K_a	symbol for acid dissociation constant
Kcal	kilocalories
L	stereochemical designation of carbohydrates and amino acids
LAH	lithium aluminum hydride, $LiAlH_4$
Me	methyl group, CH_3-
mg	milligram (0.001 g)
MHz	megahertz (10^6 cycles per second)
mL	milliliter (0.001 L)
mm	millimeter (0.001 m)
mol. wt.	molecular weight
mp	melting point
$\mu\gamma$	microgram (10^{-6} gram)
$m\mu$	millimicron (nanometer, 10^{-9} meter)
n-	normal, straight chain alkane or alkyl group
ng	nanogram (10^{-9} gram)
nm	nanometer (10^{-9} meter)
NMR	nuclear magnetic resonance
-OAc	acetate group, $-O\overset{\overset{\displaystyle O}{\|}}{C}CH_3$
PCC	pyridinium chlorochromate
Ph	phenyl group, $-C_6H_5$
pH	measure of acidity of aqueous solution
pK_a	measure of acid strength (= $-\log K_a$)
pm	picometer (10^{-12}) meter)
ppm	parts per million
n-Pr	*n*-propyl group, $CH_3CH_2CH_2-$
i-Pr	isopropyl group, $(CH_3)_2CH-$

prim	primary
R-	symbol for a generalized alkyl group
(R)	rectus, stereochemical designation of chiral centers
RNA	ribonucleic acid
(S)	sinister, stereochemical designation of chiral centers
sec-	secondary
S_N1	unimolecular substitution reaction
S_N2	bimolecular substitution reaction
tert-	tertiary
THF	tetrahydrofuran
TMS	tetramethylsilane nmr standard, $(CH_3)_4Si$

Tos   Tosylate group

UV	ultraviolet
yr	year
X-	halogen group (–F, –Cl, –Br, –I)
(Z)	zusammen, stereochemical designation of double bond geometry

chemical reaction in direction indicated

reversible chemical reaction

resonance symbol

curved arrow indicating direction of electron flow

≡	is equivalent to
>	greater than
<	less than
≈	approximately equal to

indicates that the organic fragment shown is a part of a larger molecule

single bond coming out of the plane of the paper

····· single bond receding into the plane of the paper

······ partial bond

δ^+, δ^-	partial charge
*	isotopically labeled atom
‡	denoting the transition state

Infrared Absorption Frequencies

Functional Group Class		Frequency (cm^{-1})	Text Section
Alcohol	–O–H	3300 - 3600 (s)	17.11
	>C–O–	1050 (s)	17.11
Aldehyde	–CO–H	2720, 2820 (m)	19.20
aliphatic	>C=O	1725 (s)	19.20
aromatic	>C=O	1705 (s)	19.20
Alkane	–C–H	2850 - 2960 (s)	12.11
	–C–C–	800 - 1300 (m)	12.11
Alkene	=C(H)	3020 - 3100 (m)	12.11
	>C=C<	1650 - 1670 (m)	12.11
	RCH=CH$_2$	910, 990 (m)	12.11
	R$_2$C=CH$_2$	890 (m)	12.11
Alkyne	≡C–H	3300 (s)	12.11
	–C≡C–	2100 - 2260 (m)	12.11
Alkyl bromide	–C–Br	500 - 600 (s)	12.11
Alkyl chloride	–C–Cl	600 - 800 (s)	12.11

Amine, *primary*	−N(−H)(−H) (H top, H bottom)	3400, 3500 (s)	25.211
secondary	N−H	3350 (s)	25.11
Ammonium salt	−N$^+$−H	2200 - 3000 (broad)	25.11
Aromatic ring	Ar−H	3030 (m)	15.12
monosubstituted	Ar−R	690 - 710 (s)	15.12
		730 - 770 (s)	15.12
o-disubstituted		735 - 770 (s)	15.12
m-disubstituted		690 - 710 (s)	15.12
		810 - 850 (s)	15.12
p-disubstituted		810 - 840 (s)	15.12
Carboxylic acid	−O−H	2500 - 3300 (broad)	20.10
associated	C=O	1710 (s)	20.10
free	C=O	1760 (s)	20.10
Acid anhydride	C=O	1820, 1760 (s)	21.11
Acid chloride			
aliphatic	C=O	1810 (s)	21.11
aromatic	C=O	1770 (s)	21.11

Amide, *aliphatic*	C=O	1690 (s)	21.11
aromatic	C=O	1675 (s)	21.11
N-substituted	C=O	1680 (s)	21.11
N,N-disubstituted	C=O	1650 (s)	21.11
Ester, *aliphatic*	C=O	1735 (s)	21.11
aromatic	C=O	1720 (s)	21.11
Ether	-O-C-	1050 - 1150 (s)	18.10
Ketone, *aliphatic*	C=O	1715 (s)	19.20
aromatic	C=O	1690 (s)	19.20
6-memb. ring	C=O	1715 (s)	19.20
5-memb. ring	C=O	1750 (s)	19.20
Nitrile, *aliphatic*	$\text{-C}\equiv\text{N}$	2250 (m)	21.11
aromatic	$\text{-C}\equiv\text{N}$	2230 (m)	21.11
Phenol	-O-H	3500 (s)	26.9

(s) = strong; (m) = medium intensity

Proton NMR Chemical Shifts

Type of Proton		Chemical Shift (δ)	Text Section
Alkyl, primary	$R\text{-}CH_3$	0.7 – 1.3	13.10
Alkyl, secondary	$R\text{-}CH_2\text{-}R$	1.2 – 1.4	13.10
Alkyl, tertiary	$R_3C\text{-}H$	1.4 – 1.7	13.10
Allylic	$-C{=}C\text{-}C\text{-}H$	1.6 – 1.9	13.10
Alpha to carbonyl	$-\overset{\overset{\displaystyle O}{\|\|}}{C}\text{-}C\text{-}H$	2.0 – 2.3	19.20
Benzylic	$Ar\text{-}C\text{-}H$	2.3 – 3.0	15.12
Acetylenic	$R\text{-}C{\equiv}C\text{-}H$	2.5 – 2.7	13.10
Alkyl chloride	$Cl\text{-}C\text{-}H$	3.0 – 4.0	13.10
Alkyl bromide	$Br\text{-}C\text{-}H$	2.5 – 4.0	13.10
Alkyl iodide	$I\text{-}C\text{-}H$	2.0 – 4.0	13.10
Amine	$\ddot{N}\text{-}C\text{-}H$	2.2 – 2.6	25.11
Epoxide	$\overset{\displaystyle O}{\underset{\displaystyle C-C}{\triangle}}\text{-}H$	2.5 – 3.5	18.10
Alcohol	$HO\text{-}C\text{-}H$	3.5 – 4.5	17.11
Ether	$RO\text{-}C\text{-}H$	3.5 – 4.5	18.10
Vinylic	$-C{=}C\text{-}H$	5.0 – 6.5	13.10
Aromatic	$Ar\text{-}H$	6.5 – 8.0	15.12
Aldehyde	$R\text{-}\overset{\overset{\displaystyle O}{\|\|}}{C}\text{-}H$	9.7 – 10.0	19.20
Carboxylic acid	$R\text{-}\overset{\overset{\displaystyle O}{\|\|}}{C}\text{-}O\text{-}H$	11.0 – 12.0	20.10
Alcohol	$R\text{-}O\text{-}H$	3.5 – 4.5	17.11
Phenol	$Ar\text{-}O\text{-}H$	2.5 – 6.0	26.9

Ethylene (18,740,000 tons/yr):
prepared by thermal cracking of ethane and propane during petroleum refining; used as starting material for manufacture of polyethylene, ethylene oxide, ethylene glycol, ethylbenzene, 1,2-dichloroethane, and other bulk chemicals.

Propylene (11,060,000 tons/yr):
prepared by steam cracking of light hydrocarbon fractions during petroleum refining; used as starting material for the manufacture of polypropylene, acrylonitrile, propylene oxide, and isopropyl alcohol.

1,2-Dichloroethane (Ethylene dichloride; 6,650,000 tons/yr):
prepared by addition of chlorine to ethylene in the presence of $FeCl_3$ catalyst at 50°C; used as a chlorinated solvent and as starting material for the manufacture of vinyl chloride.

Benzene (5,930,000 tons/yr):
obtained from petroleum by catalytic reforming of hexane and cyclohexane over a platinum catalyst; used as starting material for the synthesis of ethylbenzene, cumene, cyclohexane, and aniline.

Vinyl chloride (5,325,000 tons/yr):
prepared by addition of chlorine to ethylene followed by elimination of HCl; used as starting material for preparation of poly(vinyl chloride) polymers (hoses, pipes, molded objects).

Ethylbenzene (4,495,000 tons/yr):
prepared during catalytic reforming in petroleum refining and by an acid-catalyzed Friedel-Crafts alkylation of benzene with ethylene; used almost exclusively for production of styrene.

Styrene (4,008,000 tons/yr):
prepared by high-temperature catalytic dehydrogenation of ethylbenzene; used in the manufacture of polystyrene polymers (thermoplastics, packaging materials).

Methanol (3,993,000 tons/yr):
prepared by high temperature reaction of a mixture of H_2, CO, and CO_2 ("synthesis gas") over a catalyst at 100 atmospheres pressure; used as a solvent and as starting material for the manufacture of formaldehyde, acetic acid, and methyl *tert*-butyl ether.

Dimethyl terephthalate (3,845,000 tons/yr):
prepared from *p*-xylene by oxidation and esterification; used in the manufacture of polyester polymers (textiles, upholstery, recording tape, and film).

Formaldehyde (3,206,000 tons/yr):
   prepared by air oxidation of methanol over a silver or metal oxide catalyst; used in the manufacture of phenolic resins, melamine resins, and plywood adhesives.

Methyl *tert*-butyl ether (MTBE, 3,150,000 tons/yr):
   prepared by acid-catalyzed addition of methanol to isobutylene; used as an octane enhancer in gasoline.

Toluene (3,050,000 tons/yr):
   prepared during catalytic reforming of petroleum; used as a gasoline additive and as a degreasing solvent.

Ethylene oxide (2,790,000 tons/yr):
   prepared by high-temperature air oxidation of ethylene over a silver catalyst; used as starting material for the preparation of ethylene glycol and poly(ethylene glycol).

***p*-Xylene** (2,600,000 tons/yr):
   prepared by separation from the mixed xylenes that result during catalytic reforming in gasoline refining; used as starting material for manufacture of the dimethyl terephthalate needed for polyester synthesis.

Ethylene glycol (2,513,000 tons/yr):
   prepared by high-temperature reaction between water and ethylene oxide at neutral pH; used as antifreeze and as a starting material for polymers and latex paints.

Cumene (2,155,000 tons/yr):
   prepared by a phosphoric-acid-catalyzed Friedel-Crafts reaction between benzene and propylene; used primarily for conversion into phenol and acetone.

Acetic acid (1,877,000 tons/yr):
   prepared by metal-catalyzed air oxidation of acetaldehyde under pressure at 80°C and by reaction of methanol with carbon monoxide; used to make vinyl acetate polymers, ethyl acetate solvent, and cellulose acetate polymers.

Phenol (1,756,000 tons/yr):
   prepared from cumene by air oxidation to cumene hydroperoxide, followed by acid-catalyzed decomposition; used as starting material for preparing phenolic resins, epoxy resins, and caprolactam.

Propylene oxide (1,600,000 tons/yr):
   prepared from propylene by high-temperature oxidation with air and by formation of propylene chlorohydrin followed by loss of HCl; used in the manufacture of propylene glycol, polyurethanes, and polyesters.

1,3-Butadiene (1,577,000 tons/yr):
   prepared by steam cracking of gas oil during petroleum refining and by dehydrogenation of butane and butene; used primarily as a monomer component in the manufacture of styrene-butadiene rubber (SBR), polybutadiene rubber, and acrylonitrile-butadiene-styrene (ABS) copolymers.

Acrylonitrile (1,514,000 tons,yr):

    prepared by the Sohio ammoxidation process in which propylene, ammonia, and air are passed over a catalyst at 500°C; used in the preparation of acrylic fibers, nitrile rubber, and acrylonitrile-butadiene-styrene (ABS) copolymer.

Vinyl acetate (1,272,000 tons/yr):

    prepared from reaction of acetic acid, ethylene, and oxygen; used for manufacture of poly(vinyl acetate) (paint emulsions, plywood adhesives, textiles).

Cyclohexane (1,233,000 tons/yr):

    prepared by catalytic hydrogenation of benzene; used as starting material for synthesis of the caprolactam and adipic acid needed for nylon.

Acetone (1,110,000 tons/yr):

    prepared by acid-catalyzed decomposition of cumene hydroperoxide and by air oxidation of isopropyl alcohol at 300°C over a metal oxide catalyst; used as a solvent and as starting material for synthesizing bisphenol A and methyl methacrylate.

Isopropyl alcohol (690,000 tons/yr):

    prepared by direct high-temperature addition of water to propylene; used in cosmetics formulations, as a solvent and deicer, and as starting material for manufacture of acetone.

Caprolactam (689,000 tons/yr):

    prepared from phenol by conversion into cyclohexanone, followed by formation and acid-catalyzed rearrangement of cyclohexanone oxime; used as starting material for the manufacture of nylon-6.

1-Butanol (634,000 tons/yr):

    prepared in the oxo process by reaction of propylene with carbon monoxide; used as solvent and as a starting material for synthesis of butyl acetate and dibutyl phthalate.

Methyl methacrylate (603,000 tons/yr):

    prepared by acetone cyanohydrin by treatment with sulfuric acid to effect dehydration, followed by esterification with methanol; used for the synthesis of methacrylate polymers such as Lucite.

Bisphenol A (570,000 tons/yr):

    prepared by reaction of phenol with acetone; used in the manufacture of epoxy resins and adhesives, polycarbonates, and polysulfones.

Isobutylene (572,000 tons/yr):

    prepared from catalytic cracking of petroleum; used in the manufacture of methyl *tert*-butyl ether, isoprene, and butylated phenols.

Aniline (473,000 tons/yr):

    prepared by catalytic reduction of nitrobenzene with hydrogen at 350°C; used as starting material for preparing toluene diisocyanate and for the synthesis of dyes and pharmaceuticals.

o-Xylene (471,000 tons/yr):
obtained by separation from the mixed xylenes that result during catalytic reforming in petroleum refining; used as starting material for preparation of phthalic acid and phthalic anhydride.

Phthalic anhydride (470,000 tons/yr):
prepared by oxidation of o-xylene at 400°C and by oxidation of naphthalene obtained from coal tar; used for the synthesis of polyesters and plasticizers.

Propylene glycol (400,000 tons/yr):
prepared by high temperature reaction of propylene oxide with water; used in the preparation of polyesters and as an additive in the food industry.

1,1,1-Trichloroethane (Methylchloroform; 392,000 tons/yr):
prepared by addition of HCl to vinyl chloride to give 1,1-dichloroethane, followed by radical chlorination with Cl_2; used as a solvent, industrial cleaner, and metal degreaser.

Ethanolamines (339,000 tons/yr):
prepared by reaction of ammonia with ethylene oxide at 100°C; used in soaps, detergents, cosmetics, and corrosion inhibitors.

2-Ethylhexanol (298,000 tons/yr):
prepared from butanal by aldol condensation and catalytic hydrogenation (the Oxo Process); used in the manufacture of plasticizers, lubricating-oil additives, and detergents.

Ethanol (275,000 tons/yr):
prepared by direct vapor phase hydration of ethylene at 300°C over an acidic catalyst; used as a solvent, as a constituent of cleaning preparations, and as starting material for ester synthesis.

Chloromethane (Methyl chloride; 249,000 tons/yr):
prepared by reaction of methanol with HCl at 0°C and by radical chlorination of methane; used in the manufacture of silicones, synthetic rubber, and methyl cellulose.

2-Butanone (Methyl ethyl ketone; 236,000 tons/yr):
prepared by oxidation of 2-butanol over a ZnO catalyst at 400°C; used as a solvent for vinyl coatings, lacquers, rubbers, and paint removers.

Nobel Prizes in Chemistry

1901   **Jacobus H. van't Hoff** (Dutch):
"for the discovery of laws of chemical dynamics and of osmotic pressure"

1902   **Emil Fischer** (German):
"for syntheses in the groups of sugars and purines"

1903   **Svante A. Arrhenius** (Swedish):
"for his theory of electrolytic dissociation"

1904   **Sir William Ramsey** (British):
"for the discovery of gases in different elements in the air and for the determination of
their place in the periodic system"

1905   **Adolf von Baeyer** (German):
"for his researches on organic dyestuffs and hydroaromatic compounds"

1906   **Henri Moissan** (French):
"for his research on the isolation of the element fluorine and for placing at the service
of science the electric furnace that bears his name"

1907   **Eduard Buchner** (German):
"for his biochemical researches and his discovery of cell-less formation"

1908   **Ernest Rutherford** (British):
"for his investigation into the disintegration of the elements and the chemistry of
radioactive   substances"

1909   **Wilhelm Ostwald** (German):
"for his work on catalysis and on the conditions of chemical equilibrium and velocities
of chemical reactions"

1910   **Otto Wallach** (German):
"for his services to organic chemistry and the chemical industry by his pioneer work in
the field of alicyclic substances"

1911   **Marie Curie** (French):
"for her services to the advancement of chemistry by the discovery of the elements
radium and polonium"

1912   **Victor Grignard** (French):
"for the discovery of the so-called Grignard reagent, which has greatly helped in the
development of organic chemistry"

Paul Sabatier (French):
"for his method of hydrogenating organic compounds in the presence of finely divided metals"

1913   **Alfred Werner** (Swiss):
"for his work on the linkage of atoms in molecules by which he has thrown new light on earlier investigations and opened up new fields of research especially in inorganic chemistry"

1914   **Theodore W. Richards** (U.S.):
"for his accurate determinations of the atomic weights of a great number of chemical elements"

1915   **Richard M. Willstätter** (German):
"for his research on plant pigments, principally on chlorophyll"

1916   No award

1917   No award

1918   **Fritz Haber** (German):
"for the synthesis of ammonia from its elements, nitrogen and hydrogen"

1919   No award

1920   **Walther H. Nernst** (German):
"for his thermochemical work"

1921   **Frederick Soddy** (British):
"for his contributions to the chemistry of radioactive substances and his investigations into the origin and nature of isotopes"

1922   **Francis W. Aston** (British):
"for his discovery, by means of his mass spectrograph, of the isotopes of a large number of nonradioactive elements, as well as for his discovery of the whole-number rule"

1923   **Fritz Pregl** (Austrian):
"for his invention of the method of microanalysis of organic substances"

1924   No award

1925   **Richard A. Zsigmondy** (German):
for his demonstration of the heterogeneous nature of colloid solutions, and for the methods he used, which have since become fundamental in modern colloid chemistry"

1926   **Theodor Svedberg** (Swedish):
"for his work on disperse systems"

1927   **Heinrich O. Wieland** (German):
"for his research on bile acids and related substances"

1928   **Adolf O. R. Windaus** (German):
"for his studies on the constitution of the sterols and their connection with the vitamins"

1929   **Arthur Harden** (British):
Hans von Euler-Chelpin (Swedish):
"for their investigation on the fermentation of sugar and of fermentative enzymes"

1930   **Hans Fischer** (German):
"for his researches into the constitution of hemin and chlorophyll, and especially for his synthesis of hemin"

1931   **Frederich Bergius** (German):
Carl Bosch (German):
"for their contributions to the invention and development of chemical high-pressure methods"

1932   **Irving Langmuir** (U.S.):
"for his discoveries and investigations in surface chemistry"

1933   No award

1934   **Harold C. Urey** (U.S.):
"for his discovery of heavy hydrogen"

1935   **Frederic Joliot** (French):
Irene Joliot-Curie (French):
"for their synthesis of new radioactive elements"

1936   **Peter J. W. Debye** (Dutch/U.S.):
"for his contributions our knowledge of molecular structure through his investigations on dipole moments and on the diffraction of X-rays and electrons in gases"

1937   **Walter N. Haworth** (British):
"for his researches into the constitution of carbohydrates and vitamin C"

Paul Karrer (Swiss):
"for his researches into the constitution of carotenoids, flavins, and vitamins A and B"

1938   **Richard Kuhn** (German):
"for his work on carotenoids and vitamins"

1939   **Adolf F. J. Butenandt** (German):
"for his work on sex hormones"

Leopold Ruzicka (Swiss):
"for his work on polymethylenes and higher terpenes"

1940   No award

1941   No award

1942   No award

1943   **Georg de Hevesy** (Hungarian):
"for his work on the use of isotopes as tracer elements in researches on chemical processes"

1944   **Otto Hahn** (German):
"for his discovery of the fission of heavy nuclei"

1945   **Artturi I. Virtanen** (Finnish):
"for his researches and inventions in agricultural and nutritive chemistry, expecially for his fodder preservation method"

1946   **James B. Sumner** (U.S.):
"for his discovery that enzymes can be crystallized"

John H. Northrop (U.S.):
Wendell M. Stanley (U.S.):
for their preparation of enzymes and virus proteins in a pure form"

1947   **Sir Robert Robinson** (British):
"for his investigations on plant products of biological importance, particularly the alkaloids"

1948   **Arne W. K. Tiselius** (Swedish):
"for his researches on electrophoresis and adsorption analysis, especially for his discoveries concerning the complex nature of the serum proteins"

1949   **William F. Giauque** (U.S.):
"for his contributions in the field of chemical thermodynamics, particularly concerning the behavior of substances at extremely low temperatures"

1950   **Kurt Alder** (German):
Otto P. H. Diels (German):
"for their discovery and development of the diene synthesis"

1951   **Edwin M. McMillan** (U.S.):
Glenn T. Seaborg (U.S.):
"for their discoveries in the chemistry of the transuranium elements"

1952   **Archer J. P. Martin** (British):
Richard L. M. Synge (British):
"for their development of partition chromatography"

1953   **Hermann Staudinger** (German):
"for his discoveries in the field of macromolecular chemistry"

1954 **Linus C. Pauling** (U.S.):
"for his research into the nature of the chemical bond and its application to the elucidation of the structure of complex substances"

1955 **Vincent du Vigneaud** (U.S.):
"for his work on biochemically important sulfur compounds, especially for the first synthesis of a polypeptide hormone"

1956 **Sir Cyril N. Hinshelwood** (British):
Nikolai N. Semenov (U.S.S.R.):
"for their research in clarifying the mechanisms of chemical reactions in gases"

1957 **Sir Alexander R. Todd** (British):
"for his work on nucleotides and nucleotide coenzymes"

1958 **Frederick Sanger** (British):
"for his work on the structure of proteins, particularly insulin"

1959 **Jaroslav Heyrovsky** (Czechoslovakian):
"for his discovery and development of the polarographic method of analysis"

1960 **Willard F. Libby** (U.S.):
"for his method to use carbon-14 for age determination in archaeology, geology, geophysics, and other branches of science"

1961 **Melvin Calvin** (U.S.):
"for his research on the carbon dioxide assimilation in plants"

1962 **John C. Kendrew** (British):
Max F. Perutz (British):
"for their studies of the structures of globular proteins"

1963 **Giulio Natta** (Italian):
Karl Ziegler (German):
"for their work in the controlled polymerization of hydrocarbons through the use of organometallic catalysts"

1964 **Dorothy C. Hodgkin** (British):
"for her determinations by X-ray techniques of the structures of important biochemical substances, particularly vitamin B-12 and penicillin"

1965 **Robert B. Woodward** (U.S.):
"for his outstanding achievements in the 'art' of organic synthesis"

1966 **Robert S. Mulliken** (U.S.):
"for his fundamental work concerning chemical bonds and the electronic structure of molecules by the molecular orbital method"

1967   **Manfred Eigen** (German):
Ronald G. W. Norrish (British):
George Porter (British):
"for their studies of extremely fast chemical reactions, effected by disturbing the equilibrium with very short pulses of energy"

1968   **Lars Onsager** (U.S.):
"for his discovery of the reciprocal relations bearing his name, which are fundamental for the thermodynamics of irreversible processes"

1969   **Sir Derek H. R. Barton** (British):
Odd Hassel (Norwegian):
"for their contributions to the development of the concept of conformation and its application in chemistry"

1970   **Luis F. Leloir** (Argentinian):
"for his discovery of sugar nucleotides and their role in the biosynthesis of carbohydrates"

1971   **Gerhard Herzberg** (Canadian):
"for his contributions to the knowledge of electronic structure and geometry of molecules, particularly free radicals"

1972   **Christian B. Anfinsen** (U.S.):
"for his work on ribonuclease, especially concerning the connection between the amino acid sequence and the biologically active conformation"

Stanford Moore (U.S.):
William H. Stein (U.S.):
"for their contribution to the understanding of the connection between chemical structure and catalytic activity of the active center of the ribonuclease molecule"

1973   **Ernst Otto Fischer** (German):
Geoffrey Wilkinson (British):
"for their pioneering work, performed independently, on the chemistry of the organometallic sandwich compounds"

1974   **Paul J. Flory** (U.S.):
"for his fundamental achievements, both theoretical and experimental, in the physical chemistry of macromolecules"

1975   **John Cornforth** (Australian/British):
"for his work on the stereochemistry of enzyme-catalyzed reactions"

Vladimir Prelog (Yugoslavian/Swiss):
"for his work on the stereochemistry of organic molecules and reactions"

1976 **William N. Lipscomb** (U.S.):
"for his studies on the structures of boranes illuminating problems of chemical bonding"

1977 **Ilya Pregogine** (Belgian):
"for his contributions to nonequilibrium thermodynamics, particularly the theory of dissipative structures"

1978 **Peter Mitchell** (British):
"for his contribution to the understanding of biological energy transfer through the formulation of the chemiosmotic theory"

1979 **Herbert C. Brown** (U.S.):
"for his application of boron compounds to synthetic organic chemistry"

Georg Wittig (German):
"for developing phosphorus reagents, presently bearing his name"

1980 **Paul Berg** (U.S.):
"for his fundamental studies of the biochemistry of nucleic acids, with particular regard to recombinant DNA"

Walter Gilbert (U.S.)
Frederick Sanger (British):
"for their contributions concerning the determination of base sequences in nucleic acids"

1981 **Kenichi Fukui** (Japanese)
Roald Hoffmann (U.S.):
for their theories, developed independently, concerning the course of chemical reactions"

1982 **Aaron Klug** (British):
"for his development of crystallographic electron microscopy and his structural elucidation of biologically important nucleic acid - protein complexes"

1983 **Henry Taube** (U.S.):
"for his work on the mechanisms of electron transfer reactions, especially in metal complexes"

1984 **R. Bruce Merrifield** (U.S.):
"for his development of methodology for chemical synthesis on a solid matrix"

1985 **Herbert A. Hauptman** (U.S.):
Jerome Karle (U.S.):
"for their outstanding achievements in the development of direct methods for the determination of crystal structures"

1986 **John C. Polanyi** (Canadian):
"for his pioneering work in the use of infrared chemiluminescence in studying the dynamics of chemical reactions"

Dudley R. Herschbach (U.S.):
Yuan T. Lee (U.S.):
"for their contributions concerning the dynamics of chemical elementary processes"

1987 **Donald J. Cram** (U.S.):
Jean-Marie Lehn (French):
Charles J. Pedersen (U.S.):
"for their development and use of molecules with structure-specific interactions of high selectivity"

1988 **Johann Deisenhofer** (German):
Robert Huber (German):
Hartmut Michel (German):
"for their determination of the structure of the photosynthetic reaction center of bacteria"

1989 **Sidney Altman** (U.S.):
Thomas R. Cech (U.S.):
"for their discovery that RNA acts as a biological catalyst as well as a carrier of genetic information"

1990 **E. J. Corey** (U.S.):
"for his contributions to organic synthesis"

Suggested Additional Readings

Chapters 1 and 2 - Structure, Bonding, and Molecular Properties

J. R. Partington, "A History of Chemistry." Vol. I - IV, MacMillan, London, 1961–1964.

T. L. Brown, H. E. LeMay, and B. E. Bursten, "Chemistry", Prentice-Hall, Englewood Cliffs, NJ, 1991

L. Salem, "A Faithful Couple: The Electron Pair," **J. Chem. Educ., 55,** 344 (1978).

R. J. Gillespie, "The Electron-Pair Repulsion Model for Molecular Geometry," **J. Chem. Educ., 47,** 18 (1970).

R. H. Maybury, "The Language of Quantum Mechanics," **J. Chem. Educ., 39,** 367 (1962).

D. Kolb, "Acids and Bases," **J. Chem. Educ., 55,** 459 (1978).

D. Kolb, "The pH Concept," **J. Chem. Educ., 56,** 49 (1979).

Chapter 3 - The Nature of Organic Compounds: Alkanes and Cycloalkanes

J. G. Traynham, "Organic Nomenclature: A Programmed Introduction," 3rd Ed., Prentice-Hall, Englewood Cliffs, N.J., 1985.

J. H. Fletcher, O. C. Dermer, and R. B. Fox, "Nomenclature of Organic Compounds: Principles and Practice," Advances in Chemistry Series No. 126, American Chemical Society, Washington, D. C., 1974.

"Nomenclature of Organic Chemistry, Sections A, B, C, D, E, F, and H," International Union of Pure and Applied Chemistry, Pergamon Press, Oxford, 1979.

D. Kolb and K. E. Kolb, "Petroleum Chemistry," **J. Chem. Educ., 56,** 465 (1979).

Chapter 4 - Stereochemistry of Alkanes and Cycloalkanes

C. A. Kingsbury, "Conformations of Substituted Ethanes," **J. Chem. Educ., 56,** 431 (1979).

E. L. Eliel, "Stereochemistry of Carbon Compounds," McGraw-Hill, New York, 1962.

L. N. Ferguson, "Ring Strain and Reactivity of Alicycles," **J. Chem. Educ., 47,** 46 (1970).

J. B. Lambert, "The Shapes of Organic Molecules," **Scientific American,** Jan., 1970, p. 58.

Chapter 5 - An Overview of Organic Reactions

J. March, "Advanced Organic Chemistry," 3rd ed., McGraw-Hill, New York, 1985, Chapter 6.

W. H. Saunders, Jr., "Ionic Aliphatic Reactions," Prentice-Hall, Englewood Cliffs, 1965.

P. Sykes, "A Guidebook to Mechanism in Organic Chemistry," 5th ed., Longman, Green, New York 1981.

R. Breslow, "Organic Reaction Mechanisms," 2nd ed., Benjamin, Menlo Park, 1969.

Chapters 6 and 7 - Alkenes: Structure, Reactions, and Synthesis

H. Salzman, "Arthur Lapworth: The Genesis of Reaction Mechanism," **J. Chem. Educ., 49,** 750 (1972).

W. R. Dolbier Jr., "Electrophilic Additions to Alkenes," **J. Chem. Educ., 46,** 342 (1969).

J. March, "Advanced Organic Chemistry," 3rd ed., McGraw-Hill, New York, 1985, Chapters 15 and 17.

N. Isenberg and M. Grdinic, "A Modern Look at Markovnikov's Rule and the Peroxide Effect," **J. Chem. Educ., 46,** 601 (1969).

S. Patai (ed.), "Chemistry of the Alkenes, Part I," Wiley Interscience, New York, 1964.

J. Zabicky (ed.), "Chemistry of the Alkenes, Part II," Wiley Interscience, New York, 1970.

M. Jones, Jr., "Carbenes," **Scientific American,** February, 1976, p. 101.

Chapter 8 - Alkynes

S. Patai (ed.), "The Chemistry of the Carbon-Carbon Triple Bond, Parts I and II," Wiley Interscience, New York, 1978.

Chapter 9 - Stereochemistry

D. F. Mowery, Jr., "Criteria for Optical Activity in Organic Molecules," **J. Chem. Educ., 46,** 269 (1969).

R. S. Cahn, "An Introduction to the Sequence Rule," **J. Chem. Educ., 41,** 116 (1964).

E. L. Eliel, "Elements of Stereochemistry," Wiley, New York, 1969.

Chapter 10 - Alkyl Halides

W. A. Pryor, "Introduction to Free-Radical Chemistry," Prentice-Hall, Englewood Cliffs, 1966.

S. Patai (ed.), "Chemistry of the Carbon-Halogen Bond, Parts I and II," Wiley Interscience, New York, 1973.

Chapter 11 - Reactions of Alkyl Halides: Nucleophilic Substitutions and Eliminations

J. March, "Advanced Organic Chemistry," 3rd ed., McGraw-Hill, New York, 1985, Chapters 10 and 17.

A. Streitwieser, Jr., "Solvolytic Displacement Reactions," McGraw-Hill, New York, 1962.

Chapter 12 - Structure Determination: Mass Spectroscopy and Infrared Spectroscopy

F. W. McLafferty, "Interpretation of Mass Spectroscopy," 3rd ed., Benjamin, Menlo Park, 1981.

J. R. Dyer, "Applications of Absorption Spectroscopy of Organic Compounds," Prentice-Hall, Englewood Cliffs, 1965.

J. W. Cooper, "Spectroscopic Techniques for Organic Chemists," Wiley Interscience, New York, 1980.

Chapter 13 - Structure Determination: Nuclear Magnetic Resonance Spectroscopy

J. R. Dyer, "Applications of Absorption Spectroscopy of Organic Compounds," Prentice-Hall, Englewood Cliffs, 1965.

J. W. Cooper, "Spectroscopic Techniques for Organic Chemists," Wiley Interscience, New York, 1980.

F. W. Wehrli and T. Wirthlin, "Interpretation of Carbon-13 NMR Spectra," Heyden, Philadelphia, 1978.

Chapter 14 - Conjugated Dienes and Ultraviolet Spectroscopy

H. H. Jaffe and M. Orchin, "Theory and Application of Ultraviolet Spectroscopy," Wiley, New York, 1962.

J. R. Dyer, "Applications of Absorption Spectroscopy of Organic Compounds," Prentice-Hall, Englewood Cliffs, 1965.

J. W. Cooper, "Spectroscopic Techniques for Organic Chemists," Wiley Interscience, New York, 1980.

Chapters 15 and 16 - Aromatic Compounds

R. Breslow, "The Nature of Aromatic Molecules," **Sci. Am.,** August, 1972, p. 32.

D. Kolb, "The Aromatic Ring," **J. Chem. Educ., 56,** 334 (1979).

J. March, "Advanced Organic Chemistry," 3rd ed., McGraw-Hill, New York, 1985, Chapter 11.

G. A. Olah, "Friedel-Crafts Chemistry," Wiley, New York, 1973.

Chapter 17 - Alcohols and Thiols

S. Patai (ed.), "The Chemistry of the Hydroxyl Group, Parts I and II," Wiley Interscience, New York, 1971.

S. Patai (ed.), "The Chemistry of Ethers, Crown Ethers, Hydroxyl Groups, and Their Sulfur Analogues, Parts I and II", Wiley Interscience, New York, 1980.

K. L. Rinehart, "Oxidation and Reduction of Organic Compounds," Prentice-Hall, Englewood Cliffs, 1973.

Chapter 18 - Ethers and Epoxides

S. Patai (ed.), "The Chemistry of the Ether Linkage", Wiley Interscience, New York, 1967.

S. Patai (ed.), "The Chemistry of Ethers, Crown Ethers, Hydroxyl Groups, and Their Sulfur Analogues, Parts I and II", Wiley Interscience, New York, 1980.

Chapter 19 - Aldehydes and Ketones: Nucleophilic Addition Reactions

C. D. Gutsche, "The Chemistry of Carbonyl Compounds," Prentice-Hall, Englewood Cliffs, 1967.

H. Hart and M. Sasaoka, "Simple Enols: How Rare Are They?," **J. Chem. Educ., 57,** 685 (1980).

S. Patai (ed.), "The Chemistry of the Carbonyl Group, Part I." Wiley Interscience, New York, 1966.

J. Zabicky (ed.), "The Chemistry of the Carbonyl Group, Part II," Wiley Interscience, New York, 1970.

Chapters 20 and 21 - Carboxylic Acids and Their Derivatives

C. D. Gutsche, "The Chemistry of Carbonyl Compounds," Prentice-Hall, Englewood Cliffs, 1967.

S. Patai (ed.), "The Chemistry of Carboxylic Acids and Esters," Wiley Interscience, New York, 1969.

H. O. House, "Modern Synthetic Reactions," 2nd ed., Benjamin, Menlo Park, 1972, Chapter 11.

Chapter 22 - Carbonyl Alpha-Substitution Reactions

C. D. Gutsche, "The Chemistry of Carbonyl Compounds," Prentice-Hall, Englewood Cliffs, 1967.

H. O. House, "Modern Synthetic Reactions," 2nd ed., Benjamin, Menlo Park, 1972, Chapter 9.

Chapter 23 - Carbonyl Condensation Reactions

H. O. House, "Modern Synthetic Reactions," 2nd ed., Benjamin, Menlo Park, 1972, Chapter 10.

C. D. Gutsche, "The Chemistry of Carbonyl Compounds," Prentice-Hall, Englewood Cliffs, 1967.

Chapter 24 - Carbohydrates

C. S. Hudson, "Emil Fischer's Discovery of the Configuration of Glucose," **J. Chem. Educ., 18,** 353 (1941).

W. W. Pigman and D. Horton (eds.), "The Carbohydrates: Chemistry and Biochemistry," continuing series, Academic Press, New York, 1972 -.

L. Stryer, "Biochemistry," 3rd ed., Freeman, San Francisco, 1988, Chapters 14-16.

G. M. Bodner, "Glycolysis, or the Embden-Meyerhof Pathway," **J. Chem. Educ., 63,** 566, (1986).

G. M. Bodner, "The Tricarboxylic Acid (TCA), Citric Acid, or Krebs Cycle," **J. Chem. Educ., 63,** 673 (1986).

Chapters 25 and 26 - Amines and Phenols

J. M. McIntosh, "Phase-Transfer Catalysis Using Quaternary 'Onium Salts," **J. Chem. Educ., 55,** 235 (1978).

I. T. Miller and H. D. Springall, "Sidgwick's Organic Chemistry of Nitrogen," 3rd ed., Oxford Press, London, 1966.

P. A. S. Smith, "The Chemistry of Open-Chain Organic Nitrogen Compounds, Vols. I and II," Benjamin, Menlo Park, 1966.

S. Patai, "The Chemistry of the Amino Group," Wiley Interscience, New York, 1968.

Chapter 27 - Amino Acids, Peptides, and Proteins

L. Stryer, "Biochemistry," 3rd ed., Freeman, San Francisco, 1988, Chapters 2-6.

Special Issue on "The Molecules of Life," **Scientific American**, October, 1985.

Chapter 28 - Lipids

L. Stryer, "Biochemistry," 3rd ed., Freeman, San Francisco, 1988, Chapter 23.

J. H. Richards and J. B. Hendrickson, "Biosynthesis of Steroids, Terpenes, and Acetogenins," Benjamin, Menlo Park, 1964.

Chapter 29 - Heterocycles and Nucleic Acids

L. A. Paquette, "Principles of Modern Heterocyclic Chemistry," Benjamin, Menlo Park, 1968.

L. Stryer, "Biochemistry," 3rd ed., Freeman, San Francisco, 1988, Chapters 27-34.

Special Issue on "The Molecules of Life," **Scientific American**, October, 1985.

Chapter 30 - Pericyclic Reactions

R. Wollenberg and R. Belloli, "Woodward-Hoffmann Made Easy," **Chem. Brit.,** 10, 95 (1974).

T. L. Gilchrist and R. C. Storr, "Organic Reactions and Orbital Symmetry," Cambridge University Press, London, 1972.

R. E. Lehr and A. P. Marchand, "Orbital Symmetry: A Problem-Solving Approach," Academic Press, Inc., New York, 1972.

Chapter 31 - Synthetic Polymers

F. W. Harris, et.al., "State of the Art: Polymer Chemistry," **J. Chem. Educ., 58,** 836-955 (1981).

F. W. Billmeyer, Jr., "Textbook of Polymer Science," 3rd ed., Wiley, New York, 1984.